Analytic and Algebraic Geometry

Common Problems, Different Methods

Ω IAS/PARK CITY
MATHEMATICS SERIES
Volume 17

Analytic and Algebraic Geometry

Common Problems, Different Methods

Jeffery McNeal
Mircea Mustaţă
Editors

American Mathematical Society
Institute for Advanced Study

John C. Polking, Series Editor
Ezra Miller, Volume Editor
Victor Reiner, Volume Editor
Bernd Sturmfels, Volume Editor

IAS/Park City Mathematics Institute runs mathematics education programs that bring together high school mathematics teachers, researchers in mathematics and mathematics education, undergraduate mathematics faculty, graduate students, and undergraduates to participate in distinct but overlapping programs of research and education. This volume contains the lecture notes from the Graduate Summer School program on Analytic and Algebraic Geometry: Common Problems, Different Methods held in Park City, Utah in the summer of 2008.

2000 *Mathematics Subject Classification.* Primary 14E15, 14E30, 14F18, 32W05, 53C21.

Library of Congress Cataloging-in-Publication Data

Analytic and algebraic geometry : common problems, different methods / Jeffery McNeal, Mircea Mustaţă, editors.
p. cm. — (IAS/Park City mathematics series, ISSN 1079-5634 ; v. 17)
Includes bibliographical references.
ISBN 978-0-8218-4908-8 (alk. paper)
1. Geometry, Analytic. 2. Geometry, Algebraic. I. McNeal, Jeffery, D. II. Mustaţă, Mircea, 1971–

QA551.A485 2010
516.3—dc22

2010029457

♾ The paper used in this book is acid-free and falls within the guidelines
established to ensure permanence and durability.
Visit the AMS home page at `http://www.ams.org/`

10 9 8 7 6 5 4 3 2 1   15 14 13 12 11 10

Contents

Preface

The IAS/Park City Mathematics Institute (PCMI) was founded in 1991 as part of the "Regional Geometry Institute" initiative of the National Science Foundation. In mid 1993 the program found an institutional home at the Institute for Advanced Study (IAS) in Princeton, New Jersey.

The IAS/Park City Mathematics Institute encourages both research and education in mathematics and fosters interaction between the two. The three-week summer institute offers programs for researchers and postdoctoral scholars, graduate students, undergraduate students, high school teachers, undergraduate faculty, and researchers in mathematics education. One of PCMI's main goals is to make all of the participants aware of the total spectrum of activities that occur in mathematics education and research: we wish to involve professional mathematicians in education and to bring modern concepts in mathematics to the attention of educators. To that end the summer institute features general sessions designed to encourage interaction among the various groups. In-year activities at the sites around the country form an integral part of the High School Teachers Program.

Each summer a different topic is chosen as the focus of the Research Program and Graduate Summer School. Activities in the Undergraduate Summer School deal with this topic as well. Lecture notes from the Graduate Summer School are being published each year in this series. The first seventeen volumes are:

- Volume 1: *Geometry and Quantum Field Theory* (1991)
- Volume 2: *Nonlinear Partial Differential Equations in Differential Geometry* (1992)
- Volume 3: *Complex Algebraic Geometry* (1993)
- Volume 4: *Gauge Theory and the Topology of Four-Manifolds* (1994)
- Volume 5: *Hyperbolic Equations and Frequency Interactions* (1995)
- Volume 6: *Probability Theory and Applications* (1996)
- Volume 7: *Symplectic Geometry and Topology* (1997)
- Volume 8: *Representation Theory of Lie Groups* (1998)
- Volume 9: *Arithmetic Algebraic Geometry* (1999)
- Volume 10: *Computational Complexity Theory* (2000)
- Volume 11: *Quantum Field Theory, Supersymmetry, and Enumerative Geometry* (2001)
- Volume 12: *Automorphic Forms and their Applications* (2002)
- Volume 13: *Geometric Combinatorics* (2004)
- Volume 14: *Mathematical Biology* (2005)
- Volume 15: *Low Dimensional Topology* (2006)
- Volume 16: *Statistical Mechanics* (2007)
- Volume 17: *Analytic and Algebraic Geometry: Common Problems, Different Methods* (2008)

Volumes are in preparation for subsequent years.

Some material from the Undergraduate Summer School is published as part of the Student Mathematical Library series of the American Mathematical Society. We hope to publish material from other parts of the IAS/PCMI in the future.

This will include material from the High School Teachers Program and publications documenting the interactive activities which are a primary focus of the PCMI. At the summer institute late afternoons are devoted to seminars of common interest to all participants. Many deal with current issues in education: others treat mathematical topics at a level which encourages broad participation. The PCMI has also spawned interactions between universities and high schools at a local level. We hope to share these activities with a wider audience in future volumes.

John C. Polking
Series Editor
July 2010

Introduction

Jeffery D. McNeal and Mircea Mustaţă

IAS/Park City Mathematics Series
Volume 17, 2008

Introduction

Jeffery D. McNeal and Mircea Mustaţă

1. The subject

Interactions between algebraic and analytic techniques in the study of complex algebraic varieties go back to the beginnings of the twentieth century. However, over the past twenty years these connections have particularly flourished. Analytic methods and ideas around multiplier ideals have been successfully introduced in the study of higher-dimensional algebraic varieties, and tools developed as part of the Minimal Model Program have had growing impact. Taken together these new methods achieved great success, culminating in the proof of a fundamental problem in the field, the finite generation of the canonical ring of an algebraic variety. While some of the analytic techniques have already found an algebraic counterpart and vice versa, much is left to be done, and it is expected that we will see more of this interaction in the future.

The 2008 PCMI Summer School was centered around these exciting new developments at the crossroads of analytic and algebraic geometry, and in particular, on the two existing approaches to the finite generation of the canonical ring. The program had several components. There was a Graduate Program, consisting of eight mini-courses, half of which were devoted to analytic, and respectively, algebraic topics. The lecturers were Bo Berndtsson, John D'Angelo, Jean-Pierre Demailly, Christopher Hacon, János Kollár, Robert Lazarsfeld, Mircea Mustaţă, and Dror Varolin. In parallel with this there was an active Research Program, organized around one or two daily seminar talks. The research environment was enhanced by the presence of the Clay Senior Scholar Robert Lazarsfeld, and of the Program Principal Yum-Tong Siu. They each gave a public lecture, introducing some of the main questions in analytic and algebraic geometry.

This volume consists of the contributions of the eight lecturers in the Graduate Program. In addition, it contains two expository presentations introducing the finite generation of the canonical ring, from the two different perspectives.

2. Content of the volume

The mini-courses consisted of five lectures each, and three additional problem sessions run by TA's. Each sequence of four mini-courses, on the algebraic and on the analytic side, was organized as to give a gradual introduction to the recent developments. We have decided to keep the same order for the contributions in this

Department of Mathematics, Ohio State University, Columbus, Ohio 43210, USA
E-mail address: mcneal@math.ohio-state.edu
Department of Mathematics, University of Michigan, Ann Arbor, Michigan 48109, USA
E-mail address: mmustata@umich.edu

volume. At the end of each of the two sequences, we have included one expository paper that was not based on the lectures in the PCMI program. We hope that in terms of both topic and presentation, these additions fit well with the rest of the volume.

As expected, there was a lot of interaction between the two sequences of courses, and this is transparent in the lecture notes in this volume. Some key concepts (most notably, multiplier ideals) and results (such as vanishing and lifting theorems) have played an important role in the cross-fertilization of analytic and algebraic geometry, and especially in the recent applications to birational geometry. We hope that seeing these notions and results developed in both contexts, will give the reader a sense of the close interaction between these two fields.

The contributions on the analytic side are the following.

An Introduction to Things $\bar{\partial}$. These are the lecture notes from Bo Berndtsson's course at PCMI 2008. The aim of this course was to give an introduction, appropriate for a beginning student in complex analysis, to the basic analytic techniques behind weighted L^2 estimates for the $\bar{\partial}$-equation. The notes here reflect this aim by starting with the $\bar{\partial}$-equation in one-variable, carefully examining the fundamental issues in this special case, and then seeing how these issues are handled in ever greater generality, i.e. for domains in several variables, on Stein manifolds, and for bundle-valued forms over general complex manifolds. The emphasis throughout the notes is on how the basic L^2 estimates for $\bar{\partial}$ change as additional structure (primarily metrics on both the base manifold and on the bundles) is added. Several applications of L^2 estimates on $\bar{\partial}$ are also presented, including vanishing theorems of Kodaira! type and various extension results connected to the Ohsawa-Takegoshi theorem.

Real and Complex Geometry Meet the Cauchy-Riemann Equations. In his course at the Summer School, John D'Angelo discussed several analytic topics that involve an interplay between real and complex geometry. The order of contact between a complex analytic variety and a real hypersurface inside a complex manifold, and its relation to estimates for the $\bar{\partial}$ operator, forms the basis for the first half of these notes. An expanded discussion of Kohn's subelliptic multipliers for the $\bar{\partial}$-Neumann problem, which inspired the general notion of multiplier ideals studied in algebraic and analytic geometry, is also given. A detailed presentation of a version of Hilbert's 17th problem, and a sketch of its relationship with special metrics over the complex projective space concludes these notes.

Three Variations on a Theme in Complex Analytic Geometry. This chapter is based on Dror Varolin's lecture series. These notes show how L^2 methods are used to obtain three foundational results in analytic geometry: Kodaira's embedding theorem, the L^2 holomorphic extension theorem of Ohsawa-Takegoshi, and Skoda's division theorem. The connection between curvature, in several forms, and the estimates on $\bar{\partial}$ used to obtain these results is emphasized throughout the notes. Multiplier ideal sheaves are also discussed, from the analytic perspective, and proofs of Nadel's vanishing theorem and Siu's theorem on the global generation of these ideals are presented.

Structure Theorems for Projective and Kähler Varieties. These are lecture notes based on Jean-Pierre Demailly's course. They examine vanishing results for $\bar{\partial}$ cohomology on compact Kähler manifolds (or more particularly, projective manifolds) that carry line bundles of various types, e.g. numerically effective, pseudo-effective, ample, or big line bundles. The positivity conditions expressed by these types of bundles is the central theme of these notes. Recent results on the deformation theory of curves in Kähler manfolds and approximation of closed, positive $(1,1)$-currents are discussed in some detail. The notes also give an outline of Siu's analytic approach to finite generation of the canonical ring and a sketch of Păun's related non-vanishing theorem.

Lecture Notes on Rational Polytopes and Finite Generation. Mihai Păun, who was unable to attend the PCMI 2008 program, graciously contributed this chapter in his absence. The notes present an overview of an analytic approach to the Birkar-Cascini-Hacon-McKernan theorem on the finite generation of the canonical ring, using techniques pioneered by Siu, Demailly, and Shokurov. The polytopes in the pseudoeffective cone of a nonsingular projective variety were introduced by Shokurov, and they play an important role in this proof of finite generation (as well as in the original one) The technical core of the arguments relies on analytic methods, in particular on extension theorems established previously by Păun, which are discussed also in these notes. The flexibility of the analytic approach to finite generation of the canonical ring, as proposed by Siu, motivates much of the presentation.

The following are the contributions on the algebraic side.

Introduction to Resolution of Singularities. This is based on Mircea Mustaţă's lecture series. Assuming a minimum of background in algebraic geometry, it presents a proof of Hironaka's Theorem on resolution of singularities. Despite being arguably one of the most important results in algebraic geometry, for many years its proof has been understood by only a handful of experts. The combined work of many people has resulted in making the proof accessible to any algebraic geometry student. These notes give a full account of the proof, based on the recent simplifications of Włodarczyk and Kollár.

A Short Course on Multiplier Ideals. These are lecture notes of Robert Lazarsfeld's mini-course. Multiplier ideals have been central to many of the recent developments in higher-dimensional algebraic geometry. These notes give a broad overview of the algebraic theory of multiplier ideals. They cover the definition and basic properties of multiplier ideals, the connection with vanishing theorems that lies at the heart of the subject, as well as a wide range of applications. In particular, one describes the use of multiplier ideals in the context of lifting theorems, as pioneered by Siu in his proof of Invariance of Plurigenera.

Exercises in the Birational Geometry of Algebraic Varieties. In his PCMI course, János Kollár gave an introduction to the main ideas, results, and open problems in the Minimal Model Program. The contribution in this volume offers an introduction to the same topic through one hundred exercises. It contains the basic definitions, as well as the statements of and comments on the main results. The rest of the notes is devoted to illustrating through exercises and examples these results, the existing techniques in birational geometry, with their strengths as well as their shortcomings. They cover basic topics about birational maps, classical results about surfaces, the cone of curves, flips and minimal models.

Higher Dimensional Minimal Model Program for Varieties of Log General Type. These are lecture notes for the course given by Christopher Hacon during the Summer School. They present some of the recent results of Hacon-McKernan and of Birkar-Cascini-Hacon-McKernan leading to the proof of the finite generation of the canonical ring. They cover more advanced topics on multiplier ideals, and in particular, the lifting theorems that are then used in the proof of existence of flips. The notes end with a discussion of the Minimal Model Program with scaling, and with a sketch of the argument for the existence of minimal models for varieties of log general type.

Lectures on Flips and Minimal Models. In April 2007, MSRI organized the workshop "Minimal and Canonical Models in Algebraic Geometry", motivated by the recent progress in higher-dimensional birational geometry. These lecture notes of Alessio Corti, Paul Hacking, János Kollár, Robert Lazarsfeld, and Mircea Mustaţă are based on talks at this workshop, giving an informal overview of the work of Hacon-McKernan and Birkar-Cascini-Hacon-McKernan proving the finite generation of the canonical ring. In particular, these notes could be used as an introduction to the more detailed expositions of Lazarsfeld and Hacon in this volume.

3. Acknowledgments

We would like to thank all the people who have contributed to the success of this PCMI program, and implicitly, to this volume. We are grateful to the lecturers for their effort in making accessible the recent developments in this exciting field, to the TA's for running the problem sessions, and last but not least, to all the participants for creating a stimulating and vibrant environment. Special thanks go to the PCMI staff, and especially to Catherine Giesbrecht, for the invaluable help she has provided all along. We are indebted to Herb Clemens, who first envisioned this program, and to Robert Bryant, for his constant help and advice in running the program. Finally, we would like to thank John Polking for his assistance in preparing this volume.

An Introduction to Things $\overline{\partial}$

Bo Berndtsson

IAS/Park City Mathematics Series
Volume 17, 2008

An Introduction to Things $\bar{\partial}$

Bo Berndtsson

Introduction

These are the notes of a series of lectures given at the PCMI summer school 2008. They are intended to serve as an introduction to the weighted L^2-estimates for the $\bar{\partial}$-equation, by Kodaira, [**12**], Andreotti-Vesentini, [**2**] and, in the most complete form, Hörmander, [**10**].

Three lecture series on this topic was given at the school; this is the first and supposedly the most elementary of them. Therefore we start by a discussion of the one dimensional case, which I believe shows very clearly the main ideas. After that we discuss briefly the functional analytic set up which is needed in the case of higher dimensions. The third lecture presents the geometric notions needed to discuss the $\bar{\partial}$-equation for forms with values in a line bundle on a manifold, and then gives the fundamental existence theorems and L^2-estimates. The proofs differ a bit from the more common ones, that are based on use of the Kähler identities and the Kodaira-Nakano formula for twisted Laplace operators. Instead we use the so called $\partial\bar{\partial}$-Bochner-Kodaira method introduced by Siu, which in my opinion is the most elementary approach.

The remaining lectures deal with applications and generalizations of this basic material. Lecture 4 gives some basic facts about Bergman kernels associated to holomorphic line bundles and uses the $\bar{\partial}$-estimates to deduce a rudimentary asymptotic formula. Lecture 5 introduces singular metrics and generalizes the results of Lecture 3 to this setting, leading up to the Demailly-Nadel and Kawamata-Viehveg vanishing theorems. The main topic of Lecture 6 is the Ohsawa-Takegoshi extension theorem, which we prove by another application of the $\partial\bar{\partial}$-Bochner-Kodaira method. The final section discusses briefly one of the main recent applications of the L^2-theory, Siu's theorem on the 'invariance of plurigenera', which we prove following the method of Paun.

It is a pleasure to thank here Jeff McNeal and Mircea Mustata for making the summer school for making the summer school an extremely nice and memorable event. I also like to thank the other participants, in particular Ann-Katrin Herbig and Yannir Rubinstein, for a lot of feedback on the material presented here.

Department of Mathematics, CTH, 412 96 Goteborg, Sweden
E-mail address: bob@chalmers.se

LECTURE 1

The one-dimensional case

In this lecture we state and prove the basic Hörmander L^2-estimate for the $\bar{\partial}$-equation in the case of one complex variable. This case contains the two main ideas in the subject, and almost no technical difficulties, so it is a good introduction to the discussion in later lectures.

1.1. The $\bar{\partial}$-equation in one variable

Let Ω be any domain in $\mathbb{C}$, and ϕ any function satisfying

$$\frac{\partial^2 \phi}{\partial z \partial \bar{z}} =: \Delta\phi > 0.$$

The one dimensional version of the theorem we will discuss says that we can solve any inhomogeneous $\bar{\partial}$-equation

$$\frac{\partial u}{\partial \bar{z}} = f$$

with a function u satisfying

$$\int |u|^2 e^{-\phi} \leq \int \frac{|f|^2}{\Delta\phi} e^{-\phi}.$$

Even this one variable case is a very precise and useful result, and it is quite surprising that it was discovered in several variables first. Moreover, the proof when $n = 1$ is considerably more elementary than the general case.

We begin by giving the problem a dual formulation. Remember that, interpreted in the sense of distributions, the equation $\frac{\partial}{\partial \bar{z}} u = f$ means precisely that

$$-\int u \frac{\partial}{\partial \bar{z}} \alpha = \int f \alpha \tag{1.1}$$

for all $\alpha \in C_c^2(\Omega)$. To introduce the weighted L^2-norms of the theorem we substitute $\bar{\alpha} e^{-\phi}$ for α. The equality (1.1) then says

$$\int u \overline{\bar{\partial}_\phi^* \alpha} e^{-\phi} = \int f \bar{\alpha} e^{-\phi}, \tag{1.2}$$

where

$$\bar{\partial}_\phi^* \alpha =: -e^{\phi} \frac{\partial}{\partial z}(e^{-\phi}\alpha),$$

is the *formal adjoint* of the $\bar{\partial}$-operator with respect to our weighted scalar product

$$\langle f, g \rangle_\phi = \int f \bar{g} e^{-\phi}.$$

The following proposition is one of the key ideas in the subject. It reduces the proof of an existence statement to the proof of an inequality.

Proposition 1.1.1. *Given f there exists a solution, u, to $\frac{\partial}{\partial \bar z}u = f$ satisfying*

$$\int |u|^2 e^{-\phi} \leq C, \tag{1.3}$$

if and only if the estimate

$$|\int f\,\bar\alpha e^{-\phi}|^2 \leq C \int |\bar\partial^*_\phi \alpha|^2 e^{-\phi} \tag{1.4}$$

holds for all $\alpha \in C^2_c(\Omega)$. On the other hand, for a given function $\mu > 0$, (1.4) holds for all f satisfying

$$\int \frac{|f|^2}{\mu} e^{-\phi} \leq C \tag{1.5}$$

if and only if

$$\int \mu |\alpha|^2 e^{-\phi} \leq \int |\bar\partial^*_\phi \alpha|^2 e^{-\phi}, \tag{1.6}$$

holds for all $\alpha \in C^2_c(\Omega)$.

Proof. It is clear that if (1.2), and (1.3) hold, then (1.4) follows. Suppose conversely that the inequality (1.4) is true. Let

$$E = \{\bar\partial^*_\phi \alpha; \alpha \in C^2_c(\Omega)\},$$

and consider E as a subspace of

$$L^2(e^{-\phi}) = \{g \in L^2_{loc}; \int |g|^2 e^{-\phi} < \infty\}.$$

Define an antilinear functional on E by

$$L(\bar\partial^*_\phi \alpha) = \int f\bar\alpha e^{-\phi}.$$

The inequality (1.4) then says that L is (well defined and) of norm not exceeding C. By Hahn-Banach's extension theorem L can be extended to an antilinear form on all of $L^2(e^{-\phi})$, with the same norm. The Riesz representation theorem then implies that there is some element,u, in $L^2(e^{-\phi})$, with norm less than C, such that

$$L(g) = \int u\bar g e^{-\phi},$$

for all $g \in L^2(e^{-\phi})$. Choosing $g = \bar\partial^*_\phi \alpha$, we see that

$$\int u\overline{\bar\partial^*_\phi \alpha} e^{-\phi} = \int f\bar\alpha e^{-\phi},$$

so u solves $\frac{\partial}{\partial \bar z}u = f$.

The first part of the proposition is therefore proved. The second part is obvious if μ is identically equal to 1. The general case follows if we write

$$f \cdot \bar\alpha = (f/\sqrt{\mu}) \cdot \overline{(\sqrt{\mu}\alpha)}.$$

□

To complete the proof of Hörmander's theorem in the one-dimensional case it is therefore enough to prove an inequality of the form (1.4). This will be accomplished by the following integral identity. The general case of this basic identity (that we will see in Lecture 3) is the other key idea in the L^2-theory.

Proposition 1.1.2. *Let Ω be a domain in $\mathbb{C}$ and let $\phi \in C^2(\Omega)$. Let $\alpha \in C_c^2(\Omega)$. Then*

$$\int \Delta\phi|\alpha|^2 e^{-\phi} + \int |\frac{\partial}{\partial \bar{z}}\alpha|^2 e^{-\phi} = \int |\bar{\partial}_\phi^* \alpha|^2 e^{-\phi} \tag{1.7}$$

PROOF. Since α has compact support we can integrate by parts and get

$$\int |\bar{\partial}_\phi^*\alpha|^2 e^{-\phi} = \int \bar{\partial}\bar{\partial}_\phi^*\alpha \cdot \bar{\alpha} e^{-\phi}.$$

Next note that

$$\bar{\partial}_\phi^*\alpha = -\frac{\partial}{\partial z}\alpha + \phi_z \alpha,$$

so that

$$\bar{\partial}\bar{\partial}_\phi^*\alpha = -\Delta\alpha + \phi_z \frac{\partial}{\partial \bar{z}}\alpha + \Delta\phi\alpha = \bar{\partial}_\phi^* \frac{\partial}{\partial \bar{z}}\alpha + \Delta\phi\alpha.$$

Hence

$$\int |\bar{\partial}_\phi^*\alpha|^2 e^{-\phi} = \int \Delta\phi|\alpha|^2 e^{-\phi} + \int |\frac{\partial}{\partial \bar{z}}\alpha|^2 e^{-\phi}$$

and the proof is complete. □

Combining the last two propositions we now immediately conclude

Theorem 1.1.3. *Let Ω be a domain in $\mathbb{C}$ and suppose $\phi \in C^2(\Omega)$ satisfies $\Delta\phi \geq 0$. Then, for any f in $L^2_{loc}\Omega$ there is a solution u to $\frac{\partial}{\partial \bar{z}}u = f$ satisfying*

$$\int |u|^2 e^{-\phi} \leq \int \frac{|f|^2}{\Delta\phi} e^{-\phi}.$$

Note that the theorem says two things: the $\bar{\partial}$-equation can be solved, and there is a good estimate for the solution. If we disregard the first aspect we get as a corollary the following Poincaré type inequality for the $\bar{\partial}$-operator.

Corollary 1.1.4. *Suppose $\phi \in C^2(\Omega)$ satisfies $\Delta\phi \geq 0$. Let u be a C^1 function in a domain Ω such that*

$$\int_\Omega u\bar{h}e^{-\phi} = 0$$

for any holomorphic function h in $L^2(\Omega, e^{-\phi})$. Then

$$\int_\Omega |u|^2 e^{-\phi} \leq \int_\Omega \frac{|\bar{\partial}u|^2}{\Delta\phi} e^{-\phi}. \tag{1.8}$$

PROOF. The previous theorem says that the equation

$$\bar{\partial}v = \bar{\partial}u$$

has *some* solution satisfying (0.8). But, the condition that u is orthogonal to all holomorphic functions means that u is the minimal solution to this equation. Hence u satisfies the estimate as well which is what the corollary claims. □

1.2. An alternative proof of the basic identity

We have proved the basic identity using integration by parts. We will here give an alternative proof, akin to the classical *Bochner method* from differential geometry, which gives a more general statement. This method can be generalized in a rather surprising way to the case of higher dimensions and complex manifolds, and we will get back to it in Lecture 3.

The idea is to calculate the Laplacian of the weighted norm of a test function α. Again we use the complex Laplacian

$$\Delta = \frac{\partial^2}{\partial z \partial \bar z}.$$

To compute $\Delta|\alpha|^2 e^{-\phi}$ we will use the product rule in the form

$$\frac{\partial}{\partial z}(u \cdot \bar v e^{-\phi}) = \delta u \cdot \bar v e^{-\phi} + u \cdot \overline{\frac{\partial}{\partial \bar z} v} e^{-\phi},$$

where

$$\delta := e^{\phi} \frac{\partial}{\partial z} e^{-\phi} = \frac{\partial}{\partial z} - \phi_z = -\bar\partial^*_\phi,$$

and a similar computation for $\frac{\partial}{\partial \bar z}$. Applying this twice we obtain

$$\Delta|\alpha|^2 e^{-\phi} = \alpha \cdot \overline{\frac{\partial}{\partial \bar z}\delta \alpha} e^{-\phi} + \delta \frac{\partial}{\partial \bar z}\alpha \cdot \bar\alpha e^{-\phi} + |\frac{\partial}{\partial \bar z}\alpha|^2 e^{-\phi} + |\delta\alpha|^2 e^{-\phi}.$$

Then we apply the commutation rule

$$\delta \frac{\partial}{\partial \bar z} = \frac{\partial}{\partial \bar z}\delta + \Delta\phi$$

and obtain

$$\Delta|\alpha|^2 e^{-\phi} = 2\Re \frac{\partial}{\partial \bar z}\delta\alpha \cdot \bar\alpha e^{-\phi} + |\frac{\partial}{\partial \bar z}\alpha|^2 e^{-\phi} + |\delta\alpha|^2 e^{-\phi} + \Delta\phi|\alpha|^2 e^{-\phi}. \tag{1.9}$$

Integrating this over a domain containing the support of α, and recalling that $\delta = -\bar\partial^*_\phi$ we obtain the basic identity.

1.3. An application: Inequalities of Brunn-Minkowski type

In this section we will give an application of the corollary, or rather of an even simpler, real variable variant of it.

Proposition 1.3.1. *Let ϕ be a convex function on $\mathbb{R}$ of class C^2 with strictly positive second derivative. Let u be a function of class C^1 such that*

$$\int_{\mathbb{R}} u e^{-\phi} = 0.$$

Then

$$\int_{\mathbb{R}} |u|^2 e^{-\phi} \le \int_{\mathbb{R}} \frac{|u'|^2}{\phi''} e^{-\phi}.$$

Note that this is formally similar to the the statement of Corollary 1.1.4. We have replaced a subharmonic ϕ by a convex ϕ, and require u to be orthogonal to constants (elements of the kernel of d) instead of orthogonal to holomorphic functions (elements of the kernel of $\bar\partial$).

Exercise: Prove the proposition imitating the proof of Corollary 1.0.4. □

We shall now use this proposition to prove a generalization of the Brunn Minkowski theorem.

Theorem 1.3.2. *Let $\phi(t,x)$ be a convex function on $\mathbb{R}^m_t \times \mathbb{R}^n_x$. Define the function $\tilde{\phi}(t)$ by*

$$e^{-\tilde{\phi}(t)} = \int_{\mathbb{R}^n} e^{-\phi(t,x)} dx.$$

Then $\tilde{\phi}$ is convex.

PROOF. We start by a few reductions. By Fubini's theorem, we may assume that $n = 1$. Since convexity means convexity on any line, we may also assume that $m = 1$. Now

$$\tilde{\phi}(t) = -\log \int e^{-\phi(t,x)} dx.$$

Differentiating once with respect to t we get

$$\tilde{\phi}' = \frac{\int \phi'_t e^{-\phi(t,x)} dx}{\int e^{-\phi(t,x)} dx},$$

and differentiating once more

$$\tilde{\phi}'' = \frac{\int (\phi''_{t,t} - (\phi'_t)^2) e^{-\phi} \int e^{-\phi} + (\int \phi'_t e^{-\phi})^2}{(\int e^{-\phi})^2}$$

Let a be the mean value

$$a := \frac{\int \phi'_t e^{-\phi}}{\int e^{-\phi}}.$$

Then the expression for the second derivative simplifies to

$$\tilde{\phi}'' = \frac{\int (\phi''_{t,t} - (\phi'_t - a)^2) e^{-\phi}}{\int e^{-\phi}}.$$

It is now time to use the inequality in the proposition. Since $u := \phi'_t - a$ by construction has integral 0 against the weight $e^{-\phi}$, the proposition shows that

$$\tilde{\phi}'' \geq \frac{\int (\phi''_{t,t} - (\phi''_{t,x})^2 / \phi''_{x,x}) e^{-\phi}}{\int e^{-\phi}}.$$

Since ϕ is convex

$$\phi''_{t,t} - (\phi''_{t,x})^2 / \phi''_{x,x} = (\phi''_{t,t}\phi''_{x,x} - (\phi''_{t,x})^2) / \phi''_{x,x} \geq 0.$$

This completes the proof. □

Theorem 1.2.2 was first given by Prekopa, [**18**], but the proof we have given is essentially due to Brascamp and Lieb, [**6**]. The theorem is a functional form of the Brunn Minkowski inequality, which can be stated as follows.

Theorem 1.3.3. *Let D be a convex open set in $\mathbb{R}^m_t \times \mathbb{R}^n_x$, and let D_t be the slices*

$$\{x; (t,x) \in D\}.$$

Let $|D_t|$ be the Lebesgue measure of D_t. Then

$$\log \frac{1}{|D_t|}$$

is convex.

PROOF. Take ϕ to be the convex function that equals 0 in D and ∞ outside D in the previous result. Admittedly, this is not a classical convex function, but it can be written as an increasing limit of smooth convex functions, to which the proof above applies. □

There is another, perhaps more common, way of stating the Brunn Minkowski theorem. It says that if D_0 and D_1 are convex open sets in $\mathbb{R}^n$, then the volume of their Minkowski sum $tD_1 + (1-t)D_0$ satisfies

$$|tD_1 + (1-t)D_0|^{1/n} \geq t|D_1|^{1/n} + (1-t)|D_0|^{1/n},$$

if t lies between 0 and 1. This formulation can be obtained from the ("multiplicative") form we have given above in the following way.
(1) There is a convex open set D in $\mathbb{R}_t \times \mathbb{R}^n$ such that

$$tD_1 + (1-t)D_0 = D_t.$$

(2) Theorem 1.3.3 implies trivially that

$$|D_t| \geq \min(|D_0|, |D_1|).$$

(3) Hence, if A and B are convex open sets

$$|A+B|^{1/n} = |tA/t + (1-t)B/(1-t)|^{1/n} \geq \min(|A|^{1/n}/t, |B|^{1/n}/(1-t)).$$

Choosing

$$t = \frac{|A|^{1/n}}{|A|^{1/n} + |B|^{1/n}}$$

we then get

$$|A+B|^{1/n} \geq |A|^{1/n} + |B|^{1/n},$$

which is the usual ("additive") form of the inequality.

The multiplicative form has the advantage that it applies to many other measures besides Lebesgue measure. The additive statement — $-|D_t|^{1/n}$ is convex — is formally stronger, but in the presence of the natural homogeneity of Lebesgue measure the two statements are equivalent.

Note that the main point in the proof of Prekopa's theorem was the one-dimensional case — the general case then followed by a simple induction. This is clearly not the case for Brunn Minkowski; there the one-dimensional case does not say very much!

1.4. Regularity — a disclaimer

In the previous discussion we have interpreted the $\bar\partial$-equation in the weak, or distributional sense. Thus,

$$\frac{\partial u}{\partial \bar z} = f$$

has been interpreted as saying that

$$-\int u \frac{\partial}{\partial \bar z}\alpha = \int f\alpha$$

for any smooth α of compact support. If f is smooth, this implies that u is also smooth and that the $\bar\partial$-equation is also satisfied in the classical pointwise sense. This is a consequence of the classical *Weyl's lemma*, and holds also in several variables (at least if f is of bidegree $(0,1)$). We will not discuss these issues further in the notes; all the $\bar{\partial}$-equations we deal with are to be interpreted in the sense of distributions.

LECTURE 2

Functional analytic interlude

In this lecture we begin to look at the $\bar\partial$-equation in higher dimension. For the moment we shall think of Ω as a domain in $\mathbb{C}^n$, and consider the $\bar\partial$-equation when the right hand side is a $(0,1)$-form, but this is for motivational purposes only — the formalism we develop will later apply also to the case of complex manifolds and general bidegrees.

2.1. Dual formulation of the $\bar\partial$-problem

Denote by $D_{(0,1)}$ the class of $(0,1)$-forms whose coefficients are, say, of class C^2 with compact support in Ω. If f and α are $(0,1)$-forms we denote by $f \cdot \bar\alpha$ their pointwise scalar product, i e

$$f \cdot \bar\alpha = \sum f_j \bar\alpha_j.$$

The equation $\bar\partial u = f$, explicitly

$$\frac{\partial u}{\partial \bar z_j} = f_j,$$

taken in the sense of distributions, means that

$$\int f \cdot \alpha = -\int u \sum \frac{\partial \alpha_j}{\partial \bar z_j}, \tag{2.1}$$

for all $\alpha \in D_{(0,1)}$. Just like in the one-dimensional case we replace α by $\bar\alpha e^{-\phi}$ (where ϕ is a C^2-function which will later be chosen to be plurisubharmonic). The condition (2.1) is then equivalent to

$$\int f \cdot \bar\alpha e^{-\phi} = \int u \overline{\bar\partial^*_\phi \alpha} e^{-\phi} \tag{2.2}$$

for all $\alpha \in D_{(0,1)}$, where

$$\bar\partial^*_\phi \alpha = -e^{\phi} \sum \frac{\partial}{\partial z_j}(e^{-\phi}\alpha_j).$$

Assume now that we can find a solution, u, to $\bar\partial u = f$, satisfying

$$\int |u|^2 e^{-\phi} \le C.$$

Then (2.2) implies

$$|\int f \cdot \bar\alpha e^{-\phi}|^2 \le C \int |\bar\partial^*_\phi \alpha|^2 e^{-\phi}.$$

The next proposition says that the converse of this also holds.

Proposition 2.1.1. *There is a solution, u, to the equation $\bar\partial u = f$ satisfying*

$$\int |u|^2 e^{-\phi} \leq C. \tag{2.3}$$

if and only if the inequality

$$|\int f \cdot \bar\alpha e^{-\phi}|^2 \leq C \int |\bar\partial^*_\phi \alpha|^2 e^{-\phi}. \tag{2.4}$$

holds for all $\alpha \in D_{(0,1)}$.

PROOF. It only remains to prove that (2.4) implies that there is a solution to the $\bar\partial$-equation satisfying (2.3). This is done precisely as in the one-dimensional case (cf first part of Proposition 1.0.1). □

To prove inequality (2.4) one might, as in the one dimensional case, first try to prove an inequality of the form

$$\int |\alpha|^2 e^{-\phi} \leq C \int |\bar\partial^*_\phi \alpha|^2 e^{-\phi}.$$

The main problem in higher dimensions is that no such inequality can hold. Indeed, if it did, then we would be able to solve $\bar\partial u = f$, even when f does not satisfy the compatibility condition $\bar\partial f = 0$ — which is clearly not possible. Thus we must somehow feed this information, $\bar\partial f = 0$, into the method. This requires a little bit more of functional analysis.

Let us consider a linear operator T between two Hilbert spaces H_1 and H_2. Assume that T is *closed* and densely defined. The first condition means that the graph of T

$$\{(x, Tx); x \in \mathrm{Dom}(T)\}$$

is a closed linear subspace of the product $H_1 \times H_2$, or, in more concrete terms, that if x_n lie in the domain of T and converge to x, and if moreover Tx_n converge to y, then x lies in the domain of T and $Tx = y$. The main example to think of for the moment is $T = \bar\partial$,

$$H_1 = L^2(\Omega, e^{-\phi}),$$

the space of functions that are square integrable against the weight $e^{-\phi}$ and

$$H_2 = L^2_{(0,1)}(\Omega, e^{-\phi}),$$

the space of $(0,1)$-forms that are square integrable against the weight $e^{-\phi}$. In this example, the domain of T is taken to be the space of all functions u in H_1 such that $\bar\partial u$ taken in the sense of distributions lies in H_2.

Exercise: Show that in this example T is closed and densely defined. □

The next proposition is the Hilbert space version of the theorem that a linear operator between finite dimensional vector spaces is surjective if and only if its adjoint is injective.

Proposition 2.1.2. *T is surjective if and only if there is a constant $c > 0$ such that for all y in the domain of T^**

$$c|y|^2 \leq |T^* y|^2. \tag{2.5}$$

More generally, let F be a closed subspace of H_2 containing the range of T. Then T is surjective onto F if and only if (2.5) holds for y in F intersected with the domain of T^*. In that case, for any y in F there is an x in H_1 such that

$$Tx = y$$

and

$$|x|^2 \leq \frac{1}{c}|y|^2.$$

PROOF. First assume 2.5 holds for all y in the domain of T^*. Take z in H_2 and define an antilinear functional on the range of T^* by

$$L(T^*y) = (z, y).$$

Our hypothesis implies that L is well defined and of norm at most $|z|\sqrt{c}$. By the Hahn-Banach theorem, L extends to a linear operator with the same norm on all of H_1, and by the Riesz representation theorem there is an element in H_1 of norm at most $|z|\sqrt{c}$ such that

$$(z, y) = L(T^*y) = (x, T^*y).$$

Since $T^{**} = T$ (this is a nontrivial statement!) it follows that $Tx = z$.

Suppose conversely that T is surjective. By the open mapping theorem there is then a constant so that for any y in H_2 we can solve $Tx = y$ with

$$|x|^2 \leq \frac{1}{c}|y|^2.$$

If y lies in the domain of T^*

$$|y|^2 = (y, Tx) = (T^*y, x) \leq \sqrt{\frac{1}{c}}|T^*y||y|,$$

so (2.5) holds.

Let now F be a closed subspace of H_2 containing the range of T. We leave as an exercise to prove that that the adjoint of T considered as a map from H_1 to F is the restriction of T^* to F. Hence the last part of the proposition follows from the first part. □

Notice that the first part of the proof is virtually identical to the proof of Proposition 1.1.1. In the application that we have in mind the space F will of course consist of the subspace of $\bar\partial$-closed forms — which is closed since differentiation is a continuous operation in distribution theory.

What we have gained with this proposition is that to prove solvability of the $\bar\partial$-equation we need only prove an inequality of type

$$\int |\alpha|^2 e^{-\phi} \leq C \int |\bar\partial_\phi^* \alpha|^2 e^{-\phi},$$

when α is a *closed* form. We will then automatically also get estimates for the solution. But, we have also lost something. In the one dimensional case we could work all the time with α smooth and of compact support; now we have to deal with forms in the domain of a rather abstract operator: the Hilbert space adjoint of $\bar\partial$. We will later overcome this complication by approximating a general element in the domain of T^* by smooth forms with compact support. This is a somewhat delicate business which we will, following Demailly [**8**], handle by introducing complete Kähler metrics. We illustrate the issue involved in the next subsection in a very simple model example.

2.1.1. The role of completeness in a simple model example

Let us consider the differential equation

$$\frac{du}{dx} = f$$

on an open interval I in $\mathbb{R}$. To imitate the weighted estimates for the $\bar\partial$-equation that we are dealing with we will study estimates for norms of the solution in the Hilbert space H with norm

$$\|u\|^2 = \int_I u^2 e^{-x^2/2}.$$

We consider the operator

$$u \mapsto \frac{du}{dx} =: Tu$$

as a closed densely defined operator and we let the domain of T consist of all functions u in $L^2(I, e^{-x^2/2})$ such that du/dx in the sense of distributions lie in $L^2(I, e^{-x^2/2})$.

The *formal adjoint* of the differential operator d/dx is the operator ϑ defined by

$$\int_I (du/dx) v e^{-x^2/2} = \int_I u \vartheta v e^{-x^2/2}$$

for all smooth u and v of compact support. Simple integration by parts shows that

$$\vartheta v = -dv/dx + xv.$$

On the other hand, the Hilbert space adjoint of T is the operator T^* satisfying

$$(Tu, v)_H = (u, T^* v)_H$$

for all u in the domain of T. This should be interpreted as saying that v lies in the domain of T^* if there is some w in H satisfying

$$(Tu, v)_H = (u, w)_H$$

and $T^* v$ is then equal to w (which is uniquely determined since the domain of T is dense).

There are now two main cases: $I = \mathbb{R}$ (a complete manifold) or $I \neq \mathbb{R}$ (not complete). Assume first that $I = \mathbb{R}$. Then if v lies in the domain of T^* and we take u to be smooth with compact support, we see that

$$T^* v = \vartheta v$$

where the right hand side is taken in the sense of distributions. Hence ϑv must lie in H and must be equal to $T^* v$ if v lies in the domain of T^*. Conversely if v and ϑv lies in H, and $\chi_k(x) = \chi(x/k)$ is a sequence of cut-off function tending to 1, we get

$$(u, \vartheta v)_H = \lim(\chi_k u, \vartheta v)_H = \lim(\chi_k' u, v)_H + (Tu, v)_H = (Tu, v)_H,$$

(since χ_k' goes to zero).

Hence we have a precise description of T^*: $T^* v = \vartheta v$ and its domain consist of functions for which ϑv is in H. Moreover, if v lies in the domain of T^* then $\chi_k v$ tends to v and $T^* \chi_k v$ tends to $T^* v$, so to prove the crucial inequality

$$\|v\|^2 \leq C \|T^* v\|^2$$

one may assume that v has compact support. In fact one may also assume that v is smooth, since we can achieve this by taking convolutions with a sequence of smooth functions tending to the Dirac measure.

Exercise: Assuming that v is smooth with compact support, prove that

$$\|v\|^2 \leq \|\vartheta v\|^2$$

□

Let us now take $I = (0, 1)$ instead. As before we see that if v lies in the domain of the adjoint, then $T^* v = \vartheta v$ and of course this expression must lie in H then. However, this is no longer sufficient to be in the domain of T^*, not even if v is smooth up to the boundary.

Exercise: Show that if v is smooth on $[0, 1]$ then v lies in the domain of T^* if and only if $v(0) = v(1) = 0$. □

Moreover it is in this case also less evident how to approximate an element in the domain with functions of compact support, and also how to regularize by taking convolutions. Any sequence of cut-off functions with compact support in I that tend to 1 must have unbounded derivatives, and there is also the problem to take convolutions near the boundary. These problems are certainly possible to overcome (and it *is* possible to solve the equation $du/dx = f$ on the unit interval), but the fact that they arise serve to illustrate the advantage of working with complete metrics. In the case of the $\bar{\partial}$-equation we shall see later that we may introduce complete (Kähler) metrics precisely in the domains where we expect to be able to solve the $\bar{\partial}$-equation.

LECTURE 3

The $\bar{\partial}$-equation on a complex manifold

In this section we will discuss the $\bar{\partial}$-equation on an n-dimensional complex manifold X. We start by discussing the linear algebra of forms and metrics on complex manifolds, and then apply this formalism to derive the basic integral identity for $\bar{\partial}$ for forms with values in a line bundle (Proposition 3.4.1). In many texts this is done via the Kodaira-Nakano identity for associated Laplace operators, but we have chosen another route: Siu's $\partial\bar{\partial}$-Bochner-Kodaira method that generalizes the 'alternative proof' in the one dimensional case from Section 1.1. This formalism is particularly efficient for forms of bidegree (n, q). It avoids the use of the so called 'Kähler identities' completely and also gives a stronger result that will be of use later when we discuss extension of holomorphic sections from subvarieties.

3.1. Metrics

The first thing we have to do is to find convenient expressions for norms of forms that will enable us to compute the adjoint of $\bar{\partial}$. For this we need a hermitian metric on X. This is by definition a hermitian scalar product on each complex tangent space

$$T_p^{1,0}(X).$$

If $z = (z_1, ...z_n)$ are local coordinates near p, then

$$\frac{\partial}{\partial z_1}, \dots \frac{\partial}{\partial z_n}$$

at p form a basis for $T_p^{0,1}(X)$ with dual basis

$$dz_1, ...dz_n$$

for the complex cotangent space. With respect to this basis the metric is given by a hermitian metric $g = (g_{j,k})$, so that if

$$v = \sum v_j \frac{\partial}{\partial z_j},$$

then

$$|v|^2 = \sum g_{j,k} v_j \bar{v}_k.$$

We also get an induced metric on the cotangent space, such that if

$$\eta = \sum \eta_j dz_j,$$

then

$$|\eta|^2 = \sum g^{j,k} \eta_j \bar{\eta}_k.$$

The dual metric is defined so that

$$|\eta|^2 = \sup_{|v| \leq 1} |\eta(v)|^2.$$

Exercise: Show that $(g^{j,k})^* = (g_{j,k})^{-1}$. □

It is assumed that $g(p)$ varies smoothly with p. Formally,

$$g = \sum g_{j,k} dz_j \otimes d\bar z_k,$$

but it turns out to be more convenient to let the metric be represented by its *Kähler form*

$$\omega = \omega_g = i \sum g_{j,k} dz_j \wedge d\bar z_k.$$

One says that g is a Kähler metric if $d\omega = 0$. Obviously the euclidean metric with Kähler form

$$\beta = i \sum dz_j \wedge d\bar z_j$$

is Kähler. One can always choose local coordinates so that $\omega = \beta$ to order 0 at any given point. We say that the local coordinates z are normal at a point p if $\omega = \beta$ to first order at p. More precisely

$$g_{j,k} = \delta_{j,k}$$

and

$$dg_{j,k} = 0$$

at p. Clearly, if we can choose normal coordinates at any point, then $d\omega = 0$, so the metric must be Kähler.

Exercise: Verify that the converse to this also holds. (Hint: This is a bit tricky. Start from coordinates whose differentials are orthonormal at the point. Then look for new coordinates of the form

$$\zeta_j = z_j + Q_j(z)$$

where Q_j are quadratic in z.) □

A metric determines a volume element on the manifold X. This is a differential form of maximal degree that can be written

$$dV_\omega = i^n \xi_1 \wedge \bar\xi_1 \wedge ... \xi_n \wedge \bar\xi_n$$

if ξ is an orthonormal basis for the holomorphic cotangent space. If we take another orthonormal basis it is related to the first one via a unitary linear transformation. Then the expression on the right hand side above gets multiplied by the modulus square of the determinant of the unitary transformation, i.e. by 1. Hence the volume element is well defined. In terms of the Kähler form

$$\omega = i \sum g_{j,k} dz_j \wedge d\bar z_k = i \sum \xi_j \wedge \bar\xi_j,$$

$$dV_\omega = \omega^n/n!.$$

This formula is very convenient in computations. Note that there is nothing (at least nothing obvious) that corresponds to this for real manifolds. We will introduce the notation

$$\alpha_p := \alpha^p/p!$$

for expressions of this kind. If Y is a complex submanifold of dimension p, then the induced metric on Y has Kähler form ω restricted to Y, i.e. the pullback of ω to Y under the inclusion map. Hence the volume element on the submanifold is simply

$$\omega_p$$

restricted to Y.

Exercise: Show that this implies: If Y is a complex curve, i.e. a complex submanifold of dimension 1 in $\mathbb{C}^n$, then the area of Y equals the sum of the areas of the projections of Y onto the coordinate axes (counted with multiplicities).

How can this be geometrically possible? □

Exercise*: In general, the volume element on an oriented Riemannian manifold M, can be defined as

$$dV_M = \xi_1 \wedge ...\xi_n,$$

where ξ_j form an oriented orthonormal basis for the space of real one-forms. Show that if Y is a submanifold of $\mathbb{C}^n$ of real dimension 2, then

$$dV_Y \geq \omega|Y$$

with equality (if and) only if Y is a complex curve. This is *Wirtinger's inequality.* Deduce that if Y and Y' are such submanifolds with the same boundary and Y' is complex, the Y' has area smaller than or equal to that of Y. (Hint: It is enough to prove Wirtinger's inequality for a linear subspace. Choose complex coordinates on $\mathbb{C}^n$ so that $\frac{\partial}{\partial x_1}$ and $a\frac{\partial}{\partial y_1}$ + remainder form an orthonormal basis for the tangent space.) □

3.2. Norms of forms

We have already said that if $\eta = \sum \eta_j dz_j$ is a $(1,0)$-form, and dz_j are orthonormal at a point, then

$$|\eta|^2 = \sum |\eta_j|^2.$$

We would now like to define the norm of a form of arbitrary bidegree by saying that if

$$\eta = \sum \eta_{I,J} dz_I \wedge d\bar z_J,$$

then

$$|\eta|^2 = \sum |\eta_{I,J}|^2, \tag{3.1}$$

if dz_i is orthonormal. It is however not crystal clear that this definition is independent of the choice of orthonormal basis. Fortunately there is a simple way out of this, at least for forms of bidegree $(p,0)$ or $(0,q)$.

Note first that if dz_j are orthonormal and we denote by

$$dV_J = \bigwedge_{j\in J} i dz_j \wedge d\bar z_j,$$

then

$$\omega_p = \sum dV_J,$$

the sum running over multiindices J of length p. Let η be a form of bidegree $(p,0)$. Then we can define the norm of η by

$$|\eta|^2 dV_\omega = c_p \eta \wedge \bar\eta \wedge \omega_{n-p},$$

where $c_p := i^{p^2}$ is a unimodular constant chosen to make the right hand side positive.

This constant, $c_p = i^{p^2}$ will appear again and again in the sequel, so it might be good to pause and see what it comes from. Say first that η is of the form

$$\eta = a_1 \wedge a_2 \wedge ...a_p,$$

where a_j are differentials dz_j. Then

$$(-1)^{p(p-1)/2}\eta\wedge\bar\eta = a_1\bar a_1\wedge ...a_p\wedge\bar a_p.$$

But $(-1)^{p(p-1)/2} = i^{p(p-1)} = i^{p^2}i^{-p}$, so

$$i^{p^2}\eta\wedge\bar\eta = ia_1\bar a_1\wedge ...ia_p\wedge\bar a_p,$$

which is a positive form (after wedging with ω_{n-p}). For general η we get a sum of such terms, so the same choice of c_p will do.

This definition is independent of choice of basis and coincides with (3.1), so (3.1) is indeed independent of choice of basis if η is of bidegree $(p,0)$. Of course, norms on forms of bidegree $(0,q)$ are defined similarly so that

$$|\eta|^2 = |\bar\eta|^2.$$

We shall next motivate why formula (3.1) is independent of choice of ON-basis for forms of general bidegree (p,q). Fix one choice of ON-basis dz_i and define the scalar product by (3.1). Then, if $\mu=\sum\mu_J dz_j$ is of bidegree $(p,0)$ and $\xi=\sum\xi_K d\bar z_K$ is of bidegree $(0,q)$ it follows that

$$\|\mu\wedge\xi\|^2 = \sum|\mu_J|^2|\xi_K|^2 = \|\mu\|^2\|\xi\|^2,$$

and polarizing that

$$(\mu\wedge\xi,\mu'\wedge\xi') = (\mu,\mu')(\xi,\xi').$$

From this it follows that if $\eta=\sum\eta_{J,K}dw_J\wedge d\bar w_K$

$$\|\eta\|^2 = \sum\eta_{JK}\bar\eta_{LM}(dw_J,dw_L)(d\bar w_K,d\bar w_M).$$

But our previous discussion for $(p,0)$-forms shows that

$$(dw_J,dw_K) = \delta_{JK}$$

so (3.1) holds for any ON-basis.

We will later have use for a special formula for forms of bidegree (n,q).

Lemma 3.2.1. *Let η be a form of bidegree (n,q). Then there is a unique form γ_η of bidegree $(n-q,0)$ such that*

$$\eta = \gamma_\eta\wedge\omega_q.$$

This follows immediately from a computation in ON-coordinates. Note that if in ON-coordinates

$$\eta = \sum\eta_J dz\wedge d\bar z_J,$$

γ_η equals

$$\sum\epsilon_J\eta_J dz_{J^c},$$

where ϵ_J are unimodular constants. Consequently

$$|\eta|^2 = |\gamma_\eta|^2.$$

In terms of Riemannian geometry, γ_η is the Hodge-* of η, again up to a unimodular constant. We have

$$|\eta|^2 dV_\omega = c_{n-q}\eta\wedge\bar\gamma_\eta.$$

Polarizing we get

$$(\eta,\xi)dV_\omega = c_{n-q}\eta\wedge\bar\gamma_\xi, \tag{3.2}$$

a formula that will be very helpful later to compute adjoints.

3.3. Line bundles

Formally, a line bundle L over a complex manifold X is a complex manifold fibered over X

$$p : L \to X$$

in such a way that all fibers $L_x = p^{-1}(x)$ are complex lines. One also requires that the fibration be *locally trivial*, so that each point in X has a neighbourhood U with $p^{-1}(U)$ isomorphic to $U \times \mathbb{C}$ via a fiber preserving holomorphic map

$$g_U : p^{-1}(U) \to U \times \mathbb{C},$$

which is linear on each fiber. If V is another trivializing open set, we then get a map

$$g_U g_V^{-1} : U \cap V \times \mathbb{C} \to U \cap V \times \mathbb{C}.$$

Since this map is fiber preserving and linear on fibers, it must be given as

$$g_U g_V^{-1}(z, \xi) = (z, g_{UV}(z)\xi)$$

where g_{UV} is a zero free holomorphic function of z. The maps g_{UV} arising in this way are known as *transition functions* of the line bundle. Notice that they satisfy the *cocycle conditions*

$$g_{UV} g_{VU} = 1$$

and

$$g_{UV} g_{VW} g_{WU} = 1.$$

Conversely, given an open covering U_j of the base manifold X, and a corresponding collection of functions $g_{ij} = g_{U_i U_j}$ satisfying the cocycle conditions we can define a line bundle over X having these as transition functions. For this, one starts with the disjoint union

$$\bigcup_i U_i \times \mathbb{C}$$

and identifies (z, ξ) in $U_j \times \mathbb{C}$ with $(z, g_{ij}(z)\xi$ in $U_i \times \mathbb{C}$. A *section* of a line bundle is a map s from (a subset of) X to L satisfying

$$p \circ s(z) = z.$$

In terms of the local trivializations, s is given by a collection of scalar functions

$$s_i = g_{U_i} s,$$

satisfying

$$s_i = g_{ij} s_j$$

A main reason to be interested in line bundles comes from the following example:

Example: Let $\mathbb{P}^n$ be n-dimensional projective space, i.e. the set of equivalence classes

$$[z_0, z_1 ... z_n] = [z]$$

where z lies in $\mathbb{C}^{n+1} \setminus \{0\}$ under the equivalence $[z] = [\lambda z]$ for λ in $\mathbb{C} \setminus \{0\}$. Denote by π the natural projection map from $\mathbb{C}^{n+1} \setminus \{0\}$ to $\mathbb{P}^n$, $\pi(z) = [z]$. Let

$$U_i = \{[z]; z_i \neq 0\}$$

Then each U_i is isomorphic to $\mathbb{C}^n$, e g $U_0 = \{[1, \zeta]; \zeta \in \mathbb{C}^n\}$. We now define, for each integer k, the line bundles $\mathcal{O}(k)$, using the transition functions

$$g_{ij} = (z_j / z_i)^k.$$

A section of $\mathcal{O}(k)$ is then a collection of functions s_i on U_i such that

$$z_j^k s_j = z_i^k s_i$$

on overlaps. Then $h := z_i^k s_i \circ \pi$ is a globally defined k-homogeneous function on $\mathbb{C}^{n+1} \setminus \{0\}$, and working backwards we conversely see that any k-homogeneous function gives rise to a section of $\mathcal{O}(k)$. For k nonnegative it follows that the global holomorphic sections extend across the origin (they are bounded near 0) and therefore correspond to homogeneous polynomials of degree k (their Taylor expansion near 0 is homogeneous). Such a polynomial is of course uniquely determined by its restriction to the dense set U_0, where it can be thought of as a polynomial on $\mathbb{C}^n$ of degree k (put $z_0 = 1$). For k negative there are no global holomorphic sections. □

This example has much wider scope than is visible at first sight. Any line bundle over a complex manifold restricts to a line bundle over submanifolds. Hence $\mathcal{O}(k)$ restricts to a line bundle over any submanifold of $\mathbb{P}^n$, and the bundles we obtain in this way have at least the sections that come from restricting homogeneous polynomials.

Example: The *canonical bundle* K_X of a complex manifold X is a line bundle whose sections are forms of bidegree $(n, 0)$. Locally such section can be written

$$s = s_i dz_1^i \wedge ...dz_n^i =: s_i dz^i,$$

where $(z_1^i, ...z_n^i)$ are local coordinates. If

$$g_{ij} dz^i = dz^j,$$

then g_{ij} are transition functions for the canonical bundle. □

Exercise: Prove that on $\mathbb{P}^n$ the canonical bundle is equal to $\mathcal{O}(-(n+1))$. □

A (classical) metric on a line bundle is a smoothly varying norm on each fiber, which thus enables us to define the norm $\|s\|$ of any section of the bundle. Under the representation of s as a collection of local functions s_i, the metric becomes represented by a collection of smooth real valued functions ϕ_i so that

$$\|s\|^2 = |s_i|^2 e^{-\phi_i}.$$

Since $s_i = g_{ij} s_j$ on overlaps the local representatives for the metric must be related by

$$\log |g_{ij}|^2 = \phi_i - \phi_j$$

on overlaps. We will in the sequel denote metrics by e g ϕ, with the understanding that this means a collection of local functions ϕ_j related in this manner.

Note that if ϕ is a metric on L and χ is a function, then

$$\phi + \chi$$

is also a metric.

Exercise Show that if ϕ is some metric on L, all other metrics on L can be written $\phi + \chi$ where χ is a function. □

Since $\log |g_{ij}|^2$ is pluriharmonic, we see that the form

$$c(\phi) = \partial\bar\partial\phi_i$$

is globally defined (even though the ϕ_i:s are not global). By definition, $c(\phi)$ is the *curvature form* of the metric.

Example: Let $\rho_i\sqrt{-1}dz^i \wedge d\bar{z}^i$ be the local Kähler forms of a metric on a complex manifold X of dimension 1. Then

$$-\Delta \log \rho_i$$

is the classical Gaussian curvature of the corresponding Riemannian metric $\rho(dx^2 + dy^2)$, and $\partial\bar{\partial} - \log \rho$ is the curvature form. If $s = (s_i)$ is a section of K_X, then

$$|s_i|^2 dz^i \wedge d\bar{z}^i$$

is globally well defined, so $|s_i|^2/\rho_i$ is also global. Therefore

$$\rho_i = e^{\phi_i}$$

where ϕ is a metric on K_X and the curvature of this metric is the negative of the Gaussian curvature of the underlying manifold. (The minus sign comes from ϕ being a metric on K_X which is the dual of the tangent space.) In higher dimensions a Kähler metric on X also induces a metric on K_X. The curvature of this induced metric on K_X is the negative of the *Ricci curvature* of the Kähler metric on X. □

Definition: We say that a line bundle L equipped with a metric ϕ is *positive* if $ic(\phi)$ is a positive form (i.e. if all the local representatives ϕ_i are strictly plurisubharmonic). L itself is called positive (or equivalently *ample* in the algebraic geometry terminology) if it has some metric with positive curvature. □

Example/Exercise: We have seen that for the line bundles $\mathcal{O}(k)$ on $\mathbb{P}^n$ their global holomorphic sections are in one-one correspondence with functions on $\mathbb{C}^{n+1}$ that are homogeneous of degree k. Under this correspondence:

$$s \to z_i^k s_i \circ \pi := h$$

a metric on $\mathcal{O}(k)$ can be defined by

$$\|s\|^2 = |h|^2/|z|^{2k}.$$

(This defines $\|s\|$ as a function on $\mathbb{C}^{n+1}$ that is homogeneous of degree 0, i.e. a function on $\mathbb{P}^n$.) Show that if we represent this mètric by local functions ϕ_i then

$$\phi_0(1, \zeta) = k \log(1 + |\zeta|^2).$$

Hence $\mathcal{O}(k)$ is positive (ample) if $k > 0$. □

Example: Let D be a *divisor* in X. Intuitively, this is a hypersurface, every branch of which is endowed with certain multiplicities; positive or negative. D is called *effective* if all multiplicities are nonnegative. Locally, in an open set U, a divisor can be defined by a meromorphic function s_U in U which vanishes to the given multiplicity on every branch of D that intersects U. Another local holomorphic function s_V defines the same divisor if

$$g_{UV} : s_U/s_V$$

is holomorphic and zero free on $U \cap V$. Then, if U_i is a covering of X by open sets and $g_{ij} := g_{U_i U_j}$, g_{ij} the g_{ij} satisfy the cocycle condition so they form transition functions of a certain line bundle which is usually denoted (D). Any line bundle arising in this way from a divisor has a meromorphic section, and any line bundle arising from an effective divisor has a holomorphic section. □

If L and F are line bundles over X, we can form a new line bundle $L \otimes F$ by taking tensor products of the fibers. If g_{ij} and h_{ij} are the transition functions of L and F respectively (by passing to a common refinement we may assume they are defined w r t the same covering with open sets), then $g_{ij}h_{ij}$ are the transition functions of $L \otimes F$. If $\phi = (\phi_i)$ is a metric on L and $\psi = (\psi_i)$ is a metric on F, then $\phi + \psi$ is a metric on $L \otimes F$. For this reason one sometimes uses additive notation for the tensor product of vector bundles so that

$$L \otimes F := L + F.$$

We will use this convention (often) in the sequel and write in particular

$$L^{\otimes k} = kL$$

for the product of L with itself k times.

3.3.1. Forms with values in a line bundle

Recall that a section of a line bundle L can be thought of as a collection of functions (s_i), each defined in an open set U_i, where all the U_i together make up an open covering of the base manifold X. Moreover, the s_i are related by

$$s_i = g_{ij} s_j$$

where g_{ij} are the transition functions of the line bundle. What this means is that in each U_i we have a local basis element $e_i := g_{U_i}^{-1} 1$ for the fibers of L, and the section can be recovered from the local functions as

$$s = s_i \otimes e_i.$$

The norm of s with respect to a metric ϕ on L is then defined by

$$\|s\|^2 = |s_i|^2 \|e_i\|^2 = |s_i|^2 e^{-\phi_i},$$

so the local representatives ϕ_i of the metric satisfy

$$e^{-\phi_i} = \|e_i\|^2.$$

In the same way a differential form η with values in L is given by a collection of local forms η_i in U_i, such that

$$\eta_i = g_{ij} \eta_j,$$

and we can recover η from the local representatives as

$$\eta = \eta_i \otimes e_i.$$

Given a metric ϕ on L we can now define the norm of an L valued form by

$$\|\eta\|^2 = |\eta_i|^2 e^{-\phi_i},$$

where the norm of η_i is defined with respect to some metric ω on the underlying manifold. The transformation rules for η_i and ϕ_i under change of local trivialization then show that the norm is well defined. Polarizing we also get a scalar product on L-valued forms. Notice that expressions like

$$\eta_i \wedge \bar\xi_i e^{-\phi_i} =: \eta \wedge \xi e^{-\phi}$$

also become well defined for the same reason. We will write $\langle , \rangle$ for the scalar product on forms, so that for L-valued η and ξ

$$\langle \eta, \xi \rangle = (\eta_i, \xi_i) e^{-\phi_i}.$$

Note now that the $\bar\partial$-operator is well defined on sections of L and on forms with values L:

$$\bar\partial \eta = \bar\partial \eta_i \otimes e_i.$$

Since the transition functions are holomorphic this definition is independent of which local representative η_i that we choose. However we can not define the d or ∂

operator on L-valued forms in this way unless the transition functions are locally constant. Instead we define a differential operator of degree $(1,0)$ on L valued forms by

$$\delta\eta = (e^{\phi_i}\partial e^{-\phi_i}\eta_i)\otimes e_i.$$

Then the operator δ satisfies

$$\bar\partial(\eta\wedge\bar\xi e^{-\phi}) = \bar\partial\eta\wedge\bar\xi e^{-\phi} + (-1)^{\deg\eta}\eta\wedge\overline{\delta\xi}e^{-\phi}, \tag{3.3}$$

so, since the $\bar\partial$-operator is well defined, the δ-operator must be well defined as well.

Exercise: Show by hand that δ is independent of choice of local trivialization. □

Together δ and $\bar\partial$ make up the analog of the d-operator for L-valued forms

$$D = \delta + \bar\partial,$$

which is the *Chern connection* operator defined by the metric on L. Then

$$D^2 = \delta\bar\partial + \bar\partial\delta = -\bar\partial\partial\phi = c(\phi),$$

the curvature of the connection. It is instructive to compare this formalism to the calculations in Section 1.1.

3.4. Calculation of the adjoint and the basic identity

We now set out to determine the formal adjoint of the $\bar\partial$ operator on L-valued forms.We will be principally interested in the case of forms of bidegree (n,q). Recall that (Section 3.2) we have defined a map

$$\eta \to \gamma_\eta$$

mapping forms of bidegree (n,q) to forms of bidegree $(n-q,0)$, and that the scalar product on forms satisfies

$$(\eta,\xi)dV_\omega = c_{n-q}\eta\wedge\bar\gamma_\xi. \tag{3.4}$$

The formal adjoint of the $\bar\partial$ operator, $\bar\partial^*_\phi$ must, for any η of bidegree $(n,q-1)$, satisfy

$$\int\langle\bar\partial\eta,\xi\rangle dV_\omega = \int\langle\eta,\bar\partial^*_\phi\xi\rangle dV_\omega.$$

By (3.4), the left hand side equals

$$c_{n-q}\int\bar\partial\eta\wedge\overline{\gamma_\xi}e^{-\phi},$$

which by Stokes' theorem and formula (3.3) equals

$$c_{n-q}(-1)^{n-q}\int\eta\wedge\overline{\delta\gamma_\xi}e^{-\phi}.$$

The right hand side is

$$c_{n-q+1}\int\eta\wedge\overline{\gamma_{\bar\partial^*_\phi\xi}}e^{-\phi}.$$

Since $c_{n-q+1} = i(-1)^{n-q}c_{n-q}$ we see that $\bar\partial^*_\phi$ satisfies

$$\gamma_{\bar\partial^*_\phi\xi)} = i\delta\xi.$$

We will now use this description of the adjoint to derive a basic integral identity generalizing what we found in the one dimensional case. Let α be an L-valued form of bidegree (n, q). We will follow the method outlined in Section 1.1, but instead of considering the norm of the form α we will use α to define a differential form of bidegree $(n-1, n-1)$. (This method was introduced by Siu in [**19**].) In the one dimensional case an $(n-1, n-1)$-form is a function, and in that case our construction will repeat what we did in Section 1.1.

DEFINITION : Let α be a differential form of bidegree (n, q). Then

$$T_\alpha := c_{n-q}\gamma_\alpha \wedge \bar\gamma_\alpha \wedge \omega_{q-1} e^{-\phi},$$

where the unimodular constant $c_{n-q} = i^{(n-q)^2}$ is chosen so that $T_\alpha \geq 0$. □

Proposition 3.4.1. *Let α be a smooth (n, q)-form on X. Then, if ω is Kähler,*

$$i\partial\bar\partial T_\alpha = (-2\Re\langle\partial\bar\partial^*_\phi\alpha, \alpha\rangle + \|\bar\partial\gamma_\alpha\|^2 + \|\bar\partial^*_\phi\alpha\|^2 - \|\bar\partial\alpha\|^2)dV_\omega + ic(\phi) \wedge T_\alpha, \tag{3.5}$$

and, if α has compact support,

$$\int ic(\phi) \wedge T_\alpha + \int \|\bar\partial\gamma_\alpha\|^2 dV_\omega = \int \|\bar\partial\alpha\|^2 + \|\bar\partial^*_\phi\alpha\|^2 \tag{3.6}$$

PROOF. Note first that the second formula follows immediately from the first one, since the integral of the left hand side of (3.5) vanishes by Stokes' theorem and

$$\int \langle\bar\partial\bar\partial^*_\phi\alpha, \alpha\rangle dV_\omega = \int \|\bar\partial^*_\phi\alpha\|^2.$$

To prove (3.5) we shall basically use (3.3) twice, keeping in mind that $d\omega = 0$. (This is one place where we use the Kähler assumption. Notice however that for the perhaps most important case, $q = 1$, we don't need the Kähler assumption — here.) We then get

$$i\partial\bar\partial T_\alpha =$$

$$ic_{n-q}(\delta\bar\partial\gamma_\alpha \wedge \bar\gamma_\alpha \wedge \omega_{q-1} + \gamma_\alpha \wedge \overline{\bar\partial\delta\gamma_\alpha} \wedge \omega_{q-1} +$$

$$+(-1)^{n-q}\delta\gamma_\alpha \wedge \overline{\delta\gamma_\alpha} \wedge \omega_{q-1} + (-1)^{n-q+1}\bar\partial\gamma_\alpha \wedge \overline{\bar\partial\gamma_\alpha} \wedge \omega_{q-1})e^{-\phi}.$$

Now we use the commutation rule

$$\delta\bar\partial + \bar\partial\delta = c(\phi)$$

in the first term. The first two terms then combine to give the first and last two terms in (3.5). Moreover, by our formula for $\bar\partial^*_\phi$

$$ic_{n-q}(-1)^{n-q}\delta\gamma_\alpha \wedge \overline{\delta\gamma_\alpha} \wedge \omega_{q-1} = \|\bar\partial^*_\phi\alpha\|^2 dV_\omega.$$

(Again, it is useful to note that $c_{n-q+1} = i(-1)^{n-q}c_{n-q}$.) It only remains to analyse the term

$$\bar\partial\gamma_\alpha \wedge \overline{\bar\partial\gamma_\alpha} \wedge \omega_{q-1},$$

which even though it looks simple enough is actually the trickiest term. By the lemma below it equals

$$(\|\bar\partial\gamma_\alpha\|^2 - \|\bar\partial\alpha\|^2)dV_\omega,$$

since $\gamma_\alpha \wedge \omega_q = \alpha$ implies $\bar\partial\gamma_\alpha \wedge \omega_q = \bar\partial\alpha$ by the Kähler assumption (this is the only place where we use the Kähler assumption when $q = 1$). This finishes the proof.

□

Lemma 3.4.2. *Let ξ be a form of bidegree $(n-q,1)$. Then*

$$ic_{n-q}(-1)^{n-q-1}\xi\wedge\bar\xi\wedge\omega_{q-1}=(\|\xi\|^2-\|\xi\wedge\omega_q\|^2)dV_\omega.$$

PROOF. We give the proof for $q=1$, leaving the general case as an exercise for the reader. Since we are dealing with a pointwise formula we may choose a ON basis dz_j for the differentials. Write

$$\xi=\sum\xi_{jk}d\hat{z}_j\wedge d\bar{z}_k.$$

Here $d\hat{z}_j$ denotes the product of all differentials dz_i except dz_j, ordered so that

$$dz_j\wedge d\hat{z}_j=dz:=dz_1\wedge...dz_n.$$

Then one checks that

$$ic_{n-q}(-1)^{n-q-1}\xi\wedge\bar\xi\wedge\omega_{q-1}=\sum\xi_{jk}\bar\xi_{kj}dV_\omega.$$

On the other hand

$$\|\xi\wedge\omega\|^2=\sum_{j<k}|\xi_{jk}-\xi_{kj}|^2=\sum|\xi_{jk}|^2-\sum\xi_{jk}\bar\xi_{kj},$$

so

$$\sum\xi_{jk}\bar\xi_{kj}=\|\xi\|^2-\|\xi\wedge\omega\|^2.$$

□

Corollary 3.4.3. *Assume the curvature form of the metric ϕ is strictly positive so that*

$$ic(\phi)\geq c\omega$$

for some positive c. Then

$$qc\int\|\alpha\|^2dV_\omega+\int\|\bar\partial\gamma_\alpha\|^2dV_\omega\leq\int\|\bar\partial\alpha\|^2+\|\bar\partial_\phi^*\alpha\|^2 \tag{3.7}$$

PROOF. This follows from (3.6) since $ic(\phi)\geq c\omega$ implies

$$ic(\phi)\wedge T_\alpha\geq cT_\alpha\wedge\omega=qc\gamma_\alpha\wedge\bar\gamma_\alpha\wedge\omega_q=qc\|\alpha\|^2dV_\omega$$

□

3.5. The main existence theorem and L^2-estimate for compact manifolds

We are now ready to state and prove the main existence theorem for the $\bar\partial$-equation for positive line bundles over compact manifolds. Apart from the estimate for the solution obtained, this is the celebrated *Kodaira vanishing theorem*, [**12**].

Theorem 3.5.1. *Let L be a holomorphic line bundle endowed with a metric ϕ over a compact complex manifold X. Assume the metric ϕ has (strictly) positive curvature and that*

$$ic(\phi)\geq c\omega$$

where $c(\phi)=\partial\bar\partial\phi$ is the curvature of ϕ and ω is some Kähler form on X.

Let f be a $\bar\partial$-closed (n,q)-form ($q>0$) with values in L. Then there is an $(n,q-1)$-form u with values in L such that

$$\bar\partial u=f$$

and

$$\|u\|^2\leq\frac{1}{cq}\|f\|^2.$$

For the proof we use the strategy from Section 2 and consider the two Hilbert spaces

$$H_1 = L^2_{(n,q-1)}(X, L, \phi)$$

and

$$H_2 = L^2_{(n,q)}(X, L, \phi)$$

of L-valued forms in L^2 with respect to the metric ϕ, and the closed and densely defined operator $T = \bar\partial$ from H_1 to H_2. By the discussion in Section 2, the theorem follows if we can establish the dual inequality

$$cq\|\alpha\|^2 \leq \|T^*\alpha\|^2 \tag{3.8}$$

for any $\bar\partial$-closed α in the domain of the Hilbert space adjoint T^* of the operator T. Such an estimate will follow from Corollary 3.4.3 together with the following important approximation lemma.

Lemma 3.5.2. *If X is compact any smooth (n,q) form α lies in the domain of T^* and*

$$T^*\alpha = \bar\partial^*_\phi\alpha.$$

If α lies in the domain of T^* and in addition $\bar\partial\alpha$ (in the sense of distributions) lies in $H_3 := L^2_{(n,q+1)}(X, L, \phi)$ then there is a sequence α_k of smooth L-valued (n,q) forms such that

$$\|\alpha_k - \alpha\|^2,$$

$$\|\bar\partial^*_\phi\alpha_k - T^*\alpha\|^2,$$

and

$$\|\bar\partial\alpha_k - \bar\partial\alpha\|^2$$

all tend to zero.

Proof. The first statement means that if u lies in the domain of T (i.e. if $\bar\partial u$ taken in the sense of distributions lies in H_2) and α is smooth, then

$$\int \langle\bar\partial u, \alpha\rangle dV_\omega = \int \langle u, \bar\partial^*_\phi\alpha\rangle dV_\omega.$$

But this is exactly the definition of $\bar\partial u$ in the sense of distributions. The main part of the lemma is the possibility to approximate a general form by smooth forms in the way described.

We first claim that if α lies in $\mathrm{Dom}(T^*)$ and if χ is a real valued smooth function on X, then $\chi\alpha$ also lies in $\mathrm{Dom}(T^*)$. For this we need to prove that

$$|(\bar\partial u, \chi\alpha)_{H_2}| \leq C\|u\|_{H_1}.$$

But

$$(\bar\partial u, \chi\alpha)_{H_2} = (\chi\bar\partial u, \alpha)_{H_2} = (\bar\partial\chi u, \alpha)_{H_2} - (\bar\partial\chi \wedge u, \alpha)_{H_2}.$$

The first term on the right hand side satisfies the desired estimate since α lies in $\mathrm{Dom}(T^*)$, and the second term is trivially OK too.

We now choose χ to belong to an appropriate partition of unity with small supports, and decompose

$$\alpha = \sum \chi_j\alpha.$$

It is enough to approximate each term, so we see that we may assume that α has support in a coordinate neighbourhood which moreover trivializes L.

We then simply approximate α by taking the convolution with a sequence of functions

$$\chi_k(z) : k^{2n}\chi(kz)$$

where χ is smooth with compact support and has integral equal to 1. It is then a standard property of such convolutions that

$$\alpha_k := \chi_k * \alpha$$

converges to α in L^2. Since $\bar{\partial}\alpha_k = \chi_k * \bar{\partial}\alpha$ it follows that $\bar{\partial}\alpha_k$ converges to $\bar{\partial}\alpha$ as well. The remaining property, that $T^*\alpha_k$ converges to T^* is somewhat more delicate, but follows from a classical result of Friedrich, that we omit. □

Given the lemma, we can apply Corollary 3.4.3 to each α_k and conclude that

$$cq\|\alpha_k\|^2 \leq \|T^*\alpha_k\|^2 + \|\bar{\partial}\alpha_k\|^2.$$

If α lies in the domain of T^* and is $\bar{\partial}$-closed we get 3.8 by taking limits. This completes the proof of Theorem 3.5.1.

3.6. Complete Kähler manifolds

In this section we shall prove that Theorem 3.5.1 also holds for certain noncompact manifolds, namely those that carry some complete Kähler metric. We stress that we do not need to assume that the Kähler metric appearing in the final estimates is complete, only that the manifold has *some* complete metric. First we need to recall some definitions.

Definition : A complex manifold X with an hermitian metric (or a Riemannian manifold) is *complete* if there is some function

$$\chi : X \to [0, \infty)$$

which is proper and satisfies

$$\|d\chi\| \leq C.$$

□

Admittedly, this definition is not overly intuitive. The more intuitive definition is that the hermitian (or Riemannian) metric induces a structure on X as a metric space, the distance between points being the infimum of the length of paths connecting the two points. The Riemannian manifold is then complete if this metric space is complete, i.e. if any Cauchy sequence has a limit. One can prove however that this definition is equivalent to the one we have given, but we have chosen the definition above since it is more convenient in applications. (It is both easier to verify and to apply.) Notice that if we have two Hermitian metrics with Kähler forms ω_1 and ω_2 respectively, if $\omega_1 \leq C\omega_2$ for some constant C and ω_1 is complete, then ω_2 is also complete (the same function χ will do).

Recall that a domain Ω in $\mathbb{C}^n$ is pseudoconvex if it has a strictly plurisubharmonic exhaustion function. This means that there exists some function

$$\psi : X \to [0, \infty)$$

which is proper and satisfies $i\partial\bar{\partial}\psi > 0$. More generally, a complex manifold is *Stein* if it has some strictly plurisubharmonic exhaustion.

Proposition 3.6.1. *Any Stein manifold has a complete Kähler metric.*

PROOF. Let $\Psi = e^{\psi}$ where ψ is strictly plurisubharmonic and exhaustive, and let

$$\omega = i\partial\bar\partial\Psi = (i\partial\psi \wedge \bar\partial\psi + i\partial\bar\partial\psi)e^{\psi}.$$

This is then a Kähler metric (since ψ is strictly plurisubharmonic) and we claim that it is complete. For this we take

$$\chi = e^{\psi/2}.$$

Then

$$\partial\chi \wedge \bar\partial\chi = (1/4)i\partial\psi \wedge \bar\partial\psi e^{\psi} \le \omega,$$

so $|\partial\chi|^2$ is bounded. □

Theorem 3.5.1 holds verbatim for manifolds having a complete Kähler metric, and hence in particular for Stein manifolds.

Theorem 3.6.2. *Let L be a holomorphic line bundle endowed with a metric ϕ over a complex manifold X which has some complete Kähler metric. Assume the metric ϕ on L has (strictly) positive curvature and that*

$$ic(\phi) \ge c\omega$$

where $c(\phi) = \partial\bar\partial\phi$ is the curvature of ϕ and ω is some Kähler form on X.

Let f be a $\bar\partial$-closed (n, q)-form ($q > 0$) with values in L. Then there is an $(n, q-1)$-form u with values in L such that

$$\bar\partial u = f$$

and

$$\|u\|^2 \le \frac{1}{cq}\|f\|^2,$$

provided that the right hand side is finite.

To prove the theorem we first assume that the metric ω is itself complete. (Hopefully, it is useful here to refer back to Section 2.0.1 for the same argument in a model example.) As in the previous section, we then need to establish an inequality

$$cq\|\alpha\|^2 \le \|T^*\alpha\|^2 \qquad (3.9)$$

for the adjoint of the operator $T = \bar\partial$, where α is $\bar\partial$-closed. (We use the same notation for the Hilbert spaces and operators between them as in the previous section.) An (n, q)- form α lies in the domain of T^* if for some η in H_1

$$(Tu, \alpha)_{H_2} = (u, \eta)_{H_1},$$

and then $T^*\alpha = \eta$. Testing this condition for u smooth with compact support it follows that $\eta = \bar\partial^*_\phi\alpha$, the formal adjoint of $\bar\partial$ acting on α in the sense of distributions.

Lemma 3.6.3. *Let α belong to the domain of T^* and assume that $\bar\partial\alpha$ lies in H_3. Then there is a sequence of forms with compact support α_k such that*

$$\|\alpha_k - \alpha\|^2,$$

$$\|\bar\partial^*_\phi\alpha_k - T^*\alpha\|^2,$$

and

$$\|\bar\partial\alpha_k - \bar\partial\alpha\|^2$$

all tend to zero.

PROOF. Let $g(t)$ be a smooth function on $\mathbb{R}$ which equals 1 for $t \leq 0$ and 0 for $t \geq 1$. Let

$$\chi_k = g(\chi - k),$$

where χ is a smooth exhaustion with bounded differential (which exists by the completeness). Then χ_k tends to 1 on X and has uniformly bounded differential. Let $\alpha_k = \chi_k \alpha$. The three limits are then easy to check using dominated convergence. □

The proof now follows the argument in the previous section. We approximate all the α_ks by forms with compact support that are moreover smooth. For these forms we apply Corollary 3.4.3, and then get 3.9 by taking limits (recall $\bar{\partial}\alpha = 0$). The theorem then follows from Proposition 2.0.5.

For the general case we now assume only that $ic(\phi) \geq c\omega$, but that there is on X some other metric with Kähler form ω' which is complete. To simplify the presentation, we shall assume that this metric is of the form

$$\omega' = i\partial\bar{\partial}\psi$$

where ψ is a strictly plurisubharmonic function. This is not necessary for the theorem, but makes the exposition simpler, and it is certainly satisfied in the Stein case. Let

$$\omega_{(k)} = \omega + \omega'/k.$$

All metrics $\omega_{(k)}$ are then complete, and if

$$\phi_{(k)} = \phi + c\psi/k$$

then $ic(\phi_{(k)}) \geq c\omega_{(k)}$. By the case we have just discussed, there is a solution u_k to $\bar{\partial}u_k = f$ with

$$cq\|u_k\|_k^2 \leq \|f\|_k^2,$$

where $\|\cdot\|_k$ are the norms with respect to the metrics $\omega_{(k)}$ on X and $\phi_{(k)}$ on L, provided the right hand side is finite.

Lemma 3.6.4. *Let $\omega_1 \leq \omega_2$ be two Kähler forms, and let $\|\cdot\|_{1,2}$ be the corresponding norms. Then, if f is an (n,q)-form*

$$\|f\|_2^2 dV_{\omega_2} \leq \|f\|_1^2 dV_{\omega_1}.$$

PROOF. Write $f = w \wedge g$ where w is $(n,0)$ and g is $(0,q)$. Then

$$\|f\|_i^2 = \|w\|_i^2 \|g\|_i^2$$

for both norms. But for $(n,0)$-forms

$$\|w\|^2 dV = c_n w \wedge \bar{w}$$

is independent of the metric. Hence we need only prove that

$$\|g\|_2^2 \leq \|g\|_1^2.$$

Choose a basis (at a point) such that

$$\omega_1 = i\sum dz_i \wedge d\bar{z}_i$$

and

$$\omega_2 = i\sum \lambda_i dz_i \wedge d\bar{z}_i,$$

with $\lambda_i \geq 1$. Then $\sqrt{\lambda_i} dz_i$ are orthonormal for ω_2. Hence, if $g = \sum g_J dz_j$

$$\|g\|_1^2 = \sum |g_J|^2$$

while

$$\|g\|_2^2 = \sum |g_J|^2/\lambda_J,$$

with λ_J the product of all λ_i for i in J. This proves the lemma. □

From the lemma we get that

$$\|f\|_k^2 \leq \|f\|^2,$$

so if the right hand side here is finite we have a uniform bound for all the norms with respect to $\omega_{(k)}$. We therefore get a uniform estimate for the norms of all u_k and the final theorem follows by taking limits of a suitably chosen subsequence, weakly convergent on any compact part of X.

Exercise Carry out the last part of the proof in detail! □

We stress once again that Theorem 3.6.2 applies to any Kähler metric, complete or not. The assumption that there exist some complete Kähler metric is a condition on the domain, but it is not visible in the final estimate. The assumption is satisfied by any pseudoconvex domain in $\mathbb{C}^n$ as we have seen, and the theorem therefore gives existence and estimates for $\bar\partial$ in e g the ball with respect to e g the Euclidean metric. This particular case is not much easier than the general case.

3.6.1. A basic fact of life on Stein manifolds

As an application of the results above for Stein manifolds, we shall now prove a Runge-like approximation theorem. Let X be a Stein manifold, so that we know that there is some smooth strictly plurisubharmonic exhaustion function on X. If Y is a compact subset of X we say that Y is *Runge* in X if for any open neighbourhood U of Y, there is some exhaustion as above which is negative on Y, and positive outside U. A main example of this situation is $X = \mathbb{C}^n$ and Y a polynomially convex subset, i.e. a compact that can be approximated arbitrarily well from outside by sets defined by polynomial inequalities, $\{|P_j| \leq 1, j = 1, 2...N\}$.

Exercise: Verify this! □

Theorem 3.6.5. *Let F be a holomorphic line bundle on X and let Y be a compact Runge subset of X. Then any holomorphic section of F defined in some open neighbourhood of Y can be approximated arbitrarily well in the supremum norm on Y by global sections of F. In particular, F has some nontrivial global holomorphic section.*

Proof. Let $L = F - K_X$ so that $F = K_X + L$. Then sections of F are $(n, 0)$-forms with values in L, and F-valued $(0, 1)$-forms are $(n, 1)$-forms with values in L. Let h be a holomorphic section of F defined in a neighbourhood of Y. Choose some cut-off function χ which equals 1 in some open neighbourhood of Y and is supported in the open neighbourhood where h is defined. Then $v = \chi h$ and

$$f = \bar\partial v = \bar\partial \chi h$$

are global objects. We shall solve the $\bar\partial$-equation

$$\bar\partial u = f$$

in such a way that u is small on Y. Then $h' := v - u$ is a global holomorphic section of F that approximates h on Y.

By assumption there is some strictly plurisubharmonic plurisubharmonic exhaustion function on X, ψ, such that $\psi \leq 0$ on a neighbourhood V of Y and $\psi \geq \delta$ on the support of $\bar{\partial}\chi$. Let ϕ be some smooth metric on L such that $i\partial\bar{\partial}\phi \geq \omega$, where ω is some Kähler form on X. (Such a metric can always be constructed by choosing an arbitrary smooth metric on L and then adding $k \circ \psi$ where k is a sufficiently increasing convex function.) Let $\phi_m = \phi + m\psi$. By Theorem 3.6.2 we can solve

$$\bar{\partial}u = f$$

with u satisfying

$$\int |u|^2 e^{-\phi_m} \leq \int |f|^2 e^{-\phi_m}.$$

The right hand side here is bounded by $Ce^{-m\delta}$ so it follows that

$$\int_V |u|^2 \leq Ce^{-m\delta}.$$

But u is holomorphic near Y so it follows from the Cauchy estimates that u tends to 0 uniformly on Y. This completes the proof of the first part. To see that there are some compact Runge subsets on X, so that the theorem is not void, it suffices to take

$$Y = \{\psi \leq c\}.$$

If c is small enough, Y is a small neighbourhood of the minimum point of ψ, over which L is trivial, so we can find plenty of local holomorphic sections there. □

LECTURE 4

The Bergman kernel

In the first section of this lecture we will give the basic definitions and properties of Bergman kernels associated to Hilbert spaces of functions such that point evaluations are bounded linear functionals on the space. (The main example to keep in mind is an L^2 space of holomorphic functions on a domain in $\mathbb{C}^n$.) The next section discusses the analogous constructions for spaces of sections of a line bundle. This is very similar to the scalar valued case, but formally a bit different. In particular, the Bergman kernel is then no longer a function, but behaves like a metric on the line bundle. Here we will also give the simplest asymptotic estimates for Bergman kernels associated to high powers of the line bundle.

4.1. Generalities

Let (X, μ) be a measure space, and let $\mathcal{H}$ be a closed subspace of $L^2(X, \mu)$. Assume that $\mathcal{H}$ is such that each function in $\mathcal{H}$ has a pointwise value at every point, and that the map

$$h \mapsto h(x)$$

is a bounded linear functional on $\mathcal{H}$. A basic example of this situation is that X is a domain in $\mathbb{C}^n$, $d\mu = d\lambda$ is Lebesgue measure, or $d\mu = e^{-\phi}d\lambda$ where ϕ is a weight function locally bounded from above. Then, by the Riesz representation theorem, there is for any x in X a unique element k_x of $\mathcal{H}$ such that

$$h(x) = (h, k_x).$$

Definition: The function k_x is the *Bergman kernel* for the point x. The function $K(x) = k_x(x)$ is the Bergman kernel on the diagonal. □

Another way to obtain the Bergman kernel is to start from an orthonormal basis of $\mathcal{H}$, h_j. Then

$$k_x(y) = \sum h_j(y)\overline{h_j(x)}$$

and

$$K(x) = \sum |h_j(x)|^2.$$

This requires some justification:

Proposition 4.1.1. *For any $N < \infty$*

$$\sum^N |h_j(x)|^2 \leq K(x).$$

PROOF. Let $h = \sum^N a_j h_j$ with $\sum |a_j|^2 \leq 1$. Then $\|h\| \leq 1$ so

$$|h(x)|^2 = |(h, k_x)|^2 \leq (k_x, k_x) = K(x).$$

Since the coefficients a_j are here arbitrary, this implies that

$$\sum^N |h_j(x)|^2 \leq K(x).$$

□

Hence we see that

$$\sum^\infty |h_j(x)|^2$$

and therefore

$$\sum^\infty h_j(y)\overline{h_j(x)}$$

converges pointwise. Moreover

$$\sum h_j(\cdot)\overline{h_j(x)}$$

converges in L^2 to some function h_x. Then, for any l,

$$(h_l, h_x) = h_l(x)$$

so h_x must be equal to k_x.

From this expression for the Bergman kernel, we see that if we take the scalar product

$$(g, k_x) := \hat{g}(x)$$

for a function g which is not necessarily in $\mathcal{H}$, then

$$\hat{g} = \sum h_j(g, h_j)$$

is the orthogonal projection of g on $\mathcal{H}$. This gives us yet another way of looking at the Bergman kernel: it is the kernel of the orthogonal projection of a function on $\mathcal{H}$.

Another important property is the extremal characterization of the Bergman kernel on the diagonal:

Proposition 4.1.2. *The Bergman kernel on the diagonal $K(x)$ is the extremum*

$$s_x := \sup |h(x)|^2$$

over all elements h in $\mathcal{H}$ of norm at most 1. In other words, $K(x)$ is the norm of the point evaluation at x.

PROOF.

$$s_x = \sup |(h, k_x)|^2 = \|k_x\|^2 = (k_x, k_x) = K(x).$$

□

Much of the strength of the Bergman kernel comes from the interplay between those characterizations of k_x. Here is one striking example.

Proposition 4.1.3. *Let $\mathcal{H}$ be a subspace of $L^2(X, \mu)$ consisting of continuous functions, and assume that for any h in $\mathcal{H}$*

$$|h(x)|^2 \leq C(x)\|h\|^2.$$

Then

$$\dim \mathcal{H} \leq \int_X C(x) d\mu.$$

PROOF. The first condition implies that $K(x) \leq C(x)$. Hence if h_j is an ON-basis for $\mathcal{H}$

$$\dim \mathcal{H} = \int_X \sum |h_j|^2 = \int_X K(x) \leq \int_X C(x) d\mu.$$

□

We end this section with an asymptotic estimate for the Bergman kernel associated to weighted L^2-spaces of holomorphic functions. Let ϕ be a function of class C^2 in an open set Ω in $\mathbb{C}^n$. Consider the space $L^2_k := L^2(\Omega, e^{-k\phi})$ and its subspace $\mathcal{H}_k$ of holomorphic functions. Let K_k be the Bergman kernel on the diagonal for $\mathcal{H}_k$.

Proposition 4.1.4.

$$\limsup_{k \mapsto \infty} \frac{1}{k^n} K_k(z) e^{-k\phi(z)} d\lambda \leq \pi^{-n} \chi_0(z)(i\partial\bar{\partial}\phi)_n,$$

where χ_0 is the characteristic function of the open set X_0 where $i\partial\bar{\partial}\phi$ is positive.

PROOF. We will use the extremal characterization of $K_k(z)$, so let h be a holomorphic function in Ω such that

$$\int_\Omega |h|^2 e^{-k\phi} \leq 1.$$

Say $z = 0$. We need to estimate $h(0)$. Assume first that ϕ has the Taylor expansion

$$\phi(\zeta) = \phi(0) + \sum \lambda_j |\zeta_j|^2 + o(|\zeta|^2)$$

with $\lambda_j \geq 0$ near the origin. Then for any finite A

$$1 \geq \int_{|\zeta|^2 \leq A/k} |h|^2 \, e^{-k\phi} \geq (1 - \epsilon_k) e^{-k\phi(0)} \int_{|\zeta|^2 \leq A/k} |h|^2 \, e^{-k \sum \lambda_j |\zeta_j|^2},$$

where ϵ_k tends to zero.

By the mean value property of holomorphic functions the last integral dominates

$$|h(0|^2 \int_{|\zeta|^2 \leq A/k} e^{-k\sum \lambda_j |\zeta_j|^2} = |h(0)|^2 (\lambda_1 ... \lambda_n k^n)^{-1} \pi^n (1 - \delta_A)),$$

where δ_A tends to 0 as A goes to infinity. Hence

$$h(0) e^{-k\phi(0)} / k^n \leq \pi^{-n} (\lambda_1 ... \lambda_n)(1 + 2\delta_A)(1 + 2\epsilon_k)$$

for A and k large. Taking the supremum over all h of norm at most 1 we get that

$$K(0) e^{-k\phi(0)} / k^n \leq \pi^{-n} (\lambda_1 ... \lambda_n)(1 + 2\delta_A)(1 + 2\epsilon_k)$$

and the claimed estimate follows by letting first k and then A tend to infinity, since

$$\lambda_1 ... \lambda_n d\lambda = (i\partial\bar{\partial}\phi)_n.$$

If on the other hand ϕ has the same type of Taylor expansion with one of the eigenvalues λ_j negative or 0 it follows from a similar argument that $K_k e^{-k\phi}(0)$ tends to 0.

The same argument applies if ϕ has the Taylor expansion

$$\phi(\zeta) = \phi(0) + q(z, \bar{z}) + o(|\zeta|^2)$$

at the origin, where $q(z,\bar z)$ is an hermitian form, since we can then diagonalize q by a unitary change of variables. A general ϕ has the Taylor expansion

$$\phi(\zeta) = \phi(0) + q(z,\bar z) + 2\Re p(\zeta) + o(|\zeta|^2)$$

where p is a holomorphic polynomial of degree 2 and no constant term. The substitution

$$h \mapsto he^{-p}$$

reduces this case to the one we have treated. □

4.2. Bergman kernels associated to complex line bundles

When generalizing the notion of Bergman kernel to spaces of sections of a line bundle instead of scalar valued functions one runs into no serious problem — but several minor complications. The first is that point evaluations are not well defined. The value of a section at a point x is an element in the fiber over that point, so the Bergman kernel k_x will also be L-valued, or rather $\bar L$-valued, as a function of x. When restricting to the diagonal, $K(x) = k_x(x)$ will then take its values in $L \otimes \bar L$, which means that

$$\psi := \log K$$

will behave like a metric on L. To avoid going through this in detail we will instead focus directly on the Bergman kernel on the diagonal and define it using an orthonormal basis instead.

Definition: Let L be a holomorphic line bundle over a complex manifold X, and let ϕ be a metric on L. Let μ be a measure on X, and let u_j be an orthonormal basis for the space

$$H^0(X,L)$$

of global holomorphic sections of L, with respect to the scalar product

$$(u,v) = \int_X u\bar v e^{-\phi} d\mu.$$

Then

$$B := \sum |u_j|^2 e^{-\phi}$$

is the *Bergman function*, and

$$K(z) := B(z)e^{\phi}$$

is the *Bergman kernel* (on the diagonal) for L, ϕ, μ. □

It is clear that in case L is trivial so that ϕ is just a global function on X, this definition coincides with the definition we gave in the previous section. Notice that since

$$\log K = \log B + \phi$$

and $\log B$ is a function, $\log K$ is a metric on L, just like we expected. One checks that the sum defining B is convergent in much the same way as we did in the scalar valued case before. The extremal characterization of the Bergman kernel takes the following form.

Proposition 4.2.1.

$$B(z) = \sup |u(z)|^2 e^{-\phi(z)}$$

where the supremum is taken over all global holomorphic sections of L with L^2-norm at most 1.

PROOF. A global section of L^2-norm at most 1 can be written

$$u = \sum a_j u_j$$

where $\sum |a_j|^2 \leq 1$. Evaluating at z, the proposition follows immediately from this:

$$|u(z)|^2 e^{-\phi} \leq B(z)$$

and equality can occur. □

Consider now powers L^k of the line bundle L, and give them the metrics $k\phi$. For each k we then get a Bergman function, that we denote B_k. The next proposition generalizes 5.1.4.

Proposition 4.2.2. *Let μ be an arbitrary measure on X with smooth density. Then*

$$\limsup_{k \mapsto \infty} \frac{1}{k^n} B_k(z) d\mu \leq \pi^{-n} \chi_0(z)(i\partial\bar{\partial}\phi)_n,$$

where χ_0 is the characteristic function of the open set X_0 where $i\partial\bar{\partial}\phi$ is positive.

We omit the proof of this since it is virtually identical to the proof of 4.1.4 — the arguments there were entirely local, using only the submeanvalue property of holomorphic functions near z. It should however be noted explicitly that the measure μ plays a very small role in the asymptotic estimate. This may seem a bit surprising, but the reason is that the estimates are carried out in a shrinking neighbourhood of a given point. Different choices of μ (with smooth densities!) are then related by a multiplicative factor that is almost constant, and it is very easy to see that the expression

$$K d\mu$$

is unchanged if we multiply μ by a constant.

From this we can rather easily prove the next estimate on the dimension of the space of global sections to high powers of a line bundle.

Theorem 4.2.3. *Let X be compact and put*

$$h_k^0 = dim\, H^0(X, L^k).$$

Then

$$\limsup \frac{h_k^0}{k^n} \leq \pi^{-n} \int_{X_0} (i\partial\bar{\partial}\phi)_n.$$

In particular, $h_k^0 = O(k^n)$ for any line bundle over any compact manifold. Notice that the left hand side is independent of the choice of metrics.

The proof of Theorem 4.2.3 is, just like the proof of Proposition 4.1.3, based on the fact that

$$\dim H^0(X, L^k) = \int B_k.$$

(Again, this holds since

$$B := \sum |u_j|^2 e^{-\phi}$$

and each term in this sum contributes a 1 to the integral.) We now want to apply Fatou's lemma, and for this we claim first that

$$\frac{1}{k^n} B_k(z)$$

is uniformly bounded from above. To see this, we use the extremal characterization

$$B_k(z) = \sup |u(z)|^2 e^{-k\phi(z)}$$

where u is a section of L^k of unit L^2-norm. Fix the point $z = 0$ and choose a local trivialization of L near 0 in which ϕ gets the form

$$\phi(\zeta) = q(z, z) + o(|\zeta|^2)$$

(cf the end of the proof of Proposition 4.1.4). Then, if u has unit norm

$$1 \geq \int_{|\zeta|^2 \leq 1/k} |u|^2 \, e^{-k\phi} \geq C \int_{|\zeta|^2 \leq 1/k} |u|^2 \geq C'|u(0)|^2 k^n$$

by the submeanvalue property. Hence

$$|u(0)|^2 e^{-k\phi(0)} \leq A/k^n,$$

since $\phi(0) = 0$, so the uniform upper bound for B_k follows. Hence, by Fatou's lemma,

$$\limsup \int B_k/k^n \leq \int \limsup B_k/k^n \leq \pi^{-n} \int_{X_0} (i\partial\bar\partial\phi)_n$$

which proves the theorem. □

We finally note that in case the metric has positive curvature, equality actually holds in the asymptotic estimate for the bergman kernels.

Theorem 4.2.4. *Assume X is compact and $ic(\phi) > 0$. Then*

$$\lim \frac{h_k^0}{k^n} = \pi^{-n} \int_{X_0} (i\partial\bar\partial\phi)_n.$$

This follows just like before (this time using dominated convergence instead of Fatou's lemma) from the following precision of Proposition 4.2.2 for positive bundles.

Theorem 4.2.5. *Assume that X is compact and that $ic(\phi) > 0$. Define a Kähler metric on X by $\omega = ic(\phi)$ and let $dV_\omega = \omega_n$ be its volume form on X. Let B_k be the associated Bergman function. Then*

$$\lim B_k/k^n = \pi^{-n}.$$

To prove this last theorem we have to construct a global holomorphic section with a large value at a given point x. This is no longer a local estimate and we have to use $\bar\partial$-estimates to obtain it. For this we choose local coordinates z centered at x and choose a local trivialization so that our metric on L is represented by a function

$$\phi(z) = \sum \lambda_j |z_j|^2 + o(|z|^2)$$

near that point. For simplicity we can even change coordinates by a linear transformation so that

$$\phi(z) = |z|^2 + o(|z|^2).$$

Let χ be a smooth function in $\mathbb{C}^n$, supported in a ball with radius 2 centered at the origin and equal to 1 in the unit ball. Let δ_k be a sequence of numbers tending to zero (to be chosen later such that $\delta_k k^{1/2}$ goes to infinity) and put

$$\chi_k(z) = \chi(z k^{1/2} \delta_k).$$

We can consider χ_k to be a section of L supported in a shrinking neighbourhood of x. Simple estimates (using the change of variables $w = zk^{1/2}$) show that the L^2-norm of χ_k is asymptotic to $k^n\pi^n$, and the L^2 norm of $f_k := \bar\partial\chi_k$ is dominated by

$$k^{n+1}\delta_k^2 e^{-\delta_k^{-2}}.$$

We now want to apply Theorem 3.5.1 to solve the equation $\bar\partial u_k = f_k$. There is a very minor problem here that the bundle we are dealing with is L^k and does not contain the canonical bundle K_X as a summand. This is easily resolved by writing

$$L^k = K_X + F_k$$

with $F_k = L^k - K_X$. This bundle has a natural metric with curvature asymptotic to $k\omega$ so Theorem 3.5.1 does apply.

Hence we can solve $\bar\partial u_k = f_k$ with the L^2-norm of u_k controlled by $k^n\delta_k^2 e^{-\delta_k^{-2}}$. We can certainly choose δ_k so that this quantity goes to zero faster than polynomially. Since u_k is holomorphic near the origin, it then follows from the submeanvalue property that $u_k(0)$ also goes to zero.

Let $s_k = \chi_k - u_k$. Then s_k is a global holomorphic section of L^k with L^2-norm asymptotic to $k^n\phi^n$. Moreover, $|s_k(x)|^2 e^{-k\phi(x)}$ tends to 1. Hence

$$\limsup B_k(x)/k^n \geq \pi^{-n}$$

which together with Proposition 4.2.2 proves the theorem. □

This asymptotic formula for the Bergman function of a positive bundle was first obtained (in a sharper form) by Bouche, [**5**], and Tian, [**22**]. Much sharper asymptotic formulas, giving asymptotic developments for the Bergman function in powers of k, have been found later, see [**7**], [**24**] and also [**4**].

4.2.1. The field of meromorphic functions on a compact complex manifold

In this section we will, as an application of the upper estimate on the dimension of the space of global sections, give a proof of a classical result of Siegel.

Theorem 4.2.6. *Let X be a compact complex manifold of dimension n, and let $g_1, ...g_{n+1}$ be meromorphic functions on X. Then there is a polynomial P such that*

$$P(g_1, ...g_{n+1}) = 0.$$

In other words, any $n+1$ meromorphic functions are algebraically dependent.

Proof. We first claim that there is a holomorphic line bundle L over X with a holomorphic section s, such that $s_j := sg_j$ are all holomorphic (sections of L). To see this, we argue as follows. A meromorphic function function g can be written locally, in a neighbourhood U of any given point,

$$g = f/h$$

where f and h are holomorphic, and by choosing them without common factors, this factorization is unique up to units, i.e. up to nonvanishing holomorphic factors. Cover X by open sets U_j where such a representation

$$g = f_j/h_j$$

holds. Then $h_j/h_i := h_{ij}$ are nonvanishing functions satisfying the cocycle condition, so they define a line bundle L of which $s = (h_i)$ is a section. (This is the line bundle associated to the *polar divisor* of g.) Clearly, gs is then holomorphic.

This proves the claim for one single meromorphic function; the case of an N-tuple (g_j) follows by taking the product of sections, one for each g_j.

With this in our hands, we consider sections to L^k of the form

$$s_0^{k_0}...s_{n+1}^{k_{n+1}},$$

where $k_0 + ...k_{n+1} = k$ and $s_0 = s$. The number of such sections is of the order k^{n+1}. (It is equal to the number of choices of $k_1, ...k_{n+1}$ with $k_1 + ...k_{n+1} \leq k$ which is the number of lattice points in k times the unit simplex in $\mathbb{R}^{n+1}$.) By our estimate for the dimension of the space of global sections of L^k $(= O(k^n))$ they must be linearly dependent for k large. This means that there is a homogeneous polynomial in the s_j that vanishes identically. Clearing out the factor s we get a polynomial in g_j that vanishes identically. □

Exercise: Consider the meromorphic functions $1/z$ and $1/(z-1)$ on $\mathbb{P}^1$. Find a polynomial P satisfying the conclusion of the theorem. □

LECTURE 5

Singular metrics and the Kawamata-Viehweg vanishing theorem

In Lecture 3 we have defined a (smooth) metric on a line bundle L as a collection of functions ϕ_i defined on trivializing open sets U_i for L, satisfying the compatibility condition

$$\phi_i - \phi_j = \log |g_{ij}|^2$$

A *singular* metric on L is defined in the same way, but allowing ϕ_i to be non-smooth. To fix ideas one usually requires ϕ_i to be locally in L^1, but in practice ϕ_i is locally the difference between two plurisubharmonic functions and so lies in any L^p with $p < \infty$.

If L is *effective*, i.e. has a nontrivial global holomorphic section s, given by local holomorphic functions s_j, one can define a singular metric by $\phi = \log |s|^2$, i.e. $\phi_i = \log |s_i|^2$. More generally, if s^λ are global sections of L, then

$$\phi = \sum a_\lambda \log |s^\lambda|^2,$$

(for $\sum a_\lambda = 1$) and

$$\psi = \log \sum |s^\lambda|^2$$

are singular metrics on L. If the s^λs are holomorphic, then the singularities of the latter metric lies precisely on the common zero locus of the sections.

5.1. The Demailly-Nadel vanishing theorem

Just as before we can define the curvature of a singular metric by

$$c(\phi) = \partial\bar{\partial}\phi = \partial\bar{\partial}\phi_i$$

on U_i. This is of course not necessarily a smooth form anymore, but merely a current. We say that L is *pseudoeffective* if L has a metric with nonnegative curvature current, meaning precisely that the local representatives can be chosen to be plurisubharmonic. The two metrics ϕ and ψ above constructed from sections have this property if s^λ are holomorphic and ($a_\lambda > 0$), so effective bundles are pseudoeffective. The principal aim of this lecture is to generalize the main existence theorems of Lecture 3 to the setting of singular metrics. This is the content of the Demailly-Nadel vanishing theorem, see **[8]**, **[15]**. Demailly actually proved this theorem in the more general setting of manifolds with complete Kähler metrics. This used his technique of regularisation of singular metrics. Here we will restrict attention to *projective* manifolds and Stein manifolds, which is more elementary. Since we have not defined projective manifolds yet in these notes, we adopt the preliminary definition that a compact manifold X is projective if there is some ample line bundle F over X (recall that F is ample if it has some smooth metric, ψ with strictly positive curvature). Later, we shall use the Demailly-Nadel vanishing theorem to prove the Kodaira embedding theorem, which says that manifolds that carry ample line bundles are precisely those that can be embedded in projective space - hence the terminology. Projective manifolds are in particular Kähler, since $i\partial\bar\partial\psi$ is a Kähler form.

Theorem 5.1.1. *Let X be a complex manifold which is either projective or Stein. Let L be a holomorphic line bundle over X which has a possibly singular metric ϕ whose curvature satisfies*

$$i\partial\bar\partial\phi \geq \epsilon\omega,$$

where ω is a Kähler form. Let f be an L-valued $\bar\partial$-closed form of bidegree (n,q). Then there is a solution u to the equation $\bar\partial u = f$ satisfying

$$\|u\|^2_{\omega,\phi} \leq \frac{1}{\epsilon q}\|f\|^2_{\omega,\phi},$$

provided the right hand side is finite.

Notice the last proviso of the theorem. That the L^2-norm with respect to a singular metric is finite is not only a global condition, but also a local condition which imposes vanishing of f on the nonintegrability locus of the metric $e^{-\phi}$. In case $\phi = \log|s|^2$ it essentially means that f vanishes on the zero divisor of s.

The strategy of the proof is to reduce to the case of smooth metrics (we already know the theorem for smooth metrics) by regularizing the singular metric. This regularization is however a somewhat delicate business: on a compact (even projective) manifold it is in general not possible to approximate singular metrics with smooth metrics, keeping positivity of the curvature. Therefore we shall first prove the theorem in the Stein case — where regularization *is* possible — and then get the projective case from there.

Proposition 5.1.2. *Let X be a Stein manifold and let D be a relatively compact Stein subdomain of X. Let χ be a plurisubharmonic function on X. Then there is a sequence of smooth, strictly plurisubharmonic functions, χ_ν, defined in a neighbourhood of the closure of D, that decrease to χ on D. Moreover, if $i\partial\bar\partial\chi \geq \omega$ where ω is a Kähler form and $\delta > 0$ the χ_νs can be chosen so that $i\partial\bar\partial\chi_\nu \geq (1-\delta)\omega$.*

We will not give a complete proof of this result, but merely indicate some steps in the proof. First, since X is Stein, X can be (properly) embedded in $\mathbb{C}^N$, so we may assume from the start that X is a submanifold of $\mathbb{C}^N$. Let D' be a larger Stein open subset of X containing D in its interior. Then there is a neighbourhood U of D' in $\mathbb{C}^N$, such that D' is a *holomorphic retract* of U, i.e. there is a holomorphic map p from U to D' which is the identity on D' (see Forster and Ramspott...). Then $\chi' := \chi \circ p$ is plurisubharmonic in U, so by the well known technique of convolution with an approximate identity χ' can be approximated by a decreasing sequence of strictly plurisubharmonic smooth functions on any relatively compact subset. The $\partial\bar\partial$ of the approximating functions will then be bounded from below of a quantity tending to $p^*\omega$ on compact subsets. Restricting to D the proposition follows.

Using the proposition we can prove Theorem 6.0.1. Assume first that L is the trivial line bundle, and that ϕ is a global plurisubharmonic function on X. Let D

be a relatively compact Stein open set in X, and choose approximating plurisubharmonic functions ϕ_ν as in the proposition. By the Hörmander L^2-estimates, we can solve the equation $\bar\partial u = f$ with $u = u_\nu$ satisfying

$$\|u_\nu\|^2_{\omega,\phi_\nu} \leq \frac{1}{\epsilon(1-\delta)q}\|f\|^2_{\omega,\phi_\nu} \leq \frac{1}{\epsilon(1-\delta)q}\|f\|^2_{\omega,\phi}.$$

If $\nu > \nu_0$, $\|\cdot\|_{\omega,\phi_\nu} \geq \|\cdot\|_{\omega,\phi_{\nu_0}}$ Therefore we may, using a diagonal argument, choose a subsequence of the u_ν converging weakly in $L^2(e^{-\phi_{\nu_0}})$, for any ν_0, to a limit, u_D. It is easily checked that u still solves the $\bar\partial$-equation and satisfies the required estimate in D. We then let D increase to X and take weak limits again.

If X is still Stein, but we no longer assume that L is trivial, we proceed as follows. Let s be some nontrivial holomorphic section of L. (Such a section exists by Theorem 3.6.5.) Let S be the zero divisor of s. Then $X \setminus S$ is still Stein (why?) so we may apply the previous argument and solve our $\bar\partial$ equation on $X \setminus S$. The proof will then be completed by the following very important lemma.

Lemma 5.1.3. *Let M be a complex manifold and let S be a complex hypersurface in M. Let u and f be (possibly bundle valued) forms in L^2_{loc} of M satisfying $\bar\partial u = f$ outside of S. Then the same equation holds on all of M (in the sense of distributions).*

PROOF. Since the statement is local we may assume that M is an open ball in $\mathbb{C}^n$, that u and f are scalar valued, and that $S = h^{-1}(0)$ where h is holomorphic in M. Let $\xi(w)$ be smooth and nonnegative in $\mathbb{C}$, equal to 0 for $|w| \leq 1/2$ and equal to 1 outside of the unit disk. Put

$$\xi_\delta = \xi(h/\delta).$$

Then

$$\bar\partial(u\xi_\delta) = f\xi_\delta + \bar\partial\xi_\delta \wedge u.$$

We then let δ tend to 0, and it suffices to show that the last term on the right hand side goes to zero locally in L^1. For this it is by the Cauchy inequality enough to prove that

$$\int |\bar\partial\xi_\delta|^2 \leq \frac{C}{\delta^2}\int_{\delta/2\leq|h|\leq\delta} |\partial h|^2$$

is uniformly bounded. This is clear if h vanishes to order 1 on M. The general case is left as an exercise below. □

Exercise: Prove that if h is holomorphic in a ball of radius 2, then

$$\int_{\delta/2\leq|h|\leq\delta,|z|<1} |\partial h|^2 \leq C\delta^2.$$

(Hint: Prove first that if χ is a cutoff function equal to 1 on the ball with radius 1 and supported in the bigger ball, then

$$\int \chi\Delta\log(|h|^2+\delta^2)$$

is bounded.) □

This completes the proof of the theorem in the Stein case: Our L^2 estimate for the solution u shows in particular that u is locally in (unweighted!) L^2 outside of S, so the lemma implies that u actually solves the $\bar\partial$-equation across S too.

A very similar argument is used to prove the projective case of Theorem 6.0.1. If X is projective there is some positive line bundle F over X. A sufficiently high power of F then has a nontrivial holomorphic section. The complement of the zero divisor of that section is then Stein, so by the Stein case of the theorem we can solve our $\bar\partial$-equation there. But, then the lemma shows that we have actually solved the $\bar\partial$-equation in all of X.

5.2. The Kodaira embedding theorem

Let L be a line bundle over a compact complex manifold X, and let $s_0, ...s_N$ be a basis for the space, E, of global holomorphic sections of L (we assume that there are such sections so that the dimension of E is not zero). The *Kodaira map* associated to L is a holomorphic map from X to N-dimensional projective space, defined as follows :

$$\mathcal{K}(x) = [s_0(x), ...s_N(x)].$$

Some comments are in order. The s_js are sections to a line bundle so the values in the right hand side needs some interpretation. Choose a local trivialisation of L and let the right hand side mean the values of the sections s_j with respect to that trivialization. If we change to another trivialization, all s_j get multiplied by the same quantity, so the corresponding point in projective space is the same. Thus $\mathcal{K}$ is well defined - and of course holomorphic.

With the definition that we have just given, the Kodaira map is dependent of the choice of basis. Somewhat more elegantly we can consider the Kodaira map as a map from X to the projectivization of the dual space, E^*, of E: A point x in X is mapped to the element of E^* that is evaluation in that point with respect to some local trivialization near x, and then to the corresponding point in $\mathbb{P}(E^*)$. Again, when we take projectivization of E^* the final map will be independent of the choice of local trivialization. In terms of the basis above any section can be written

$$s = \sum a_j s_j.$$

Thus $a = (a_j)$ are the coordinates of s with respect to the basis and the evaluation map is given by

$$s(x) = \langle a, (s_j(x))\rangle,$$

so our previous definition gives the coordinates of the evaluation map in the dual basis. The Kodaira embedding theorem says that if L is sufficiently positive, i.e. if the curvature of L is sufficiently large, then the Kodaira map is an embedding.

Theorem 5.2.1. *Let X be a compact complex manifold. Let L be a holomorphic line bundle over X, having a (smooth) metric ϕ of positive curvature. Then the Kodaira map for $K_X + L^k$ is an embedding if k is large enough.*

PROOF. We will prove only that the Kodaira map is injective, leaving the injectivity of its differential as an exercise. We need to prove that if x_1 and x_2 are distinct points in X, then there is a section of kL that vanishes at x_2 but not at x_1. Take local coordinates near x_1, z, centered at x_1 and put

$$\xi_1 = \chi_1(z) n \log |z|^2$$

where χ_1 is a cut-off function with compact support in the coordinate neighbourhood that equals 1 near 0. Define ξ_2 in a similar way, and let $\xi = \xi_1 + \xi_2$ Then

$$i\partial\bar\partial\xi \geq -Ci\partial\bar\partial\phi,$$

where C can be taken independent of the choice of the x_is. Consider the bundle kL with k larger than $C + 1$, and give kL the (singular) metric

$$\psi = k\phi + \xi,$$

so that $i\partial\bar{\partial}\psi \geq \omega = i\partial\bar{\partial}\phi$. We define a smooth section to $K_X + kL$ by

$$s' = \chi dz_1 \wedge ...dz_n,$$

where χ is again a cut-off function equal to 1 near 0, and we also arrange things so that χ vanishes near x_2 Put $f = \bar{\partial}s'$. We now apply the Demailly-Nadel vanishing theorem and find a solution u to $\bar{\partial}u = f$ such that

$$\|u\|^2_{\omega,\psi} \leq \|f\|^2.$$

Note that f vanishes near the points where $e^{-\psi}$ is not integrable (i.e. near x_1 and x_2), so the right hand side is finite and the theorem does apply. The left hand side is also finite so u must vanish at x_1 and x_2. Hence $s := s' - u$ is holomorphic, vanishes at x_2 but not at x_1. □

Exercise: Use a similar technique to find a holomorphic section of $K_X + kL$ with prescribed first order Taylor expansion at a given point x in X. Show that this means that the Kodaira map has nondegenerate differential if k is large enough. □

Thus we see that any compact X which carries a positive line bundle is actually biholomorphic to a smooth submanifold of some projective space. The converse to this statement is also true - since projective space carries the positive bundle $\mathcal{O}(1)$ we get a positive bundle on any submanifold by taking restrictions.

Finally a word on terminology. A line bundle is said to be *very ample* if its Kodaira map is an embedding. It is said to be *ample* if some power of it is very ample. Thus Kodaira's embedding theorem can be rephrased as saying that positive bundles are ample. Conversely, very ample bundles must be positive since they are isomorphic to the pullback of $\mathcal{O}(1)$ under the Kodaira embedding.

Exercise: Prove this!

Hence ample bundles are also positive, so positivity and ampleness are equivalent.

5.3. The Kawamata-Viehweg vanishing theorem

A line bundle L over a compact manifold is said to be *numerically effective* if for any $\epsilon > 0$ there is a smooth metric on L with

$$i\partial\bar{\partial}\phi > -\epsilon\omega,$$

where ω is any given Kähler form. This is not the original definition, which says that L is numerically effective if for some (and hence any) smooth metric ψ on L the integral

$$\int_C i\partial\bar{\partial}\psi \geq 0,$$

for any curve C in X. If X is projective, the *Kleiman criterion* says that this definition coincides with the one we have given. In a non-projective manifold the definition we have chosen (which originates with Demailly) is the only one possible, since there may not be any curves C to test the integral condition on. Moreover, because of its more analytic character it fits much better with the methods of these notes.

We already know from Lecture 5 that if d_k is the dimension of the space of global sections of kL, then d_k grows with k at most at the rate

$$d_k \leq Ck^n.$$

We say that L is *big* if this maximal growth rate is actually attained, so that for k large

$$d_k \geq \epsilon k^n.$$

By the Bergman kernel asymptotics in the previous lecture any positive bundle is big. Bigness is however more general than positive. While, as we have seen in the previous section, "positive" means that the Kodaira map for kL is eventually an embedding, one can prove that bigness means that the image of the Kodaira map has maximal dimension ($= n$).

Using a method by Demailly ([**9**]) we shall now see that the Demailly-Nadel theorem implies a famous result of Kawamata [**11**] and Viehweg, [**23**].

Theorem 5.3.1. *Let X be a compact manifold with a holomorphic line bundle L that is both numerically effective and big. Then*

$$H^{n,q}(X, L) = 0,$$

for $q > 0$.

In order to prove the theorem it is enough to construct a singular metric, ϕ, on L, satisfying the hypotheses of the Demailly-Nadel vanishing theorem, which moreover is such that $e^{-\phi}$ is locally integrable. Then any smooth form has finite norm with respect to this metric, so theorem 5.1.1 implies that it is exact if it is closed. The first step is the following criterion of bigness, due to Kodaira.

Proposition 5.3.2. *A line bundle L on a projective manifold X is big (if and) only if some multiple of it can be written*

$$kL = A + E,$$

where A is ample and E is effective.

Proof. Since X is projective it carries some ample bundle A. Possibly after replacing A by some multiple we can assume that A has a nontrivial holomorphic section s. We can also arrange thing so that the zero divisor, S of s is smooth. (This will be generically true if A is very ample, as follows from *Bertini's theorem*, a variant of Sard's lemma.) Now consider the short sequence of maps

$$H^0(X, kL - A) \mapsto H^0(X, kL) \mapsto H^0(X, kL|_S).$$

The first map here is taking products with s; the second is restricting to S. By the Bergman kernel asymptotics of Lecture 5, the dimension of the last space is at most Ck^{n-1}, and since L is big, the dimension of the middle space is of order k^n. Hence the kernel of the last map is nontrivial if k is large, and we let s_k be some element in the kernel. Then s_k can be written

$$s_k = st_k,$$

where t_k is some global holomorphic section of $E := kL - A$. Therefore E is effective, so we have proved the "only if" part. For the "if" part we note that if $kL = A + E$, then for any large integer m $H^0(X, mkL)$ has large dimension (since $H^0(X, mA)$ has large dimension and mE has at least some section). Hence

$$\liminf d_p/p^n > 0.$$

This actually implies that

$$\lim d_p/p^n > 0,$$

(since the limit exists), but we shall not prove it (nor use it). □

The next lemma is almost trivial.

Lemma 5.3.3. *If L is numerically effective and A is ample, then $L + A$ is ample.*

PROOF. Let ψ be a metric on A with strictly positive curvature. By assumption, L has a metric ϕ with

$$i\partial\bar{\partial}\phi > -i\partial\bar{\partial}\psi.$$

Hence $\phi + \psi$ is a metric on $L + A$ with strictly positive curvature. □

We can now conclude the proof of the Kawamata-Viehweg theorem. Write

$$kL = A + E,$$

with A ample and E effective. By the lemma

$$kL + L = A_1 + E$$

where A_1 is still ample. Iterating, we see that

$$(k+m)L = A_m + E,$$

where A_m is ample and E is the same effective bundle (it is important that E does not change with m). We equip $(k+m)L$ with the (singular) metric

$$\phi_m + \log|t|^2,$$

where ϕ_m has strictly positive curvature and t is one fixed section of E. Then

$$\phi_L := \frac{1}{k+m}(\phi_m + \log|t|^2)$$

is a metric on L with strictly positive (albeit small!) curvature current. Its only singularities come from the last term, so if m is large enough, $e^{-\phi_m}$ is integrable. This concludes the proof. □

Exercise: Justify the last part of the proof by showing that if h is a holomorphic function defined in a neighbourhood of a closed ball B, then

$$\int_B \frac{1}{|h|^\delta} < \infty$$

if δ is small enough. □

LECTURE 6

Adjunction and extension from divisors

In this lecture we will discuss the Ohsawa-Takegoshi extension theorem, which deals with the extension from a divisor to the ambient space of holomorphic sections of a line bundle. The statement of the theorem is more natural if we consider *adjoint bundles* i.e. line bundles of the form

$$L + K_X,$$

where K_X is the canonical bundle of the manifold. This is of course only a matter of convenience, since any bundle can be put in this form, and we will afterwords also translate the theorem back to the non adjoint case. In the first section we discuss the relation between the canonical bundle of the divisor and the canonical bundle of the ambient space, which is described in the *adjunction formula.* There we also discuss some basic formulas describing the current of integration on a divisor that we will need in the sequel. In the second section we prove the extension theorem of Ohsawa-Takegoshi, see [**16**] for the first version of this theorem that has many variants. Apart from proving the possibility to extend sections under optimal conditions, this theorem gives sharp L^2-estimates for the extension. The hypotheses in the extension theorem involve an inequality between the curvatures of the line bundle L and the line bundle defined by the divisor. In the simple case of line bundles over the Riemann sphere the sections of these line bundles are polynomials, and the inequalities translate to an inequality between the number of points in the divisor, and the degree of an interpolating polynomial. In this way we see that the hypotheses are in fact sharp, already in the simplest case.

6.1. Adjunction and the currents defined by divisors

Let S be a (smooth) hypersurface of X. Then S defines a divisor, that in turn defines an associated line bundle (S), having a global holomorphic section s which vanishes to degree 1 exactly on S. The adjunction formula expresses the canonical bundle of the hypersurface in terms of the canonical bundle of the ambient space and (S).

Theorem 6.1.1. *Let S be a smooth hypersurface in a complex manifold X. Then the canonical bundle of S satisfies*

$$K_S = (K_X + (S))|_S.$$

PROOF. Let u be any local section of K_S, i.e. a locally defined $n-1$-form in an open set in S. Then

$$ds \wedge u$$

is a form of bidegree $(n, 0)$ with values in (S) (ds is not globally defined on X but its restriction to S is a well defined $(1, 0)$-form with values in (S) (why?)). Thus the map that takes u to $ds \wedge u$ is well defined and it is easily checked that it is an isomorphism at any point. □

In the next section we will discuss 'extension' of sections u of $K_S + L|_S$ to sections U of $K_X + (S) + L$ over all of X. Extension is here to be taken in the sense of the adjunction theorem so that

$$U = ds \wedge u$$

on S. The method of proof we will use (see [**1**], [**3**]) reduces this extension problem to the problem of solving a $\bar\partial$-equation

$$\bar\partial v = f,$$

where

$$f = u \wedge [S],$$

$[S]$ being the *current of integration on* S. We therefore first discuss such currents of integration in an attempt to make the main lines of the proof understandable also to readers without much knowledge of the theory of currents.

By the *Lelong-Poincaré formula* the current of integration on a hypersurface S is given by

$$[S] = \frac{i}{2\pi}\partial\bar\partial \log |h|^2$$

if h is any (local) holomorphic function vanishing to degree one precisely on S. This means that if α is any compactly supported form of bidegree $(n-1, n-1)$ then

$$\int_S \alpha = \frac{1}{2\pi}\int_X \log |h|^2 i\partial\bar\partial\alpha.$$

Exercise: Prove this formula! (Hint: Since h vanishes to degree 1 and the formula is local, you may choose coordinates so that $h = z_1$. Use that, in one complex variable,

$$\frac{\partial^2}{\partial\zeta\partial\bar\zeta}\frac{1}{2\pi}\log|\zeta|^2 = \delta_0,$$

a Dirac mass at $\zeta = 0$.) □

We will also have use for the closely related formula

$$[S] = \frac{i}{2\pi}\partial h \wedge \bar\partial\frac{1}{h}.$$

This can again be proved by choosing local coordinates so that $h = z_1$. It also follows from our previous expression for $[S]$ since

$$\partial\bar\partial \log|h|^2 = -\bar\partial\frac{\partial h}{h} = \partial h \wedge \bar\partial\frac{1}{h}.$$

The estimates that we will find for the extension U is in term of the L^2-norms

$$\int_X c_n U \wedge \bar U e^{-\phi-\psi},$$

(where ϕ and ψ are metrics on L and (S) respectively) and

$$\int_S c_{n-1} u \wedge \bar u e^{-\phi}.$$

These norms are well defined if U and u are $(n, 0)$ and $(n-1, 0)$ forms with values in $L + (S)$ and L respectively. They can also be expressed in terms of L^2

norms with respect to the volume elements on X and S. For this we need to use explicitly the Kähler form ω on X.

This form ω induces a norm on $(n,0)$ forms on X by

$$c_n U \wedge \bar{U} = |U|^2_\omega \omega_n.$$

We can also write

$$|U|^2_\omega = |U|^2 e^{-\phi_\omega}$$

to emphasise that this way ω induces a metric ϕ_ω on the canonical bundle of X. Then

$$\int_X c_n U \wedge \bar{U} e^{-\phi-\psi} = \int_X |U|^2_\omega e^{-\phi-\psi}\omega_n = \int_X |U|^2 e^{-\phi-\psi-\phi_\omega}\omega_n$$

To express the integral over S in a similar way we need to use dS, the volume (or surface) measure on the hypersurface S. This is a measure on X (concentrated on S of course), and can hence be viewed as a current of bidegree (n,n). We define a corresponding current of bidegree $(0,0)$ by

$$*dS\omega_n = dS.$$

We will use a representation of the current of integration on S that we state as a separate lemma.

Lemma 6.1.2.

$$[S] = \frac{ids \wedge d\bar{s}}{|ds|^2_\omega} * dS, \tag{6.1}$$

where the right hand side is defined by taking any local representative for the section s.

PROOF. The formula means that if α is any form of degree $2n-2$ on X then

$$\int_S \alpha$$

equals the integral of α against the right hand side. Since booth sides depend only on the restriction of α to S, and since on S any form of maximal degree is a multiple of the volume form, it is enough to verify this for $\alpha = g\omega_{n-1}$ where g is some function. But since $|ds|^2_\omega$ satisfies

$$|ds|^2_\omega \omega_n = ids \wedge d\bar{s} \wedge \omega_{n-1}$$

we get

$$\int g\omega_{n-1} \wedge \frac{ids \wedge d\bar{s}}{|ds|^2_\omega} * dS = \int g\omega_n * dS = \int gdS = \int_S g\omega_{n-1},$$

by the definition of the volume form on S as ω_{n-1}. This proves (6.1). □

Exercise*: This exercise outlines the corresponding formulas in the setting of real manifolds. It is not used in the sequel.

a. Let M be a smooth hypersurface in an n-dimensional Riemannian manifold X. Let M be defined locally by an equation $\rho = 0$, where $d\rho \neq 0$ on M, and let dV_M be the Riemannian volume element on M (so that dV_M is a form of degree $n-1$ on M). Let γ be a (locally defined) form of degree $n-1$ on X. Then the restriction of γ to M equals dV_M if and only if

$$\gamma \wedge d\rho/|d\rho| = dV_X,$$

the Riemannian volume element of X.

b. Let dM be the surface measure on M considered as a measure on X concentrated on M, and define $*dM$ by

$$*dM dV_X = dM.$$

Then the current of integration on M is given by

$$[M] = \frac{d\rho}{|d\rho|} * dM.$$

(In both (a) and (b) we assume that M is appropriately oriented.) □

6.2. The Ohsawa-Takegoshi extension theorem

We can now state the "adjunction version" of the Ohsawa-Takegoshi extension theorem. We first deal with the case of a compact manifold and smooth metrics.

Theorem 6.2.1. *Let X be a compact Kähler manifold and let S be a smooth hypersurface in X, defined by a global holomorphic section s of the line bundle (S). Let L be a complex line bundle over all of X. Assume that L and (S) have smooth metrics, ϕ and ψ respectively, satisfying the curvature assumptions*

$$i\partial\bar\partial\phi \geq 0$$

and

$$i\partial\bar\partial\phi \geq \delta i\partial\bar\partial\psi,$$

with $\delta > 0$. Assume moreover that s is normalized so that

$$|s|^2 e^{-\psi} \leq e^{-1/\delta}.$$

Let finally u be a global holomorphic section of $K_S + L|_S$.

Then there is a global holomorphic section U of $K_X + (S) + L$ such that

$$U = ds \wedge u$$

on S and such that U satisfies the estimate

$$\int_X c_n U \wedge \bar U e^{-\phi-\psi} \leq C_\delta \int_S c_{n-1} u \wedge \bar u e^{-\phi}. \tag{6.2}$$

Here C_δ is a constant depending only on δ.

One very interesting feature of the statement is that the constant C_δ depends only on δ- this is one reason why the theorem is so useful. Even more remarkably, if $i\partial\bar\partial\psi = 0$ (or ≤ 0), then the only assumption on ϕ is that $i\partial\bar\partial\phi \geq 0$. Thus we get a uniform constant even if we don't have a strict lower bound for the curvature. Let now u be a (local or global) section of $K_S + L|_S$. Suppose U is some section of $K_X + (S) + L$ such that

$$U = ds \wedge u$$

on S. Put $v' = -(i/2\pi)U/s$ so that v is a section of $K_X + L$. Then

$$\bar\partial v' = -(i/2\pi)\bar\partial\frac{1}{s} \wedge U = (i/2\pi) ds \wedge \bar\partial\frac{1}{s} \wedge u = u \wedge [S],$$

since $\bar\partial(1/s)$ vanishes outside of S. Conversely, suppose we are able to solve the $\bar\partial$-equation

$$\bar\partial v = u \wedge [S]. \tag{6.3}$$

Then $v - v' = h$ is holomorphic and hence in particular smooth. Therefore sv is also holomorphic and satisfies

$$2\pi i s v = ds \wedge u$$

on S, so any such v gives us a solution to the extension problem. The extension problem is therefore completely equivalent to the problem of solving (6.3).

We should point out that if we assume that we have *strict* inequality in the curvature assumption for ϕ so that $i\partial\bar{\partial}\phi > 0$, and if we don't care about estimates for the solution, then the possibility of solving (6.3) follows from the Kodaira vanishing theorem. Indeed, the Kodaira theorem says that we can solve any equation $\bar{\partial}v = f$ if f is a smooth L-valued $(n, 1)$-form and from this it follows that one can solve such an equation even if f is not smooth, but just a current. This is a well known fact — the cohomology defined with currents, and the cohomology defined with smooth forms are isomorphic — but we will not prove it.

To solve (6.3) we follow the method used in the proof of the Hörmander L^2-estimates, but a new twist is needed since the right hand side is now a form with measure coefficients instead of a form in L^2. Put

$$f := u \wedge [S],$$

so that f is a form of bidegree $(n, 1)$ with values in L. Indeed, u is a form of bidegree $(n-1, 0)$ on S and we can extend it smoothly in an arbitrary way to a form of the same bidegree on X. The wedge product $u \wedge [S]$ is independent of the choice of extension since $[S]$ is supported on S and contains a factor ds.

As in the proof of Hörmander's theorem we need to estimate the "scalar product" between f and a smooth compactly supported form α of bidegree $(n, 1)$. Write

$$\alpha = \gamma \wedge \omega,$$

where $\gamma = \gamma_\alpha$ is $(n-1, 0)$ and ω is the Kähler form. As in Lecture 3 the scalar product then becomes

$$\int_X f \wedge \bar{\gamma} e^{-\phi}$$

which equals

$$\int u \wedge \bar{\gamma} e^{-\phi} \wedge [S] = \int_S u \wedge \bar{\gamma} e^{-\phi}.$$

By Cauchy's inequality this can be estimated

$$|\int_S u \wedge \bar{\gamma} e^{-\phi}|^2 \leq \int c_{n-1} \gamma \wedge \bar{\gamma} e^{-\phi},$$

if we normalize things so that the L^2-norm of u over S equals 1.

To estimate this quantity we use again Proposition 3.4.1, but this time we multiply the formula for $i\partial\bar{\partial}T_\alpha$ by a certain function $w \geq 0$ before integrating. After applying Stokes' formula this introduces an extra term $i\partial\bar{\partial}w$ which is the key to the estimate. Take

$$w = -r \log |s|^2 e^{-\psi},$$

where $0 < r < 1$. We formulate the basic estimate as a separate lemma.

Lemma 6.2.2.

$$|\int_S c_{n-1} \gamma \wedge \bar{\gamma} e^{-\phi}|^2 \leq C_\delta\left(\int_X e^w c_n \partial_\phi \gamma \wedge \overline{\partial_\phi \gamma} e^{-\phi} + \int_X (w+1) |\bar{\partial}\alpha|^2 e^{-\phi} \omega_n\right)$$

PROOF. By our size estimate on s, $w \geq r/\delta$. Proposition 3.4.1 then implies after integration by parts that

$$c_{n-1}\Big(\int wi\partial\bar\partial\phi \wedge \gamma \wedge \bar\gamma e^{-\phi} - \int i\partial\bar\partial w \wedge \gamma \wedge \bar\gamma e^{-\phi}\Big) \leq$$

$$\leq 2c_{n-1}\int w\bar\partial\partial_\phi\gamma \wedge \bar\gamma e^{-\phi} + \int w|\bar\partial\alpha|^2 e^{-\phi}\omega_n. \tag{6.4}$$

By the Lelong-Poincaré formula

$$i\partial\bar\partial w = ri\partial\bar\partial\psi - r[S] \leq (r/\delta)i\partial\bar\partial\phi - r[S].$$

The first term here gives a negative contribution to the second integral in the left hand side of (6.4). Since $w \geq r/\delta$ this term is however controlled by the first integral. Hence

$$r\int_S c_{n-1}\gamma \wedge \bar\gamma e^{-\phi} \leq 2c_{n-1}\int w\bar\partial\partial_\phi\gamma \wedge \bar\gamma e^{-\phi} + \int w|\bar\partial\alpha|^2 e^{-\phi}\omega_n. \tag{6.5}$$

Apply Stokes' formula to the first term in the right hand side. We then get

$$\int w\bar\partial\partial_\phi\gamma \wedge \bar\gamma e^{-\phi} = \int w\partial_\phi\gamma \wedge \overline{\partial_\phi\gamma} e^{-\phi} + \int \bar\partial w \wedge \partial_\phi\gamma \wedge \bar\gamma e^{-\phi}.$$

The first term in the right hand side here is OK as it stands, but the second one needs some extra work. By the Cauchy inequality it is dominated by

$$|\int \bar\partial w \wedge \partial_\phi\gamma \wedge \bar\gamma e^{-\phi}| \leq \frac{1}{2}(c_n\int e^w\partial_\phi\gamma \wedge \overline{\partial_\phi\gamma}e^{-\phi} + e^{-w}\partial w \wedge \bar\partial w \wedge \gamma \wedge \bar\gamma e^{-\phi})$$

The first term on the right hand side is exactly what we want. The second term again contains γ, but it is less singular than the integral over S that we have just estimated, so it is at least intuitively clear that it should not cause any serious trouble.

To estimate it we use Proposition 3.4.1 once more. This time we multiply $i\partial\bar\partial T_\alpha$ by $W = 1 - e^{-w} \geq 1 - e^{-r/\delta}$ before integrating. Note that

$$i\partial\bar\partial W = r(i\partial\bar\partial\psi - i\partial w \wedge \bar\partial w)e^{-w} \leq \frac{r}{\delta}e^{-w}i\partial\bar\partial\phi - e^{-w}i\partial w \wedge \bar\partial w.$$

We then get

$$c_n\int (W - re^{-w}/\delta)i\partial\bar\partial\phi \wedge \gamma \wedge \bar\gamma e^{-\phi} + c_n\int e^{-w}\partial w \wedge \bar\partial w \wedge \gamma \wedge \bar\gamma e^{-\phi} \leq$$

$$c_n\int W\bar\partial\partial_\phi\gamma \wedge \bar\gamma e^{-\phi} + \int W|\bar\partial\alpha|^2 e^{-\phi}\omega_n.$$

If δ is small enough $W - re^{-w}/\delta = 1 - (1 + r/\delta)e^{-w} \geq 0$ (since $w \geq r/\delta$). Hence we can neglect the first term on the left hand side. We then repeat the same procedure as above and apply Stokes' to the first integral in the right hand side. This again produces a good term plus an undesired term containing $\bar\partial W = \bar\partial we^{-w}$. Now however this term can be absorbed in the left hand side. The result is (since $e^{-w} \leq 1$ and $W \leq 1$) that

$$c_{n-1}\int e^{-w}i\partial w \wedge \bar\partial w \wedge \gamma \wedge \bar\gamma e^{-\phi} \leq C_\delta(\int c_n\partial_\phi\gamma \wedge \overline{\partial_\phi\gamma}e^{-\phi} + |\bar\partial\alpha|^2 e^{-\phi}\omega_n).$$

Inserting this in our previous estimate we finally get

$$\int_S c_{n-1}\gamma \wedge \bar{\gamma} e^{-\phi} \leq C_\delta \Big(\int e^w c_n \partial_\phi \gamma \wedge \overline{\partial_\phi \gamma} e^{-\phi} + w|\bar{\partial}\alpha|^2 \Big), \tag{6.6}$$

which proves the lemma. □

Let us now see how we get an existence theorem for $\bar{\partial}$ from this a priori estimate. The argument follows basically the reasoning in Lecture 3 but it is complicated by two things. First, $f = u \wedge [S]$ is no longer a form in L^2. Therefore we can not use the same functional analysis set up as before. Second, we can not hope to get a solution in L^2 either. Remember that our solution v will basically be U/s where U solves the extension problem. Hence v will never be in L^2; the estimate we are looking for is an L^2 estimate for sv.

Normalize so that

$$c_{n-1} \int u \wedge \bar{u} e^{-\phi} = 1.$$

The next lemma replaces the estimate for the scalar product $\langle f, \alpha \rangle$ from the standard L^2-theory.

Lemma 6.2.3.

$$|\int_S u \wedge \bar{\gamma} e^{-\phi}|^2 \leq C_\delta \int e^w c_n \partial_\phi \gamma \wedge \overline{\partial_\phi \gamma} e^{-\phi}, \tag{6.7}$$

To understand the significance of the lemma, recall that $\partial_\phi \gamma = \bar{\partial}^* \alpha$. Hence we again estimate the 'scalar product' between f and α by $\bar{\partial}^* \alpha$. The additional weight factor e^w in the right hand side is the price we pay for not having f in L^2.

PROOF. To prove (6.7) we decompose $\alpha = \alpha_1 + \alpha_2$ where α_1 is $\bar{\partial}$-closed and α_2 is orthogonal to the space of $\bar{\partial}$-closed forms, and put $\alpha_j = \gamma_j \wedge \omega$. Since $f = u \wedge [S]$ is $\bar{\partial}$-closed we claim that

$$\int f \wedge \gamma e^{-\phi} = \langle f, \alpha \rangle = \langle f, \alpha_1 \rangle = \int f \wedge \gamma_1 e^{-\phi}.$$

This would have been completely evident if f had been a form in L^2; as it is now it requires a small argument that we will only sketch. The main point is that due to elliptic regularity, α_1 and α_2 are still smooth. Moreover, when X is compact, the ranges of $\bar{\partial}$ and $\bar{\partial}^*$ are closed. Since α_2 is orthogonal to the kernel of $\bar{\partial}$, this implies that $\alpha_2 = \bar{\partial}^* \chi$ for some smooth χ. Hence $\langle f, \alpha_2 \rangle = \langle \bar{\partial} f, \chi \rangle = 0$.

The claim (6.7) now follows since we may replace γ by γ_1 in the left hand side. This gives

$$|\int_S u \wedge \bar{\gamma} e^{-\phi}|^2 = |\int_S u \wedge \bar{\gamma}_1 e^{-\phi}|^2 \leq c_{n-1} \int_S \gamma_1 \wedge \bar{\gamma}_1 e^{-\phi}$$

which by Lemma 6.2.3 is dominated by

$$C_\delta \int c_n e^w \partial_\phi \gamma_1 \wedge \overline{\partial_\phi \gamma_1} e^{-\phi}$$

since $\bar{\partial}\alpha_1 = 0$. On the other hand $\partial_\phi \gamma_1 = \partial_\phi \gamma$ since $\partial_\phi \gamma_2 = \bar{\partial}^* \alpha_2 = 0$ since α_2 is orthogonal to closed forms. This proves the lemma. □

The theorem now follows in essentially the same way as we proved the standard L^2-estimate for $\bar\partial$. The Riesz representation theorem implies that there is some $(n,0)$-form η with

$$c_n \int \eta \wedge \bar\eta e^{w} e^{-\phi} \leq C_\delta$$

and

$$\int f \wedge \bar\gamma e^{-\phi} = \int_S u \wedge \bar\gamma e^{-\phi} = \int e^{w} \eta \wedge \overline{\partial_\phi \gamma} e^{-\phi},$$

for all smooth compactly supported $(n-1,0)$-forms γ. It might be appropriate to point out that this is the point where we use that we have chosen $w = -r \log |s|^2 e^{-\psi}$ with r smaller than 1. Then e^w is integrable so smooth forms lie in $L^2(e^w)$. Then $v = e^w \eta$ solves $\bar\partial v = f$ and

$$c_n \int v \wedge \bar v e^{-w} e^{-\phi} \leq C_\delta.$$

Concretely this means that

$$c_n \int v \wedge \bar v e^{-\phi} |s|^{2r} e^{-r\psi} \leq C_\delta.$$

Hence, since $U = sv$ and $|s|^2 e^{-\psi} \leq 1$,

$$\int_X c_n U \wedge \bar U e^{-\phi-\psi} \leq \int_X c_n v \wedge \bar v e^{-\phi} |s|^{2r} e^{-r\psi} \leq C_\delta,$$

which proves the theorem.

Let us now translate this form of the Ohsawa-Takegoshi theorem to the non adjoint case. Let dS be the surface (or volume) measure on the hypersurface S induced by the Kähler metric ω.

Let $F := K_X + (S) + L$, and put $\phi_F = \phi_\omega + \phi + \psi$. Recall that

$$c_n U \wedge \bar U = |U|^2 e^{-\phi_\omega}.$$

Hence, by lemma 6.1.2 our estimate for U reads

$$\int_X |U|^2 e^{-\phi_F} \omega_n \leq C_\delta \int_S |U|^2 e^{-\phi_F} \frac{dS}{|ds|^2 e^{-\psi}}.$$

Hence we have the following form of Theorem 6.2.1.

Theorem 6.2.4. *Let F be a holomorphic line bundle over X and let S be a divisor in X. Assume F and (S) have metrics ϕ_F and ψ respectively satisfying*

$$i\partial\bar\partial\phi_F \geq (1+\delta) i\partial\bar\partial\psi + \phi_{K_X}$$

where ϕ_{K_X} is some smooth metric on K_X. Assume also that

$$i\partial\bar\partial\phi_F \geq i\partial\bar\partial\psi + \phi_{K_X}.$$

Then any holomorphic section U_0 of F over S extends holomorphically to a section of the same bundle over X satisfying

$$\int_X |U|^2 e^{-\phi_F} \omega_n \leq C_\delta \int_S |U_0|^2 e^{-\phi_F} \frac{dS}{|ds|^2 e^{-\psi}}.$$

Strictly speaking we have proved this theorem only in the case when $\phi_{K_X} = \phi_\omega$ is a metric on K_X coming from some metric ω on X. The general case however follows from this. An arbitrary metric on K_X differs from ϕ_ω by some smooth function χ, and we can write

$$\phi_{K_X} + \phi + \psi = \phi_\omega + (\chi + \phi) + \psi,$$

so changing ϕ_ω to ϕ_{K_X} is equivalent to changing ϕ to $\phi + \chi$.

Note that in this formulation the curvature assumption on the bundle F involves comparison with the canonical bundle. In the case when the canonical bundle is nonpositive (meaning that the Ricci curvature of X is nonnegative) this makes the hypotheses easier to fulfill, cf the exercise below.

Exercise: Let X be the Riemann sphere and let S be a divisor consisting of n points. Prove that there are sections of $\mathcal{O}(n-1)$ over X that attains arbitrary given values at the points, but that this in general is not true for $\mathcal{O}(n-2)$. (You need a polynomial of degree $n-1$ to interpolate at n points!). Check that this means that the curvature assumption in the Ohsawa-Takegoshi extension theorem is sharp — it is not possible to take $\delta = 0$. (Recall that $K_{\mathbb{P}^1} = \mathcal{O}(-2)$.) □

6.2.1. A more general version involving nonsmooth metrics and noncompact manifolds

It is often important to be able to relax the assumptions in Theorem 6.2.1 as we have stated it. We will next give a version that allows for singular metrics and also noncompact manifolds. I do not know if the theorem holds as it stands for any complete Kähler manifold and line bundles with singular metrics. A reasonably general situation is the condition from [**14**] of a variety that becomes Stein after removal of some divisor. This certainly includes projective manifolds and of course Stein manifolds as well.

Theorem 6.2.5. *Let X be a complex manifold. Assume that X contains a divisor D such that $X \setminus D$ is Stein.*

Let S be a smooth hypersurface in X, defined by a global holomorphic section s of the line bundle (S), and let L be a complex line bundle over all of X. Assume that L and (S) have not necessarily smooth metrics, ϕ and ψ respectively, satisfying the curvature assumptions

$$i\partial\bar\partial\phi \geq 0$$

and

$$i\partial\bar\partial\phi \geq \delta i\partial\bar\partial\psi,$$

with $\delta > 0$. Assume moreover that s is normalized so that

$$|s|^2 e^{-\psi} \leq e^{-1/\delta}.$$

Let u be a global holomorphic section u of $K_S + L|_S$ such that

$$I := \int_S c_{n-1} u \wedge \bar{u} e^{-\phi} < \infty.$$

Then there is a global holomorphic section U of $K_X + (S) + L$ such that

$$U = ds \wedge u$$

on S and such that U satisfies the estimate

$$\int_X c_n U \wedge \bar{U} e^{-\phi-\psi} \leq C_\delta I. \tag{6.8}$$

Here C_δ is a constant depending only on δ.

Note that we have assumed that the L^2-norm of the section we wish to extend is finite. Just like in the Demailly-Nadel vanishing theorem this is a nontrivial condition even in the case of a compact manifold, since our line bundle metrics may have nonintegrable singularities. If it is not satisfied there is no guarantee that we can extend sections from S, with or without L^2-estimates!

We will not give a detailed proof of Theorem 6.2.5, but merely make a few remarks. First, it is enough to prove the theorem in the Stein case. The general case then follows from Lemma 5.1.3, which implies in particular that any holomorphic section on $X \setminus D$ extends holomorphically across D (given the L^2-condition). Hence, if we can extend u to $X \setminus D$, we automatically have an extension to all of X.It is then also enough to consider the case of smooth metrics, by the same arguments as in Lecture 5: Exhaust $X \setminus D$ by a sequence of relatively compact subdomains, on each of which we can approximate nonsmooth metrics with smooth ones. The basic estimate Lemma 6.2.2 is proved in exactly the same way on a noncompact manifold, provided we assume from the start that α and hence γ have compact support. The lemma then follows for not necessarily compactly supported forms if we first equip our Stein manifold with a complete Kähler metric. Then apply the lemma to $\chi_k \gamma$ where χ_k is an exhausting sequence of compactly supported cutoff functions having uniformly bounded gradients.

The only serious complication in the analysis is Lemma 6.2.3. In the proof of that lemma we decompose $\alpha = \alpha_1 + \alpha_2$ where α_1 is $\bar{\partial}$-closed and α_2 is orthogonal to the space of $\bar{\partial}$-closed forms. Then we write $\alpha_j = \gamma_j \wedge \omega$ and need to prove that

$$\int_S u \wedge \bar{\gamma}_2 e^{-\phi} = 0.$$

This is the "scalar product" between f and α_2, so it should be zero since f is $\bar{\partial}$-closed and α_2 is orthogonal to $\bar{\partial}$-closed forms, but again the crux is that f is not in L^2. So we need to approximate $f = u \wedge [S]$ in a suitable way by forms f_ϵ in L^2.

Again, we work on a relatively compact Stein subdomain of $X \setminus D$. Since this latter space is Stein, we can find a global holomorphic $(n-1, 0)$-form that restricts to u on S, in the sense that its pullback under the inclusion map equals u. We denote this global form by u too, and note that on our relatively compact subdomain u is bounded. We will approximate $f = u \wedge [S]$ by L^2-forms $f_\epsilon = u \wedge [S]_\epsilon$, where $[S]_\epsilon$ is a smooth approximation to $[S]$. For this we take

$$w_\epsilon = -\log(|s|^2 e^{-\psi} + \epsilon) = -\log(e^{-w} + \epsilon),$$

$$[S]_\epsilon = i\partial\bar{\partial} w_\epsilon - i\partial\bar{\partial}\psi,$$

and

$$f_\epsilon = u \wedge [S]_\epsilon.$$

Then f_ϵ is for positive ϵ a $\bar{\partial}$-closed form in L^2. We claim that

$$\int f \wedge \bar{\gamma}_2 e^{-\phi} = \lim \int f_\epsilon \wedge \bar{\gamma}_2 e^{-\phi}. \tag{6.9}$$

For this, note first that $\bar{\partial}\alpha_2 = \bar{\partial}\alpha$ and $\partial_\phi \gamma_2 = 0$. Hence it follows from Lemma 6.2.2 (or rather the noncompact version of that lemma) that

$$\int c_{n-1}\gamma_i \wedge \bar{\gamma}_i e^{-\phi} \wedge [S] < \infty.$$

A similar estimate is satisfied uniformly with $[S]$ replace by $[S]_\epsilon$ — it is proved in the same way, replacing w by w_ϵ To verify (6.9) we now decompose the integrals into two pieces, one close to the boundary and the remaining part. The first part is uniformly small if we are sufficiently close to the boundary since u is uniformly bounded. The other part converges as ϵ tends to zero since γ_2 is smooth. This completes the proof. □

Exercise: Let ϕ be a plurisubharmonic function defined in a neighbourhood of the origin in $\mathbb{C}^n$. Use the Ohsawa-Takegoshi theorem to prove the following: If for some $\epsilon > 0$

$$\int_{|\zeta|<\epsilon} e^{-\phi(\zeta,0,\ldots 0)} dm(\zeta) < \infty,$$

then fore some other ϵ' the integral in n variables is also finite:

$$\int_{|z|<\epsilon'} e^{-\phi(z)} dm(z) < \infty.$$

Try to prove this without using the Ohsawa-Takegoshi extension theorem! □

LECTURE 7

Deformational invariance of plurigenera

In this lecture we will use the Ohsawa-Takegoshi theorem to prove a celebrated result of Siu, **[20]**,**[21]**, on the "invariance of plurigenera".

Recall that for a compact Riemann surface, the *genus* is the dimension of the space of holomorphic one-forms, i.e. the space of global holomorphic sections of the canonical bundle. In the same way the dimension of the space of global holomorphic sections of mK_X is called the m-genus, also for compact manifolds of any dimension. Collectively these are referred to as the *plurigenera* of the manifold. In the one dimensional case, the genus is a topological invariant, hence in particular invariant under deformations of the manifold. Siu's theorem is that in any dimension, all the plurigenera are invariant under deformations. The main point in Siu's proof is to show that the dimensions do not jump down when we perturb the manifold, and this is accomplished by an extension theorem. We will give here a simplification, due to Paun, **[17]**, of Siu's original argument.

7.1. Extension of pluricanonical forms

In this section we will prove Siu's theorem on the extension of sections of multiples of the canonical bundle from the central fiber of a projective family to the ambient space. First we need a few basic definitions.

Let X be a complex manifold together with a holomorphic map p from X to the unit disk. We will assume that p defines a smooth fibration with compact fibers. By this we mean that the differential of p is surjective everywhere and that the fibers $X_t = p^{-1}(t)$ are compact manifold. We can then think of the fibers as forming a family of complex manifolds and we say that this family is projective if there is a positive line bundle A over the total space X. The fiber X_0 is a smooth hypersurface defined by an equation $p = 0$.

In the notation of the previous lecture $S = X_0$, $s = p$ and (S) is now a *trivial* line bundle. The adjunction theorem says in this case that $K_X|_{X_0} = K_{X_0}$ the isomorphism being given by

$$u \mapsto U = dp \wedge u.$$

Abusing notation slightly we will identify u and $U|_{X_0}$. The main result of Siu, **[21]**, is the following theorem.

Theorem 7.1.1. *Let u be a section of $mK_X|_{X_0}$. Then there is a holomorphic section U of mK_X over all of X that extends u.*

We give immediately the main corollary.

Corollary 7.1.2. *Let for any natural number m and any t in the disk $g_m(t)$ be the m-genus of the fiber X_t,*

$$g_m(t) = dimH^0(X_t, mK_{X_t}).$$

Then $g_m(t)$ is independent of t.

PROOF. The previous result shows that for any N the sets where $g_m(t)$ is at least N is open. A simple argument with normal families shows that it is also closed. The maximal value of N for which it is nonempty is the plurigenus. □

For the proof of Theorem 7.1.2 we will follow the method of Paun, [**17**], which simplified the original proof of Siu considerably. Notice first that the theorem follows immediately from Theorem 6.2.1 in case $m = 1$. In this case L and (S) are both trivial and ϕ and ψ are both zero, so by Theorem 6.2.1 u extends. For general m we write

$$(m-1)K_X =: L$$

so that $mK_X = K_x + L$. The crux of the matter is to find a metric ϕ on L over all of X with semi positive curvature current such that the section u that we want to extend satisfies

$$\int_{X_0} c_{n-1} u \wedge \bar{u}\, e^{-\phi} < \infty.$$

Then we can apply Theorem 6.2.1 again and get an extension of u. (We don't need any strict positivity of the curvature since the bundle (S) here is trivial!)

Over X_0 we can easily find such a metric. Since $\psi = \log |u|^2$ is a metric on mK_{X_0}, $\phi' = (1 - 1/m)\psi$ is a metric on $L = (m-1)K_X|_{X_0}$, and moreover

$$\int_{X_0} c_{n-1} u \wedge \bar{u}\, e^{-\phi'} = \int_{X_0} |u|^{2/m} < \infty.$$

The proof consists in finding an extension of ψ (and hence ϕ') as a metric with positive curvature current. This is simpler than extending u as a holomorphic section.

Let B be a line bundle over X that is sufficiently positive so that the following two conditions hold:

1. Any section of $pK_X + B$ over X_0 extends holomorphically to all of X if $p \le m-1$.

2. Still for $p \le m-1$, $pK_X + B$ is *base point free* over X_0, i.e. there is no point on X_0 where all sections to this bundle vanishes.

The first of these conditions is easy to achieve. Start with any smooth metric on K_X and some smooth metric of positive curvature on A. Let $B = lA$ for some l and take the induced metric on $pK_X + B$. This will have positive curvature if l is large enough and $p \le m-1$. Hence, Theorem 6.2.1 implies again that any holomorphic section extends.

The second condition will also hold if l is large enough, since by the proof of the Kodaira embedding theorem we can then find sections of $pK_X + B$ with prescribed (i.e. nonzero) values at any point in the compact X_0.

Choose for $p \le m-1$ a basis $(s_j^{(p)})$ for the space of global sections of $pK_X + B$ over the central fiber X_0.

Lemma 7.1.3. *For $k = 0, 1, \ldots$ and $p \le m-1$ any section*

$$u^k s_j^{(p)}$$

of $(mk+p)K_X + B$ over X_0 extends holomorphically to all of X.

PROOF. We prove this by induction over $l = mk+p$, and we know by hypothesis that the statement holds for $l < m$, i.e. $k = 0$ and $p \leq m-1$. The first nontrivial step is therefore to extend $us_j^{(0)}$. Put

$$h_{m-1} = \sum_j |\widetilde{s_j^{(m-1)}}|^2$$

where $\tilde{s}$ means an extension of s. Then $h_{m-1} = e^{\phi_{m-1}}$ where ϕ_{m-1} is a metric on $(m-1)K_X + B$. Since this bundle is base point free this metric is actually smooth and

$$\int_{X_0} |us_j^{(0)}|^2 e^{-\phi_{m-1}}$$

is thus finite. By Theorem 6.2.1 we can find extensions of $us_j^{(0)}$ satisfying

$$\int_X |\widetilde{us_j^{(0)}}|^2 e^{-\phi_{m-1}} \leq C \int_{X_0} |us_j^{(0)}|^2 e^{-\phi_{m-1}}. \tag{7.1}$$

Put

$$h_m = \sum_j |\widetilde{us_j^{(0)}}|^2$$

and define ϕ_m so that $e^{\phi_m} = h_m$. The new metric ϕ_m is no longer smooth, but the only singularities come from u so we have that

$$\int_{X_0} |us_j^{(1)}|^2 e^{-\phi_m} < \infty$$

Then we can iterate the argument again and continuing this way the lemma follows.

□

Notice also that during the proof of the lemma we obtain a sequence of metrics on $lK_X + B$,

$$h_l = \sum_j |\widetilde{u^k s_j^{(p)}}|^2$$

for $l = km + p$. These metrics satisfy good estimates namely

$$\int_X h_{l+1}/h_l \leq C \int_{X_0} h_{l+1}/h_l. \tag{7.2}$$

To verify this for $l = m-1$ we just sum over j in (7.1), and since all the metrics are constructed in a similar way all of the metrics satisfy (7.2). The integral in the right hand side here is

$$\int_{X_0} \sum |u^k s_j^{(p)}|^2 / \sum |u^k s_j^{(p-1)}|^2$$

if $p > 0$, and

$$\int_{X_0} \sum |u^k s_j^{(0)}|^2 / \sum |u^{k-1} s_j^{(m-1)}|^2$$

if $p = 0$. Hence they are bounded by a fixed constant, depending only on the choice of u and the choice of bases $s_j^{(p)}$, so we get

$$\int_X h_{l+1}/h_l \leq C.$$

By Hölder's inequality this implies that (define h_l to be some arbitrary smooth metric for $l < m-1$)

$$\int_X h_l^{1/l} = \int_X (h_l/h_{l-1})^{1/l}(h_{l-1}/h_{l-2})^{1/l}...h_1^{1/l} \leq$$
$$(\int_X h_l/h_{l-1})^{1/l}(\int_X h_{l-1}/h_{l-2})^{1/l}...(\int_X h_1)^{1/l} \leq C.$$

Take in particular $l = mk$ (the reason for this will be seen below). Thus by the submeanvalue property of plurisubharmonic functions

$$\phi_\infty := \limsup \frac{1}{mk}\phi_{mk}$$

is finite everywhere. Since ϕ_{mk} is a metric on $mkK_X + B$, ϕ_∞ is a metric on K_X (B disappears in the limit!).

After taking the upper semicontinuous regularization we get a metric ψ on K_X with semipositive curvature current which is greater than or equal to ϕ_∞ everywhere. On X_0

$$h_{km} = |u|^{2k}h$$

where h is a smooth positive function. Hence $e^\psi \geq e^{\phi_\infty} = |u|^{2/m}$ on X_0, so

$$\int_{X_0} |u|^2 e^{-(m-1)\psi} \leq \int_{X_0} |u|^{2/m} < \infty.$$

This completes the proof of the theorem. $\square$

Bibliography

[1] E AMAR: *Extension de fonctions holomorphes et courants*, Bull. Sciences Mathmatiques, 2me srie, 107 (1983), 25-48.

[2] A ANDREOTTI AND E VESENTINI: *Carleman estimates for the Laplace-Beltrami equation on complex manifolds.* , Inst. Hautes tudes Sci. Publ. Math. No. 25 1965 81–130.

[3] B BERNDTSSON: *The extension theorem of Ohsawa-Takegoshi and the theorem of Donnelly-Fefferman*, Annales de l'institut Fourier, 46 no. 4 (1996), p. 1083-1094 .

[4] R BERMAN, B BERNDTSSON AND J SJÖSTRAND: *A direct approach to Bergman kernel asymptotics for positive line bundles.* , Ark. Mat. 46 (2008), no. 2, 197–217.

[5] TH BOUCHE: *Convergence de la mtrique de Fubini-Study d'un fibr linaire positif.*, Ann. Inst. Fourier (Grenoble) 40 (1990), no. 1, 117–130.

[6] H J BRASCAMP AND E H LIEB: *On extensions of the Brunn-Minkowski and Prkopa-Leindler theorems, including inequalities for log concave functions, and with an application to the diffusion equation.*, J. Functional Analysis 22 (1976), no. 4, 366–389..

[7] CATLIN, D: *The Bergman kernel and a theorem of Tian*, Analysis and geometry in several complex variables (Katata, 1997), 1–23, Trends Math., Birkhuser Boston, Boston, MA, 1999. .

[8] J-P DEMAILLY: *Estimations L^2 pour l'oprateur $\bar\partial$ d'un fibré vectoriel holomorphe semi-positif au-dessus d'une varit kählrienne complète.*, Ann. Sci. cole Norm. Sup. (4) 15 (1982), no. 3, 457–511.

[9] J-P DEMAILLY: *Une généralisation du théorème d'annulation de Kawamata-Viehweg*, C. R. Acad. Sci. Paris Sr. I Math. 309 (1989), no. 2, 123–126.

[10] L. HÖRMANDER: *L^2-estimates and existence theorems for the $\bar\partial$-equation* , Acta Math., 113 (1965).

[11] Y KAWAMATA: *A generalization of Kodaira - Ramanujam's vanishing theorem*, Math. Ann. 261 (1982), no. 1, 43–46.

[12] K KODAIRA: *On a differential-geometric method in the theory of analytic stacks.* , Proc. Nat. Acad. Sci. U. S. A. 39, (1953). 1268–1273. .

[13] K KODAIRA: *On Kähler varieties of restricted type (an intrinsic characterization of algebraic varieties*, Ann. of Math., 60 (1954), 28-40.

[14] J McNeal and D VarolinAnalytic inversion of adjunction: L^2 extension theorems with gain. Ann. Inst. Fourier (Grenoble) 57 (2007), no. 3, 703–718

[15] A M NADEL: *Multiplier ideal sheaves and Khler-Einstein metrics of positive scalar curvature*, Ann. of Math. (2) 132 (1990), no. 3, 549–596.

[16] T OHSAWA AND K TAKEGOSHI: *On the extension of L^2 holomorphic functions*, Math. Z. 195 (1987) pp. 197-204.

[17] M PAUN: *Siu's invariance of plurigenera: a one-tower proof.*, J. Differential Geom. 76 (2007), no. 3, 485–493. .

[18] A PREKOPA: *On logarithmic concave measures and functions.*, Acad Sci Math (Szeged) 34 (1973) 335-343..

[19] Y-T SIU: *Complex-Analyticity at harmonic maps, vanishing and Lefschetz theorems*, J. Diff. Geom., 17 (1982), 55-138 .

[20] Y-T SIU: *Invariance of Plurigenera*, Inv. Math. 134 (1998) 661–673 .

[21] Y-T SIU: *Extension of twisted pluricanonical sections with plurisubharmonic weight and invariance of semipositively twisted plurigenera for manifolds not necessarily of general type*, Complex geometry (G"ottingen, 2000), 223–277, Springer, Berlin, 2002 .

[22] G TIAN: *On a set of polarized Khler metrics on algebraic manifolds.*, J. Differential Geom. 32 (1990), no. 1, 99–130..

[23] E Viehweg: *Vanishing theorems*, J. Reine Angew. Math. 335 (1982), 1–8.

[24] Zelditch, S.: *Szeg kernels and a theorem of Tian.*, Internat. Math. Res. Notices 1998, no. 6, 317–331..

Real and Complex Geometry meet the Cauchy-Riemann Equations

John P. D'Angelo

IAS/Park City Mathematics Series
Volume 17, 2008

Real and Complex Geometry meet the Cauchy-Riemann Equations

John P. D'Angelo

Preface

The principal themes in these notes arise from studying the interaction between real and complex geometry. Several of the specific questions come from studying the Cauchy-Riemann equations in several variables as a first-order system of PDE. Other questions arise naturally from positivity conditions in complex analysis. The same issues form the foundations of all these subjects. We need to analyze information arising from second derivatives in situations when that information degenerates in some manner. The simplest such question asks under what conditions is a degenerate critical point of a smooth function a local extremum? We therefore begin by discussing the difficulties inherent in this question for functions of several variables. We then go on to investigate real hypersurfaces with degenerate Levi forms; the Levi form comes from second derivatives and the parallels will become quite transparent. We will make the connection between degenerate Levi forms and Kohn's work on subelliptic multipliers.

These notes are a slightly expanded and edited version of what was provided to the students at PCMI in July 2008; while not all of the material here was covered in the talks given there, nearly all the material did appear in that earlier written form. The material here is divided into seven Lectures. Lecture 1 consists of introductory material and Lecture 7 states eleven open problems. Lectures 2-6 roughly correspond to what was presented in the talks given at PCMI, although only Lectures 2-4 were discussed in detail. The final talk at PCMI carefully covered Theorems 6.2 and 6.3; it also presented Proposition 6.3.

The intended audience for these lectures consists of graduate students in both complex analysis and algebraic geometry as well as research mathematicians. Finding a common intersection of interests is not simple. The logical development in the exposition does not seek maximal economy of thought. The author hopes that the occasional repetition of ideas will be useful rather than annoying. He feels that well-chosen examples facilitate the understanding of related abstract statements; some readers might tire of too many formulas, and the author apologizes to those who find some of the examples too trivial.

More than one hundred exercises appear throughout the text. Many of these exercises are routine, some are a bit harder, and a few are open problems which are stated again in Lecture 7. The lengthy bibliography might be useful to some

Dept. of Mathematics, University of Illinois, 1409 W.Green St., Urbana IL 61801
E-mail address: jpda@math.uiuc.edu

readers; the author's interest in several complex variables does seem to create an imbalance in the references, for which he apologizes.

Lecture 1 summarizes some standard facts and notations. We determine in Lecture 2 when an algebraic real hypersurface contains non-constant holomorphic curves. One of the main techniques in these notes is to analyze functions of several variables by investigating their restrictions to various curves in one dimension. This idea leads in Lecture 3 to the notion of *point of finite type* on a real hypersurface in complex Euclidean space; much of our discussion concerns motivation, basic properties, and applications of this concept. We will see a beautiful combination of ideas from complex analysis and elementary algebraic geometry. Lecture 4 investigates Kohn's method of subelliptic multipliers in a situation simple enough to be understandable but subtle enough to reveal the depths of the connections between PDE and Algebraic Geometry. Lecture 5 sketches without proofs various difficult analytic results; the discussion attempts to unify the ideas from the earlier lectures. In particular it explains why finite type is a biholomorphic invariant. Finally in Lecture 6 we investigate anew the basic techniques, by studying positivity conditions for Hermitian symmetric polynomials and real-analytic functions. We discuss there a complex variables analogue of Hilbert's seventeenth problem. The ideas useful in studying the geometry of real hypersurfaces turn out to help understand basic questions of positivity but new ideas emerge as well.

The author acknowledges support from NSF Grant DMS-07-53978. He thanks Jeff McNeal and Mircea Mustata for organizing the PCMI program for 2008 and for inviting him to give this course. Catherine Giesbrecht and Dena Vigil do a superb job in running PCMI. Many of the participants at PCMI made useful comments on the preliminary version; Yunus Zeytuncu and Andre Minor provided particularly useful suggestions. The author and Dror Varolin exchanged comments on each other's lecture notes. The author's colleague Tom Nevins provided several useful suggestions and clarifications. The author appreciates John Polking's efforts in turning the preliminary version into this final form. The author thanks Joe Kohn for many years of discussions on much of this material; especially in the discussion of subelliptic estimates Kohn's influence is more than evident.

LECTURE 1

Background material

1. Complex linear algebra

Let $\mathbf{C}^n$ denote n-dimensional complex Euclidean space. The inner product of z and w is given by

$$\langle z, w\rangle = \sum_{j=1}^{n} z_j \overline{w}_j \tag{1}$$

and the Euclidean norm is given by $||z||^2 = \langle z, z\rangle$. The Euclidean topology on $\mathbf{C}^n$ agrees with the usual Euclidean topology on $\mathbf{R}^{2n}$; we will use standard terms from point-set topology without comment. On occasion we will use facts from Hermitian linear algebra. We therefore recall the notion of a complex inner product space and related ideas. Of course complex Euclidean space is the basic model for complex inner product space.

A *complex inner product space* is a complex vector space V equipped with an inner product. An inner product assigns to each pair z, w of elements in V a complex number $\langle z, w\rangle$, called the inner product of z and w, satisfying the following axioms:

1) (positive definiteness) For all nonzero $z \in V$, $\langle z, z\rangle > 0$.

2) (linearity in the first slot). For all $z, \zeta, w \in V$ and for all $c \in \mathbf{C}$, we have

$$\langle cz, w\rangle = c\langle z, w\rangle$$

$$\langle z + \zeta, w\rangle = \langle z, w\rangle + \langle \zeta, w\rangle.$$

3) (Hermitian symmetry) For all $z, w \in V$, we have $\langle z, w\rangle = \overline{\langle w, z\rangle}$.

It follows from these axioms that the inner product is *conjugate linear* in the second slot. In the physics literature complex inner products are often assumed to be linear in the second slot and conjugate linear in the first slot. The inner product defines a norm via the formula $||z||^2 = \langle z, z\rangle$.

Let V be a complex inner product space and let $L : V \to V$ be a linear transformation. In particular L is continuous. Its *adjoint* L^* is defined by the formula

$$\langle Lz, w\rangle = \langle z, L^* w\rangle.$$

A linear transformation U from V to itself is *unitary* if $\langle Uz, Uw\rangle = \langle z, w\rangle$ for all z and w; equivalently U is invertible and $U^{-1} = U^*$. A linear transformation L from V to itself is *Hermitian* or *self-adjoint* if $L = L^*$. Finally a linear transformation is *nonnegative definite* if $\langle Lz, z\rangle \geq 0$ for all z, and *positive definite* if there is a positive c such that $\langle Lz, z\rangle \geq c||z||^2$ for all z. Such mappings (see Exercise 1.3) are necessarily Hermitian. A mapping is *positive semidefinite* if it is nonnegative definite but not positive definite.

We briefly discuss polarization now before saying much more in Lecture 2. Let V be a complex inner product space with squared norm given by $||z||^2 = \langle z, z\rangle$. We can recover the inner product $\langle z, w\rangle$ from certain squared norms. Let L be a linear transformation on V. Fix an integer m at least 3 and let η be a primitive m-th root of unity. The following formula (P1) holds for all z, w and L:

$$\langle Lz, w\rangle = \frac{1}{m}\sum_{j=0}^{m-1} \eta^j \langle L(z+\eta^j w), z+\eta^j w\rangle. \tag{P1}$$

When L is the identity, (P1) shows how to recover the inner product from the squared norm:

$$\langle z, w\rangle = \frac{1}{m}\sum_{j=0}^{m-1} \eta^j ||z+\eta^j w||^2. \tag{P}$$

When $m = 4$ formula (P) is known as the *polarization identity.*

Exercise 1.1. Prove the Cauchy-Schwarz inequality (CS) and the triangle inequality (T) in a complex inner product space:

$$|\langle z, w\rangle| \le ||z||\;||w|| \tag{CS}$$

$$||z+w|| \le ||z|| + ||w||. \tag{T}$$

Exercise 1.2. Prove (P1) and hence (P). What are the analogues when $m = 2$?

Exercise 1.3. Show that there are real inner product spaces and nonzero linear transformations T satisfying $\langle Tx, x\rangle = 0$ for all x. Show in the complex case that $\langle Tz, z\rangle = 0$ for all z implies that $T = 0$. Using this fact, verify that L is Hermitian if and only if $\langle Lz, z\rangle$ is real for all z, and hence note that a nonnegative definite linear map is automatically Hermitian.

Exercise 1.4. Let $\{x_j\}$ be a finite collection of distinct positive numbers. Consider a square matrix whose entries are $\frac{1}{x_j+x_k}$. Prove that this matrix is positive definite. Here is one possible approach. Show first the following fact: a finite-dimensional matrix whose jk entry is the inner product $\langle \zeta_j, \zeta_k\rangle$ of linearly independent vectors must be positive definite. Then define an inner product (using integration) on a specific infinite-dimensional space of functions to obtain the result.

2. Differential forms

Next we discuss differential forms; as is customary in differential geometry, we will use upper indices when expressing differential forms in coordinates. The interaction between the real and the complex viewpoints dominates the discussion. Coordinates $(z^1, ..., z^n)$ on $\mathbf{C}^n$ can be expressed as usual in terms of real coordinates by writing $z^j = x^j + iy^j$, where of course $i^2 = -1$. Hence the differentials dz^j and $d\overline{z}^j$ can be expressed by the formula

$$\begin{aligned} dz^j &= dx^j + i dy^j \\ d\overline{z}^j &= dx^j - i dy^j. \end{aligned} \tag{2}$$

The formulas (2) lead to formulas for the coordinate vector fields. We have

$$\begin{aligned} \frac{\partial}{\partial z^j} &= \frac{1}{2}\frac{\partial}{\partial x^j} - \frac{i}{2}\frac{\partial}{\partial y^j} \\ \frac{\partial}{\partial \overline{z}^j} &= \frac{1}{2}\frac{\partial}{\partial x^j} + \frac{i}{2}\frac{\partial}{\partial y^j}. \end{aligned} \tag{3}$$

Exercise 1.5. Students sometimes believe that the formulas in (3) are definitions, but in fact they are consequences of (2) and the invariance of the exterior derivative d. Verify that the formulas in (2) imply the formulas in (3).

Let Ω be an open, connected set in $\mathbf{C}^n$. We write $C^\infty(\Omega)$ for the ring of smooth complex-valued functions on Ω and we write $A(\Omega)$ for the subring of holomorphic functions on Ω. A smooth function $f:\Omega\to\mathbf{C}$ lies in $A(\Omega)$ if and only if it satisfies the Cauchy-Riemann equations. We can express this system of PDE efficiently by using the notation of differential forms. We write the Cauchy-Riemann equations in the form

$$0=\overline{\partial} f=\sum_{j=1}^{n}\frac{\partial f}{\partial \overline{z}^j}d\overline{z}^j. \tag{4}$$

Equivalently, f satisfies the Cauchy-Riemann equations if for each j we have

$$\frac{\partial f}{\partial \overline{z}^j}=0.$$

The expression on the right-hand side of (4) is called a $(0,1)$ form, or a differential form of *type* $(0,1)$. The right-hand side of (4) is a differential 1-form whose differentials involve only the $d\overline{z}^j$. A 1-form could also involve the differentials dz^j. More generally we consider differential forms of *type* (p,q); another word for type is *bi-degree*. A differential form is of type (p,q) if it is a sum of forms each involving wedge products of p of the dz^j and q of the $d\overline{z}^j$. It is convenient to use multi-index notation; thus we write for example dz^I instead of $dz^{i_1}\wedge\ldots\wedge dz^{i_k}$ when $I=(i_1,\ldots,i_k)$ is a multi-index.

We extend the $\overline{\partial}$ operator to differential forms of all bi-degrees in the usual way. Let

$$u=\sum u_{IJ}dz^I\wedge d\overline{z}^J$$

be a (p,q) form. We define a $(p,q+1)$ form $\overline{\partial}u$ by

$$\overline{\partial}u=\sum\frac{\partial u_{IJ}}{\partial\overline{z}^k}d\overline{z}^k\wedge dz^I\wedge d\overline{z}^J. \tag{5}$$

Using the definition of wedge product one can rewrite (5) in the form

$$\overline{\partial}u=\sum_{|I|=p,|K|=q+1}G_{IK}dz^I\wedge d\overline{z}^K$$

for appropriate functions G_{IK}.

Exercise 1.6. Using the equality of mixed partial derivatives verify that $\overline{\partial}^2=0$.

Exercise 1.7. Let $\phi=\sum\phi_k d\overline{z}^k$ be a $(0,1)$ form. What are the functions g_{jk} such that

$$\overline{\partial}\phi=\sum_{j<k}g_{jk}d\overline{z}^j\wedge d\overline{z}^k?$$

Then find $\sum_{j,k}|g_{jk}|^2$ in terms of the derivatives of the ϕ_j.

Many problems in complex analysis involve the $\overline{\partial}$ operator. As in classical de Rham theory, it is natural to consider the operator $\overline{\partial}$ on differential forms of all degrees. In the complex case we consider the operator $\overline{\partial}$ as a mapping from forms of type (p,q) to forms of type $(p,q+1)$. In these lecture notes there will be no loss in generality if we assume that $p=0$ in this context.

Complex Analysis in one variable has become a standard subject, with nearly all books proceeding in a similar order and manner, based on the Cauchy theory. Complex Analysis in Several Variables has not yet reached such a stage. One approach to the subject uses sheaf theory ([Gun1], [GR]); a second approach relies on the methods of PDE ([H], [FK]), and a third approach emphasizes integral formulas [Range]. The subject of CR Geometry also allows for different perspectives. Furthermore there is a wealth of information about specific topics such as function theory in the unit ball (See [Ru]) that fits into none of these categories.

In the PDE approach to complex analysis, one thinks of holomorphic functions as solutions to the first-order system of PDE called the Cauchy-Riemann equations. Let f be a continuously differentiable function on an open connected set Ω in $\mathbf{C}^n$. Then f is holomorphic if and only if $\overline{\partial} f = 0$. One basic idea, dating from the work of Riemann, is to study solutions to the homogeneous equation $\overline{\partial} f = 0$ by considering the inhomogeneous equation

$$\overline{\partial} u = \alpha. \tag{6}$$

In (6) the right hand side is a $(0,1)$ form. Since $\overline{\partial}^2 = 0$, there can be a solution only if $\overline{\partial}\alpha = 0$. On the other hand, if u is a solution to (6), and f is holomorphic, $u + f$ is also a solution. Thus in general there are many solutions to (6); part of the philosophy is to study (6) as a system of PDE. Given α with some property, we ask whether we can find a solution u with a similar property. See [H] and [FK] for several compelling examples of this approach.

Some of the big advances in the theory of the $\overline{\partial}$ equation developed from considering (6) on domains with smooth boundary, and studying the regularity of solutions up to the boundary. While we will not redevelop that theory, most of our geometric questions will be motivated by considerations from it.

3. Solving the Cauchy-Riemann equations

In this section we briefly discuss the formalism for solving the Cauchy-Riemann equations in L^2 spaces. First recall that a Hilbert space is a complete complex inner product space. For us the relevant Hilbert spaces will be spaces of square integrable differential forms, possibly with respect to a weight function, on a domain Ω in $\mathbf{C}^n$. See Lectures 4 and 5.

Let $\mathcal{H}_1$, $\mathcal{H}_2$, and $\mathcal{H}_3$ be Hilbert spaces. We write $||\ ||$ to denote the norm on each of these spaces. Consider linear operators $T : \mathcal{H}_1 \to \mathcal{H}_2$ and $S : \mathcal{H}_2 \to \mathcal{H}_3$ such that $ST = 0$. These operators will not be continuous (bounded); they will be defined only on dense subspaces rather than on all of $\mathcal{H}_1$ and $\mathcal{H}_2$. The operators will be *closed*, and hence their adjoints are defined on appropriate dense domains as well.

We think of both T and S as the $\overline{\partial}$ operator, defined on forms of consecutive degrees. Since $\overline{\partial}^2 = 0$, we must have $ST = 0$. To solve the equation $Tu = \alpha$, we therefore require $S\alpha = 0$. The equation $Tu = \alpha$ is in general over-determined; if $Tf = 0$, then $T(u + f) = Tu = \alpha$ as well. A nice formalism enables us to choose a unique canonical solution in many cases. Consider the operator $L = TT^* + S^*S : \mathcal{H}_2 \to \mathcal{H}_2$. With its natural domain, L is self-adjoint.

Suppose for some positive constant C we can establish the inequality

$$||f||^2 \leq C(||T^* f||^2 + ||Sf||^2) \tag{7}$$

for all f in the intersection of the domain of T^* with the domain of S. Then L is invertible on its domain. To verify the invertibility notice that $0 = (TT^* + S^*S)f$ implies

$$0 = \langle (TT^* + S^*S)f, f\rangle = ||T^*f||^2 + ||Sf||^2. \tag{8}$$

If (7) holds then we see from (8) that $f = 0$. Thus L is injective; since it is self-adjoint it is also invertible on its domain.

Assuming (7) we write $N = L^{-1}$. Then we have the Hodge decomposition

$$\alpha = TT^*N\alpha + S^*SN\alpha. \tag{9}$$

If we assume $S\alpha = 0$, then it follows (see Exercise 1.8) that $SN\alpha = 0$, and we obtain $\alpha = T(T^*N\alpha)$.

Put $u = T^*N\alpha$. Then u is the unique solution to $Tu = \alpha$ orthogonal to the null-space of T. It is therefore the solution of minimal norm.

Exercise 1.8. Show that $S\alpha = 0$ implies $SN\alpha = 0$. Hint: Use (9) to first get $SS^*SN\alpha = 0$. Then take the inner product with $SN\alpha$. Repeat the idea.

These formal considerations indicate how one solves the Cauchy-Riemann equations in L^2 spaces; we are interested in solving the equations in spaces of smooth forms, and hence various approximation methods also come into play. See the lectures by Berndtsson in this volume.

LECTURE 2

Complex varieties in real hypersurfaces

We will be studying whether a given real hypersurface in complex Euclidean space contains the image of a nonconstant holomorphic curve. We will see that the answer is no when the hypersurface is strongly pseudoconvex; answering the question in general will require analyzing weakly pseudoconvex hypersurfaces. The condition of weak pseudoconvexity at a point is roughly analogous to the condition of being a degenerate critical point for a smooth function at a local minimum. We therefore start with elementary calculus.

1. Degenerate critical points of smooth functions

We begin with a short discussion about the one variable case. Let $f : \mathbf{R} \to \mathbf{R}$ be a smooth function, and suppose $f(0) = 0$. What are the possibilities for the behavior of f near 0? If f does not vanish identically near 0, then f could have a local minimum there, a local maximum there, or neither. In case of a local extremum, we naturally ask additionally whether the extremum is strict. How do we decide which of these possibilities holds? For f to have a local minimum or maximum at 0 it is of course necessary that $f'(0) = 0$, that is, the origin must be a *critical point.* Assume the origin is a critical point. If $f''(0) \neq 0$, then we have a strict local minimum when $f''(0) > 0$ and a strict local maximum when $f''(0) < 0$. When $f''(0) = 0$ we have a degenerate critical point, and we need to investigate further. The following result is quite easy to prove:

Lemma 2.1. *Let $f : \mathbf{R} \to \mathbf{R}$ be a smooth function with $f(0) = 0$, and suppose that f vanishes to finite order at 0. Let k be the smallest integer for which $f^{(k)}(0) \neq 0$. If k is odd, then f takes on both positive and negative values in every neighborhood of 0. If k is even, then f has a strict local minimum at 0 when $f^{(k)}(0) > 0$ and a strict local maximum at 0 when $f^{(k)}(0) < 0$.*

PROOF. The conclusion follows from expanding f in a Taylor series about 0 and estimating the remainder term. □

Exercise 2.1. Complete the details of the proof of Lemma 2.1.

There is no similar test in several variables. One problem is that the lowest order terms in the Taylor expansion for a smooth function f do not govern the situation. While one can decide whether a homogeneous polynomial has a minimum at 0 by investigating its behavior along lines, there is no good algorithm to check all lines. Hence there is no good test for deciding whether a homogeneous polynomial in several variables is non-negative. In Lecture 6 we will say more about this problem.

The following remarkable example of Peano illustrates why degenerate critical points are so difficult to analyze.

Example 2.1. (Peano) Consider the polynomial f in two real variables given by

$$f(x,y) = (y - x^2)(y - 4x^2). \tag{7}$$

It is evident that $f(0,0) = 0$ and that $f(x,y) < 0$ in the set defined by $x^2 < y < 4x^2$. Thus f does not have a minimum at $(0,0)$. But the restriction of f to each line through the origin has a strict local minimum there! Consider the line $L_{a,b}$ given by $t \to (at, bt)$, where $(a,b) \neq (0,0)$. We have $f(at,bt) = (bt - a^2t^2)(bt - 4a^2t^2) = b^2t^2+$ higher order terms. Thus, for $b \neq 0$, the restriction to the line $L_{a,b}$ has a strict non-degenerate local minimum at the origin. When $b = 0$ the restriction to the line $L_{a,b}$ is $4a^4t^4$ which also has a strict local minimum there, albeit a degenerate minimum. Thus the restriction of f to every line but one has a nondegenerate strict local minimum, the restriction of f to the remaining line has a strict local minimum, and yet f does not have a minimum at $(0,0)$.

We next consider the following related example. Define a homogeneous polynomial q in three variables by

$$q(x,y,z) = (yz - x^2)(yz - 4x^2) + z^4. \tag{8}$$

Again this function is negative somewhere in every neighborhood of the origin. For example $q(1,2,1) = -1$ and hence $q(t,2t,t) = -t^4 < 0$ for $t \neq 0$. We homogenized f and added the positive term z^4. Yet now there are lines along which the resulting polynomial q is never positive and hence the restriction to such a line has a strict maximum of 0 at 0.

We note the following simple general fact.

Lemma 2.2. *Let p be a homogeneous polynomial of even degree in $x = (x_1, ..., x_n)$. Then the restriction p to each line through $0 \in \mathbf{R}^n$ has a minimum at $0 \in \mathbf{R}$ if and only if p has a minimum at 0.*

PROOF. For $0 \neq v \in \mathbf{R}^n$, we consider the line parametrized by $t \to tv$. By homogeneity $p(tv) = t^{2d}p(v)$. The result follows. □

We consider (7) from the point of view of Lemma 2.2. After homogenizing we restrict to the line $t \to (at, bt, ct)$ for $c \neq 0$. Without loss of generality we may assume that $c = 1$. We then need to consider the expression

$$p(at, bt, t) = t^4 p(a, b, 1) = t^4 f(a, b). \tag{9}$$

To decide whether this restriction has a minimum at $t = 0$ requires knowing whether $f(a,b)$ is nonnegative; deciding whether we have a minimum for all (a,b) is then equivalent to deciding whether the Peano example itself has a minimum!

Even in the homogeneous case, it therefore follows that restricting to lines doesn't really help us decide anything. To determine whether there is a line along which the homogeneous polynomial fails to have a minimum requires us to determine what happens at a critical point of an inhomogeneous polynomial. We are back where we started.

We give another example to illustrate why things are so difficult in several variables. Let $p(x,y) = (x^2 - y^3)^2 + \phi(x,y)$ where ϕ vanishes to infinite order at the origin. The restriction of p to each line through the origin vanishes to finite order (either 4 or 6) and the restriction has a local minimum at the origin. Whether p itself has a local minimum depends on how ϕ behaves along the curve given by

$(x(t), y(t)) = (t^3, t^2)$. There is no finite order condition one can check, even though p vanishes to finite order in every direction.

Later we will consider polynomials in several complex variables, and we will restrict these polynomials to more complicated curves. The local behavior will be closely related to important considerations in the theory of the Cauchy-Riemann equations. Nonetheless dealing with degenerate critical points will not be easy.

Exercise 2.2. In the Peano example there is a single direction in which the function has a degenerate minimum. Prove (or look up in an advanced calculus book) that if f is a smooth function of n variables, $df(0) = 0$, and the restriction of f to *every* line through 0 has a nondegenerate minimum there, then f has a strict local minimum at 0.

Exercise 2.3. (Harder) Suppose that f is smooth near 0, $df(0) = 0$, and that $D^2 f(0)$ is nondegenerate. Assume it has k positive eigenvalues and $n - k$ negative eigenvalues. Prove that there is a local coordinate system $(x_1, ..., x_n)$ such that $f(x) = \sum_{j=1}^{k} x_j^2 - \sum_{j=k+1}^{n} x_j^2$.

2. Hermitian symmetry and polarization

Let p be a polynomial (with real or complex coefficients) on $\mathbf{R}^{2n}$; by substituting the formulas for x_j and y_j in terms of z_j and $\overline{z}_j$ we obtain a polynomial in the complex variables z and $\overline{z}$. We may treat these variables as independent. Let $R(z, \overline{w})$ be a polynomial in the $2n$ complex variables $(z_1, ..., z_n)$ and $(\overline{w}_1, ..., \overline{w}_n)$. We call such a polynomial R *Hermitian symmetric* if

$$R(w, \overline{z}) = \overline{R(z, \overline{w})}. \tag{10}$$

It is easy to see that R is Hermitian symmetric if and only if the polynomial $R(z, \overline{z})$ is real-valued. At the opposite extreme, we call R *holomorphic* if it is independent of $\overline{w}$. In general we will write, using multi-index notation,

$$R(z, \overline{w}) = \sum_{|a|,|b|=0}^{d} c_{ab} z^a \overline{w}^b. \tag{11}$$

Generalizing (11) we also consider real-analytic real-valued functions r defined in a neighborhood of the origin in $\mathbf{C}^n$. Thus r is given near the origin 0 by a convergent power series (in multi-index notation)

$$r(z, \overline{z}) = \sum_{|a|,|b|=0}^{\infty} c_{ab} z^a \overline{z}^b. \tag{12}$$

In this case we call (c_{ab}) the *underlying matrix of coefficients* of r. We may polarize by treating z and $\overline{z}$ as independent variables. The condition that r be real-valued is equivalent to the Hermitian symmetry condition (10) for r. This condition is equivalent to the statement

$$c_{ba} = \overline{c_{ab}} \tag{13}$$

for all pairs (a, b) of multi-indices. When r is a polynomial, the underlying matrix of coefficients is finite-dimensional.

Exercise 2.4. When r satisfies (11) or (12) and (c_{ab}) is non-negative definite, it is obvious that $r(z,\overline{z}) \geq 0$ for all z. Give an example where $r(z,\overline{z}) \geq 0$ for all z but (c_{ab}) has at least one negative eigenvalue. See Lecture 6 for much more in this direction.

Definition 2.1. Let D be an open subset of $\mathbf{C}^n$. We denote by D^* the complex conjugate domain; $D^* = \{\overline{z} : z \in D\}$. Let R be holomorphic on $D \times D^*$. We say that R is *Hermitian symmetric* if $D = D^*$ (as sets) and (10) holds for all $(z,\overline{w}) \in D \times D^*$.

Polarization is among the most powerful ideas in complex analysis; if r is Hermitian symmetric, then we can recover the values of $r(z,\overline{w})$ from the values of $r(z,\overline{z})$. In this sense we are treating z and $\overline{z}$ as independent variables. We state several forms of this idea as a Proposition whose proof is left to the reader:

Proposition 2.1. *(Polarization) Let r be a real-analytic real-valued function defined in a connected neighborhood D of the origin such that $D = D^*$. Then there is a unique Hermitian symmetric holomorphic function R defined on $D \times D^*$ such that $R(z,\overline{z}) = r(z,\overline{z})$ for all $z \in D$. Equivalently, if $(z,\overline{w}) \to R(z,\overline{w})$ is Hermitian symmetric on $D \times D^*$, and $R(z,\overline{z}) = 0$ for all $z \in D$, then $R(z,\overline{w}) = 0$ for all $(z,\overline{w}) \in D \times D^*$.*

Exercise 2.5. Prove Proposition 2.1.

Exercise 2.6. Let L be a linear mapping on a finite-dimensional inner product space. Assume that $||Lz||^2 = ||z||^2$ for all z. Prove, without using Proposition 2.1, that $\langle Lz, Lw\rangle = \langle z, w\rangle$ for all z and w. Suggestion: Use a *polarization identity*.

Exercise 2.7. Let $f : \mathbf{C}^n \to \mathbf{C}$ be holomorphic. Identify $\mathbf{C}^n$ with $\mathbf{R}^{2n}$. Assume that f vanishes on a subspace $V \subset \mathbf{R}^{2n}$ of dimension n. Under what condition on V must f vanish identically? What does this exercise have to do with Proposition 2.1?

Exercise 2.8. Consider a harmonic function $U(x,y)$ defined for $(x,y) \in \mathbf{R}^2$. Given z_0 and ζ in $\mathbf{C}$, we seek a holomorphic F whose real part is U and such that $F(z_0) = \zeta$. Find a formula for F in terms of U. Avoid differentiation or integration. Generalize your result to higher dimensions. What is the appropriate analogue of harmonic in this setting?

3. Holomorphic decomposition

We will study the geometry of the zero set of a smooth function of several complex variables by considering the zero sets of its polynomial truncations, and by using algebraic means to study these algebraic zero sets. The geometric interaction between the real and complex aspects will be crucial. Our first step is to write Hermitian symmetric functions in terms of holomorphic functions. In Theorem 2.3 we answer one of our fundamental questions: when does the zero-set of a real-valued polynomial contain non-constant holomorphic curves?

Holomorphic decomposition for polynomials and real-analytic functions

We first consider the polynomial case and then we study the real-analytic case. While this approach is a bit redundant, it helps clarify the issues; furthermore the polynomial case plays a major role in our development whereas the real-analytic case does not.

Lemma 2.3. *(Holomorphic decomposition for polynomials) Let r be a Hermitian symmetric polynomial. There are linearly independent holomorphic polynomials f_j and g_j such that*

$$r(z,\overline{z}) = \sum_{j=1}^{k} |f_j(z)|^2 - \sum_{j=1}^{l} |g_j(z)|^2 = ||f(z)||^2 - ||g(z)||^2. \tag{14.1}$$

Furthermore there is a holomorphic polynomial h and linearly independent holomorphic polynomials F_j and G_j such that

$$r(z,\overline{z}) = 2\mathrm{Re}(h(z)) + \sum_{j=1}^{K} |F_j(z)|^2 - \sum_{j=1}^{L} |G_j(z)|^2 = 2\mathrm{Re}(h(z)) + ||F(z)||^2 - ||G(z)||^2. \tag{14.2}$$

Finally, if $r(0,0) = 0$, then we may choose all the functions in (14.2) to vanish at 0 as well.

Proof. We begin with (14.1). There are several easy proofs. One way to prove (14.1) is to diagonalize the Hermitian matrix (c_{ab}). The advantage of this proof is that we obtain the linear independence of the set of holomorphic polynomials directly. One can also prove (14.1) directly as follows. For each multi-index b we have

$$|\sum_a c_{ab}z^a + z^b|^2 - |\sum_a c_{ab}z^a - z^b|^2 = 4\mathrm{Re}\,(\sum_a c_{ab}z^a\overline{z}^b). \tag{15}$$

Summing (15) on the indices b we obtain

$$4r(z,\overline{z}) = 4\mathrm{Re}\,(r(z,\overline{z})) = \sum_b \left(|\sum_a c_{ab}z^a + z^b|^2 - |\sum_a c_{ab}z^a - z^b|^2 \right). \tag{16}$$

Put $f_b(z) = \frac{1}{2}(\sum_a c_{ab}z^a + z^b)$ and put $g_b(z) = \frac{1}{2}(\sum_a c_{ab}z^a - z^b)$. Then (14.1) follows from (16). This second approach thus immediately gives a holomorphic decomposition, but without linear independence. Theorem 2.2 and its Corollary reveal how to ensure that we can rewrite (16) with linearly independent f's and g's. We suggest that the reader fill in this point after reading Theorem 2.3.

To prove (14.2) we first put $h(z) = r(z,0) = \sum_a c_{a0}z^a$. Put $s(z,\overline{z}) = r(z,\overline{z}) - 2\mathrm{Re}(h(z))$ and apply (14.1) to s. We obtain F and G for which (14.2) holds. If also $r(0,0) = 0$, then $h(0) = 0$. By (16) each component of F and G vanishes there as well. □

Exercise 2.9. Assume that (14.2) is known. Prove (14.1). In other words, write $2\mathrm{Re}(h(z))$ in the form $|f_0(z)|^2 - |g_0(z)|^2$.

Exercise 2.10. Put $r = |f_1|^2 + |f_2|^2 - |f_1 + f_2|^2$, where f_1 and f_2 are linearly independent. Show that the rank of the underlying matrix of coefficients is 2 by writing r is the form $r = |F|^2 - |G|^2$ for linearly independent F and G.

We write (14.1) as $r(z,\overline{z}) = ||f(z)||^2 - ||g(z)||^2$. The linear algebra proof via diagonalization reveals that we may choose k and l in (14.1) to be the numbers of positive and negative eigenvalues of the Hermitian matrix (c_{ab}). In some contexts we will prefer the decomposition (14.2), because it better captures the underlying CR geometry. We call either (14.1) or (14.2) a *holomorphic decomposition* of r.

We now extend holomorphic decomposition to the real-analytic case. Let r be a Hermitian symmetric real-analytic function defined near $0 \in \mathbf{C}^n$ with $r(0) = 0$. We write

$$r(z, \overline{z}) = \sum_{a,b} c_{ab} z^a \overline{z}^b. \tag{17}$$

We call a term $c_{ab} z^a \overline{z}^b$ in the power series expansion of r *pure* if either a or b is zero; in other words, if it is either holomorphic or its conjugate is holomorphic. All other terms are called *mixed*. In the proof of (14.2) we isolated the pure terms in r and called them 2Re $(h(z))$. We are witnessing a simple example of polarization, as we treated z and $\overline{z}$ as independent variables and wrote

$$h(z) = \sum_a c_{a0} z^a = r(z, 0). \tag{18}$$

Formula (18) also applies in the real-analytic case. We wish to find holomorphic vector-valued mappings f and g such that

$$r(z, \overline{z}) = 2\mathrm{Re}(h(z)) + ||f(z)||^2 - ||g(z)||^2. \tag{19}$$

Unlike in the polynomial case, the mappings f and g need not take values in a finite-dimensional space. We will obtain (19); because of convergence issues it might be possible on only a small neighborhood of a given point.

Theorem 2.1. *Let r be a Hermitian symmetric real-analytic function with $r(0,0) = 0$. There is a neighborhood D of the origin, a holomorphic function h defined in D, and sequences of holomorphic functions f_j and g_j such that the following hold:*
1) All these functions vanish at the origin.
2) $\sum_j |f_j(z)|^2$ and $\sum_j |g_j(z)|^2$ converge in D.

$$3)\ r(z, \overline{z}) = 2\mathrm{Re}(h(z)) + \sum_j |f_j(z)|^2 - \sum_j |g_j(z)|^2 = 2\mathrm{Re}(h(z)) + ||f(z)||^2 - ||g(z)||^2. \tag{20}$$

PROOF. First we put $h(z) = r(z, 0)$ as in (18). Then let D be a polydisk in $\mathbf{C}^n$ whose closure lies within the region of convergence of $r(z, \overline{z})$. We follow the proof of Lemma 2.3, with a small variation. First we choose δ such that $\delta \in D \subset \mathbf{C}^n$ and all its components are non-zero real numbers. For each nonzero multi-index b we let $\mu_b = \frac{1}{2}(\frac{z}{\delta})^b$ and then put

$$f_b(z) = \frac{1}{2}\left(\sum_{1 \le |a|} c_{ab} \delta^b z^a + \delta^{-b} z^b\right) = \zeta_b + \mu_b$$

$$g_b(z) = \frac{1}{2}\left(\sum_{1 \le |a|} c_{ab} \delta^b z^a - \delta^{-b} z^b\right) = \zeta_b - \mu_b. \tag{21}$$

It follows from (21) that

$$|f_b(z)|^2 - |g_b(z)|^2 = \mathrm{Re}\left(\sum_{1 \le |a|} c_{ab} z^a \overline{z}^b\right),$$

and hence that

$$\sum_b \left(|f_b(z)|^2 - |g_b(z)|^2\right) = r(z,\overline{z}) - 2\mathrm{Re}(h(z)). \tag{22}$$

The functions f_b and g_b are defined by convergent series and hence are holomorphic near 0. Furthermore the sums in 22) converge. To see the convergence, note that $\sum |\mu_b|$ is summable by the geometric series if $|z_j| < |\delta_j|$ for each j. Since also $\sum_b \zeta_b$ converges absolutely for z close to 0 (it is essentially the Taylor series of r) we can use the estimate

$$\sum_b |\zeta_b \pm \mu_b|^2 \leq 2\left(\sum_b |\zeta_b|^2 + \sum_b |\mu_b|^2\right) \tag{23}$$

to obtain convergence of the sums on b determining $||f||^2$ and $||g||^2$. □

In the polynomial case we need not worry about convergence and we recover (14.2) from Lemma 2.3. We will use the holomorphic decomposition (14.2) many times in what follows.

We pause to ask under what circumstances we can write r as a *squared norm*; that is, $r(z,\overline{z}) = ||f(z)||^2$ for some holomorphic mapping f. We discuss this matter in more detail in Lecture 6. The next Lemma gives one simple way of deciding whether a real-analytic r is the squared norm of a holomorphic mapping f. If r is a polynomial (in z and $\overline{z}$), then the components of f will be polynomials in z but independent of $\overline{z}$.

Lemma 2.4. *Let r be a Hermitian symmetric real-analytic function. Its underlying matrix of coefficients is nonnegative definite if and only if there is a sequence of holomorphic functions f_j such that*

$$r(z,\overline{z}) = \sum_{j=1}^{\infty} |f_j(z)|^2 = ||f(z)||^2. \tag{24}$$

PROOF. Note that (c_{ab}) is nonnegative definite if and only if there are vectors F_a such that $c_{ab} = \langle F_a, F_b\rangle$. Plugging this relationship into r gives

$$r(z,\overline{z}) = \sum_{a,b} \langle F_a, F_b\rangle z^a \overline{z}^b = ||\sum_a F_a z^a||^2. \tag{25}$$

Conversely the functions f_j in (24) determine the vectors for which (25) holds. □

We pause to discuss the role of the function h. The zero set of r defines a real submanifold of real codimension one when $dr(0) \neq 0$. In this case we find that $dh(0) \neq 0$, and we may choose local holomorphic coordinates such that $h(z) = z_n$. In CR geometry the variable z_n plays a different role from the variables $z_1, ..., z_{n-1}$. The author finds it interesting that the naive difference of pure versus mixed is closely connected with deep aspects of the anisotropic behavior of the tangent spaces on the zero set of r. We will see more of this geometry when we pull r back to holomorphic curves.

We wish to discuss the extent to which the holomorphic decomposition of r is unique. The following result is interesting on its own and plays a major role in later developments. We emphasize that the linear map L appearing in it is independent of z.

Theorem 2.2. *Let Ω be an open ball containing the origin in $\mathbf{C}^n$. Let $f : \Omega \to \mathbf{C}^N$ and $g : \Omega \to \mathbf{C}^K$ be holomorphic mappings such that*

$$||f(z)||^2 = ||g(z)||^2 \tag{26}$$

on Ω. Then there is a linear mapping $L : \mathbf{C}^K \to \mathbf{C}^N$ such that $f = Lg$. If in addition $N = K$, then we may choose L to be unitary.

PROOF. On Ω we may write $f(z) = \sum f_a z^a$ for $f_a \in \mathbf{C}^N$ and $g(z) = \sum g_a z^a$ for $g_a \in \mathbf{C}^K$. After equating Taylor coefficients, condition (26) becomes

$$\langle f_a, f_b \rangle = \langle g_a, g_b \rangle \tag{27}$$

for all pairs of multi-indices a and b.

We then define L by setting $Lg_a = f_a$ for a maximal linearly independent set of the g_a. The compatibility conditions (27) show that L is well-defined and that $Lg_a = f_a$ for all indices. When the dimensions are equal we observe that L preserves all inner products, and hence it can be extended to be unitary. □

Corollary 2.1. *Suppose that*

$$||f||^2 - ||g||^2 = ||A||^2 - ||B||^2 \tag{28}$$

on a ball Ω. Then there is a linear map L such that

$$f \oplus B = L(g \oplus A).$$

PROOF. Rewriting (28) gives $||f||^2 + ||B||^2 = ||g||^2 + ||A||^2$ and hence the claim follows from Theorem 2.3. □

We can obtain analogues of the Peano example for Hermitian symmetric functions of complex variables in several ways:

$$r_1(z, \overline{z}) = (|z_2|^2 - |z_1|^4)(|z_2|^2 - 4|z_1|^4) = |z_2|^4 - 5|z_2|^2|z_1|^4 + 4|z_1|^8 \tag{29}$$

$$r_2(z, \overline{z}) = |z_2|^2 - 5\mathrm{Re}(z_2)|z_1|^2 + 4|z_1|^4. \tag{30}$$

Neither of these functions has a local minimum at the origin, yet a naive analysis using complex lines might suggest otherwise.

We make some simple remarks about calculus and complex linear algebra. Let r be a smooth real-valued function on a neighborhood of p in $\mathbf{C}^n$; then p is a critical point for r if and only if $dr(p) = 0$, which holds if and only if both $\partial r(p) = 0$ and $\overline{\partial} r(p) = 0$. For such r the analogue of the second derivative matrix is the complex Hessian, defined by $H(r) = \left(r_{z_i \overline{z}_j}\right)$. Notice that this matrix does not include all second derivatives of r, as pure second derivatives do not arise. The following simple statements hold. If p is a critical point, and r has a local minimum at p, then $H(r)$ is non-negative definite. If p is a critical point, the pure second derivatives vanish at p, and $H(r)$ is positive definite at p, then r has a strict local minimum at p.

Exercise 2.11. Let r be a smooth real-valued function defined near p in $\mathbf{C}^n$. Show that if $dr(p) = 0$, the pure second derivatives of r vanish at p, and $H(r)(p)$ is positive definite at p, then r has a strict local minimum at p. Show that if $dr(p) = 0$ and $H(r)(p)$ has at least one negative eigenvalue, then r does not have a strict local minimum at p.

Example 2.2. Consider the quadratic function r defined on $\mathbf{C}$ by

$$r(z,\overline{z}) = \frac{c}{2}(z^2 + \overline{z}^2) + |z|^2.$$

Here c is a real constant. Then the complex Hessian of r is 1, but r has a minimum at 0 only if the constant c satisfies $|c| \leq 1$. This example indicates why one must assume the pure second derivatives vanish in order to obtain a test for a local minimum.

Exercise 2.12. Express the function r from Example 2.2 in terms of x, y and verify that the condition $|c| \leq 1$ is necessary and sufficient for r to have a minimum at 0.

Suppose that $r(z,\overline{z}) = ||f(z)||^2$. Then a simple computation shows that the complex Hessian $H(r)$ can be written

$$H(r) = r_{z_i \overline{z}_j} = \langle f_{z_i}, f_{z_j} \rangle.$$

More interesting is a formula for the determinant of the complex Hessian:

$$\det(H(r)) = \sum_I |J(f_{i_1}, ..., f_{i_n})|^2. \tag{31}$$

In (31) the sum is taken over all n-tuples of component functions of f, and J denotes the Jacobian determinant. There is an analogous formula when $r(z,\overline{z}) = ||f(z)||^2 - ||g(z)||^2$ whose derivation we leave as an exercise to the reader.

Exercise 2.13. Find a formula for $\det(H(r))$ when $r(z,\overline{z}) = ||f(z)||^2 - ||g(z)||^2$. Each term in the sum should be $\pm$ the squared modulus of a Jacobian.

Remark 2.1. Smooth functions whose complex Hessians are nonnegative definite at each point are called *plurisubharmonic* and play a major role in complex analysis.

Holomorphic decomposition and holomorphic curves

Let $r : \mathbf{C}^n \to \mathbf{R}$ be a function. As usual we write $V(r)$ for its zero set. We are most often interested in the case when $V(r)$ is a smooth real hypersurface; thus we will often assume that r is smooth and that $dr(p) \neq 0$ when $r(p) = 0$. In the particular situation when r is an Hermitian symmetric polynomial or real-analytic function we sometimes write $r(z,\overline{z})$ rather than $r(z)$ in order to emphasize the possibility of polarization.

For later convenience we pause to discuss the concept of *germs* of functions. Consider the collection of smooth functions defined in some unspecified neighborhood of a given point p in $\mathbf{R}^n$ or on a smooth manifold M. We call two functions equivalent at p if there is some neighborhood of p on which these functions are identical. This relation defines an equivalence relation; a germ of a function is its equivalence class in this relation. Although we cannot evaluate a germ at a general point, it does make sense to evaluate a germ at the base point; it is thus sensible to write $f(p)$ when f is a germ at p. The collection of germs of smooth functions at p defines a ring C_p^∞ (also written $\mathcal{E}_p$) under the usual operations; we choose representatives, add and multiply as usual, and then take equivalence classes.

In a similar fashion we have the ring of germs of holomorphic functions $\mathcal{O}_p$ at a point $p \in \mathbf{C}^n$ or on a complex manifold X. A germ that does not vanish at p is a *unit* in the ring $\mathcal{O}_p$; the non-units in $\mathcal{O}_p$ form the unique maximal ideal $\mathcal{M}_p$ and hence $\mathcal{O}_p$ is a regular local ring. We can also consider *germs of sets.* We write $(\mathbf{C}^n, 0)$ to denote the germ of the set $\mathbf{C}^n$ at 0. We then write $f : (\mathbf{C}^n, 0) \to (\mathbf{C}^N, 0)$ to mean that f is a holomorphic mapping, defined on some unspecified neighborhood of $0 \in \mathbf{C}^n$, taking values in $\mathbf{C}^N$, and with $f(0) = 0$.

Exercise 2.14. Show that the ring $\mathcal{O}_p$ is isomorphic to the ring of convergent power series $\mathbf{C}\{z_1 - p_1, ..., z_n - p_n\}$.

Germs of holomorphic maps from $(\mathbf{C}, 0)$ to $(\mathbf{C}^n, 0)$ will play a particular role for us. We generally write these maps as $t \to z(t)$; thus t is now a complex variable. We are assuming that these mappings are holomorphic; they do not depend on $\bar{t}$, and we call them (germs of) holomorphic curves. As is customary, we define the curve to be the mapping, but we usually think of the curve as the image of the mapping. When $r : \mathbf{C}^n \to \mathbf{R}$ is a function we consider the pullback function z^*r defined by $z^*r(t) = r(z(t))$. When r is an Hermitian symmetric polynomial or real-analytic function we will write $z^*r(t) = r(z(t), \overline{z(t)})$.

In Section I.5.2 we introduce complex-analytic varieties and their germs. These varieties are subsets of $\mathbf{C}^n$ locally defined by the simultaneous vanishing of finitely many holomorphic functions. We give a simple illustration of the kind of thing we will be doing.

Consider the set V of points in $\mathbf{C}^2$ defined by $z_1^2 = z_2^3$. This set V can also be described as the image of $\mathbf{C}$ under the map $z : \mathbf{C} \to \mathbf{C}^2$ defined by $z(t) = (t^3, t^2)$. Near any point of V except the origin $(0, 0)$, the geometry is easy to understand. Near the origin, however, things are difficult. The origin is a *singularity*. One of our main ideas will be finding geometric conditions on a real hypersurface M at a point p that preclude the existence of complex-analytic varieties containing p and lying in M. Things become subtle because the variety could be singular at p.

At this stage we can only sketch the proof of the following fundamental result.

Lemma 2.5. *Let r be a real-valued function defined in a neighborhood of 0 in $\mathbf{C}^n$. Then $V(r)$ contains a complex-analytic variety through 0 if and only if there is a (germ of a) nonconstant holomorphic curve $z : (\mathbf{C}, 0) \to (\mathbf{C}^n, 0)$ such that z^*r is identically zero.*

PROOF. (Sketch) Suppose that such a curve exists. Its image is a complex one-dimensional variety contained in $V(r)$. Suppose conversely that $(X, 0) \subset (V(r), 0)$ is the germ of a complex variety. We may choose an irreducible branch of a one-dimensional subvariety of $(X, 0)$. By standard considerations in basic algebraic geometry this irreducible branch is the image of (the germ of) a holomorphic curve. □

We now develop a method for deciding whether there is a nonconstant map z such that z^*r vanishes identically. First we study the case when r is a polynomial. We assume that r has been written in the form (14.2). Suppose that $z^*r = 0$. We obtain

$$0 = 2\mathrm{Re}(z^*h) + ||z^*f||^2 - ||z^*g||^2. \tag{32}$$

We may regard the unwritten variables t and $\bar{t}$ in (32) as independent. It follows that the pure terms and mixed terms in (32) must vanish separately, and hence

$$\begin{aligned} z^*h &= 0 \\ ||z^*f||^2 &= ||z^*g||^2. \end{aligned} \tag{33}$$

We may assume, by including additional zero components, that f and g map into the same dimensional space. Thus we abbreviate $f \oplus 0$ by f and $g \oplus 0$ by g. By Theorem 2.2 there is a unitary mapping U such that $z^*(f) = U(z^*g)$. Hence we obtain a unitary mapping U for which

$$z^*h = 0$$

$$z^*(f - Ug) = 0.$$

Let us write $U = (U_{jk})$. We are now in the setting of algebraic geometry; we have a collection of holomorphic polynomials

$$h, f_1 - \sum_k U_{1k} g_k, \ldots, f_N - \sum_k U_{Nk} g_k; \tag{34}$$

each polynomial in (34) vanishes along the curve $t \to z(t)$.

Thus the variety defined by these functions is positive dimensional. We obtain the following result:

Theorem 2.3. *Let r be a real-valued polynomial on $\mathbf{C}^n$, written as in (14.2). If V is an irreducible one-dimensional complex-analytic variety contained in $V(r)$, then there is a unitary matrix U such that V is a subvariety of the variety defined by $V(h, f - Ug)$. Conversely, for each unitary U, the variety $V(h, f - Ug)$ is contained in $V(r)$.*

Example 2.3. Put $r(z, \overline{z}) = 2\mathrm{Re}(z_3) + |z_1^2 - z_2^3|^2 + |z_1^4|^2 - |z_2^6|^2$. Then the holomorphic curve defined by $t \to (t^3, t^2, 0)$ lies in $V(r)$. Furthermore, if γ is a holomorphic curve lying in $V(r)$, then there is unitary matrix of constants $U = (U_{jk})$ such that γ is a subvariety of the variety defined by the equations

$$(z_3, z_1^2 - z_2^3 - U_{11} z_2^6, z_1^4 - U_{21} z_2^6).$$

If the zero set of these equations is nontrivial, then we must have $U_{11} = 0$ and $U_{21} = 1$. A unitary U exists with these entries. We discover the variety V defined by the functions $(z_3, z_1^2 - z_2^3, z_1^4 - z_2^6)$; here V is precisely the image of the curve $t \to (t^3, t^2, 0)$.

Example 2.4. Let $f_1, \ldots f_k$ be holomorphic functions defined near the origin in $\mathbf{C}^n$ and vanishing there. For r of the form

$$r(z) = 2\mathrm{Re}(z_n) + \sum_{j=1}^{K} |f_j(z)|^2, \tag{35}$$

the only complex subvarieties contained in $V(r)$ are subvarieties of the variety given by the equations

$$z_n = f_1(z) = \cdots = f_K(z) = 0.$$

In this example, the mapping g from (20) vanishes and the geometry simplifies.

Hypersurfaces defined by r of the form (35) play a big role for several reasons. The geometry is simpler than in general, but it is sufficiently complicated to illustrate many of the main points. See especially Lecture 4. For now, note that the zero set of a squared norm of a holomorphic mapping f defines the variety $V(f)$. Theorem 2.3 generalizes this simple idea by providing a method to decide whether the germ of an algebraic real hypersurface contains the germ of a complex variety of positive dimension.

Exercise 2.15. Put $R = (|h_1|^2 - |h_2|^2)^2 + |h_3|^2$ for holomorphic polynomials h_1, h_2, h_3, Give necessary and sufficient conditions on the h_j such that the zero set of R is a complex variety.

For convenience we discuss the usage of the term *defining function*. Let r be a smooth function defined near a point p in real Euclidean space $\mathbf{R}^N$. Suppose $dr(p) \neq 0$. There is then a neighborhood of p on which the zero set of r defines a hypersurface (submanifold of codimension one) in $\mathbf{R}^N$. We say that r is a local defining function for M near p. Next let Ω be an open subset of $\mathbf{R}^N$; we call r a defining function for Ω if the following hold: $\Omega = \{x : r(x) < 0\}$, $b\Omega = \{x : r(x) = 0\}$, and $dr(x) \neq 0$ on $b\Omega$. Most of the time in these lectures Ω will be an open set in $\mathbf{C}^n$ and $b\Omega$ will be a real hypersurface in $\mathbf{C}^n$. While the notion of a defining function is the same in the real and complex settings, the allowable local coordinate changes differ.

We close this section with a simple comment. If r is a defining function for a real hypersurface in $\mathbf{C}^n$ containing 0, then by definition $dr(0) \neq 0$, and the origin is not a critical point. On the other hand, if the pure term $2\mathrm{Re}(z_n)$ in (35) were not there, then we would have a squared norm $||f||^2$ of a holomorphic mapping, and the origin would be a critical point for this function. The function $||f||^2$ has a strict local minimum at 0 if and only if the (germ of the) variety defined by the components of f consists of the origin alone. There are many algorithms in commutative algebra and complex analysis for deciding whether a variety is a single point. These algorithms enable us to handle a special kind of singularity. The double inequality in Theorem 3.4 will help us understand a more general situation.

4. Real analytic hypersurfaces and subvarieties

Let M be a real-analytic subvariety of $\mathbf{C}^n$ containing the point p. By definition M is locally given by the common zeroes of finitely many real-analytic real-valued functions. In fact we may assume that M is given near p by the vanishing of a single real-analytic real-valued function r; if M were defined by the vanishing of several such real functions, say $r_1, ..., r_k$, then M could also be defined by the zeroes of the single function $\sum r_j^2$. Assume M is the zero set of a single function r. When M is of higher codimension than one at p, or when M is of codimension one but not smooth at p, then $dr(p)$ must vanish. When $dr(p) \neq 0$, then M is a smooth real hypersurface near p, and hence of real dimension $n - 1$. While our primary interest is in the geometry of real hypersurfaces, many of the techniques apply for real subvarieties.

Let r be a real-analytic Hermitian symmetric function defined near p. We may suppose that it is defined near p by $r(z, \overline{z}) = 2\mathrm{Re}(h(z)) + ||f(z)||^2 - ||g(z)||^2$, where f and g are Hilbert space valued holomorphic functions defined near p. The analogue of Theorem 2.3 holds in this setting; if V is the (germ of a) complex-analytic variety passing through p and lying in M, then there is a holomorphic curve $t \to z(t)$ such that $z^*h = 0$ and $||z^*f||^2 = ||z^*g||^2$. By a generalization of Theorem 2.2, it follows that there is a constant linear map U such that $z^*(f - Ug) = 0$. Using this result (and assuming Theorem 3.5) we will establish in Theorem 2.4 a fundamental fact about compact real analytic subvarieties.

By Theorem 3.5 and Remark 3.7, when M is real-analytic, the set W of points $p \in M$ for which there is a (germ of a) complex-analytic variety through p and lying in M is a closed subset of M. Furthermore we will express this set precisely in terms of the holomorphic decomposition of r. Assuming this result for the moment, we derive a basic result of Diederich-Fornaess [DF1]: if M is a compact real-analytic subvariety, then M contains no germs of complex-analytic varieties; thus W is empty when M is compact.

Theorem 2.4. *(Diederich-Fornaess) A compact real analytic subvariety of $\mathbf{C}^n$ contains no positive-dimensional complex-analytic subvarieties.*

PROOF. Let M be the given compact real analytic subvariety, and let W be the set of $p \in M$ for which there is a positive dimensional ambient complex variety passing through p and lying in M. We want to show that W is empty. By Theorem 3.5 and Remark 3.7, W is a closed subset of M, and hence W is a compact subset of $\mathbf{C}^n$. Suppose W is not empty. The function $z \to ||z||^2$ is continuous on W, and hence achieves a maximum at some point $p \in W$. Thus p is the point of W farthest from the origin.

Let ϕ be the rational function given $\phi(z) = \frac{1}{1-\langle z,p\rangle+||p||^2}$. Then $\phi(p) = 1$. Also ϕ is holomorphic except where the denominator vanishes, and hence near each point of W. By the Cauchy-Schwarz inequality, the restriction of ϕ to W achieves its maximum at p. Theorem 2.3 provides an explicit complex variety $V = V(h, f - Ug)$ containing p and lying in M. Since $V \subset W$, the holomorphic function ϕ achieves a maximum on V, which contradicts the maximum principle for holomorphic functions on a variety. Alternatively we could use Lemma 2.5 to find a holomorphic curve whose image is in W, and then apply the maximum principle in one dimension. Hence W must be empty. □

Corollary 2.2. *Let Ω be a bounded domain in $\mathbf{C}^n$ with real-analytic boundary. Then $b\Omega$ contains no non-constant holomorphic curves.*

5. Complex varieties, local algebra, and multiplicities

We will require some elementary commutative algebra in order to provide a quantitative measurement of how closely complex-analytic curves can touch a real hypersurface. Our rather intuitive account centers around multiplicities.

Orders of vanishing

We have glimpsed the importance of holomorphic mappings and their squared norms. We next develop some algebra to help understand ideas such as orders of vanishing and multiplicities. We start with a simple but decisive fact from complex analysis. If f is holomorphic in a connected neighborhood of 0 in $\mathbf{C}$, and $f(0) = 0$, then either f is identically zero or f has a zero of finite order at 0. There are many ways to compute the order of vanishing m of f when f is not identically zero. We list four such methods; later we will see how to generalize these methods to higher dimensions.

1) We can use an integral formula:

$$m = \frac{1}{2\pi i} \int_\gamma \frac{f'(z)}{f(z)} dz,$$

where the integral is taken over a contour γ with winding number 1 about 0.

2) We can use algebra. Let $\mathcal{O}$ denote the convergent power series ring at 0 and let $\mathcal{M} \subset \mathcal{O}$ denote the maximal ideal of functions vanishing at 0. When $n = 1$, the ring $\mathcal{O}$ is a principal ideal domain; its ideals are the full ring, the zero ideal, and the ideals (z^k) for positive integers k. This simple description applies only in one dimension. Thus, in one dimension, $(f) = (z^m)$ for some integer m, and m is the dimension of the complex vector space $\mathcal{O}/(f)$. This quotient space is an algebra; we may take $1, z, z^2, ..., z^{m-1}$ as a basis.

3) We can use topology. The generic number of inverse images of a point w under f is m; this integer is the topological degree of the (germ of the) map f.

4) Here is another interesting integral formula. We claim that $\frac{1}{m}$ is the supremum of the set of c such that

$$\int \frac{1}{|f|^{2c}} dV < \infty.$$

The integral is taken over a small deleted neighborhood of 0 on which $f(z) \neq 0$.

Exercise 2.16. Verify that each of the above four methods computes the order of vanishing of f. Find several additional methods to determine m.

We now turn to several-variable analogues of these ideas. First suppose that $f : (\mathbf{C}^n, 0) \to (\mathbf{C}, 0)$ is the germ of a holomorphic mapping. There is an obvious notion of the order of vanishing of f, namely the smallest m for which there is a term in the Taylor expansion of f of order m. This single number is inadequate for many purposes. Consider the following simple example. For a complex number a let $f(z_1, z_2) = z_1^2 - az_2^3$. Then the order of vanishing at f at the origin is two. When $a = 0$, f is reducible, and the number 2 measures the multiplicity. When $a \neq 0$, however, f is irreducible, and its zero set is a complicated complex variety. More information is needed. One naive approach would be to associate with f the pair of numbers $(2, 3)$ representing the order of vanishing of f in the coordinate directions. This approach is also inadequate, as we shall see.

The ideas from one dimension extend to higher dimensions in various ways; one of the nicest situations arises from *finite analytic mappings*. Let $f : (\mathbf{C}^n, 0) \to (\mathbf{C}^N, 0)$ be the germ of a holomorphic mapping preserving the origin; f is called a finite analytic mapping if $f^{-1}(0) = \{0\}$. One basic issue which distinguishes complex geometry from real geometry concerns the dimensions of zero-sets. The simultaneous zero-set of fewer than n holomorphic functions f must be a complex-analytic variety of positive dimension. Therefore f can be a finite mapping only if $N \geq n$. When f is a finite mapping, things more closely resemble the one-dimensional case. For example, one can associate with f various measurements of the *multiplicity* of the zero-dimensional variety $V(f)$. In one dimension these measurements will all be the same, whereas they will differ in higher dimensions. In order to make these things precise we must discuss preliminary local algebra.

Complex-analytic varieties

First we discuss complex subvarieties and their germs. We will then need to compare various rings of germs of functions. See [GH], [Gun1], [Gun2], [Gun3] for considerable information on complex-analytic varieties and [Nar] for information about real-analytic spaces.

Definition 2.2. (Complex-analytic subvariety) Let Ω be an open subset of $\mathbf{C}^n$. A complex-analytic subvariety of Ω is a closed subset V locally given by the vanishing of holomorphic functions; thus, each point $p \in V$ has a neighborhood U_p such that $V \cap U_p$ is the set of common zeroes of finitely many holomorphic functions on U_p.

In these notes we will not generally mention the ambient open set Ω. Furthermore we will typically care only about local behavior of V. We usually say only *complex-analytic variety* or even just *variety* if the meaning is clear. For us a variety is the set of points; it will be important for us to consider the possibility that the same point set could be defined multiple times (especially when the variety is a single point), but we do not introduce additional terminology such as *reduced complex-analytic variety* or *reduced analytic space.*

Let V be a complex subvariety of $\Omega \subset \mathbf{C}^n$. Suppose, for each point $p \in V$, the set V is locally defined by the vanishing of l holomorphic functions $f_1, ..., f_l$ for which the differentials $df_1(p), ..., df_l(p)$ are linearly independent. Then, by the holomorphic version of the implicit function theorem, V is a *complex manifold* of dimension $n - l$ or of *codimension* l. It is easy to see that *most* points of a variety V are *smooth* points; p is a smooth point if it has a neighborhood on which V is a complex manifold. A complex variety V has dimension k if k is the largest integer for which V is a k-dimensional complex manifold at some smooth point.

A complex-analytic variety V is *reducible* if it can be written as the union of two proper complex-analytic varieties, and *irreducible* otherwise. A variety can be written as an irredundant union of irreducible varieties. At each smooth point of an irreducible variety the dimension is the same. See [Wh] or [GR] for these facts.

Exercise 2.17. Let Ω be an open neighborhood of 0 in $\mathbf{C}^n$. Suppose $f : \Omega \to \mathbf{C}^k$ is a holomorphic mapping with $f(0) = 0$. Show that the dimension over $\mathbf{C}$ of $V(f)$ is at least $n - k$. For $n \geq 2$, show that $V(f)$ cannot be a compact subset of Ω. Suggestion: First assume $k = 1$; consider the reciprocal of f and apply Hartog's extension theorem.

Local algebra

Let R denote one of the following rings of germs of functions at the point p in $\mathbf{C}^n$:

$\mathcal{O}$, the ring of germs of holomorphic functions at p.

$\mathcal{A}$, the ring of germs of real-analytic functions at p.

$\mathcal{E}$, the ring of germs of smooth functions at p.

We write R_n when we wish to emphasize the underlying dimension. In each case the ring is *local*; we let $\mathcal{M}$ denote the unique maximal ideal in the local ring; elements of $\mathcal{M}$ are those germs vanishing at p. These ideas arise in our work on real hypersurfaces bounding domains; a defining function r for $b\Omega$ can be interpreted as a generator of the principal ideal in $\mathcal{E}$ of (germs of) functions vanishing on $b\Omega$.

In working with the ring $\mathcal{O}$, two classical results of Weierstrass enable us to perform the induction step in proofs by induction on the dimension. We state these results here; versions of them can be found in many sources. First it is convenient to introduce the term *Weierstrass polynomial.* Write the variable $(z_1, ..., z_n)$ as (z, w), where $z \in \mathbf{C}^{n-1}$. A Weierstrass polynomial is a monic polynomial in w of degree m, whose coefficients c_j for $0 \leq j < m$ are (germs of) holomorphic functions of z and which vanish at the origin.

Theorem 2.5 (Weierstrass Preparation Theorem). *Let f be holomorphic in some neighborhood of the origin in $\mathbf{C}^n$. Suppose, with coordinates as above, that the function $w \to f(0, w)$ vanishes to finite order m at 0. Then there are a unique unit*

$u \in \mathcal{O}$ and a unique Weierstrass polynomial p of degree m such that $f = up$. Thus

$$f(z,w) = u(z,w)\left(w^m + a_{m-1}(z)w^{m-1} + \ldots + a_0(z)\right). \tag{36}$$

Theorem 2.6 (Weierstrass Division Theorem). *Let ϕ be a Weierstrass polynomial of degree m, and let f be holomorphic near the origin. Then there is a unique expression $f = \phi q + r$, where $q \in \mathcal{O}$ and r is a polynomial of degree less than m in w with holomorphic coefficients in z (but not necessarily a Weierstrass polynomial).*

Exercise 2.18. Easy! Give an example (with notation as in the division theorem) where $f = \phi q + r$ and r is not a Weierstrass polynomial.

Exercise 2.19. Easy! Suppose f is holomorphic near the origin in $\mathbf{C}^n$ and that f is not identically 0. Show that there is a linear change of coordinates $z \to \zeta$ such that $f(0, \zeta_n)$ vanishes to finite order at 0.

Exercise 2.20. Suppose $f_1, \ldots f_k$ are holomorphic near the origin in $\mathbf{C}^n$, each vanishes there, and none of the f_k is identically 0. Show that there is a linear change of coordinates $z \to \zeta$ such that all the $f_k(0, \zeta_n)$ vanish to finite order at 0.

Exercise 2.21. Verify the uniqueness parts of both Weierstrass theorems.

Exercise 2.22. Let $f \in \mathcal{O}$ be a polynomial in w with holomorphic coefficients in z. Assume that f is a multiple of a Weierstrass polynomial ϕ; thus $f = g\phi$ for some $g \in \mathcal{O}$. Show that g must be a polynomial in w. Give an example where ϕ is not a Weierstrass polynomial and g is not a polynomial.

Exercise 2.23. Look up the proofs of these theorems in various sources and see how many fundamentally different proofs you can find.

Exercise 2.24. Give a proof of the preparation theorem that works in the formal power series setting.

Exercise 2.25. Using the Weierstrass division theorem and induction on the dimension, prove that $\mathcal{O}$ is Noetherian and that it is a unique factorization domain.

The Weierstrass preparation theorem provides the basic information about the geometry of complex-analytic varieties. Using the notation of this theorem, we note that the germ at 0 of the variety $V(f)$ is the same as the germ of the variety $V(p)$. Because p is a polynomial in the w variable, its zero set is more tractable.

Exercise 2.26. The polynomial $z + z^2$ is *reducible* in the polynomial ring $\mathbf{C}[z]$, yet it is irreducible in the local ring $\mathcal{O}$. Why?

Exercise 2.27. Give an example of a holomorphic f in three variables such that f is irreducible at 0 but such that the variety $V(f)$ is reducible at points arbitrarily close to 0.

In the ring $\mathcal{E}$, a function can vanish to infinite order without being identically zero. A related point is that this ring is not Noetherian. Furthermore the geometry of zero sets has no meaningful structure. For example, every closed subset of real Euclidean space can be expressed as the zero-set of a smooth function.

The zero sets in the other two cases are much more restricted. We are especially interested in equations whose simultaneous zero set is a single point. We first recall the notion of *radical* of an ideal. Let W be a commutative ring and let I be an ideal in W. The radical of I, written $\mathrm{rad}(I)$, is the set of all $w \in W$ for which there is an integer m such that $w^m \in I$. It is easy to verify that rad(I) is itself an ideal. For an ideal I in the polynomial ring $\mathbf{C}[z_1, ..., z_n]$, the famous Hilbert Nullstellensatz states that $p \in \text{rad}(I)$ if and only if $p(z) = 0$ whenever $g(z) = 0$ for all $g \in I$. In the ring $\mathcal{O}$, the analogous Rückert Nullstellensatz holds. We pause to express these results in a simple notation. Given an ideal I, we write $V(I)$ for the variety defined by the elements in I. Given a variety V, we write $I(V)$ for the ideal of functions vanishing on V. These two versions of the Nullstellensatz take the same elegant form using this notation:

$$I(V(I)) = \text{rad}(I). \tag{37}$$

In these notes we will emphasize the special case when $V(I)$ consists of a single point. In each of the rings R above, the maximal ideal $\mathcal{M}$ at p consists of those germs vanishing at p. Let $I \subset \mathcal{M}$ be an ideal. By (37) the (germ of a) variety $V(I)$ is zero-dimensional if and only if

$$\text{rad}(I) = \mathcal{M}. \tag{38}$$

We mention corresponding language from algebra. We say that I is $\mathcal{M}$-primary if there an integer k such that

$$\mathcal{M}^k \subset I \subset \mathcal{M}. \tag{39}$$

The left-hand containment in (39) guarantees that any germ vanishing to order at least k lies in I. Hence there are finitely many multi-indices α for which z^α is not in I. Putting these remarks together gives the following special case of the Rückert Nullstellensatz.

Theorem 2.7. *(Nullstellensatz for zero-dimensional varieties in $\mathcal{O}$) The following properties of an ideal I in $\mathcal{O}$ are equivalent:*

1) I is $\mathcal{M}$-primary.

2) $\text{rad}(I) = \mathcal{M}$.

3) $V(I)$ is a single point.

4) The quotient algebra $\mathcal{O}/I$ is a finite-dimensional complex space.

Let I be an ideal in $\mathcal{O}$ satisfying any of the equivalent properties from this theorem. By (39), there is an integer k for which $\mathcal{M}^k \subset I \subset \mathcal{M}$. We could take the smallest such k as a numerical measurement of the singularity. By property 4), we could take the dimension of the quotient algebra as a measurement. When $n = 1$ these numbers are equal; in general they are not the same. We write $\mathbf{D}(I)$ for the dimension of $\mathcal{O}/I$ over $\mathbf{C}$; we will investigate its properties in Section 3.5.

Exercise 2.28. Verify that the radical of an ideal is itself an ideal.

Exercise 2.29. In n dimensions, find $\mathbf{D}(\mathcal{M}^k)$ as a function of n and k.

Exercise 2.30. Let I be the ideal generated by $z_1^{m_1}, ..., z_n^{m_n}$. Compute $\mathbf{D}(I)$.

Exercise 2.31. For $n = 2$ let I be generated by $z_1^a - z_2^b$ and $z_1^c z_2^d$. Assume the exponents are positive. Compute $\mathbf{D}(I)$. A harder problem: Do the same for $I = (z_1^a - z_2^b, z_1^c - z_2^d)$. First determine the conditions on the exponents for $\mathbf{D}(I)$ to be finite.

Exercise 2.32. Let I be $\mathcal{M}$-primary in n dimensions, and let $\kappa(I)$ denote the smallest k for which (39) holds. Prove that $\kappa(I) \leq \mathbf{D}(I) \leq \binom{n-1+\kappa(I)}{\kappa(I)-1}$.

Next we briefly consider a more general notion of radical appropriate in $\mathcal{A}$. For $\mathcal{A}$,the Lojasiewicz inequality plays the role of the Nullstellensatz. Let germs $g_1, ...g_k$ generate an ideal (g) in $\mathcal{M} \subset \mathcal{A}$. Consider the real-analytic variety $V(g)$ defined by them. The following result provides the analogue of the Nullstellensatz in this setting:

Theorem 2.8. *(Lojasiewicz) A germ f in $\mathcal{A}$ vanishes on $V(g)$ if and only if there is an integer N and a constant C such that $|f|^N \leq C \sum_j |g_j|$.*

Based on this theorem, one therefore defines the real radical $\mathrm{rad}_{\mathbf{R}}(g)$ of an ideal in $\mathcal{A}$ or $\mathcal{E}$ to be the set of germs satisfying such an inequality. See [Nar] for detailed information. We indicate why inequalities are used rather than equalities. The real function x vanishes on the zero-set of the function $x^2 + y^2$, yet we cannot possibly have an integer m for which x^m is a multiple of $x^2 + y^2$. On the other hand, we do have $x^2 \leq x^2 + y^2$. This simple example illustrates the idea.

Things are much more complicated in the smooth case. For example, let ϕ be a smooth function with a zero of infinite order at 0 in $\mathbf{R}$. Then x vanishes on the zero set of ϕ, but there are no C and N such that $|x|^N \leq c\phi(x)$.

In Lecture 4, we will discuss Kohn's results on subelliptic multipliers. The special case when an ideal in $\mathcal{A}$ is generated by the squared norm of a holomorphic mapping will be especially important there. Kohn's method then reduces to an algorithm in $\mathcal{O}$ and leads to an unusual but fascinating method for showing that a collection of germs defines a zero-dimensional variety. This method is not well known to algebraists and hence we will discuss it some detail.

LECTURE 3

Pseudoconvexity, the Levi form, and points of finite type

We introduce the notion of pseudoconvexity for a real hypersurface M in $\mathbf{C}^n$. We are especially interested in a geometric notion called *finite type*. This notion serves as an intermediate condition between *strong pseudoconvexity* and Levi *flatness*. The mathematics is somewhat analogous to that of a critical point for a smooth function f of several real variables; strong pseudoconvexity at p is similar to p being a nondegenerate local minimum point; Levi flatness at p is similar to f vanishing to infinite order at p, and finite type is similar to f having a strict local minimum there. After discussing pseudoconvexity, we define finite type by measuring the order of contact of holomorphic curves with the real hypersurface. We use commutative algebra to compute the type of a point. One of the main results of this lecture is that *type* is a locally bounded function and hence the set of points of finite type is an open subset of M.

1. Euclidean convexity

A subset S of real Euclidean space is *convex* if, for each pair of points $p, q \in S$, the line segment between them lies in S. Thus, for $0 \leq t \leq 1$, $tp + (1-t)q \in S$. Let us recall from calculus several tests for convexity. These real variable ideas motivate our approach to *pseudoconvexity*, the more elusive analogous notion appropriate in complex analysis.

The simplest situation to visualize arises in freshman calculus. A function on the real line is called *convex* if the following inequality holds for all $p, q \in \mathbf{R}$ and for all $t \in [0,1]$:

$$f(tp + (1-t)q) \leq tf(p) + (1-t)f(q). \tag{1}$$

For each $x \in [p,q]$, the point $(x, f(x))$ is below or on the line segment from $f(p)$ to $f(q)$.

The following statement is both obvious geometrically and easy to prove: f is a convex function if and only if the subset S of $\mathbf{R}^2$ defined by $S = \{(x,y) : y > f(x)\}$ is a convex set. Furthermore, if f is twice differentiable, then f is convex if and only if f'' is a nonnegative function. One can define convex function of several variables as well. A twice differentiable function of several real variables is convex on a domain if and only if its matrix of second derivatives is non-negative definite at each point of the domain.

Exercise 3.1. Verify that a differentiable function on $\mathbf{R}$ is convex if and only if its derivative is nondecreasing. Then verify that a twice differentiable function is convex if and only if its second derivative is nonnegative.

Exercise 3.2. Prove that $S = \{(x, y) : y > f(x)\}$ is a convex set if and only if f is convex. Suggestion: show first that it suffices to verify the line segment condition for convexity assuming p and q are on the boundary.

Exercise 3.3. Suppose that f is continuous and satisfies (1) for all p, q when $t = \frac{1}{2}$. Show that f satisfies (1) for all t and hence is convex.

Suppose next that S is a subset of $\mathbf{R}^n$ with a twice continuously differentiable *defining function* r. Thus $r : \mathbf{R}^n \to \mathbf{R}$ is of class C^2, $dr(x) \neq 0$ when $r(x) = 0$, and S is the set of points where $r < 0$. We write M for the boundary bS; then M is a real submanifold of $\mathbf{R}^n$ of codimension one. If $x \in M$, then the tangent space T_xM at x is the set of $v \in \mathbf{R}^n$ such that $\langle dr(x), v\rangle = 0$. Each tangent space is a real vector space of dimension $n - 1$.

It is convenient to choose coordinates and notation in which calculations simplify. We pause to describe one such choice. Suppose $p \in M$. By the implicit function theorem, we can find a neighborhood Ω of p, a local coordinate system, and a smooth function g such that such that the following hold on Ω:

1) p is the origin in $\mathbf{R}^n$.
2) $M \cap \Omega = \{x : x_n = g(x_1, ..., x_{n-1})\}$.
3) $g(0) = 0$ and $dg(0) = 0$.

We will sometimes use the notation (x', x_n) for these coordinates.

The function r given by $r(x) = g(x_1, ..., x_{n-1}) - x_n = g(x') - x_n$ is then a local defining function for M near p, analogous to the function $f(x) - y$ above. We naturally ask for conditions on g equivalent to convexity of $\{x_n > g(x')\}$. The answer is that g must be a convex function of its $n - 1$ variables. When g is twice differentiable, it is convex if and only if its matrix of second derivatives, or (real) Hessian, is non-negative definite at each point. For a twice differentiable function r we let $D^2r(x)$ denote its Hessian at x.

Let r be a defining function of class C^2 for a set S. What is the condition for convexity? Proposition 3.1 answers this question and also motivates our work on pseudoconvexity. We show that S is convex if and only if, at each point x of M, the restriction of $D^2r(x)$ to T_xM is non-negative definite. In other words, if $\langle dr, X\rangle(p) = 0$, then $D^2r(p)(v, v) \geq 0$. In case $r(x) = g(x') - x_n$, the condition of tangency for a vector $v = (v', v_n)$ at x becomes

$$v_n = \sum_{j=1}^{n-1} g_{x_j}(x')v_j = \langle dg(x'), v'\rangle, \tag{2}$$

and the condition for convexity is simply that $D^2g(x)$ be non-negative definite on $\mathbf{R}^{n-1}$. We do not need to restrict a Hessian to a subspace when using these coordinates.

Proposition 3.1. *With the notation in the above paragraphs, assume for all $x \in M$ that $D^2r(x)$ is nonnegative definite on T_xM. Then S is convex. Conversely, if S is convex, then for each $x \in M$, $D^2r(x)$ is nonnegative definite on T_xM.*

PROOF. Suppose first that $D^2r(x)$ is actually positive definite on T_xM. Given distinct points $p, q \in S$, let $v = p - q$. Consider the line segment $tp + (1 - t)q$ connecting them. We want to show that $tp + (1 - t)q \in S$ for all $t \in [0, 1]$. As in the proof of Exercise 3.2, we may without loss of generality assume that p and q are boundary points.

We pull back to the line segment connecting them. Thus we consider the smooth function ϕ defined by $t \to r(tp + (1-t)q) = \phi(t)$. Then $\phi(0) = \phi(1) = 0$. Let W be the subset of $[0,1]$ for which $\phi(t)$ is in the closure of S; W contains 0 and 1. If W isn't $[0,1]$, then ϕ achieves a positive maximum for some $t \in (0,1)$, and hence there is some point where $\phi'(t) = 0$ and $\phi''(t) \leq 0$. Using subscripts for partial derivatives, we then have

$$0 = \sum r_{x_j}(\phi(t))(p_j - q_j) = \langle dr(\phi(t)), v \rangle \tag{3.1}$$

$$0 \geq \phi''(t) = \sum_{j,k} r_{x_j x_k}(tp + (1-t)q)(p_j - q_j)(p_k - q_k) = D^2 r(\phi(t))(v,v). \tag{3.2}$$

Since $v \neq 0$, we contradict the definiteness of the Hessian. When $D^2 r$ is only semi-definite, we may add a small convex term $\epsilon ||x||^2$ to r, and then use a limiting argument to reduce to the case where $D^2 r$ is positive definite. Thus the nonnegativity of $D^2 r$ guarantees convexity.

The converse statement is proved by contrapositive and a Taylor series argument. We assume that the Hessian has a negative eigenvalue at some point x in M, and we will show that S is not convex. Without loss of generality we may choose coordinates as above such that x is the origin, r has the form $g(x') - x_n$, that $dg(0) = 0$, but for some v, $D^2 g(0)(v,v) < 0$. Put $p(t) = tv - \epsilon e_n = (tv, -\epsilon)$ for sufficiently small positive ϵ. Note that the vector e_n is the inner unit normal at the origin and that $p(0)$ is not in S. We expand r in a Taylor series to order two in t:

$$r(p(t)) = r(tv - \epsilon e_n) = g(tv) + \epsilon = \epsilon + \frac{t^2}{2} D^2 g(0)(v,v) + \ldots$$

Since the Hessian is negative we can find t such that $r(p(t)) < 0$, and hence by continuity we can proceed from $p(0)$ in both the $\pm v$ directions to obtain t_1 and t_2 where $r(p(t_j)) = 0$. Thus we have a line segment connecting two boundary points of S and also containing the point $p(0)$, which is outside of S. Therefore S is not convex. □

Exercise 3.4. Fill in the details of the proof of Proposition 3.1.

2. The Levi form

Pseudoconvexity is the complex analogue of Euclidean convexity. The notion of pseudoconvexity is however considerably more elusive than that of Euclidean convexity. For example, every open set in $\mathbf{C}$ is pseudoconvex! See [H] or [Kr] for detailed discussion about pseudoconvexity in general. Perhaps the most important result from twentieth century complex analysis was the solution of the Levi problem, which characterized pseudoconvex domains. Here we will need only to understand when a domain with smooth boundary is pseudoconvex. The answer then involves the Levi form, which is the complex analogue of the real Hessian from Section 0.

What do we mean by a domain in $\mathbf{C}^n$ with smooth boundary? A domain is an open and connected set; a domain has smooth boundary if its boundary is a smooth real manifold (necessarily of real codimension one). Let Ω be a domain in $\mathbf{C}^n$ with smooth boundary, and assume that $p \in b\Omega$. We may suppose that there is a neighborhood of p on which $b\Omega$ is given by the vanishing of a smooth defining function r with $dr(p) \neq 0$. By convention $r < 0$ on Ω.

The condition of pseudoconvexity can be expressed in terms of the first and second derivatives of a defining function r. We will continue to write partial derivatives as subscripts, but we also try to use more aesthetic notation when possible. Let $\langle\ ,\ \rangle$ denote the contraction of forms and vector fields.

First we observe that a complex vector field X on $\mathbf{C}^n$ can be expressed as a smooth combination of the first order derivative operators:

$$X = \sum_{j=1}^{n} a_j \frac{\partial}{\partial z^j} + \sum_{j=1}^{n} b_j \frac{\partial}{\partial \overline{z}^j}.$$

We say that X is a $(1,0)$ vector field on $\mathbf{C}^n$ if each b_j vanishes identically. Let $T^{1,0}b\Omega$ be the bundle whose sections are the $(1,0)$ vector fields tangent to $b\Omega$. In coordinates, a vector field $L = \sum_{j=1}^{n} a_j \frac{\partial}{\partial z_j}$ is a local section of $T^{1,0}b\Omega$ if, on $b\Omega$

$$\langle \partial r(z), L(z) \rangle = \sum_{j=1}^{n} a_j(z) r_{z_j}(z) = 0. \tag{4}$$

Then $b\Omega$ is pseudoconvex at p if, whenever (4) holds at p, we have

$$D^2 r(p)(a,a) = \sum_{j,k=1}^{n} r_{z_j \overline{z}_k}(p) a_j(p) \overline{a_k(p)} \geq 0. \tag{5}$$

We wish to express (5) more invariantly. The bundle $T^{1,0}(b\Omega)$ is a subbundle of $T(b\Omega) \otimes \mathbf{C}$. The intersection of $T^{1,0}(b\Omega)$ with its complex conjugate bundle is the zero bundle, and their direct sum has fibers of codimension one in $T(b\Omega) \otimes \mathbf{C}$. Let η be a non-vanishing purely imaginary one form annihilating this direct sum. Then (4) and (5) together become

$$\lambda(L, \overline{L}) = \langle \eta, [L, \overline{L}] \rangle \geq 0 \tag{6}$$

on $b\Omega$ for all local sections of $T^{1,0}(b\Omega)$. The left-hand side of (6) defines a Hermitian form λ on $T^{1,0}(b\Omega)$ called the Levi form. The Levi form is defined only up to a multiple, but this ambiguity makes no difference in what we will do. The boundary $b\Omega$ is called *pseudoconvex* if at each point of $b\Omega$ all nonzero eigenvalues of the Levi form have the same sign. In this case, we multiply by a constant to ensure that the Levi form is *nonnegative* definite. We collect this information in the next definition.

Definition 3.1. A real hypersurface M is *pseudoconvex* if its Levi form is nonnegative definite on $T^{1,0}M$. A real hypersurface is *strongly pseudoconvex* at p if its Levi form is positive definite at p. A domain with smooth boundary is (Levi) pseudoconvex if its Levi form is nonnegative definite at each boundary point and *strongly pseudoconvex* if its Levi form is positive definite at each boundary point.

If a real hypersurface M is the boundary of a bounded domain, then we must have at least one point where the Levi form is positive definite; take the point p farthest from the origin. Then M will osculate a sphere to second order at p, and hence its Levi form will be positive definite there.

We provide some coordinate free formulas for clarifying the several points of view we have introduced. Let M be a a real hypersurface in $\mathbf{C}^n$, and suppose that M is locally given by the vanishing of a smooth function r such that $dr \neq 0$ on M. A vector field X on $\mathbf{C}^n$ is tangent to M if and only if $X(r) = \langle dr, X \rangle = 0$ on

M. If X is also a $(1,0)$ vector field, then $\langle \overline{\partial} r, X\rangle = 0$, and this tangency condition becomes $\langle \partial r, X\rangle = 0$ on M. We may choose the differential form η to be

$$\eta = \frac{1}{2}(\partial - \overline{\partial})(r). \tag{7}$$

We then have $d\eta = -\partial\overline{\partial} r$. Let L and K be vector fields. By the Cartan formula for the exterior derivative of a one-form we obtain

$$\langle d\eta, L \wedge \overline{K}\rangle = L\langle \eta, \overline{K}\rangle - \overline{K}\langle \eta, L\rangle - \langle \eta, [L, \overline{K}]\rangle. \tag{8}$$

When L and K are $(1,0)$ vector fields tangent to M, two of these terms vanish, giving

$$\lambda(L, \overline{K}) = \langle \eta, [L, \overline{K}]\rangle = \langle -d\eta, L \wedge \overline{K}\rangle = \langle \partial\overline{\partial} r, L \wedge \overline{K}\rangle. \tag{9}$$

Note that the $(1,1)$ form $\partial\overline{\partial} r$ is essentially the complex Hessian of r. Thus the Levi form can be regarded as the restriction of the complex Hessian of r to the space of $(1,0)$ vector fields tangent to M.

Exercise 3.5. Let r be a defining function for a strongly pseudoconvex hypersurface M. Show for sufficiently large real λ that $e^{\lambda r} - 1$ is a strongly plurisubharmonic defining function for M.

Remark 3.1. According to the previous exercise, all strongly pseudoconvex domains admit strongly plurisubharmonic defining functions. Not all pseudoconvex domains admit plurisubharmonic defining functions, however. See [DF2], [DF3]. Diederich-Fornaess introduced a class of smoothly bounded pseudoconvex domains in C^2 with remarkable properties. The closure of such a domain does not admit a Stein neighborhood basis, and there is no differentiable plurisubharmonic defining function. The boundaries of these *worm* domains are strictly pseudoconvex except on a certain annulus.

Exercise 3.6. Easy. Show that the domain defined by $\sum_{j=1}^{n} |z_j|^2 - 1 < 0$ (the unit ball) and the domain defined by $2\mathrm{Re}(z_n) + \sum_{j=1}^{n} |z_j|^2 < 0$ are biholomorphically equivalent.

Exercise 3.7. Show that the unit ball and the Siegel half space (the domain given by $\mathrm{Re}(z_n) + \sum_{j=1}^{n-1} |z_j|^2 < 0$) are biholomorphically equivalent.

The analogy between domains with degenerate Levi forms and functions with degenerate critical points provides a unifying theme in these lectures. In both cases we need to take higher derivatives, and in both cases there is no easy naive way to do so and still preserve the simplicity of the one-dimensional case. We saw in Lecture 2 that degenerate critical points are harder to handle in two or more real dimensions than they are in one real dimension. In the same way, a degenerate Levi form is harder to handle for hypersurfaces in three or more complex dimensions than it is in two complex dimensions.

Next we consider a context in which strongly pseudoconvex points behave like nondegenerate critical points. Let g be a smooth convex function defined near $0 \in \mathbf{R}^{n-1}$ with $dg(0) = 0$. The origin is a nondegenerate critical point if and only if $D^2 g(0)$ is positive definite, in which case g has a strict local minimum there. If g has a strict local minimum, then the subset of $\mathbf{R}^n$ defined by $\{x_n = g(x')\}$ contains no line segment through the origin. This geometric condition can occur even when the Hessian is degenerate. The analogous situation for us is whether a smooth hypersurface contains any nonconstant holomorphic curves. One difference arises because we allow these curves to have singularities. We begin by proving that strongly pseudoconvex hypersurfaces contain no nonconstant holomorphic curves.

Theorem 3.1. *Let M be a real hypersurface in $\mathbf{C}^n$, and assume that M is strongly pseudoconvex at p. Then every holomorphic curve lying in M and passing through p is a constant.*

PROOF. If z is a nonconstant holomorphic curve, then we may assume that $z(t) = p + t^m v + ...$, where v is a nonzero vector, $m \geq 1$, and the dots denote higher order terms. Choose a defining function r for M near p. If z lies in M, then $r(z(t)) = 0$. Hence (10) and (11) hold for all t:

$$0 = \left(\frac{\partial}{\partial t}\right)^m r(z(t)) \tag{10}$$

$$0 = \left(\frac{\partial}{\partial t}\right)^m \left(\frac{\partial}{\partial \overline{t}}\right)^m r(z(t)), \tag{11}$$

After routine computation with the chain rule and evaluation at $t = 0$, (10) tells us that $v \in T_p^{1,0}(M)$ and (11) tells us that the Levi form on v vanishes. By strong pseudoconvexity $v = 0$. We obtain a contradiction unless z is a constant map. □

We can strengthen Theorem 3.1 by making it quantitative. After we define *order of contact* we will discover (for $n \geq 2$) that M is strongly pseudoconvex at p if and only if the order of contact of every holomorphic curve with M at p is at most two.

Exercise 3.8. Put $n = 1$ and set $z = x + iy$. What is the differential operator $\frac{\partial}{\partial z}\frac{\partial}{\partial \overline{z}}$ in terms of x and y? Use it to prove that a holomorphic function (defined on a connected open set) whose absolute value is constant must itself be constant.

Exercise 3.9. Give an example of a real hypersurface $M \subset \mathbf{C}^3$ such that $0 \in M$, such that M has a polynomial defining equation whose Hessian has eigenvalues $(1, 1, -1)$ at every point, and such that there are complex lines through 0 and lying in M. Then give a second example where $0 \in M$, the Hessian has eigenvalues $(1, 1, -1)$ at every point, and M is strongly pseudoconvex at 0. Hence this M contains no positive dimensional complex varieties through 0.

Exercise 3.10. Give an example of a hypersurface M in $\mathbf{C}^n$ containing p for which there are no nonsingular holomorphic curves through p lying in M but for which there are such nonconstant singular holomorphic curves.

Exercise 3.11. Suppose that $r = \mathrm{Re}(z_n) + \sum_{j=1}^K |f_j(z)|^2$ for holomorphic functions f_j. Verify that the level sets of r are pseudoconvex. Find a formula for the determinant of the Levi form when the f_j are independent of z_n.

Exercise 3.12. Find a defining equation for a smooth pseudoconvex real hypersurface M that is strongly pseudoconvex at no points. Can you find such an M that contains no complex-analytic varieties?

Exercise 3.13. Find a defining equation for a smooth real hypersurface M that is strongly pseudoconvex at all points except one.

Exercise 3.14. Let Ω be a strongly pseudoconvex domain with $p \in b\Omega$. Show that there is a linear function L with $L(p) = 0$ but whose other nearby zeroes lie outside the closed domain. Thus $\frac{1}{L}$ is holomorphic at points of Ω near p but it blows up at the boundary point p. In fact there is a function f holomorphic on all of Ω such that f blows up at p, but this fact is harder to prove. See [FK], [H], [Kr], or [Range] for various approaches to this idea, which verifies that a strongly pseudoconvex domain is a *domain of holomorphy.*

3. Higher order commutators

In Lecture 2 we considered a smooth function of one variable, vanishing to finite order. We found an easy generalization of the second derivative test to determine the nature of the critical point. We needed only to differentiate enough times, keep track of the number of times we differentiated, and determine the sign of the first non-vanishing derivative. For us the analogous easy situation will be a point of finite type on a hypersurface in $\mathbf{C}^2$ with degenerate Levi form. All reasonable definitions of *point of finite type* correspond in this setting, and the type of a point can be calculated by computing iterated commutators of vector fields. We therefore begin with the notion of type for a single vector field. We include an interesting open problem in the exercises in this section.

Let M be a real hypersurface in $\mathbf{C}^n$, with $p \in M$. Let L denote a $(1,0)$ vector field tangent to M that does not vanish at p. Let T be a vector field, tangent to M and nonzero at p, which lies outside the span of the $(1,0)$ and $(0,1)$ vector fields. This vector field T is sometimes called the *missing direction* or the *bad direction*, for reasons relating to estimates for $\overline{\partial}$. We may assume that T is purely imaginary and that $\langle \eta, T \rangle = 1$. The Levi form captures the component of the commutator $[L, \overline{L}]$ in the T direction:

$$\lambda(L, \overline{L}) = \langle \eta, [L, \overline{L}] \rangle. \tag{12}$$

The situation when $\lambda(L, \overline{L})(p) = 0$ provides our analogue of a degenerate critical point.

How do we take higher order derivatives? One method is to take higher order commutators; one may regard the commutator (or Lie bracket) $[X, Y]$ of vector fields X and Y as the Lie derivative of Y with respect to X. Hence higher order brackets amount to higher order derivatives.

Fix a $(1,0)$ vector field L with $L(p) \neq 0$. We write $\mathcal{L}_m$ to denote any vector field defined by

$$\mathcal{L}_m = [...[[L_1, L_2], L_3], ..., L_m], \tag{13}$$

where each L_j is either L or $\overline{L}$.

Definition 3.2. The *type* at p of the $(1,0)$ vector field L is the smallest m for which there is an $\mathcal{L}_m$ with

$$\langle \eta, \mathcal{L}_m \rangle (p) \neq 0.$$

We write $\text{type}(L, p) = m$. When no such m exists we say that L has *infinite type* or that the type of L at p is infinite.

Note that the type of L at p equals 2 if and only if $\lambda(L, \overline{L})(p) \neq 0$. Higher commutators amount to taking higher derivatives. We can also take higher derivatives in another fashion, leading to a different concept.

Definition 3.3. Let M be a real hypersurface in $\mathbf{C}^n$ with $p \in M$. Let L be a $(1,0)$ vector field on M with $L(p) \neq 0$. Then
$c(L,p) = 2 + \min\{k : L^a \overline{L}^b \lambda(L, \overline{L})(p) \neq 0 \text{ for some } a, b \text{ with } a + b = k\}$.

Notice that $c(L,p) = 2$ if and only if $\text{type}(L,p) = 2$ if and only if $\lambda(L,\overline{L})(p) \neq 0$. In general these numbers are not equal, but the conjecture (proved in many cases) from Exercise 3.20 states that pseudoconvexity is equivalent to the equality of these numbers for all $(1,0)$ vector fields.

Exercise 3.15. Suppose $M = \{r = 0\}$, where $r_{z_n}(p) \neq 0$. For $1 \leq j \leq n-1$ define L_j by

$$L_j = \frac{\partial}{\partial z_j} - \frac{r_{z_j}}{r_{z_n}} \frac{\partial}{\partial z_n}.$$

Show that these vector fields span the local sections of $T^{1,0}(M)$ near p. Show also that they commute. Find the Levi matrix $\lambda(L_j, \overline{L}_k)$ with respect to this basis.

Exercise 3.16. Suppose that M is a pseudoconvex hypersurface in $\mathbf{C}^2$ and that L is a $(1,0)$ vector field of finite type at p. Show that $\text{type}(L,p)$ is an even integer.

Exercise 3.17. Let M be defined by $2\text{Re}(z_3) + |z_1|^4 + |z_2|^2$. Let $L = L_1 + \overline{z}_1 L_2$. Compute the expressions $\langle \eta, [[[L, \overline{L}], L], \overline{L}] \rangle$ and $L\overline{L}\lambda(L,\overline{L})$ and evaluate them at the origin.

Exercise 3.18. Assume that M is pseudoconvex, and that L and K are $(1,0)$ vector fields. Prove (easy!) that

$$|\lambda(L,\overline{K})|^2 \leq \lambda(L,\overline{L})\ \lambda(K,\overline{K}). \tag{14}$$

Then (harder) use (14) to show that, if M is pseudoconvex, then $c(L,p) = 4$ if and only if $\text{type}(L,p) = 4$.

Exercise 3.19. Let M be a hypersurface and assume there are $(1,0)$ vector fields L and K with $\lambda(L,\overline{L})(p) > 0$ and $\lambda(K,\overline{K})(p) < 0$. Show that there is a $(1,0)$ vector field A with $c(A,p) \neq \text{type}(A,p)$.

Exercise 3.20. Prove the following conjecture. (See [D1] and [D8] for more information). Show that a real hypersurface in $\mathbf{C}^n$ is pseudoconvex at p if and only if $c(L,p) = \text{type}(L,p)$ for all $(1,0)$ vector fields L with $L(p) \neq 0$.

Things simplify when $M \subset \mathbf{C}^2$. In particular this conjecture is easy; in higher dimensions different directions interact with each other. For example, when finding an expression such as (13), the $(1,0)$ part of the vector field $[L,\overline{L}]$ arises, and the Levi form on it matters as well.

Let us continue by assuming that $M \subset \mathbf{C}^2$. We leave most of the work as exercises. Let L denote a $(1,0)$ vector field that does not vanish at p. The complexified tangent space at p is then spanned by the three vectors $L(p)$, $\overline{L(p)}$, and $T(p)$. When M is contained in $\mathbf{C}^2$, it is possible to fully analyze the situation by taking higher order commutators.

Example 3.1. Let $M = \{z : r(z) = 0\}$, where $r(z) = 2\text{Re}(z_2) + |z_1|^{2m}$. The single vector field L, defined by

$$L = \frac{\partial}{\partial z_1} - \frac{\partial r}{\partial z_1}\frac{\partial}{\partial z_2},$$

spans the space of $(1, 0)$ vectors tangent to M. The type of L at 0 is $2m$. The situation is virtually the same if M is the zero set of r, where

$$r(z) = 2\mathrm{Re}(z_2) + p_{2m}(z_1, \overline{z}_1) + \text{higher order terms},$$

and p_{2m} is a homogeneous polynomial.

Exercise 3.21. Suppose that ψ is a smooth function vanishing at the origin and M is defined by the vanishing of the function r given by

$$r(z_1, z_2) = 2\mathrm{Re}(z_2) + \psi(z_1, \mathrm{Im}(z_2)).$$

Let L be a $(1, 0)$ vector field, nonvanishing at the origin, and tangent to M. Show that the type of L at 0 equals the order of vanishing of the function $\psi(z_1, 0)$ there.

Exercise 3.22. For $p \in M \subset \mathbf{C}^2$, show that $\Delta(M, p) = \mathrm{type}(L, p)$, where L is a $(1, 0)$ vector field not zero at p. The number $\Delta(M, p)$ is defined in the next section.

4. Points of finite type

The notion of *type* for a complex vector field is important in many places, such as sub-Riemannian geometry and harmonic analysis on nilpotent Lie groups. For our purposes, however, we will need more subtle ideas in order to relate the singularities arising from a degenerate Levi form with the theory of $\overline{\partial}$. In some sense the notion of type of a vector field exhibits the same drawbacks as did restricting to lines in the Peano example from Lecture 2. We therefore introduce the notion of *point of finite type*, and the ideas will be based on the interplay between the local rings of smooth and holomorphic functions.

Our definition of point of finite type on a real hypersurface arises from making quantitative the notion that a real hypersurface contains no complex-analytic varieties. Let (M, p) be the germ of a smooth real hypersurface at $p \in \mathbf{C}^n$; let r be a local defining function for M. We write $j_{k,p}r$ for the k-th order Taylor polynomial of r at p. Let z be (the germ of) a holomorphic curve with $z(0) = p$. We write $\nu(z)$ for the order of vanishing at 0 of the function $t \to z(t) - p$. We note some simple facts about orders of vanishing we will use.

Lemma 3.1. *If ζ_1 and ζ_2 are $\mathbf{C}$-valued formal or convergent power series in $t, \overline{t}$, then*

$$\nu(\zeta_1 + \zeta_2) \geq \min\left(\nu(\zeta_1), \nu(\zeta_2)\right)$$
$$\nu(\zeta_1 \zeta_2) = \nu(\zeta_1) + \nu(\zeta_2).$$

PROOF. The proof is left as an exercise to the reader. □

Recall that if the curve z is singular at p, then $\nu(z) > 1$. Earlier we saw the singular curve z defined by $z(t) = (t^2, t^3, 0)$. For this curve we have $\nu(z) = 2$. On the other hand the curve w defined by $w(t) = (t^2, t^2 + t^4, 0)$ also has order 2, but its image is smooth.

Exercise 3.23. Prove Lemma 3.1. Suppose more generally that ζ_1 and ζ_2 take values in an inner-product space. What can you say about $\nu(\langle \zeta_1, \zeta_2 \rangle)$?

We are now prepared to define *order of contact* and *point of finite type*.

Definition 3.4. Let M be a real hypersurface containing p and let z be a holomorphic curve with $z(0) = p$. We define the *order of contact* $\Delta(M, p, z)$ of z with M at p by $\Delta(M, p, z) = \frac{\nu(z^*r)}{\nu(z)}$. This number is independent of the choice of r. We define the 1-type or type, written $\Delta(M, p)$, of M at p by $\Delta(M, p) = \sup_z \Delta(M, p, z)$, where the supremum is taken over all nonconstant holomorphic curves.

Remark 3.2. Note that we divide by $\nu(z)$ in the definition of order of contact. Of course $\nu(z) = \nu(\zeta)$ if ζ is a nonsingular reparametrization of z. What happens under a singular reparametrization? If we replace t by $t^m u(t)$ for some u with $u(0) \neq 0$, then we multiply both the numerator and denominator of $\Delta(M, p, z)$ by m, and the ratio is unchanged.

Definition 3.5. Let M be a real hypersurface in $\mathbf{C}^n$. Assume $p \in M$; then p is a point of finite type if $\Delta(M, p)$ is finite.

Exercise 3.24. Assume that M is a smooth hypersurface in $\mathbf{C}^n$ for $n \geq 2$. Show that $\Delta(M, p) = 2$ if and only if M is strongly pseudoconvex at p.

Exercise 3.25. Give an example of a hypersurface M with nondegenerate Levi form at p for which $\Delta(M, p) = 4$.

The following example illustrates why iterated commutators are inadequate to understand finite type.

Example 3.2. Let M be the zero set of r, where

$$r(z, \overline{z}) = 2\mathrm{Re}(z_3) + |z_1^2 - z_2^3|^2.$$

Then the type of each nonzero $(1, 0)$ vector field at 0 is either 4 or 6, but $\Delta(M, 0) = \infty$. For most points $p \in M$ near 0 the type of each nonzero $(1, 0)$ vector field at p is either 2 or ∞. Catlin [C2] assigns to the origin in this example the multi-type $(1, 4, 6)$. At most nearby points the multi-type is $(1, 2, \infty)$.

Exercise 3.26. Verify that the type of L is either 4 or 6 in Example 3.2.

We next show that the function $p \to \Delta(M, z)$ is not semi-continuous from either side. We remark that the failure of lower semi-continuity is to be expected, but that the failure of upper semi-continuity is more surprising.

Example 3.3. Let M be the zero set of r, where

$$r(z, \overline{z}) = 2\mathrm{Re}(z_3) + |z_1^2 - z_2 z_3|^2 + |z_2|^4.$$

Then $\Delta(M, 0) = 4$. For a real but nonzero put $p = (0, 0, ia)$. Then $\Delta(M, p) = 8$; the curve z given by $z(t) = (t, \frac{t^2}{ia}, ia)$ satisfies $\Delta(M, p, z) = 8$ and this value is maximal. Most other points p near 0 are strongly pseudoconvex points, at which $\Delta(M, p) = 2$. Thus $p \to \Delta(M, z)$ is neither upper nor lower semi-continuous. The maximum order of contact at the origin is 4; there are nearby points where the maximum order of contact is 2 and others where it is 8.

Exercise 3.27. Verify the statements in Example 3.3.

In order to use the ideas from Lecture 2, we will approximate M by hypersurfaces M_k. Here M_k is the zero set of the k-th order Taylor polynomial $j_{k,p}r$ of r at p. Thus $p \in M_k$ for all k. Finite type means more than $\Delta(M_k, p)$ being finite for some k. Proposition 3.2 tells us however that we can determine whether p is finite type by studying $\Delta(M_k, p)$ for sufficiently large k.

Example 3.4. Put $r(z) = 2\mathrm{Re}(z_3) + |z_1^2 - z_2^3|^2 + |z_1^4|^2$. Then 0 is a point of finite type with $\Delta(M, 0) = 12$. If we added the term $-|z_2|^{12}$ to r, then the resulting function defines a hypersurface M' for which 0 is of infinite type, as the curve $(t^3, t^2, 0)$ lies in M'.

Exercise 3.28. Prove that the type at 0 in Example 3.4 is 12.

Exercise 3.29. Give an example of a hypersurface M with the following property. There is a point $p \in M$ such that $\Delta(M_k, p)$ is finite for an infinite sequence of integers k, but $\Delta(M, p)$ is infinite.

Proposition 3.2. *Let M be a real hypersurface with $p \in M$. Then p is a point of finite type if and only if there is an integer k for which $\Delta(M_k, p) \leq k$.*

PROOF. For each integer k we may write $r = j_{k,p}r + E_{k,p}r$, where $E_{k,p}$ is the remainder term from Taylor's formula. Since $E_{k,p}r$ vanishes to order $k+1$ at p, for each z we have $\nu(z^*(E_{k,p}r)) \geq \nu(z)(k+1)$.

Suppose first that $\Delta(M_k, p) \leq k$; then for each z

$$\nu(z^* j_{k,p}r) \leq k\nu(z) < \nu(z^*(E_{k,p}r)), \tag{15}$$

and hence $\nu(z^*r) = \nu(z^* j_{k,p}r) \leq k\nu(z)$ for all z. Thus $\Delta(M, p, z) = \Delta(M_k, p, z) \leq k$ for all z and hence $\Delta(M, p) = \Delta(M_k, p) \leq k$.

Conversely, if $\Delta(M, p)$ is finite, choose some integer k larger than $\Delta(M, p)$. As above it follows for each z that $\nu(z^* j_{k,p}r) = \nu(z^*r) \leq k\nu(z)$. Hence $\Delta(M_k, p) \leq k$ as well. □

Thus the study of finite type for smooth real hypersurfaces reduces to its study for algebraic real hypersurfaces, i. e., those defined by polynomial equations. We will therefore proceed by combining the methods of Lecture 2 with some quantitative measurements of the singularities defined by $\mathcal{M}$-primary ideals.

Let r be a smooth function with $r(p) = 0$ and $dr(p) \neq 0$. As above $j_{k,r}$ denotes the k-th order Taylor polynomial at p and let $M_k = \{j_{k,p}r = 0\}$. Following the ideas in Lecture 2 we write

$$j_{k,p}r(z, \overline{z}) = 2\mathrm{Re}(h_{k,p}(z)) + ||f_{k,p}(z)||^2 - ||g_{k,p}(z)||^2 \tag{16}$$

where

$$h_{k,p}(z) = \sum_{|\alpha| \leq k} \frac{(D^\alpha r)(p)}{\alpha!}(z-p)^\alpha, \tag{17}$$

and the components of $f_{k,p}$ and $g_{k,p}$ are given by

$$(f_{k,p})_\beta(z) = \frac{1}{2}\left(\sum_{|\alpha| \leq k} \frac{(D^\alpha r)(p)}{\alpha!}\frac{\overline{(D^\beta r)(p)}}{\beta!}(z-p)^\alpha + (z-p)^\beta\right) \tag{18.1}$$

$$(g_{k,p})_\beta(z) = \frac{1}{2}\left(\sum_{|\alpha| \leq k} \frac{(D^\alpha r)(p)}{\alpha!}\frac{\overline{(D^\beta r)(p)}}{\beta!}(z-p)^\alpha - (z-p)^\beta\right) \tag{18.2}$$

These formulas come from the proof of Lemma 2.3. We have

$$j_{k,p}r(z, \overline{z}) = 2\mathrm{Re}(h_{k,p})(z)) + ||f_{k,p}(z)||^2 - ||g_{k,p}(z)||^2. \tag{19}$$

In (19) the function $h_{k,p}$ and the mappings $f_{k,p}$ and $g_{k,p}$ vanish at p. Furthermore it is evident from the previous formulas that the coefficients of the Taylor series of these polynomials for a fixed k depend smoothly on the base point p. Based on our ideas from Lecture 2 we let $I(U,k,p)$ denote the ideals in $\mathcal{O}_p$ generated by $h_{k,p}$ and the components of $f_{k,p} - Ug_{k,p}$. For p be a point of finite type, we require each of these ideals $I(U,k,p)$ to be $\mathcal{M}$-primary, and we will also need quantitative information.

In Theorem 3.4 we will establish two-sided bounds on $\Delta(M_k,p)$ in terms of the orders of contact of ideals in $\mathcal{O}_p$. The order of contact of an ideal I is defined by

$$\mathbf{T}(I) = \sup_z \inf_{w \in I} \frac{\nu(z^* w)}{\nu(z)}.$$

This invariant $\mathbf{T}$ indicates how the commutative algebra and the analysis will interact, allowing us to glimpse the tip of an iceberg.

5. Commutative algebra

The ideas so far have revealed how to study the local geometry of a real algebraic hypersurface by studying ideals in the ring of germs of holomorphic functions. We need more precise quantitative information to extend these ideas to the smooth case. We therefore temporarily turn toward commutative algebra.

We will work with the ring $\mathcal{O}_p$ of germs of holomorphic functions at a point p in $\mathbf{C}^n$. We will not need to include the dimension n in the notation, although much of what we say is trivial when $n = 1$ because the ring is then a principal ideal domain. We will also drop p from the notation, thus denoting the ring by $\mathcal{O}$ when, as usual, p is the origin. In Section 3.6 however we will let p vary.

The ring of germs of holomorphic functions

The ring $\mathcal{O}$ is a Noetherian local ring with identity. The maximal ideal $\mathcal{M}$ consists of the non-units. An ideal I is called *primary* to $\mathcal{M}$ when there is an integer k such that $\mathcal{M}^k \subset I \subset \mathcal{M}$. By the Nullstellensatz I is primary to $\mathcal{M}$ if and only if the variety $V(I)$ consists of the origin alone. In this case the quotient vector space $\mathcal{O}/I$ is finite-dimensional over $\mathbf{C}$; its dimension $\mathbf{D}(I)$ is called the *codimension, length,* or *multiplicity* of I. We give a simple example. Suppose $a, b \geq 2$ and $I = (z_1^a, z_1 z_2, z_2^b)$. We find that $\mathbf{D}(I) = a + b - 1$, by listing a basis for the quotient algebra $\mathcal{O}/I$:

$$\mathcal{O}/I = \{1, z_1, ..., z_1^{a-1}, z_2, ..., z_2^{b-1}\}. \tag{20}$$

Exercise 3.30. Compute $\mathbf{T}(I)$ for the above ideal, and compute both $\mathbf{D}(I)$ and $\mathbf{T}(I)$ when $I = (z_1^a - z_2^b, z_1 z_2)$.

The number $\mathbf{D}(I)$ measures the singularity of I, in the following sense. Although I defines a single point, it defines the point more than once. The number $\mathbf{D}(I)$ tells us how many times. More generally in commutative algebra one uses lengths of modules (a generalization of dimension for vector spaces) to count the number of times a given collection of equations defines an object. These ideas are made precise in many algebraic geometry texts such as [Shaf]. The book [D1] discusses how and why this concept arises in considerations of finite type.

We make the following analogy. Strongly pseudoconvex points correspond to the maximal ideal $\mathcal{M}$, and pseudoconvex points of finite type correspond to ideals primary to $\mathcal{M}$. We can measure the singularity of an ideal in several ways; see

Theorem 3.3 and [D1] for some inequalities relating these invariants. Similarly we can measure finite type in several different ways. One of the key observations will be that while $\mathbf{D}(I)$ and $\mathbf{T}(I)$ are not equal, they provide equivalent information.

In Lectures 4 and 5 we will discuss the relationship between points of finite type and subelliptic estimates. Strongly pseudoconvex points correspond to a subelliptic estimate called the $\frac{1}{2}$-estimate. Points of finite type will correspond to a subelliptic estimate of order ϵ, for some ϵ with $0 < \epsilon \leq \frac{1}{2}$.

Remark 3.3. A useful generalization of these ideas arises when studying subellipticity on pseudoconvex domains for forms of type $(0, q)$. The analogue of strong pseudoconvexity is that the Levi form have at least $n - q$ positive eigenvalues; the analogue of point of finite type is that the order of contact at p of all q-dimensional varieties with the boundary be finite. See [K4], [C2], [C3]. The analogue of $\mathcal{M}$-primary for an ideal I is that the variety defined by I has dimension at most $q - 1$.

We briefly consider the situation when $I = (f)$ is generated by precisely n elements and $\mathbf{D}(I)$ is finite. In the language of commutative algebra, $f_1, ..., f_n$ form a *regular sequence* of length n in a ring of n dimensions. In this situation $\det(df)$ is not a member of I. Hence the ideal $I + \det(df)$ has smaller codimension than I does. On the other hand, for each $g \in \mathcal{M}$, we have $g \det(df) \in I$. One says that $\det(df)$ generates the *socle* of (f). From our perspective, $\det(df)$ is as close as possible to being in I without being in it. See [D1] or [EL] for more information. This simple idea is an algebraic analogue of Kohn's notion of subelliptic multipliers [K4]. The determinant $\det(df)$ is analogous to the determinant of the Levi form and therefore plays the role of the subelliptic multiplier that gets Kohn's process going. See Lecture 4.

The beautiful idea in the preceding paragraph requires additional considerations when I is not generated by the right number of functions. Suppose for example that $n = 2$ and $I = (z_1^2, z_1 z_2, z_2^2)$. For each of the three choices of pairs of generators of I we compute Jacobian determinants, and we get nothing new. In Kohn's process, however, one is allowed to take the radical of the ideal J generated by the determinants, and in this case the radical of J is $\mathcal{M}$. See Lecture 4.

Exercise 3.31. Put $f(z) = (z_1^2 - z_2^3, z_1 z_2)$. Find $\det(df)$ and prove that it is not in the ideal (f). Prove that $z_1 \det(df)$ and $z_2 \det(df)$ do lie in the ideal (f).

Exercise 3.32. (Easy) Suppose $n = 1$ and $f \in \mathcal{M}$. Prove that f' is not in (f), but that $zf' \in (f)$. Thus f' generates the socle of (f). (Harder) Prove the analogous result in higher dimensions.

The associativity formula

We will require one formula from commutative algebra that relates the invariants $\mathbf{D}(I)$ and $\mathbf{T}(I)$; hence we develop these ideas anew. We begin with the situation in dimension 2, beautifully discussed in [Ful]. See [Shaf] for an accessible treatment of the general case.

Let f and g be polynomials in two complex variables, and suppose each vanishes at the origin. We would like to associate with this intersection a number $\iota(f, g)$ measuring the multiplicity with which the variety $V(f)$ intersects $V(g)$ at the origin. The approach in [Ful] first considers properties describing how we want this number to depend on f and g, and then proves that there is a unique number $\iota(f, g)$ satisfying the list of desired properties. The fundamental results states that $\iota(f, g) = \mathbf{D}(I) = \dim_{\mathbb{C}} \mathcal{O}/I$, where I is the ideal in $\mathcal{O}$ generated by f and g.

Let us recall the ideas from [Ful] in an informal fashion. Let f and g be germs at the origin of holomorphic functions in two variables, and let (f, g) denote the ideal they generate in $\mathcal{O}$. We will assign to this pair a number (possibly infinity) $\iota(f, g)$ depending on only the ideal (f, g). In particular we have $\iota(f, g) = \iota(g, f)$. First we consider the three values $0, 1, \infty$. We want $\iota(f, g) = 0$ if and only if $(f, g) = \mathcal{O}$; this choice is natural because the varieties intersect zero times at the origin in this case. We want $\iota(f, g) = 1$ if and only if $(f, g) = \mathcal{M}$. This choice is natural because the intersection of the varieties is transverse in this case. By the holomorphic inverse function theorem, the intersection number is therefore 1 if and only if we can use f and g as local coordinates preserving the origin. We want $\iota(f, g) = \infty$ if and only $V(f, g)$ is of positive dimension; that is, if and only if the (germs of the) varieties $V(f)$ and $V(g)$ have a common component through 0.

What does it mean for $\iota(f, g)$ to be a positive integer greater than 1? We want $\iota(f, g)$ to measure the order of tangency of the possibly singular curves. It is natural to demand that $\iota(f, g^m) = m\iota(f, g)$, and more generally,

$$\iota(f, gh) = \iota(f, g) + \iota(f, h). \tag{21}$$

Formula (21) is crucial for the development; the proof that $\mathbf{D}$ works uses a strong induction argument. Imagine that we have computed the number $\iota(f, g)$ in all cases when it is less than some value. Given a pair (f, g) for which we have not yet found $\iota(f, g)$, we use the given properties to find F, G, H such that $\iota(f, g) = \iota(F, GH)$. Using (21) we reduce to two situations where the intersection number is smaller, thereby allowing an inductive step.

The number $\mathbf{D}(f, g) = \dim_{\mathbf{C}} \mathcal{O}/(f, g)$ has the requisite properties. The proof uses the following result, whose proof is left to the reader.

Proposition 3.3. *For f, g, h elements of $\mathcal{M}$, assume that $V(f, gh)$ consists of the origin alone. Then there is an exact sequence of vector spaces*

$$0 \to (h)/(gh) \to (\mathcal{O}/(f))/(gh) \to (\mathcal{O}/(f))/(h) \to 0. \tag{22}$$

When $0 \to U \to V \to W \to 0$ is an exact sequence of finite-dimensional vector spaces, elementary linear algebra gives $\dim(V) = \dim(U) + \dim(W)$. Thus (22) yields

$$\dim (\mathcal{O}/(f))/(gh) = \dim ((h)/(gh)) + \dim ((\mathcal{O}/(f))/(h)). \tag{23}$$

When h is not a divisor of zero in the ring $(\mathcal{O}/(f))/(g)$, multiplication by h determines an isomorphism of $(\mathcal{O}/(f))/(g)$ with $(h)/(gh)$. That h is not a divisor of zero follows from the Nullstellensatz. Therefore in (23) we may replace $\dim((h)/(gh))$ by $\dim((\mathcal{O}/(f))/(g))$. Hence (23) implies two crucial results:

$$\mathbf{D}(f, gh) = \mathbf{D}(f, g) + \mathbf{D}(f, h), \tag{24.1}$$

$$\mathbf{D}(f, g^m) = m\mathbf{D}(f, g). \tag{24.2}$$

Based on these facts we define $\iota(f, g)$ to be $\mathbf{D}(f, g)$. In Proposition 3.5 we generalize them to higher dimensions.

Exercise 3.33. Prove Proposition 3.3.

There is a remarkable formula for computing $\mathbf{D}(f, g)$ in two dimensions. Its generalization to higher dimensions will be called the *associativity formula*. Let us consider the simplest case first.

Proposition 3.4. *Associativity formula for* $n = 2$. *Let* f, g *be germs in* $\mathcal{M}$ *such that* $V(f,g)$ *consists of the origin alone. Suppose* $V(f)$ *is irreducible and that the curve* $t \to z(t)$ *of minimal order parametrizes* $V(f)$. *Then*

$$\mathbf{D}(f,g) = \nu(z^*g). \tag{25}$$

More generally, suppose that $f = \prod_{j=1}^{K} f_j^{m_j}$, *where the* f_j *are irreducible. Let* z_j *be a minimal parametrization of* $V(f_j)$. *Then*

$$\mathbf{D}(f,g) = \sum_{j=1}^{K} m_j \nu(z_j^* g). \tag{26}$$

PROOF. Note that (26) follows from (25) and repeated application of (24.1). To prove (25) requires some algebra that we omit. See [Ful] or [Shaf]. □

We now turn to the ring $\mathcal{O}$ in n-dimensions. We are interested in collections of germs at the origin which define the origin alone. By Exercise 2.17, when an ideal I is generated by fewer than n elements, its variety is positive dimensional. We therefore consider ideals generated by at least n functions; the simplest situation is when an ideal I is generated by precisely n elements $f_1, ..., f_n$. There are several reasons why this situation is much simpler. Some of these reasons are algebraic; for example, in this setting the dimension of the variety $V(f_1, ..., f_k)$ is $n - k$ for each k. Furthermore, for each k the germ f_{k+1} is not a zero-divisor in the ring $\mathcal{O}/(f_1, ..., f_k)$. The same holds for any permutation of the germs. One says that $f_1, ..., f_n$ form a *regular sequence* of length n in $\mathcal{O}$.

Proposition 3.5. *Let* $(f_1, ..., f_n)$ *be an* $\mathcal{M}$*-primary ideal in* $\mathcal{O}$ *in dimension* n. *Suppose that* $f_n = gh$. *Then*

$$\mathbf{D}(f_1, ..., f_{n-1}, gh) = \mathbf{D}(f_1, ..., f_{n-1}, g) + \mathbf{D}(f_1, ..., f_{n-1}, h).$$

PROOF. The proof is almost identical to the proof of (24.1). The sequence (22) remains exact if we replace the ideal (f) in (22) by $(f_1, ..., f_{n-1})$. Then (23) holds with the same notational change. By the above remarks $f_1, .., f_{n-1}, f_n$ form a regular sequence and hence so do $f_1, ..., f_{n-1}, h$. In particular h is not a zero divisor in the ring $\mathcal{O}/(f_1, ..., f_{n-1})$. Hence the argument that $(\mathcal{O}/(f))/(g)$ is isomorphic to $(h)/(gh)$ remains valid as well. □

The number $\mathbf{D}(I)$ has many geometric and analytic interpretations when I is generated by a regular sequence of length n. For example, $\mathbf{D}(I)$ equals the generic number of inverse images of a point close to the origin under the mapping f. Also $\mathbf{D}(I)$ can be expressed using integral formulas. Our intuition from the one variable discussion carries over precisely. Furthermore, $\mathbf{D}(I)$ equals the *Hilbert-Samuel multiplicity* in this situation. See [Mat].

The next method for computing $\mathbf{D}$ will be useful because it involves pulling back to holomorphic curves.

Theorem 3.2. *Associativity formula. Let* $f_1, ..., f_n$ *be elements of* $\mathcal{M}$ *in* n *dimensions. Suppose that* $V(f) = \{0\}$. *Let* $W = V(f_1, ..., f_{n-1})$ *be the* 1*-dimensional variety defined by the first* $n-1$ *functions. Let* $W = \sum_{j=1}^{K} n_j W_j$ *denote the decomposition of* W *into irreducible one-dimensional branches, where the positive integers*

$n_j = n_j(f_1, ..., f_{n-1})$ *denote the number the times (multiplicity) the equations define* W_j. *Let* z_j *be a curve of minimal order that parametrizes* W_j. *The following formula holds:*

$$\mathbf{D}(f) = \sum_{j=1}^{K} n_j(f_1, ..., f_{n-1})\nu(z_j^*(f_n)). \tag{27}$$

PROOF. See [Shaf], page 190. (The notation there differs considerably, but the result is the same.) □

Exercise 3.34. Compute $\mathbf{D}(I)$ using the associativity formula if the variables are (a, b, c) in $\mathbf{C}^3$ and

$$I = (a^3 - bc, b^3 - ac, c^3 - ab).$$

$$I = (a^3 - bc, (b^3 - ac)^2, c^3 - ab).$$

Theorem 3.3. *Let* I *be an* $\mathcal{M}$*-primary ideal in* $\mathcal{O}$ *in* n *dimensions. Assume that* I *contains* q *independent linear forms. Then*

$$\mathbf{T}(I) \leq \mathbf{D}(I) \leq (\mathbf{T}(I))^{n-q}. \tag{28}$$

PROOF. The first inequality follows easily because

$$\mathbf{T}(I) \leq \kappa(I) \leq \mathbf{D}(I), \tag{29}$$

where $\kappa(I)$ is the smallest k such that $\mathcal{M}^k \subset I$. See Exercise 2.32.

The second inequality is harder; it relies on the associativity formula. Given I, we first choose n elements $f_1, ..., f_n$ in I for which $\mathbf{D}(f_1, ..., f_n)$ is minimal. A generic choice of n elements will have this property, and furthermore precisely q of these functions will be linear. If there are any linear forms, then we list these first. Since $(f) \subset I$, we have

$$\mathbf{D}(I) \leq \mathbf{D}(f_1, ..., f_n). \tag{30}$$

To verify (28) we will show that $\mathbf{D}(f_1, ..., f_n) \leq \mathbf{T}(I)^{n-q}$. The functions $f_1, ..., f_n$ define a regular sequence; hence the variety $V = V(f_1, ..., f_{n-1})$ is one-dimensional. We decompose it into irreducible branches and we obtain $\mathbf{D}(f_1, ..., f_n)$ by the associativity formula:

$$\mathbf{D}(f_1, ..., f_n) = \sum n_j(f_1, ..., f_{n-1})\nu(z_j^* f_n) \tag{31}$$

Now $\nu(z_j^* f_n) \leq \mathbf{T}(I)v(z_j)$ by definition of the invariant $\mathbf{T}$. Therefore

$$\mathbf{D}(f_1, ..., f_n) \leq \mathbf{T}(I) \sum_j n_j(f_1, ..., f_{n-1})\nu(z_j). \tag{32}$$

For a generic linear function g_1, $\nu(z_j) = \nu(z_j^* g_1)$. Using (32) and the formula again,

$$\mathbf{D}(f_1, ..., f_n) \leq \mathbf{T}(I) \sum n_j(f_1, ..., f_{n-1})\nu(z_j^* g_1) = \mathbf{T}(I)\mathbf{D}(f_1, ..., f_{n-1}, g_1). \tag{33.1}$$

Next we apply the associativity formula to the ideal $(g_1, f_1, ..., f_{n-1})$. Repeating the above procedure, we replace f_{n-1} by a generic linear function g_2 and we have

$$\mathbf{D}(f_1, ..., f_n) \leq \mathbf{T}(I)\mathbf{D}(g_1, f_1, ..., f_{n-1}) \leq (\mathbf{T}(I))^2\mathbf{D}(g_1, g_2, f_1, ..., f_{n-2}). \qquad (33.2)$$

Continuing in this fashion we replace nonlinear elements of the list $f_1, ..., f_n$ by linear forms. Each such replacement creates a factor of $\mathbf{T}(I)$. Since $\mathbf{D}(g_1, ..., g_n) = 1$ if the g_j are independent linear forms. we obtain the desired result. □

Remark 3.4. At the PCMI lectures M. Mustata showed the author an alternative definition of $T(I)$ involving integral closures from which the harder inequality in (28) follows.

Exercise 3.35. Let $I = (z_1^3 - z_2^3, z_2^3 - z_3^3, z_1z_2z_3)$. Carry out the proof of Theorem 3.2 and prove from it that $\mathbf{D}(I) = (\mathbf{T}(I))^3 = 27$.

Exercise 3.36. Assume that $I = (f_1, ..., f_n)$, that $V(f) = \{0\}$, and that each f_j is homogeneous of degree d. Prove that $\mathbf{D}(I) = d^n$ and $\mathbf{T}(I) = d$.

Exercise 3.37. (Not easy) Assume that $f_1, ..., f_n \in \mathcal{O}$ and $f_j(0) = 0$. We can then find a matrix $A = a_{jk}$ of elements in $\mathcal{O}$ such that $f_j(z) = \sum a_{jk}(z)z_k$. Let J denote the Jacobian determinant of df. Prove that $J - \mathbf{D}(f)\det(A)$ lies in the ideal (f).

6. A return to finite type

We now unify the discussion about points of finite type on a smooth real hypersurface in $\mathbf{C}^n$. We repeat certain points in order to make this section self-contained.

Let r be a smooth defining function for a real hypersurface M in $\mathbf{C}^n$, and let $p \in M$. Let C_p^∞ denote the ring of germs of smooth real-valued functions at p. Since $dr(p) \neq 0$ we may consider the germ of r at p as the generator of the principal ideal in C_p^∞ of smooth functions vanishing on M near p. Although this particular ideal is principal, general ideals in the ring C_p^∞ are not finitely generated, as $C_p^\infty$is not Noetherian. At this stage we are therefore far from the nice situation that applies in $\mathcal{O}$.

In order to employ the algebraic ideas we have developed, we consider the k-th order Taylor polynomials of r at p. We study the approximating hypersurfaces M_k defined near p by these real-valued polynomials. By Proposition 3.2, finite type is characterized by the condition that there is an integer k for which $\Delta(M_k, p) \leq k$.

We recall and generalize the ideas from Section I.2. Let V be a complex-analytic subvariety of $\mathbf{C}^n$ containing 0. Let W be an irreducible one-dimensional branch of V through 0. We can parametrize W by a Puiseaux series; in more modern language, the normalization of W is a neighborhood D of the origin in $\mathbf{C}$ together with a surjective holomorphic mapping $z : D \to W$. We think of $t \to z(t)$ as a parametrized holomorphic curve in $\mathbf{C}^n$.

Let R be a Hermitian symmetric polynomial; in particular R might be $j_{k,p}r$ for some k. As usual we consider the holomorphic decomposition $R = 2\mathrm{Re}(h) + ||f||^2 - ||g||^2$. If the complex-analytic variety V were contained in $\{R = 0\}$, then there would be a holomorphic curve z for which the function z^*r vanishes identically. If $z^*R = 0$, then $z^*h = 0$ and $||z^*f||^2 = ||z^*g||^2$. By including components equal to zero if necessary we may assume that f and g have the same number of component functions, say N. It follows that there is an N-by-N unitary matrix U of constants such that

$$z^*h = 0 \qquad (34.1)$$

$$z^*(f - Ug) = 0. \tag{34.2}$$

This discussion leads to the collection of ideals I_U in $\mathcal{O}_p$ generated by h and the components of $f - Ug$, and therefore to the corresponding complex-analytic varieties $V(I_U)$. Each ideal is contained in $\mathcal{M}_p$ and $p \in V(I_U)$. Theorem 2.3 summarized the key idea; each $V(I_U)$ lies in the algebraic surface defined by $R = 0$, and any irreducible complex variety lying in this surface is a subvariety of $V(I_U)$ for some U. In particular this algebraic hypersurface is of finite type at p if and only if each of the varieties $V(I_U)$ is the single point p. By the Nullstellensatz, this property holds if and only if each ideal I_U is primary to the maximal ideal. Therefore an algebraic M is of finite type at p if and only if there is no germ of a complex-analytic variety passing through p and lying in M.

To understand the smooth case we will need to make this algebraic information quantitative. Recall the definition of *order of contact* for an ideal of holomorphic functions.

Definition 3.6. Let I be a proper ideal in $\mathcal{O}$, the ring of germs of holomorphic functions at 0 in $\mathbf{C}^n$. We define the *order of contact* $\mathbf{T}(I)$ by the formula

$$\mathbf{T}(I) = \sup_z \inf_{g \in I} \frac{\nu(z^*g)}{\nu(z)}.$$

By Theorem 3.3, $\mathbf{T}(I)$ is finite if and only if I is $\mathcal{M}$-primary and

$$\mathbf{T}(I) \leq \mathbf{D}(I) \leq \mathbf{T}(I)^n.$$

Hence the order of contact measures the singularity in a manner consistent with the multiplicity. These inequalities will be used to verify that the function $p \to \Delta(M, p)$ is locally bounded on M. Local boundedness then implies that the set of points of finite type is an open subset of M.

We now prove a decisive quantitative result in the algebraic case.

Theorem 3.4. *Let M be an algebraic real hypersurface containing p. Let I_U denote the ideals in $\mathcal{O}_p$ defined above. The following double inequality holds for order of contact:*

$$\sup_U \mathbf{T}(I_U) + 1 \leq \Delta(M, p) \leq 2\sup_U \mathbf{T}(I_U). \tag{35}$$

PROOF. We may assume that p is the origin and that $r = 2\mathrm{Re}(h) + ||f||^2 - ||g||^2$. By definition I_U is generated by h and the components of $f - Ug$. Let z be a holomorphic curve. The supremum over z in the definitions of both $\mathbf{T}(I_U)$ and $\Delta(M, 0)$ occurs for some z with $z^*h = 0$. We may therefore assume that $z^*h = 0$ and work with $f - Ug$. If there is a U for which $z^*f = Uz^*g$, then all expressions in (35) are infinite, and the double inequality holds. Otherwise choose some unitary U for which the supremum is attained. For a given unitary U we write $A = z^*f$ and $B = z^*Ug$. The series A and B are vector-valued. We must then verify that

$$\nu(A - B) + 1 \leq \nu(||A||^2 - ||B||^2) \leq 2\nu(A - B). \tag{36}$$

Assume that l is the smallest integer for which the coefficients of t^l in A and B differ. Put $A = \phi + A_1$ and $B = \phi + B_1$ where $\nu(A_1 - B_1) = l$. Since A and B vanish at 0, we have $1 \leq \nu(\phi) < l$. For vector-valued functions ζ_1 and ζ_2, it follows by Exercise 3.23 after Lemma 3.1 that $\nu(\langle \zeta_1, \zeta_2 \rangle) \geq \nu(\zeta_1) + \nu(\zeta_2)$.

Therefore $\nu(\langle\phi, A_1 - B_1\rangle) \geq \nu(\phi) + l$. Also $\nu(||A_1||^2 - ||B_1||^2 \geq 2l$, and hence $\nu(\phi) + l < \nu(||A_1||^2 - ||B_1||^2)$. Putting these facts together yields

$$\nu(||A||^2 - ||B||^2) = \nu(||\phi + A_1||^2 - ||\phi + B_1||^2) =$$

$$\nu\left(2\mathrm{Re}(\langle\phi, A_1 - B_1\rangle) + ||A_1||^2 - ||B_1||^2\right) \geq \nu(\phi) + l \geq 1 + \nu(A - B). \tag{37}$$

Since $\mathbf{T}$ and Δ are computed by taking the supremum over curves z, we obtain the left-hand inequality in (35).

The second inequality is a bit subtle. If $A(t) = A_a t^a + ...$ and $B(t) = B_b t^b + ...$ and $a \neq b$, then we have $\nu(||A||^2 - ||B||^2) = 2\min(a, b)$ and $\nu(A - B) = \min(a - b)$. In this case the desired result holds. If $a = b$ however, then the inequality $\nu(||A||^2 - ||B||^2) \leq 2\nu(A - B)$ need not hold. See Exercise 3.38 below. In this potential problem case we have cancellation of the coefficient of $|t|^{2a}$ in $||A(t)||^2 - ||B(t)||^2$ if also $||A_a||^2 = ||B_a||^2$. This equality of norms guarantees however that there is a unitary V such that $VB_a = A_a$. Put $C = VB = z^*VUg$; then the product VU has the property that $\nu(z^*(f - VUg)) > a = \nu(z^*(f - Ug))$. Hence the original assumption that U was chosen for which the supremum is attained must be violated. We conclude that the other inequality $\Delta(M, p) \leq \sup_U \mathbf{T}(I_U)$ also holds. □

Neither inequality in (35) is sharp in general.

Exercise 3.38. (Easy) Give examples of $\mathbf{C}^n$-valued power series $A(t)$ and $B(t)$ such that

$$\nu(||A||^2 - ||B||^2) > 2\nu(A - B).$$

Exercise 3.39. For $m \geq 2$, put $r(z, \overline{z}) = 2\mathrm{Re}(z_2) + |z_1 + z_1^m|^2 - |z_1 - z_1^m|^2$ and let $M = \{r = 0\}$. What is $\Delta(M, 0)$? What is $2\sup_U \mathbf{T}(I_U)$?

Exercise 3.40. Put $r = 2\mathrm{Re}(h) + ||f||^2$. Show that $\Delta(M, 0) = 2\sup_U \mathbf{T}(I_U))$. (Harder) Show that the same equality holds for pseudoconvex algebraic hypersurfaces.

Let now M be a smooth real hypersurface with defining equation r. For each positive integer k and each point p in M, we considered the algebraic hypersurface M_k defined by $j_{k,p}r$. The holomorphic decomposition of a polynomial changes from point to point. We recall formula (16) for $j_{k,p}r$:

$$j_{k,p}r(z, \overline{z}) = 2\mathrm{Re}(h_{k,p}(z)) + ||f_{k,p}(z)||^2 - ||g_{k,p}(z)||^2, \tag{16}$$

where the notation emphasizes the dependence on the truncation integer k and on the base point p. We saw that the coefficients of (the holomorphic mappings) $h_{k,p}$, $f_{k,p}$, and $g_{k,p}$ depend smoothly on p. We then follow the above procedure and define ideals $I(U, k, p)$ and varieties $V(U, k, p)$.

Proposition 3.2 characterizes finite type by seeking a positive integer k with a certain stability property. First M_k must contain no positive-dimensional complex-analytic variety passing through p, but this property alone is insufficient, as Example 3.5 reveals. We also require the inequality $\Delta(M_k, p) \leq k$. It follows, for any integer l with $l \geq k$, that $\Delta(M_l, p) = \Delta(M_k, p) < \infty$. We obtain a finite determination result; not only does M_k contain no complex-analytic varieties through p, but the same is true no matter how we perturb the higher order Taylor coefficients.

The truncation process requires care. Our next example and the corresponding discussion reveal why we need the inequality $\Delta(M_k, p) \leq k$ rather than simply the finiteness of $\Delta(M_k, p)$ for some k.

Example 3.5. Let M be given by the defining equation

$$r(z) = 2\mathrm{Re}(z_3) + |z_1^2 - z_2^3|^2 + |z_1^4|^2. \tag{38}$$

We consider the situation at the origin. For each unitary U, the ideal I_U is given by

$$I_U = (z_3, z_1^2 - z_2^3, z_1^4),$$

and this ideal is evidently primary to $\mathcal{M}_p$. Thus M is finite type at 0. Here $\mathbf{T}(I_U) = 6$ for all U, and $\Delta(M, 0) = 12$.

Observe that M is finite type at 0 and is defined by a polynomial of degree eight. There are hypersurfaces tangent to M to order *eleven* that are not finite type at 0, such as the surface defined by

$$s(z) = 2\mathrm{Re}(z_3) + |z_1^2 - z_2^3|^2 + |z_1^4|^2 - |z_2^6|^2. \tag{39}$$

In this case the ideals V_U depend on the two-by-two unitary matrix U:

$$I_U = (z_3, z_1^2 - z_2^3 - U_{12}z_2^6, z_1^4 - U_{22}z_2^6). \tag{40}$$

Choosing $U_{12} = 0$ and $U_{22} = 1$, we find that there is a particular unitary U for which

$$I_U = (z_3, z_1^2 - z_2^3, z_1^4 - z_2^6).$$

This ideal evidently defines the one-dimensional variety parametrized by the holomorphic curve $t \to (t^3, t^2, 0)$. On the other hand, a smooth hypersurface M' that is tangent to M to order at least twelve contains no nonconstant holomorphic curves through the origin.

Thus some Taylor polynomials of defining functions are better than others. The strongly pseudoconvex case is deceptively simple; the second-order Taylor polynomial gives us the information we need. In the general finite type case we get the information we need from an appropriate Taylor polynomial, but its degree must be sufficiently (and perhaps surprisingly) high. This phenomenon also arises in the classification of degenerate critical points of smooth functions of several real variables.

We recall the definition of point of finite type on a smooth hypersurface M.

Definition 3.7. Let M be a smooth real hypersurface in $\mathbf{C}^n$ containing p with defining function r. Then M is called *of finite type* at p if and only if there is a constant C such that, for all nonconstant holomorphic curves z with $z(0) = p$, we have

$$\nu(z^*r) \leq C\nu(z). \tag{41}$$

The type of M at p, written $\Delta(M, p)$, is the supremum of $\Delta(M, p, z) = \frac{\nu(z^*r)}{\nu(z)}$ over all nonconstant holomorphic curves z with $z(0) = p$. Equivalently, the type is the infimum of the set of C for which (41) holds.

The definition does not depend upon the choice of defining equation for M near p. The ideal in C_p^∞ of germs of smooth functions vanishing on M near p is principal, and the germ of r may be used as its generator. It follows that the existence of the constant C in (41) is independent of the choice of defining function. We mention that the type need not be an integer.

Example 3.6. Put $r(z) = 2\text{Re}(z_3) + |z_1^a - z_2^b|^2 + |z_1^c|^2$, where $a < b \leq c$ and a, b are relatively prime. The type at 0 is then $\frac{2bc}{a}$.

Exercise 3.41. Verify Example 3.6.

Exercise 3.42. (A Bezout type theorem) (not easy) Let M be a real algebraic hypersurface in $\mathbf{C}^n$ defined by a polynomial equation of degree d. There is a number $m = m(n, d)$ with the following property: If there is a complex-analytic curve V with order of contact at least m at some point in M, then V must lie in M. See [D4] for a sharp formula for $m(n, d)$.

The original definition from [K3] of point of finite type on a hypersurface M in two dimensions involved commutators. We saw in Lecture 3 that finite type in higher dimensions is not equivalent to every $(1, 0)$ vector field being of finite type. Nonetheless there is an approach to finite type via iterated commutators.

Suppose that V is a complex-analytic curve contained in M and passing through p. When V is the image of the holomorphic mapping $t \to z(t)$, we naturally consider the holomorphic tangent vector $z'(t)$. We can find a vector field L on M such that $L = z'(t)$ along V, and all iterated commutators of L and its conjugate $\overline{L}$ must vanish at p. If L itself does not vanish at p, then L is a nonzero vector field of *infinite type* at p. Thus the assumption that all vector fields be of finite type at p precludes the existence of a *nonsingular* holomorphic curve through p that lies in M. If, however, V is singular at p, then L vanishes at p, and as Example 3.2 reveals, there need not be a nonzero vector field of infinite type there. The definition of point of finite type takes this possibility into account, by considering singular curves. We illustrate in an example how to do the same by allowing vector fields with nonsmooth coefficients.

Example 3.7. Put $r(z) = 2\text{Re}(z_3) + |z_1^2 - z_2^3|^2$, and let p be the origin. Here M is of infinite type at p, yet every $(1, 0)$ vector field is of finite type there. Let L_1 and L_2 be the usual local basis of tangential vector fields of type $(1, 0)$; thus $L_j = \frac{\partial}{\partial z_j} - r_{z_j}\frac{\partial}{\partial z_3}$. Finally let L be the vector field

$$L = 3\frac{z_1}{z_2}L_1 + 2L_2.$$

Although L has does not have smooth coefficients, the restriction of L to the holomorphic curve $(t^3, t^2, 0)$ is simply $3tL_1 + 2L_2$, and t is well behaved along the curve. We may thus take iterated commutators of L and $\overline{L}$ along V, and we discover that L has infinite type. This phenomenon occurs because V is not a *normal* variety.

It is possible in general to introduce a class of vector fields with nonsmooth coefficients, and to prove that p is a point of finite type if and only if all such vector fields have finite type at p. Furthermore one can express the type of a point in terms of the types of these vector fields. Such an approach closely models the situation in two dimensions, but the author feels that pulling back to parametrized curves expresses the ideas in a more appealing fashion.

7. The set of finite type points is open

We show in this section that *finite type* is an open condition. The number $\Delta(M,p)$ measures the maximum order of contact of holomorphic curves with M at p. Example 3.3 shows that $\Delta(M,p)$ does not vary semicontinuously (it is neither upper nor lower semicontinuous) as we change the base point p. The failure of lower semicontinuity is to be expected; the failure of upper semicontinuity is more interesting. The viability of the concept of finite type depends on the information that the function $p \to \Delta(M,p)$ is nonetheless locally bounded; see Theorem 3.5. See [D1] and [D2] for proofs of sharp local bounds and many examples.

Remark 3.5. The author believes that another number, written $B(M,p)$ in [D1], might be a better numerical measurement of the singularity than is $\Delta(M,p)$. The two numbers are equivalent in the sense that they are related by inequalities going both ways, but the number $B(M,p)$ is more closely related to the codimension of an ideal and it depends upper semicontinuously on p. This number has also been used in work on boundary orbit accumulation sets. See pages 19-22 of [IK].

We make the simple observation that the set S of strongly pseudoconvex points on a real hypersurface M is an open subset of M. The reason is that S is the complement of the zero set of the determinant of the Levi form; this function $\det(\lambda)$ is smooth, and hence its zero set is closed in M; therefore S is open. We note as an aside that the same argument would also apply if M were merely of class C^2. For smooth M in general, there is no obvious candidate for a continuous function whose zero set defines the complement of the points of finite type. Instead we will compute the type and prove that it is a locally bounded function of the base point.

The model for strongly pseudoconvex domains is the unit ball or its unbounded version, the Siegel half-space. In Exercise 3.17 you were asked to show that they are biholomorphically equivalent. The defining equation for the Siegel half-space in $\mathbf{C}^{n+1}$ is given by

$$r(z) = 2\mathrm{Re}(z_{n+1}) + \sum_{j=1}^{n} |z_j|^2. \tag{42}$$

A simple model for pseudoconvex domains of finite type, appropriate in some contexts, is given by those domains Ω defined by

$$r(z) = 2\mathrm{Re}(z_{n+1}) + \sum_{j=1}^{K} |f_j(z)|^2. \tag{43}$$

We write M for $b\Omega = \{r = 0\}$. The origin is a point of finite type if and only the variety $V(f)$ defined by the f_j consists of the origin alone. By an earlier exercise,

$$\det(\lambda) = \sum_I |J(f_{I_1}, ... f_{I_n})|^2, \tag{44}$$

where J denotes the Jacobian determinant, and the sum is taken over all choices of n of the component functions. Therefore the zero set of $\det(\lambda)$ is the intersection with M of the complex-analytic variety defined by all these Jacobians. In Lecture 4 we will see how to use these Jacobians to decide whether $V(f)$ is a single point.

Exercise 3.43. Assume $M = \{r = 0\}$, where r has the form (43). In each case give explicit f such that the following hold:

1) For $0 \leq q \leq n-1$, there is a q-dimensional complex variety through the origin lying in M but not such a $q+1$ dimensional complex variety.

2) There is a positive dimensional variety lying in M which is singular at 0 but no smooth variety lying in M that is smooth at 0.

3) There is a positive dimensional variety lying in M containing 0 but for which $\det(\lambda)$ is not identically zero near the origin.

Exercise 3.44. Let M be a real hypersurface in $\mathbf{C}^n$, and suppose M contains a complex variety V of dimension $n-1$. Show that V is smooth.

When M is defined by an equation of the form (43) we see that the origin is of finite type if and only if $V(f)$ is the origin alone. In Lecture 2 we gave a method for deciding whether a hypersurface defined by a polynomial equation contained complex-analytic varieties. We use that method to establish openness for smooth domains in general.

Let r be smooth near the origin. For each positive integer k and each point p with $r(p) = 0$, we consider the Taylor polynomial $j_{k,p}$ of order k at p. Let M_k be the hypersurface defined by $j_{k,p}r$. We obtain the ideals

$$I(U,k,p) = (h_{k,p}, f_{k,p} - Ug_{k,p}). \tag{45}$$

Each $I(U,k,p)$ lies in $\mathcal{M}_p$. We showed that M is of finite type at p if and only if there is an integer k such that

$$\Delta(M_k, p) \leq k.$$

We will show that the set of points of finite type is an open subset of M by showing that $p \to \Delta(M,p)$ is a locally bounded function. In fact, if p is near p_0 then

$$\Delta(M,p) \leq 2(\Delta(M,p_0) - 1)^{n-1}. \tag{46}$$

The bound (46) has the following useful consequences:

Theorem 3.5. *Let M be a smooth real hypersurface in $\mathbf{C}^n$. The set of points of finite type is open in M.*

PROOF. The result follows immediately from the bound (46). We prove (46) after stating a Corollary and making an important remark. □

Corollary 3.1. *Let M be a smooth compact real hypersurface, and assume each point $p \in M$ is a point of finite type. Then there is a constant C such that $\Delta(M,p) \leq C$ for all $p \in M$.*

PROOF. A locally bounded function on a compact space is globally bounded. □

Exercise 3.45. Prove using the open cover definition of compactness that a locally bounded real-valued function on a compact topological space is globally bounded.

Let M be a smooth hypersurface and denote by W the set of points p for which there is a complex variety of positive dimension through p and lying in M. In general there are three possibilities; $p \in W$, or p is not in W but $\Delta(M,p) = \infty$, or $\Delta(M,p) < \infty$ and p is a point of finite type.

Remark 3.6. Let M be real-analytic near p. One can show by our methods that only two of these possibilities can occur: either $p \in W$ or p is a point of finite type. See [Lem3] or [D1]. It follows from Theorem 3.5 that when M is real-analytic, W is a closed subset of M.

PROOF OF OPENNESS. To complete the proof of openness we must establish the local bound (46) for order of contact. Suppose that $\Delta(M, p_0)$ is finite. By Proposition 3.2 there is an integer k such that $\Delta(M_k, p_0) \leq k$. We have

$$\Delta(M,p) \leq 2\text{sup}_U \mathbf{T}(I(U,k,p)) \leq 2\text{sup}_U \mathbf{D}(I(U,k,p)) \leq$$
$$2\text{sup}_U \mathbf{D}(I(U,k,p_0)) \leq 2\text{sup}_U \mathbf{T}(I(U,k,p_0))^{n-1}) \leq 2(\Delta(M,p_0)-1)^{n-1}. \quad (47)$$

In passing from p to p_0 in the middle we used the upper semi-continuity of $\mathbf{D}$ with respect to parameters. We have also used the inequalities

$$\text{sup}_U \mathbf{T}(I(U,k,p)) + 1 \leq \Delta(M,p) \leq 2\text{sup}_U \mathbf{T}(I(U,k,p)) \quad (48)$$

$$\mathbf{T}(I) \leq \mathbf{D}(I) \leq \mathbf{T}(I)^{n-1}. \quad (49)$$

Inequality (49) follows from Theorem 3.3. We can use the exponent $n-1$ instead of n because the ideal $I(U)$ always contains at least one linear form, as $h_{k,p} \in I(U,k,p)$ and $dh_{k,p}(p) \neq 0$.

Consider the inequalities in (48). The parameters k and p are now fixed, so we may drop them from the notation, writing h, f, g. Thus we may assume that we have a polynomial defining function of the usual form

$$2\text{Re}(h) + ||f||^2 - ||g||^2$$

and p is the origin. Put $I_U = (h, f - Ug)$ as usual; (48) is precisely what we proved in Theorem 3.4. Therefore both (47) and (48) hold and hence (46) holds. We have proved the Theorem. □

Exercise 3.46. What does (46) tell us when $\Delta(M, p_0) = 2$?

Exercise 3.47. Suppose we happen to know that $\Delta(M,p) = 2\text{sup}_U \mathbf{T}(I_U)$ at each of the various steps in (47). What estimate replaces (46)? (Comment: This improved version holds in the pseudoconvex case. See [D1])

LECTURE 4

Kohn's algorithm for subelliptic multipliers

1. Introduction

In conjunction with proving subelliptic estimates for the Cauchy-Riemann equations, Kohn [K4] invented an algorithm for finding subelliptic multipliers. This work helped lead to the development of multiplier ideal sheaves, which have had considerable influence in algebraic geometry. See the notes to the other lectures at this PCMI meeting and [Siu1] for these connections and many references; we mention here that [ELSV] and [DFEM] include algebraic work related to some of the ideas in this lecture.

This lecture has two purposes. After precisely defining Kohn's algorithm, we will discuss its *effectiveness*. This topic is of some recent interest. See [N] and [Siu2]. We describe Kohn's algorithm from [K4] in terms of a sequence of pairs $(\mathbf{M}_k, I_k)$, where $\mathbf{M}_k$ is a module of $(1,0)$ forms and I_k is an ideal in the ring of germs $\mathcal{E}$ of smooth functions. The algorithm stabilizes if some I_k is the full ring $\mathcal{E}$, in which case a subelliptic estimate holds. Effectiveness means that there is a positive lower bound on the ϵ obtained in (4), depending on only the dimension and the type of the point. Proposition 4.3 gives an example where effectiveness fails.

We introduce a simplified situation, called a *triangular system*, where we can modify the algorithm to make it even better than effective. The original algorithm is not effective even for general triangular systems. The modified version in Theorem 4.1 provides a finite list of subelliptic multipliers, with the length of the list equal to the multiplicity of the original ideal, all roots are *under control*, and such that the last item is a unit in the ring. In dimension n we take no root of order more than n. Recently Catlin and Cho [CC] have determined sharp subelliptic estimates for some pseudoconvex domains whose defining equations come from triangular systems.

Several examples here will include computations of what is known in algebraic geometry as the log canonical threshold [ELSV]. It is interesting to relate this concept, which plays a key role in the kind of multiplier ideals used in commutative algebra, with the original ideas concerning subelliptic estimates. The full impact of the ideas behind these estimates has not yet been applied in the theory of multiplier ideals, but it suggests a direction for future research.

See [K1], [K2], [K3], [K4] for Kohn's work on the Cauchy-Riemann equations; [K4] introduced the method of subelliptic multipliers. Crucial early results on subelliptic estimates and their consequences such as local regularity in the $\overline{\partial}$-Neumann problem appear in [KN]. See [K5] and [DK] for surveys. Catlin's papers [C1], [C2], and [C3] give necessary and sufficient conditions for subelliptic estimates in the smooth case. Nicoara's preprint [N] connects Catlin's work with Kohn's work. The posted paper [Siu2] gives a modified version of Kohn's algorithm which effectively terminates.

2. Subelliptic estimates

Let Ω be a bounded pseudoconvex domain in $\mathbf{C}^n$ with smooth boundary, and assume that $p \in b\Omega$. We suppose that $b\Omega$ is given by the vanishing of a smooth function r with $dr(p) \neq 0$. We may further assume without loss of generality that ∂r has Euclidean length one on $b\Omega$.

We work in $L^2(\Omega)$. We will require L^2 and Sobolev norms. See [Ta] or [SR] for material on Sobolev spaces. We will write either $\langle f, g\rangle$ or $\langle f, g\rangle_{L^2}$ (we use this second notation to avoid confusion with the inner product on $\mathbf{C}^n$) to denote the L^2 inner product of functions or forms f and g. Also $||u||^2$ denotes the squared L^2-norm of u, whether u is a function or a differential $(0, q)$ form. Thus for example if u is a function and $\phi = \sum \phi_j d\overline{z}^j$ is a $(0,1)$-form, we have

$$||u||^2 = \int_\Omega |u|^2 dV$$

$$||\phi||^2 = \sum_{j=1}^n \int_\Omega |\phi_j|^2 dV.$$

We will need to understand the L^2-adjoint of $\overline{\partial}$. Since differential operators are only densely defined on L^2, one has to be careful in defining the domains of an operator and its adjoint. A function or form f lies in the domain of $\overline{\partial}$ if $\overline{\partial} f$, defined in the sense of distributions, lies in L^2. We state the condition for a smooth $(0,1)$ form being in the domain of the adjoint. A smooth differential $(0,1)$ form $\sum_{j=1}^n \phi_j d\overline{z}^j$, defined near p, lies in the domain of $\overline{\partial}^*$ if the vector field $\sum_{j=1}^n \phi_j \frac{\partial}{\partial z_j}$ lies in $T_z^{1,0} b\Omega$ for z near p. Thus $\sum \phi_j r_{z_j} = 0$ on $b\Omega$. See [FK] for a detailed explanation and for the definition in general. To motivate this definition of domain for the adjoint, let f be a function and let g be a $(0,1)$ form. Assume that on $b\Omega$

$$\sum_{j=1}^n g_j r_{z_j} = 0. \tag{1}$$

Using the divergence theorem together with our assumption on ∂r we obtain

$$\langle \overline{\partial} f, g\rangle = \sum_j \int_\Omega f_{\overline{z}_j} \overline{g}_j dV = -\sum_j \int_\Omega f \overline{(g_j)_{z_j}} dV + \int_{b\Omega} f \overline{\sum_j g_j r_{z_j}} dS. \tag{2}$$

The boundary integral vanishes when (1) holds on $b\Omega$, and we obtain

$$\langle \overline{\partial} f, g\rangle = \langle f, \overline{\partial}^* g\rangle,$$

where $\overline{\partial}^* g = -\sum (g_j)_{z_j}$.

Before defining subelliptic estimates, we pause to mention one of their principal consequences [KN]. Consider the inhomogeneous Cauchy-Riemann equation $\overline{\partial} u = \alpha$ for an unknown function u on a smoothly bounded pseudoconvex domain Ω. Here α is a $(0,1)$ form with L^2-coefficients and of course $\overline{\partial}\alpha = 0$. The $\overline{\partial}$-Neumann solution to this equation is the unique solution u that is orthogonal to the holomorphic functions; by standard Hilbert space reasoning, this solution must be in the range of the adjoint $\overline{\partial}^*$. We write $u = \overline{\partial}^* N\alpha$. The question of *local regularity* asks whether this particular u must be smooth on any open subset Ω_0 of the closure of Ω whenever α is smooth on Ω_0. Not all solutions u can satisfy this property; to note this fact, imagine that u and α are smooth. If f is holomorphic on Ω but blows up at a boundary point p (such a function exists for any boundary point of any pseudoconvex domain), then

$$\overline{\partial}(u+f) = \overline{\partial}u = \alpha,$$

but $u+\alpha$ is not smooth at p even though α is. To repeat, local regularity considers whether the unique solution u in the range of the adjoint must be smooth wherever α is smooth. Local regularity holds whenever there is a subelliptic estimate at each boundary point of Ω. More generally, if there is a subelliptic estimate in a neighborhood of a boundary point, then local regularity holds for forms supported in that neighborhood.

We write $||u||_\epsilon$ for the Sobolev ϵ norm of a function or form u. For $0 < \epsilon < 1$ this norm measures fractional derivatives of u. Subelliptic estimates imply local Sobolev estimates for the solution u discussed above. For each s, if α has s derivatives in L^2 (on some open set), and a subelliptic estimate holds, then u has $s+\epsilon$ derivatives in L^2. There is a constant C (depending on s but not on α) such that

$$||u||_{s+\epsilon} \leq C||\alpha||_s. \tag{3.1}$$

Estimates such as (3.1) generalize *elliptic estimates* such as

$$||u||_{s+m} \leq C||Lu||_s, \tag{3.2}$$

where L is an elliptic linear PDE of order m. In the subelliptic case we gain derivatives, but the gain is less than the order of the operator. See [K7] for a stunning result where operators that *lose* derivatives are nonetheless hypoelliptic.

Definition 4.1. A subelliptic estimate holds on $(0,1)$ forms at p if there is a neighborhood U of p and positive constants C and ϵ such that (4) holds for all smooth $(0,1)$ forms ϕ supported in U and in the domain of $\overline{\partial}^*$:

$$||\phi||_\epsilon^2 \leq C\left(||\overline{\partial}\phi||^2 + ||\overline{\partial}^*\phi||^2 + ||\phi||^2\right). \tag{4}$$

Remark 4.1. The term $||\phi||^2$ can be estimated in terms of $||\overline{\partial}\phi||^2 + ||\overline{\partial}^*\phi||^2$ and hence it is not necessary to include it in (4).

Exercise 4.1. Put $\phi = \sum_{j=1}^n \phi_j d\overline{z}^j$. Show that $\overline{\partial}^*\phi = -\sum(\phi_j)_{z_j}$. Then compute explicitly the right-hand side of (4) if $\phi = \sum_{j=1}^n \phi_j d\overline{z}^j$. The answer will involve sums of integrals. Integrate by parts to obtain a boundary integral involving the Levi form.

Exercise 4.2. Let Ω be a bounded strongly pseudoconvex domain. Assume that ϕ is a $(0,1)$ form with $\overline{\partial}\phi = 0$ and $\overline{\partial}^*\phi = 0$. Prove that $\phi = 0$. Suggestion: Use Exercise 4.1 to show that the components of ϕ are holomorphic and also vanish on $b\Omega$.

Exercise 4.3. Suppose that ϕ is supported in Ω. Show that (4) holds with $\epsilon = 1$. Note that all you need to do is to estimate the L^2 norms of each $(\phi_j)_{z_k}$ and $(\phi_j)_{\overline{z_k}}$.

We next describe Kohn's approach [K4] to proving subelliptic estimates. Instead of estimating $||\phi||_\epsilon^2$ directly, we consider functions f for which we can estimate $||f\phi||_\epsilon^2$ for some ϵ. The collection of (germs of) such functions, called *subelliptic multipliers*, has remarkable structure.

Definition 4.2. Let f be the germ of a smooth function at p. We say that f is a *subelliptic multiplier* at p if there are positive constants C and ϵ and a neighborhood U such that

$$||f\phi||_{\epsilon}^{2} \leq C\left(||\overline{\partial}\phi||^{2} + ||\overline{\partial}^{*}\phi||^{2} + ||\phi||^{2}\right) \tag{5}$$

for all smooth forms ϕ supported in U and in the domain of $\overline{\partial}^{*}$.

We write $Q(\phi,\phi)$ for $||\overline{\partial}\phi||^{2} + ||\overline{\partial}^{*}\phi||^{2} + ||\phi||^{2}$. A subelliptic estimate holds at p if and only if 1 is a subelliptic multiplier at p. The constant C is not important. The constant ϵ is important; a fundamental question not completely understood is the precise relationship between the largest possible value of this parameter and the geometry of $b\Omega$. If $b\Omega$ is strongly pseudoconvex at p, then one can take $\epsilon = \frac{1}{2}$ in (4). A stronger statement appears in (7) from Proposition 4.1 below.

Exercise 4.4. Let Λ^{ϵ} denote the usual pseudodifferential operator defining the ϵ norm: $||f||_{\epsilon}^{2} = ||\Lambda^{\epsilon}f||^{2}$. Assume the standard result that, for any operator T of order zero, the commutator $[T, \Lambda^{2\epsilon}]$ is a pseudodifferential operator of order $2\epsilon - 1$. Prove the following fact. If f is a subelliptic multiplier, and $|g| \leq f$, then g is a subelliptic multiplier with the same ϵ. You may assume that $\epsilon \leq \frac{1}{2}$.

The collection of subelliptic multipliers is a non-trivial ideal in $\mathcal{E}$ closed under taking radicals. Furthermore, the defining function r and the determinant of the Levi form $\det(\lambda)$ are subelliptic multipliers. We state these results of Kohn:

Proposition 4.1. *The collection I of subelliptic multipliers is a radical ideal in $\mathcal{E}$; in particular, if $f^{N} \in I$ for some N, then $f \in I$. Also, r and $\det(\lambda)$ are in I.*
If $||f^{N}\phi||_{\epsilon}^{2} \leq CQ(\phi,\phi)$, then $||f\phi||_{\frac{\epsilon}{N}}^{2} \leq C'Q(\phi,\phi)$. Also

$$||r\phi||_{1}^{2} \leq CQ(\phi,\phi) \tag{6}$$

$$||\det(\lambda)\phi||_{\frac{1}{2}}^{2} \leq CQ(\phi,\phi). \tag{7}$$

Kohn's algorithm starts with these two subelliptic multipliers and constructs additional ones. In order to do so, we need also to discuss modules of $(1,0)$ forms. Consider a $(1,0)$ form $\sum f_{j}dz_{j}$. We call this $(1,0)$ form *allowable* if there are positive constants C and ϵ such that, for all ϕ as in the definition of subelliptic estimate,

$$||\sum_{j} f_{j}\phi_{j}||_{\epsilon}^{2} \leq CQ(\phi,\phi). \tag{8}$$

We think of the components of f as a row in a matrix, and we say that an n-tuple $(f_{1}, ..., f_{n})$ of germs of functions is an *allowable row* (or *vector multiplier*) if (8) holds.

The sum of allowable rows is an allowable row, and for any germ g and allowable row f, gf is an allowable row. Thus the collection of allowable rows is a module. By combining (6) and Proposition 4.2, the components of ∂r are an allowable row with $\epsilon = \frac{1}{2}$. Furthermore (See [K4] or [D1]), for each j, the j-th row of the Levi form is an allowable row:

$$||\sum r_{z_{i}\overline{z}_{j}}\phi_{i}||_{\frac{1}{2}}^{2} \leq CQ(\phi,\phi). \tag{9}$$

The next result relates allowable rows and subelliptic multipliers:

Proposition 4.2. *Let f be a subelliptic multiplier such that*

$$||f\phi||_{2\epsilon}^2 \leq Q(\phi, \phi). \tag{10}$$

Then the n-tuple of functions $(\frac{\partial f}{\partial z^1}, ..., \frac{\partial f}{\partial z^n})$ is an allowable row, and we have:

$$||\sum_j \frac{\partial f}{\partial z^j}\phi_j||_\epsilon^2 \leq CQ(\phi, \phi). \tag{11}$$

Conversely, consider any $n \times n$ matrix (f_{ij}) whose rows are allowable for some ϵ. Then $\det(f_{ij})$ is a subelliptic multiplier for the same value of ϵ.

PROOF. See [K4] or [D1]. □

By combining these results, Kohn obtains a method for proving (4). We start with ∂r and the rows of the Levi form as the initial collection of allowable rows. From Proposition 4.2 we obtain (7). By Proposition 4.1 all elements of the real radical of the ideal in $\mathcal{E}$ generated by r and $\det(\lambda)$ satisfy (5) for some ϵ. Proposition 4.2 yields new allowable rows. Form all possible determinants of n allowable rows, consider the ideal they generate, and then take its real radical. Its elements are multipliers and their derivatives define allowable rows. Keep on going. A subelliptic estimate holds for some ϵ if we obtain the function 1 as a subelliptic multiplier in finitely many steps. Speaking loosely we call this process *Kohn's algorithm.* We define the algorithm precisely below.

When $b\Omega$ is real-analytic and pseudoconvex near p, Kohn proved using geometric results from [DF1] that the process stabilizes in finitely many steps and reaches 1 if and only if there is no germ of a complex-analytic variety of positive dimension containing p and lying in $b\Omega$. Recall that the type $\Delta(b\Omega, p)$ of a point measures the maximum order of contact of complex-analytic varieties with $b\Omega$ at p. Catlin proved ([C1],[C2],[C3]), when $b\Omega$ is smooth and pseudoconvex, that there is a subelliptic estimate at p on $(0, 1)$ forms if and only if $\Delta(b\Omega, p)$ is finite. The proof of necessity shows that the largest value of ϵ in a subelliptic estimate at p is at most the reciprocal of $\Delta(b\Omega, p)$. Examples from [C1] and our Example 3.3 show that this value is not always possible. The precise connection between the type and the parameter ϵ remains an open problem. We remark however that the definition of subelliptic estimate forces the supremum of the set of ϵ that work at p to be lower-semi continuous as a function of p; hence its reciprocal must be upper semi-continuous.

3. Kohn's algorithm

We next describe Kohn's algorithm precisely. Let $\mathcal{E}$ denote the ring of germs of smooth functions at a point p in $\mathbf{C}^n$. Let Ω be a bounded pseudoconvex domain in $\mathbf{C}^n$ and assume that $p \in b\Omega$. Let r be a local defining function; thus r is a generator of the principal ideal in $\mathcal{E}$ of functions vanishing on $b\Omega$ near p.

Kohn's algorithm consists of a sequence of pairs $(\mathbf{M}_k, I_k)$, where each $\mathbf{M}_k$ is a module of $(1, 0)$ forms and each I_k is an ideal in $\mathcal{E}$ closed under taking real radicals. The starting pair $(\mathbf{M}_0, I_0)$ is known; it is built from r and its derivatives up to order two. Namely $\mathbf{M}_0$ is generated by ∂r and the rows of the Levi form, and I_0 is the radical of the ideal generated by r and $\det(\lambda)$.

Given $(\mathbf{M}_k, I_k)$, the algorithm defines the next pair as follows: $\mathbf{M}_{k+1}$ is the module generated by $\mathbf{M}_k$ and the forms ∂h for $h \in I_k$, and I_{k+1} is the real radical of the ideal generated by I_k and determinants of n by n matrices whose rows are the components of elements of $\mathbf{M}_k$. Thus the sequence of ideals is increasing. We say that Kohn's algorithm stabilizes if there is k such that $I_j = I_k$ for $j \geq k$.

In the real-analytic case the sequence of ideals I_j stabilizes; in fact $I_k = I_n$ for $k \geq n$. Furthermore I_n is the full ring (and thus 1 is a subelliptic multiplier) if and only if there is no germ of a complex-analytic variety of positive dimension in $b\Omega$ and containing p. Thus under this assumption there is a subelliptic estimate of order ϵ on $(0,1)$ forms for some positive ϵ. We prove in Proposition 4.3 that Kohn's process is not effective; it is not possible to bound ϵ away from 0 in terms of the dimension and the type of the point using the algorithm.

4. Kohn's algorithm for holomorphic and formal germs

Consider a pseudoconvex domain Ω in $\mathbf{C}^{n+1}$ with smooth boundary $b\Omega$. Suppose $0 \in b\Omega$ and that there is a local defining function of the form

$$r(z) = 2\mathrm{Re}(z_{n+1}) + \sum_{j=1}^{K} |h_j(z)|^2, \tag{12}$$

where the function h_j are holomorphic near 0 and vanish there. For simplicity we also assume that they are independent of z_{n+1}. Kohn's algorithm then simplifies to an interesting procedure that stays within the category of germs of holomorphic functions. A similar version applies even to formal power series. [Cho].

Consider the domain defined by (12). It is pseudoconvex near 0, and it is of finite type there if and only if the variety $V(h)$ in $\mathbf{C}^n$ consists of 0 alone. We follow Kohn's algorithm for finding subelliptic multipliers; the first multiplier is the determinant of the Levi form, which in this case is a sum of squared absolute values of various Jacobian determinants of the h_j (by formula (31) from Lecture 2). Because the collection of subelliptic multipliers is closed under taking real radicals, we conclude that each of these Jacobians is itself a subelliptic multiplier. In a similar fashion each dh_j as an allowable row. We stay in the category of germs of holomorphic functions and carry out steps analogous to those in the general case. The procedure yields 1 as a subelliptic multiplier if and only if 0 is a point of finite type, which is in turn equivalent to $V(h)$ consisting of 0 alone. In particular we obtain an interesting procedure for deciding whether or not $V(h)$ is trivial.

Let $R = R_n$ be either the convergent or formal power series ring in n variables over the complex numbers $\mathbf{C}$. In either case R is a commutative local Noetherian ring with maximal ideal $\mathcal{M}$. Let $h = (h_1, ..., h_N)$ denote a finite number of elements of $\mathcal{M}$; in the convergent case we can think of h as the germ of a mapping $h : (\mathbf{C}^n, 0) \to (\mathbf{C}^N, 0)$ and we are interested in the variety $V(h)$ it defines. In the convergent case, by the Rückert Nullstellensatz, $h^{-1}(0) = \{0\}$ if and only if the ideal (h) is $\mathcal{M}$-primary; that is, the radical of (h) is $\mathcal{M}$. Even in the formal power series setting, sometimes we think of $h^{-1}(0)$ as a formal variety and we say $V(h) = \{0\}$ if and only (h) is $\mathcal{M}$-primary. There are many well-known methods for determining whether an ideal (h) in R is $\mathcal{M}$-primary; several were discussed in Lecture 3. Kohn's algorithm for finding subelliptic multipliers provides an unusual and intriguing method for deciding this matter.

The starting point is the collection of differential 1-forms $\{dh_j\}$; these play the role of *allowable rows* discussed above. Let $\mathbf{M}_0$ denote the module generated by the $\{dh_j\}$; we can think of each 1-form in $\mathbf{M}_1$ as a row vector. Let I_0 denote the radical of the ideal generated by all determinants of n by n matrices formed from these rows. Let $\mathbf{M}_1$ denote the module generated by the union of $\mathbf{M}_0$ and $\{dg : g \in I_0\}$. Let I_1 denote the radical of the ideal generated by determinants of all n by n matrices formed from the rows of $\mathbf{M}_1$. Proceeding inductively, we obtain a sequence of pairs $(\mathbf{M}_k, I_k)$, where $\mathbf{M}_{k+1}$ is the module of $(1,0)$ forms generated by $\mathbf{M}_k$ and $\{dg : g \in I_k\}$, and I_{k+1} is the radical of the ideal generated by I_k and all determinants of n by n matrices formed from rows of $\mathbf{M}_k$. By definition

$$I_1 \subset I_2 \subset \ldots$$

Since the ring R is Noetherian, this sequence of ideals stabilizes.

When R_n is the convergent power series ring, Kohn [K4] proved that $I_k = I_{n+1}$ whenever $k \geq n+1$, and that I_{n+1} is the full ring R_n if and only if the initial ideal (h) is $\mathcal{M}$-primary. Cho [Cho] generalized these two results to other rings, including the formal power series ring in n variables over $\mathbf{C}$. Cho's proof is algebraic whereas Kohn's proof uses information at nearby points (coherence). The increasing sequence of ideals above always stabilizes; when (h) is not $\mathcal{M}$-primary the stabilized ideal cannot be the full ring. We show this simple fact in Proposition 4.4. Thus the stabilized ideal is the full ring if and only if (h) is $\mathcal{M}$-primary.

Remark 4.2. There is an analogous process, involving q by q minor determinants, for determining whether the (formal) variety defined by (h) has dimension $n-q$. We do not consider this generalization here.

Before proceeding we note several subtleties. First of all, Kohn's algorithm depends on the choice of generators for the ideal (h); whether or not the stabilized ideal equals the full ring is independent of the choice of generators, but issues such as the number of roots taken in the steps does depend on the generators. Here is a second subtlety. At no stage of the algorithm do we take the radical of the initial ideal (h). We begin instead with the differentials dh_j; the first ideal of which we take the radical is the ideal generated by the Jacobian determinants of each choice of n elements of $\mathbf{M}_0$. Simple examples such as Example 4.1 show that we must take radicals (or do something similar) for the process to continue.

We clarify this last point. Suppose (h) is $\mathcal{M}$-primary. If we were allowed to take the radical of (h) at the outset, then we would obtain $\mathcal{M}$ in the first step. The next step would give the full ring. Kohn's algorithm differs. In the language of subelliptic multipliers and allowable rows, the starting information consists of the allowable rows dh_j rather than the h_j themselves. If we knew that the h_j were subelliptic multipliers, then we would be done in one step, because the radical of (h) is $\mathcal{M}$.

In case an $\mathcal{M}$-primary ideal (h) is generated by n functions $h_1, ..., h_n$ in R_n, the determinant $J = \det(\frac{\partial h_j}{\partial z_k})$ is not in the ideal (h), although it does generate the socle of (h). We stated these facts in Lecture 3.5; see also [D1], [EL], [GH]. In this case the 1-form dJ therefore provides new information. On the other hand, given K elements $h_1, ..., h_K$ where $K > n$ the corresponding statement fails. Although the Jacobian determinant of some n of these K elements is not zero in the ring R_n (See [BER] Theorem 5.1.37), it is possible (Example 4.1) that all such determinants are in (h). In this case the process would halt unless we are allowed to take roots of some kind.

Example 4.1. Put $n=2$, and let h denote the three functions (z^2, zw, w^2). Then the ideal J_0 of Jacobians is generated by the three functions $(2z^2, 2w^2, 4zw)$. If we tried to use J_0 instead of its radical I_0, then the algorithm would get stuck. We elaborate; the functions z^2, zw, w^2 are not known to be subelliptic multipliers at the start. Since they arise as Jacobians, their differentials are allowable. Taking determinants again then shows that they are multipliers, and hence we are allowed to take the radical. This strange phenomenon illustrates one of the subtleties in Kohn's algorithm.

Let I be an $\mathcal{M}$-primary ideal. There are several different numerical ways to measure the singularity; see Lecture 3 and [D1] for considerable discussion and inequalities relating them. We recall several possibilities; the order of contact, the index of primality, and the multiplicity or length. All of these measurements make sense in both the convergent and the formal power series setting:

$$\mathbf{T}(I) = \sup_z \inf_{g \in I} \frac{\nu(z^* g)}{\nu(z)} \tag{13.1}$$

$$\kappa(I) = \min\{k : \mathcal{M}^k \subset I\} \tag{13.2}$$

$$\mathbf{D}(I) = \dim_{\mathbf{C}} R_n/(I). \tag{13.3}$$

It is easy to check (see Exercise 2.32) that

$$\mathbf{T}(I) \leq \kappa(I) \leq \mathbf{D}(I) \leq \binom{n-1+\kappa(I)}{\kappa(I)-1}. \tag{14}$$

In Theorem 3.3 we showed that

$$\mathbf{D}(I) \leq (\mathbf{T}(I))^n. \tag{15}$$

By (14) and (15) these measurements, though unequal for $n \geq 2$, provide equivalent measurements of the singularity in all dimensions. See [C1] for the role order of contact plays in subelliptic estimates, and see [D3] for the consequences of its failure of upper semi-continuity.

We naturally ask how these measurements relate to the algorithm. The surprising answer is that we have no control on the parameter ϵ in terms of these measurements. See Proposition 4.3.

4.1. Examples of Kohn's algorithm and log canonical thresholds

We give in this section three illustrative examples of Kohn's algorithm. For two of them we also compute the log canonical threshold $\mathcal{L}$ from [ELSV] for purposes of comparison. We recall the definition of $\mathcal{L}$. Let $(f_1, ..., f_k)$ generate an ideal in $\mathcal{M} \subset \mathcal{O}$. Given a single function $g \in \mathcal{M}$ we let $\mathcal{L}(g)$ denote the supremum of the set of c for which the integral

$$\int \frac{dV(z)}{|g(z)|^{2c}} \tag{16}$$

converges in small deleted neighborhood of the origin. For the ideal $f = (f_1, ..., f_k)$ we define $\mathcal{L}(f)$ by putting

$$\mathcal{L}(f) = \mathcal{L}(\sum \lambda_i f_i) \tag{17}$$

where the sum $\sum \lambda_i f_i$ is a generic linear combination of the generators.

Example 4.2 (Kohn Algorithm). Suppose $h_j(z) = z_j^{m_j}$; assume each $m_j \geq 2$. Here $\mathbf{D}(h) = \prod m_j$. One can easily check that

$$I_0 = \left(\prod_{j=1}^{n} z_j\right) \tag{18}$$

and I_1 is generated by all products of $n-1$ of the coordinate functions. Furthermore, if $1 \leq k \leq n$, then I_{k-1} is generated by all products of $n-k+1$ coordinate functions. In particular I_{n-1} is the maximal ideal, and $I_n = R = \mathcal{O}$. We can also keep track of how many roots are extracted when taking radicals. For convenience we may assume that $m_1 \geq m_2 ... \geq m_n \geq 2$. The first time we take a root of order $m_1 - 1$. In each of the next $n-1$ steps we take a root of order m_1. The total number of roots required is $nm_1 - 1$. Note that we also took derivatives; in the theory of subelliptic estimates we need to account for these as well.

Example 4.3. (Log canonical threshold). With h as in Example 4.2, $\mathcal{L}(h) = \sum_{j=1}^{n} \frac{1}{m_j}$.

PROOF. It suffices to determine the supremum of the set of c for which

$$\int \frac{dV(z)}{|\sum z_j^{m_j}|^{2c}} \tag{19}$$

is finite. Here the integral is taken over a small deleted neighborhood of the origin. We do not effect convergence if we make the finite-to-one change of variables

$$(z_1^{m_1}, ..., z_n^{m_n}) = (w_1, ..., w_n) \tag{20}$$

and hence

$$(z_1, ..., z_n) = (w_1^{\frac{1}{m_1}}, ..., w_n^{\frac{1}{m_n}}). \tag{21}$$

We have

$$dz_1 \wedge ... \wedge dz_n = c \prod_j w_j^{\frac{1}{m_j}-1} dw_1 \wedge ... \wedge dw_n. \tag{22}$$

Hence the integral in (19) converges if and only if (with a different deleted neighborhood of the origin) the integral in (23) converges:

$$\int \frac{\prod_j |w_j|^{2(\frac{1}{m_j}-1)}}{|\sum w_j|^{2c}} dV(w) < \infty. \tag{23}$$

Now change variables again (blow up) by putting $z_1 = u_1$ and $z_j = u_1 u_j$ for $j \geq 2$. In terms of the u_j variables we get an integrand of the form

$$|u_1|^{2n-2+\sum_{j=1}^{n} 2(\frac{1}{m_j}-1)-2c} \prod_{j=2}^{n} |u_j|^{2(\frac{1}{m_j}-1)} h(u), \tag{24}$$

where h does not vanish at 0. After introducing polar coordinates in the first variable, the condition for convergence becomes

$$2n - 2 + \sum_{j=1}^{n} 2(\frac{1}{m_j} - 1) - 2c + 1 > -1, \tag{25}$$

from which we obtain $c < \sum \frac{1}{m_j}$. $\square$

The processes in Examples 4.2 and 4.3 both take all the m_j into account. For completeness we recall well-known information about subelliptic estimates for this example. Define a pseudoconvex domain Ω by the equation

$$r(z) = \mathrm{Re}(z_{n+1}) + \sum |z_j^{m_j}|^2 < 0; \tag{26}$$

the origin lies in $b\Omega$. The n-tuple of these integers (ignoring a factor of 2 and setting $m_{n+1} = 1$) gives Catlin's multi-type (see [C2],[C3] and Lecture 5). We may assume that $m_1 \geq m_2 ... \geq m_n$. For each q with $1 \leq q \leq n$ there is a subelliptic estimate on $(0, q)$ forms; the value of epsilon in (4) can be taken as large as the reciprocal of $2m_{n-q+1}$. The value of epsilon on $(0, n)$ forms is the reciprocal of twice the minimum of the numbers; the value of epsilon in on $(0, 1)$ forms is the reciprocal of twice the maximum of the numbers. The maximum order of contact T_q of q-dimensional complex varieties with $b\Omega$ at 0 is the number $2m_{n-q+1}$. The multiplicity of the point ([D1]) is twice the product of the m_j. The log canonical threshold is the sum of the reciprocals of the m_j. The point of these comments is that there are many ways to measure the singularity of the Levi form at the origin. Different measurements matter in different situations.

Example 4.4 (Kohn algorithm)**.** We investigate in detail the specific ideal $I = (z^a - w^b, zw)$ where $1 \leq a \leq b$. First we find $\mathbf{T}(I)$ and $\mathbf{D}(I)$. Note that the candidates for curves with maximal contact are given by $(z(t), w(t)) = (0, t)$ and by $(z(t), w(t)) = (t^b, t^a)$. Hence $\mathbf{T}(I) = \max(b, \frac{a+b}{a})$. Unless $a = 1$ the maximum is b. We have $\mathbf{D}(I) = a + b$, by using either (24.1) or (25) from Lecture 3. The allowable rows in $\mathbf{M}_0$ are generated by

$$\begin{pmatrix} az^{a-1} & -bw^{b-1} \\ w & z \end{pmatrix} \tag{27.1}$$

The Jacobian J is $az^a + bw^b$. Adding dJ to the list of allowable rows shows that $\mathbf{M}_1$ is generated by

$$\begin{pmatrix} az^{a-1} & -bw^{b-1} \\ w & z \\ a^2 z^{a-1} & b^2 w^{b-1} \end{pmatrix}. \tag{27.2}$$

Applying row operations and dividing by non-zero constants shows that the rows $(z^{a-1} \ \ 0)$ and $(0 \ \ w^{b-1})$ are in $\mathbf{M}_1$. Taking determinants again yields the subelliptic multipliers z^a and w^b. Taking radicals shows that z and w are subelliptic multipliers. Therefore the rows $(1 \ 0)$ and $(0 \ 1)$ are in M_3, and hence their determinant, namely 1, is a subelliptic multiplier. The only root we took was of order b, and hence is controlled by $\mathbf{D}(I)$.

Example 4.5. Let $n = 2$, and write the variables as (z, w). Let $h(z, w) = (z^M, w^N + wz^k)$, where $M \geq 2$, $N \geq 3$, and $k > M$. Here, for all k, $\mathbf{D}(h) = MN$. We will show in Proposition 4.3 that the number of roots required in the process depends on k, and hence that Kohn's algorithm is not effective.

Example 4.6. Let h be as in Example 4.5. We compute the log canonical threshold. (This number depends only on the ideal and not on the choice of generators; we do the calculation without assuming this fact.) We must take a generic linear combination of the two generators; there is no loss in generality in assuming both constants equal 1. We therefore need to determine the supremum of those c for which the integral

$$\int \frac{dV}{|z^M + w^N + wz^k|^{2c}}, \tag{28}$$

converges in some neighborhood of the origin. Note that $k > m$, and hence that

$$z^M + wz^k = z^M(1 + wz^{k-M}) = z^M u(z, w),$$

where u is a unit. We make a local biholomorphic change of variables $\zeta_1 = zu(z, w)^{\frac{1}{m}}$ and $\zeta_2 = w$. The Jacobian does not vanish near the origin. We therefore need to decide when

$$\int \frac{dV(\zeta)}{|\zeta_1^M + \zeta_2^N|^{2c}} \tag{29}$$

converges. By Example 4.3, the condition for convergence is $c < \frac{1}{M} + \frac{1}{N}$. In particular the log canonical threshold depends on M and N but it is independent of k. On the other hand the Kohn algorithm depends on N, M, and k.

5. Failure of effectiveness for Kohn's algorithm

In this section R_n again denotes either the convergent or formal power series ring. Let $h_1, ..., h_K$ be a finite number of nonzero elements in $\mathcal{M} \subset R_n$. We call h the *initial set*. Let $\mathbf{M}_0$ denote the module generated by the $\{dh_j\}$; we think of each 1-form in $\mathbf{M}_0$ as a row vector. Let J_0 denote the collection of determinants of n by n matrices formed from these rows. (Using the Hodge * operator, we can also express J_0 as the collection of elements $*(dg_1 \wedge ... \wedge dg_n)$ for $dg_j \in \mathbf{M}_0$.) Let I_0 be the radical of the ideal generated by J_0. For $k \geq 0$, let $\mathbf{M}_{k+1}$ denote the module generated by the union of $\mathbf{M}_k$ and $\{dg : g \in I_k\}$. Let J_{k+1} denote the collection of determinants of n by n matrices formed from the rows of $\mathbf{M}_{k+1}$; again $J_{k+1} = \{*(dg_1 \wedge ... \wedge dg_n)\}$ such that $dg_j \in \mathbf{M}_{k+1}$. Let I_{k+1} denote the radical of the ideal generated by the elements of J_{k+1}. Each I_k is a radical ideal and

$$I_1 \subset I_2 \subset ... \subset I_k \subset ...$$

Since the ring R_n is Noetherian, this increasing sequence of ideals stabilizes. Kohn [K4] (in the convergent case) and Cho [Cho] (in the formal case) have proved that $I_j = I_n$ holds in R_n for $j \geq n$

In dimension at least 2, Kohn's algorithm depends on the initial set h, not simply on the ideal h defines. The next result shows that one cannot take radicals in a controlled fashion unless one revises the algorithm.

Proposition 4.3. *Kohn's algorithm is not effective.*

PROOF. We consider the triangular system defined by $z^m, w^n + wz^k$ in two variables. Let (h) be the ideal they generate. The exponents are positive integers; we assume that $k > m \geq 2$ and that $n \geq 3$. By any of several methods from Lecture 3, or by Lemma 4.1, $\mathbf{D}(h) = mn$. We denote the second function by g and in this proof we use subscripts to denote its partial derivatives.

The starting module $\mathbf{M}_0$ of *allowable rows* is generated by (30):

$$\begin{pmatrix} z^{m-1} & 0 \\ g_z & g_w \end{pmatrix} \tag{30}$$

Therefore the starting ideal is given by (31):

$$I_0 = \mathrm{rad}(z^{m-1}g_w) = (zg_w). \tag{31}$$

Therefore $\mathbf{M}_1$ is generated by $\mathbf{M}_0$ and $d(zg_w) = (zg_{wz} + g_w)dz + zg_{ww}dw$. Using the row notation as before we see that $\mathbf{M}_1$ is generated by (32):

$$\begin{pmatrix} z^{m-1} & 0 \\ g_z & g_w \\ zg_{wz} + g_w & zg_{ww} \end{pmatrix}. \tag{32}$$

The ideal generated by zg_w and the two new determinants is

$$J_1 = (zg_w, z^m g_{ww}, zg_z g_{ww} - zg_w g_{zw} - g_w^2). \tag{33}$$

It is easy to see that

$$I_1 = \mathrm{rad}(J_1) = \mathcal{M}. \tag{34}$$

We claim that we cannot bound the root taken in the definition of radical, in passing from J_1 to I_1, in terms of m and n. To do so we show that z^{k-1} is not an element of J_1. In other words, $\kappa(J_1) \geq k$, and hence the number of roots taken must be at least k. Since k can be independently of m and n and also arbitrarily large, there is no bound on the number of roots taken in terms of the dimension 2 and the intersection number $\mathbf{D}(h) = mn$.

It remains to prove the claim. If $z^{k-1} \in J_1$, then we could write

$$z^{k-1} = a(z,w)zg_w + b(z,w)z^m g_{ww} + c(z,w)(zg_z g_{ww} - zg_{zw} - g_w^2) \tag{35}$$

for some a, b, c. We note that $g_{ww}(z,0) = 0$, that $g_w(z,0) = z^k$, and $g_{zw}(z,0) = kz^{k-1}$. Using this information we set $w = 0$ in (35) and obtain

$$z^{k-1} = a(z,0)z^k + b(z,0)0 + c(z,0)(-zkz^{k-1} + 0). \tag{36}$$

It follows from (36) that z^{k-1} is divisible by z^k; this contradiction proves that z^{k-1} is not in J_2, and hence that passing to I_2 requires at least k roots. (It is easy to show, but the information is not needed here, that taking k roots suffices.) □

Corollary 4.1. *Assume $k > m \geq 2$ and $n \geq 3$. Let Ω be a pseudoconvex domain in $\mathbf{C}^3$ with defining function near 0 given by*

$$\mathrm{Re}(z_3) + |z_1^m|^2 + |z_2^n + z_1^k z_2|^2. \tag{37}$$

Then, although $\Delta(b\Omega, 0) = 2mn$ and is independent of k, the ϵ in a subelliptic estimate determined by Kohn's algorithm depends on k and tends to 0 as k tends to infinity.

6. Triangular systems

Two computational difficulties in Kohn's algorithm are finding determinants and determining radicals of ideals. We describe a nontrivial class of examples for which finding determinants is easy. At each stage we require only determinants of triangular matrices. Furthermore for this class of examples we never need to take a root of order larger than the underlying dimension.

We call this special class of examples *triangular systems*. The author introduced a version of these examples in [D5], using the term *regular coordinate domains*, but the calculations there give a far from optimal value of the parameter ϵ in a subelliptic estimate. The version in this section improves the work from [D5]. Catlin and Cho [CC] have recently established sharp subelliptic estimates in some specific triangular systems. The crucial point in this section is that triangular systems enable one to choose allowable rows in Kohn's algorithm, one at a time and with control on all radicals. In Theorem 4.1 we establish a decisive result on effectiveness for triangular systems.

Definition 4.3 (Triangular Systems). Let $\mathcal{H}$ be a collection of nonzero elements of $M \subset R_n$. We say that $\mathcal{H}$ is a *triangular system of full rank* if, possibly after a linear change of coordinates, there are elements, $h_1, ..., h_n \in \mathcal{H}$ with the following properties:

1) For each i with $1 \leq i \leq n$, we have $\frac{\partial h_i}{\partial z_j} = 0$ whenever $j > i$. In other words, h_i depends on only the variables $z_1, ..., z_i$.

2) For each i with $1 \leq i \leq n$, $h_i(0, z_i) \neq 0$. Here $(0, z_i)$ is the i-tuple $(0, ..., 0, z_i)$.

It follows from 1) that the derivative matrix $dh = (\frac{\partial h_i}{\partial z_j})$ for $1 \leq i, j \leq n$ is lower triangular. (All the entries above the main diagonal vanish identically.) It follows from 2) that $\frac{\partial h_i}{\partial z_i}(0, z_i) \neq 0$. By combining these facts we see that $J = \det(dh)$ is not identically zero. Our modified procedure makes no use of the other elements of $\mathcal{H}$. Hence there is little lost if we consider a triangular system of full rank simply to be this ordered set of n functions $h_1, ..., h_n$.

Any ideal defining a zero-dimensional variety contains a triangular system of full rank; we are assuming here that the gradients of such functions define the initial allowable rows.

Remark 4.3. Triangular systems of rank less than n are useful for understanding the generalization of the algorithm where we consider q by q minors. We do not consider these systems here, and henceforth we drop the phrase *of full rank*, assuming henceforth that our triangular systems have full rank.

Let $\mathcal{H}$ be a triangular system. After renumbering, we may assume that h_1 is a function of z_1 alone, h_2 is a function of (z_1, z_2), and so on. Note that $h_1(z_1) = z_1^{m_1} u_1(z_1)$ for a unit u_1, that $h_2(z_1, z_2) = z_2 u_2(z_2) + z_1 g_2(z_1, z_2)$ for a unit u_2, and so on. After changing coordinates again we may assume that these units are constant. For example $z_1^{m_1} u_1(z_1) = \zeta_1^{m_1}$, where ζ_1 is a new coordinate. We may therefore assume that a triangular system includes functions $h_1, ..., h_n$ as follows:

$$h_1(z) = z_1^{m_1} \tag{38.1}$$

$$h_2(z) = z_2^{m_2} + z_1 g_{21}(z_1, z_2) \tag{38.2}$$

$$h_3(z) = z_3^{m_3} + z_1 g_{31}(z_1, z_2, z_3) + z_2 g_{32}(z_1, z_2, z_3) \tag{38.3}$$

$$h_n(z) = z_n^{m_n} + \sum_{j=1}^{n-1} z_j g_{nj}(z_1, ..., z_{n-1}). \tag{38.n}$$

In (38) the holomorphic germs g_{kl} are arbitrary. Our approach will work uniformly without regard to what these functions are, but the original Kohn algorithm depends upon them.

Each h_j depends upon only the first j variables and has a pure monomial in z_j. A useful special case is where each h_j is a Weierstrass polynomial of degree m_j in z_j whose coefficients depend upon only the first $j-1$ variables.

Example 4.7. Write the variables (z, w) in two dimensions. The pair of functions

$$h(z, w) = (h_1(z, w), h_2(z, w)) = (z^m, w^n + zg(z, w)), \tag{39}$$

where g is any element of R_2, form a triangular system.

Lemma 4.1. *Let $h_1, ..., h_n$ define a triangular system in R_n and let (h) denote the ideal generated by them. Then*

$$\mathbf{D}(h) = \prod_{j=1}^{n} m_j \tag{40}$$

PROOF. There are many possible proofs. One is to compute the vector space dimension of $R_n/(h)$ by listing a basis of this algebra. The collection $\{z^\alpha\}$ for $0 \le \alpha_i \le m_i - 1$ is easily seen to be a basis. □

Theorem 4.1. *There is an effective algorithm for subelliptic estimates for triangular systems. That is, let $h_1, ..., h_n$ define a triangular system with $L = \mathbf{D}(h) = \prod m_j$. The following hold:*

1) There is a finite sequence of pairs of multipliers $(B_1, A_1), ..., (B_L, A_L)$ such that $B_1 = A_1 = \det(\frac{\partial h_i}{\partial z_j})$ and $B_L = A_L = 1$.

2) Each B_j divides a power of A_j. The power depends on only the dimension and not on the functions h_j. In fact, in dimension n we never require any power larger than n.

3) The length L of the sequence equals the multiplicity $\mathbf{D}(h) = \prod m_j$.

PROOF. We induct on the dimension, writing out the cases $n = 1$ and $n = 2$ in full for clarity. When $n = 1$ we never need to take radicals. When $n = 1$ we may assume $h_1(z_1) = z_1^{m_1}$. We set $B_1 = A_1 = (\frac{\partial}{\partial z_1})h_1$, and we set $B_j = A_j = (\frac{\partial}{\partial z_1})^j h_1$. Then B_1 is a subelliptic multiplier, and each B_{j+1} is the derivative of B_j and hence also a subelliptic multiplier; it is the determinant of the one-by-one matrix given by dB_j. By construction $B_j = A_j$ and hence 2) holds. Since h_1 vanishes to order m_1 at the origin, the function B_{m_1} is a non-zero constant. Hence 3) holds. Here $L = m_1$ and hence 1) holds. Thus the Theorem is proved when $n = 1$.

We next write out the proof when $n = 2$. The initial allowable rows are dh_1 and dh_2, giving a lower triangular two-by-two matrix, because $\frac{\partial h_1}{\partial z_2} = 0$. We set

$$B_1 = A_1 = \det(\frac{\partial h_i}{\partial z_j}) = Dh_1 Dh_2, \tag{41}$$

where we use the following convenient notation:

$$Dh_k = \frac{\partial h_k}{\partial z_k}. \tag{42}$$

For $1 \le k \le m_2$ we set

$$B_j = (Dh_1)^2 \ D^j h_2 \tag{43.1}$$

$$A_j = Dh_1 \ D^j h_2. \tag{43.2}$$

Each B_{j+1} is a subelliptic multiplier, obtained by taking the determinant of the allowable matrix whose first row is dh_1 and second row is dA_j. Since $A_1 = B_1$, and otherwise B_j divides A_j^2, each A_{j+1} is a subelliptic multiplier. By the assumption on h_2, we find that A_{m_2} is a unit times Dh_1. We therefore obtain Dh_1 as a subelliptic multiplier. We now use $d(Dh_1)$ as the first allowable row, and repeat the process. After m_2 more steps we obtain $D^2(h_1)$ as a subelliptic multiplier. We then repeat the process a total of m_1 times, obtaining $B_{m_1 m_2} = A_{m_1 m_2}$ and this expression is a unit. Furthermore, except in the few cases where $A_j = B_j$ we have B_j divides A_j^2. In particular, no power more than 2 arises. Thus 1), 2), and 3) hold when $n = 2$.

We mention why we needed to take square roots, that is, replacing B_i with A_i. After establishing that $A_j = Dh_1 D^j h_2$ is a multiplier, we use dA_j as an allowable row. The next determinant becomes $(Dh_1)^2 D^{j+1} h_2$. We replace this multiplier with $Dh_1 D^{j+1} h_2$ in order to avoid having both Dh_1 and $D^2 h_1$ appearing.

Now we make the induction hypothesis: we assume that $n \geq 2$, and that $h_1, ..., h_n$ defines a triangular system. We assume that 1),2), and 3) hold for all triangular systems in $n-1$ variables. We set

$$B_1 = A_1 = \det(\frac{\partial h_i}{\partial z_j}) = Dh_1 Dh_2 \cdots Dh_n. \tag{44}$$

We replace the last allowable row by dA_n and take determinants, obtaining

$$B_2 = Dh_1 Dh_2 \cdots Dh_{n-1}\ Dh_1 Dh_2 \cdots Dh_{n-1} D^2 h_n \tag{45}$$

as a subelliptic multiplier. Taking a root of order two, we obtain

$$A_2 = Dh_1 Dh_2 \cdots Dh_{n-1} D^2 h_n \tag{46}$$

as a subelliptic multiplier. Repeating this process m_n times we obtain

$$A_{m_n} = Dh_1 Dh_2 \cdots Dh_{n-1} \tag{47}$$

as a subelliptic multiplier. We use its differential dA_{m_n} as the $n-1$-st allowable row, and use dh_n as the n-th allowable row. Taking determinants shows that

$$A_{m_n+1} = Dh_1 Dh_2 \cdots Dh_{n-2} Dh_1 Dh_2 \cdots D^2 h_{n-1} Dh_n \tag{48}$$

is a subelliptic multiplier.

What we have done? We are in the same situation as before, but we have differentiated the function h_{n-1} one more time, and hence we have taken one step in resolving the singularity. We go through the same process $m_{n-1} m_n$ times and we determine that $A_{m_n m_{n-1}}$ is a subelliptic multiplier which depends upon only the first $n-2$ variables. We then use its differential as the $n-2$-nd allowable row. We obtain, after $m_n m_{n-1} m_{n-2}$ steps, a nonzero subelliptic multiplier independent of the last three variables. By another induction, after $\prod m_j$ steps, we obtain a non-zero constant. Thus 3) holds. Each determinant is the product of n diagonal elements. At any stage of the process we can have a fixed derivative of h_1 appearing as a factor to at most the first power in each of the diagonal elements. Similarly a derivative of Dh_2 can occur as a factor only in the last $n-1$ diagonal elements. It follows that we never need to take more than n-th root in passing from the B_k (which is a determinant) to the A_k. Thus 1) and 2) hold. $\square$

Corollary 4.2. *Let $h_1, ..., h_n$ be a triangular system in $\mathbf{C}^n$. Let Ω be a domain in $\mathbf{C}^{n+1}$ defined near the origin by*

$$\mathrm{Re}(z_{n+1}) + \sum_{j=1}^{n} |h_j(z)|^2.$$

Then there is a single ϵ such that a subelliptic estimate of order ϵ holds at 0 for all choices of the functions g_{jk} in (38).

The algorithm used in the proof of Theorem 4.1 differs from the Kohn algorithm. At each stage we choose a single function A with two properties. Some power of A is a determinant of a matrix of allowable rows, and the differential dA provides a new allowable row. The number of steps the algorithm takes equals $\mathbf{D}(h)$.

Example 4.8. We illustrate the difference when $n = 1$. Let I be an ideal with $I \subset \mathcal{M}$. Since the ring is a principal ideal domain in this case, $I = (h)$ for some h. Of course $\mathbf{D}(h)$ is the order of vanishing of h. We know that Dh is a multiplier; hence its derivative D^2h also is a multiplier; it is the determinant of a one-by-one matrix of allowable rows. Therefore each function in the sequence of derivatives $Dh, D^2h, ..., D^m h$ is a subelliptic multiplier, and the last is a unit. Notice that we never took a radical here; we simply differentiated. In Kohn's algorithm we would take the radical of the ideal (Dh) at the start, obtaining $I_0 = \mathcal{M}$. Then $I_1 = (1)$. We avoided taking a radical by taking more derivatives.

The multiplicity $\mathbf{D}(h)$ is the dimension over $\mathbf{C}$ of the quotient space $R/(h)$. The process in Theorem 4.1 seems to be moving through basis elements for this algebra; the multipliers B_j might be in the ideal however. We give a simple example.

Example 4.9. Let $h(z, w) = (z^2, w^2)$. The multiplicity is 4. We have $(A_1, B_1) = (zw, zw)$. We have $(A_2, B_2) = (z, z^2)$. We have $(A_3, B_3) = (w, w^2)$, and finally $(A_4, B_4) = (1, 1)$. Notice that the A_j give the basis for the quotient algebra, whereas two of the B_j lie in the ideal (h).

7. Additional remarks

We state and prove a simple result which explains what happens if the initial set does not define an $\mathcal{M}$-primary ideal.

Proposition 4.4. *Let $h_j \in \mathcal{M}$ for each j, and suppose $(h_1, ..., h_K)$ is not $\mathcal{M}$-primary. Then the stabilized ideal from the algorithm is not the full ring R_n.*

PROOF. Since the (analytic or formal) variety defined by the h_j is positive dimensional, we can find a (convergent or formal) nonconstant n-tuple of power series in one variable t, written $z(t)$, such that $h_j(z(t)) = 0$ in R_1 for all j. Differentiating yields

$$\sum \frac{\partial h_j}{\partial z_k}(z(t))z_k'(t) = 0. \tag{49}$$

Hence the matrix $\frac{\partial h_j}{\partial z_k}$ has a nontrivial kernel, and so each of its n by n minor determinants J vanishes after substitution of $z(t)$. Since $J(z(t)) = 0$,

$$\sum \frac{\partial J}{\partial z_k}(z(t))z_k'(t) = 0. \tag{50}$$

Hence including the 1-form dJ does not change the collection of vectors annihilated by a matrix of allowable rows. Continuing we see that $z'(t)$ lies in the kernel of all new matrices we form from allowable rows, and hence $g(z(t))$ vanishes for all functions g in the stabilized ideal. Since $z(t)$ is not constant, we conclude that the variety of the stabilized ideal is positive dimensional, and hence the stabilized ideal is not R_n. $\square$

Exercise 4.5. Using appropriate minor determinants, describe the version of Kohn's algorithm that decides whether the dimension of the variety defined by $h_1, ..., h_K$ has dimension at most $n - q$.

Exercise 4.6. (continuation of the previous exercise) What does Kohn's algorithm say when $q = n - 1$?

We have mentioned the log canonical threshold. Consider the ideal in two dimensions by $(z^2 - w^3, z^m)$ where $m > 2$. The log canonical threshold is $\frac{5}{6}$, and is thus independent of m. The m matters for subellipticity. Consider the defining equation

$$\text{Re}(z_3) + |z^2 - w^3|^2 + |z|^{2m} \tag{51}$$

for a domain Ω. The origin is a boundary point, Ω is a domain of finite type, and $b\Omega$ is real-analytic. The type of the origin is $3m$. By either the work of Kohn or Catlin, there is a subelliptic estimate at 0. As m tends to infinity, the largest epsilon one can use in the estimate (4) tends to 0. This simple example suggests that one needs to define new analogues of the log canonical threshold. More precisely, for each integer k with $0 \leq k \leq n-1$ there should be a number $\mathcal{L}_k$; the special case when $k = n - 1$ corresponds to the log canonical threshold. Sharp subelliptic estimates would then be related to all these numbers. Such a theory would further solidify connections between PDE and Commutative Algebra anticipated by Kohn's work.

LECTURE 5

Connections with partial differential equations

The main purpose of this Lecture is to unify what we have done so far. We will discuss some of the relevant literature rather than develop the necessary mathematics in detail. The material here is more advanced than in the rest of the lectures. We begin by discussing why finite type conditions arise. We mention various difficult results about the Cauchy-Riemann equations. We briefly discuss the method of L^2-estimates in Section 4. We conclude by briefly explaining why *finite type* is a biholomorphic invariant.

1. Finite type conditions

Let Ω be a domain in $\mathbf{C}^n$. Researchers in several complex variables have long found success by relating the geometry of $b\Omega$ to function-theoretic properties of Ω. Perhaps the quintessential example of this approach dates back to Levi, around 1906. Levi pseudoconvexity, defined by the condition that $-\log(\text{distance}, b\Omega)$ be plurisubharmonic, is a local property of $b\Omega$. The notion of domain of holomorphy is a global property of Ω; there is no part of $b\Omega$ past which all holomorphic functions on Ω extend. Levi observed that a domain of holomorphy is Levi pseudoconvex; the converse assertion that pseudoconvex domains are domains of holomorphy was finally established, via the work of many mathematicians, in the 1950's. See [H], [Kr], [Range] for proofs and historical information.

Beginning in the 1960's methods of partial differential equations have been used for passing from geometric information on the boundary to function theoretic information on the domain. We therefore suppose that the boundary of Ω is a smooth real submanifold M of $\mathbf{C}^n$. Under this assumption Ω is Levi pseudoconvex if and only if the *Levi form* of M is nonnegative definite. The PDE approach to several complex variables focuses on the Cauchy-Riemann operator $\overline{\partial}$. Kohn's approach requires establishing regularity results for the $\overline{\partial}$ equation up to the boundary. He first proved such results (including the $\frac{1}{2}$ estimate) in the strongly pseudoconvex case. The natural generalizations led to the notion of *point of finite type.* So far we have described how the notion of finite type relates to ideas from geometry and commutative algebra. In this Lecture we will describe the PDE issues from which the concept of finite type was introduced and developed.

The boundary M of a domain Ω is strongly pseudoconvex at p when the Levi form is positive definite there. Strong pseudoconvexity is a natural condition for two important reasons. First, the set of strongly pseudoconvex points is an *open* subset of the boundary. Second, the notion of strong pseudoconvexity is *finitely determined.* If M is strongly pseudoconvex at p, and M' is any hypersurface tangent to second order to M at p, then M' is also strongly pseudoconvex at p. At a given strongly pseudoconvex point, one can osculate M to second order by a sphere.

The observations in the previous paragraph partially explain why many theorems in several complex variables are first proved for strongly pseudoconvex domains. One then seeks more general geometric conditions on the boundary for which these theorems remain valid. Often these theorems fail for Levi flat boundaries, and hence one is led to seek intermediate conditions guaranteeing the same conclusions. Different problems lead to different intermediate conditions, but many of the same motifs apply. Our definition of *finite type* yields a finitely determined open condition. One fundamental difference however is that there is no single *model* hypersurface in the finite type case analogous to the unit sphere (or Heisenberg group) in the strongly pseudoconvex case.

The $\overline{\partial}$-Neumann problem nicely illustrates the discussion in the previous paragraph. Kohn solved the $\overline{\partial}$-Neumann problem for strongly pseudoconvex domains in 1962, and the subelliptic $\frac{1}{2}$-estimate played a crucial role. See [K1], [K2], [FK], or [CS] for example. Let α be a differential form of bidegree $(0,q)$, with square-integrable coefficients. The $\overline{\partial}$-Neumann problem constructs an operator N such that $u = \overline{\partial}^* N\alpha$ solves the inhomogeneous Cauchy-Riemann system $\overline{\partial}u = \alpha$ when $\overline{\partial}\alpha = 0$. When N is *pseudolocal*, the solution u is smooth wherever α is smooth. This property is called *local regularity*. Kohn and Nirenberg [KN] proved that an *a priori* subelliptic estimate for $(0,q)$ forms on a smoothly bounded pseudoconvex domain guarantees local regularity for the particular solution $\overline{\partial}^* N\alpha$. Under what geometric conditions do these estimates hold?

In 1972 [K3] introduced *points of finite type* for boundary points of pseudoconvex domains in two complex dimensions. This definition involved iterated commutators of tangential vector fields to the boundary; the type of a point p was defined to be the type at p of a nonzero $(1,0)$ vector field L. See Lecture 3. This concept led to many developments and generalizations to higher dimensions. One possible notion of finite type for a boundary point in higher dimensions could be that there is some $(1,0)$ vector field of finite type there. This condition turns out to be equivalent to subelliptic estimates for $(0,n-1)$ forms and it is appropriate in many contexts in CR geometry, including the holomorphic extension of CR functions. See [Tre]. From the point of view of local regularity for $\overline{\partial}$, however, the more elusive notion of subelliptic estimates on $(0,1)$ forms needed to be understood.

Generalizations of the finite type condition involving types of vector fields occur in many problems. One might consider the following notion; each $(1,0)$ vector field is of finite type at p. This notion fails to be an open condition (see Example 3.2), and hence it is not viable. For issues related to subellipticity, conditions involving iterated commutators of smooth vector fields are inadequate, because they do not account for the possibility of singular complex varieties lying in the boundary. We mentioned in Lecture 3 a way to express finite type by using vector fields with nonsmooth coefficients. See [D1] for lengthy discussion on various equivalent definitions of point of finite type in two dimensions and how these definitions generalize to higher dimensions.

Kohn [K3] proved that finite type in two dimensions implied subellipticity on $(0,1)$ forms and Greiner [Gr] established the converse. Rothschild and Stein [RS], again in two dimensions, used the theory of nilpotent Lie algebras to establish the precise relationship between the type of a point and the largest ϵ that works in a subelliptic estimate. Kohn posed the problem of finding necessary and sufficient conditions for subelliptic estimates on $(0,q)$ forms for pseudoconvex domains in arbitrary dimensions. See [DK] for a survey of this work and related developments. See [K4] and [C1], [C2], and [C3] for the main results.

Why do these delicate analytic considerations have anything to do with commutative algebra? To indicate the answer and to make peace with our work in Lecture 3, we briefly consider a second question where intermediate conditions arise. Let D be a neighborhood of the origin in $\mathbf{C}^n$, and let $f : D \to \mathbf{C}^{\mathrm{N}}$ be a holomorphic mapping with $f(0) = 0$. The function $z \to ||f(z)||^2$ has a minimum at the origin. When the rank of the Jacobian $df(0)$ is n, the complex Hessian of $||f||^2$ is positive definite at the origin, and we conclude by elementary calculus that f attains a *strict* minimum at the origin. We naturally ask "Under what general conditions is the origin a strict minimum for $||f||^2$?"

It is evident that $||f||^2$ has a strict minimum at the origin if and only if the origin is an isolated point in the set of common zeroes of the component functions of f. By the Rückert Nullstellensatz, the (germ of the) variety $V(f)$ consists of the origin alone if and only if the ideal (f) in the ring $\mathcal{O}$ of germs of holomorphic functions at the origin is primary to the maximal ideal $\mathcal{M}$. Lecture 3 provided many equivalent statements and Lecture 4 related this situation to subelliptic estimates.

The circumstance where $||f||^2$ has a *strict minimum* at the origin provides a simple example of a finite type condition. This situation generalizes the case where the second derivative test applies, and it exhibits two crucial properties of a non-degeneracy condition. First, by standard commutative algebra, the condition that (f) be primary to $\mathcal{M}$ is finitely determined. Second, when the Taylor coefficients of f depend continuously on some parameter λ, the set of λ for which (f) is primary to $\mathcal{M}$ is an open subset of the parameter space.

2. Local regularity for $\overline{\partial}$

Let Ω be a bounded pseudoconvex domain with smooth boundary. Let α be a $(0, q)$ form with L^2 coefficients on the closed domain and suppose $\overline{\partial}\alpha = 0$. Consider the inhomogeneous Cauchy-Riemann system of equations

$$\overline{\partial}u = \alpha. \tag{1}$$

Equation (1) is overdetermined; adding any $\overline{\partial}$-closed $(0, q-1)$ form to u gives another solution. The $\overline{\partial}$-Neumann solution to (1), also called the Kohn solution or canonical solution, is the unique solution orthogonal to the nullspace of $\overline{\partial}$. In terms of the operator N it can be written $\overline{\partial}^* N\alpha$, where $\overline{\partial}^*$ is the adjoint of $\overline{\partial}$. One determines the domain of the adjoint via integration by parts, so forms in its domain must satisfy a boundary condition. For example, a smooth $(0, 1)$ form $\sum a_j d\overline{z}^j$ is in the domain of $\overline{\partial}^*$ when the vector field $\sum a_j \frac{\partial}{dz^j}$ is tangent to $b\Omega$.

Local regularity asks whether there is some solution u to (1) such that u is smooth wherever α is smooth. When a subelliptic estimate holds on $(0, q)$ forms, then the $\overline{\partial}$-Neumann solution has this property. Local regularity does not hold on general bounded weakly pseudoconvex domains; the existence of complex-analytic varieties in the boundary leads to a propagation of singularities. In [C1,C2,C3] Catlin proved that subelliptic estimates hold on $(0, 1)$ forms near p if and only if p is a point of finite type in the sense described in Lecture 3. He also proved analogous results for forms of degree $(0, q)$.

The definition of subelliptic estimates for $(0, 1)$ forms appears in Lecture 4. We state the natural generalization to $(0, q)$ forms.

Definition 5.1. Let Ω be a smoothly bounded pseudoconvex domain in $\mathbf{C}^n$, and let p be a point in the closure $\overline{\Omega}$. The $\overline{\partial}$-Neumann problem is *subelliptic* on $(0,q)$ forms at p if there is a neighborhood U of p in $\overline{\Omega}$ and positive constants C and ϵ such that the *a priori* estimate

$$||\phi||_\epsilon^2 \leq C(||\overline{\partial}\phi||^2 + ||\overline{\partial}^*\phi||^2 + ||\phi||^2) \tag{2}$$

holds for all smooth $(0,q)$ forms ϕ, supported in U, and in the domain of $\overline{\partial}^*$.

It is not necessary to include the term $||\phi||^2$ on the right-hand side of (2), as it can be estimated in terms of the other two terms. See Section 5.4 for the case when $q=1$. When p is in the interior, (2) holds with $\epsilon = 1$. This conclusion follows because of the ellipticity of the problem in the interior. When p is a strongly pseudoconvex boundary point, (2) holds for $(0,1)$ forms with $\epsilon = \frac{1}{2}$. More generally, if the Levi form has at least $n-q$ positive eigenvalues at a boundary point p, then (2) holds for $(0,q)$ forms with $\epsilon = \frac{1}{2}$.

Kohn [K4] introduced subelliptic multipliers and combined them with geometric results of Diederich-Fornaess [DF1] to establish subelliptic estimates on $(0,q)$ forms at p on bounded pseudoconvex domains with real-analytic boundary, assuming that the boundary contains no q-dimensional complex-analytic variety containing p.

Catlin's proof of subellipticity [C2,C3] for smoothly bounded domains differs considerably. Catlin shows that subelliptic estimates for $(0,1)$ forms at a boundary point of a pseudoconvex domain follow from the existence of bounded smooth plurisubharmonic functions with large Hessians. More precisely, suppose there is a neighborhood of U of p such that for all $\delta > 0$ there is a smooth real-valued function Φ_δ satisfying the following properties:

$$|\Phi_\delta| \leq 1 \text{ on } U$$

$$(\Phi_\delta)_{z_i\overline{z_j}} \geq 0 \text{ on } U$$

$$\sum_{i,j=1}^{n} (\Phi_\delta)_{z_i\overline{z_j}} a_i \overline{a}_j \geq c\frac{||a||^2}{\delta^{2\epsilon}} \text{ on } U \cap \{-\delta < r \leq 0\}. \tag{3}$$

Then there is a subelliptic estimate of order ϵ at p. To help construct the functions ϕ_δ, which are used as weight functions in L^2 estimates, Catlin introduces an n-tuple of rational numbers ($+\infty$ is also allowed) called the multi-type of p. We give a short discussion of an easier version of these ideas in Section 5.4.

3. Hypoellipticity, global regularity, and compactness

In order to develop a feeling for subelliptic estimates and finite type we briefly consider a simpler situation, namely subelliptic estimates for second-order self-adjoint scalar operators on real Euclidean space. We mention a result of Fediĭ [F] on hypoellipticity; the setting exhibits subellipticity as a natural intermediate finite order condition between ellipticity and subellipticity.

Consider the second-order linear partial differential operator defined by:

$$L = \frac{\partial^2}{\partial x^2} + a(x)\frac{\partial^2}{\partial y^2}. \tag{4}$$

Here a is a smooth even function that is positive away from the origin. By Fediĭ's theorem, the operator L is hypoelliptic for all such functions a. It is evident that

L is elliptic if and only if $a(0) > 0$, and it is easy to show that L is subelliptic if and only if A vanishes to finite order at 0.

Hypoellipticity for second-order operators is not well-understood, as it depends on delicate analytic considerations. Suppose we work in three dimensions with variables x, y, z and we add the term $\frac{\partial^2}{\partial z^2}$ to L; hypoellipticity then holds if and only if $\lim_{x \to 0} x\log(a(x)) = 0$. This result was proved by Kusuoka and Stroock using stochastic methods; see [Ch3] for references and an alternative proof. Determination of conditions for hypoellipticity of second-order operators is a subject of considerable current research. See [K6] and [Ch3] for information concerning situations where hypoellipticity holds but subellipticity fails.

The Fedii example exhibits subellipticity as a natural intermediate, finite-order condition between ellipticity and hypoellipticity. The situation roughly corresponds to that for $\overline{\partial}$ in two complex dimensions. In all cases subellipticity in the $\overline{\partial}$-Neumann problem requires the analysis of an intermediate finite-order condition.

Subelliptic estimates in the $\overline{\partial}$-Neumann problem yield local regularity. The problem of *global regularity* was also answered by Kohn. Let α be smooth on the closure of a smoothly bounded pseudoconvex domain. Is there some solution u to (1) that is smooth everywhere on the closed domain? The answer is yes. The proof passes through a weighted version of the $\overline{\partial}$-Neumann problem, where the Euclidean metric is multiplied by the factor $e^{-t||z||^2}$. It is often useful to think of ordinary forms as L-valued forms, where L is a trivial line bundle, but with a Hermitian metric given by $e^{-t||z||^2}$ times the standard metric. See [Gre] and also Berndtsson's lectures in this volume for lucid explanations in terms of curvature and line bundles.

It is natural to ask whether the $\overline{\partial}$-Neumann solution $\overline{\partial}^* N\alpha$ enjoys the global regularity property of the preceding paragraphs. This particular property is sometimes phrased *global regularity* for the $\overline{\partial}$-Neumann problem. For a long time complex analysts believed that global regularity would always hold on pseudoconvex domains; this conclusion turns out to fail. In 1996, Christ [Ch2], building on work of Barrett [Ba1,Ba2] showed that global regularity for the $\overline{\partial}$-Neumann problem fails on worm domains. Worm domains, discovered by Diederich-Fornaess [DF2], have many strange properties, even though they are smoothly bounded and pseudoconvex. Barrett [Ba1] had given in 1984 an example of a smoothly bounded but nonpseudoconvex domain, for which the Bergman projection fails to preserve the space $C^\infty(\overline{\Omega})$. (Condition R of Bell fails; see [Be1] and [Be2]). In [Ba2] Barrett showed that the Bergman projection on worm domains failed to preserve certain Sobolev spaces; Christ used this result in a crucial way in establishing the failure of global regularity. See [Ch4] for detailed discussion of these ideas.

Thus the condition for global regularity of the $\overline{\partial}$-Neumann solution of (1) is an intermediate condition, distinct from finite type, whose precise geometric formulation is currently an open problem. Compactness of N, to be discussed in Section 3, implies global regularity. Boas-Straube have given other positive results; for example, for domains with a defining function that is plurisubharmonic on the boundary, global regularity of the $\overline{\partial}$-Neumann problem holds. See [BS] and [Ch4] for extended discussion on global regularity and additional references.

The main point for us is that subellipticity implies compactness which in turn implies global regularity, but that neither reverse implication holds. Necessary and sufficient conditions for compactness and for global regularity are not known. It is known that the existence of complex varieties in the boundary of a smoothly bounded domain prevents compactness, but even this conclusion fails when the boundary is not smooth.

Compactness of the N operator on smoothly bounded pseudoconvex domains is a delicate matter. For each such domain, and for all q with $1 \leq q \leq n$, the $\overline{\partial}$-Neumann operator N_q on $(0,q)$ forms is *bounded* on the Hilbert space of square-integrable $(0,q)$ forms. It need not be compact. Subelliptic estimates on $(0,q)$ forms yield compactness for N_q, but compactness holds more generally. One can show [FuS] that compactness for N_q is equivalent to compactness for both operators $\overline{\partial}^* N_q$ and $\overline{\partial}^* N_{q+1}$, and it is also equivalent to an *a priori* estimate. ([KN] or [C4]). Catlin [C4] established compactness when the boundary of Ω satisfies *property P*; later Sibony [Si] called the same concept B-regularity, and he gave many equivalent formulations and applications. McNeal [Mc3] introduced a more general sufficient condition for compactness which has additional applications. See [FuS] for a thorough survey of compactness in the $\overline{\partial}$-Neumann problem. See [CF] for a result relating compactness on Hartog's domains to physics. In Section 4 we recall some of the techniques from [C4] and we state McNeal's result.

Subelliptic estimates and compactness estimates involve L^2 norms. Estimates for solutions of the Cauchy-Riemann equations and related operators are known in the strongly pseudoconvex case for many other function spaces, such as Hölder and Lipschitz spaces. Also Fornaess and Sibony [FS] wrote an extended article on estimates for $\overline{\partial}$ in L^p spaces. Many of these results follow from explicit integral formulas for solving the $\overline{\partial}$ equation. Such integral formulas are not known in the weakly pseudoconvex case, and generalizations of the estimates to that case are therefore difficult and a subject of much current research. See [FeK], [FKM] and the references there. In the elliptic case Hölder and elliptic estimates are equivalent; this fact might suggest some simple connection between subelliptic estimates and Hölder estimates. Guan [Gu] showed however that there exist second-order subelliptic operators for which Hölder estimates fail completely. A complete understanding of Hölder estimates for $\overline{\partial}$ in weakly pseudoconvex domains is far away.

4. An introduction to L^2-estimates

We begin with a useful result about Hessians and exponentials. When λ is a smooth real-valued function, we write $H(\lambda) \geq c$ if, for all points, the minimum eigenvalue of the Hessian $H(\lambda)$ is at least c. Alternatively,

$$\sum_{j,k=1}^{n} \lambda_{z_j \overline{z}_k}(p) f_j \overline{f}_k \geq c \sum |f_j|^2.$$

Exponentiating helps make a Hessian larger. One can also compose a real-valued function with a convex increasing function on $\mathbf{R}$ and play similar tricks.

Lemma 5.1. *Assume that λ is smooth, and put $\phi = e^{\lambda}$. Then*

$$|\langle \partial\phi, f\rangle|^2 = e^{2\lambda} |\langle \partial\lambda, f\rangle|^2, \tag{5}$$

$$H(\phi)(f,f) = e^{\lambda}(H(\lambda)(f,f) + |\langle \partial\lambda, f\rangle|^2). \tag{6}$$

PROOF. Note that both calculations are several variable analogues of elementary calculus. In one variable we have the following. If $\phi = e^{\lambda}$, then $\phi' = e^{\lambda}\lambda'$ and $\phi'' = e^{\lambda}(\phi'' + (\phi')^2)$. Statements (5) and (6) follow in the same manner. $\square$

Definition 5.2. (Property P) Let Ω be a smoothly bounded pseudoconvex domain; Ω satisfies *Property P* if, for every positive number C there is a smooth plurisubharmonic function λ on the closed domain such that

1) $0 \leq \lambda \leq 1$
2) $H(\lambda) \geq C$.

Catlin established that Property P implies a compactness estimate for the Cauchy-Riemann equations. His methods combine aspects of the approaches of both Kohn and Hörmander for proving regularity results. Property P is weaker than finite type. To obtain subelliptic estimates near a point of finite type one must find plurisubharmonic functions as in Property P, but also whose Hessians grow at a prescribed rate as one approaches the boundary. In order to understand these ideas we first consider a compactness estimate.

Definition 5.3. Let $\mathcal{H}$ be a Hilbert space with norm $|| \; ||$ and let $L : \mathcal{H} \to \mathcal{H}$ be a linear operator. Then L is called compact if the following holds. For each $\epsilon > 0$, there is an operator T with finite-dimensional range such that $||L - T|| < \epsilon$. (Thus L is the limit in norm of *finite rank operators.*)

The following self-referential but useful characterization of compactness helps clarify the concept: L is compact if and only if, for each $\epsilon > 0$ there is a compact operator K_ϵ such that

$$||Lf||^2 \leq \epsilon ||f||^2 + ||K_\epsilon f||^2. \tag{7}$$

The formulation (7) suggests why the estimate (17) below is called a compactness estimate and it leads to easy proofs of basic facts about compact operators.

Exercise 5.1. Suppose L is compact and T is bounded. Show that TL and LT are compact.

Exercise 5.2. Prove that (7) characterizes compactness. Use (7) to show that the adjoint of a compact operator is compact.

Let Ω be a smoothly bounded pseudoconvex domain. We will study the equation $\overline{\partial} u = \alpha$, where α is a $(0,1)$ form with L^2 coefficients and $\overline{\partial}\alpha = 0$. Rather than working in L^2, we will work in L^2 with respect to a weight function. Thus we write $L^2(\Omega, \phi)$ to denote the space of functions f for which

$$\int_\Omega |f|^2 e^{-\phi} dV < \infty. \tag{8}$$

Similarly we write $L^2_{0,q}(\Omega, \phi)$ for the corresponding space of $(0,q)$ forms.

Following Hörmander's approach, we consider the diagram

$$0 \to \; L^2(\Omega, \phi_1) \to \; L^2_{0,1}(\Omega, \phi_2) \to \; L^2_{0,2}(\Omega, \phi_3) \to \; 0. \tag{9}$$

We define the domain of the operator $T : L^2(\Omega, \phi_1) \to \; L^2_{0,1}(\Omega, \phi_2)$ to be the set of functions f for which $\overline{\partial} f$, defined in the sense of distributions, is in $L^2_{0,1}(\Omega, \phi_2)$, and on this domain $Tf = \overline{\partial} f$. Similarly we define the domain of the operator $S : L^2_{0,1}(\Omega, \phi_1) \to \; L^2_{0,2}(\Omega, \phi_2)$. Both T and S are closed and densely defined; hence they have adjoints T^* and S^* whose domains we ignore for the moment. Also $ST = 0$.

Let $f = \sum f_j d\overline{z}^j$ be a $(0,1)$ form. When $\phi_1 = \phi_2 = 0$, we saw earlier that $\overline{\partial}^* f = -\sum (f_j)_{z_j}$. We now write $\mathcal{D}f = -\sum (f_j)_{z_j}$. The quadratic form Q discussed in Lecture 4 is defined by

$$Q(f,f) = ||\overline{\partial} f||^2 + ||\mathcal{D}f||^2 + ||f||^2. \tag{10}$$

There are no weight functions in (10). Why are they useful? First we express T^* in terms of the functions ϕ_1 and ϕ_2.

Lemma 5.2. *For smooth f in the domain of T^*, we have $T^*f = e^{\phi_1}\mathcal{D}(e^{-\phi_2}f)$.*

PROOF. Let g be a smooth function and f a smooth $(0,1)$ form. Integrate by parts in the integral $\langle f, Tg\rangle_{\phi_2}$ and assume the boundary integral vanishes to obtain

$$\langle f, Tg\rangle_{\phi_2} = \sum_j \int_\Omega f_j \overline{g_{\overline{z}_j}} e^{-\phi_2} dV = -\sum_j \int_\Omega \frac{\partial}{\partial z_j}(e^{-\phi_2} f)\overline{g} dV$$

$$= -\int_\Omega e^{\phi_1} \sum_j \frac{\partial}{\partial z_j}(e^{-\phi_2} f)\overline{g} e^{-\phi_1} dV = \langle e^{\phi_1}\mathcal{D}(e^{-\phi_2}f), g\rangle_{\phi_1} = \langle T^*f, g\rangle_{\phi_1}. \tag{11}$$

□

For an appropriate choice of weight functions there is a constant C such that

$$||f||^2_{\phi_2} \leq C(||Sf||^2_{\phi_3} + ||T^*f||^2_{\phi_1}) \tag{12}$$

for all f in the domains of S and T^*. Inequality (12) guarantees existence of solutions to the Cauchy-Riemann equations. We need stronger estimates to obtain regularity. We state a basic computation from [C4] and indicate how it leads to the basic inequality (15) below.

Theorem 5.1. *Assume that $\phi_3 = \phi$, $\phi_2 = \phi - \psi$ and $\phi_1 = \phi - 2\psi$. The following estimate holds:*

$$\int_\Omega H(\psi)(f,f)e^{-\psi} dV + \int_{b\Omega} H(r)(f,f)e^{-\psi} dS + \sum_{i,j=1}^n \int_\Omega |(f_i)_{\overline{z}_j}|^2 e^{-\psi} dV$$

$$\leq ||Sf||^2_{\phi_3} + 2||T^*f||^2_{\phi_1} + 2\int_\Omega |\langle \partial\psi, f\rangle|^2 e^{-\psi} dV. \tag{13}$$

The second and third terms on the right-hand side of (13) are not part of the usual expression $Q(f,f)$. If we make the right choices of ϕ and ψ, however, we can estimate these terms in terms of $Q(f,f)$. To do so, first set $\phi_2 = \phi - \psi = 0$. By Lemma 5.2 the term T^*f becomes a smooth multiple of $\mathcal{D}(f)$, and the second term can be estimated by $Q(f,f)$. To handle the third term we set $\phi = ce^\lambda$ for a small constant c with the purpose of controlling it by the Hessian on the left.

After these choices and some computations we discover, for any smooth λ with $0 \leq \lambda \leq 1$, that we get the estimate

$$\int_\Omega H(\lambda)(f,f) dV \leq ||\overline{\partial} f||^2 + ||\mathcal{D}f||^2. \tag{14}$$

The estimate (14) has a beautiful consequence. If we choose $\lambda = c||z||^2$, we obtain

$$||f||^2 \leq C(||\overline{\partial} f||^2 + ||\mathcal{D}f||^2), \tag{15}$$

where the constant C in (15) depends only upon the diameter of Ω.

Now consider (14) for general λ. To get something out of it we need $0 \leq \lambda \leq 1$ and also for the Hessian of λ to be large. If we can make $H(\lambda) \geq \frac{1}{\epsilon}$, while keeping $0 \leq \lambda \leq 1$, then we obtain

$$\frac{1}{\epsilon}||f||^2 \leq \int_\Omega H(\lambda)(f,f)dV \leq ||\overline{\partial} f||^2 + ||\mathcal{D}f||^2. \tag{16}$$

In general circumstances we can obtain (16) up to a compact error. Let χ be a smooth function with compact support in Ω. The estimate

$$||f||^2 \leq \epsilon(||\overline{\partial} f||^2 + ||\mathcal{D}f||^2) + ||\chi f||^2_{-1} \tag{17}$$

implies compactness for the canonical solution operator, and (17) holds whenever Property P holds.

For $(0,1)$ forms f supported in Ω, it is possible to estimate all first derivatives in terms of $Q(f,f)$. The barred derivatives are included in the term $||\overline{\partial} f||^2$, and two integrations by parts show that the unbarred derivatives have the same L^2 norm:

Lemma 5.3. *Suppose that f is smooth and compactly supported in Ω. For each j,*

$$||f_{z_j}||^2 = ||f_{\overline{z}_j}||^2.$$

As a consequence, if each $f_{\overline{z_j}} \in L^2$, then f is in the Sobolev space W^1.

To finish the compactness estimate we need an estimate of the form

$$||\chi_\epsilon f||^2 \leq C_1(||\overline{\partial} f||^2 + ||\mathcal{D}f||^2) + C_2||f||^2 + ||\chi_\epsilon f||^2_{-1}. \tag{18}$$

Then one obtains (7) because the terms with negative Sobolev norms involve compact operators. See [C4] for details.

Property P thus enables us to prove a compactness estimate for N on $(0,1)$ forms. If Catlin's multi-type, discussed briefly in the next subsection, is finite in a neighborhood of p, then Property P holds. See [Si1] for many equivalent conditions. Next we state McNeal's general result on compactness of the operator N on $(0,q)$ forms. See [He] for a related result about subellipticity.

Theorem 5.2. *Let Ω be a smoothly bounded pseudoconvex domain in $\mathbf{C}^n$. Let $1 \leq q \leq n$ and suppose properties 1) and 2) hold. Then the $\overline{\partial}$-Neumann operator N is compact on $(0,q)$ forms.*

1) For all $C > 0$, there is a smooth plurisubharmonic function λ such that, for all $p \in \overline{\Omega}$, the sum of any q eigenvalues of the Hessian $H(\lambda)(p)$ is at least C.

2) The function λ satisfies (19) for all p and for all $a \in \mathbf{C}^n$:

$$|\langle \partial\lambda(p), a\rangle|^2 \leq H\lambda(p)(a,a). \tag{19}$$

Multi-types

Let $b\Omega$ be the boundary of a pseudoconvex domain in $\mathbf{C}^n$. We briefly consider the notion of *holomorphic dimension* for a real manifold. Let X be a real submanifold of $\mathbf{C}^n$ for which $T_p^{1,0}X$ has constant dimension. Assume X is a submanifold of $b\Omega$. We say that X has *holomorphic dimension zero* in $b\Omega$ if the Levi form on $b\Omega$ is positive on every $(1,0)$ vector tangent to X. Otherwise it has *positive holomorphic dimension.* A complex submanifold lying in $b\Omega$, when thought of as a real manifold in $b\Omega$, will have positive holomorphic dimension. Thus this concept is related to the existence of complex varieties in $b\Omega$. See [C1], [DF1], [K4], and [Si2].

Let r be a local defining function for $b\Omega$, and suppose $p \in b\Omega$. Catlin assigns to p an n-tuple of positive integers or plus infinity $m(p) = (m_1, m_2, ..., m_n)(p)$ with the following properties:

1) $m_n \leq m_{n-1} \leq ... \leq m_1$.

2) With respect to the lexicographic ordering, the function $p \to m(p)$ is upper semicontinuous.

3) There is a neighborhood of p and a submanifold M of $b\Omega$ of holomorphic dimension 0 such that the set where the multi-type equals $m(p)$ is a subset of M.

4) Assume p is the origin. Then $D^\alpha \overline{D}^\beta r(0) = 0$ if $\sum \frac{\alpha_i + \beta_i}{m_i} < 1$, but there is some α, β such that $\sum \frac{\alpha_i + \beta_i}{m_i} = 1$ and this derivative isn't zero.

5) $m_1 \leq \Delta(M, p)$.

Example 5.1. Put $r(z) = 2\mathrm{Re}(z_3) + |z_1^2 - z_2^3|^2$. Let p be the origin. Then $m(p) = (1, 4, 6)$. At nearby points on $\{r = 0\}$ where $z_1 \neq 0$, we have $m(p) = (1, 2, \infty)$.

Example 5.2. If $\Omega \subset \mathbf{C}^2$, and $p \in b\Omega$, then the multi-type at p is the pair $(1, m_2)$, where m_2 is the type $\Delta(b\Omega, p)$.

Lemma 5.4. *If $m_1(p)$ is finite, then there are only a finite number of possible values for the n-tuple $m(p)$.*

We omit the proof of this Lemma. Recall that Ω is bounded. If $b\Omega$ is finite type at each point, Corollary 3.1 gives a global bound on the type. By the Lemma the set of all possible multi-types at all points is a finite set. List these multi-types as $(1, 2, ..., 2) = M_0 < M_1 ... < M_N$. Consider the sets S_j where $M(z) \geq M_i$; each S_j is compact. We obtain a stratification of the boundary into sets where if $z \in S_i - S_{i+1}$, then z is in a submanifold of holomorphic dimension zero.

We briefly remark on the difference between Property P and subellipticity. For Property P one needs to find, for each positive C, a plurisubharmonic function λ whose values are bounded between 0 and 1, but whose Hessian satisfies $H(\lambda) \geq C$. For subellipticity, one works in the set defined by $-\delta < r \leq 0$.To prove that there is a subelliptic estimate of order ϵ one must find a plurisubharmonic function whose Hessian satisfies $H(\lambda) \geq c\delta^{-2\epsilon}$.

We close this lecture with an important comment concerning the behavior of *type* under biholomorphic maps. Bell [Be1] proved the following result:

Theorem 5.3. *Let $f : \Omega_1 \to \Omega_2$ be a biholomorphic mapping between smoothly bounded pseudoconvex domains. Assume that the $\overline{\partial}$-Neumann problem is globally regular on Ω_1. Then f extends smoothly to $b\Omega_1$, and the extended map defines a CR diffeomorphism of the boundaries.*

Corollary 5.1. *Let Ω_1 be a smoothly bounded pseudoconvex domain such that $\Delta(b\Omega, p) < \infty$ for all $p \in b\Omega$. Let $f : \Omega_1 \to \Omega_2$ be a biholomorphic mapping. Then*

$$\Delta(b\Omega_2, f(p)) = \Delta(b\Omega_1, p).$$

PROOF. Finite type implies subellipticity which implies global regularity. Bell's Theorem applies. □

LECTURE 6

Positivity conditions

1. Introduction

The complex numbers are not an ordered field and any inequalities we use must therefore ultimately involve real-valued functions. In this Lecture we discuss various positivity conditions for real-analytic real-valued functions on complex Euclidean space and the relationships among them.

To get started we mention Hilbert's 17-th problem. Hilbert asked whether a nonnegative polynomial on real Euclidean space was necessarily the sum of squares of rational functions. The problem was solved in the affirmative by Artin in the 1920's. See [PD], [R], and [HLP] for various aspects of its history and recent related developments. We are interested in analogous situations in complex analysis. Consider a nonnegative polynomial p on $\mathbf{R}^{2n}$; we may think of it as an Hermitian symmetric polynomial on $\mathbf{C}^n$. By invoking Artin's result and putting things over a common denominator, there exist real polynomials $g_1, ..., g_k$ and q such that $q^2p = \sum g_j^2$. We naturally ask whether we can find *holomorphic* q and G such that $|q|^2p = ||G||^2 = \sum |g_j|^2$? The answer is easily seen to be no. More generally we can ask whether we can find a vector-valued holomorphic polynomial q such that $||q||^2p = ||G||^2$? The answer again is no; simple counterexamples appear in Example 6.3. On the other hand, Theorem 6.2 of this Lecture provides a fairly general situation in which real-valued polynomials must indeed be quotients of squared norms of holomorphic mappings. Furthermore we include several interesting applications of this Theorem. We aim to provide a general discussion about positivity conditions in complex analysis from a perspective consistent with our work so far.

We now turn to the complex case, where we introduce a family of positivity conditions. Let $\mathcal{P}_0$ denote the set of Hermitian symmetric entire real-analytic functions on $\mathbf{C}^n$, and suppose that $R \in \mathcal{P}_0$. We say that $R \in \mathcal{P}_1$ if $R(z, \overline{z}) \geq 0$ for all z. More generally we say that $R \in \mathcal{P}_k$ if, for every choice of k points $z_1, ..., z_k \in \mathbf{C}^n$, the $k \times k$ matrix with entries $R(z_i, \overline{z}_j)$ is non-negative definite. Evidently for each k, $\mathcal{P}_{k+1} \subset \mathcal{P}_k$; even if we restrict our consideration to polynomials, each of these containments is strict. See Example 6.1. We therefore obtain an interesting filtration of the collection of Hermitian symmetric functions.

In this Lecture we will discuss these positivity conditions in detail and relate them to concepts such as squared norms, quotients of squared norms, and plurisubharmonicity. Our applications include a result about rational proper mappings between balls and an interpretation of Theorem 6.2 as an isometric embedding theorem for holomorphic line bundles over complex projective space.

2. The classes $\mathcal{P}_k$

Fix the underlying dimension n. We will be considering Hermitian symmetric polynomials and real-analytic functions on $\mathbf{C}^n \times \mathbf{C}^n$. We will continue to write such objects as $R(z, \overline{w})$. As above we say that $R \in \mathcal{P}_k$ if, for all $a \in \mathbf{C}^k$ and all $z_1, ... z_k \in \mathbf{C}^n$ we have

$$0 \leq \sum_{i,j=1}^{k} R(z_i, \overline{z}_j) a_i \overline{a}_j. \tag{1}$$

For a k by k matrix to be nonnegative definite it is necessary that each smaller principal minor be nonnegative definite, and hence $\mathcal{P}_{j+1} \subset \mathcal{P}_j$ holds for each j.

Definition 6.1. $\mathcal{P}_\infty = \cap_{j=0}^{\infty} \mathcal{P}_j$.

Exercise 6.1. Assume there are entire holomorphic functions $f_1, f_2, ...$ such that

$$R(z, \overline{z}) = \sum_{j=1}^{\infty} |f_j(z)|^2 = ||f(z)||^2.$$

Show that $f \in \mathcal{P}_k$ for all k.

We call a Hermitian symmetric function R a *squared norm* if there is a Hilbert space $\mathcal{H}$ and a holomorphic $\mathcal{H}$-valued function such that $R(z, \overline{z}) = ||f(z)||^2$. By polarization we have $R(z, \overline{w}) = \langle f(z), f(w) \rangle$. We next sketch a proof of the following standard result in functional analysis characterizing $\mathcal{P}_\infty$. See for example [AM] for considerable discussion of this result, although the terminology there differs.

Theorem 6.1. *$\mathcal{P}_\infty$ consists precisely of squared norms of Hilbert space valued holomorphic functions.*

Proof. (Sketch) It is easy to see (Exercise 6.1) that squared norms are in $\mathcal{P}_k$ for each k and hence in $\mathcal{P}_\infty$. We verify the opposite containment. Suppose $R \in \mathcal{P}_\infty$. Let V be the vector space of $\mathbf{C}$-valued functions of finite support. Let δ_z be the element of V that is 1 at z and 0 elsewhere. Define a pseudo-inner product $\langle u, v \rangle_R$ on V by

$$\langle u, v \rangle_R = \sum u(z) \overline{v(w)} R(z, \overline{w}). \tag{2}$$

The sum is finite by the support condition. Since $R \in \mathcal{P}_\infty$, we have $\langle u, u \rangle_R \geq 0$ for all u. Linearity in the first slot and Hermitian symmetry are evident. Formula (2) therefore has all the properties of being an inner product (see Lecture 1) except positive definiteness. Let W be the set of u for which $\langle u, u \rangle_R = 0$. We claim that W is a subspace of V. It is obvious that if $u \in W$, then $cu \in W$. The subtle point is that if $u, v \in W$, then $u + v \in W$. To verify this point, consider $u + \lambda v$ for $\lambda \in \mathbf{C}$. Assume $u, v \in W$. We then have

$$0 \leq \langle u + \lambda v, u + \lambda v \rangle_R = 2\mathrm{Re}(\langle \lambda v, u \rangle_R) = 2\mathrm{Re}(\lambda \langle v, u \rangle_R). \tag{3}$$

If we now choose λ in (3) to be $-\langle u, v \rangle_R$ we obtain a contradiction unless $\langle u, v \rangle_R = 0$. Thus this cross term vanishes. Hence, if $\langle u, u \rangle_R = \langle v, v \rangle_R = 0$, then $\langle u+v, u+v \rangle_R = 0$. Thus W is a subspace.

It follows that $\langle u, v \rangle_R$ induces an inner product on the quotient space V/W. Complete this quotient space to a Hilbert space $\mathcal{H}$. Let $f(z)$ denote the image in

$\mathcal{H}$ of the delta function $\delta_z \in V$. Since R is real-analytic, we obtain a holomorphic map $f : \mathbf{C}^n \to \mathcal{H}$ for which $R(z, \overline{z}) = ||f(z)||^2$. □

Remark 6.1. If r is a polynomial then the only Hilbert spaces that arise in Theorem 6.1 are complex Euclidean spaces $\mathbf{C}^N$, for some N.

Exercise 6.2. Fill in the details of the proof of Theorem 6.1.

If all the classes $\mathcal{P}_k$ were the same, then things would be dull. We next give an example showing that the classes are distinct.

Example 6.1. Let m be a positive integer and $a \in \mathbf{R}$. Consider the two parameter family of polynomials given by

$$R_{m,a}(z, \overline{z}) = (|z_1|^2 + |z_2|^2)^{2m} - a|z_1|^{2m}|z_2|^{2m}.$$

When $m = 1$, completing the square shows that $R_{m,a} \in \mathcal{P}_1$ for $a \leq 4$ and $R_{m,a} \in \mathcal{P}_\infty$ for $a \leq 2$. For $m = 1$ and $k \geq 2$, $R_{m,a} \in \mathcal{P}_k$ if only if $R_{m,a} \in \mathcal{P}_\infty$. Thus there are two critical values of a, $a = 2$ and $a = 4$. As m increases the number of critical values increases, and more of the classes become distinct. Letting m tend to infinity shows that all the classes are distinct. We state some specific information. See [DV] for details.

1) For each fixed m we have $R_{m,a} \in \mathcal{P}_1$ if and only if $a \leq 2^{2m}$.

2) For each fixed m we have $R_{m,a} \in \mathcal{P}_2$ if and only if $a \leq 2^{2m-1}$.

3) For each fixed m we have $R_{m,a} \in \mathcal{P}_\infty$ if and only if $a \leq \binom{2m}{m}$.

4) For each fixed m and $k > m$, we have $R_{m,a} \in \mathcal{P}_k$ if and only if $R_{m,a} \in \mathcal{P}_\infty$.

5) For each fixed m there are a finite number of critical values before things stabilize. See [DV] for a general result on stabilization.

Exercise 6.3. Check the details of Example 6.1.

Exercise 6.4. Give another proof that the classes are distinct.

3. Intermediate conditions

We are also interested in other conditions and properties intermediate between $\mathcal{P}_1$ and $\mathcal{P}_\infty$. One such property is that R is the quotient of squared norms of Hilbert space valued holomorphic mappings. Other properties include that R is plurisubharmonic, that $\log(R)$ is plurisubharmonic, and so on.

We begin by introducing two necessary conditions for being a quotient of squared norms. First it is evident that if $R(z, \overline{z}) = \frac{||A(z)||^2}{||B(z)||^2}$, then the zero set of R must be a complex-analytic variety. The second condition is more interesting.

Definition 6.2. A real-analytic function r (or formal series) in t and $\overline{t}$ lies in $\mathcal{B}$ if $r(t, \overline{t}) = 0$ or if r vanishes to even order $2m$ at 0 and there is a positive c such that

$$j_{2m,0}r(t, \overline{t}) = c|t|^{2m}. \tag{4}$$

Thus if r vanishes to finite order, then the order must be even and the initial form can have only the term in (4). It follows that r has a strict local minimum at 0. We say that r has a *good jet* at 0.

Example 6.2. 1) $r \in \mathcal{B}$ if $r(0, 0) > 0$.

2) $r \in \mathcal{B}$ if $r(t, \overline{t}) = |t|^{2m}$.

3) If $r_c = |t|^2 + c(t^2 + \overline{t}^2)$, then $r_c \in \mathcal{B}$ only when $c = 0$.

4) If $r_c = |t|^4 + c(t\overline{t}^4 + t^4\overline{t})$, then $r_c \in \mathcal{B}$ for all c.

We have pulled back to various curves throughout these notes. Let us continue to do so. Let R be a Hermitian symmetric real analytic function (or formal series) defined near 0 in $\mathbf{C}^n$. We say that $R \in \mathcal{J}$ if, for every germ of a holomorphic curve z (or formal series) with $z(0) = 0$, $z^*R \in \mathcal{B}$. We say informally that R has the *good jet pullback property.*

Lemma 6.1. *Let r be real-analytic (or a formal series) near $0 \in \mathbf{C}^n$. Suppose there are holomorphic maps (or formal maps in z alone) A and B taking values in a Hilbert space such that $r(z, \overline{z}) = \frac{||A(z)||^2}{||B(z)||^2}$. Then $r \in \mathcal{J}$.*

PROOF. For any holomorphic curve z we have $z^*r = \frac{||z^*A||^2}{||z^*B||^2}$. If $z^*A = 0$ then $z^*r \in \mathcal{B}$. Otherwise we may write $z^*A = vt^m + ...$ and $z^*B = wt^k + ...$ where $m \geq k$, $v \neq 0$,and $w \neq 0$. Then there is a unit u such that

$$z^*r(t, \overline{t}) = |t|^{2m-2k} \frac{||v + ...||^2}{||w + ...||^2} = |t|^{2m-2k} u(t, \overline{t}). \tag{5}$$

By (5) the lowest order part of z^*r is a positive constant times $|t|^{2m-2k}$ and the result follows. □

We pause to define the term bihomogeneous polynomial.

Definition 6.3. A *bihomogeneous polynomial* is a polynomial $R(z, \overline{z})$ of even degree $2d$ such that $R(tz, \overline{tz}) = |t|^{2d} R(z, \overline{z})$.

Although the definition allows for $R(z, \overline{z})$ to be complex-valued, we will usually assume that R is Hermitian symmetric and hence that $R(z, \overline{z})$ is real-valued. A bihomogeneous polynomial is homogeneous of the same degree in z and $\overline{z}$. For a polynomial r in one variable, the condition $r \in \mathcal{B}$ implies that the initial form of r is bihomogeneous. We next use Lemma 6.1 to give some examples of elements in $\mathcal{P}_1$ that are not quotients of squared norms.

Example 6.3. The following bihomogeneous polynomials are non-negative but cannot be written as quotients of squared norms:

$$p(z, \overline{z}) = (|z_1|^2 - |z_2|^2)^2 \tag{6}$$

$$h(z, \overline{z}) = (|z_1 z_2|^2 - |z_3|^4)^2 + |z_1|^8. \tag{7}$$

The polynomial p from (6) is not a quotient of squared norms; its zero set is three real dimensional, and hence not a complex variety. Alternatively the necessary condition of Lemma 6.1 fails, we can choose $z(t) = (1 + t, 1)$. Then

$$p(z(t), \overline{z(t)}) = (t + \overline{t} + |t|^2)^2 \tag{8}$$

and the initial form prevents z^*p from being in $\mathcal{B}$.

The zero set of the non-negative bihomogeneous polynomial h in (7) is the complex variety defined by $z_1 = z_3 = 0$, and thus a copy of $\mathbf{C}$. Yet h is not a quotient of squared norms; it doesn't satisfy the necessary condition of Lemma 6.1. Consider the mapping given by $z(t) = (t^2, 1 + t, t)$. The pullback $h(z(t), \overline{z(t)})$ vanishes to order ten at the origin, but the terms of order ten include $t^4\overline{t}^6$.

Exercise 6.5. Verify the calculations in Example 6.3.

Exercise 6.6. Assume $n = 1$. Give a necessary and sufficient condition for a polynomial $r(z,\overline{z})$ to be a quotient of squared norms. See [D11].

Another interesting condition is that there is a positive integer N such that R^N is a squared norm of a holomorphic mapping. Thus there is a Hilbert space $\mathcal{H}$ and a holomorphic mapping $f : \mathbf{C}^n \to \mathcal{H}$ such that

$$r(z,\overline{z})^N = ||f(z)||^2 \tag{9}$$

We say that $r \in \text{Rad}(\mathcal{P}_\infty)$ if r satisfies (9) for some N. See [D6] and [DV] for information about these conditions beyond what we say in this lecture.

Exercise 6.7. Assume $n = 1$ and put $r(z,\overline{z}) = |z|^4 - |z|^2 + 1$. Show that r can be written as a quotient of squared norms of holomorphic polynomial mappings, but that r cannot be written as a squared norm.

Exercise 6.8. Show that the function $2 - |z|^2$ agrees with a squared norm on no open subset of $\mathbf{C}$.

Exercise 6.9. Give an example of polynomials r, s such that each is in $\text{Rad}(\mathcal{P}_\infty)$ but such that their sum is not. (Hint: there is an example with $s = 1$.)

We naturally also consider positivity conditions on specific sets. For example, consider the following problem. Suppose that $r(z,\overline{z})$ is a Hermitian symmetric polynomial, and that $r(z,\overline{z}) \geq 0$ on the unit sphere. Must r agree with a squared norm of a holomorphic polynomial on the sphere? The answer is no, but the answer would be yes if we had assumed that r was strictly positive on the sphere. We discuss this result and give some applications later in this lecture.

Exercise 6.10. (Open problem). Let M be strongly pseudoconvex and algebraic; thus M is defined by a polynomial equation. Let $r(z,\overline{z})$ be a polynomial with $r > 0$ on M. Must r equal a squared norm of a holomorphic polynomial mapping on M?

4. The global Cauchy-Schwarz inequality

The positivity class $\mathcal{P}_2$ turns out to be particularly interesting and hence we discuss it in some detail. The two conditions for $r \in \mathcal{P}_2$ are that $r(z,\overline{z}) \geq 0$ for all z and that r satisfies the global Cauchy-Schwarz inequality (10). For all z and w,

$$r(z,\overline{z})r(w,\overline{w}) \geq |r(z,\overline{w})|^2. \tag{10}$$

The inequality (10) has many consequences. It implies that $r(z,\overline{z})$ achieves only one sign; we therefore without loss of generality usually assume $r(z,\overline{z}) \geq 0$ when we state (10). It is easy to give examples functions in $\mathcal{P}_2$.

Lemma 6.2. $\text{Rad}(\mathcal{P}_\infty) \subset \mathcal{P}_2$. *Thus roots of squared norms satisfy (10).*

PROOF. Suppose that $r^N = ||f||^2$. By the usual Cauchy-Schwarz inequality,

$$(r(z,\overline{z})r(w,\overline{w}))^N = ||f(z)||^2||f(w)||^2 \geq |\langle f(z), f(w)\rangle|^2 = |r(z,\overline{w})|^{2N} \tag{11}$$

Since $N > 0$, we may take N-th roots of both sides of (11) and preserve the direction of the inequality. □

We next recall the definition of plurisubharmonicity and discuss its relationship with the global Cauchy-Schwarz inequality. See [Cal] for related geometric ideas. We also introduce bordered Hessians.

Definition 6.4. A C^2 function is *plurisubharmonic* if for all $z, a \in \mathbf{C}^n$, we have

$$\sum_{i,j=1}^{n} r_{z_i \overline{z}_j}(z,\overline{z}) a_i \overline{a}_j = Hr(z,\overline{z})(a,\overline{a}) \geq 0 \tag{12}$$

Thus r is plurisubharmonic when its complex Hessian is nonnegative, and plurisubharmonicity is the complex analogue of convexity. When $n = 1$ the condition is that the Laplacian of r be nonnegative; that is, r is *subharmonic*.

In addition to the Hessian of a function, the *bordered Hessian* often arises. Given a C^2 function r on $\mathbf{C}^n$, we define its *bordered Hessian* to be the $n+1$ by $n+1$ matrix

$$\begin{pmatrix} r_{z_1\overline{z}_1} & r_{z_1\overline{z}_2} & \dots & r_{z_1\overline{z}_n} & r_{z_1} \\ r_{z_2\overline{z}_1} & r_{z_2\overline{z}_2} & \dots & r_{z_2\overline{z}_n} & r_{z_2} \\ \dots & \dots & \dots & \dots & \dots \\ r_{z_n\overline{z}_1} & \dots & \dots & r_{z_n\overline{z}_n} & r_{z_n} \\ r_{\overline{z}_1} & \dots & \dots & r_{\overline{z}_n} & r \end{pmatrix} = \begin{pmatrix} H(r) & \partial r \\ \overline{\partial} r & r \end{pmatrix}.$$

We leave several facts about bordered Hessians as part of an extended exercise. See [D6] or [D10] for details.

Exercise 6.11. Show that the logarithm of a smooth nonnegative function r is plurisubharmonic if and only if the bordered Hessian of r is nonnegative definite.

Exercise 6.12. Show that the determinant of the bordered Hessian of a bihomogeneous polynomial vanishes identically. Suggestion: Start with $|t|^{2d} R(z,\overline{z}) = R(tz, \overline{tz})$. Differentiate twice to discover a nonzero vector in the null space of the bordered Hessian.

Exercise 6.13. Show that a nonnegative bihomogeneous polynomial is plurisubharmonic if and only if its logarithm also is.

Discussion: One direction is easy, as $x \to e^x$ is convex and increasing. The other direction can be proved as follows. By Exercise 6.11, it suffices to show that the bordered Hessian of r is nonnegative definite. To do so, first show that the Hessian of $||z||^{2d}$ is positive definite away from 0. Given that r is plurisubharmonic and bihomogeneous, consider $u_\epsilon = r + \epsilon ||z||^{2d}$. The n by n principal minor of its bordered Hessian is positive definite, and by Exercise 6.12 the determinant of its bordered Hessian is zero. We conclude that there are n positive eigenvalues and 1 vanishing eigenvalue. Letting ϵ tend to zero shows that the bordered Hessian of r is nonnegative definite.

Proposition 6.1. *Suppose* $r \in \mathcal{P}_2$. *Then both* $\log(r)$ *and* r *are plurisubharmonic.*

Proof. If $\log(r)$ is plurisubharmonic, then r is also, because $x \to e^x$ is a convex increasing function on $\mathbf{R}$. Instead of proving things this way, however, we compute the complex Hessian for r to see how (10) gets used. Note that $r \geq 0$. Put $w = z + ta$ for $t \in \mathbf{C}$ and $a \in \mathbf{C}^n$. Then by (10) we have

$$0 \leq r(z,\overline{z}) r(z+ta, \overline{z+ta}) - |r(z, \overline{z+ta})|^2 = h(t,\overline{t}). \tag{13}$$

Obviously h vanishes at $t = 0$, so 0 is a local minimum point for the function h. Hence its Laplacian (complex Hessian in this case) $h_{t\overline{t}}(0)$ is nonnegative. Computing the Laplacian by the chain rule and evaluating at $t = 0$ gives

$$0 \leq h_{t\bar{t}}(0) = r(z,\overline{z}) \sum r_{z_j \overline{z}_k}(z,\overline{z}) a_j \bar{a}_k - |\sum r_{z_j}(z,\overline{z}) a_j|^2 \tag{14}$$

Restating (14) in alternative notation gives

$$r(z,\overline{z})\ Hr(z,\overline{z})(a,a) \geq |\partial r(z,\overline{z})(a)|^2 \geq 0. \tag{15}$$

In case r is positive at z, we can divide by it and see that r is also plurisubharmonic there. If r vanishes at z, then z is a local minimum point for r and its complex Hessian is nonnegative there as well.

In fact (15) is the stronger condition for $\log(r)$ to be plurisubharmonic. In case r can be 0, as is customary we allow the value $-\infty$ for $\log(r)$. We have

$$(\log(r))_{z_j \overline{z}_k} = \frac{r r_{z_j \overline{z}_k} - r_{z_j} r_{\overline{z}_k}}{r^2}. \tag{16}$$

Combining (15) and (16) shows that $\log(R)$ is plurisubharmonic. □

If $\log(r)$ is plurisubharmonic, then r need not be in $\mathcal{P}_2$. We give a simple example here (from Lecture 2) and a more subtle one in Proposition 6.3.

Example 6.4. Let $n = 1$, and put $r(z,\overline{z}) = |z|^2 + c(z^2 + \overline{z}^2)$. Then r is subharmonic for all choices of the constant c, but r changes sign when $c > 1$ in which case (10) fails. In fact (10) fails for any positive c. Evaluating at $z = 1$ and $w = i$ shows that

$$|r(1,i)|^2 = 1 > 1 - 4c^2 = r(1,1) r(i,-i)$$

Thus, even when r is a nonnegative, plurisubharmonic, homogeneous polynomial, we cannot conclude that (10) holds.

In order to develop a feeling for the Cauchy-Schwarz inequality (10), let r be a real-valued polynomial r on $\mathbf{C}^n$. By Lemma 2.3 there are holomorphic polynomial mappings f and g, taking values in a finite-dimensional space, such that

$$r(z,\overline{z}) = ||f(z)||^2 - ||g(z)||^2. \tag{17.1}$$

We may assume that the components of f and g are linearly independent. We next interpret some of our positivity conditions in the setting of (17.1).

Of course $r \in \mathcal{P}_1$ if and only if $||f(z)||^2 \geq ||g(z)||^2$ for all z, and r is a squared norm if and only if we may choose $g = 0$. It is plurisubharmonic if and only if the Hessian of $||f||^2$ is greater than or equal to (as a matrix) the Hessian of $||g||^2$. All of these assertions are obvious. We have the following general formula characterizing when the global Cauchy-Schwarz inequality holds for $||f||^2 - ||g||^2$.

Assume f, g are Hilbert space valued holomorphic functions. As is customary in multilinear algebra we write $(f \wedge g)(z,w) = f(z) \otimes g(w) - f(w) \otimes g(z)$.

Proposition 6.2. *Let $\mathcal{H}$ be a Hilbert space, and assume that f and g are holomorphic mappings to $\mathcal{H}$. Put*

$$r(z,\overline{w}) = \langle f(z), f(w) \rangle - \langle g(z), g(w) \rangle \tag{17.2}$$

Then (10) holds if and only if, for every pair of points z and w, either of the equivalent inequalities (18.1) or (18.2) holds:

$$||f(z) \otimes g(w) - f(w) \otimes g(z)||^2$$
$$\leq ||f(z)||^2 ||f(w)||^2 - |\langle f(z), f(w) \rangle|^2 + ||g(z)||^2 ||g(w)||^2 - |\langle g(z), g(w) \rangle|^2 \tag{18.1}$$

$$||(f \wedge g)(z,w)||^2 \leq ||(f \wedge f)(z,w)||^2 + ||(g \wedge g)(z,w)||^2. \tag{18.2}$$

PROOF. Substitute (17.2) into (10) and expand. Use the identity $\langle u_1 \otimes v_1, u_2 \otimes v_2\rangle = \langle u_1, u_2\rangle\langle v_1, v_2\rangle$, collect terms, and simplify. It follows that inequality (18.1) is equivalent to (10). We leave the more elegant form (18.2) to the reader. □

Exercise 6.14. Verify that (18.1) and (18.2) are equivalent and give a geometric interpretation in terms of area.

Observe that each term on the right side of (18.1) is nonnegative by the usual Cauchy-Schwarz inequality. Version (10) of the Cauchy-Schwarz inequality demands more; their sum must bound an explicit nonnegative expression that reveals the symmetry of the situation.

5. A complicated example

We next give an example from [DV] illustrating the various positivity conditions. We set $m = 2$ in Example 6.1 and discuss all the positivity conditions. Some of the calculations are quite complicated.

Proposition 6.3. *The following hold for the bihomogeneous polynomial r_a, defined for $a \in \mathbf{R}$ by*

$$r_a(z,\overline{z}) = (|z_1|^2 + |z_2|^2)^4 - a|z_1 z_2|^4.$$

1) $r_a \in \mathcal{P}_1$ if and only if $a \leq 16$.

2) r_a is the quotient of squared norms of holomorphic polynomial mappings if and only if $a < 16$.

3) r_a is plurisubharmonic if and only if $a \leq 12$.

4) $\log(r_a)$ is plurisubharmonic if and only if $a \leq 12$.

5) $r_a \in \mathcal{P}_2$ if and only if $a \leq 8$.

6) $r_a \in \mathrm{Rad}(\mathcal{P}_\infty)$ if and only if $a < 8$.

7) $r_a^2 \in \mathcal{P}_\infty$ if and only if $a \leq 7$.

8) $r_a \in \mathcal{P}_\infty$ if and only if $a \leq 6$.

9) The underlying matrix of coefficients for r_a is positive definite if and only if $a < 6$ and nonnegative definite if and only if $a \leq 6$.

PROOF. Since $r_a(z,\overline{z})$ depends upon only the absolute values of the variables, it is convenient to write $|z_1|^2 = x$ and $|z_2|^2 = y$. We obtain

$$r_a(z,\overline{z}) = (x+y)^4 - ax^2y^2 = x^4 + 4x^3y + (6-a)x^2y^2 + 4xy^3 + y^4. \tag{19}$$

Statements 9) and 8) are easy. The underlying matrix of coefficients is diagonal; its eigenvalues are 1, 4, $6-a$, 4, and 1. Thus 9) holds and 8) follows from this information and Theorem 6.1. Statement 7) is similar; one expands r_a^2 in terms of x, y and easily sees that all the coefficients are nonnegative if and only if $a \leq 7$.

Statement 1) also follows from (19). By homogeneity it suffices to find the condition for the minimum of r_a to be nonnegative when $x + y = 1$. This minimum happens when $x = y = \frac{1}{2}$ and we obtain $1 - \frac{a}{16} \geq 0$.

Statement 2) relies on Theorem 6.2. Note that r_a is positive away from the origin if and only if $a < 16$. Theorem 6.2 guarantees that r_a is a quotient of squared norms when $a < 16$. When $a = 16$ the zero set of r_a is not a complex variety, and hence r_a is not a squared norm.

Statements 3) and 4) are equivalent because r_a is bihomogeneous. We establish 3) by finding the Hessian explicitly. Writing λ_{jk} for the partial derivative $\frac{\partial}{\partial z_j}\frac{\partial}{\partial \overline{z}_k}(r_a)$ we obtain

$$\lambda_{11} = 4(|z_1|^2 + |z_2|^2)^3 + 12(|z_1|^2 + |z_2|^2)^2|z_1|^2 - 4a|z_1|^2|z_2|^4 \tag{20}$$

$$\lambda_{12} = \overline{z}_1 z_2(12(|z_1|^2 + |z_2|^2)^2 - 4a|z_1|^2|z_2|^2) \tag{21}$$

$$\lambda_{22} = 4(|z_1|^2 + |z_2|^2)^3) + 12(|z_1|^2 + |z_2|^2)^2|z_2|^2 - 4a|z_1|^4|z_2|^2. \tag{22}$$

By the bihomogeneity it suffices to determine the positive definiteness of the Hessian matrix on the unit sphere. We then obtain

$$\lambda_{11} = 4 + 12|z_1|^2 - 4a|z_1|^2|z_2|^4 \tag{23}$$

$$\lambda_{12} = \overline{z}_1 z_2(12 - 4a|z_1|^2|z_2|^2) \tag{24}$$

$$\lambda_{22} = 4 + 12|z_2|^2 - 4a|z_1|^4|z_2|^2. \tag{25}$$

We compute $\det\lambda = \lambda_{11}\lambda_{22} - |\lambda_{12}|^2$ and eliminate $|z_2|^2$ to get

$$\det\lambda = 64 - 64a|z_1|^2 + 256a|z_1|^4 - 384a|z_1|^6 + 192a|z_1|^8. \tag{26}$$

We know that r_a is plurisubharmonic if and only if both λ_{11} and the determinant $\det\lambda$ are nonnegative. When $a = 12$ one can verify that (26) equals

$$64(1 - 6|z_1|^2 + 6|z_1|^4)^2 \tag{27}$$

and that $\lambda_{11} > 0$ for $0 \leq |z_1|^2 \leq 1$. The expression (27) vanishes at two values in the unit interval. The expression for λ_{11} is a cubic in $|z_1|^2$ with only one real zero, and it lies outside the unit interval. It follows that r_a is plurisubharmonic when $a = 12$; it follows easily that the same is true when $a \leq 12$. When $a > 12$, expression (26) is negative for points close to the roots of (27). Thus $a = 12$ is the cut off point for plurisubharmonicity.

It remains to discuss statements 5) and 6). By Lemma 6.2, $\mathrm{Rad}(\mathcal{P}_\infty) \subset \mathcal{P}_2$. Of course $r_a \in \mathcal{P}_2$ if and only if (10) holds. We first show that (10) fails when $a > 8$, and then r_a is not in $\mathcal{P}_2$ and hence not in $\mathrm{Rad}(\mathcal{P}_\infty)$. To do so we evaluate both sides of (10) at the points $z = (\frac{1}{\sqrt{2}}, \frac{1}{\sqrt{2}})$ and $w = (\frac{1}{\sqrt{2}}, \frac{-1}{\sqrt{2}})$. We obtain $r_a(z, \overline{w}) = \frac{-a}{16}$ and $r_a(z, \overline{z}) = r_a(w, \overline{w}) = 1 - \frac{a}{16}$. Plugging this into the inequality (10) gives the condition

$$|\frac{-a}{16}|^2 \leq (1 - \frac{a}{16})^2 \tag{28}$$

which implies that $1 - \frac{a}{8} \geq 0$. Hence (10) implies $a \leq 8$. In fact (10) holds for $a = 8$, but the proof is quite technical and we omit it. See [DV]. Finally, verifying that 6) holds for $a < 8$ relies on a Theorem from [CD3], and we omit it also. Using (19) and Mathematica one can easily check that there is an N for which r_a^N has all positive coefficients for values of a such as 7.98. □

This example generalizes to n dimensions. The corresponding function is

$$r_a(z, \overline{z}) = ||z||^{4n} - a|\prod z_j|^4. \tag{29}$$

Then $r \in \mathcal{P}_1$ if and only if $a \leq n^{2n}$. Also $r_a \in \mathcal{P}_2$ if and only if $a \leq \frac{n^{2n}}{2}$. Finally $r_a \in \mathcal{P}_\infty$ if and only if $a \leq \frac{(2n)!}{2^n}$.

Exercise 6.15. Let r_a be as in Proposition 6.3. Use a symbolic package such as Mathematica to verify that there is an N for which r_a^N is a squared norm for $a < 7.99$. Try to find a short proof that (10) holds for r_a if and only if $a \leq 8$.

6. Stabilization in the bihomogeneous polynomial case

The results in this section provide a beautiful application of L^2-methods for proving concrete statements about polynomials. Let Ω be a bounded domain in $\mathbf{C}^n$, and let $A^2(\Omega)$ denote the square integrable holomorphic functions on Ω. Then $A^2(\Omega)$ is a closed subspace of $L^2(\Omega)$. The Bergman projection P from $L^2(\Omega)$ to $A^2(\Omega)$ will be a key tool in this section. We will use it when Ω is the unit ball B_n to provide an analogue of Hilbert's 17th problem for Hermitian symmetric polynomials. See Theorems 6.2 and 6.3.

Even for bihomogeneous polynomials, nonnegativity does not imply any of our other positivity conditions. If, however, r is a bihomogeneous polynomial and it is *strictly positive* away from the origin, then r is a quotient of squared norms of holomorphic polynomial mappings. One can choose the denominator to be $||z^{\otimes d}||^2$ for sufficiently large d. This result was proved in [Q] and [CD1]; it is the main point of Theorem 6.2. For generalizations when r is nonnegative see [D12] and [V].

Let R be a polynomial with $R(0,0) = 0$:

$$R(z, \overline{z}) = \sum_{1 \leq |\alpha| + |\beta| \leq 2m} c_{\alpha\beta} z^\alpha \overline{z}^\beta \tag{30}$$

We have observed in Lecture 2 that $R(z, \overline{z})$ will be real for all z if and only if the *underlying matrix of coefficients* $C = (c_{\alpha\beta})$ is Hermitian symmetric. If C is non-negative definite, then R will take on non-negative values, and if C is positive definite, then R will be positive away from the origin. Let N denote the number of multi-indices possible in (30). The polynomial R can be considered as the restriction of the Hermitian form in N variables

$$\sum_{\alpha,\beta=1}^{N} c_{\alpha\beta} \zeta_\alpha \overline{\zeta}_\beta$$

to a Veronese variety given by the image under the parametric equations $\zeta_\alpha(z) = z^\alpha$.

We write V_m for the complex vector space of homogeneous holomorphic polynomials of degree m, and we can thus identify a bihomogeneous R, via its underlying matrix of coefficients, with an Hermitian form on V_m. In this section on polynomials we reserve the term *squared norm* for a *finite* sum of squared absolute values of holomorphic polynomials. Note that a power of the squared Euclidean norm is itself a squared norm, as $||z||^{2d} = ||z^{\otimes d}||^2$.

We recall some facts about the Bergman projection P on a bounded domain Ω; we use this information only when Ω is the unit ball B_n. For $f \in L^2(\Omega)$, we have the formula

$$Pf(z) = \int_\Omega B(z,\overline{w})f(w)dV(w),$$

where $B(z,\overline{w})$ is called the Bergman kernel function of Ω. The kernel $B(z,\overline{w})$ is known explicitly for a few domains; the unit ball is one of them, and

$$B(z,\overline{w}) = \frac{n!}{\pi^n}(1 - \langle z, w\rangle)^{-n-1}. \tag{31}$$

Note that $B(z,\overline{w}) = \sum c_j \langle z, w\rangle^j$ where each c_j is a positive number.

Recall that a bounded operator T on L^2 is called *compact* if, for every $\epsilon > 0$, there is an operator K with finite-dimensional range such that $||T - K|| < \epsilon$. The collection of compact operators forms a two-sided ideal in the algebra of bounded operators. In other words, if K is compact and T is bounded, then KT and TK are compact. See Exercise 5.1. The Bergman projection P on a bounded domain is not compact; it is the identity operator on an infinite dimensional subspace. On the other hand, for every bounded multiplication operator T, the commutator $[P, T]$ is compact. This result can be proved directly for the ball. It follows also from a general result; if the $\overline{\partial}$-Neumann operator on Ω is compact, then $[P, T]$ is compact for every bounded multiplication operator T. See [CD1], [CD2], and [D6].

We have the following elementary but crucial Lemma.

Lemma 6.3. *Let $R(z,\overline{w})$ be a bihomogeneous polynomial of degree (m, m) with underlying matrix of coefficients $(E_{\mu\nu})$. The following are equivalent:*

1) $(E_{\mu\nu})$ is positive definite.

2) $R = ||h||^2$ is the squared norm of a holomorphic homogeneous polynomial mapping h whose components for a basis for V_m.

3) The integral operator K_R defined by

$$K_R g(z) = \int_{B_n} R(z,\overline{w})g(w)dV(w)$$

is positive definite on V_m. In other words, there is a positive c such that

$$\langle K_R g, g\rangle_{L^2} \geq c||g||_{L^2}^2 \tag{Pos}$$

for every $g \in V_m$.

PROOF. We leave the details as an exercise. Note that the equivalence of 1) and 2) is elementary linear algebra. The equivalence of 3) follows from computing the integrals in (Pos) and using the orthogonality of distinct monomials. □

Exercise 6.16. Prove that distinct monomials in $A^2(B_n)$ are orthogonal.

Exercise 6.17. For the unit ball, prove that $B(z,\overline{w}) = \sum_\alpha c_\alpha (z\overline{w})^\alpha$. Here c_α^2 is the reciprocal of the squared L^2 norm of z^α, and the sum is taken over all multi-indices α. Explicitly sum the series to obtain the formula for B given in (31).

Exercise 6.18. Prove Lemma 6.3.

Theorem 6.2. *Let $R(z,\overline{z})$ be a real-valued bihomogeneous polynomial. The following are equivalent:*

1) R achieves a positive minimum value on the sphere.

2) There is an integer d such that

$$||z||^{2d}R(z,\overline{z}) = \sum E_{\mu\nu} z^\mu \overline{z}^\nu \tag{32}$$

and $(E_{\mu\nu})$ *is positive definite.*

3) There is an integer d such that the integral operator T_{m+d} *defined by the kernel* $k_d(z,\zeta) = \langle z,\zeta\rangle^d R(z,\overline{\zeta})$ *is a positive operator from* $V_{m+d} \subset A^2(B_n)$ *to itself.*

4) There is an integer d and a holomorphic homogeneous vector-valued polynomial g of degree $m+d$ *such that* $\mathbf{V}(g) = \{0\}$ *and such that* $||z||^{2d}R(z,\overline{z}) = ||g(z)||^2$*. In particular, R is a quotient of squared norms.*

5) Write $R(z,\overline{z}) = ||P(z)||^2 - ||N(z)||^2$ *for holomorphic homogeneous vector-valued polynomials P and N of degree m. Then there is an integer d and a linear transformation L such that the following are true:*

5.1) $I - L^*L$ *is nonnegative definite.*

5.2) $z^{\otimes d} \otimes N = L(z^{\otimes d} \otimes P)$

5.3) $\mathbf{V}(\sqrt{I - L^*L}(z^{\otimes d} \otimes P)) = \{0\}$.

PROOF. We will omit the proof of the equivalence of statement 5). By using Lemma 6.3, it is obvious that statements 2) and 3) are equivalent and that either implies statements 4) and 1). The main point is that 1) implies 2) or 3). We will show that 1) implies 3).

We will write M_g for the operator given by multiplication by g. To show that 1) implies 3) we consider the integral operator T defined on $L^2(B_n)$ whose integral kernel is given by $B(z,\overline{w})R(z,\overline{w})$. We choose a smooth nonnegative function χ such that $\chi(0) > 0$ and χ has compact support. ¿From the explicit form of the Bergman kernel for the ball, we see that $B(z,\overline{w})\chi(w)$ is continuous with compact support and hence is the kernel of a compact operator PM_χ. We write

$$B(z,\overline{w})R(z,\overline{w}) =$$
$$B(z,\overline{w})\left(R(z,\overline{w}) - R(z,\overline{z})\right) + B(z,\overline{w})\left(R(z,\overline{z}) + \chi(w)\right) - B(z,\overline{w})\chi(w))$$

to obtain $T = T_1 + T_2 + T_3$, where the kernels of the operators are given by the corresponding terms. We showed above that T_3 is compact on $L^2(B_n)$. Next we note that T_2 is positive on $A^2(B_n)$. Because P is a self-adjoint projection and because $\chi(z) + R(z,\overline{z})$ is a positive function, for $f \in A^2(B_n)$ we obtain

$$\langle T_2 f, f\rangle_{L^2} = \langle P(M_R + M_\chi)f, f\rangle_{L^2} = \langle (M_R + M_\chi)f, f\rangle \geq c||f||^2_{L^2}.$$

Finally we analyze T_1. Note that

$$R(z,\overline{w}) - R(z,\overline{z}) = \sum c_{ab} z^a(\overline{w}^b - \overline{z}^b).$$

Let L_b denote the bounded operator given by multiplication by $\sum_a c_{ab} z^a$, and let M_b denote multiplication by the monomial $\overline{\zeta}^b$. Then T_3 can be written in the form $T_3 = \sum_b L_b[P, M_b]$. The commutator $[P, M_b]$ is compact and L_b is bounded on L^2. The sum defining T_3 is finite, and hence T_3 is also compact. Thus T is the sum of a compact operator and an operator T_2 that is positive on $A^2(B_n)$. After discarding a finite dimensional subspace of $A^2(B_n)$ it follows that the restriction of T to the complement of that subspace will be positive. Finally we expand $B(z,\overline{w})$ as a series $\sum c_j \langle z, w\rangle^j$, written in terms of the Euclidean inner product. For sufficiently large j, it follows that the operator with kernel $c_j\langle z, w\rangle^j R(z,\overline{w})$ will be positive on V_{m+j}. Thus 1) implies 3), and the main point has been proved. □

This proof uses the Bergman kernel function on the unit ball B_n and facts about compact operators on $L^2(B_n)$, whereas the proof by Quillen uses a priori estimates and Gaussian integrals on all of $\mathbf{C}^n$. In both cases the orthogonality of the monomials makes things much easier. In Theorem 6.6, following [CD2], we reinterpret Theorem 6.2 in terms of isometric embedding for holomorphic line bundles over complex projective space. See [CD3] for a general result about isometric embedding of vector bundles over compact complex manifolds.

For these notes the main point of Theorem 6.2 is the equivalence of 1) and 4). A bihomogeneous polynomial R is positive on the sphere if and only if there is a d and g such that $R = \frac{||g||^2}{||z^{\otimes d}||^2}$ and g vanishes only at the origin. What happens when R is nonnegative on the sphere? Example 6.1 provides two simple counterexamples for nonnegative bihomogeneous polynomials. It is possible for a nonnegative bihomogeneous polynomial to be a quotient of squared norms. See [D11] for an elegant necessary and sufficient condition in the one-dimensional case and [D12] for the general analogue of statement 5). Using blow-ups, Varolin [V] found a decisive necessary and sufficient condition in terms of the holomorphic decomposition of R. In particular he proved that a bihomogeneous polynomial is a quotient of squared norms if and only if its restriction to each rational curve is a good jet; one must slightly generalize the definition (4), by allowing m to be negative, to deal with infinity.

Exercise 6.19. Put $R = (|h_1|^2 - |h_2|^2)^2 + |h_3|^2$, for holomorphic polynomials h_1, h_2, h_3. Give necessary and sufficient conditions on the h_j such that R is a quotient of squared norms. If you cannot find such conditions, give an example where the h_j are linearly independent and R is a quotient of squared norms, and give another example where the h_j are linearly independent and R is not a quotient of squared norms.

To gain some feeling for Theorem 6.2, we next describe a version of it in one real variable which dates back to Poincare. A version due to Polya [HLP] holds in $\mathbf{R}^n$. See [D6] for extensive discussion including a proof of Proposition 6.4 along the lines suggested here. See [R] for extensive discussion of Polya's theorem and related ideas. See [TY] for a proof of Theorem 6.2 using these ideas.

Proposition 6.4. *Let p be a polynomial in one real variable. Then $p(t) > 0$ for all $t \geq 0$ if and only if there is an integer d such that the polynomial given by $(1+t)^d p(t)$ has only positive coefficients. The minimum such d can be arbitrarily large for polynomials of fixed degree.*

PROOF. (sketch) Factor p into linear and quadratic factors. It suffices to verify the conclusion for each factor. The result is trivial for linear factors. It therefore suffices to check it for the quadratic polynomial $p(t) = t^2 + bt + c$. We may assume $b < 0$ and $c > \frac{b^2}{4}$. One can check this special case directly by finding formulas for the coefficients of $(1+t)^d p(t)$ and showing that the coefficients are positive for d large enough. See [D6]. □

Proposition 6.4 is a special case of Theorem 6.2. To see why, replace the variable t in p by $|\frac{z}{w}|^2$ and clear denominators. The result will be a bihomogeneous polynomial R which will be positive away from the origin. Theorem 6.2 guarantees that $(|z|^2+|w|^2)^d R(z, w, \overline{z}, \overline{w})$ will be a polynomial in $|z|^2$ and $|w|^2$ whose coefficients are all positive. Express it in terms of t and the result follows.

Exercise 6.20. Fill in the details of the sketched proof of Proposition 6.4.

Exercise 6.21. Prove that Proposition 6.4 follows from Theorem 6.2. Then find the analogue (Polya's Theorem) of Proposition 6.4 in n real variables. See [HLP] for a beautiful proof of this result.

Exercise 6.22. Show that Proposition 6.4 fails for real-analytic functions on $\mathbf{R}$ that are positive for $t \geq 0$.

Theorem 6.3 provides a nice application of Theorem 6.2. Its proof uses a bihomogenization argument; we therefore first consider the technique of bihomogenization. Let $R(z, \overline{z})$ be Hermitian symmetric and of degree m in z; hence it is also of degree m in $\overline{z}$. Its total degree could be anything from m to $2m$. The bihomogenization r of R is the polynomial $r(z, t, \overline{z}, \overline{t})$ defined by

$$r(z, t, \overline{z}, \overline{t}) = |t|^{2m} R(\frac{z}{t}, \frac{\overline{z}}{\overline{t}}) \tag{33}$$

for $t \neq 0$ and by continuity when $t = 0$. We can recover R from r, because $r(z, 1, \overline{z}, 1) = R(z, \overline{z})$. Note that r is bihomogeneous in the $n+1$ variables (z, t), that it is of degree m in both the unbarred and barred variables in $\mathbf{C}^{n+1}$, and that it is of total degree $2m$. We use the technique of bihomogenization to prove the next beautiful result.

Theorem 6.3. *Let $R(z, \overline{z})$ be a (not necessarily bihomogeneous) real-valued polynomial such that $R(z, \overline{z}) > 0$ on the unit sphere in $\mathbf{C}^n$. Then there is an integer N and a (holomorphic) polynomial mapping $g : \mathbf{C^n} \to \mathbf{C^N}$ such that $R(z, \overline{z}) = ||g(z)||^2$ on the unit sphere.*

PROOF. Suppose that R is of degree m in z. If m is even we leave R alone. If m is odd we multiply R by $||z||^2$, getting a polynomial R' satisfying the same hypotheses as R and agreeing with R on the sphere. We therefore without loss of generality may assume that the degree m of R in z is even. We bihomogenize R, obtaining as above a bihomogeneous polynomial $r(z, t, \overline{z}, \overline{t})$ such that $r(z, 1, \overline{z}, 1) = R(z, \overline{z})$. Then r is bihomogeneous of total degree $2m$, where m is even.

For $C > 0$, define a bihomogeneous polynomial F_C by the formula

$$F_C(z, t, \overline{z}, \overline{t}) = r(z, t, \overline{z}, \overline{t}) + C(||z||^2 - |t|^2)^m. \tag{34}$$

Suppose we can choose C such that F_C is positive on the unit sphere in $n+1$ variables. By Theorem 6.2 there is an integer d and a holomorphic polynomial mapping g for which

$$(||z||^2 + |t|^2)^d F_C(z, t, \overline{z}, \overline{t}) = ||g(z, t)||^2. \tag{35}$$

Set $t = 1$ and then $||z||^2 = 1$ in (35), and use (34). Since $r(z, 1, \overline{z}, 1) = R(z, \overline{z})$, we obtain

$$2^d R(z, \overline{z}) = ||g(z, 1)||^2, \tag{36}$$

which gives the conclusion of the Theorem.

It remains to show that we can choose C sufficiently large to make F_C positive on the unit sphere S in $n+1$ variables; by homogeneity it suffices to find C such that F_C is strictly positive on the sphere S' defined by $||z||^2 + |t|^2 = 2$.

Let ϵ denote the positive minimum of R on the unit sphere in $\mathbf{C}^n$. Thus there is some (z_0, t_0) with $||z_0||^2 = |t_0|^2 = 1$ with $r(z_0, t_0, \overline{z}_0, \overline{t}_0) = \epsilon$. By continuity there is a $\delta > 0$ such that $|\ ||z||^2 - |t|^2\ | < \delta$ implies $r(z, t, \overline{z}, \overline{t}) \geq \frac{\epsilon}{2}$. For $|\ ||z||^2 - |t|^2\ | < \delta$

we then have $F_C \geq \frac{\epsilon}{2}$. Thus $F_C > 0$ when $||z||$ and $|t|$ are approximately equal and $||z||^2 + |t|^2 = 2$, no matter what positive C is chosen.

On the other hand, suppose $|\ ||z||^2 - |t|^2\ | \geq \delta$, and $||z||^2 + |t|^2 = 2$. The function $r(z, t, \overline{z}, \overline{t})$ achieves a minimum η on S'. We then have, because m is even,

$$F_C \geq \eta + C(||z||^2 - |t|^2)^m \geq \eta + C\delta^m. \tag{37}$$

If we choose C large enough, the right-hand side of (37) is positive, and F_C is strictly positive on the set where both $|\ ||z||^2 - |t|^2\ | \geq \delta$ and $||z||^2 + |t|^2 = 2$. Hence F_C is strictly positive on all of S', and by homogeneity also on the unit sphere. Thus Theorem 6.2 applies. □

Exercise 6.23. Give an example of an Hermitian symmetric polynomial of odd degree that is positive on the unit sphere.

Exercise 6.24. Suppose that R is positive on the unit sphere. Show that R must contain some monomial of even degree.

Exercise 6.25. Assume $n = 1$. What is the bihomogenization of $z\overline{z}^4 + z^4\overline{z}$? What is the bihomogenization of $z\overline{z}^4 + z^4\overline{z} + |z|^{10}$?

Exercise 6.26. Look up Herglotz's or Bochner's theorem. Then give a necessary and sufficient condition for a trigonometric polynomial to be nonnegative.

7. Squared norms and proper mappings between balls

Theorem 6.2 has a nice application to proper holomorphic mappings between balls. We show the following in Theorem 6.4. Let $q : \mathbf{C}^n \to \mathbf{C}$ be a holomorphic polynomial that does not vanish on the closed unit ball. Then there is an integer N and a holomorphic polynomial mapping $p : \mathbf{C}^n \to \mathbf{C}^N$ such that $\frac{p}{q}$ is a rational proper mapping between B_n and B_N, and $\frac{p}{q}$ is reduced to lowest terms. The reader should observe that this conclusion is easy when $n = 1$, where one can take $N = 1$ also. Even for $n = 2$ and q of degree two, however, the minimum target dimension N can be arbitrarily large.

Remark 6.2. We want $\frac{p}{q}$ to be in lowest terms, or else we have the trivial example where $p(z) = q(z)(z_1, ..., z_n)$.

We recall a few facts about proper mappings between domains in (perhaps different dimensional) complex Euclidean spaces. Let Ω and Ω' be bounded domains in $\mathbf{C}^n$ and $\mathbf{C}^N$. A holomorphic mapping $f : \Omega \to \Omega'$ is proper if $f^{-1}(K)$ is compact in Ω whenever K is compact in Ω'. In case such a mapping extends to be continuous on $b\Omega$, it will map $b\Omega$ to $b\Omega'$. Let $f : B_n \to B_N$ be a holomorphic mapping between balls. Then f is proper if and only if

$$\lim_{||z||^2 \to 1} ||f(z)||^2 = 1.$$

A proper map $f : B_1 \to B_1$ is necessarily a finite Blaschke product. Thus there are finitely many points a_j in B_1 and positive integer multiplicities m_j such that

$$f(z) = e^{i\theta} \prod_{j=1}^{K} \left(\frac{z - a_j}{1 - \overline{a}_j z}\right)^{m_j}. \tag{38}$$

Note that (38) and the fundamental theorem of algebra show that there is no restriction on the denominator. Every polynomial q that is not zero on the closed ball is a constant times a denominator of the form in (38), and hence q arises as the denominator of a rational function reduced to lowest terms.

When $n = N > 1$, a well-known result of Alexander [A] and Pinchuk [P] states that a proper holomorphic self map of a ball is necessarily an automorphism. In particular the map is a linear fractional transformation and the denominator is $1 - \langle z, a \rangle$. To obtain analogues of finite Blaschke products, one must replace multiplication by the tensor product. Since the tensor product of complex vector spaces of dimensions n and k is of dimension nk, the target dimension increases. It is thus natural to consider proper maps from a given B_n to B_N for all N when asking what denominators are possible. The answer is provided by the following consequence of Theorems 6.2 and 6.3.

Theorem 6.4. *Let $q : \mathbf{C}^n \to \mathbf{C}$ be a holomorphic polynomial, and suppose that q does not vanish on the closed unit ball. Then there is an integer N and a holomorphic polynomial mapping $p : \mathbf{C}^n \to \mathbf{C}^N$ such that $\frac{p}{q}$ is a rational proper mapping between B_n and B_N and $\frac{p}{q}$ is reduced to lowest terms.*

PROOF. The result is trivial when q is a constant. It is easy when $n = 1$. When the degree d of q is positive, we define p by $p(z) = z^d \overline{q}(\frac{1}{z})$ and we easily see that this p works. Such a proof cannot work in higher dimensions! The minimum integer N can be arbitrarily large even when $n = 2$ and the degree of q is also two.

Now assume $n \geq 2$. Suppose that $q(z) \neq 0$ on the closed ball. Let g be an arbitrary polynomial such that q and g have no common factor. Then there is a constant c for which

$$|q(z)|^2 - |cg(z)|^2 > 0 \tag{39}$$

for $||z||^2 = 1$. We set $p_1 = cg$.

By Theorem 6.3 a polynomial $R(z, \overline{z})$ that is positive on the unit sphere agrees with a squared norm of a holomorphic polynomial mapping on the sphere. Therefore there is an integer N and polynomials $p_2, ..., p_N$ such that

$$|q(z)|^2 - |p_1(z)|^2 = \sum_{j=2}^{N} |p_j(z)|^2 \tag{40}$$

on the sphere. It then follows that $\frac{p}{q}$ does the job. □

Related theorems due to Lempert ([Lem1], [Lem2]) and to Løw [Lw] provide conditions under which positive functions on the boundaries of strongly pseudoconvex domains agree with squared norms of holomorphic mappings there. In Lempert's work, the boundary is real-analytic, the given positive function is real-analytic, and the holomorphic mapping takes values in an infinite-dimensional space. In Løw's work, the domain has C^2 boundary, the given positive function is continuous, the mapping takes values in a finite-dimensional space, but it is holomorphic on only the interior of the domain. Theorem 6.3 applies only for the sphere, but it draws the stronger conclusion that the holomorphic mapping is a polynomial.

These ideas show also that one can choose various components p of a proper holomorphic polynomial mapping arbitrarily, assuming only that they satisfy the necessary condition $||p(z)||^2 < 1$ on the sphere.

Theorem 6.5. *Let p be a vector-valued polynomial on $\mathbf{C}^{\mathbf{n}}$ with $||p(z)||^2 < 1$ on the unit sphere. Then there is a positive integer $N-k$ and a polynomial mapping $g : \mathbf{C}^n \to \mathbf{C}^{N-k}$ such that $p \oplus g : \mathbf{C}^n \to \mathbf{C}^N$ defines a proper holomorphic mapping between balls.*

PROOF. Note that $1 - ||p(z)||^2$ is a polynomial that is positive on the sphere. Hence we can find a holomorphic polynomial mapping g such that

$$1 - ||p(z)||^2 = ||g(z)||^2$$

on the sphere. We may assume that not both p and g are constant. Then $p \oplus g$ is a non-constant holomorphic polynomial mapping whose squared norm $||p||^2 + ||g||^2$ equals unity on the sphere. By the maximum principle $p \oplus g$ is the required mapping. □

Exercise 6.27. Consider the polynomial $p_\lambda = \lambda z_1 z_2$. Determine the condition on the constant λ such that p_λ is a component of a proper polynomial mapping between balls. For $\lambda = \sqrt{3}$, show that one can choose $N = 3$ in Theorem 6.5. For each allowable λ, determine the minimum N guaranteed in Theorem 6.5.

8. Holomorphic line bundles

We will reinterpret Theorem 6.2 in terms of line bundles over complex projective space. We refer to [CD3] and [D9] for generalizations to vector bundles over compact complex manifolds and to [V] for developments regarding singular metrics.

We first consider complex projective space $\mathbf{P_N}$, the complex manifold of lines through the origin in $\mathbf{C^{N+1}}$. Let $(z_0, ..., z_N)$ be a point in $\mathbf{C}^{N+1}$. If $z \neq 0$, then we identify it as usual with the line $t \to tz$. The standard open covering $\mathcal{U}$ of $\mathbf{P_N}$ is given for $0 \leq j \leq N$ by the open sets U_j, where U_j is the set of lines tz for which $z_j \neq 0$. Let $\mathcal{O}(-m)$ denote the m-th power of the universal line bundle over $\mathbf{P_N}$. This bundle is defined by the transition functions $(\frac{z_j}{z_k})^m$. A metric on a line bundle with transition functions g_{jk} consists of a family of positive function f_k defined on U_k such that $f_k = |g_{kj}|^2 f_j$ on $U_j \cap U_k$. We can obtain such metrics from positive bihomogeneous polynomials.

Let R be a bihomogeneous polynomial that is positive away from the origin in $\mathbf{C^{N+1}}$. We can construct a metric from R as follows: In U_j we define f_j by

$$f_j(z, \overline{z}) = \frac{R(z, \overline{z})}{|z_j|^{2m}}. \tag{41}$$

On the overlap $U_j \cap U_k$ these functions transform via the rule

$$f_k = |(\frac{z_j}{z_k})^m|^2 f_j. \tag{42}$$

Since $(\frac{z_j}{z_k})^m$ are the transition functions for $\mathcal{O}(-m)$, the functions f_j determine an Hermitian metric on $\mathcal{O}(-m)$. We call a metric obtained from a bihomogeneous polynomial in this fashion a *special* metric. We write (L, g) when L is a line bundle over $\mathbf{P_N}$ and g is a special metric on L. In case R is only nonnegative we obtain a *degenerate metric* on $\mathcal{O}(-m)$.

We can express the ideas of this Lecture in the language of these special metrics. First we note that the standard Euclidean metric on $\mathcal{O}(-1)$ is the same as the special metric defined by the bihomogeneous polynomial $||\zeta||^2 = \sum_{j=0}^N |\zeta_j|^2$.

Let R be a bihomogeneous polynomial of degree $2m$ on $\mathbf{C}^n$. It defines via (41) a metric on $\mathcal{O}(-m)$ over $\mathbf{P}_{n+1}$ if and only if it is positive as a function away from the origin, that is, if and only if the origin is a strict minimum for R. This metric $(\mathcal{O}(-m), R)$ is the holomorphic pullback of $(\mathcal{O}(-1), ||\zeta||^2)$ over some $\mathbf{P}_N$ if and only if R is a squared norm. Thus $R = ||g||^2$ if and only if

$$(\mathcal{O}(-m), R) = g^*(\mathcal{O}(-1), ||\zeta||^2).$$

Similarly, the d-th tensor power of $\mathcal{O}(-m)$ is a pullback of the universal bundle if and only if $R^d = ||g||^2$ for some d:

$$(\mathcal{O}(-m), R)^{\otimes d} = g^*(\mathcal{O}(-1), ||\zeta||^2).$$

The other positivity conditions have interpretations as well. The Cauchy-Schwarz inequality implies plurisubharmonicity. Since R is bihomogeneous, R is plurisubharmonic if and only $\log(R)$ also is. Logarithmic plurisubharmonicity is equivalent to negativity of the curvature of the bundle, or to pseudoconvexity of the unit disk bundle.

This discussion applies in particular to the function r_a from Proposition 6.3. When $a < 16$, this bihomogeneous polynomial is strictly positive away from the origin, and hence defines a metric on $\mathcal{O}(-4)$ over $\mathbf{P_1}$. By varying the parameter a we see that the various positivity properties of bundle metrics are also distinct. Note by Proposition 6.3 that the Cauchy-Schwarz inequality gives an intermediate condition between being a squared norm and having negative curvature.

We next restate Theorem 6.2 in this language.

Theorem 6.6. *Let $(\mathcal{O}(-m), R)$ denote the m-th power of the universal line bundle over $\mathbf{P_n}$ with special metric defined by R. Then there is an integer d such that $(\mathcal{O}(-m-d), ||z||^{2d} R(z, \overline{z}))$ is a (holomorphic) pullback $g^*(\mathcal{O}(-1), ||L(\zeta)||^2)$ of the standard metric on the universal bundle over $\mathbf{P_N}$. The mapping $g : \mathbf{P_n} \to \mathbf{P_N}$ is a holomorphic (polynomial) embedding and L is an invertible linear mapping.*

$$(\mathcal{O}(-m), R) \otimes (\mathcal{O}(-d), ||z||^{2d}) = (\mathcal{O}(-m-d), ||z||^{2d} R(z, \overline{z})) = (\mathcal{O}(-m-d), ||g(z)||^2)$$

We have the bundles and metrics

$$\pi_1 : (\mathcal{O}(-m), R) \to \mathbf{P_n}$$
$$\pi_2 : (\mathcal{O}(-m-d), ||z||^{2d} R) \to \mathbf{P_n}$$
$$\pi_3 : (\mathcal{O}(-1), ||L(\zeta)||^2) \to \mathbf{P_N}$$

Thus π_1 is not an isometric pullback of π_3, but, for sufficiently large d, π_2 is.

LECTURE 7

Some open problems

We have described how considerations of finite type arose in PDE, and how the language of algebra has been useful for understanding the PDE. Before giving an extensive bibliography on these kinds of problems we mention several open problems.

Open Problem 7.1. (Peak Points) Let Ω be a smoothly bounded pseudoconvex domain, and suppose each boundary point is of finite type. Then, for each point $p \in b\Omega$, there is a holomorphic function f, smooth up to the boundary, and peaking at p.

Problem 1 has been solved by Bedford-Fornaess [BF] in two dimensions. The author believes that the use of pullbacks to holomorphic curves in the definition of finite type could perhaps be modified to reduce the n-dimensional case to the 2-dimensional case. See also [FM] for results on peak points related to estimates for the Bergman projection.

Open Problem 7.2. (Finite ideal-type) It is known ([C2] and [C3]) that finite type is equivalent to subelliptic estimates on $(0,1)$ forms. Call p of *finite ideal-type* if Kohn's algorithm generates the function 1 in finitely many steps. Thus finite ideal-type implies subelliptic estimates. The logical circle is not complete; does finite type imply finite ideal-type? See [N].

Open Problem 7.3. (Global regularity and compactness) Give necessary and sufficient geometric conditions for global regularity and for compactness in the $\overline{\partial}$-Neumann problem.

See [BS], [Ch4],[CF], and [FuS] for surveys and recent progress.

Open Problem 7.4. (Sharp subelliptic estimates) Suppose that a subelliptic estimate holds at p. Can one express the largest possible ϵ in terms of the geometry? If this isn't possible, can we always choose ϵ to be the reciprocal of $B(M,p)$?

Open Problem 7.5. *Hölder estimates.* Extend the results of Fefferman, Kohn, Machedon ([FeK] and [FKM]) to domains of finite type.

Open Problem 7.6. Describe precisely the boundary behavior of the Bergman kernel function at a point of finite type. In particular, describe the Bergman kernel on a finite type domain defined by $||f||^2 < 1$ if f is a holomorphic polynomial mapping. See [BFS] for explicit computations and [NRSW] for what happens near points of finite type in two dimensions.

Open Problem 7.7. Find necessary and sufficient conditions for subellipticity on $(0,q)$ forms on smoothly bounded domains that are not pseudoconvex. See [Ho] for example.

Open Problem 7.8. (See Lecture 6) Give a precise necessary and sufficient condition for a bihomogeneous polynomial to be in $\mathrm{Rad}(\mathcal{P}_\infty)$.

Open Problem 7.9. (See Lecture 6 and [V].) Prove directly (without using blow-ups) that $R \in \mathcal{J}$ if and only if R is a quotient of squared norms.

Open Problem 7.10. Prove that if a CR manifold is pseudoconvex at p, and L is a $(1,0)$ vector field, then $C(L,p) = \mathrm{type}(L,p)$.

See Lecture 3, [D1], [D8], and [Bl] for this problem and related matters. This result is claimed in [Si1] but the proof there is incorrect.

Bibliography

[AM] Agler, Jim, and McCarthy, John E., Pick Interpolation and Hilbert Function Spaces, Graduate Studies in Mathematics, Volume 44, American Mathematical Society, Providence, Rhode Island, 2000.

[A] Alexander, H., Proper holomorphic mappings in C^n. Indiana Univ. Math. J. 26 (1977), no. 1, 137–146.

[BER] Baouendi, M. Salah; Ebenfelt, Peter; Rothschild, Linda Preiss, Real submanifolds in complex space and their mappings, Princeton Mathematical Series, 47. Princeton University Press, Princeton, NJ, 1999.

[Ba1] Barrett, D. E., Irregularity of the Bergman projection on a smooth bounded domain in $\mathbf{C}^2$, Annals of Math (2) 119 (1984), 431-436.

[Ba2] ____, Behavior of the Bergman projection on the Diederich-Fornaess worm, Acta Math 168:1-2 (1992), 1-10.

[BF] Bedford, E. and Fornaess, J. E., A construction of peak functions on weakly pseudoconvex domains, Annals of Math 107(1978), 555-568.

[Be1] Bell, S., Biholomorphic mappings and the $\overline{\partial}$ problem, Annals of Math 114 (1981), 103-113.

[Be2] ____, Mapping problems in complex analysis and the $\overline{\partial}$ problem, Bulletin AMS, 22(2) (1990), 233-259.

[Bl] Bloom, T., On the contact between complex manifolds and real hypersurfaces in $\mathbf{C}^3$, Trans. Am. Math Soc. 263(1981), 515-529.

[BS] Boas, Harold P. and Straube, Emil J., Global regularity of the $\overline{\partial}$-Neumann problem: A survey of the L^2-Sobolev theory, pages 79-111 in Several Complex Variables (M. Schneider and Y. T. Siu, eds.), Math Sci. Res. Inst. Publ. 37, Cambridge Univ. Press, 1999.

[BFS] Boas, Harold P., Fu, Siqi, and Straube, Emil, The Bergman kernel function: explicit formulas and zeroes, Proc. Amer. Math. Soc., Vol. 127 (1999), 805-811.

[Cal] Calabi, Eugenio, Isometric imbedding of complex manifolds, Annals of Math 58 (1953), 1-23.

[C1] Catlin, D., Necessary conditions for subellipticity of the $\overline{\partial}$-Neumann problem, Annals of Math 117(1983), 147-171.

[C2] ____, Boundary invariants of pseudoconvex domains, Annals of Math. 120(1984), 529-586.

[C3] ____, Subelliptic estimates for the $\overline{\partial}$-Neumann problem on pseudoconvex domains, Annals of Math 126(1987), 131-191.

[C4] ____, Global regularity for the $\overline{\partial}$-Neumann problem, pages 39-49 in Complex Analysis of Several Variables (Yum-Tong Siu ed.), Proc. Symp. Pure Math 41, Amer. Math Soc., Providence, RI, 1984.

[CC] Catlin, D. ; Cho, Jae-Seong, personal communication.

[CD1] Catlin, David W. and D'Angelo, John P., A stabilization theorem for Hermitian forms and applications to holomorphic mappings, Math Research Letters 3, 149-166 (1996).

[CD2] Catlin, David W. and D'Angelo, John P., Positivity conditions for bihomogeneous polynomials, Math Research Letters 4, 1-13 (1997).

[CD3] Catlin, David W. and D'Angelo, John P., An isometric imbedding theorem for holomorphic bundles, Math Research Letters 6, 1-18 (1999).

[CNS] Chang, D. C., Nagel, A., and Stein, E. M., Estimates for the $\overline{\partial}$-Neumann problem in pseudoconvex domains of finite type in $\mathbf{C^2}$, Acta Math 169(1992), 153-228.

[CS] Chen, So-Chin and Shaw, Mei-chi, Partial differential equations in complex analysis, Amer. Math Soc./International Press, 2001.

[Cho] Cho, Jae-Seong, An algebraic version of subelliptic multipliers, Michigan Math J., Vol. 54 (2006), 411-426.

[Ch1] Christ, M., Regularity properties of the $\overline{\partial}_b$ operator on weakly pseudoconvex CR manifolds of dimension 3, J. Amer. Math Soc. 1 (1988), 587-646.

[Ch2] ____, Global C^∞ irregularity in the $\overline{\partial}$-Neumann problem for worm domains, J. Amer. Math Soc. 9 (1996), 1171-1185.

[Ch3] ____, Hypoellipticity in the infinitely degenerate regime, Pages 59-84 in Complex Analysis and Geometry, edited by Jeffery D. McNeal, Ohio State Univ. Math Res. Inst. Publ 9, Walter de Gruyter, Berlin, 2001.

[Ch4] ____, Remarks on global irregularity in the $\overline{\partial}$-Neumann problem, pages 161-198 in Several Complex Variables (M. Schneider and Y. T. Siu, eds.), Math Sci. Res. Inst. Publ. 37, Cambridge Univ. Press, 1999.

[CF] Christ, Michael; Fu, Siqi, Compactness in the $\overline{\partial}$-Neumann problem, magnetic Schrödinger operators, and the Aharonov-Bohm effect. Adv. Math. 197 (2005), no. 1, 1–40.

[D1] D'Angelo, J. P., Several Complex Variables and the Geometry of Real Hypersurfaces, CRC Press, Boca Raton, 1992.

[D2] ____, Real hypersurfaces, orders of contact, and applications, Annals of Math 115(1982), 615-637.

[D3] ____, Subelliptic estimates and failure of semi-continuity for orders of contact, Duke Math. J., Vol. 47(1980), 955-957.

[D4] ____, A Bezout type theorem for real hypersurfaces, Duke Math. J., Vol. 50(1983), 197-201.

[D5] ____, Finite type conditions and subelliptic estimates, Pp. 63-78 in Modern Methods in Complex Analysis, Annals of Math Studies 137, Princeton Univ. Press, Princeton, 1995.

[D6] ____, Inequalities from Complex Analysis, Carus Mathematical Monograph No. 28, Mathematics Association of America, 2002.

[D7] ____, Kohn's algorithm for subelliptic multipliers, preprint 2007.

[D8] ____, Iterated commutators and derivatives of the Levi form. 103–110, Lecture Notes in Math., 1268, Springer, Berlin, 1987.

[D9] ____, Proper holomorphic mappings, positivity conditions, and isometric imbedding, J. Korean Math Society, May 2003, 1-30.

[D10] ____, Bordered complex Hessians, Journal of Geometric Analysis, Vol. 11, Number 4 (2001), 559-569.

[D11] ____, An Analogue of Hilbert's Seventeenth Problem in One Complex Dimension, Contemporary Mathematics, Volume 305 (2006), 47-58.

[D12] ____, Complex variables analogues of Hilbert's Seventeenth Problem, International Journal of Mathematics, Vol. 16, No. 6 (2005), 609-627.

[DK] D'Angelo, J. and Kohn, J. J., Subelliptic estimates and finite type, pages 199-232 in Several Complex Variables (M. Schneider and Y. T. Siu, eds.), Math Sci. Res. Inst. Publ. 37, Cambridge Univ. Press, 1999.

[DV] D'Angelo, J. and Varolin, Dror, Positivity conditions for Hermitian symmetric functions, Asian Journal of Mathematics, Volume 7, No. 4, December 2003.

[DFEM] de Fernex, Tommaso; Ein, Lawrence; Mustaţă, Mircea Multiplicities and log canonical threshold. J. Algebraic Geom. 13 (2004), no. 3, 603–615.

[DF1] Diederich K., and Fornaess, J.E., Pseudoconvex domains with real analytic boundary, Annals of Math (2) 107(1978), 371-384.

[DF2] ____, A pseudoconvex domain with nontrivial Nebenhülle, Math Annalen 225:3 (1977), 275-292.

[DF3] ____, A strange bounded smooth domain of holomorphy, Bull. Amer. Math. Soc. 82 (1976), no. 1, 74–76.

[EL] Eisenbud, David; Levine, Harold I., An algebraic formula for the degree of a C^∞ map germ. With an appendix by Bernard Teissier, "Sur une ingalit la Minkowski pour les multiplicits". Ann. Math. (2) 106 (1977), no. 1, 19–44.

[ELSV] Ein, Lawrence; Lazarsfeld, Robert; Smith, Karen E.; Varolin, Dror, Jumping coefficients of multiplier ideals. Duke Math. J. 123 (2004), no. 3, 469–5.

[F] Fediĩ, V. S., On a criterion for hypoellipticity, Math. USSR-Sb. 14 (1971), 15-45.

[Fe1] Fefferman, C., The Bergman kernel and biholomorphic mappings of pseudoconvex domains, Inventiones Math 26 (1974), 1-65.

[Fe2] ____, On Kohn's microlocalization of $\overline{\partial}$ problems, Pp. 119-133 in Modern Methods in Complex Analysis, Annals of Math Studies 137, Princeton Univ. Press, 1995.

[Fe3] ____, Monge-Ampere equations, the Bergman kernel, and geometry of pseudoconvex domains, Annals of Mathematics, 103 (1976), 395-416.

[FeK] Fefferman, C. and Kohn, J. J. , Hölder estimates on domains of complex dimension two and on three-dimensional CR manifolds, *Adv. of Math.*, 69, 1988, 223-303.

[FKM] Fefferman, C. and Kohn, J. J., and Machedon, M., Hölder estimates on CR manifolds with a diagonalizable Levi form, *Adv. of Math.*, 84 (1990), 1-90.

[FK] Folland, G. B., and Kohn, J. J., The Neumann problem for the Cauchy-Riemann complex, *Annals of Math. Studies*, 75, Princeton University Press, 1972.

[FM] Fornaess, John Erik and McNeal, Jeffery D., A construction of peak functions on some finite type domains, Amer. J. Math. 116 (1994), no. 3, 737–755.

[FS] Fornaess, J. E., and Sibony, N., On L^p estimates for $\overline{\partial}$, Pp. 129-163 in Proceedings of Symposia in Pure Mathematics, Vol. 52, Part 3, Several Complex Variables and Complex Geometry, (1991) American Math. Society, Providence, Rhode Island.

[FuS] Fu, S. and Straube, E., Compactness in the $\overline{\partial}$-Neumann problem, Pages 141-160 in Complex Analysis and Geometry, edited by Jeffery D. McNeal, Ohio State Univ. Math Res. Inst. Publ 9, Walter de Gruyter, Berlin, 2001.

[Ful] Fulton, William, Algebraic curves. An introduction to algebraic geometry. Mathematics Lecture Notes Series. W. A. Benjamin, Inc., New York-Amsterdam, 1969.

[Gre] Greene, Robert E., The geometry of complex manifolds: an overview, Pp. 1-22 in Proceedings of Symposia in Pure Mathematics, Vol. 52, Part 2, Several Complex Variables and Complex Geometry, (1991) American Math. Society, Providence, Rhode Island.

[Gr] Greiner, P., Subellipticity estimates of the $\bar{\partial}$-Neumann problem, J. Differential Geometry 9 (1974), 239-260.

[GH] Griffiths, Phillip; Harris, Joseph, Principles of algebraic geometry, Pure and Applied Mathematics, Wiley-Interscience ,John Wiley and Sons, New York, 1978.

[Guan] Guan, Pengfei, Hölder regularity of subelliptic pseudodifferential operators, Duke Math J. 60(1990), 563-598.

[Gun1] Gunning, Robert C., Introduction to holomorphic functions of several variables. Volumes. I,II, III. Function theory. The Wadsworth and Brooks/Cole Mathematics Series, Pacific Grove, CA, 1990.

[Gun2] ____, Lectures on complex analytic varieties: finite analytic mappings. Mathematical Notes, No. 14. Princeton University Press, Princeton, N.J.; University of Tokyo Press, Tokyo, 1974.

[Gun3] ____, Lectures on complex analytic varieties: The local parametrization theorem. Mathematical Notes, Princeton University Press, Princeton, N.J.; University of Tokyo Press, Tokyo, 1970.

[GR] Gunning, Robert C.; Rossi, Hugo, Analytic functions of several complex variables. Prentice-Hall, Inc., Englewood Cliffs, N.J. 1965

[HLP] Hardy, G. H., Littlewood, J. E., and Polya, G., Inequalities, Cambridge Univ. Press, London, 1934.

[He] Herbig, Anne-Katrin, A sufficient condition for subellipticity of the $\bar{\partial}$-Neumann operator. J. Funct. Anal. 242 (2007), no. 2, 337–362.

[Ho] Ho, L., Subelliptic estimates for the $\bar{\partial}$-Neumann problem for $(n-1)$ forms, Trans. Am. Math Soc. 325(1991), 171-185.

[H] Hörmander, L., An introduction to complex analysis in several variables, Van Nostrand, Princeton, 1966.

[IK] Isaev, A. V. and Krantz, S. G., Domains with Non-Compact Automorphism Group: A Survey, Advances in Math 146, 1-38 (1999).

[K1] Kohn, J. J. Harmonic integrals on strongly pseudo-convex manifolds, I, *Annals of Mathematics*, 78, 1963, 112-148.

[K2] ____ Harmonic integrals on strongly pseudo-convex manifolds, II, *Annals of Mathematics*, 79, 1964, 450-472.

[K3] ____ Boundary behavior of $\bar{\partial}$ on weakly pseudo-convex manifolds of dimension two, *Journal of Differential Geometry*, 6, 1972, 523-542.

[K4] ____ Subellipticity of the $\bar{\partial}$-Neumann problem on pseudo-convex domains: sufficient conditions, *Acta Mathematica*, 142, March 1979.

[K5] ____ A Survey of the $\bar{\partial}$-Neumann Problem, *Proc. of Symposia in Pure Math.*, Amer. Math. Soc. 41, 1984, 137-145.

[K6] ____ Hypoellipticity at non-subelliptic points, pages 161-170 in Complex Analysis and Geometry, edited by Jeffery D. McNeal, Ohio State Univ. Math Res. Inst. Publ 9, Walter de Gruyter, Berlin, 2001.

[K7] ____ Hypoellipticity and loss of derivatives, With an appendix by Makhlouf Derridj and David S. Tartakoff, Ann. of Math. (2) 162 (2005), no. 2, 943–986.

[KN] Kohn, J. J. and Nirenberg, L. Non-coercive boundary value problems, *Comm. Pure Appl. Math.*, 18, 1965, 443-492.

[KN2] Kohn, J. J. and Nirenberg, L. A pseudo-convex domain not admitting a holomorphic support function, *Mathematische Annalen*, 201, 1973, 265-268.

[Kr] Krantz, Steven G., Function Theory of Several Complex Variables, Wadsworth and Brooks/Cole, 1992.

[Lem1] Lempert, László, Imbedding Cauchy-Riemann manifolds into a sphere, International Journal of Math 1 (1990), 91-108.

[Lem2] ____, Imbedding pseudoconvex domains into a ball, American Journal of Math 104 (1982), 901-904.

[Lem3] ____, On the boundary behavior of holomorphic mappings, In *Contributions to Several Complex Variables in honor of Wilhelm Stoll*, A. Howard and P. M. Wong, eds. pp. 193-215, Friedr., Vieweg and Sons, Braunschweig/Weisbaden, 1986.

[Lw] Løw, Eric, Embeddings and proper holomorphic maps of strictly pseudoconvex domains into polydiscs and balls, Math Z. 190 (1985), 401-410.

[Mat] Matsumura, Hideyuki, Commutative algebra. Second edition. Mathematics Lecture Note Series, 56. Benjamin/Cummings Publishing Co., Inc., Reading, Mass., 1980.

[Mc1] McNeal, J., Convex domains of finite type, J. Functional Anal., 108 (1992), 361-373.

[Mc2] McNeal, J, Boundary behavior of the Bergman kernel function in $\mathbf{C}^2$, Duke Math J. 58 (1989), 499-512.

[Mc3] McNeal, J, A sufficient condition for compactness of the $\bar{\partial}$-Neumann operator. J. Funct. Anal. 195 (2002), no. 1, 190–205.

[McS] McNeal, J. and Stein, E. M., Mapping properties of the Bergman projection on convex domains of finite type, Duke Math J. 73 (1994), 177-199.

[NRSW] Nagel. A, Rosay, J. P., Stein, E. M., and Wainger, S., Estimates for the Bergman and Szegö kernels in $\mathbf{C}^2$, Annals of Math 129 (1989), 113-149.

[Nar] Narasimhan, Raghavan, Introduction to the theory of analytic spaces. Lecture Notes in Mathematics, No. 25 Springer-Verlag, Berlin-New York 1966.

[N] Nicoara, A., Equivalence of types and Catlin boundary systems, Preprint 2007.

[P] Pinchuk, S. I., Proper holomorphic mappings of strictly pseudoconvex domains, (Russian) Dokl. Akad. Nauk SSSR 241 (1978), no. 1, 30–33.

[PD] Prestel, Alexander and Delzell, Charles N., Positive Polynomials: From Hilbert's 17th Problem to Real Algebra, Springer-Verlag, Berlin, 2001.

[Q] Quillen, Daniel G., On the representation of Hermitian forms as sums of squares, Inventiones math. 5 (1968), 237-242.

[Range] Range, R. Michael, Holomorphic functions and integral representations in several complex variables. Graduate Texts in Mathematics, 108. Springer-Verlag, New York, 1986.

[R] Reznick, Bruce, Some concrete aspects of Hilbert's 17th Problem, Contemporary Mathematics, Volume 253 (2000), 251-272.

[RS] Rothschild, L. P. , and Stein, E. M. , Hypoelliptic operators and nilpotent Lie groups, Acta Math. 137(1976), 247-320.

[Ru] Rudin, Walter, Function Theory in the Unit Ball of $\mathbf{C}^n$, Springer-Verlag, New York, 1980.

[SR] Saint Raymond, Xavier, Elementary Introduction to the Theory of Pseudodifferential Operators, CRC Press, Boca Raton, 1991.

[Shaf] Shafarevich, I. R., Basic Algebraic Geometry, Springer-Verlag, Berlin, 1977.

[Sh] Shaw, M.C., L^2 estimates and existence theorems for the Cauchy-Riemann complex, Inventiones Math 82(1985), 133-150.

[Si1] Sibony, N. Une classe de domaines pseudoconvexes, Duke Math. J. 55 (1987), no. 2, 299–319.

[Si2] ____, Some aspects of weakly pseudoconvex domains, Pp. 199-232 in Proceedings of Symposia in Pure Mathematics, Vol. 52, Part 1, Several Complex Variables and Complex Geometry, (1991) American Math. Society, Providence, Rhode Island.

[Siu1] Fujita's conjecture and multiplier ideal sheaves of Kohn and Nadel, pages 171-191 in Complex Analysis and Geometry, edited by Jeffery D. McNeal, Ohio State Univ. Math Res. Inst. Publ 9, Walter de Gruyter, Berlin, 2001.

[Siu2] Siu, Yum-Tong, Effective Termination of Kohn's Algorithm for Subelliptic Multipliers, preprint, arXiv:0706.4113v1.

[Ta] Taylor, Michael E., Partial Differential Equations: Basic Theory, Springer, New York, 1996.

[Tre] Trepreau, J. M., Sur le prolongement holomorphe des fonctions CR definies sur une hypersurface reelle de classe C^2 dans $\mathbf{C^n}$, Inventiones Math 83(1986), 583-592.

[TY] Wing-Keung To and Sai-Kee Yeung, Effective isometric embeddings for certain Hermitian holomorphic line bundles., J. London Math. Soc. (2), 73(3):607-624, 2006.

[V] Varolin, Dror, Geometry of Hermitian algebraic functions: quotients of squared norms, preprint 2006.

[Wh] Whitney, Hassler, Complex analytic varieties, Addison-Wesley Publishing Co., Reading, Mass.-London-Don Mills, Ont., 1972.

[Yu] Yu, Jiye, Peak functions on weakly pseudoconvex domains, Indiana Univ. Math J., 43(1994), 1271-1295.

Three Variations on a Theme in Complex Analytic Geometry

Dror Varolin

IAS/Park City Mathematics Series
Volume 17, 2008

Three Variations on a Theme in Complex Analytic Geometry

Dror Varolin

The idea of treating polynomials as holomorphic functions with additional restrictions is probably as old as complex analysis itself. The growth condition obeyed by polynomials, which may be expressed as saying that the reciprocal of a polynomial p is uniformly bounded on the complement of any neighborhood of the zeros of p, leads via the Cauchy Estimates (through an intermediate result known as Liouville's Theorem) to a simple and elegant proof of the fundamental theorem of algebra; a proof everyone learns early in a first course in complex analysis.

Another natural growth constraint on holomorphic functions is square-integrability, possibly with respect to a weight. The extraordinary compatibility of square-integrability conditions with algebraic and complex geometry, and especially in connection with birational geometry, is to my mind one of the most beautiful and miraculous symbioses in mathematics. Unfortunately, this relatively simple — if slightly more technical — circle of ideas is often not taught to students of several complex variables or algebraic geometry, with the exceptions frequently perceived as esoteric specialists. I hope these notes contribute to a change in perception, with the long term wish that mathematicians will find complex analytic L^2 methods as useful as I find them, while being only slightly more involved than basic complex analysis.

Some of the earliest fundamental contributions to the L^2 method in several complex variables were due to C. B. Morrey and J. J. Kohn. Kohn [**Kohn-1963, Kohn-1964**] solved Spencer's $\bar{\partial}$-Neumann problem, thereby establishing the Hodge Theorem for bounded strictly pseudoconvex domains in a Hermitian manifold. The extension of Kohn's result to minimal boundary positivity, namely finite type boundary, arose from the work of D'Angelo, who defined finite type [**D'Angelo-1979**] and subsequently established that finite type is an open condition [**D'Angelo-1982**]. The equivalence of subelliptic estimates and finite type was proved for domains with smooth real analytic boundary by Kohn [**Kohn-1979**] using work of Diederich and Fornæss [**Diederich-Fornæss-1978**], and for arbitrary smoothly bounded domains by Catlin [**Catlin-1983, Catlin-1987**]. The notes of D'Angelo in this volume address finite type, subelliptic estimates, and many other ideas.

A number of fundamental lines of inquiry into the L^2 method, from a number of authors, followed Kohn's first papers. Three of these foundational contributions, which start in the mid 1960's and are continuing until now, make up the bulk of

Department of Mathematics, Stony Brook University, Stony Brook NY, 11794-3651
E-mail address: dror@math.sunysb.edu
Partially supported by the NSF

the lectures. Specific references will be furnished shortly. The recent resurgence of interest in these works was certainly initiated and sustained by a vision, and consequently a stream of spectacular results, of J.-P. Demailly and Y.-T. Siu. This stream continues to run, and in fact the banks are overflowing. In these notes we can only give a few examples of the applications of these applications of the L^2 method, since the central goal of the notes is to provide a sort of introduction to L^2 methods in complex analytic and algebraic geometry.

The first foundational result we discuss is the theorem on solution of the $\bar{\partial}$ equation with L^2 estimates, often referred to as Hörmander's Theorem, which was proved independently by Hörmander [**Hörmander-1965**] and by Andreotti-Vesentini [**Andreotti-Vesentini-1965**]. We demonstrate the utility and power of Hörmander's Theorem by using it to prove the Kodaira Embedding Theorem and Nadel's Coherence and Vanishing Theorems.

The second result we discuss is the L^2 extension theorem, versions of which were proved first by Ohsawa and Takegoshi [**Ohsawa-Takegoshi-1987**] and later extended by Ohsawa in a still-continuing series of papers, by Berndtsson [**Berndtsson-1996**], by Manivel [**Manivel-1996**], by McNeal [**McNeal-1996a**], and by Siu [**Siu-1996, Siu-2002**]. Another, very useful version was later proved by Demailly in [**Demailly-2000**]. The proof presented here can be found in [**Varolin-2007**], and is an adaptation of one written by McNeal and me in [**McNeal-Varolin-2007**] to treat a very general collection of extension theorems. We apply the extension theorem to prove Siu's theorem on the deformation invariance of plurigenera [**Siu-1998, Siu-2002**].

The third result is the division theorem of Skoda [**Skoda-1972**]. The proof presented here is a direct adaptation of Skoda's original proof to the setting of so-called essentially Stein manifolds. A beautiful and powerful geometric extension of Skoda's Theorem, due to Skoda and more generally Demailly, can be found in [**Demailly-2001**]. Skoda's Theorem is then used to prove Siu's Theorem on the global generation of multiplier ideal sheaves. Finally we prove Siu's result on the equivalence of the finite generation of the canonical ring and the finite achievement of singularities of a certain metric for the canonical bundle called the Siu metric.

There are other sources where a student can learn some of this material, aside from the original research papers in which it was written. I learned almost all of the material of the present notes from the lectures and papers of Yum-Tong Siu (as well as countless private discussions) and the many notes of Jean-Pierre Demailly. The present notes have significant overlap with small parts of the book [**Demailly-Book**] and the notes [**Demailly-2001**] of Demailly. Much of the geometry of Lecture 0 can be found, with rather different presentation style, in the well-known book [**Griffiths-Harris-1978**].

SUMMATION CONVENTION: Throughout the notes we use the complex version of Einstein's summation convention; we sum over any repeated indices, one upper and the other lower. If one of the indices is conjugated, the other must be conjugated as well.

I owe a debt of gratitude to a number of people. I am indebted to Yum-Tong Siu for his teachings, his generosity with mathematical ideas, and his inspiring and electric personality. I am grateful to Jeff McNeal for giving me the opportunity to lecture to such a large group of very bright students, and for support before, during, after and besides the lectures. John D'Angelo and I have been in contact about PCMI from the early stages of planning, and I have benefited from that contact, as well as from suggestions for the lectures and the notes. Stas Ostrovsky was a wonderful TA at PCMI. Joe Kohn's support during the presentation of my lectures was inspiring and will remain with me long after these notes are published. The participants of the graduate program at PCMI made many helpful comments during and after my lectures. I have tried to incorporate as many of these suggestions as I could. Finally, thanks also go to Catherine Giesbrecht and the PCMI staff for helping to create a wonderful mathematical research and education experience, and to John Polking for making the publication of these notes such a simple matter.

Dror Varolin
Brooklyn, NY

LECTURE 0

Basic notions in complex geometry

This lecture is not part of the course proper. The primary reason for its inclusion is to serve as a basic and rapid review of some of the material that is prerequisite to the course. The reader who finds the concepts of this lecture highly unfamiliar is likely to have a hard time with the rest of the material to be presented in later lectures. A second purpose of this lecture is to fix some notation and convention for what follows.

We assume a basic familiarity with the definition of a holomorphic function of several complex variables, and with some of the most basic properties of these functions, such as the maximum principle, the Cauchy estimates, and other properties that can easily be obtained from the Cauchy Integral Formula.

We also assume familiarity with basic ideas in differential topology, such as the notions of manifolds, vector bundles, differential forms and tensors. We do not assume a knowledge of geometry of Hermitian vector bundles; we will develop what we need in due course.

1. Complex manifolds

1.1. Definition of a complex manifold

Recall that a function of several complex variables is said to be holomorphic if it is holomorphic in each variable separately. Equivalently, the function restricts to a holomorphic function of one complex variable on every complex line.

DEFINITION 1.1. *A complex manifold X is a manifold satisfying the following additional requirement: there is an open cover $\{U_j\}_{j\in\mathscr{J}}$ of X, and homeomorphisms $z_j : U_j \to V_j \subset \mathbb{C}^n$, $j \in \mathscr{J}$, such that for all $j, k \in \mathscr{J}$, the map $z_k \circ z_j^{-1}$ is holomorphic on its domain $z_j(U_j \cap U_k)$.*

Simply stated, one can choose an atlas with holomorphic transition functions. Thus one can define on a complex manifold all objects whose definition is locally invariant under biholomorphic maps.

PROPOSITION 1.2. *Every complex manifold is orientable.*

PROOF. Since the transition functions $\varphi_{\alpha\beta}$ are holomorphic, their real Jacobian is

$$\det D\varphi_{\alpha\beta} = \left|\det\left(\frac{\partial z^i_\alpha}{\partial z^j_\beta}\right)\right|^2,$$

which is positive. Thus the atlas given by holomorphic charts is oriented. □

EXAMPLE (Domain in $\mathbb{C}^n$). Every open subset of $\mathbb{C}^n$ is a complex manifold of dimension n. Indeed, one can cover the domain by a single chart, using the global coordinates that $\mathbb{C}^n$ is already endowed with.

EXAMPLE (Zero sets of non-degenerate holomorphic mappings in $\mathbb{C}^n$). Suppose $f_1, ..., f_k : \mathbb{C}^n \to \mathbb{C}$ are k holomorphic functions. Let

$$Z = \{x \in \mathbb{C}^n \ ; \ f_j(x) = 0 \text{ for all } 1 \le j \le k\}.$$

Assume moreover that the differential 1-forms $df_1, ..., df_k$ are linearly independent at each point of Z. Then the set Z is a complex manifold of dimension $n-k$, as one can easily show from the implicit function theorem applied to the mapping $F = (f_1, ..., f_k) : \mathbb{C}^n \to \mathbb{C}^k$.

REMARK. The implicit function theorem for holomorphic maps follows from the implicit function theorem for smooth maps. More precisely, one uses the smooth implicit function theorem to prove that locally Z is the graph of $n-k$ complex-valued functions $h_1, ..., h_{n-k}$. Because the functions $f_1, ..., f_k$ are holomorphic, the tangent space of Z is invariant under multiplication by $\sqrt{-1}$ (in $\mathbb{C}^n$, with the origin being the origin of the tangent space, which can be arranged after translation). The $\sqrt{-1}$-invariance of the tangent space is equivalent to the statement that the functions $h_1, ..., h_{n-k}$ satisfy the Cauchy-Riemann equations.

REMARK. Closed submanifolds of $\mathbb{C}^n$ are called *Stein manifolds.* The manifolds of this example are also cut out by the minimum number k of functions needed to produce a codimension-k submanifold. Such manifolds, called *complete intersections*, do not constitute all possible closed submanifolds of $\mathbb{C}^n$ for n and k sufficiently large.

EXAMPLE (Projective space). Consider the set $\mathbb{C}^{n+1} - \{0\}$, with coordinates

$$z = (z^0, z^1, ..., z^n).$$

We say that $x \in \mathbb{C}^{n+1} - \{0\}$ is equivalent to $y \in \mathbb{C}^{n+1} - \{0\}$, and write $[x] = [y]$, if there is a complex number λ such that $y = \lambda x$. The relation $[\ ]$ is clearly an equivalence relation. We define the projective space $\mathbb{P}_n$ to be the set of all equivalence classes of $[\ ]$. The map

$$[\ ] : \mathbb{C}^{n+1} - \{0\} \to \mathbb{P}_n$$

induces a topology on $\mathbb{P}_n$: we say that $U \subset \mathbb{P}_n$ is open if and only if $[\ ]^{-1}(U)$ is open.

Observe that (the complement of) the zero set of a homogeneous polynomial on $\mathbb{C}^{n+1} - \{0\}$ gives rise to a well defined subset of $\mathbb{P}_n$. We thus define the subsets

$$U_i := \{[x] \in \mathbb{P}_n \ ; \ z^i(x) = x^i \neq 0\}, \quad i = 0, ..., n.$$

The map

$$\varphi_i : U_i \ni [x^0, ..., x^n] \mapsto \frac{1}{x^i}(x^0, ..., x^{i-1}, x^{i+1}, ..., x^n) \in \mathbb{C}^n$$

is clearly one-to-one and onto. In other words

$$\zeta_i^\mu = \begin{cases} \frac{z^\mu}{z^i} & \mu < i \\ \\ \frac{z^{\mu+1}}{z^i} & \mu \ge i \end{cases}$$

are coordinates on U_i, and $\varphi_i = (\zeta_i^1, \ldots, \zeta_i^n)$.

Now suppose we want to pass from $\varphi_i(U_i \cap U_j)$ to $\varphi_j(U_i \cap U_j)$. We consider the case $i < j$; the case $i \geq j$ is similar. Then one easily computes the following.

$$\zeta_i^\mu = \tfrac{z^j}{z^i}\zeta_j^\mu = \zeta_j^\mu/\zeta_j^i \qquad \mu < i \text{ or } \mu \geq j$$

$$\zeta_i^\mu = \tfrac{z^j}{z^i}\zeta_j^{\mu+1} = \zeta_j^{\mu+1}/\zeta_j^i \qquad i \leq \mu < j.$$

Thus $\varphi_i \circ \varphi_j^{-1}$ is a holomorphic diffeomorphism, and we have shown that $\mathbb{P}_n$ is a complex manifold of dimension n.

Example (Zero locus of homogeneous polynomial mapping). Suppose we have homogeneous holomorphic polynomials $F_1, ..., F_k : \mathbb{C}^{n+1} \to \mathbb{C}$, possibly with different degrees of homogeneity. On the set $[\ ]^{-1}(U_j)$, we may write

$$F_\ell = (z^j)^{d_\ell} f_{\ell,j}(\zeta_j), \quad 1 \leq \ell \leq k.$$

It follows that on $\mathbb{C}^{n+1} \cap [\ ]^{-1}(U_j) \cap \{F_\ell = 0\}$,

$$dF_\ell = (z^j)^{d_\ell} df_{j,\ell}(\zeta_j).$$

Let Z denote the common zero locus of $F_1, ..., F_k$ in $\mathbb{C}^{n+1} - \{0\}$ and $[Z]$ the image of Z in $\mathbb{P}_n$. Then the calculation above shows that $[Z]$ is a smooth submanifold of $\mathbb{P}_n$ if and only if Z is a smooth submanifold of $\mathbb{C}^{n+1} - \{0\}$. Both manifolds have codimension k. Thus submanifolds of $\mathbb{P}_n$ can be defined as the common zero loci of homogeneous polynomials on $\mathbb{C}^{n+1}$ whose differentials are independent on the part of this zero locus that lies away from the origin.

Remark. Note that we must remove the origin. For example, in $\mathbb{C}^2$ the common zero set of $z^0 = z^1 = 0$ is not a manifold; this set is a pair of transverse lines through the origin, and is therefore not smooth at the origin. On the other hand, the induced set in $\mathbb{P}_1$ is a pair of distinct points.

Not all codimension-k submanifolds of $\mathbb{P}_n$ are complete intersections, i.e., cut out by exactly k homogeneous polynomials. Nevertheless, it is a fact that all submanifolds of $\mathbb{P}_n$ are cut out by homogeneous polynomials.

Remark. Submanifolds of $\mathbb{P}_n$ are called projective manifolds. Later on we will prove Kodaira's Embedding Theorem, which characterizes all projective manifolds as precisely those complex manifolds that support a holomorphic line bundle having a smooth metric of positive curvature.

Example (Complex Tori). Let Λ be a lattice in $\mathbb{C}^n$ (Recall that a lattice in a real vector space is a collection of vectors that is closed under addition and whose convex hull is the whole vector space.) We say that $x \in \mathbb{C}^n$ is equivalent to $y \in \mathbb{C}^n$, and write $x \sim y$, if there exists $\lambda \in \Lambda$ such that $y = x + \lambda$. The set of equivalence classes is denoted $\mathbb{T}^n(\Lambda)$, and we have a map

$$\pi : \mathbb{C}^n \to \mathbb{T}^n(\Lambda)$$

sending x to its equivalence class. Again we define a topology on $\mathbb{T}^n(\Lambda)$ by requiring π to be continuous.

We define coordinate charts on $\mathbb{T}^n(\Lambda)$ as follows. Let $x \in \mathbb{C}^n$ and let U be a neighborhood of x such that for all

$$U \cap (U + \lambda) = \emptyset \text{ for all } \lambda \in \Lambda - \{0\}.$$

Then $\pi|_U$ is 1-to-1, and is clearly a homeomorphism. We let

$$\varphi_U := (\pi|_U)^{-1} : \pi U \to U$$

be our coordinate neighborhood. Clearly the set of all such coordinate charts covers $\mathbb{T}^n(\Lambda)$, and if any two such charts intersect on $\mathbb{T}^n(\Lambda)$, then their images in $\mathbb{C}^n$ intersect after a translation. Thus the transition functions are holomorphic.

DEFINITION 1.3. *The manifold $\mathbb{T}^n(\Lambda)$ is called a complex torus.*

It is easy to see that complex tori are all diffeomorphic to the Cartesian product of $2n$ circles. However, among all of these mutually diffeomorphic manifolds there are many different complex structures. For example, it turns out that some complex tori can be realized as submanifolds of projective space, but that most tori cannot. (Tori that are projective manifolds are called *Abelian varieties.*) In fact, the generic torus does not even have proper complex submanifolds.

Though complex tori appear as fundamental objects in complex and algebraic geometry, they will not figure prominently in the present notes.

EXAMPLE (Matrix groups). Consider the set $GL(n, \mathbb{C})$ of $n \times n$ matrices with complex entries, whose determinant is non-zero. Clearly $GL(n, \mathbb{C})$ is an open subset of the set of $n \times n$ matrices, and the latter may be identified with $\mathbb{C}^{n^2}$. Thus $GL(n, \mathbb{C})$ is a complex manifold.

By using the implicit function theorem, one can show that the subgroups

$$SL(n, \mathbb{C}) = \{A \in GL(n, \mathbb{C}) \ ; \ \det A = 1\},$$

$$Sp(2n, \mathbb{C}) = \{A \in GL(2n, \mathbb{C}) \ ; \ AJ + JA = 0\} \quad J = \begin{pmatrix} 0 & -Id \\ Id & 0 \end{pmatrix}$$

and

$$O(n, \mathbb{C}) = \{A \in GL(n, \mathbb{C}) \ ; \ A^{\mathrm{T}} A = I\}$$

are closed complex submanifolds of $GL(n, \mathbb{C})$. There are other such closed complex subgroups. On the other hand, the unitary group

$$\mathbb{U}(n) = \{A \in GL(n, \mathbb{C}) \ ; \ A^{\dagger} A = I\}$$

is not a complex submanifold. Indeed, its tangent space at the identity is the subspace of $n \times n$ matrices ξ satisfying

$$\xi^{\dagger} = -\xi,$$

and this subspace is not invariant under multiplication by $\sqrt{-1}$.

1.2. Almost complex structure

1.2.1. Almost complex manifolds.

DEFINITION 1.4. *An almost complex structure on a manifold M is a section J of the vector bundle $Hom(T_M, T_M)$, such that $J^2 = -I$.*

PROPOSITION 1.5. *If a manifold M has an almost complex structure, then the dimension of M is even.*

PROOF. Let m be the dimension of M. Looking at a fixed tangent space, we see that $J : T_{M,p} \to T_{M,p}$ is an invertible linear transformation. We then calculate that

$$(\det J)^2 = \det(J^2) = \det(-I) = (-1)^m.$$

Thus m is even. □

As the name suggests, every complex manifold X admits an (almost) complex structure. Indeed, if we choose local holomorphic coordinates $z = (z^1, ..., z^n)$ on an open subset $U \subset X$ and write $z^i = x^i + \sqrt{-1}y^i$, then the endomorphism $J_o : T_X|_U \to T_X|_U$ defined by

$$J_o \frac{\partial}{\partial x^i} = \frac{\partial}{\partial y^i} \quad \text{and} \quad J_o \frac{\partial}{\partial y^i} = -\frac{\partial}{\partial x^i}, \quad 1 \leq i \leq n$$

satisfies $J_o^2 = -I$. We leave it to the reader to check that J_o is well-defined precisely because the changes of coordinates we are using satisfy the Cauchy-Riemann equations. Note that the almost complex structure J_o has the additional property that in some coordinate chart it is locally constant.

DEFINITION 1.6. *An almost complex structure J on a manifold M is said to be integrable if there is an atlas for M such that the local expression for J with respect to any chart in this atlas is constant, i.e., with respect to a local coordinate system $x = (x^1, ..., x^{2n})$,*

$$J \frac{\partial}{\partial x^j} = a^i_j \frac{\partial}{\partial x^i}$$

for some constant matrix $A = (a^i_j)$.

It is an exercise in linear algebra to show that if an almost complex structure is integrable, then the underlying manifold has an atlas with holomorphic transition functions, i.e., it is a complex manifold.

1.2.2. **The splitting.** The purpose of the endomorphism J is to act as multiplication by $\sqrt{-1}$ on the tangent spaces. We now make this idea more precise.

All of the eigenvalues of an almost complex structure J are $\pm\sqrt{-1}$. Indeed, if v is an eigenvector with eigenvalue λ, then

$$-v = J^2 v = \lambda^2 v.$$

Thus to diagonalize J, we must complexify T_M. Moreover, J is a real linear operator on each tangent space, and thus every eigenvalue is the complex conjugate of some other eigenvalue. It follows that $T_M \otimes \mathbb{C}$ splits as a direct sum of the two (isomorphic) eigenbundles $T^{1,0}_M$ and $T^{0,1}_M$ for the eigenvalues $\sqrt{-1}$ and $-\sqrt{-1}$, respectively:

$$T_M \otimes \mathbb{C} = T^{1,0}_M \oplus T^{0,1}_M.$$

The operation of complex conjugation maps $T^{1,0}_M$ to $T^{0,1}_M$ and vice versa. Since conjugation is an involution, we see that $T^{1,0}_M$ and $T^{0,1}_M$ are isomorphic as real vector bundles.

Next, we note that the composite map

$$s^{1,0} : T_M \hookrightarrow T_M \otimes \mathbb{C} = T^{1,0}_M \oplus T^{0,1}_M \to T^{1,0}_M,$$

where the second arrow is the projection onto the first factor, is an isomorphism sending J to multiplication by $\sqrt{-1}$. We claim that for a real vector v, $s^{1,0}v = \frac{1}{2}(v - \sqrt{-1}Jv)$. Indeed, the first arrow $\hookrightarrow$ says we should now think of v as a complex vector that happens to lie in the generating real subspace. Then v decomposes uniquely as

$$v = v^{1,0} + v^{0,1}$$

with $Jv^{1,0} = \sqrt{-1}v^{1,0}$ and $Jv^{0,1} = -\sqrt{-1}v^{0,1}$. But as $\bar{v} = v$, we see that $v^{0,1} = \overline{v^{1,0}}$, and thus $v = v^{1,0} + \overline{v^{1,0}}$. On the other hand, $Jv = \sqrt{-1}(v^{1,0} - \overline{v^{1,0}})$. We find that

$$s^{1,0}v = \tfrac{1}{2}(v - \sqrt{-1}Jv),$$

as claimed. (Note that equivalently, $(s^{1,0})^{-1} = 2\mathrm{Re}$.)

EXAMPLE. For the almost complex structure J_o defined on a complex manifold X, a choice of holomorphic coordinates $z = (z^1, ..., z^n)$ shows that bases for $T_X^{1,0}$ and $T_X^{0,1}$ are respectively given by

$$\frac{\partial}{\partial z^1}, ..., \frac{\partial}{\partial z^n} \quad \text{and} \quad \frac{\partial}{\partial \bar{z}^1}, ..., \frac{\partial}{\partial \bar{z}^n},$$

where as usual,

$$\frac{\partial}{\partial z^j} = \frac{1}{2}\left(\frac{\partial}{\partial x^j} - \sqrt{-1}\frac{\partial}{\partial y^j}\right),$$

and $\frac{\partial}{\partial \bar{z}^j}$ is the complex conjugate of $\frac{\partial}{\partial z^j}$.

The splitting $T_M \otimes \mathbb{C} = T_M^{1,0} \oplus T_M^{0,1}$ induces a splitting on all other vector bundles obtained from M by multilinear operations, such as the cotangent bundle, the various other exterior bundles, the tensor bundles, etc. Below we look carefully at the case of differential forms, and discuss the convention of so-called (p, q)-forms.

1.2.3. A few remarks on integrability. We begin with the following definition.

DEFINITION 1.7. *An almost complex manifold (M, J) is said to be involutive if the set of local sections of the subbundle $T_M^{1,0}$, also called $(1, 0)$ complex vector fields, is closed under (the complexification of) Lie brackets, i.e.,*

$$[\Gamma(M, T_M^{1,0}), \Gamma(M, T_M^{1,0})] \subset \Gamma(M, T_M^{1,0}).$$

A famous result of Newlander-Nirenberg states that a sufficiently regular almost complex manifold (M, J) is integrable if and only if it is involutive. If M has a real-analytic structure and the almost complex structure is real analytic with respect to this structure, then the Newlander-Nirenberg Theorem reduces to a complexified version of the Fröbenius Theorem on integrability of tangent distributions.

If M is a surface (i.e. $\dim_{\mathbb{R}}(M) = 2$) then every almost complex structure is easily seen to be involutive, and therefore integrable by Newlander-Nirenberg. This case of the Newlander-Nirenberg Theorem predates the higher dimensional case, and was first proved by Korn and Lichtenstein. A more elementary proof was given by S.-S. Chern.

In 1-dimensional complex dynamics, one makes use of much less regular almost complex structures. One requires only measurability of the almost complex structure, but there is an additional 'eccentricity' hypothesis. This less regular result was proved by Ahlfors and Bers.

Although we will not do so, the methods that will be developed in this course can be easily modified to give a proof, due to Kohn, of the Newlander-Nirenberg Theorem. The Ahlfors-Bers result requires deeper harmonic analysis, namely Calderon-Zygmund Theory, and still lacks an appropriate analog in higher dimensions.

1.3. Differential (p, q)-forms on complex manifolds

Let X be a complex manifold. The splitting of the complexified tangent bundle $T_X \otimes \mathbb{C}$ obtained from the almost complex structure of X induces a splitting on any bundle obtained from the tangent bundle by an operation from multi-linear algebra. We now elaborate on this splitting when the space in question is $\Lambda^r(T_X^*)$.

The splitting $T_X^* \otimes \mathbb{C} = T_X^{*1,0} \oplus T_X^{*0,1}$ induces the splitting

$$\Lambda^r(T_X^*) = \bigoplus_{p+q=r} \Lambda^{p,q}(T_X^*),$$

where

$$\Lambda^{p,q}(T_X^*) := \underbrace{T_X^{*1,0} \bigwedge \cdots \bigwedge T_X^{*1,0}}_{p \text{ copies}} \bigwedge \underbrace{T_X^{*0,1} \bigwedge \cdots \bigwedge T_X^{*0,1}}_{q \text{ copies}}.$$

The sections of $\Lambda^{p,q}(T_X^*)$ are called differential forms of bi-degree (p,q), or simply (p,q)-forms.

REMARK. Although (p,q)-forms are defined for almost complex manifolds in the obvious way, we will stick to the case of complex manifolds, since this is the only case we will use. The modification to the almost complex case is not difficult, and is left to the interested reader.

In terms of a local coordinate system z, the forms

$$dz^I \wedge d\bar{z}^J = dz^{i_1} \wedge \ldots \wedge dz^{i_p} \wedge d\bar{z}^{j_1} \wedge \ldots \wedge d\bar{z}^{j_q}, \tag{0.1}$$

where $I = (i_1, \ldots, i_p) \in \{1, \ldots, n\}^p$ and $J = (j_1, \ldots, j_q) \in \{1, \ldots, n\}^q$, thus provide a local basis of sections for $\Lambda^{p,q}(T_X^*)$. In particular, one can write any (p,q)-form α as

$$\alpha = \alpha_{I\bar{J}} dz^I \wedge d\bar{z}^J,$$

where the collection of functions $\alpha_{I\bar{J}}$ is skew symmetric with respect to the re-arrangement of the indices $i_1, \ldots, i_p$ and $j_1, \ldots, j_q$.

Using the sections (0.1) to locally trivialize $\Lambda^{p,q}(T_X^*)$ over each coordinate chart of a complex atlas, one can use the chain rule to compute the transition functions for $\Lambda^{p,q}(T_X^*)$.

1.3.1. **The exterior operators.** We define the operators ∂ and $\bar{\partial}$ on local sections of the complexified cotangent bundle by

$$\partial = \Pi^{1,0} \circ d \qquad \text{and} \qquad \bar{\partial} := \Pi^{0,1} \circ d,$$

where $\Pi^{1,0}$ and $\Pi^{0,1}$ are the projections from $T_X^* \otimes \mathbb{C}$ to $(T_X^*)^{1,0}$ and $(T_X^*)^{0,1}$ respectively, and by d we actually mean the complexification of d. Similarly, one has projections $\Pi^{p,q} : \Lambda^{p+q}(T_X^* \otimes \mathbb{C}) \to \Lambda^{p,q}(T_X^*)$, and we may define ∂ and $\bar{\partial}$ on (p,q)-forms respectively by the formulas

$$\partial = \Pi^{p+1,q} \circ d \quad \text{and} \quad \bar{\partial} = \Pi^{p,q+1} \circ d.$$

On a complex manifold X, it is easy to give a formula for the action of ∂ and $\bar{\partial}$. We leave it to the reader to show that

$$\partial(f dz^\alpha \wedge d\bar{z}^\beta) = \frac{\partial f}{dz^j} dz^j \wedge dz^\alpha \wedge d\bar{z}^\beta \quad \text{and}$$

$$\bar{\partial}(f dz^\alpha \wedge d\bar{z}^\beta) = \frac{\partial f}{d\bar{z}^k} d\bar{z}^k \wedge dz^\alpha \wedge d\bar{z}^\beta.$$

REMARK (The Fröbenius condition in terms of $\bar{\partial}$). Let (M, J) be an almost complex manifold and let $\bar{\partial}_J$ be the associated $(0,1)$-exterior operator acting on 1-forms. Then $T_M^{1,0}$ is closed under Lie brackets if and only if

$$\bar{\partial}_J \bar{\partial}_J = 0.$$

Note that if f is a function, then for any almost complex structure J, whether integrable or not,

$$df = \partial_J f + \bar{\partial}_J f.$$

However, if α is a k-form for $k \geq 1$, then in general the equality

$$d\alpha = \partial_J \alpha + \bar{\partial}_J \alpha$$

does not hold. (We leave it to the reader to show that the relation $df = \partial_J f + \bar\partial_J f$ holds for k-forms with $k \ge 1$ if and only if it holds for 1-forms.)

1.4. Hermitian manifolds

Let X be a complex manifold and denote by J_o the almost complex structure induced by the complex structure of X.

1.4.1. Definition of Hermitian metric.

DEFINITION 1.8. *A Hermitian metric on X is a Riemannian metric h on X such that $J_o^* h = h$.*

1.4.2. Complex form for Hermitian metric. The condition of J_o-invariance puts strong restrictions on the form of the metric; conditions that are best seen in the complexified tangent space. In terms of the splitting $T_X \otimes \mathbb{C} = T_X^{1,0} \oplus T_X^{0,1}$, let us write (the complexification of) our metric as

$$h = \begin{pmatrix} A & B \\ C & D \end{pmatrix},$$

i.e.,

$$h((v_1, \bar w_1), (v_2, \bar w_2)) = v_1 \cdot A v_2 + v_1 \cdot B \bar w_2 + \bar w_1 \cdot C v_2 + \bar w_1 \cdot D \bar w_2.$$

We shall now analyze the matrix $\left(\begin{smallmatrix} A & B \\ C & D \end{smallmatrix}\right)$.

First, if $v, w \in T_X^{1,0}$, then

$$\begin{aligned} h((v,0),(w,0)) &= h(J_o(v,0), J_o(w,0)) \\ &= h(\sqrt{-1}(v,0), \sqrt{-1}(w,0)) \\ &= -h((v,0),(w,0)), \end{aligned}$$

and thus $A = 0$. Similarly, $D = 0$.

Now, even though h is a complex metric, it is just the complexification of a real metric, and thus remains invariant under conjugation. We thus have

$$\bar v \cdot \overline{B} w = \overline{h((v,0),(0,\overline{w}))} = h((0,\overline{v}),(w,0)) = \bar v \cdot C w.$$

It follows that

$$\overline{B} = C.$$

Finally, since h is symmetric, $B^T = C = \overline{B}$, and therefore $B^\dagger = B$. It is standard to write

$$h = (h_{\alpha\bar\beta}),$$

in place of

$$h = \begin{pmatrix} 0 & (h_{\alpha\bar\beta}) \\ \overline{(h_{\alpha\bar\beta})} & 0 \end{pmatrix}.$$

Note that the object $(h_{\alpha\bar\beta})$ is actually a fiberwise Hermitian metric for the complex vector bundle $T_X^{1,0} \to X$, which is one reason for using this notion for a J_o-invariant Riemannian metric.

REMARK. Another reason for the notation $h = (h_{\alpha\bar{\beta}})$ comes from the previously discussed identification of the $(1,0)$-tangent space with the real tangent space. Indeed, for real vectors v and w, one has

$$\begin{aligned} 2h(s^{1,0}x, \overline{s^{1,0}y}) &= 2h(\tfrac{1}{2}(x - \sqrt{-1}J_o x), \tfrac{1}{2}(y + \sqrt{-1}J_o y)) \\ &= \frac{1}{2}\left(h(x,y) + \sqrt{-1}h(x, J_o y) - \sqrt{-1}h(J_o x, y) + h(J_o x, J_o y)\right) \\ &= h(x,y) + \sqrt{-1}h(x, J_o y), \end{aligned}$$

so that

$$h(x,y) = 2\mathrm{Re}\ h(s^{1,0}x, \overline{s^{1,0}y}).$$

We also define

$$\omega(x,y) := \frac{1}{2}h(x, J_o y).$$

Note that

$$\omega(y,x) = h(y, J_o x) = h(J_o y, J_o^2 x) = -h(J_o y, x) = -\omega(x,y),$$

so that ω is a 2-form. But in fact, the relation

$$\omega(x,y) = \frac{1}{2}\mathrm{Im}\ h(s^{1,0}x, s^{0,1}y)$$

shows that ω is a $(1,1)$-form.

One can also approach the description of a Hermitian metric as follows. The complexification of a real metric is a section of the bundle

$$\mathrm{Sym}^2(\mathbb{C} \otimes T_X^*) := \mathbb{C} \otimes T_X^* \otimes T_X^* / \sim,$$

where $\sim$ is the linear equivalence relation defined by

$$a \otimes b \sim b \otimes a.$$

If we denote the image of $a \otimes b$ in $\mathrm{Sym}^2(\mathbb{C} \otimes T_X^*)$ by $a \cdot b$, then our work above just says that, locally, one can write

$$h = h_{\alpha\bar{\beta}} dz^\alpha \cdot d\bar{z}^\beta$$

for some positive definite Hermitian symmetric matrix

$$\left(h_{\alpha\bar{\beta}}\right)_{\alpha,\beta=1}^n.$$

Since the metric $h = h(z)_{\alpha\bar{\beta}} dz^\alpha \cdot d\bar{z}^\beta$ is a global object, if we change coordinates then, with $w = w(z)$, we have

$$h(z)_{\alpha\bar{\beta}} = h(w)_{\gamma\bar{\delta}} \frac{\partial w^\gamma}{\partial z^\alpha} \overline{\left(\frac{\partial w^\delta}{\partial z^\beta}\right)}.$$

It follows that the $(1,1)$-form

$$\omega = \frac{\sqrt{-1}}{2} h_{\alpha\bar{\beta}} dz^\alpha \wedge d\bar{z}^\beta$$

is well defined. The $(1,1)$-form ω is called the metric form associated to the Hermitian metric h.

1.4.3. Volume form associated to a Hermitian metric. Recall that in Riemannian geometry, one defines a volume form associated to a Riemannian metric on an orientable manifold: if the metric is given by (g_{ij}) in the charts of an oriented atlas, then the volume is

$$\Omega = \sqrt{\det(g_{ij})} dx^1 \wedge ... \wedge dx^m.$$

If we apply this formula to the Hermitian metric h on a complex manifold X, we obtain the formula

$$\Omega = \frac{1}{n!}\omega^n.$$

Indeed, one has

$$\det\left((h_{ij})_{i,j=1}^{2n}\right) = (-1)^n(\det(h_{\alpha\bar{\beta}}))^2,$$

and thus

$$\begin{aligned}
\omega^n &= \frac{\sqrt{-1}^n}{2^n} h_{\alpha_1\bar{\beta}_1}...h_{\alpha_n\bar{\beta}_n} dz^{\alpha_1} \wedge d\bar{z}^{\beta_1} \wedge ... \wedge dz^{\alpha_1} \wedge d\bar{z}^{\beta_1} \\
&= \frac{(-1)^{n^2/2}}{2^n} h_{\alpha_1\bar{\beta}_1}...h_{\alpha_n\bar{\beta}_n} dz^{\alpha_1} \wedge ... \wedge dz^{\alpha_n} \wedge d\bar{z}^{\beta_1} \wedge ... \wedge d\bar{z}^{\beta_n} \\
&= \left(\sum_{\alpha_1,...,\bar{\beta}_n} \operatorname{sgn}\begin{pmatrix} 1 & ... & n & \bar{1} & ... & \bar{n} \\ \alpha_1 & ... & \alpha_n & \bar{\beta}_1 & ... & \bar{\beta}_n \end{pmatrix} h_{\alpha_1\bar{\beta}_1}...h_{\alpha_n\bar{\beta}_n}\right) \\
&\qquad \times \frac{(-1)^{n^2/2}}{2^n} dz^1 \wedge ... \wedge dz^n \wedge d\bar{z}^1 \wedge ... \wedge d\bar{z}^n \\
&= n! \det(h_{\alpha\bar{\beta}}) \frac{d\bar{z}^1 \wedge dz^1}{2\sqrt{-1}} \wedge ... \wedge \frac{d\bar{z}^n \wedge dz^n}{2\sqrt{-1}}.
\end{aligned}$$

REMARK. It is useful that the computation of the volume form of a Hermitian manifold involves only the application a multi-linear algebra operation to the metric form.

Let us consider a complex submanifold Y of our Hermitian manifold X. We can endow Y with a Hermitian metric simply by restricting the metric from the ambient space.

Now choose local coordinates $z^1, ..., z^n$ on X in such a way that $z^1, ..., z^k$ are coordinates on the submanifold Y. The tangent spaces of Y are then defined by the vanishing of the differentials $dz^{k+1}, ..., dz^n$. In other words, in these coordinates, the Hermitian metric for Y is given by

$$h_Y = h_{\alpha\bar{\beta}} dz^\alpha \cdot d\bar{z}^\beta,$$

where this time the summation is carried out for α and β running only through 1 to k. Equivalently, metric h_Y is simply the restriction to Y of the metric h, i.e., $h_Y = h|_Y$. In particular, one can carry out all of the above calculations on the submanifold and obtain the following useful proposition.

PROPOSITION 1.9. *If X is a Hermitian manifold with Hermitian metric form ω and Y is a k dimensional complex submanifold endowed with the relative metric, then the associated Hermitian volume form of Y is*

$$\Omega_Y = \frac{1}{k!}(\omega|_Y)^k = \frac{1}{k!}\omega^k|_Y.$$

2. Connections

2.1. Basic definition

Let $\pi : V \to M$ be a vector bundle of rank r, i.e., each fiber is diffeomorphic to $\mathbb{R}^r$. We denote by $\Gamma(M, V)$ the smooth sections of π.

DEFINITION 2.1. *A connection D on $V \to M$ is a linear map*

$$D : \Gamma(M, V) \to \Gamma(M, T^*M \otimes V)$$

that satisfies the Leibniz rule:

$$D(fs) = df \otimes s + fDs.$$

2.1.1. Connection matrix. Recall that a frame for a vector bundle is a collection of sections that define a basis of each fiber of the vector bundle. The local triviality of a vector bundle is equivalent to the existence of a frame locally.

Given a connection D and a local frame $e_1, ..., e_r$ for $V \to M$, the matrix of 1-forms ω_i^j defined by

$$De_i = \omega_i^j e_j$$

is called a connection matrix.

The Leibniz rule affords a full description of D in terms of a connection matrix. Indeed, we have

$$D(s^i e_i) = (ds^i + s^j \omega_j^i) e_i.$$

REMARK. One often sees the notation $Ds = ds + s\omega$.

Next we determine how the connection matrix transforms when we change frames. To this end, a second frame $\tilde{e}_1, ..., \tilde{e}_r$ over $\tilde{U}$ is related to $e_1, ..., e_r$ over $U \cap \tilde{U}$ by a function $G = (g_i^j) : U \cap \tilde{U} \to GL(n)$:

$$\tilde{e}_i = g_i^j e_j, \quad 1 \leq i \leq n.$$

We then have

$$\begin{aligned} D(s^i e_i) &= ds^i e_i + s^i \omega_i^j e_j \\ &= d(\tilde{s}^j g_j^i) e_i + \tilde{s}^j g_j^k \omega_k^\ell (g^{-1})_\ell^i \tilde{e}_i \\ &= d(\tilde{s}^j) \tilde{e}_j + \tilde{s}^j \left(dg_j^k (g^{-1})_k^i + g_j^k \omega_k^\ell (g^{-1})_\ell^i \right) \tilde{e}_i. \end{aligned}$$

Thus, in matrix notation, the transformation rule is

$$\tilde{\omega} = (dG)G^{-1} + G\omega G^{-1}.$$

2.2. Connections with additional symmetry

2.2.1. Metric Compatibility. Let the vector bundle V be endowed with a metric g.

DEFINITION 2.2. *We say that D is compatible with the metric g if*

$$d(g(s,t)) = g(Ds, t) + g(s, Dt).$$

A connection that is compatible with a given metric is by no means unique. Indeed, in terms of some fixed frame, we have

$$d(g_{ij} s^i t^j) = dg_{ij} s^i t^j + g_{ij}(ds^i t^j + s^i dt^j).$$

On the other hand,

$$g(Ds, t) + g(s, Dt) = g_{ij}(ds^i t^j + s^i dt^j) + g_{ij} s^l \omega_l^i t^j + g_{ij} s^i \omega_l^j t^l.$$

Thus, symmetry gives a relation that can be expressed in matrix form as follows:

$$dg - g\omega - \omega^T g = 0. \tag{0.2}$$

The equation (0.2) has many solutions. To get an idea of how many, we observe that all solutions are of the form $\omega = \frac{1}{2}g^{-1}dg + g^{-1}A$, where A is a matrix of 1-forms satisfying $A^T = -A$.

2.2.2. Levi-Čivita connections. Given a Riemannian manifold (M, γ), there is an induced metric g on the vector bundle T^*_M. Let $D : \Gamma(M, T^*_M) \to \Gamma(M, T^*_M \otimes T^*_M)$ be a connection. In view of the splitting

$$T^*_M \otimes T^*_M = S^2(T^*_M) \oplus \Lambda^2(T^*_M),$$

we can write

$$D = D^S + D^\Lambda.$$

REMARK. Note that D^S and D^Λ are themselves not connections.

THEOREM 2.3 (Levi-Čivita). *There exists a unique connection D such that*

$$d(g(s, s')) = g(Ds, s') + g(s, Ds') \quad \textit{and} \quad D^\Lambda = d.$$

In terms of the connection matrix, Levi-Čivita's theorem says that one can choose (in exactly one way) a connection that is compatible with the metric and whose connection matrix has no anti-symmetric part.

In order to prove Levi-Čivita's Theorem, let us examine this condition more carefully using coordinates. On the cotangent bundle, a natural basis comes from a choice of local coordinates $(x^1, ..., x^m)$ on M, namely, $dx^1, ..., dx^m$. Then, writing $\alpha = \alpha_i dx^i$, we put

$$\omega\ \alpha = \omega^i{}_{jk}\alpha_i dx^j \otimes dx^k.$$

Thus Levi-Civita's Theorem states that there is a unique solution $\omega^i{}_{jk}$ to the system of equations

$$\omega^i{}_{jk} = \omega^i{}_{kj} \quad \text{and} \quad g^{ia}\omega^j{}_{ak} + g^{jb}\omega^i{}_{bk} = \partial_k g^{ij}.$$

PROOF OF LEVI-CIVITA'S THEOREM. Things become a little more transparent if we define

$$\omega^{ijk} = g^{ja}g^{kb}\omega^i{}_{ab}.$$

Then the equations we must solve are

$$\omega^{ijk} = \omega^{ikj} \quad \text{and} \quad \omega^{ijk} + \omega^{jik} = g^{k\ell}\partial_\ell g^{ij} =: \partial^k g^{ij}.$$

Since the right hand side of the second equation does not depend on ω^{ijk}, the difference $\theta^{ijk} = \omega^{ijk} - \tilde{\omega}^{ijk}$ of two solutions must satisfy the homogeneous equations

$$\theta^{ijk} = \theta^{ikj} \quad \text{and} \quad \theta^{ijk} + \theta^{jik} = 0.$$

But then

$$\theta^{ijk} = \theta^{ikj} = -\theta^{kij},$$

and thus

$$\theta^{ijk} = -\theta^{kij} = \theta^{jki} = -\theta^{ijk}.$$

Therefore $\theta^{ijk} = 0$.

For the existence, simply take

$$\omega^{ijk} = \frac{1}{2}\left(\partial^k g^{ij} + \partial^j g^{ik} - \partial^i g^{jk}\right).$$

The proof is complete. □

REMARK. By transforming back, and using the rule $\partial_s(g^{ij}g_{jk}) = 0$, it is easy to show that

$$\omega^i{}_{jk} = \frac{1}{2}\left(g_{jm}\partial_k g^{mi} + g_{km}\partial_j g^{mi} + g^{is}\partial_s g_{jk}\right).$$

The numbers $\omega^i{}_{jk}$ are called the *Christoffel symbols* of the connection D, with respect to the chosen coordinates on M.

REMARK. The Riemannian metric sets up a correspondence between the tangent and cotangent bundle. Thus a connection $D : \mathscr{C}^\infty_M(T^*_M) \to \mathscr{C}^\infty_M(T^*_M \otimes T^*_M)$ has associated to it a dual connection $\nabla : \mathscr{C}^\infty_M(T_M) \to \mathscr{C}^\infty_M(T_M \otimes T^*_M)$. It is then possible to show that in the frame $\frac{\partial}{\partial x^1}, ..., \frac{\partial}{\partial x^m}$ the Christoffel symbols for this dual connection are given by

$$\Gamma_i{}^j{}_k = g^{j\ell}\omega^m{}_{\ell k}g_{mi} + g^{j\ell}\partial_k g_{\ell i},$$

and thus, that

$$\Gamma_i{}^j{}_k = \frac{1}{2}g^{j\ell}\left(\partial_i g_{\ell k} + \partial_k g_{\ell i} - \partial_\ell g_{ik}\right).$$

In particular, $\Gamma_i{}^j{}_k = \Gamma_k{}^j{}_i$.

2.2.3. **Hermitian connection.** The notion of a connection on a complex vector bundle is the same as that of a connection on a real vector bundle, except that by *linear* one means *complex linear.* Thus if D is a such a connection, then we have

$$D(cs) = cD(s) \text{ for all } c \in \mathbb{C}.$$

A Hermitian metric on a complex vector bundle is just a continuous family of Hermitian inner products on the fibers.

Given a Hermitian vector bundle $(V, k) \to M$ on a (not necessarily complex) manifold M, one can define a special connection.

DEFINITION 2.4. *A connection D is said to be compatible with the Hermitian metric k if*

$$d(k(v, \bar{w})) = k(Dv, \bar{w}) + k(v, \overline{Dw}).$$

2.2.4. **Chern connection.** On a complex manifold X we have a splitting $T^*_X \otimes \mathbb{C} = T^{*1,0}_X \oplus T^{*0,1}_X$. It follows that for a complex vector bundle $V \to X$, a connection D splits as

$$D = D^{1,0} + D^{0,1}.$$

REMARK. Note that $D^{1,0}$ and $D^{0,1}$ are not connections.

If the vector bundle is also holomorphic, then there is a special differential operator acting on sections, and whose $(1,0)$-part is identically 0. If $e_1, ..., e_r$ is a holomorphic frame, then this differential operator is given by the following formula:

$$s = s^i e_i \mapsto (\bar{\partial} s^i) \otimes e_i.$$

Note that if $\tilde{e}_j$ is another holomorphic frame, then

$$\tilde{e}_j = g^i_j e_i$$

for some holomorphic functions g^i_j, and thus

$$s^j = \tilde{s}^i g^j_i.$$

We thus obtain

$$(\bar\partial \tilde s^i)\otimes \tilde e_i = \bar\partial(\tilde s^j)\otimes g^i_j e_i = \bar\partial(\tilde s^j g^i_j)\otimes \tilde e_i = (\bar\partial s^i)\otimes e_i.$$

Thus it is reasonable (as well as common) to denote this differential operator by $\bar\partial$.

DEFINITION 2.5. *A connection D for a holomorphic vector bundle $V \to X$ is said to be Chern if in terms of the splitting above,*

$$D = D^{1,0} + \bar\partial$$

2.2.5. The fundamental theorem of holomorphic Hermitian geometry. The basic result about connections for holomorphic Hermitian vector bundles is the following analogue of Levi-Čivita's theorem.

THEOREM 2.6. *On a holomorphic Hermitian vector bundle there exists a unique Chern connection compatible with the Hermitian metric.*

PROOF. We begin with uniqueness. Since a connection has the form $D = d+\omega$, the difference of two connections is locally a 0-th order differential operator, i.e., a matrix multiplier. Moreover, if $D^{0,1} = \bar\partial$, this multiplier is a matrix of $(1,0)$-forms.

Suppose, then, that D_1 and D_2 are two Chern connections that are compatible with the given Hermitian metric h. Then writing $\theta = D_2 - D_1$, we have that for all vectors u, v, $h(\theta u, v) + h(u, \theta v) = 0$. Defining $\theta^\dagger$ by $h(\theta^\dagger u, v) = h(u, \theta v)$, we have the equation

$$\theta + \theta^\dagger = 0. \tag{0.3}$$

But $\dagger$ maps matrices of $(1,0)$-forms to matrices of $(0,1)$-forms, and thus $\theta = 0$.

To prove existence, we simply calculate the connection. To this end, let s, t be local sections of $V \to X$. Fixing a frame $e_1, ..., e_r$, we write $s = s^\alpha e_\alpha$, $t = t^\beta e_\beta$ and $h_{\alpha\bar\beta} := h(e_\alpha, e_\beta)$. Then

$$\begin{aligned}
& d(h(s,t)) \\
= \;& d\left(h_{\alpha\bar\beta} s^\alpha \bar t^\beta\right) \\
= \;& ds^\alpha \bar t^\beta h_{\alpha\bar\beta} + \overline{dt^\beta} s^\alpha h_{\alpha\bar\beta} + s^\alpha \bar t^\beta dh_{\alpha\bar\beta} \\
= \;& \left(\partial s^\alpha + s^\gamma h^{\alpha\bar\delta}\partial h_{\gamma\bar\delta} + \bar\partial s^\alpha\right)\bar t^\beta h_{\alpha\bar\beta} + s^\alpha\left(\overline{\partial t^\beta + t^\mu h^{\beta\bar\nu}\partial h_{\mu\bar\nu} + \bar\partial t^\beta}\right) h_{\alpha\bar\beta} \\
= \;& h(Ds,t) + h(s,Dt),
\end{aligned}$$

provided we define D by the formula

$$Ds = \left(\partial s^\alpha + s^\gamma h^{\alpha\bar\delta}\partial h_{\gamma\bar\delta} + \bar\partial s^\alpha\right) e_\alpha. \tag{0.4}$$

Clearly $D = D^{1,0} + \bar\partial$, and the proof is thus complete. □

2.3. The Kähler condition

2.3.1. The definition of Kähler metric. The cotangent bundle T^*_X of a Hermitian complex manifold X with Hermitian metric g (for the cotangent bundle) now carries two natural connections. One connection is the Hermitian connection for the holomorphic Hermitian vector bundle (T^*_X, g), and the other connection is the Levi-Čivita connection for (T^*_X, g). (Here g is viewed both as a Hermitian metric on $T^{1,0}_X$ and as a Riemannian metric on T_X, via the correspondence discussed in Section 1.4.)

DEFINITION 2.7. *A Hermitian metric g for which the holomorphic connection agrees with the Levi-Čivita connection is said to be Kähler. A complex manifold admitting a Kähler metric is called a Kähler manifold.*

2.3.2. Equivalent formulations.

PROPOSITION 2.8. *A Hermitian metric g with metric form ω is Kähler if and only if the $(1,1)$-form ω associated to g is closed.*

PROOF. Considering first the tangent bundle as a Hermitian holomorphic vector bundle, we get that the Christoffel symbols are

$$\Gamma^{\beta}_{\alpha\gamma} = -g_{\alpha\bar{\delta}}\partial_\gamma g^{\beta\bar{\delta}} = g^{\beta\bar{\delta}}\partial_\gamma g_{\alpha\bar{\delta}},$$

where in the second equality we have used the fact that $g_{\alpha\bar{\mu}}g^{\beta\bar{\mu}} = \delta^{\beta}_{\alpha}$. The connection is then the Levi-Čivita connection if and only if the Christoffel symbols are symmetric in α and γ, i.e., if and only if

$$\partial_\gamma g_{\alpha\bar{\delta}} = \partial_\alpha g_{\gamma\bar{\delta}}.$$

Now

$$\begin{aligned} d(g_{\alpha\bar{\delta}}dz^\alpha \wedge d\bar{z}^\delta) &= \partial_\gamma(g_{\alpha\bar{\delta}})dz^\gamma \wedge dz^\alpha \wedge d\bar{z}^\delta + \partial_{\bar{\mu}}(g_{\alpha\bar{\delta}})dz^\alpha \wedge d\bar{z}^\mu \wedge d\bar{z}^\delta \\ &= \sum_{\gamma<\alpha} \left(\partial_\gamma g_{\alpha\bar{\delta}} - \partial_\alpha g_{\gamma\bar{\delta}}\right) dz^\gamma \wedge dz^\alpha \wedge d\bar{z}^\delta \\ &\quad - \sum_{\mu<\delta} \overline{(\partial_\mu g_{\delta\bar{\alpha}} - \partial_\delta g_{\mu\bar{\alpha}})} dz^\alpha \wedge d\bar{z}^\mu \wedge d\bar{z}^\delta. \end{aligned}$$

By type considerations, we see that $d\omega = 0$ if and only if g is Kähler. This completes the proof. □

Another useful formulation of the Kähler condition is contained in the following theorem.

THEOREM 2.9. *The metric g is Kähler if and only if there exist coordinates z on X so that*

$$g = \sum_\alpha dz^\alpha \cdot \, d\bar{z}^\alpha + O(|z|^2).$$

The coordinates referred to in the theorem are called *Kähler coordinates.*

PROOF. Observe that if two $(1,1)$-forms ω_1 and ω_2, defined in a neighborhood of 0 in $\mathbb{C}^n$, have the same Taylor expansion up to second order, then one has

$$(d\omega_1)_0 = (d\omega_2)_0.$$

Thus g is Kähler if it is locally Euclidean to second order.

Conversely, suppose g is Kähler. Let ω be the metric $(1,1)$-form associated to g, and let z be local coordinates such that

$$h_{i\bar{j}}(0) = \delta_{i\bar{j}} \tag{0.5}$$

(Such coordinates, called *normal coordinates,* are easy to find.) Then the Taylor expansion of ω with respect to z is

$$\omega = \frac{\sqrt{-1}}{2}\left(\delta_{i\bar{j}} + a_{i\bar{j}k}z^k + a_{i\bar{j}\bar{k}}\bar{z}^k + O(2)\right) dz^i \wedge d\bar{z}^j.$$

The two properties of the Taylor coefficients $a_{i\bar{j}k}, a_{i\bar{j}\bar{k}}$ are

a. $g_{i\bar{j}} = \overline{g_{j\bar{i}}} \Rightarrow a_{i\bar{j}\bar{k}} = \overline{a_{j\bar{i}k}}$,
b. $d\omega = 0 \Rightarrow a_{i\bar{j}k} = a_{k\bar{j}i}$.

Since we are making an assertion about the second Taylor coefficients, it is reasonable to seek a quadratic change of variables. Thus we seek coordinates w such that

$$z^k = w^k + \frac{1}{2} b^k_{\ell m} w^\ell w^m$$

(which do not modify condition (0.5)) such that

$$\omega = \frac{\sqrt{-1}}{2} \left(\delta_{i\bar{j}} + O(2) \right) dw^i \wedge d\bar{w}^j.$$

Let us further choose b^i_{jk} so that $b^i_{jk} = b^i_{kj}$. Then

$$dz^k = dw^k + b^k_{\ell m} w^\ell dw^m,$$

and we have

$$\begin{aligned}
\frac{2}{\sqrt{-1}} \omega &= \delta_{i\bar{j}} \left(dw^i + b^i_{\ell m} w^\ell dw^m \right) \wedge \left(d\bar{w}^j + \overline{b^j_{rs}} \bar{w}^r d\bar{w}^s \right) \\
&\quad + \left(a_{i\bar{j}k} w^k + a_{i\bar{j}\bar{k}} \bar{w}^k \right) dw^i \wedge d\bar{w}^j + O(2) \\
&= \left(\delta_{i\bar{j}} + (a_{i\bar{j}k} + b^j_{ki}) w^k + (a_{i\bar{j}\bar{k}} + \overline{b^i_{kj}}) \bar{w}^k + O(2) \right) dw^i \wedge d\bar{w}^j.
\end{aligned}$$

Thus, if we set $b^j_{ki} = -a_{i\bar{j}k}$, then

$$b^j_{ki} = -a_{i\bar{j}k} = -a_{k\bar{j}i} = b^j_{ik}$$

and

$$\overline{b^i_{kj}} = -\overline{a_{j\bar{i}k}} = -a_{i\bar{j}\bar{k}},$$

so that g is indeed Euclidean to second order in the coordinate system w. □

2.4. Connections for product bundles

2.4.1. The general definition of product connections. Given a pair of vector bundles $V_i \to X$, $i = 1, 2$, equipped with connections D_1 and D_2 respectively, there are natural definitions of connections for various products of V_1 and V_2.

In seeking to satisfy the Leibniz rule, one is naturally led to the connection formula

$$D\left(s_1 \times s_2\right) = \left(D_1 s_1\right) \times s_2 + s_1 \times \left(D_2 s_2\right)$$

where $\times$ denotes whatever product one has in mind. For example, one can consider tensor, symmetric or wedge products.

By induction, one can pass to any finite product of vector bundles.

EXAMPLE. We can think of metrics of a given vector bundle $V \to M$ as sections of the vector bundle $V^* \otimes \overline{V^*} \to M$ that in addition satisfy a symmetry condition (as tensors). If $V \to M$ is equipped with a connection D whose connection form is ω in some frame, then the induced action of D on a section g of $V \otimes V^* \to M$ is given in the associated frame by

$$Dg = dg - g\omega - \omega^T g. \tag{0.6}$$

We can thus interpret (0.2) above as follows: connections compatible with the metric are simply those connections that annihilate the metric.

2.4.2. **Induced connections for determinant bundles.** Let $V \to X$ be a complex vector bundle of rank r and let a connection D_V for V be given. Consider the complex line bundle

$$\det V \to X$$

whose transition functions are just the determinants of the corresponding transition functions for V. Fix a frame $e_1, ..., e_r$ for V. Then $e_1 \wedge ... \wedge e_r$ is a frame for $\det V$, and we have

$$e_1 \wedge ... \wedge De_j \wedge ... \wedge e_r = e_1 \wedge ... \wedge \omega_j^k e_k \wedge ... \wedge e_r = \omega_j^k \delta_{jk} e_1 \wedge ... \wedge e_j \wedge ... \wedge e_r.$$

Thus by the definition of product connection,

$$D_{\det V}(e_1 \wedge ... \wedge e_r) = \omega_j^j e_1 \wedge ... \wedge e_r,$$

i.e., the connection matrix for $D_{\det V}$ is the trace of the connection matrix for D_V.

2.4.3. **The Chern connection of the determinant bundle.** If $V \to X$ is a complex vector bundle with Hermitian metric h, then $\det V \to X$ has a metric $\det h$ defined as follows. If $e_1, ..., e_r$ is a frame for $V \to X$, then

$$\det h(fe_1 \wedge ... \wedge e_r, ge_1 \wedge ... \wedge e_r) = f\bar{g} \det(h(e_\alpha, e_\beta))^n_{\alpha,\bar{\beta}=1}.$$

Assume now that X is complex, $V \to X$ is holomorphic, and $e_1, ..., e_r$ is a holomorphic frame for V. Then the unique holomorphic Hermitian connection for the metric $\det h$ of the determinant bundle $\det V \to X$ has connection (1×1) matrix

$$\omega = \frac{1}{\det h} \partial(\det h).$$

In view of the previous paragraph, we also have

$$\omega = \mathrm{trace}(\partial h h^{-1}).$$

We thus have the identity

$$\mathrm{trace}(\partial h h^{-1}) = \frac{1}{\det h} \partial(\det h). \tag{0.7}$$

REMARK. The identity (0.7) can also be established directly as follows. First, note that for a Hermitian matrix A, the identity

$$\det e^A = e^{\mathrm{trace} A}$$

holds. Indeed, if we let $H_t = \det e^{tA}$, then $H_0 = 1$, $H_1 = \det e^A$, and $H_t H_s = H_{t+s}$. Moreover,

$$\left.\frac{d}{dt}\right|_{t=0} H_t = \mathrm{trace}(A).$$

Let $K_t = e^{t \cdot \mathrm{trace} A}$. Then $K_0 = 1$, $K_t K_s = K_{t+s}$ and

$$\left.\frac{d}{dt}\right|_{t=0} K_t = \mathrm{trace}(A).$$

It follows that $H_t = K_t$ for all t.

Now, since h is smooth, we can write $h = e^{A(x)}h_o$ for some local coordinates x on X, where $A(0) = 0$. We thus have

$$\begin{aligned}\left.\frac{\partial \det h}{\det h}\right|_{x=0} &= \left.\partial e^{\mathrm{trace} A(x)}\right|_{x=0} \\ &= \left.\partial \mathrm{trace} A(x)\right|_{x=0} \\ &= \left.\mathrm{trace} \partial A(x)\right|_{x=0} \\ &= \left.\mathrm{trace}(\partial h h^{-1})\right|_{x=0}.\end{aligned}$$

2.5. Induced connection on twisted 1-forms

Let $V \to M$ be a vector bundle with connection D. We can define a twisted version of the exterior derivative for sections of

$$\Gamma(M, V \otimes \Lambda^k(T^*_M)),$$

or *V-valued k-forms.* This twisted exterior derivative should produce a V-valued $(k+1)$-form.

Definition 2.10. *Let $V \to M$ be a vector bundle with connection D. We define the twisted exterior derivative*

$$D = D_k : \Gamma(M, V \otimes \Lambda^k(T^*_M)) \to \Gamma(M, V \otimes \Lambda^{k+1}(T^*_M))$$

to be the linear differential operator given locally as follows. If $e_1, ..., e_r$ is a frame for V and $x^1, ..., x^m$ is a local coordinate system on M, then for a section $\sigma \in \Gamma(M, V \otimes \Lambda^k(T^*_M))$ given locally by

$$\sigma = \sigma^\mu_I e_\mu \otimes dx^I,$$

we set

$$\begin{aligned} D\sigma &= \frac{\partial(\sigma^\mu_J)}{\partial x^i} e_\mu \otimes dx^i \wedge dx^I + \sigma^\mu_i e_\mu \otimes \omega^\mu_\nu \wedge dx^I \\ &= \frac{\partial(\sigma^\mu_J)}{\partial x^i} e_\mu \otimes dx^i \wedge dx^I + (-1)^k \sigma^\mu_i e_\mu \otimes dx^I \wedge \omega^\mu_\nu. \end{aligned}$$

Informally, we write

$$D\sigma = d\sigma + (-1)^k \sigma \wedge \omega.$$

Remark. Note that when the vector bundle V is trivial of rank 1, we simply recover the definition for the exterior derivative of a k-form on M.

The formula for D does not automatically make sense; we need to verify that if s is a V-valued k-form, then Ds is a V-valued $(k+1)$-form. We shall verify this fact using matrix notation. Let g be a change of frame, and let $\tilde{\omega} = -dgg^{-1} + g^{-1}\omega g$ be the connection matrix in the new frame. We have

$$\begin{aligned} D(\sigma g) &= d(\sigma g) + (-1)^k \sigma g \wedge \omega \\ &= d\sigma g + (-1)^k \sigma((dg)g^{-1})g + (-1)^k \sigma \wedge g\omega g^{-1} g \\ &= (d\sigma + (-1)^k \sigma \wedge \tilde{\omega})g = (D\sigma)g. \end{aligned}$$

3. Curvature

The name *curvature* is both suggestive and vague. And indeed, there are many ways in which curvature gets used in mathematics.

When we turn to L^2 methods, our main interest will be to see how curvature affects changing the order of differentiation for mixed (covariant) partial derivatives. We will take advantage of the situation when this commuting of the order of differentiation gives rise to a positive operator.

In this section, we will define curvature of a connection on a vector bundle and develop its basic properties.

3.1. Definition

We start with the most general object.

DEFINITION 3.1. *Let $V \to M$ be a vector bundle with connection D and, in terms of some frame, connection matrix ω. The curvature of (V, D) is the operator*

$$DD : \Gamma(M, V) \to \Gamma(M, V \otimes \Lambda^2(T^*_M)),$$

where the connection on the left is the connection induced on V-valued 1-forms as in Definition 2.10.

Computing, we see that

$$\begin{aligned} DDs &= D(ds + s\omega) \\ &= d(ds + s\omega) - (ds + \omega) \wedge \omega \\ &= ds \wedge \omega + sd\omega - ds \wedge \omega - s\omega \wedge \omega \\ &= s(d\omega - \omega \wedge \omega). \end{aligned}$$

Thus curvature is not a differential operator, but rather a multiplier. Note that if the connection is trivial ($D = d$) the curvature vanishes, which is consistent with $d^2 = 0$.

3.2. Transformation formula

Let us see how the matrix of 2-forms

$$\Omega := d\omega - \omega \wedge \omega$$

transforms under a change G of local basis. In fact, because the operator D is globally defined, there is almost no need for calculation. Writing $s = \tilde{s}G$, $\tilde{\omega} = (dG)G^{-1} + G\omega G^{-1}$ and $\tilde{\Omega} = d\tilde{\omega} - \tilde{\omega} \wedge \tilde{\omega}$, we find that $(s\Omega)G = \tilde{s}G\tilde{\Omega}$. Therefore

$$\tilde{\Omega} = G\Omega G^{-1}.$$

3.3. Chern classes

It is beyond the scope of these notes to get into a thorough description of Chern classes. Instead, we content ourselves with a statement, devoid of proof, of the main result of the so-called Chern-Weil Theory.

THEOREM 3.2. *Let $V \to M$ be a complex vector bundle of rank r with a connection having curvature form Ω.*

(1) *The coefficients $P_j(\Omega)$ of the Chern polynomial*

$$\det\left(I + t\frac{\sqrt{-1}}{2\pi}\Omega\right) = 1 + \sum_{j=1}^{r} P_j(\Omega)t^j,$$

are closed $2j$-forms on M.

(2) *Moreover, for each j the cohomology class of $P_j(\Omega)$ is independent of the connection on $V \to M$.*

DEFINITION 3.3. *With the notation of theorem 3.2, the cohomology class*

$$c_i(V) := [P_i(\Omega)] \in H^{2i}(M, \mathbb{C})$$

is called the ith Chern class of V.

3.4. Curvature of holomorphic Hermitian connection

In this section, fix a holomorphic Hermitian vector bundle $(V, h) \to X$. We want to compute the curvature of the unique holomorphic Hermitian connection, in terms of the metric. To this end, fix a holomorphic frame $\{e_\alpha\}$ for V, and let

$$h_{\alpha\bar{\beta}} = h(e_\alpha, e_\beta).$$

According to the proof of Theorem 2.6, the unique holomorphic Hermitian connection is given by $d + \omega$, where the matrix ω of $(1,0)$ forms is given by

$$\omega^\alpha_\beta = h^{\alpha\bar{\mu}} \partial h_{\beta\bar{\mu}}.$$

Thus, since the curvature Ω is given by

$$\Omega = d\omega - \omega \wedge \omega,$$

we have the following proposition.

PROPOSITION 3.4. *The curvature of the unique holomorphic Hermitian connection of $(V, h) \to M$ is given by the formula*

$$\Omega^\alpha_\beta = \bar{\partial}(h^{\alpha\bar{\mu}} \partial h_{\beta\bar{\mu}}).$$

PROOF. For ease of calculation, we use matrix notation; we write $H = h_{\alpha\bar{\beta}}$. Our connection matrix is then simply

$$\omega = \partial H H^{-1}.$$

Recall that $\partial(H^{-1}) = -H^{-1}\partial H H^{-1}$. We then have

$$\begin{aligned} \Omega &= d\omega - \omega \wedge \omega \\ &= (\partial + \bar{\partial})(\partial H H^{-1}) - \partial H H^{-1} \wedge \partial H H^{-1} \\ &= \bar{\partial}(\partial H H^{-1}) - \partial H \wedge \partial(H^{-1}) - \partial H H^{-1} \wedge \partial H H^{-1} \\ &= \bar{\partial}(\partial H H^{-1}). \end{aligned}$$

This completes the proof. □

REMARK. We leave it to the reader to check that if $\Omega_{\alpha\bar{\beta}} = h_{\gamma\bar{\beta}} \Omega^\gamma_\alpha$ then

$$\Omega_{\alpha\bar{\beta}} = -\left(\partial\bar{\partial} h_{\alpha\bar{\beta}} - h^{\lambda\bar{\mu}} \partial h_{\alpha\bar{\mu}} \wedge \bar{\partial} h_{\lambda\bar{\beta}}\right).$$

3.5. Curvature of determinant bundles

PROPOSITION 3.5. *Let $(V, D) \to M$ be a vector bundle of rank r with connection, and let* $(\det V, \mathrm{trace}(D)) \to M$ *be its determinant line bundle. Then*

$$\Omega(\mathrm{trace} D) = \mathrm{trace}(\Omega(D)).$$

PROOF. Let $e_1, ..., e_r$ be a frame for V. Then

$$
\begin{aligned}
(\mathrm{trace}(D))^2(e_1 \wedge ... \wedge e_r) &= \mathrm{trace}(D)\left(\sum_{j=1}^{r} e_1 \wedge ... \wedge De_j \wedge ... \wedge e_r\right) \\
&= \sum_{j=1}^{r}\sum_{k=1}^{j} e_1 \wedge ... \wedge De_k \wedge ... \wedge De_j \wedge ... \wedge e_r \\
&\quad - \sum_{j=1}^{k}\sum_{k=j+1}^{r} e_1 \wedge ... \wedge De_j \wedge ... \wedge De_k \wedge ... \wedge e_r \\
&\quad + \sum_{j=1}^{r} e_1 \wedge ... \wedge DDe_j \wedge ... \wedge e_r \\
&= \mathrm{trace}(D^2)(e_1 \wedge ... \wedge e_r).
\end{aligned}
$$

This completes the proof. □

3.5.1. Holomorphic hermitian vector bundles. Let $(V, h) \to X$ be a holomorphic Hermitian vector bundle of rank r. We have already mentioned the determinant bundle $\det V \to X$, and observed that $\det h$ is a metric for the line bundle $\det V \to X$, whose (unique) Chern connection is

$$\omega = \frac{1}{\det h}\partial(\det h).$$

It follows that the curvature matrix of $\det h$ is

$$\Omega = d\omega - \omega \wedge \omega = \bar{\partial}(\partial \log \det h) = \partial\bar{\partial}(-\log \det h).$$

With calculations similar to those we used in the study of the connection, one can easily see that

$$\Omega = \mathrm{trace}\ \bar{\partial}(\partial h h^{-1}),$$

a fact we already know from Proposition 3.5

3.5.2. The canonical bundle. Given a complex manifold X of complex dimension n, the *canonical bundle* K_X of X is the line bundle $\det T_X^{*1,0}$. The local sections K_X are holomorphic n-forms.

If X is a Hermitian complex manifold with Hermitian metric h, Theorem 3.5 tells us that the curvature of $(K_X, \det(h^{-1}))$ is just the negative of the trace of the curvature of (X, h).

If the metric h for T_X is furthermore Kähler, the curvature of $\det(h^{-1})$ agrees with the negative of the so-called *Ricci curvature* of h. In the notes, we take the curvature of $\det(h)$ to be the definition of the Ricci curvature of h:

$$\mathrm{Ricci}(h) := \mathrm{trace}(\Omega(h)).$$

3.6. Symmetry of Levi-Čivita and Kähler curvatures

Since the Levi-Čivita connection is determined by vanishing of a certain part of the connection matrix, one can expect that less numerical data is needed to determine its curvature. In the Kähler case, where one has two vanishing conditions that have been required to hold simultaneously, one can expect even less numerical data to determine the curvature of the Kähler connection. These constraints on the curvature matrices of the Levi-Čivita and more generally the Kähler connection are expressed in terms of symmetries in the curvature tensor. We now write down some of these symmetries.

In this section, we agree to let Greek letters run through $\{1, ..., n\}$ and Latin letters through $\{1, ..., n, \bar{1}, ..., \bar{n}\}$.

First, the Christoffel symbols are defined by the relation

$$(\nabla \partial_{z^i})(\partial_{z^j}) = \Gamma^k_{ij}(\partial_{z^k}),$$

and by taking complex conjugates we obtain that

$$\Gamma^{\bar{k}}_{\bar{i}\bar{j}} = \overline{\Gamma^k_{ij}} \tag{0.8}$$

Next we observe that, since the Kähler connection matrix has only a $(1,0)$-component, its Christoffel symbols must satisfy

$$\Gamma^{\alpha}_{i\bar{\gamma}} = \Gamma^{\bar{\alpha}}_{i\gamma} = 0 \tag{0.9}$$

Let us introduce the following notation. Let R^i_j be the curvature tensor with respect to the local frame $\partial_1, ..., \partial_n, \partial_{\bar{1}}, ...\partial_{\bar{n}}$ for $T_X \otimes \mathbb{C}$ and write

$$R_{i\bar{j}} = R^k_i g_{k\bar{j}}.$$

The matrix entries R_{ij} are $(1,1)$-forms, and so we write

$$R_{ij} = R_{ij\alpha\bar{\beta}} dz^{\alpha} \wedge d\bar{z}^{\beta}.$$

We extend the indices α, β to all $\{1, ..., n, \bar{1}, ..., \bar{n}\}$ by defining

$$R_{ij\alpha\beta} = R_{ij\bar{\alpha}\bar{\beta}} = 0 \quad \text{and} \quad R_{ij\bar{\alpha}\beta} = -R_{ij\beta\bar{\alpha}}.$$

Now, since the Kähler connection is complex, we have

$$R_{\alpha\beta ij} = R_{\bar{\alpha}\bar{\beta}ij} = 0. \tag{0.10}$$

Next, the curvature for the Levi-Čivita connection satisfies

$$R_{ijlk} = R_{jikl} = -R_{klij} \tag{0.11}$$

as well as the so-called *First Bianchi Identity*:

$$R_{ijkl} + R_{iklj} + R_{iljk} = 0. \tag{0.12}$$

Combining this with the Kähler condition, we also have

$$R_{\alpha\bar{\beta}i\bar{j}} = R_{i\bar{j}\alpha\bar{\beta}} = R_{i\bar{\beta}\alpha\bar{j}} \tag{0.13}$$

The last equality uses the Kähler condition $d\omega = 0$.

There is some symmetry to the Ricci curvature of a Kähler manifold as well. One has that the Ricci curvature tensor $\text{Ricci}(R)_{ij} = R_i{}^k{}_{kj}$ satisfies

$$\text{Ricci}(R)_{\alpha\beta} = \text{Ricci}(R)_{\bar{\alpha}\bar{\beta}} = 0 \quad \text{and} \quad \text{Ricci}(R)_{\alpha\bar{\beta}} = \overline{\text{Ricci}(R)_{\alpha\bar{\beta}}}. \tag{0.14}$$

In fact, as we have already mentioned, one has the formula

$$\text{Ricci}(R)_{\alpha\bar{\beta}} = -\partial_{z^{\alpha}}\partial_{\bar{z}^{\beta}} \log\det\left(h_{\mu\bar{\nu}}\right). \tag{0.15}$$

4. Holomorphic line bundles

Great simplification arises in the study of line bundles simply because 1×1 matrices are numbers, and so commute. In particular, the curvature is a genuine, global 2-form and line bundles form a group.

4.1. Curvature of line bundles

Let $L \to X$ be a Hermitian line bundle with metric h. A frame ξ_i for h on a neighborhood U_i defines a smooth function $\varphi_i := -\log h(\xi_i, \xi_i)$. Now, any $v \in L|_{U_i}$ can be written $v = v_i \xi_i$ for some $v_i \in \mathbb{C}$, and thus

$$h(v, v) = |v_i|^2 e^{-\varphi_i}.$$

We observe next that the connection matrix is a 1-form ω_i, and the associated curvature matrix on U_i is the 2-form

$$\Omega_i = d\omega_i - \omega_i \wedge \omega_i = d\omega_i.$$

From the transformation rule for curvature that we derived earlier, we have

$$\Omega_j = g_{ji} \Omega_i g_{ij} = \Omega_i,$$

where the second equality follows because g_{ij} are 1×1 matrices. We have thus proved the following proposition.

PROPOSITION 4.1. *The curvature form of a line bundle is globally defined.*

If the line bundle is holomorphic, we see that the curvature is the $(1,1)$-form

$$\Omega = \partial \bar{\partial} \varphi_i \quad \text{on } U_i.$$

One often encounters the sometimes confusing notation

$$h(v, v) = |v|^2 e^{-\varphi},$$

for the metric of a holomorphic line bundle, and its companion notation $h = e^{-\varphi}$, which is equally confusing. In these notes, whenever we use this notation, we are implicitly assuming that the line bundle in question is holomorphic, and that we are using a holomorphic local frame ξ to determine the function $\varphi := -\log h(\xi, \xi)$. The formula

$$\Omega = \partial \bar{\partial} \varphi$$

for the curvature form of h is then independent of the frame ξ, and the latter fact is the main reason for the notation. After some familiarity with the subject, this notation causes no confusion.

4.2. Chern class

In the case of a line bundle, we have just seen that the curvature matrix has only one entry. Thus for a line bundle L, the Chern class is just the cohomology class of $(\sqrt{-1}/2\pi)\Omega$:

$$c_1(L) = \left[\frac{\sqrt{-1}}{2\pi} \Omega \right].$$

One sometimes encounters the notation $c(L)$ for the Chern class of a line bundle.

4.3. Divisors

In analytic geometry holomorphic line bundles often arise from divisors. Especially in the algebraic setting, the correspondence between line bundles and divisors is very important and useful.

4.3.1. **Definition of divisors.** Let $\mathrm{Div}(X)$ be the Abelian group of locally finite sums of of hypersurfaces: $D \in \mathrm{Div}(X)$ is a formal sum

$$D = \sum_{\alpha} a_\alpha V^\alpha,$$

where $a_\alpha \in \mathbb{Z}$, V^α are hypersurfaces, and given any point $p \in X$, there is a neighborhood U such that the number of hypersurfaces V^α intersecting U is finite. Thus if, for example, X is compact, $\mathrm{Div}(X)$ consists of finite formal sums of hypersurfaces:

$$\mathrm{Div}(X) := \bigoplus_{V^\alpha} \mathbb{Z} V^\alpha.$$

REMARK. This definition of divisors is due to Weil, and in algebraic geometry such divisors are called *Weil divisors.*

DEFINITION 4.2. *A divisor $D = \sum a_\alpha V_\alpha$ with $a_\alpha \geq 0$ for all α is said to be effective. We write $D \geq 0$.*

4.3.2. **Line bundles associated to divisors.** Let D be a divisor on X, and let U_1 and U_2 be two sufficiently small neighborhoods of p in X. By definition, there are a finite number of hypersurfaces $V^1, ..., V^j$ among the generators defining D, and functions $f_k^1, ..., f_k^j \in \mathcal{O}(U_k)$, $k = 1, 2$, such that

$$U_k \cap V^1 \cap ... \cap V^j = (f_k^1 = ... = f_k^j = 0).$$

Suppose that, for $1 \leq \ell \leq j$, the coefficient of V^ℓ in D is a_ℓ. Then the function

$$g_{12}(q) = \left(\frac{f_1^1(q)}{f_2^1(q)} \right)^{a_1} \cdots \left(\frac{f_1^j(q)}{f_2^j(q)} \right)^{a_j}, \qquad q \in U_1 \cap U_2$$

are well-defined and nonvanishing holomorphic functions on $U_1 \cap U_2$. In this way, if X is covered by a family of sufficiently small neighborhoods $\{U_j\}$, we can use the functions g_{jk} to define transition functions for a line bundle. The line bundle thus obtained is denoted L_D. We will often denote the associated Chern class for this line bundle by $c_1(D)$ instead of $c_1(L_D)$.

Now, by its very definition, the line bundle L_D has a section. If D is given by meromorphic function f_j over open sets U_j, then the line bundle, being locally $U_j \times \mathbb{C}$ with transition functions $g_{ij} = f_i / f_j$, has a meromorphic section s_D locally given by $s_D = (p, f_j(p))$ over U_j.

DEFINITION 4.3. *The section s_D is called the canonical section of L_D associated to D.*

Note that the canonical section s_D is holomorphic if and only if the divisor D is effective.

REMARK. In a more algebraic setting, divisors are in general not defined locally by the vanishing of quotients of regular functions. In such a setting, divisors with this property are called *Cartier divisors,* and are in general a proper subset of the set of Weil divisors.

4.4. Examples

EXAMPLE. Consider the line bundle on $\mathbb{P}_n$ corresponding a divisor associated to a hyperplane H with multiplicity 1. It is possible to choose homogeneous coordinates in $\mathbb{P}_n$ so that H is the zero locus of $z^0 = 0$. Thus on the set $U_j := \{[z^0, ..., z^n] \in \mathbb{P}_n \; ; \; z^j \neq 0\}$ H can be described as the zero set of the function f_j defined by

$$f_j = \begin{cases} x_j^0 = z^0/z^j & j \neq 0 \\ 1 & j = 0 \end{cases}$$

The transition functions for the associated line bundle L_H, sending the local trivialization on U_i to that on U_j, are then given, in homogeneous coordinates on $U_i \cap U_j$, by the formula

$$g_{ji} = f_j/f_i = z^i/z^j.$$

It follows that the functions

$$\varphi_i := \log\left(\frac{|z^0|^2 + ... + |z^n|^2}{|z^i|^2}\right) \quad \text{on } U_i$$

define a metric h_{FS} for L_H, called the Fubini-Study metric. The Chern form of this metric is the well known Fubini-Study Kähler form

$$\omega_{FS} := \frac{\sqrt{-1}}{2\pi}\partial\bar{\partial}\varphi_i \quad \text{on } U_i.$$

It turns out that every line bundle on $\mathbb{P}_n$ is an integral tensor power of L_H (where negative powers means positive powers of the dual).

EXAMPLE. Let us look more closely at the Fubini-Study Kähler form $\omega_{FS} = \sqrt{-1}\partial\bar{\partial}\varphi_i$ discussed in the previous example. Consider the affine chart $U_0 = \{[z] \in \mathbb{P}_n \; ; \; z^0 \neq 0\}$ and the affine coordinates $\zeta = (\zeta^1, ..., \zeta^n)$ where $z^j = z^j/z^0$. Then

$$\begin{aligned} \omega_{FS} &= \frac{\sqrt{-1}}{2\pi}\partial\bar{\partial}\log(1 + |\zeta|^2) \\ &= \frac{\sqrt{-1}}{2\pi}\partial\left(\frac{\zeta \cdot \overline{\partial\zeta}}{1 + |\zeta|^2}\right) \\ &= \frac{\sqrt{-1}}{2\pi}\left(\frac{(1 + |\zeta|^2)(\partial\zeta \cdot \overline{\partial\zeta}) - ((\bar{\zeta} \cdot \partial\zeta) \wedge (\zeta \cdot \overline{\partial\zeta}))}{(1 + |\zeta|^2)^2}\right). \end{aligned}$$

By the Cauchy-Schwarz inequality we see that

$$\omega_{FS} \geq \frac{\sqrt{-1}}{2\pi}\frac{\partial\zeta \cdot \overline{\partial\zeta}}{(1 + |\zeta|^2)^2}$$

and thus that ω_{FS} is strictly positive.

We also note that, with $\omega_o = \frac{\sqrt{-1}}{2}\partial\zeta \cdot \overline{\partial\zeta}$ denoting the Euclidean Kähler form on $U_0 = \mathbb{C}^n$,

$$\frac{\omega_{FS}^n}{n!} = \frac{1}{(\pi)^n}\left(\frac{1}{(1 + |\zeta|^2)^n} - \frac{|\zeta|^2}{(1 + |\zeta|^2)^{n+1}}\right)\frac{\omega_o^n}{n!} = \frac{dV(\zeta)}{(1 + |\zeta|^2)^{n+1}},$$

where

$$dV = \frac{\omega_o^n}{n!}$$

is the usual Euclidean volume in $\mathbb{C}^n$. The key point in this calculation is that $(\bar{\zeta} \cdot \partial\zeta \wedge \overline{\bar{\zeta} \cdot \partial\zeta})^k = 0$ for $k \geq 2$. Indeed, if θ is a $(1,0)$-form, then $\theta \wedge \theta = 0$ by skew symmetry, and thus

$$(\theta \wedge \bar{\theta})^2 = -\theta \wedge \theta \wedge \bar{\theta} \wedge \bar{\theta} = 0.$$

PROPOSITION 4.4. *The sections of $L_H \to \mathbb{P}_n$ are in 1-to-1 correspondence with linear functionals on $\mathbb{C}^{n+1}$.*

PROOF. Let $z^0, ..., z^n$ be homogeneous coordinates on $\mathbb{P}_n$. By definition, z^j is a linear functional on $\mathbb{C}^{n+1}$ that assigns to a vector v its jth coordinate $v^j = z^j(v)$. Consider the linear functional $\lambda = c_j z^j$. Let $\sigma_j : U_j \to \mathbb{C}$ be defined by

$$\sigma_j = \frac{\lambda}{z^j}.$$

Then

$$g_{ij}\sigma_j = \frac{z^j}{z^i}\frac{\lambda}{z^j} = \sigma_i,$$

and thus λ defines a section σ of L_H.

Conversely, suppose given a section σ of L_H. Since z^j is a non-zero section of L_H over U_j, we have that $\sigma = s^j z^j$ on U_j for some holomorphic function s^j. We claim that s^j is a polynomial of degree at most 1 in the affine coordinates ζ_j on U_j. To see this, we endow $L_H \to \mathbb{P}_n$ with the Fubini-Study metric h_{FS} described above. We then have

$$h_{FS}(\sigma, \sigma) = \frac{|s^j|^2 |z^j|^2}{|z^0|^2 + ... + |z^n|^2} = \frac{|s^j|^2}{1 + |\zeta_j|^2}.$$

We then have

$$\int_{\mathbb{C}^n} \frac{|s^j|^2}{(1+|\zeta|^2)^{n+2}} dV(\zeta) \leq \int_{\mathbb{P}_n} h_{FS}(\sigma, \sigma)\frac{\omega_{FS}^n}{n!} < +\infty,$$

where the last inequality follows from the compactness of $\mathbb{P}_n$. It follows from Theorem an L^2-version of Liouville's Theorem that s^j is an affine linear function of ζ. Thus σ corresponds to a linear functional on $\mathbb{P}_n$, and the proof is complete. □

Since $\mathbb{P}_n$ is composed of lines, one obtains by tautology a line bundle $\mathbb{U} \to \mathbb{P}_n$, which assigns to each $\ell \in \mathbb{P}_n$ the set of points comprising the line ℓ.

PROPOSITION 4.5. *The dual L_{-H} of L_H is the tautological line bundle.*

PROOF. Consider a general vector $v \in \mathbb{C}^{n+1}$. Then any linear functional on V is a linear combination of the coordinate functions on $\mathbb{C}^{n+1}$. Since we have already shown that the latter are global sections of L_H, it follows that the tautological bundle is dual to L_H. □

EXAMPLE. Let us compute the canonical bundle of projective space in terms of the hyperplane divisor. If we fix homogeneous coordinates $z^0, ..., z^n$ on $\mathbb{P}_n$, let $U_j = \{z^j \neq 0\}$ be the cover of $\mathbb{P}_n$ by affine spaces, and let

$$\zeta_j^i := \begin{cases} z^{i+1}/z^j & i < j \\ z^i/z^j & i > j \end{cases}$$

be affine coordinates on U_j. Then one computes that

$$\partial\zeta_0^1 \wedge ... \wedge d\zeta_0^n = \frac{(-1)^j}{(z^0/z^j)^{n+1}} d\zeta_j^1 \wedge ... \wedge d\zeta_j^n.$$

It follows that

$$(-1)^j(z^j)^{n+1} \otimes d\zeta_j^1 \wedge ... \wedge d\zeta_j^n \quad \text{on} \quad U_j$$

defines a global section of $L_{(n+1)H} \otimes K_{p_n}$ that is zero-free. Thus $L_{(n+1)H} \otimes K_{p_n}$ is trivial, i.e.,

$$K_{\mathbb{P}_n} = L_{-(n+1)H}.$$

EXAMPLE. Let X be a compact complex manifold and $V \subset X$ a smooth analytic hypersurface. The normal bundle N_V is the quotient line bundle $T_X^{1,0}/T_V^{1,0}$. We define the conormal bundle N_V^* to be its dual. Unlike the normal bundle, which is not a subbundle of anything, the conormal bundle of V is the subbundle of $T_X^{*1,0}$ consisting of all those (1,0)-forms that vanish on $T_V^{1,0}$.

Suppose that $\{U_j\}$ is an open cover of X such that V is given by $(f_j = 0)$ on U_j. Now, the family of differentials df_j transform by the rule

$$df_i = d\left(\frac{f_i}{f_j} f_j\right) = f_j dg_{ij} + g_{ij} df_j.$$

It follows that, on the set V (where $f_j = 0$), $\{df_j\}$ defines a section of the line bundle $N *_V \otimes [V]$. If V is smooth, then this section vanishes nowhere and thus $N_V^* \otimes L_V$ is trivial. It follows that for a smooth hypersurface $V \subset X$,

$$L_V|_V = N_V \quad \text{or} \quad L_{-V}|_V = N_V^*.$$

This isomorphism is called the *Adjunction Formula.*

As a nice application, we can calculate the canonical bundle of any smooth submanifold $V \subset X$ in terms of the canonical bundle of the ambient manifold. As already pointed out, we have the exact sequence

$$0 \to N_V^* \to T_X^{*1,0} \to T_V^{*1,0} \to 0.$$

It follows from linear algebra that

$$\Lambda^n(T_X^{*1,0})|_V = \Lambda^{n-1}(T_V^{*1,0}) \otimes N_V^*,$$

or in other words

$$K_X|_V = K_V \otimes L_{-V}.$$

Thus we get the formula

$$K_V = K_X|_V \otimes L_V,$$

which is sometimes also referred to as the adjunction formula.

REMARK. By embedding $\mathbb{P}_{n-1} \subset \mathbb{P}_n$ in the usual way, i.e., $z_0 = 0$, we can derive the formula for $K_{\mathbb{P}_n}$ inductively, provided we know $K_{\mathbb{P}_1}$. The latter can be determined (in essentially the same way as above) as follows: let z be the complex variable in the affine part of $\mathbb{P}_1$. Then $1/z$ is a coordinate at infinity, we we have

$$d(1/z) = -z^{-2}dz.$$

Thus dz extends to a meromorphic section of $K_{\mathbb{P}_1}$ with no poles or zeros in the affine part, and a double pole at infinity. Thus

$$K_{\mathbb{P}_1} = L_{-2\cdot\infty} = L_{-2H}.$$

LECTURE 1

The Hörmander theorem

In this lecture, we prove Hörmander's Theorem on the solution of the $\bar{\partial}$ equation with L^2 estimates, using Hilbert space methods. The results we present were originally proved simultaneously in the papers [**Hörmander-1965**] in the case of domains in $\mathbb{C}^n$, and in [**Andreotti-Vesentini-1965**] on more general complex manifolds. The proof we present here is only slightly different, combining elements of the work of Kohn [**Kohn-1963, Kohn-1964**] with the aforementioned papers.

The $\bar{\partial}$ operator is defined initially for smooth forms. Unfortunately, in the topology induced by the L^2-norm, the space of smooth forms incomplete. In order to be able to apply Hilbert space methods, we use the topology of L^2-convergence. The domain of the operator $\bar{\partial}$ can be extended from smooth forms to all square integrable forms by applying $\bar{\partial}$ in the sense of currents, but then we run into the new problem that if α is a square integrable (p,q)-form, then $\bar{\partial}\alpha$ is in general not square-integrable. Thus we must take a slightly smaller domain for $\bar{\partial}$. We will say that a square integrable form is in the domain of $\bar{\partial}$ if its image is square integrable.

Smooth forms constitute a dense subspace of L^2, so the domain of any extension of $\bar{\partial}$ is dense. Thus we will employ so-called *densely defined operators*. Fortunately, smooth forms also constitute dense subsets of some other dense subsets of the set of all square integrable forms, and the density of smooth forms in one particular subset— the set of measurable forms satisfying the *$\bar{\partial}$-Neumann condition*, and whose *graph norm is finite*— constitutes the most technical point underlying the proof of Hörmander's Theorem.

We begin by describing densely defined linear operators on Hilbert space, and proving a functional analysis lemma that is used to invert densely defined linear equations. The functional analysis lemma says that the densely defined linear operator in question can be inverted if and only if the adjoint of this operator satisfies a certain estimate. To achieve this estimate, we then obtain a certain integral identity, called the Bochner-Kodaira Identity, for smooth forms. We then establish the density of smooth forms mentioned above, and this density allows us to extend the Bochner-Kodaira Identity to the domains of the densely defined linear operators in our problem. A positivity of curvature hypothesis in Hörmander's Theorem then allows us to obtain from the Bochner-Kodaira Identity the estimate needed to apply the functional analysis lemma, and we are able to solve the $\bar{\partial}$ equation with estimates, provided we are working in L^2 spaces defined with smooth weights. We then discuss the process of passing to singular weights defined by quasi-plurisubharmonic functions.

The lecture concludes with two applications: Kodaira's Embedding Theorem and Nadel's Vanishing Theorem. In both of these applications, we follow inspirit the presentation of [**Demailly-2001**], though some of our proofs are more adapted to the philosophy of the present notes.

1. Functional analysis

We begin by establishing some basic but crucial functional analysis on Hilbert spaces. The results we prove here will be used in a fundamental way in the proof of the main result of this lecture and the next, and a slight modification will play a similar role in the main result of the last lecture.

1.1. Densely defined operators

Let H_1 and H_2 be Hilbert spaces. We are interested in linear maps from subspaces of H_1 into H_2. Such maps will be called linear operators, or simply operators. Two important things to keep in mind about operators are

- each operator $T : H_1 \to H_2$ comes with its own domain of definition $\mathrm{Domain}(T) \subset H_1$, and
- $T : \mathrm{Domain}(T) \to H_2$ need not be continuous.

It may sometimes be possible to extend an operator $T : H_1 \to H_2$ to a larger domain. If this is so, and S is such an extension, then the Graph

$$\mathrm{Graph}(T) := \{(x, Tx) \ ; \ x \in \mathrm{Domain}(T)\} \subset H_1 \times H_2$$

of T is a subspace of the graph $\mathrm{Graph}(S)$ of S.

DEFINITION 1.1. *An operator $T : H_1 \to H_2$ is said to be*

(1) *densely defined if* $\mathrm{Domain}(T)$ *is a dense subspace of H_1, and*
(2) *closed if* $\mathrm{Graph}(T)$ *is a closed subspace of $H_1 \times H_2$.*

Our next objective is to define the adjoint $T^* : H_2 \to H_1$ of a densely defined operator $T : H_1 \to H_2$. We begin by letting $\mathrm{Domain}(T^*)$ consist of all $\eta \in H_2$ such that

$$\mathscr{L}_\eta : \xi \mapsto \langle T\xi, \eta \rangle_2$$

is a continuous linear functional on $\mathrm{Domain}(T)$. By continuity and the density of $\mathrm{Domain}(T)$, one has a unique extension of $\mathscr{L}_\eta$ to $H_1 = \overline{\mathrm{Domain}(T)}$. It follows from the Riesz Representation Theorem that $\mathscr{L}_\eta$ is represented by a unique element $T^*\eta$ of H_1 satisfying

$$\langle T\xi, \eta \rangle_2 = \langle \xi, T^*\eta \rangle_1, \quad \xi \in \mathrm{Domain}(T).$$

The linearity of T^* is clear.

PROPOSITION 1.2. *Suppose $T : H_1 \to H_2$ is a closed, densely defined operator between two Hilbert spaces. Then $T^* : H_2 \to H_1$ is a closed, densely defined operator.*

PROOF. Let $F : H_1 \times H_2 \to H_2 \times H_1$ be given by

$$F(\xi, \eta) = (\eta, -\xi).$$

Observe that $(\eta, \xi) \perp F(\mathrm{Graph}(T))$ if and only if

$$\langle x, \xi \rangle_1 = \langle Tx, \eta \rangle_2$$

for all $x \in \mathrm{Domain}(T)$. Since the assignment $x \mapsto \langle x, \xi \rangle_1$ is continuous, it follows from the density of $\mathrm{Domain}(T)$ that, for $(\eta, \xi) \perp F(\mathrm{Graph}(T))$, $\eta \in \mathrm{Domain}(T^*)$ and $\xi = T^*\eta$. Thus

$$F(\mathrm{Graph}(T))^\perp = \mathrm{Graph}(T^*),$$

and hence T^* is closed. Suppose now that $\eta \perp \mathrm{Domain}(T^*)$. Then as T is closed, the vector $(\eta, 0)$ lies in $\mathrm{Graph}(T^*)^\perp = F(\overline{\mathrm{Graph}(T)}) = F(\mathrm{Graph}(T))$. It follows that $\eta = T0 = 0$. Thus T^* is densely defined. $\square$

PROPOSITION 1.3. *If T is a closed, densely defined operator, then $T^{**} = T$.*

PROOF. We know that T^* is also closed and densely defined, and we have

$$\langle x, T^*\eta\rangle = \langle Tx, \eta\rangle$$

whenever $x \in \text{Domain}(T)$ and $\eta \in \text{Domain}(T^*)$. But since $\eta \mapsto \langle Tx, \eta\rangle$ is continuous, it follows that $x \in \text{Domain}(T^{**})$ and $T^{**}x = Tx$. Thus $\text{Graph}(T) \subset \text{Graph}(T^{**})$. Since T is closed and $\text{Domain}(T)$ is dense, $\text{Graph}(T) = \text{Graph}(T^{**})$. The proof is complete. □

1.2. The functional analysis lemma

Suppose we have a closed, densely defined linear operator $T : H_1 \to H_2$ between two Hilbert spaces. In this paragraph, we seek to solve the following problem.

PROBLEM: Given $\alpha \in H_2$, find conditions that guarantee the existence of some $u \in H_1$ such that $Tu = \alpha$. Moreover, find a solution u with estimates for $|u|$ in terms of $|\alpha|$?

The answer to this problem is contained in the following lemma.

LEMMA 1.4 (Functional Analysis Lemma). *Let $T : H_1 \to H_2$ be a closed, densely defined operator. Suppose that for some $\alpha \in H_2$ there is a constant $C > 0$ such that for all $\beta \in \text{Domain}(T^*)$,*

$$|(\alpha, \beta)|^2 \leq C||T^*\beta||^2. \tag{1.1}$$

Then there exists $u \in H_1$ such that

$$Tu = \alpha \quad \textit{and} \quad ||u||^2 \leq C.$$

PROOF. Consider the anti-linear functional

$$\mathscr{L}(T^*\beta) = (\alpha, \beta).$$

By (1.1), if $T^*\beta = 0$ then $\beta \perp \alpha$ and thus $\mathscr{L}$ is well-defined. Again by (1.1), $\mathscr{L}$ is continuous on (the closure of) the subspace $\text{Image}(T^*) \subset H_1$. By extending $\mathscr{L}$ trivially in the orthogonal complement of $\text{Image}(T^*)$, we may assume that $\mathscr{L}$ is a continuous anti-linear functional on H_1.

By the Riesz Representation Theorem, there exists $u \in H_1$ such that for all $v \in H_1$,

$$\mathscr{L}v = (u, v) \quad \text{and} \quad ||u||^2 \leq C.$$

Restricting to $\text{Image}(T^*)$, we have

$$(u, T^*\beta) = (\alpha, \beta).$$

Since $T^{**} = T$ and $\text{Domain}(T^*)$ is dense, we have that $Tu = \alpha$. The proof is complete. □

2. The Bochner-Kodaira identity

We fix a Kähler manifold (X, g) and a holomorphic line bundle $H \to X$ with Hermitian metric $e^{-\kappa}$. We let ω denote the metric form of g, and denote by $\Omega = \partial\bar{\partial}\kappa$ the curvature of the unique Chern connection associated to $(H, e^{-\kappa})$ and by R the curvature operator of the Kähler connection associated to (X, g). These curvature operators induce multiplier-type operators on H-valued (p, q)-forms. The definitions of these multiplier-type operators will be apparent from the derivation.

2.1. The Hilbert spaces

Let φ be a (p,q)-form with values in H. If we choose local coordinates z and a local frame ξ for H, then we may write

$$\varphi = \varphi_{I\bar{J}}\xi \otimes dz^I \wedge d\bar{z}^J.$$

Let

$$g_{i\bar{j}} := g(\partial_i, \partial_{\bar{j}}).$$

As usual, $g^{i\bar{j}}$ denotes the inverse matrix, which is the matrix of the metric induced on the cotangent bundle of X, with respect to the frame $dz^1, d\bar{z}^1, ..., dz^n, d\bar{z}^n$.

It follows immediately that the quantity

$$\langle \varphi, \psi \rangle_{\kappa,g} := \varphi_{I\bar{J}}\overline{\psi_{K\bar{L}}}e^{-\kappa}g^{i_1\bar{k}_1} \cdot ... \cdot g^{i_p\bar{k}_p}g^{\ell_1\bar{j}_1} \cdot ... \cdot g^{\ell_q\bar{j}_q}$$

is independent of all the frames that were chosen. When the metrics $e^{-\kappa}$ and g are clear from the context, we might simply write $\langle \varphi, \psi \rangle$ and omit the subscripts referring to the metrics.

REMARK. It is sometimes convenient to employ the notation

$$g^{J\bar{L}} := g^{j_1\bar{\ell}_1} \cdot ... \cdot g^{j_q\bar{\ell}_q} \quad \text{and} \quad \psi^{J\bar{I}} := \psi_{K\bar{L}}g^{K\bar{I}}g^{J\bar{L}}.$$

In this case, $\langle \varphi, \psi \rangle = \varphi_{I\bar{J}}\overline{\psi^{J\bar{I}}}e^{-\kappa}$.

DEFINITION 2.1. *Let dA be a volume element on X, i.e., a smooth (n,n)-form with no zeros.*

(1) *For two compactly supported H-valued (p,q)-forms φ and ψ, we define the inner product*

$$(\varphi, \psi) = (\varphi, \psi)_{\kappa,g,dA} = \frac{1}{p!q!}\int_X \langle \varphi, \psi \rangle_{\kappa,g} dA.$$

With this inner product, the space of compactly supported smooth H-valued (p,q)-forms is an inner product space.

(2) *We let $L^2_{p,q}(\kappa, g, dA)$ denote the Hilbert space completion of the inner product space of smooth H-valued (p,q)-forms, with the inner product $(\cdot,\cdot)_{\kappa,g,dA}$.*

(3) *If the volume element is $dA_g := \frac{\omega_g^n}{n!}$ where ω_g is the Kähler form of g, then we will write*

$$L^2_{p,q}(\kappa, g, dA_g) = L^2_{p,q}(\kappa, g).$$

We will often consider adjoints of operators on $L^2(\kappa, g, dA)$. One could try to calculate these adjoints using smooth compactly supported forms. But since integration by parts is often employed in the calculation of adjoints, we lose information when we use compactly supported functions. Thus the formula for an adjoint when computed using only compactly supported forms has a special name.

DEFINITION 2.2. *Let $T : L^2_{p,q}(\kappa, g, dA) \to L^2_{p',q'}(\kappa', g', dA')$ be a densely defined operator whose domain contains all smooth, compactly supported H-valued (p,q)-forms. The formal adjoint of T is an operator*

$$T^* : L^2_{p',q'}(\kappa', g', dA') \to L^2_{p,q}(\kappa, g, dA)$$

satisfying

$$(T\varphi, \psi)' = (\varphi, T^*\psi)$$

for all smooth compactly supported H-valued (p,q)-forms φ and all smooth compactly supported H'-valued (p',q')-forms ψ.

REMARK. Since smooth compactly supported forms are dense in the Hilbert spaces in question, the formal adjoint is unique.

2.2. The action of curvature on H-valued (p,q)-forms

In order to make our formulation of the Bochner-Kodaira Identity more conceptual, we will now define a sense in which curvature of a Hermitian line bundle H and the Kähler curvature on a Kähler manifold operate on H-valued (p,q)-forms.

DEFINITION 2.3. *Let (X,g) be a Kähler manifold, $H \to X$ a holomorphic line bundle with curvature Ω. Let R denote the Kähler curvature of g and* Ricci(R) *the curvature induced on the anticanonical bundle $-K_X$, i.e., on the determinant of the tangent bundle. Then we define*

$$\begin{aligned}(T_g(R+\mathrm{Ricci}(R)+\Omega)\varphi)_{i_1\dots i_p\bar{j}_1\dots\bar{j}_q} &:= \sum_{k=1}^{q}\sum_{\nu=1}^{p} R_{i_\nu}{}^{r\bar{\ell}}{}_{\bar{j}_k}\varphi_{i_1\dots(r)_\nu\dots i_p\bar{j}_1\dots(\bar{\ell})_k\dots\bar{j}_q} \\ &\quad+\sum_{k=1}^{q}\mathrm{Ricci}(R)^{\bar{s}}{}_{\bar{j}_k}\varphi_{I_p\bar{j}_1\dots(\bar{s})_k\dots\bar{j}_q} \\ &\quad+\sum_{k=1}^{q} g^{i\bar{\ell}}\Omega_{i\bar{j}_k}\varphi_{I_p\bar{j}_1\dots(\bar{\ell})_k\dots\bar{j}_q}.\end{aligned}$$

where ξ is a frame and $\varphi = \varphi_{I\bar{J}}\xi\otimes dz^I\wedge d\bar{z}^J$.

We also have the following useful observation.

PROPOSITION 2.4. *The action of curvature on (n,q) forms is given by*

$$(T_g(R+\mathrm{Ricci}(R)+\Omega)\varphi)_{1\dots n\bar{j}_1\dots\bar{j}_q} = \sum_{k=1}^{q} g^{i\bar{\ell}}\Omega_{i\bar{j}_k}\varphi_{1\dots n,\bar{j}_1\dots(\bar{\ell})_k\dots\bar{j}_q}.$$

PROOF. Since $(i_1,\dots,i_n)=(1,\dots,n)$, in the summand

$$R_{\nu}{}^{r\bar{\ell}}{}_{\bar{j}_k}\varphi_{1\dots(r)_\nu\dots n\bar{j}_1\dots(\bar{\ell})_k\dots\bar{j}_q}$$

we obtain nontrivial terms only if $r=\nu$. Thus the sum gives us a trace of R which, by the symmetries of the Kähler curvature, cancels out the term coming from the Ricci curvature of R. The proof is complete. □

2.3. The formal identity

We write

$$\nabla_i := \nabla_{\frac{\partial}{\partial z^i}} \quad\text{and}\quad \nabla_{\bar{j}} := \nabla_{\frac{\partial}{\partial \bar{z}^j}}$$

for the directional covariant derivatives along the directions $\frac{\partial}{\partial z^i}$ and $\frac{\partial}{\partial \bar{z}^j}$ respectively. We also assume that the volume form on X is the usual one obtained from the Kähler metric. Our main goal in this section is to prove the following formula, known as the Bochner-Kodaira Identity.

THEOREM 2.5. *Let (X,g) be a Kähler manifold equipped with its Kähler volume, and $H \to X$ a holomorphic line bundle with smooth Hermitian metric $e^{-\kappa}$ and* *curvature* $\Omega = \partial\bar{\partial}\kappa$. *Let* $\bar{\partial}^*$ *be the formal adjoint of* $\bar{\partial}$, *and let* $\Box = \bar{\partial}\bar{\partial}^* + \bar{\partial}^*\bar{\partial}$. *Then one has the formal identity*

$$\Box \quad = \quad -\mathrm{Trace}(\nabla\bar{\nabla}) + T_g(R + \mathrm{Ricci}(R) + \Omega), \tag{1.2}$$

where

$$\mathrm{Trace}(\nabla\bar{\nabla}) := g^{i\bar{j}}\nabla_i\nabla_{\bar{j}}.$$

REMARK. (i) The phrase *formal identity* means that the identity holds on all smooth H-valued (p, q)-forms with compact support in X.
(ii) As mentioned earlier, the curvature operators in equation (1.2) represent the natural action of the various curvature operators on H-valued (p, q)-forms. The definition of this action was dictated by the calculations in this section.

The exterior derivative, as well as its complex counterparts ∂ and $\bar{\partial}$, are local operators, and so we can calculate them from local data. To this end, we fix a nowhere zero local section ξ of $H \to X$.

PROPOSITION 2.6. *Let* $\varphi = \varphi_{I\bar{J}}\xi \otimes dz^I \wedge d\bar{z}^J$ *be an* H*-valued* (p, q)*-form. Then*

$$\left(\bar{\partial}\varphi\right)_{I\bar{j}_0\bar{j}_1\dots\bar{j}_q} = (-1)^p \sum_{k=0}^{q} (-1)^k \nabla_{\bar{j}_k}\varphi_{I\bar{j}_0\dots\hat{\bar{j}}_k\dots\bar{j}_q}.$$

PROOF. By definition of $\bar{\partial}$ we have $\bar{\partial}(\varphi_{I\bar{J}}\xi \otimes dz^I \wedge d\bar{z}^J) = \xi \otimes (\bar{\partial}\varphi_{I\bar{J}}dz^I \wedge d\bar{z}^J)$. We then have

$$\bar{\partial}(\varphi_{I\bar{J}}dz^I \wedge d\bar{z}^J) = (-1)^p \partial_{\bar{j}}\varphi^{\alpha}_{I\bar{J}}dz^I \wedge d\bar{z}^j \wedge d\bar{z}^J.$$

It follows that

$$\begin{aligned} (\bar{\partial}\varphi)_{I\bar{j}_0\dots\bar{j}_q} &= (-1)^p \sum_{k=0}^{q} (-1)^k \partial_{\bar{j}_k}\varphi_{I\bar{j}_0\dots\hat{\bar{j}}_k\dots\bar{j}_q} dz^I \wedge d\bar{z}^{j_0} \wedge \dots \wedge d\bar{z}^{j_q} \\ &= (-1)^p \sum_{k=0}^{q} (-1)^k \nabla_{\bar{j}_k}\varphi_{I\bar{j}_0\dots\hat{\bar{j}}_k\dots\bar{j}_q} dz^I \wedge d\bar{z}^{j_0} \wedge \dots \wedge d\bar{z}^{j_q}, \end{aligned}$$

where the last equality holds since $\nabla^{0,1} = \bar{\partial}$ for the Kähler connection. □

Next we calculate the formal adjoint of $\bar{\partial}$:

PROPOSITION 2.7. *Let* $\varphi = \varphi_{I\bar{J}}\xi \otimes dz^I \wedge d\bar{z}^J$ *be an* H*-valued* (p, q)*-form. Then*

$$\left(\bar{\partial}^*\varphi\right)_{I\bar{j}_1\dots\bar{j}_{q-1}} = (-1)^{p+1} g^{\bar{j}i}\nabla_i\varphi_{I\bar{j}\bar{j}_1\dots\bar{j}_{q-1}}.$$

PROOF. One has

$$\begin{aligned} &(\bar{\partial}^*\varphi, \psi) = (\varphi, \bar{\partial}\psi) \\ &= \frac{1}{p!q!}\int_X \varphi_{I\bar{j}_1\dots\bar{j}_q}(-1)^p \sum_{k=1}^{q} \overline{(-1)^{k+1} g^{j_k\bar{j}'_k}\nabla_{\overline{j'_k}}\psi^{j_1\dots\hat{j}_k\dots j_q\bar{I}}} e^{-\kappa}dA \\ &= \frac{1}{p!q!}\int_X (-1)^{p+1}\sum_{k=1}^{q} g^{j'_k\bar{j}_k}\left(\nabla_{j'_k}(-1)^{k+1}\varphi_{I\bar{j}_1\dots\bar{j}_q}\right)\overline{\psi^{j_1\dots\hat{j}_k\dots j_q\bar{I}}} e^{-\kappa}dA \\ &= \frac{1}{p!(q-1)!}\int_X \left((-1)^{p+1}g^{i\bar{j}}\nabla_i\varphi^{\alpha}_{I\bar{j}\bar{j}_1\dots\bar{j}_{q-1}}\right)\overline{\psi^{j_1\dots j_{q-1}\bar{I}}} e^{-\kappa}dA, \end{aligned}$$

Where the second-to-last inequality follows from the metric compatibility of the connection. This completes the proof. □

Combining Propositions 2.6 and 2.7, we obtain the following lemma.

LEMMA 2.8. *Let φ be an H-valued (p,q)-form. Then*

$$(\Box\varphi)_{I\bar{J}} = -g^{i\bar{j}}\nabla_i\nabla_{\bar{j}}\varphi_{I\bar{J}} - \sum_{k=1}^{q} g^{i\bar{\ell}}\left[\nabla_{\bar{j}_k}, \nabla_i\right]\varphi_{I\bar{j}_1\ldots\overline{(\ell)_k}\ldots\bar{j}_q}, \tag{1.3}$$

where $(\ell)_k$ indicates that ℓ is in the slot of the kth index.

PROOF. To make the proof easier to follow, it may be helpful to keep in mind that in order to apply the lemmas giving formulas for $\bar{\partial}$ and $\bar{\partial}^*$, some reindexing of the formulas of those lemmas is required. With that in mind, we have

$$\begin{aligned}
(\bar{\partial}^*\bar{\partial}\varphi)_{I\bar{j}_1\ldots\bar{j}_q} &= (-1)^{p+1}g^{i\bar{j}}\nabla_i(\bar{\partial}\varphi)_{I\bar{j}\bar{j}_1\ldots\bar{j}_q}\\
&= (-1)^{p+1}g^{i\bar{j}}\nabla_i\Big((-1)^p\nabla_{\bar{j}}\varphi_{I\bar{j}_1\ldots\bar{j}_q}\\
&\qquad +(-1)^p\sum_{k=1}^{q}(-1)^k\nabla_{\bar{j}_k}\varphi_{I\bar{j}\bar{j}_1\ldots\widehat{\bar{j}_k}\ldots\bar{j}_q}\Big)\\
&= -g^{i\bar{j}}\nabla_i\nabla_{\bar{j}}\varphi_{I\bar{j}_1\ldots\bar{j}_q} + \sum_{k=1}^{q} g^{i\bar{j}}\nabla_i\nabla_{\bar{j}_k}\varphi_{I\bar{j}_1\ldots(\bar{j})_k\ldots\bar{j}_q},
\end{aligned}$$

and

$$\begin{aligned}
(\bar{\partial}\bar{\partial}^*\varphi)_{I\bar{j}_1\ldots\bar{j}_q} &= (-1)^p\sum_{k=1}^{q}(-1)^{k+1}\nabla_{\bar{j}_k}(\bar{\partial}^*\varphi)_{I\bar{j}_1\ldots\widehat{\bar{j}_k}\ldots\bar{j}_q}\\
&= (-1)^p\sum_{k=1}^{q}(-1)^{k+1}\nabla_{\bar{j}_k}\left((-1)^{p+1}g^{i\bar{j}}\nabla_i\varphi_{I\bar{j}\bar{j}_1\ldots\widehat{\bar{j}_k}\ldots\bar{j}_q}\right)\\
&= -\sum_{k=1}^{q}\nabla_{\bar{j}_k}\left(g^{i\bar{j}}\nabla_i\varphi_{I\bar{j}_1\ldots(\bar{j})_k\ldots\bar{j}_q}\right)\\
&= -\sum_{k=1}^{q} g^{i\bar{j}}\nabla_{\bar{j}_k}\nabla_i\varphi_{I\bar{j}_1\ldots(\bar{j})_k\ldots\bar{j}_q},
\end{aligned}$$

where the last equality comes from the fact that, since ∇ is the metric connection, $\nabla g^{i\bar{j}} = 0$. The proof is then completed by summing these two calculations. □

Our next task is to show that the commutator $[\nabla_i, \nabla_{\bar{j}}]$ is a zeroth order operator whose action is the same as the action of the curvature tensor. We begin with the following lemma.

LEMMA 2.9. *If X is Kähler with Kähler curvature R and ξ is a section of the holomorphic line bundle $H \to X$ whose curvature is Ω, then*

$$[\nabla_i, \nabla_{\bar{j}}](\xi) = \Omega_{i\bar{j}}\xi \tag{1.4}$$

$$[\nabla_i, \nabla_{\bar{j}}](\eta_\ell dz^\ell) = R_{k\ \,i\bar{j}}^{\ \ell}\eta_\ell dz^k \tag{1.5}$$

$$[\nabla_i, \nabla_{\bar{j}}](\theta_{\bar{k}} d\bar{z}^k) = R^{\bar{k}}_{\ \bar{\ell}i\bar{j}}\theta_{\bar{k}} d\bar{z}^\ell \tag{1.6}$$

In particular, $[\nabla_i, \nabla_{\bar{j}}]$ is a 0th order operator.

REMARK. Of course, in each of these formulas the connection is different. We are using the connection associated to the vector bundle in question: respective to the formulas (1.4)-(1.6), these bundles are H, $(T^*_X)^{1,0}$ and $(T^*_X)^{0,1}$. As a consequence, we also know the value of the operator $[\nabla_i, \nabla_{\bar{j}}]$ on H-valued (p,q)-forms.

PROOF OF LEMMA 2.9. Since the derivations for (1.5) and (1.6) are analogous, we will only prove formula (1.4). To this end, $\nabla_{\bar{j}}\nabla_i s = \nabla_{\bar{j}}(\partial_i s + \omega_i s) = \partial_{\bar{j}}\partial_i s + (\partial_{\bar{j}}\omega_i)s + \omega_i\partial_{\bar{j}} s$. Similarly one has $\nabla_i\nabla_{\bar{j}} s = \partial_i\partial_{\bar{j}} s + \omega_i\partial_{\bar{j}} s$. Subtracting these two expressions and recalling the formula for the curvature of the Chern connections gives us the desired result. □

PROOF OF THEOREM 2.5. To compute the covariant derivative on a product of vector bundles with connection, one has to apply the Leibniz rule together with covariant differentiation of the various factors. This fact, together with lemmas 2.9 and 2.8 will be used in our computation below.

To proceed, let $\varphi = \varphi_{I\bar{J}}\xi \otimes dz^I \wedge d\bar{z}^J$. Then

$$\begin{aligned}
\nabla_i\nabla_{\bar{j}}(\xi \otimes dz^I \wedge d\bar{z}^J) &= \nabla_i\left((\nabla_{\bar{j}}\xi) \otimes dz^I \wedge d\bar{z}^J\right) \\
&= (\nabla_i\nabla_{\bar{j}}\xi) \otimes dz^I \wedge d\bar{z}^J \\
&\quad + \sum_{\nu=1}^{p} (\nabla_{\bar{j}}\xi) \otimes dz^{i_1} \wedge \dots \wedge \nabla_i dz^{i_\nu} \wedge \dots \wedge dz^{i_p} \wedge d\bar{z}^J \\
&\quad + \sum_{\mu=1}^{q} (\nabla_{\bar{j}}\xi) \otimes dz^I \wedge d\bar{z}^{j_1} \wedge \dots \wedge \nabla_i d\bar{z}^{j_\mu} \wedge \dots \wedge d\bar{z}^{j_q}.
\end{aligned}$$

Similarly,

$$\begin{aligned}
\nabla_{\bar{j}}\nabla_i(\xi \otimes dz^I \wedge d\bar{z}^J) &= \nabla_{\bar{j}}\Big((\nabla_i\xi) \otimes dz^I \wedge d\bar{z}^J \\
&\quad + \sum_{\nu=1}^{p} \xi \otimes dz^{i_1} \wedge \dots \wedge \nabla_i dz^{i_\nu} \wedge \dots \wedge dz^{i_p} \wedge d\bar{z}^J \\
&\quad + \sum_{\mu=1}^{q} \xi \otimes dz^I \wedge d\bar{z}^{j_1} \wedge \dots \wedge \nabla_i d\bar{z}^{j_\mu} \wedge \dots \wedge dz^{j_q}\Big) \\
&= (\nabla_{\bar{j}}\nabla_i\xi) \otimes dz^I \wedge d\bar{z}^J \\
&\quad + \sum_{\nu=1}^{p} (\nabla_{\bar{j}}\xi) \otimes dz^{i_1} \wedge \dots \wedge \nabla_i dz^{i_\nu} \wedge \dots \wedge dz^{i_p} \wedge d\bar{z}^J \\
&\quad + \sum_{\mu=1}^{q} (\nabla_{\bar{j}}\xi) \otimes dz^I \wedge d\bar{z}^{j_1} \wedge \dots \wedge \nabla_i d\bar{z}^{j_\mu} \wedge \dots \wedge d\bar{z}^{j_q} \\
&\quad + \sum_{\nu=1}^{p} \xi \otimes dz^{i_1} \wedge \dots \wedge \nabla_{\bar{j}}\nabla_i dz^{i_\nu} \wedge \dots \wedge dz^{i_p} \wedge d\bar{z}^J \\
&\quad + \sum_{\mu=1}^{q} \xi \otimes dz^I \wedge d\bar{z}^{j_1} \wedge \dots \wedge \nabla_{\bar{j}}\nabla_i d\bar{z}^{j_\mu} \wedge \dots \wedge dz^{j_q}.
\end{aligned}$$

By subtracting these two calculations and applying Lemma 2.9, we have

$$\begin{aligned}
&[\nabla_i, \nabla_{\bar{j}}](\varphi_{I\bar{J}}\xi \otimes dz^I \wedge d\bar{z}^J) \\
= \;& \varphi_{I\bar{J}}[\nabla_i, \nabla_{\bar{j}}](\xi \otimes dz^I \wedge d\bar{z}^J) \\
= \;& \varphi_{I\bar{J}}\Omega_{i\bar{j}}\xi \otimes dz^I \wedge d\bar{z}^J \\
& + \sum_{\nu=1}^{p} \varphi_{I\bar{J}} R_k{}^{i_\nu}{}_{i\bar{j}}\xi \otimes dz^{i_1} \wedge \ldots \wedge (dz^k)_\nu \wedge \ldots \wedge dz^{i_p} \wedge d\bar{z}^J \\
& + \sum_{\mu=1}^{p} \varphi_{I\bar{J}} R^{\bar{j}_\mu}{}_{\bar{\ell}i\bar{j}}\xi \otimes dz^I \wedge d\bar{z}^{j_1} \wedge \ldots \wedge (d\bar{z}^\ell)_\mu \wedge \ldots \wedge d\bar{z}^{j_q},
\end{aligned}$$

where $(\theta)_s$ means that θ appears in the sth factor in the product. This formula can be rewritten as follows.

$$\left([\nabla_i, \nabla_{\bar{j}}]\varphi\right)_{I\bar{J}} = \Omega_{i\bar{j}}\varphi_{I\bar{J}} + \sum_{\nu=1}^{p} R_{i_\nu}{}^{k}{}_{i\bar{j}}\varphi_{i_1\ldots(k)_\nu\ldots i_p\bar{J}} + \sum_{\mu=1}^{q} R^{\bar{\ell}}{}_{\bar{j}_\mu i\bar{j}}\varphi_{I\bar{j}_1\ldots(\bar{\ell})_\mu\ldots\bar{j}_q}.$$

Substituting this into equation (1.3), one has

$$\begin{aligned}
(\Box\varphi)_{I\bar{J}} \;=\;& -g^{i\bar{j}}\nabla_i\nabla_{\bar{j}}\varphi_{I\bar{J}} + \sum_{k=1}^{q} g^{i\bar{\ell}}\Omega_{i\bar{j}_k}\varphi_{I_p\bar{j}_1\ldots(\bar{\ell})_k\ldots\bar{j}_q} \\
& + \sum_{k=1}^{q}\sum_{\nu=1}^{p} R_{i_\nu}{}^{r\bar{\ell}}{}_{\bar{j}_k}\varphi_{i_1\ldots(r)_\nu\ldots i_p\bar{j}_1\ldots(\bar{\ell})_k\ldots\bar{j}_q} \\
& + \sum_{k=1}^{q}\sum_{\mu\neq k} R^{\bar{s}}{}_{\bar{j}_\mu}{}^{\bar{\ell}}{}_{\bar{j}_k}\varphi_{I\bar{j}_1\ldots(\bar{s})_\mu\ldots(\bar{\ell})_k\ldots\bar{j}_q} \\
& + \sum_{k=1}^{q} g^{i\bar{\ell}} R^{\bar{s}}{}_{\bar{j}_k i\bar{\ell}}\varphi_{I\bar{j}_1\ldots(\bar{s})_k\ldots\bar{j}_q} \\
=\;& -g^{i\bar{j}}\nabla_i\nabla_{\bar{j}}\varphi_{I\bar{J}} + \sum_{k=1}^{q} g^{i\bar{\ell}}\Omega_{i\bar{j}_k}\varphi_{I\bar{j}_1\ldots(\bar{\ell})_k\ldots\bar{j}_q} \\
& + \sum_{k=1}^{q}\sum_{\nu=1}^{p} R_{i_\nu}{}^{r\bar{\ell}}{}_{\bar{j}_k}\varphi_{i_1\ldots(r)_\nu\ldots i_p\bar{j}_1\ldots(\bar{\ell})_k\ldots\bar{j}_q} \\
& + \sum_{k=1}^{q} \mathrm{Ricci}(R)^{\bar{s}}{}_{\bar{j}_k}\varphi^{\alpha}_{I\bar{j}_1\ldots(\bar{s})_k\ldots\bar{j}_q},
\end{aligned}$$

where the last equality follows from the definition of Ricci curvature (the minus sign on the last term comes from the symmetry of the curvature), together with the fact that, by the symmetry of curvature and the skew symmetry of differential forms,

$$\sum_{k=1}^{q}\sum_{\mu\neq k} R^{\bar{s}}{}_{\bar{j}_\mu}{}^{\bar{\ell}}{}_{\bar{j}_k}\varphi_{I_p\bar{j}_1\ldots(\bar{s})_\mu\ldots(\bar{\ell})_k\ldots\bar{j}_q} = 0.$$

This completes the proof. □

REMARK. The action of the curvature term R on $(0,q)$-forms is trivial. Indeed, the $(0,1)$-part of the Kähler curvature is $\bar{\partial}$, and thus satisfies $\bar{\partial}^2 = 0$.

3. Manifolds with boundary

We will be interested in solving the $\bar\partial$ equation on a smoothly bounded domain in our complex manifold. Consequently, we will need some information about the complex geometry of the boundary of the domain in question. There is a standard geometric positivity condition for domains in complex manifolds; this is the notion of pseudoconvexity. The role of pseudoconvexity in solving the $\bar\partial$ equation is revealed via the so-called *Morrey trick*.

3.1. Manifolds we will consider

For various reasons, we are interested in the following situation. Let X be Kähler with holomorphic Hermitian line bundle $H \to X$. Suppose given a continuous function $\rho : X \to \mathbb{R}$ with the following properties.

(1) For every interval $[a,b]$, $\rho^{-1}([a,b])$ is compact. (That is to say, ρ is proper.)
(2) The set $Y : \{x \in X \ ; \ \rho(x) < 0\}$ is open and ρ is smooth in a neighborhood of the (compact) boundary ∂Y of Y.
(3) The differential $d\rho$ is nowhere vanishing on ∂Y.

(It follows from Property (3) that ∂Y is a smooth real hypersurface whose normal bundle in X is trivial.) Such a function is called a *defining function* for Y.

By rescaling ρ with a positive function if necessary, we may assume that ρ is a *Levi* defining function, which by definition means that

$$|d\rho| \equiv 1 \quad \text{on} \quad \partial Y.$$

3.2. The operators T, S and T^*

We would like now to extend $\bar\partial$ to L^2 and calculate the domain of the adjoint of this extension. As we already mentioned, $\bar\partial$ is defined on L^2 in the sense of currents, and maps smooth (p,q)-forms to smooth $(p,q+1)$ forms.

DEFINITION 3.1. *Let* Domain(T_q) *consist of all those* $(p,q)-$*forms* φ *in* $L^2_{p,q}(\kappa, g)$ *such that* $\bar\partial\varphi$, *computed in the sense of currents, is represented by an element of* $L^2_{p,q+1}(\kappa, g)$. *For all* $\varphi \in$ *Domain*(T_q), *we set*

$$T_q\varphi := \bar\partial\varphi.$$

In the rest of this lecture, we adopt the following convention: We fix $q \in \{1, ..., n\}$, write

$$T = T_q,$$

and let

$$S = T_{q+1} \ \text{ if } \ q < n, \quad \text{and} \quad S = 0 \ \text{ if } \ q = n.$$

3.3. The domain of T^*

PROPOSITION 3.2. *A smooth* H*-valued* (p,q)*-form* φ *on* $\overline{Y}$ *belongs to* Domain(T^*) *if and only if*

$$g^{s\bar t}(\partial_s\rho)\varphi_{I\bar t\bar J} = 0. \tag{1.7}$$

PROOF. Let φ be any smooth H-valued form. From the integration-by-parts formula, we can expect to have $(\varphi, \bar\partial\psi) = (\bar\partial^*\varphi, \psi) +$ a boundary integral. We can use ρ as the radial coordinate (since it is a local coordinate near the boundary of Y. Thus the volume form, restricted to ∂Y, is

$$dA = d\rho \wedge d\sigma_{\partial Y}$$

for some smooth $2n-1$-form $d\sigma_{\partial Y}$ on ∂Y.

REMARK. This restriction can be a little confusing. We think of dA is a section of a real line bundle, namely the determinant of the real cotangent bundle of Y, and we are restricting that line bundle to ∂Y. Since $d\rho$ is non-zero and annihilates the tangent bundle of ∂Y, this restricted line bundle is naturally identified with the determinant of the real cotangent bundle of ∂Y, of which $d\sigma_{\partial Y}$ is a section.

We can thus compute as follows:

$$\begin{aligned}(\varphi, \bar{\partial}\psi) &= \int_Y \varphi_{I\bar{J}}(-1)^{p+1}\sum_{k=1}^{q}(-1)^{k-1}\overline{\nabla^{j_k}\psi^{\bar{I}j_1\ldots\hat{j}_k\ldots j_q}}dA \\ &= (\bar{\partial}^*\varphi, \psi) + \frac{(-1)^{p+1}}{p!(q-1)!}\int_{\partial Y}\varphi_{I\bar{t}J}g^{s\bar{t}}\overline{\psi^{\bar{I}J}}\,(\partial_s\rho)\,d\sigma_{\partial Y}.\end{aligned}$$

Now suppose $\varphi \in \text{Domain}(T^*)$. Then by definition, the linear functional

$$\text{Domain}(T) \ni \psi \mapsto (\varphi, T\psi) \tag{1.8}$$

is continuous. Let $\varepsilon > 0$ and let χ_ε be a smooth function taking values in $[0,1]$, that is $\equiv 1$ on $\{\rho < -\varepsilon\}$ and has compact support in Y. We claim that $\chi_\varepsilon\psi \in \text{Domain}(T)$. Indeed, in the sense of distributions,

$$T(\chi_\varepsilon\psi) = \bar{\partial}\chi_\varepsilon \wedge \psi + \chi_\varepsilon T\psi,$$

which is clearly L^2. But $\chi_\varepsilon\psi$ has compact support, so

$$(\varphi, T(\chi_\varepsilon\psi)) = (\bar{\partial}^*\varphi, \chi_\varepsilon\psi).$$

Since trivially $\chi_\varepsilon\psi \to \psi$ in L^2, the continuity of the functional (1.8) implies that for all $\psi \in \text{Domain}(T)$,

$$\int_{\partial Y}\varphi_{I\bar{t}J}g^{s\bar{t}}\overline{\psi^{\bar{I}J}}\,(\partial_s\rho)\,d\sigma_Y = 0.$$

But since smooth H-valued (p,q)-forms ψ on a neighborhood of $\overline{Y}$ are contained in Domain(T), the result follows. □

We are now in a position to identify the domain of T^*. Indeed, the calculation in the proof of Proposition 3.2 holds even if $\varphi \in L^2_{p,q+1}(\kappa, g)$, provided we replace $\bar{\partial}$ by T. Thus we have the following proposition.

PROPOSITION 3.3. Domain(T^*) *consists of those forms* $\varphi \in L^2_{p,q+1}(\kappa, g)$ *such that* $\bar{\partial}^*\varphi \in L^2_{p,q}(\kappa, g)$ *and* (1.7) *holds.*

3.4. C.B. Morrey's trick

Our next goal to to extend the formal Bochner-Kodaira identity to the case of smooth forms whose support is not necessarily compact. We have the following theorem.

THEOREM 3.4. *Let* $\varphi \in$ *Domain*(T^*) *be a smooth, H-valued (p,q)-form. Then the following integral identity holds.*

$$\begin{aligned}||\bar{\partial}\varphi||^2_Y + ||\bar{\partial}^*\varphi||^2_Y &= (T_g(R + \text{Ricci}(R) + \Omega)\varphi, \varphi)_Y \\ &\quad + ||\bar{\nabla}\varphi||^2_Y + \frac{1}{p!(q-1)!}\int_{\partial_Y} g^{s\bar{\ell}}\,(\partial_s\partial_{\bar{t}}\rho)\,\varphi_{I\bar{\ell}\bar{J}}\overline{\varphi^{\bar{I}tJ}}d\sigma_{\partial Y}\end{aligned}$$

PROOF. Let $J = (j_1 ... j_q)$. Then

$$
\begin{aligned}
||\bar\partial\varphi||^2_Y &= \frac{(-1)^p}{p!(q+1)!}\int_Y (\bar\partial\varphi)_{I\bar j_0 \bar J}\sum_{k=0}^q (-1)^k \overline{\nabla^{j_k}\varphi^{j_0...\widehat{j_k}...j_q \bar I}}dA \\
&= (\bar\partial^*\bar\partial\varphi,\varphi)_Y \\
&\quad + \frac{(-1)^p}{p!(q+1)!}\sum_{k=0}^q\int_{\partial Y}(\nabla^{\bar j_k}\rho)(-1)^k(\bar\partial\varphi)_{I\bar j_0 ...\bar j_k ... \bar j_q}\overline{\varphi^{j_0...\widehat{j_k}...j_q\bar I}}d\sigma_{\partial Y} \\
&= (\bar\partial^*\bar\partial\varphi,\varphi)_Y + \frac{(-1)^p}{p!q!}\int_Y (g^{s\bar j}\partial_s\rho)(\bar\partial\varphi)_{I\bar j\bar J}\overline{\varphi^{JI}}d\sigma_{\partial Y}.
\end{aligned}
$$

Next, since we assume that $\varphi \in \mathrm{Domain}(T^*)$,

$$
||\bar\partial^*\varphi||^2_Y = (\bar\partial\bar\partial^*\varphi,\varphi)_Y.
$$

Combining, we have

$$
\begin{aligned}
||\bar\partial\varphi||^2_Y + ||\bar\partial^*\varphi||^2_Y &= (\Box\varphi,\varphi)_Y \\
&\quad + \frac{(-1)^p}{p!q!}\int_{\partial Y}(g^{s\bar j}\partial_s\rho)(\bar\partial\varphi)_{I\bar j\bar J}\overline{\varphi^{JI}}d\sigma_{\partial Y}.
\end{aligned}
$$

Now, from the Bochner-Kodaira identity we have

$$
\begin{aligned}
(\Box\varphi,\varphi)_Y &= (-\mathrm{Trace}(\nabla\bar\nabla)\varphi,\varphi)_Y + (T_g(R+\mathrm{Ricci}(R)+\Omega)\varphi,\varphi)_Y \\
&= (T_g(R+\mathrm{Ricci}(R)+\Omega)\varphi,\varphi)_Y + ||\bar\nabla\varphi||^2_Y \\
&\quad - \frac{1}{p!q!}\int_{\partial Y}(g^{s\bar j}\partial_s\rho)\nabla_{\bar j}\varphi_{I\bar J}\overline{\varphi^{JI}}d\sigma_{\partial Y}.
\end{aligned}
$$

The last term comes again from integration by parts. Putting these two calculations together, we have

$$
\begin{aligned}
||\bar\partial^*\varphi||^2_Y + ||\bar\partial\varphi||^2_Y &= (T_g(R+\mathrm{Ricci}(R)+\Omega)\varphi,\varphi)_Y + ||\bar\nabla\varphi||^2_Y \\
&\quad + \frac{1}{p!q!}\int_{\partial Y}(g^{s\bar j}\partial_s\rho)\left((-1)^p(\bar\partial\varphi)_{I\bar j\bar J}\overline{\varphi^{JI}} - \nabla_{\bar j}\varphi_{I\bar J}\overline{\varphi^{JI}}\right)d\sigma_{\partial Y}.
\end{aligned}
$$

From the formula for $\bar\partial$ given by Proposition 2.6, we have

$$
(-1)^p(\bar\partial\varphi)_{I\bar j\bar J} = \nabla_{\bar j}\varphi_{I\bar J} - \sum_{k=1}^q \nabla_{\bar j_k}\varphi_{I\bar j_1...\overline{(j)_k}...\bar j_q}.
$$

Thus

$$
\begin{aligned}
&\int_{\partial Y}(g^{s\bar j}\partial_s\rho)\left((-1)^p(\bar\partial\varphi)_{I\bar j\bar J}\overline{\varphi^{JI}} - \nabla_{\bar j}\varphi_{I\bar J}\overline{\varphi^{JI}}\right)d\sigma_{\partial Y} \\
&= -\sum_{k=1}^q\int_{\partial Y}(g^{s\bar j}\partial_s\rho)\left(\nabla_{\bar j_k}\varphi_{I\bar j_1...\overline{(j)_k}...j_q}\right)\overline{\varphi^{JI}}d\sigma_{\partial Y}.
\end{aligned}
$$

Now, by Lemma 3.2,

$$
g^{s\bar j}(\partial_s\rho)\varphi_{I\bar j_1...\overline{(j)_k}...\bar j_q}
$$

vanishes on ∂Y, and thus one can write, on $\overline{Y}$,

$$
g^{s\bar j}(\partial_s\rho)\varphi_{I\bar j_1...\overline{(j)_k}...\bar j_q} = \rho\psi_{I\bar j_1...\widehat{\bar j_k}...\bar j_q}
$$

for some smooth H-valued (p,q)-form ψ. If we now apply to this identity the operator

$$\sum_{k=1}^{q} \overline{\varphi^{J\bar{I}}} \nabla_{\bar{j}_k},$$

we obtain, using the fact that ∇ is the Kähler connection, that

$$\begin{aligned}&\sum_{k=1}^{q}(g^{s\bar{j}}\nabla_{\bar{j}_k}\partial_s\rho)\varphi_{I\bar{j}_1\ldots\overline{(j)_k}\ldots j_q}\overline{\varphi^{J\bar{I}}}+\sum_{k=1}^{q}(g^{s\bar{j}}\partial_s\rho)\left(\nabla_{\bar{j}_k}\varphi_{I\bar{j}_1\ldots\overline{(j)_k}\ldots j_q}\right)\overline{\varphi^{J\bar{I}}}\\ &=\sum_{k=1}^{q}\overline{\varphi^{J\bar{I}}}\left(\nabla_{\bar{j}_k}\rho\right)\psi^{I_p\bar{j}_1\ldots\widehat{\bar{j}_k}\ldots\bar{j}_q}+\sum_{k=1}^{q}\rho\overline{\varphi^{J\bar{I}}}\nabla_{\bar{j}_k}\psi^{I_p\bar{j}_1\ldots\widehat{\bar{j}_k}\ldots\bar{j}_q},\end{aligned}$$

and the last two sums vanish on ∂Y since $\rho = 0$ on ∂Y and $\varphi_{I_p\bar{J}_q}\left(\nabla^{\bar{j}_k}\rho\right)$ vanishes for all $\varphi \in \text{Domain}(T^*)$ according to Lemma 3.2. It follows, upon a simple reindexing, that

$$-\sum_{k=1}^{q}\int_{\partial Y}(g^{s\bar{j}}\partial_s\rho)\left(\nabla_{\bar{j}_k}\varphi_{I\bar{j}_1\ldots\overline{(j)_k}\ldots j_q}\right)\overline{\varphi^{J\bar{I}}}d\sigma_{\partial Y} = q\int_{\partial Y} g^{\bar{j}s}\left(\partial_{\bar{t}}\partial_s\rho\right)\varphi_{I\bar{j}\bar{J}'}\overline{\varphi^{tJ'\bar{I}}},$$

where $J' = (j_1 \ldots j_{q-1})$. The proof is complete. □

4. Density of smooth forms in the graph norm

The identity

$$\begin{aligned}||\bar{\partial}\varphi||_Y^2 + ||\bar{\partial}^*\varphi||_Y^2 &= (T_g(R + \text{Ricci}(R) + \Omega)\varphi, \varphi)_Y \\ &\quad + ||\bar{\nabla}\varphi||_Y^2 + \frac{1}{p!(q-1)!}\int_{\partial_Y} g^{s\bar{\ell}}\left(\partial_s\partial_{\bar{t}}\rho\right)\varphi_{I\bar{\ell}\bar{J}}\overline{\varphi^{ItJ}}d\sigma_{\partial Y}\end{aligned} \tag{1.9}$$

is proved in Theorem 3.4 only for smooth forms. We know that smooth forms are dense in $L^2_{*,*}(\kappa, g)$, but we do not know that this is the case for the images of the set of smooth forms under T^* or S, nor do we know that the set of restrictions of smooth forms to the boundary is dense.

In order to apply the Hilbert space method the (1.9), we would need to extend the latter to Domain$(T^*)\cap$Domain(S), and also to prove that the set of restrictions of smooth forms to the boundary of Y satisfying the boundary condition (1.7) is dense in the set of L^2-forms on the boundary satisfying the boundary condition (1.7).

We will not take exactly this approach. Instead, in the next section we will impose a geometric condition on the boundary ∂Y that guarantees the non-negativity of the boundary term in (1.9). To handle the remaining terms, we now establish the following theorem.

THEOREM 4.1. *In terms of the graph norm*

$$\varphi \mapsto ||\varphi||_{p,q;\kappa,g} + ||T^*\varphi||_{p,q-1;\kappa,g} + ||S\varphi||_{p,q+1;\kappa,g},$$

the subspace of smooth H-valued (p,q)-forms on $\overline{Y}$ that lie in Domain$(T^*)\cap$ Domain(S) *is dense.*

I. Friedrichs-type lemmas. By convolution with mollifiers, i.e., smooth functions with compact support and total integral 1, one establishes the standard fact that on domains in $\mathbb{R}^n$ one can approximate L^2-functions by smooth, compactly supported functions. This construction is easily adapted to manifolds by localizing in elements of an open cover and using a different mollifier in each element of the cover.

We are going to employ this localization procedure again below, in a slightly more delicate situation. In this section we will prove the needed local results for approximation not just in L^2, but also in norms involving certain derivatives. The types of derivatives we want to consider should include those appearing in the graph norm in the statement of Theorem 4.1.

Given a compactly supported function f defined on a domain in $\mathbb{R}^m$ containing the origin, and such that $\int_{\mathbb{R}^m} f dV = 1$ (such functions are called *mollifiers*), we write $f_\varepsilon(x) := \varepsilon^{-m} f(\varepsilon^{-1}x)$. Given another function g on $\mathbb{R}^m$, we write $g^\varepsilon = g * f_\varepsilon$.

Let us begin with the following lemma.

LEMMA 4.2. *Let h be a smooth function in $\mathbb{R}^m$ and denote by h^ε be a regularization of h by convolution with a mollifier φ. Let f be a smooth function on $\mathbb{R}^n$. Then for any $v \in L^2(\mathbb{R}^m)$ with compact support,*

$$f(\partial_{x^i} v * h^\varepsilon) - (f\partial_{x^i} v) * h^\varepsilon \to 0 \quad \text{in } L^2(\mathbb{R}^m) \text{ as } \varepsilon \to 0.$$

PROOF. We have

$$\begin{aligned} f(x)(\partial_{x^i} v * h^\varepsilon)(x) &= \int_{\mathbb{R}^m}\int_{\mathbb{R}^m} f(x)\partial_{x^i} v(x-y)\varphi_\varepsilon(y-z)h(z)dV(z)dV(y) \\ &= -\int_{\mathbb{R}^m}\int_{\mathbb{R}^m} f(x)\partial_{y^i} v(x-y)\varphi_\varepsilon(y-z)h(z)dV(z)dV(y) \\ &= \int_{\mathbb{R}^m}\int_{\mathbb{R}^m} f(x)v(x-y)\partial_{y^i}\varphi_\varepsilon(y-z)h(z)dV(z)dV(y) \\ &= -\int_{\mathbb{R}^m}\int_{\mathbb{R}^m} f(x)v(x-y)\partial_{z^i}\varphi_\varepsilon(y-z)h(z)dV(z)dV(y) \\ &= \int_{\mathbb{R}^m}\int_{\mathbb{R}^m} f(x)v(x-y)\varphi_\varepsilon(y-z)\partial_{z^i}h(z)dV(z)dV(y). \end{aligned}$$

Similar calculations show that

$$\begin{aligned} (f\partial_{x^i} v) * h^\varepsilon(x) &= \int_{\mathbb{R}^m}\int_{\mathbb{R}^m} \partial_{x^i} f(x-y)v(x-y)\varphi_\varepsilon(y-z)h(z)dV(z)dV(y) \\ &\quad + \int_{\mathbb{R}^m}\int_{\mathbb{R}^m} f(x-y)v(x-y)\varphi_\varepsilon(y-z)\partial_{z^i}h(z)dV(z)dV(y). \end{aligned}$$

Standard estimates from real analysis then show that

$$||(f\partial_{x^i} v) * h^\varepsilon - f(\partial_{x^i} v * h^\varepsilon)||_{L^2} \le CMm||v||_{L^2},$$

where M involves a bound for f and the derivative of f on the support of V, and m involves the derivative of h on the support of v. We are now in a position to apply the dominated convergence theorem, from which the result follows. □

Lemma 4.2 will be used to handle the interior of Y. The boundary points require a more subtle smoothing technique, to which we now turn.

For an unknown J-tuple $u = (u^1, ..., u^J)$ of functions, we consider the system of first order linear PDE

$$Au + Bu = f, \tag{1.10}$$

on the closed lower half-space

$$U^- := \{x \in \mathbb{R}^m \ ; \ x^m \leq 0\},$$

where for $1 \leq k \leq K$

$$(Au)^k = a_j^{ik} D_i u^j, \quad (Bu)^k = b_j^k u^j \quad \text{and} \quad f = (f^1, ..., f^K).$$

We will assume that the functions a_j^{ik}, b_j^k, $(1 \leq i \leq m, 1 \leq j \leq J, 1 \leq k \leq K)$ are smooth, while the functions $f^k \in L^2(U^-)$, $1 \leq k \leq K$, have compact support in U^-. Finally, we also set

$$U^+ = \{x \in \mathbb{R}^m \ ; \ x^m \geq 0\}.$$

Then we have the following lemma.

LEMMA 4.3. *Suppose we have a solution u of* (1.10) *such that the components of u and f are in $L^2(U^-)$ and vanish outside a compact subset U of U^-. Then there is a sequence $\{u_j\} \subset \mathscr{C}^\infty(U^-)$ vanishing outside a fixed compact subset of U^-, such that*

$$||u_j - u||_{(L^2(U^-))^J} + ||Au_j + Bu_j - f||_{(L^2(U^-))^K} \to 0 \quad \textit{as} \quad j \to \infty.$$

Moreover, if u vanishes on ∂U^-, in the sense that if we define f and u to be 0 in U^+ then u still satisfies (1.10)*, then u_j can be chosen with compact support in $\{x^m < 0\}$.*

PROOF. Extend u to be 0 in U – interior(U^-). Then

$$Au + Bu = f + g$$

where g is supported on ∂U^-. (In the case where the solution u satisfies the stated boundary conditions, $g = 0$ by hypothesis.) Let φ be a smooth function with compact support in the set

$$\Gamma = \{x \in \mathbb{R}^m \ ; \ |x - (0, ..., 0, 1)| < 1\}$$

and total integral 1, and let $\varepsilon > 0$ be smaller than the distance between Supp(φ) and the boundary of Γ. We use φ as a mollifier. Then $u * \varphi_\varepsilon$ has components in $\mathscr{C}_0^\infty(U)$ and by Lemma 4.2 we have

$$A(u * \varphi_\varepsilon) + B(u * \varphi_\varepsilon) - f * \varphi_\varepsilon - g * \varphi_\varepsilon \to 0 \quad \text{in} \quad L^2(U)$$

as $\varepsilon \to 0$. Moreover, upon inspection one sees that $g * \varphi_\varepsilon = 0$ on U^-. Since $f * \varphi_\varepsilon \to f$ in $L^2(U)$ as $\varepsilon \to 0$, we see that $u_j := u * \varphi_{1/j}$, $j >> 0$ has the required properties for the case without boundary conditions.

To handle the case where u vanishes in ∂U^- in the sense of the last statement of the lemma, we take instead $-1 << \varepsilon < 0$. Then the support of $u * \varphi_\varepsilon$ is compact and lies in the interior of U^-, and as we already mentioned $g = 0$ by hypothesis. Thus in this case $u_j := u * \varphi_{-1/j}$ does the job. The proof is complete. □

In the next lemma, we apply the notation used above. We also define A^o, B^o and f^o by

$$(A^o u)^k = a_j^{ik} D_i u^j, \quad (B^o u)^k = b_j^k u^j \quad \text{and} \quad f^o = (f^1, ..., f^{K_o})$$

for $1 \leq k \leq K_o$, where $K_o \leq K$ is an integer. In other words, we choose the first K_o equations. We want to investigate what happens if the boundary conditions we consider are determined by these first K_o equations in (1.10).

LEMMA 4.4. *Let $K_o \le K$ be an integer. Assume that the coefficients of A are smooth, as are those of B, and that the matrices C and C^o, with components*

$$C_j^k = a_j^{mk} \quad j = 1, ..., J, \quad k = 1, ..., K$$

and

$$(C^o)_j^k = a_j^{mk} \quad j = 1, ..., J, \quad k = 1, ..., K_o,$$

have constant ranks r and r_o, respectively. In addition, assume that if u and f^0 are extended to be zero outside the interior U_o^- of U^- then the equation $A^o u + B^o u = f^o$ holds. Then there is a sequence $u_j \in \mathscr{C}^r(U^-)$ vanishing outside a fixed compact subset U of U^-, such that

$$||u_j - u||^2_{L^2(U_o^-)} + ||Au_j + Bu_j - f||_{L^2(U_o^-)} \to 0 \quad as \quad j \to \infty,$$

and $C^o u = 0$ on $\{x^m = 0\}$.

PROOF. We are going to reduce to a normal form first. To this end, let us make a change of the unknown function: we set

$$\tilde{u}^k = a_j^{mk} u^k.$$

Then the equations (1.10) assume the form

$$\tilde{A}\tilde{u} + \tilde{B}\tilde{u} = f$$

where the coefficients of $\tilde{A}$ and $\tilde{B}$ have the same properties of those of A and B. Moreover, we have

$$\tilde{a}_j^{mk} = \delta_j^k \quad \text{for} \quad k = 1, ..., r_o, j = 1, ..., J$$

and

$$\tilde{a}_j^{mk} = 0 \quad \text{for} \quad j > r_o, k \le K_o.$$

Observe that by the rank hypothesis, the matrix whose kj^{th} coefficient is $\tilde{a}_j^{mk}$, for $j > r_o$ and $k > K_o$, must have constant rank $r - r_o$. Thus by a linear change of variables applied to $(\tilde{u}^{r_o+1}, ..., \tilde{u}^J)$, we may assume that the bottom right hand block of size $(r - r_o) \times (r - r_o)$ in the matrix of $(\tilde{a}_j^{mk})_{j,k}$ is the identity. In other words, we may assume from the outset that C^o is the $J \times K_o$-matrix having the $r_o \times r_o$-identity in its upper left hand corner and with zeros everywhere else, and that C is the $J \times K$ matrix with the $r_o \times r_o$-identity in its upper left hand corner, the $(r - r_o) \times (r - r_o)$-identity in its lower right-hand corner, and zeros everywhere else.

Now define u and f to be 0 in U^+ and let u_ε be the regularization of u in the first $m-1$ variables. We then apply Lemma 4.2 to deduce that all the components of u_ε and their first derivatives with respect to all but the m-th variable x^m are in L^2. Moreover

$$(1.11)\quad Au_\varepsilon + Bu_\varepsilon - f \to 0 \text{ in } L^2(U_o^-) \quad \text{and} \quad A^o u_\varepsilon + B^o u_\varepsilon - f^o \to 0 \text{ in } L^2(\mathbb{R}^m),$$

as $\varepsilon \to 0$. Since the m-th variable of u_ε is independent of ε, the limits (1.11) prove that u_ε satisfies the boundary conditions in the last statement of the lemma. One also sees from (1.11) that

$$\frac{\partial u^j}{\partial x^m} \in L^2(U) \text{ for } j \le r_o \quad \text{and} \quad \frac{\partial u^j}{\partial x^m} \in L^2(\text{interior}(U)) \text{ for } j > J - r + r_o.$$

These are the only x^m derivatives appearing in A.

Now let us take two mollifiers φ^+ and φ^-, with total integral 1 and compact support in the interior of U^+ and U^- respectively. We set

$$u^j_{\delta,\varepsilon} = u^j_\varepsilon * \varphi^-_\delta, \quad 1 \le j \le K_o, \quad u^j_{\delta,\varepsilon} = u^j_\varepsilon * \varphi^+_\delta \quad K_o < j \le K.$$

(The reason we must use two mollifiers lies in Lemma 4.3; some of the components of u satisfy zero boundary conditions in the coordinates we have chosen, while others do not. This is the nature of the condition of being in the domain of $\bar{\partial}^*$. Our choice of coordinates was made to handle this problem exactly.) Then $u_{\varepsilon,\delta} \in \mathscr{C}^\infty_0(\mathbb{R}^n)$ for sufficiently small ε and δ, and the support of the j-th component of $u_{\varepsilon,\delta}$ is contained in the open lower half space when $j \le K_o$. Moreover, as $\delta \to 0$, $\partial_{x^i} u_{\varepsilon,\delta} \to \partial_{x^i} u_\varepsilon$ when $i < m$ in $L^2(\mathbb{R}^m)$, and $\partial_{x^m} u^j_{\varepsilon,\delta} \to \partial_{x^m} u^j_\varepsilon$ in $L^2(\mathbb{R}^m)$ for $j \le K_o$ and in $L^2(\text{interior}(U^-))$ if $j > J - r + r_o$. If we define u_n to be $u_{\varepsilon,\delta}$ for $0 < \varepsilon = \varepsilon(n) << 1$, and then $0 < \delta = \delta(\varepsilon) << 1$, then we will get the estimate

$$||u_n - u||_{L^2(U^-)} + ||Au_n + Bu_n - f||_{L^2(U^-)} < \frac{1}{n}.$$

The proof is complete. □

II. Localization. Since our manifold-with-boundary Y is compact, we can cover it by a finite number of neighborhoods $U_1, ...U_N$ with the following properties:

(1) Each U_j is diffeomorphic to the ball of radius 2 and center 0 in $\mathbb{R}^{2n}$. Let us denote by $\varphi_j : U_j \to B^{2n}(2)$ the chosen diffeomorphism.
(2) The open sets $V_j := \varphi^{-1}(B^{2n}(1))$ also cover $\overline{Y}$.
(3) Among the open sets U_j in the cover, relabel by $W_\ell := U_{i_\ell}$ those that intersect the boundary ∂Y. Then we assume that

$$\varphi_{i_\ell}(Y \cap W_\ell) = \{(x^1, ..., x^{2n}) \in B^{2n}(2) \ ; \ x^{2n} > 0\}.$$

(Such φ_{i_ℓ} are guaranteed by the implicit function theorem.)

Let $\{\chi_m\}$ be a partition of unity subordinate to $\{V_j\}$, i.e., each χ_m is supported in some V_j, every compact subset of $\bigcup V_j$ supports at most a finite number of the χ_m, and $\sum_m \chi_m \equiv 1$. We may also assume, perhaps after renormalizing our partition of unity, that $\overline{Y}$ is covered by a finite number of the χ_m.

III. Proof of Theorem 4.1. We now have enough to complete the main goal of this section.

Suppose that χ_m is one of the elements of our partition of unity, and let $\varphi \in \text{Domain}(T^*) \cap \text{Domain}(S)$. Consider $\chi_m\varphi$.

Note first that $\chi_m\varphi \in \text{Domain}(S)$ and that

$$||S(\chi_m\varphi) - \chi_m S\varphi|| \le C||\varphi||.$$

Since only a finite number of the χ_m need be considered, the constant C is independent of m. (We note that the continuity of the weight is crucial here.) Also,

$$|(\chi_m\varphi, T\psi) - (\varphi, T(\chi_m\psi))| \le C||\psi||, \quad \psi \in \text{Domain}(T).$$

Thus $\chi_m\varphi \in \text{Domain}(T^*)$ and

$$||T^*(\chi_m\varphi) - \chi_m T^*\varphi|| \le C||\varphi||.$$

(Here we are using the continuity of the first derivative of the weight.) Thus we may assume that the φ in question is compactly supported in one of the charts U_j above.

Now, the formulas for S and T^*, and the boundary conditions defining elements in the domain of T^*, show that locally these operators, the elements φ we are considering, and the types of charts we use near ∂Y, satisfy the hypotheses of Lemmas 4.2, 4.3 and 4.4. The conclusions of those Lemmas imply that there is a sequence φ_j of smooth forms in Domain(S)∩Domain(T^*) such that

$$\varphi_j \to \varphi, \quad S\varphi_j \to S\varphi \quad \text{and} \quad T^*\varphi_j \to T^*\varphi$$

in L^2. This is what we wanted to show. □

5. Hörmander's theorem

The goal of this section is to apply the Functional Analysis Lemma and Theorem 3.4 (the Morrey Trick) to obtain a solution to the equation

$$\bar{\partial} u = \varphi \tag{1.12}$$

with L^2-estimates on u in terms of those on φ. In order to obtain from the Morrey Trick the inequality needed in the Functional Analysis Lemma, we will need to impose some positivity conditions. This positivity is going to come from two places: (i) the curvature of the metrics in question, and (ii) the curvature of the boundary. The non-negativity from (ii) is the notion of pseudoconvexity, which we now discuss.

5.1. Pseudoconvexity

Let X be a complex manifold and $Y \subset\subset X$ an open subset whose boundary ∂Y is smooth and has real codimension 1. For each $x \in \partial Y$ there is a neighborhood U of x in X and a smooth function $\rho : \overline{U} \to \mathbb{R}$ such that $U \cap \partial Y = \{\rho = 0\}$ and $d\rho|_{U\cap\partial Y}$ is nowhere zero.

DEFINITION 5.1. *The complex tangent space to ∂Y at $x \in \partial Y$ is the collection of all vectors $v \in T_{X,x}$ such that $v \in T_{\partial Y,x}$ and $J_o v \in T_{\partial Y,x}$, where J_o is the almost complex structure on X associated to the complex structure. We write*

$$v \in T^{1,0}_{\partial Y,x}.$$

Note that if $v \in T^{1,0}_{\partial Y,x}$ then $d\rho(x)v = 0$ and $d\rho(x)J_o v = 0$, and thus

$$\partial\rho(x)v = 0.$$

Conversely, if $\partial\rho(x)v = 0$ then $\partial\rho(x)J_o v = \sqrt{-1}\partial\rho(x)v = 0$, and thus we see that

$$T^{1,0}_{\partial Y,x} = \text{Kernel } \partial\rho(x).$$

REMARK. Note that the condition of a vector belonging to the complex tangent space of ∂Y strongly resembles the condition for an H-valued (p,q)-form to be in the domain of T^*.

Next we pursue a notion of curvature of the boundary that is appropriate in complex geometry. With this pursuit in mind, consider the $(1,1)$-form $\partial\bar{\partial}\rho(x)$ on the boundary ∂Y of Y.

DEFINITION 5.2. *We say that the point $x \in \partial Y$ is a pseudoconvex boundary point if for all $v \in T^{1,0}_{\partial Y,x}$,*

$$\partial\bar{\partial}\rho(x)(v,\bar{v}) \geq 0. \tag{1.13}$$

Observe that if ρ is replaced by $h\rho$ for some smooth positive function h, then on ∂Y we have

$$\partial\bar{\partial}(h\rho) = h\partial\bar{\partial}\rho + \partial\rho \wedge \overline{\partial h} + \partial h \wedge \overline{\partial \rho}.$$

It follows that for $v, w \in T^{1,0}_{\partial Y, x}$,

$$\partial\bar{\partial}(h\rho)(x)(v, \bar{w}) = h(x)\partial\bar{\partial}\rho(x)(v, \bar{w}).$$

Thus the notion of pseudoconvexity does not depend on the choice of the defining function ρ.

The form

$$\mathscr{L}_x(v, \bar{w}) := \partial\bar{\partial}\rho(x)(v, \bar{w})$$

is called the Levi form. When one talks about the Levi form, one usually cares about positivity properties of its restriction to the complex tangent space, and thus the function ρ is only important up to scaling by a positive function.

REMARK. If the Levi form is strictly positive definite at x, one says that x is a strictly pseudoconvex boundary point. While this condition is important in many problems in complex analysis, it will not play much of a role in the applications considered in these notes. As we mentioned in the introduction, the notes of D'Angelo in this volume address different positivity conditions for boundaries of domains in complex manifolds.

5.2. Proof of Hörmander's theorem

We begin with the following *a priori* estimate.

THEOREM 5.3. *Let Y be pseudoconvex domain in a Kähler manifold (X, g) and let $H \to X$ be a holomorphic line bundle with Hermitian metric having curvature Ω. Then for all H-valued (p, q)-forms $\varphi \in$ Domain$(T^*) \cap$ Domain(S) one has the estimate*

$$||T^*\varphi||^2 + ||S\varphi||^2 \geq (T_g(-R + \mathrm{Ricci}(R) + \Omega)\varphi, \varphi).$$

In particular, if the curvature operator $T_g(R + \mathrm{Ricci}(R) + \Omega)$ is bounded below by a positive constant c, in the sense that for any $\xi \in H \otimes \Lambda^{p,q}(T^*_X)$,

$$\langle T_g(R + \mathrm{Ricci}(R) + \Omega)\xi, \xi\rangle \geq c||\xi||^2,$$

then we have the estimate

$$||T^*\varphi||^2 + ||S\varphi||^2 \geq c||\varphi||^2. \tag{1.14}$$

PROOF. For smooth forms, this is an immediate consequence of Morrey's Trick and the definition of pseudoconvexity. The general case follows from Theorem 4.1 on the density of smooth forms in the graph norm. □

From the Functional Analysis Lemma, we thus obtain the following theorem.

THEOREM 5.4 (Hörmander's Theorem). *Let Y be pseudoconvex domain in a Kähler manifold (X, g) and let $H \to X$ be a holomorphic line bundle with Hermitian metric $e^{-\kappa}$ having curvature Ω. Suppose that the curvature operator $T_g(R+\mathrm{Ricci}(R)+\Omega)$ is bounded below by c in the sense of Theorem 5.3. Then for each H-valued (p, q)-form φ such that*

$$\int_Y |\varphi|^2_{\kappa,g}\omega^n < +\infty \quad \textit{and} \quad \bar{\partial}\varphi = 0$$

in the sense of distributions, there exists an H-valued $(p, q-1)$-form u such that

$$\bar{\partial} u = \varphi \quad \text{and} \int_Y |u|^2_{\kappa,g}\omega^n \leq \frac{1}{c}\int_Y |\varphi|^2_{\kappa,g}\omega^n.$$

PROOF. Note first that the image of T lies in the closed subspace of $L^2_{p,q+1}(\kappa, g)$ defined by the kernel of S, and hence that Kernel$(T^*) \supset (\text{Kernel}(S))^{\perp}$. Thus we may assume that the image space of T is the Kernel of S.

For all $\psi \in \text{Domain}(T^*) \cap \text{Kernel}(S)$ we have

$$|(\varphi, \psi)|^2 \leq ||\varphi||^2||\psi||^2 \leq \frac{||\varphi||^2}{c}||T^*\psi||^2,$$

where the last inequality follows from Theorem 5.3. Applying the Functional Analysis Lemma to the operator $T : L^2_{p,q-1}(\kappa, g) \to \text{Kernel}(S)$ completes the proof. □

REMARK. Our proof of Hörmander's Theorem could introduce some confusion. To avoid such confusion, we emphasize that the estimate (1.14) is established on the whole Hilbert space first, and not just on the kernel of S. It is necessary to do things in this order because we did not prove that smooth forms satisfying the $\bar{\partial}$-Neumann condition and also lying in the kernel of S are dense.

6. Singular Hermitian metrics for line bundles

In many applications it is convenient to have a version of Hörmander's Theorem for line bundles with a singular version of a Hermitian metric. In this section we discuss the modifications needed to pass to such a singular case.

6.1. The definition of a singular Hermitian metric

DEFINITION 6.1. *Let X be a Kähler manifold and let $L \to X$ be a holomorphic line bundle. A singular metric on L is a section $e^{-\kappa}$ of the real line bundle $|L^*|^2 \to X$ such that κ is $L^1_{\ell oc}$.*

REMARK. We continue to assume that the local expression $e^{-\kappa}$ is always obtained by using a local holomorphic frame.

The main reason for requiring such regularity for $e^{-\kappa}$ rests in the following definition.

DEFINITION 6.2. *The curvature of of a singular metric $e^{-\kappa}$ is the current*

$$\Theta_\kappa := \partial\bar{\partial}\kappa.$$

We say that $e^{-\kappa}$ has non-negative (resp. positive) curvature current if Θ_κ is a non-negative (resp. positive) $(1,1)$-current, or equivalently, the local representatives κ are plurisubharmonic (resp. strictly plurisubharmonic).

As in the case of smooth metrics, the curvature current of a singular metric is globally defined, and is independent of the holomorphic frame used to trivialize the line bundle L.

REMARK. Unfortunately, the term *singular hermitian metric* also encompasses the notion of smooth Hermitian metric. The better name *possibly singular hermitian metric* often appears in the literature, but we will nevertheless retain our confusing convention.

EXAMPLE. Let $V \subset X$ be an analytic hypersurface with local defining function f_j on U_j.

$$\kappa_j := \log |f_j|^2.$$

Then for each $\gamma \in \mathbb{Z}$, $e^{-\gamma\kappa} := \{e^{-\gamma\kappa_j}\}$ defines a singular metric on the line bundle γL_V, where L_V is the line bundle associated to the divisor V (or equivalently, having transition functions $g_{ij} = f_i/f_j$). It follows from the Lelong-Poincaré formula that the curvature current of this metric is equal to $-2\pi\sqrt{-1}\gamma[V]$, where $[V]$ is the current of integration over V.

EXAMPLE. Let $L \to X$ be a holomorphic Hermitian line bundle, $m \in \mathbb{Z}$, and let $s^1, ..., s^N$ be sections of mL. We can define a singular metric on L as follows: if e_α is a holomorphic section trivializing L, then we may write $s^j = s^j_\alpha e_\alpha$ for some functions s^j_α

$$\kappa_\alpha = \frac{1}{m} \log \left(|s^1_\alpha|^2 + ... + |s^m_\alpha|^2 \right).$$

This singular metric blows up exactly on the common zero set of the sections $s^1, ..., s^N$. Moreover, if $m > 0$ and all of the sections s^j are holomorphic, then the curvature of $e^{-\kappa}$ is non-negative.

6.2. Hörmander's theorem for singular metrics

THEOREM 6.3. *Let (X, g) be a Kähler manifold and let $L \to X$ be a holomorphic line bundle with singular metric $e^{-\kappa}$. Let $Y \subset X$ be a bounded pseudoconvex domain with smooth boundary. Fix integers p and q with $0 \le p \le n$ and $1 \le q \le n$. Assume that for some $c > 0$, the curvature operator*

$$T_g(R - Ricci(R) + \partial\bar{\partial}\kappa) \ge c$$

in the sense of currents and of Theorem 5.4. Assume also that there exist smooth metrics $e^{-\kappa_\varepsilon}$ for L and positive constants c_ε such that

$$e^{-\kappa_\varepsilon} \nearrow e^{-\kappa}, \quad T_g(R - Ricci(R) + \partial\bar{\partial}\kappa_\varepsilon) \ge c_\varepsilon \quad \text{and} \quad \lim_{\varepsilon \to 0} c_\varepsilon = c,.$$

Then for each L-valued (p, q)-form φ such that

$$\int_Y |\varphi|^2_g e^{-\kappa} \omega^n < +\infty \quad \text{and} \quad \bar{\partial}\varphi = 0$$

in the sense of currents, there exists an L-valued $(p, q-1)$-form u such that

$$\bar{\partial}u = \varphi \quad \text{and} \int_Y |u|^2_g e^{-\kappa}\omega^n \le \frac{1}{c} \int_Y |\varphi|^2_g e^{-\kappa}\omega^n.$$

EXAMPLE. On a Stein manifold, every singular Hermitian metric of non-negative curvature (and more generally of curvature bounded below by the curvature form of any smooth metric) can be approximated by smooth metrics in the manner stated in the hypotheses of Theorem 6.3. Indeed, If we fix any smooth metric $e^{-\gamma}$ for L, then the function $\kappa - \gamma$ is quasi-plurisubharmonic. Thus there exist smooth quasi-plurisubharmonic functions $\kappa_\varepsilon - \gamma$ with $\sqrt{-1}\partial\bar{\partial}\kappa_\varepsilon \ge 0$ such that $\gamma - \kappa_\varepsilon \nearrow \gamma - \kappa$. (We refer to J.-P. Demailly's book [**Demailly-Book**] for the details of how to do this approximation.) Then we obtain smooth metrics $e^{-\kappa_\varepsilon} \nearrow e^{-\kappa}$.

EXAMPLE. Metrics locally of the form

$$\frac{e^{-\gamma}}{(\sum |f_j|^2)^c}$$

where γ is smooth and c is a positive constant, can be regularized in the manner stated in the hypotheses of Theorem 6.3. The required regularization may be taken to be

$$\frac{e^{-\gamma}}{(\sum |f_j|^2 + \varepsilon)^c}$$

PROOF OF THEOREM 6.3. By Theorem 5.4, there exist sections u_ε such that

$$\bar\partial u_\varepsilon = \varphi \quad \text{and} \quad \int_Y |u_\varepsilon|_g^2 e^{-\kappa_\varepsilon}\omega^n \le \frac{1}{c_\varepsilon}\int_Y |\varphi|_g^2 e^{-\kappa_\varepsilon}\omega^n.$$

Now, *a priori* all of the solutions we have lie in different L^2-spaces. However, since $e^{-\kappa_\varepsilon} \le e^{-\kappa}$ and $c_\varepsilon \to c$, we have for a given $\delta \in (0, c)$ that for all sufficiently small $\varepsilon > 0$,

$$\int_Y |u_\varepsilon|_g^2 e^{-\kappa_\varepsilon}\omega^n \le \frac{1}{c-\delta}\int_Y |\varphi|_g^2 e^{-\kappa}\omega^n.$$

Thus $\{u_\varepsilon e^{-\kappa_\varepsilon/2}\}$ is bounded and hence a subsequence converges weakly in $L^2_{\ell oc}$. The limit $u_\delta = e^{\kappa/2}\lim u_\varepsilon$ satisfies

$$\int_Y |u_\delta|_g^2 e^{-\kappa}\omega^n \le \frac{1}{c-\delta}\int_Y |\varphi|_g^2 e^{-\kappa}\omega^n.$$

By the same reasoning, we then obtain that for a sequence of positive numbers $\delta_j \to 0$, $(1-\delta_j/c)^{1/2}u_{\delta_j} \to u$ in L^2, and thus u satisfies

$$\int_Y |u|_g^2 e^{-\kappa}\omega^n \le \frac{1}{c}\int_Y |\varphi|_g^2 e^{-\kappa}\omega^n.$$

Since $\bar\partial u = \varphi$ by the continuity of $\bar\partial$ in the sense of currents, we are done. □

With only a minor modification, we obtain the following result.

THEOREM 6.4. *Let (X, g) be a Stein manifold and let $L \to X$ be a holomorphic line bundle with singular metric $e^{-\kappa}$. Let $Y \subset X$ be a bounded pseudoconvex domain with smooth boundary. Fix integers p and q with $0 \le p \le n$ and $1 \le q \le n$. Assume that for some continuous function $\varepsilon : X \to (0, 1]$, the curvature operator*

$$T_g(R + \mathrm{Ricci}(R) + \partial\bar\partial\kappa) \ge \varepsilon$$

in the sense of distributions and Theorem 5.4. Assume also that there exist smooth metrics $e^{-\kappa_\varepsilon}$ for L and positive constants c_ε such that

$$e^{-\kappa_\varepsilon} \nearrow e^{-\kappa}, \quad T_g(R + \mathrm{Ricci}(R) + \partial\bar\partial\kappa_\varepsilon) \ge c_\varepsilon \quad \textit{and} \quad \lim_{\varepsilon\to 0} c_\varepsilon = c, .$$

Then for each L-valued (p, q)-form φ such that

$$\int_Y \frac{|\varphi|_g^2}{\varepsilon} e^{-\kappa}\omega^n < +\infty \quad \textit{and} \quad \bar\partial\varphi = 0$$

in the sense of distributions, there exists an L-valued $(p, q-1)$-form u such that

$$\bar\partial u = \varphi \quad \textit{and} \int_Y |u|_g^2 e^{-\kappa}\omega^n \le \int_Y \frac{|\varphi|_g^2}{\varepsilon} e^{-\kappa}\omega^n.$$

PROOF. The proof is the same as that of Theorem 6.3, except that in the a priori estimate, we proceed as follows.

$$|\langle f, \varphi\rangle|^2 = \left|\left\langle \sqrt{\varepsilon} f, \tfrac{\varphi}{\sqrt{\varepsilon}}\right\rangle\right|^2 \le \varepsilon|f|^2 \cdot \tfrac{|\varphi|^2}{\varepsilon}.$$

The remainder of the proof is exactly the same. □

7. Application: Kodaira embedding theorem

Kodaira's Theorem gives necessary and sufficient conditions for embedding a compact complex manifold in complex projective space. The precise statement is as follows.

THEOREM 7.1 (Kodaira's Embedding Theorem). *Let X be a compact complex manifold. Assume there is a holomorphic line bundle $L \to X$ with a smooth Hermitian metric of positive curvature. Then there is a holomorphic embedding $X \hookrightarrow \mathbb{P}_N$ of X into a complex projective space of some dimension N.*

REMARK. Conversely, if X is a submanifold of $\mathbb{P}_N$ then X has a holomorphic line bundle with a positively curved smooth Hermitian metric. Indeed, the hyperplane bundle with its Fubini-Study metric is positive, and thus its restriction to X gives a positive line bundle on X. When we prove Kodaira's Theorem, we will actually embed X in $\mathbb{P}_N$ so that some power of the positive line bundle on X maps to the restriction to X of the hyperplane line bundle. However, our metric, taken to the same power, will not map isometrically to the Fubini-Study metric.

7.1. Projective maps

Let X be as in Kodaira's Theorem. For a holomorphic line bundle $F \to X$, consider its space of global holomorphic sections

$$H^0(X, \mathcal{O}_X(F)) = \Gamma_{\mathcal{O}}(X, F).$$

It can be shown, for example by using the Hodge Theorem, that the vector space $H^0(X, \mathcal{O}_X(F))$ is finite dimensional.

Let us attempt to define a map from X into $\mathbb{P}(\Gamma_{\mathcal{O}}(X, F)^*)$. For $x \in X$, we let

$$\phi_{|F|}(x) := \{s \in \Gamma_{\mathcal{O}}(X, F) \ ; \ s(x) = 0\}.$$

If there is a section $s \in \Gamma_{\mathcal{O}}(X, F)$ such that $s(x) \neq 0$, then $\phi_{|F|}(x)$ is a hyperplane in $\Gamma_{\mathcal{O}}(X, F)$, i.e., a line in through the origin in the dual space $\Gamma_{\mathcal{O}}(X, F)^*$, i.e., a point in $\mathbb{P}(\Gamma_{\mathcal{O}}(X, F)^*)$.

REMARK. A less intrinsic way to write down the map $\phi_{|F|}$ is as follows. Let $s^0, ..., s^N$ be a basis for $\Gamma_{\mathcal{O}}(X, F)$. Then we assign to $x \in X$ the point

$$[s^0(x), ..., s^N(x)] \in \mathbb{P}_N.$$

To see that we obtain essentially the same map, note that $s^i \in \Gamma_{\mathcal{O}}(X, F) = (\Gamma_{\mathcal{O}}(X, F)^*)^*$, and thus $s^0, ..., s^N$ serve as homogeneous coordinates for $\Gamma_{\mathcal{O}}(X, F)^*$. Take $\alpha_j \in \Gamma_{\mathcal{O}}(X, F)^*$ such that $\langle \alpha_j, s^k \rangle = \delta_j^k$. A typical point $\xi \in \Gamma_{\mathcal{O}}(X, F)^*$ may then be written

$$\xi = \langle \xi, s^j \rangle \alpha_j.$$

Now, the point evaluation ξ_x at x is a linear function on $\Gamma_{\mathcal{O}}(X, F)$ with values in F_x, the fiber of F over x, and thus there exist $\lambda^j(x) \in F_x$ such that

$$\xi_x = \lambda^j(x) \otimes \alpha_j.$$

But then

$$\lambda^j(x) = \langle \xi_x, s^j \rangle = s^j(x).$$

DEFINITION 7.2. *The set $\mathbb{B}\mathrm{s}(F)$ of points $x \in X$ such that $\phi_{|F|}(x) = \mathbb{P}(\Gamma_{\mathcal{O}}(X, F)^*)$ is called the (set theoretic) base locus of F. A holomorphic line bundle is said to be free or base-point free if $\mathbb{B}\mathrm{s}(F)$ is empty.*

REMARK. Equivalently, a base point of F is a point at which each element of $\Gamma_{\mathcal{O}}(X, F)$ vanishes, and F is base-point free if at each point $x \in X$ there is a section $s \in \Gamma_{\mathcal{O}}(X, F)$ with $s(x) \neq 0$.

Thus we obtain a well-defined map

$$\phi_{|F|} : X - \mathbb{B}\mathrm{s}(F) \to \mathbb{P}(\Gamma_{\mathcal{O}}(X, F)^*).$$

So in order to make maps into projective space, we would like to be able to produce sections that do not vanish at a given point.

7.2. Properties of embeddings

An embedding has two important properties: it is injective, and it is an immersion. (If the manifold to be embedded is non-compact, one typically must add the requirement of properness: the inverse image of any compact set is compact.)

In order to separate two points $x, y \in X$ via the map $\phi_{|F|}$, it would suffice to construct a global holomorphic section s of F that does not vanish at either x or y, and a second section that vanishes at x but not at y. Note that once x and y have been separated, then some neighborhoods of x and y have also been separated.

The requirement that the map $\phi_{|F|}$ be an immersion is an infinitesimal version of point separation, as the points get closer together. To obtain the property that $\phi_{|F|}$ is an immersion at a point x, one must produce a section s^0 that does not vanish at x, and n sections $s^1, ..., s^n$, $n = \dim_{\mathbb{C}}(X)$, such that the map

$$U \ni y \mapsto \left(\frac{s^1(y)}{s^0(y)}, ..., \frac{s^n(y)}{s^0(y)}\right) \in \mathbb{C}^n$$

which is holomorphic in a neighborhood of y in X, is non-degenerate, i.e., its differential is invertible. Once again, if $\phi_{|F|}$ is an immersion at a point x, then it is an immersion in a neighborhood of X.

If we are able to construct the sections required for the point separation and immersion properties at arbitrary points of $X \times X$ and X respectively, then we obtain open covers of $X \times X$ and X in which there are sections doing the required work. It then follows that the map $\phi_{|F|}$ is an embedding, since up to a choice of homogeneous coordinates, $\phi_{|F|}$ is of the form $[s^0, ..., s^N]$ for any basis of $\Gamma_{\mathcal{O}}(X, F)$.

In fact, Theorem 7.1 will be a consequence of the following result.

THEOREM 7.3 (Kodaira's Embedding Theorem Again). *Given a positive line bundle $F \to X$, there is an integer $m >> 0$ such that the map $\phi_{|mF|}$ is an embedding.*

7.3. Separation of jets

Let us fix the line bundle $F \to X$ and a smooth hermitian metric $e^{-\kappa}$ of positive curvature on F.

Let $\{U_j\}$ be an open cover of X by coordinate neighborhoods such that

a) for each j, $F|_{U_j}$ is trivial, and
b) there are open sets V_j and W_j such that $V_j \subset\subset W_j \subset\subset U_j$ and $\{V_j\}$ is also an open cover of X.

By compactness, we may assume that the cover $\{U_j\}$ is finite. Fix local coordinates z_j on U_j, and smooth functions χ_j such that $\chi_j|_{W_j} \equiv 1$ and $\mathrm{Supp}(\chi_j) \subset\subset U_j$. Consider the functions

$$\psi_{j,x}(y) := \chi_j(y) \log |z_j(y) - z_j(x)|^{2(n+1)}, \qquad x \in V_j.$$

Then $\psi_{j,x}$ is plurisubharmonic in W_j and there in an integer n_j such that for all $x \in V_j$,

$$T_g(\text{Ricci}(R) + \partial\bar{\partial}(n_j\kappa + \psi_{j,x})) \geq \varepsilon_j \omega_g$$

for some positive number ε_j. (The choice of Hermitian metric can be made *a priori* and does not matter.) Let

$$m := 2 \max_j n_j.$$

This is going to be our m.

7.3.1. Construction of a section that does not vanish at $x \in X$. Let $x \in V_j$. Since $F|_{U_j}$ is trivial, there is a holomorphic section σ_o of L over U_j such that $\sigma_o(x) \neq 0$. Let $\sigma = \chi_j \cdot \sigma_o$. Then σ is a globally defined smooth section that is holomorphic on W_j. Consider the F-valued $(0,1)$-form

$$\varphi = \bar{\partial}\sigma.$$

This form is supported in $W_j - U_j$, and thus

$$\int_X |\varphi|^2 e^{-(m\kappa + \psi_{j,x})} \omega^n < +\infty.$$

By Hörmander's Theorem 6.3 for singular metrics, there is a section u of $F \to X$ such that

$$\bar{\partial}u = \alpha \quad \text{and} \quad \int_X |u|^2 e^{-(m\kappa+\psi_{j,x})} \omega^n < +\infty.$$

(Note that the singular metrics used here are of the form discussed in the second example following the statement of Theorem 6.3.) But because of the singularity of $\psi_{j,x}$, u vanishes to second order at x. It follows that $s = \sigma - u$ is a holomorphic section of $mF \to X$ that does not vanish at x. This shows that $mF \to X$ is base-point free.

7.3.2. Construction of a section that vanishes at $y \in X$ but not at $x \in X - \{y\}$. Let $x, y \in X$ be distinct points. We are now going to show that there is a section $s \in \Gamma_{\mathcal{O}}(X, mF)$ such that $s(x) = 0$ and $s(y) \neq 0$. To this end, suppose $x \in U_j$ and $y \in U_k$. (j and k need not be distinct.) Since $F|_{U_j}$ and $F|_{U_k}$ are trivial, there is a smooth global section σ of mF that is holomorphic on a neighborhood of $\{x, y\}$, has compact support in $U_j \cup U_k$, and vanishes at x but not at y. Then the mF-valued $(0,1)$-form

$$\varphi = \bar{\partial}\sigma$$

is smooth and supported on a relatively compact subset of $U_j \cup U_k - \{x, y\}$, and thus

$$\int_X |\varphi|^2 e^{-(m\kappa+\psi_{j,x}+\psi_{k,y})} \omega^n < +\infty.$$

We deduce from Hörmander's Theorem that there is a section u of $mF \to X$ such that

$$\bar{\partial}u = \varphi \quad \text{and} \quad \int_X |u|^2 e^{-(m\kappa+\psi_{j,x}+\psi_{k,y})} \omega^n < +\infty. \tag{1.15}$$

The estimate (1.15) also implies that u vanishes at x and y. We obtain a global holomorphic section

$$s = \sigma - u$$

with the desired properties.

If we now go through the same construction, but with σ non-vanishing at both x and y, we obtain a global holomorphic section s' that is non-zero at both x and y. The pair s, s' separate the points x and y.

7.3.3. Construction of a section with prescribed differential at $x \in X$. Returning to the construction of a section that is non-zero at $x \in X$, we note that the holomorphic correction u vanishes to second order. Thus the derivative of the section s at x is the same as that of σ. By choosing sections $\sigma^1, ..., \sigma^n$ such that the functions σ^j/σ (with σ being any non-vanishing section at x) have independent differentials at x, we obtain sections $s, s^1, ..., s^n$ such that the holomorphic map

$$\left(\tfrac{s^1}{s}, \ldots, \tfrac{s^j}{s}\right)$$

defined on a neighborhood of x is an immersion near x. Thus Kodaira's embedding theorem is proved.

8. Multiplier ideal sheaves and Nadel's Theorems

An important application of the version of Hörmander's Theorem involving singular Hermitian metrics is to the construction of sections with prescribed vanishing orders at certain points. We saw an example of the usefulness of such metrics in the proof of Kodaira's Embedding Theorem.

To have a systematic approach to such constructions, it is very useful to consider multiplier ideal sheaves, a notion introduced by Nadel, and which has its roots in earlier works of J.J. Kohn and of H. Skoda.

8.1. The definition of multiplier ideals

DEFINITION 8.1. (Multiplier Ideal sheaf)

(1) *Let $U \subset X$ be an open subset, and let φ be a locally integrable function on U. We define*

$$\mathscr{I}_\varphi(U) := \left\{ f \in \mathcal{O}_X(U) \ ; \ |f|^2 e^{-\varphi} \in L^1(U) \right\}.$$

(2) *Let $\{U_j\}$ be an open cover of X, and let $\varphi := \{\varphi_j \in L^1_{\ell oc}(U_j)\}$ be a collection of locally integrable functions such that $\varphi_i - \varphi_j$ is bounded on $U_i \cap U_j$. Then $\mathscr{I}_{\varphi_i}(U_i \cap U_j) = \mathscr{I}_{\varphi_j}(U_i \cap U_j)$, so we have a globally defined sheaf of ideals $\mathscr{I}_\varphi \subset \mathcal{O}_X$ defined by*

$$\mathscr{I}_{\varphi,x} = \mathscr{I}_{\varphi_j,x} \quad \text{for } x \in U_j.$$

This ideal sheaf is called the multiplier ideal sheaf associated to φ.

EXAMPLE. We can associated to a positively curved singular Hermitian line bundle $(L, e^{-\varphi})$ a multiplier ideal sheaf associated to the plurisubharmonic family $\{\varphi\}$ of local potentials of the singular Hermitian metric.

REMARK. As we will see later, $\mathscr{I}_\varphi$ has interesting properties when the function (resp. metric) φ is plurisubharmonic (resp. has semi-positive curvature current).

Important properties of multiplier ideal sheaves are captured in two results of A. Nadel: Nadel's Coherence Theorem and Nadel's Vanishing Theorem.

8.2. Lelong numbers of plurisubharmonic functions

The zero variety of the ideal sheaf $\mathscr{I}_\varphi$ is the set of points where $e^{-\varphi}$ is not locally integrable. Such points only occur where φ has poles of sufficiently high order. If $\varphi = \frac{1}{m}\log(|s_1|^2 + ... + |s_N|^2)$ as in example 6.1, then one has a notion of pole order, namely the $\frac{1}{m}$ times the multiplicity of the variety of common zeros of $s_1, ..., s_N$. In general, the pole orders are defined using so-called Lelong numbers.

DEFINITION 8.2. *Let X be a complex manifold and φ a plurisubharmonic function in a neighborhood U of $x \in X$. Fix a coordinate chart U near x, and let z be a system of local coordinates vanishing at x. The Lelong number of φ is the number*

$$\nu(\varphi, x) := \liminf_{z \to x} \frac{\varphi(z)}{\log|x - z|^2}.$$

We also set

$$E_c(\varphi) := \{x \in X \ ; \ \nu(\varphi, x) \geq c\}.$$

REMARK. It is not at all clear that Lelong numbers are well-defined, or that they are independent of the coordinate system chosen. This is, however, the case, as established in a famous paper of Siu, which also shows that the sets $E_c(\varphi)$ are complex analytic sets. A proof of Siu's Theorem, due to Demailly, will be outlined in the exercises at the end of Lecture 2.

Lelong numbers can be described in other ways.

PROPOSITION 8.3. *The Lelong number is given by the integral formulas:*

$$\nu(\varphi, x) = \lim_{r \searrow 0} \frac{1}{(2\pi r^2)^{n-1}} \int_{B(x,r)} \tfrac{1}{\pi}\sqrt{-1}\partial\bar\partial\varphi \wedge (\sqrt{-1}\partial\bar\partial|z - x|^2)^{n-1},$$

and

$$\nu(\varphi, x) = \lim_{r \searrow 0} \frac{1}{(2\pi)^n \log r} \int_{|z|=1} \varphi(rz) d^c|z|^2 \wedge (dd^c|z|^2)^{n-1}.$$

The proof of this proposition, as well as that of the next two results, will be omitted in these notes. The reader seeking more details may consult J.-P. Demailly's notes [**Demailly-2001**] for complete proofs and much more.

THEOREM 8.4. *Let φ be a plurisubharmonic function.*

(1) *(Siu) The Lelong number $\nu(\varphi, x)$ is invariant under holomorphic change of coordinates.*
(2) *(Lelong) $\nu(\varphi, x) = \sup\{\gamma \geq 0 \ ; \ \varphi(z) \leq \gamma \log|z - x|^2 + O(1)$ near $x\}$. In particular, if $u = \log|f|^2$ for some holomorphic function f, then $\nu(\log|f|^2, x) = \mathrm{Ord}_x(f)$.*
(3) *(Thie) If A is an analytic variety and the ideal of germs of holomorphic functions vanishing on A near x is generated by $g_1, ..., g_k$, then the Lelong number of the function $\varphi := \log(\sum |g_j|^2)$ is*

$$\nu(\varphi, x) = \mathrm{Mult}_x(A),$$

the multiplicity of A at x (which can be defined, for instance, as the number of sheets of a generic orthogonal projection of A onto an affine subspace whose dimension is the dimension of A).

LEMMA 8.5 (Skoda's Lemma). *Let φ be a plurisubharmonic function on an open set U in X containing x.*

(1) *If* $\nu(\varphi, x) < 1$ *then* $e^{-\varphi}$ *in integrable in a neighborhood of* x. *In particular,* $\mathscr{I}_{\varphi,x} = \mathcal{O}_x$.
(2) *If* $\nu(\varphi, x) \geq n+s$ *for some* $s \in \mathbb{N}$, *then the estimate* $e^{-\varphi} \geq C|z-x|^{-2(n+s)}$ *holds in a neighborhood of* x. *In particular, one obtains that* $\mathscr{I}_{\varphi,x} \subset \mathfrak{m}_x^{s+1}$, *where* $\mathfrak{m}_x$ *is the maximal ideal of* $\mathcal{O}_x$.
(3) *The zero variety* $V(\mathscr{I}_\varphi)$ *of* $\mathscr{I}_\varphi$ *satisfies*

$$E_n(\varphi) \subset V(\mathscr{I}(\varphi)) \subset E_1(\varphi).$$

8.3. $\mathbb{Q}$-line bundles

In this section, we define the notion of a $\mathbb{Q}$-line bundles and their multi-sections. The main use of multi-sections is in the construction of singular Hermitian metrics for holomorphic line bundles.

DEFINITION 8.6 ($\mathbb{Q}$-line bundles). *A $\mathbb{Q}$-line bundle L on a complex manifold X is an open cover $\{U_j\}$ of X together with a collections of functions $\{g_{ij} : U_i \cap U_j \to \mathbb{C}^*\}$, called transition functions, such that for some integer m, $\{g_{ij}^m\}$ form the transition functions for a line bundle. The line bundle obtained from $\{g_{ij}^m\}$ is denoted mL. If it is possible to choose transition functions so that $m = 1$, we say the $\mathbb{Q}$-line bundle is integral. If the functions g_{ij} are holomorphic, we say that L is a holomorphic $\mathbb{Q}$-line bundle.*

The tensor product of two $\mathbb{Q}$-line bundles is obtained by multiplying their transition functions. It is clear that this product is also a $\mathbb{Q}$-line bundle. Similarly the dual of a $\mathbb{Q}$-line bundle L is a $\mathbb{Q}$-line bundle whose transition functions are the reciprocals of those of L.

With some hesitation, we introduce next the notion of multi-sections of $\mathbb{Q}$-line bundles. Conceptually, multi-sections are meant to be roots of sections of (integral) line bundles. But taking roots of a holomorphic function with zeros creates problems, and thus should be avoided if possible. Still, the language of multi-sections does appear in the literature, and can sometimes be handy.

DEFINITION 8.7. *Let $L \to X$ be a $\mathbb{Q}$-line bundle. A multi-section s of L is an equivalence class of pairs (m, σ) where $m > 0$ is an integer such that mL is integral and $\sigma \in H^0(X, mL)$. The equivalence of two such pairs is defined as follows:*

$$(m, \sigma) \sim (\tilde{m}, \tilde{\sigma}) \iff e^{\sqrt{-1}\theta}\sigma^{\tilde{m}} = \tilde{\sigma}^m$$

for some $\theta \in \mathbb{R}$. A the class $[(m, \sigma)]$ may be denoted $\sigma^{1/m}$.

Finally, we have the notion of singular Hermitian metrics for $\mathbb{Q}$-line bundles.

DEFINITION 8.8. *Let $L \to X$ be a holomorphic $\mathbb{Q}$-line bundle. Fix an open cover $\{U_j\}$ of X together with a collections of transition functions $\{g_{ij}\}$ for L. A singular Hermitian metric for L is a collection of functions $e^{-\kappa_i}$ on U_i such that*

(1) $\kappa_i \in L^1_{\ell oc}(U_i)$ *and*
(2) $e^{-\kappa_i} = e^{-\kappa_j}|g_{ij}|^{-2}$ *on* $U_i \cap U_j$.

Note that, unlike sections and $\mathbb{Q}$-line bundles, there is no need for the integer m in the definition of singular Hermitian metric for a $\mathbb{Q}$-line bundle. Moreover, if $e^{-\kappa}$ is a singular Hermitian metric for L, then $e^{-q\kappa}$ defines a singular Hermitian metric for qL and vice versa. Finally, the curvature current

$$\sqrt{-1}\partial\bar{\partial}\kappa$$

of a singular Hermitian metric $e^{-\kappa}$ for a $\mathbb{Q}$-line bundle is well defined, and Lelong numbers and multiplier ideals are well-defined for singular Hermitian metrics of $\mathbb{Q}$-line bundles whose curvature current is non-negative.

EXAMPLE. Let $s_1, \ldots, s_N$ be multisections of the $\mathbb{Q}$-line bundle L. If we define

$$\kappa := \log\left(\sum_{j=1}^{N} |s_j|^2\right),$$

then $e^{-\kappa}$ defines a singular Hermitian metric for L.

8.4. Nadel's theorems

8.4.1. Nadel's coherence theorem.

THEOREM 8.9 (Nadel Coherence). *Let Ω be an open set in a Kähler manifold X, and let φ be a plurisubharmonic function on Ω. Then the multiplier ideal sheaf $\mathscr{I}_\varphi$ is a coherent sheaf of ideals. Moreover, if Ω is a bounded Stein domain, then $\mathscr{I}_\varphi$ is generated by any Hilbert basis of the space*

$$\mathscr{A}^2(\Omega, \varphi) := \left\{ f \in \mathcal{O}(\Omega) \ ; \ \int_\Omega |f|^2 e^{-\varphi} dV < +\infty \right\},$$

where dV denotes the Kähler volume.

PROOF. Since the coherence of $\mathscr{I}_\varphi$ is a local property, we may assume throughout that Ω is a bounded Stein domain in X. Let $h : \Omega \to (0, 1)$ be a function such that for some positive constant ε, $\sqrt{-1}\partial\bar{\partial}h \geq \varepsilon\omega$, where ω is the Kähler form of X.

By the strong Nötherian property property of coherent sheaves, the family of sheaves generated by finite subsets of $\mathscr{A}^2(\Omega, \varphi)$ has a maximal element on each compact subset of Ω. It follows that $\mathscr{A}^2(\Omega, \varphi)$ generates a coherent ideal sheaf $\mathscr{J} \subset \mathcal{O}_\Omega$. Evidently $\mathscr{J} \subseteq \mathscr{I}_\varphi$.

To establish equality requires a little more work. First, we establish equality at the level of jets. To this end, fix $x \in \Omega$ and let $f \in \mathscr{I}_{\varphi,x}$. Suppose f is holomorphic on a neighborhood U_x and let $B_x(2) \subset\subset U_x$ be a relatively compact open set biholomorphic to the ball of radius 2 in $\mathbb{C}^n$ ($n = \dim_{\mathbb{C}} X$) with coordinates $z = (z^1, \ldots, z^n)$. Denote by B_x the ball $|z| < 1$ in X. Let χ be a smooth function taking values in $[0, 1]$ such that $\chi|_{B_x} \equiv 1$ and $\mathrm{supp}(\chi) \subset B_x(2)$. Define

$$\alpha = \bar{\partial}\chi f.$$

Fix an integer $N > 0$. Then

$$\int_\Omega |\alpha|^2 \frac{e^{-(\varphi+h)}}{|z|^{2(n+N)}} \leq C \int_{B_x(2)} |f|^2 e^{-\varphi} < +\infty,$$

where the first inequality follows because $\alpha \equiv 0$ in B_x, and the finiteness is by definition of $\mathscr{I}_\varphi$. It follows from Hörmander's Theorem that there is a function u_N on Ω satisfying

$$\bar{\partial}u_N = \alpha$$

and

$$\int_\Omega |u_N|^2 \frac{e^{-\varphi}}{|z|^{2(n+N)}} \leq e \int_\Omega |u|^2 \frac{e^{-(\varphi+h)}}{|z|^{2(n+N)}} \leq \frac{Ce}{\varepsilon} \int_{B_x(2)} |f|^2 e^{-\varphi} < +\infty.$$

We thus see that the function

$$g_N = \chi f - u_N$$

lies in $\mathscr{A}^2(\Omega, \varphi)$ and agrees with f to order at least N at x.

Now observe that on B_x, u_N is holomorphic and $\chi \equiv 1$. It follows that $f = g_N + u_N$, i.e.

$$\mathscr{I}_{\varphi,x} \subseteq \mathscr{J}_x + \mathscr{I}_{\varphi,x} \cap \mathfrak{m}_x^N \subseteq \mathscr{I}_{\varphi,x},$$

where $\mathfrak{m}_x$ is the maximal ideal at x. Taking quotients, we have

$$\mathscr{I}_{\varphi,x}/\mathscr{I}_{\varphi,x} \cap \mathfrak{m}_x^N = \mathscr{J}_x/\mathscr{I}_{\varphi,x} \cap \mathfrak{m}_x^N.$$

It follows from Nakayama's Lemma that $\mathscr{J}_x = \mathscr{I}_{\varphi,x}$, and thus $\mathscr{I}_\varphi$ is a coherent ideal sheaf.

Finally, the ideal $\mathscr{J}$ is generated by any Hilbert space basis because it is generated by $\mathscr{A}^2(\Omega, \varphi)$, and the space of sections of $\mathscr{J}$ over an open subset of Ω, being a Fréchet space, is closed under L^2 convergence. □

8.4.2. Nadel's vanishing theorem.

THEOREM 8.10 (Nadel Vanishing). *Let X be a weakly pseudo-convex Kähler manifold with Kähler form ω, and let $F \to X$ be a holomorphic line bundle with singular metric $h = e^{-\varphi}$ such that $\sqrt{-1}\partial\bar{\partial}\varphi \geq \varepsilon\omega$ for some continuous function $\varepsilon > 0$. Assume there exists an increasing sequence of smooth metrics $e^{-\varphi_j} \nearrow e^{-\varphi}$ such that $\sqrt{-1}\partial\bar{\partial}\varphi_j \geq (\varepsilon - o(1))\omega$ locally uniformly as $j \to \infty$. Then*

$$H^q(X, \mathcal{O}_X(K_X + F) \otimes \mathscr{I}_\varphi) = 0 \text{ for all } q \geq 1.$$

PROOF. Let $q \geq 1$ and fix an (n, q)-form u such that $\bar{\partial}u = 0$ and for any compact set $K \subset\subset X$,

$$\int_K |u|^2_\omega e^{-\varphi} dV_\omega < +\infty.$$

Let $\psi : X \to [0, \infty)$ be plurisubharmonic exhaustion function, which exists by the hypothesis of weak pseudoconvexity. By replacing ψ with $\chi \circ \psi$ for some sufficiently rapidly increasing convex function χ, we may assume that

$$\int_X \frac{|u|^2_\omega}{\varepsilon} e^{-\varphi} e^{-\psi} dV_\omega < +\infty.$$

Since for (n, q)-forms the Ricci curvature is canceled out by the Kähler curvature (see Proposition 2.4) and

$$\sqrt{-1}\partial\bar{\partial}(\varphi + \psi) \geq \varepsilon\omega,$$

Hörmander's Theorem (version 6.4) implies that there is an $(n, q-1)$-form f such that

$$\bar{\partial}f = u \quad \text{and} \quad \int_X |f|^2_\omega e^{-\varphi} e^{-\psi} dV_\omega \leq \int_X \frac{|u|^2_\omega}{\varepsilon} e^{-\varphi} e^{-\psi} dV_\omega < +\infty.$$

It follows that

$$[u] = 0 \quad \text{in} \quad H^q(X, \mathcal{O}_X(K_X + F) \otimes \mathscr{I}_\varphi),$$

and the proof is complete. □

8.5. Realizing jets

Nadel's Vanishing Theorem is particularly useful when the zero variety of the multiplier ideal sheaf has isolated points. In fact, we have the following corollary.

COROLLARY 8.11. *Let X be a Kähler manifold of complex dimension n with Kähler form ω, $F \to X$ a holomorphic line bundle with singular Hermitian metric $e^{-\varphi}$ such that for some continuous positive function $\varepsilon : X \to (0,\infty)$, $\sqrt{-1}\partial\bar\partial\varphi \geq \varepsilon\omega$. Suppose that $x_1, \dots, x_N$ are isolated points in the zero variety $V(\mathscr{I}_\varphi)$ of the multiplier ideal sheaf associated to $e^{-\varphi}$. Assume there exists an increasing sequence of smooth metrics $e^{-\varphi_j} \nearrow e^{-\varphi}$ such that $\sqrt{-1}\partial\bar\partial\varphi_j \geq (\varepsilon - o(1))\omega$ locally uniformly as $j \to \infty$. Then there is a surjective map*

$$H^0(X, K_X + F) \twoheadrightarrow \bigoplus_{1\leq j\leq N} \mathcal{O}(K_X + F)_{x_j} \otimes (\mathcal{O}_X/\mathscr{I}_\varphi)_{x_j}.$$

In particular, if $\nu(\varphi, x_i) \geq n + s_j$ for some integers $s_j \geq 0$ and $1 \leq j \leq N$, then $H^0(X, K_X + F)$ simultaneously generates all of order s_j at the point x_j, $1 \leq j \leq N$.

PROOF. Consider the short exact sequence of coherent analytic sheaves

$$0 \to \mathscr{I}_\varphi \to \mathcal{O}_X \to \mathcal{O}_X/\mathscr{I}_\varphi \to 0.$$

Clearly $\mathcal{O}_X/\mathscr{I}_\varphi$ is supported on the zero variety of $\mathscr{I}_\varphi$. Thus

$$\mathcal{O}_X/\mathscr{I}_\varphi = \mathscr{F} \oplus \bigoplus_{j=1}^N (\mathcal{O}_X/\mathscr{I}_\varphi)_{x_j}$$

where the support of $\mathscr{F}$ contains none of the points $x_1, \dots, x_N$. Tensoring with $\mathcal{O}_X(K_X + F)$ and passing to the long exact sequence in sheaf cohomology, we have

$$H^0(X, K_X + F) \to H^0(X, \mathcal{O}_X(K_X + F) \otimes (\mathcal{O}_X/\mathscr{I}_\varphi)) \to H^1(X, \mathcal{O}_X(K_X + F) \otimes \mathscr{I}_\varphi).$$

The right-most space is zero by Nadel Vanishing. Since

$$0 \to \bigoplus_{j=1}^N (\mathcal{O}_X/\mathscr{I}_\varphi)_{x_i} \to \mathscr{F} \oplus \bigoplus_{j=1}^N (\mathcal{O}_X/\mathscr{I}_\varphi)_{x_i},$$

We find that

$$\begin{aligned}\bigoplus_{j=1}^N (\mathcal{O}_X/\mathscr{I}_\varphi)_{x_j} &= H^0\Big(X, \mathcal{O}_X(K_X + F) \otimes \bigoplus_{j=1}^N (\mathcal{O}_X/\mathscr{I}_\varphi)_{x_j}\Big) \\ &\subset H^0\Big(X, \mathcal{O}_X(K_X + F) \otimes \mathscr{F} \oplus \bigoplus_{j=1}^N (\mathcal{O}_X/\mathscr{I}_\varphi)_{x_j}\Big).\end{aligned}$$

The first statement of the corollary thus follows. The last statement of the corollary then follows since, by Skoda's Lemma 8.5, the Lelong number hypothesis implies that

$$(\mathscr{I}_\varphi)_{x_j} \subset \mathfrak{m}_{x_j}^{s_j+1}.$$

□

9. Exercises

9.1. Seip's theorem

This first exercise outlines a proof of a theorem of Seip [**Seip-1992**]. The method we used is strongly based on the paper [**Berndtsson-Ortega Cerdà-1995**].

Let $\Gamma \subset \mathbb{C}$ be a discrete set of points such that

$$\inf\{|\gamma - \gamma'| \ ; \ \gamma, \gamma' \in \Gamma, \ \gamma \neq \gamma'\} > 0. \tag{1.16}$$

We consider the following problem:

DEFINITION 9.1 (Interpolation problem for Γ). *We say that the interpolation problem is solvable for Γ if for any collection $\{a_\gamma \ ; \ \gamma \in \Gamma\} \subset \mathbb{C}$ of complex numbers such that*

$$\sum_{\gamma \in \Gamma} |a_\gamma|^2 e^{-|\gamma|^2} < +\infty, \tag{1.17}$$

there is an entire function f such that

$$f(\gamma) = a_\gamma \quad \text{and} \quad \int_{\mathbb{C}} |f(z)|^2 e^{-|z|^2} dA(z) < +\infty. \tag{1.18}$$

K. Seip proved the following theorem.

THEOREM 9.2 (Seip's Theorem). *The following are equivalent.*

(1) *The discrete set Γ satisfies* (1.16), *and*

$$D(\Gamma) := \limsup_{r \to \infty} \sup_{z \in \mathbb{C}} \frac{\#\{\gamma \in \Gamma \ ; \ |z - \gamma| < r\}}{r^2} < 1.$$

(2) *For any set of complex numbers $\{a_\gamma \ ; \ \gamma \in \Gamma\}$ satisfying* (1.17) *there is an entire function f satisfying* (1.18).

In this exercise we are going to prove one direction of Seip's Theorem, namely that if $D(\Gamma) < 1$ then Γ has the interpolation property. The proof is not Seip's original proof, but rather one due to Berndtsson and Ortega Cerdà.

(1) Let $r > 0$. Show that the function

$$s_r(z) := \sum_{\gamma \in \Gamma} \left(\log|z - \gamma|^2 - \frac{1}{\pi r^2} \int_{|z - \zeta| < r} \log|\zeta - \gamma|^2 dA(\zeta) \right)$$

has the following properties:

(a) $s_r : \mathbb{C} \to [-\infty, \infty)$ is everywhere well-defined and satisfies $s_r(z) \leq 0$ for all z.

(b) In the sense of currents,

$$\sqrt{-1}\partial\bar{\partial} s_r(z) \geq -\frac{\#\{\gamma \in \Gamma \ ; \ |z - \gamma| < r\}}{r^2} dA(z).$$

(c) e^{-s_r} is not locally integrable at any point of Γ.

(d) There is a constant $C_{r,\varepsilon} > 0$ such that if $\gamma \in \Gamma$ and $\varepsilon/2 < |z - \gamma| < \varepsilon$ then $s_r(z) \geq -C_{r,\varepsilon}$.

(2) Show that the function $\psi(z) := |z|^2 + s_r(z)$ is strictly plurisubharmonic for all $r > 0$ sufficiently large if and only if $D(\Gamma) < 1$.

(3) Let $\chi : [0,1) \to [0,1]$ be a smooth function with compact support such that $\chi|_{[0,1/2)} \equiv 1$. Show that if $\{a_\gamma \ ; \ \gamma \in \Gamma\}$ satisfies (1.17) then, for sufficiently small $\varepsilon > 0$, the function

$$h_\varepsilon(z) := \sum_{\gamma\in\Gamma} e^{\bar{\gamma}(z-\gamma)} a_\gamma \chi(|z-\gamma|^2/\varepsilon^2)$$

satisfies

$$h_\varepsilon(\gamma) = a_\gamma, \quad \gamma \in \Gamma$$

as well as the estimates

$$\int_{\mathbb{C}} |h_\varepsilon|^2 e^{-|z|^2} dA(z) \le C_\varepsilon \sum_{\gamma\in\Gamma} |a_\gamma|^2 e^{-|\gamma|^2}$$

and

$$\int_{\mathbb{C}} |\bar{\partial} h_\varepsilon|^2 e^{-\psi} dA \le C'_\varepsilon \sum_{\gamma\in\Gamma} |a_\gamma|^2 e^{-|\gamma|^2}.$$

Hint: Notice that $e^{-|z|^2} = e^{-|\gamma|^2} e^{-|z-\gamma|^2 - 2\mathrm{Re}\ ((z-\gamma)\bar{\gamma})} \le e^{-2\mathrm{Re}\ \bar{\gamma}(z-\gamma)} e^{-|\gamma|^2}$.

(4) Use Hörmander's Theorem to find a function f satisfying (1.18).

LECTURE 2

The $\mathbf{L}^2$ extension theorem

As we saw in the proof of Kodaira's Embedding Theorem, Hörmander's Theorem can be used to construct sections of a sufficiently positive line bundle L that have a given value (or given Taylor polynomial) at a point. In this lecture, we will extend sections from submanifolds using L^2 methods similar to those used to prove Hörmander's Theorem. We will also try to optimize (which means minimize) the positivity needed to do the extension.

The method we use has come to be called the *Ohsawa-Takegoshi Technique.* It was first developed by Ohsawa and Takegoshi in the mid 1980's, and has been extended by a number of mathematicians since that time. (See the introduction for references.)

1. $\mathbf{L}^2$ extension

In this section we state and prove the main L^2 extension theorem of these notes.

1.1. Essentially Stein manifolds

DEFINITION 1.1. *A Kähler manifold Y is said to be essentially Stein if there exists an analytic hypersurface $V \subset Y$ such that $Y - V$ is Stein. Thus there are relatively compact subsets $\Omega_j \subset\subset Y - V$ with smooth pseudoconvex boundary such that*

$$\Omega_j \subset\subset \Omega_{j+1} \quad \textit{and} \quad \bigcup_j \Omega_j = Y - V.$$

EXAMPLE. (1) Of course, Stein manifolds are essentially Stein.

(2) Projective manifold are also essentially Stein: simply embed the manifold in projective space and take V to be the intersection of X with a hyperplane.

(3) Let $\pi : X \to \mathbb{D}$ is a holomorphic immersion of a complex manifold X to the unit disk such that each fiber $X_t := \pi^{-1}(t)$, $t \in \mathbb{D}$ is a compact manifold. Assume there is a positive line bundle $L \to X$. Then X is essentially Stein.

1.2. Statement of the main theorem

To state our main theorem, we recall some notation from the introductory lecture, this time around stated slightly differently. Let Y be a complex manifold and $H \to Y$ a holomorphic line bundle with singular Hermitian metric h. Let s be a section of $H + K_Y$, not necessarily holomorphic. We fix a nowhere-zero section ξ of H and holomorphic coordinates $z^1, ..., z^n$ on a sufficiently small open set $U_{\xi,z}$ in Y, and in terms of these coordinates, there exist functions $f^{(z,\xi)}$ and $\varphi^{(\xi)}$ such that

$$h(\xi, \xi) = e^{-\varphi^{(\xi)}} \quad \text{and} \quad s = f^{(z,\xi)} \xi \otimes dz^1 \wedge ... \wedge dz^n.$$

Consider the expression

$$\alpha^{(z,\xi)} := |f^{(z,\xi)}|^2 e^{-\varphi^{(\xi)}} \left(\tfrac{\sqrt{-1}}{2}\right)^n dz^1 \wedge d\bar{z}^1 \wedge \ldots \wedge dz^n \wedge d\bar{z}^n.$$

Then a simple calculation shows that if η is another nowhere-zero local section of H and w another local coordinate system on an open set $U_{\eta,w} \subset Y$, then

$$\alpha^{(w,\eta)} = \alpha^{(z,\xi)} \quad \text{on } U_{\xi,z} \cap U_{\eta,w}.$$

We thus define the global (n,n)-form $|s|^2 e^{-\varphi}$ by

$$|s|^2 e^{-\varphi} := \alpha^{(z,\xi)} = |f^{(z,\xi)}|^2 e^{-\varphi^{(\xi)}} \left(\tfrac{\sqrt{-1}}{2}\right)^n dz^1 \wedge d\bar{z}^1 \wedge \ldots \wedge dz^n \wedge d\bar{z}^n \quad \text{on } U_{\xi,z}.$$

We can now state the first main result of this lecture.

THEOREM 1.2. *Let Y be an essentially Stein manifold, and suppose $V \subset Y$ is the hypersurface such that $Y - V$ is Stein. Let $Z \subset Y$ a smooth hypersurface not contained in V. Suppose given a holomorphic line bundle $H \to Y$ with a singular Hermitian metric $e^{-\kappa}$, and a singular Hermitian metric $e^{-\varphi_Z}$ for the line bundle associated to the divisor Z, such that the following properties hold.*

(i) *The restrictions $e^{-\kappa}|_Z$ and $e^{-(\kappa+\varphi_Z)}|_Z$ are singular metrics for H and $Z + H$, respectively.*
(ii) *There is a global holomorphic section $T \in H^0(Y, Z)$ such that*

$$Z = \{T = 0\} \quad \textit{and} \quad \sup_Y |T|^2 e^{-\varphi_Z} = 1.$$

(iii) $\sqrt{-1}\partial\bar{\partial}\kappa \geq 0$ *and* $\mu\sqrt{-1}\partial\bar{\partial}\kappa \geq \sqrt{-1}\partial\bar{\partial}\varphi_Z$ *for some integer* $\mu > 0$.

Then for every $s \in H^0(Z, K_Z + H|_Z)$ such that

$$\int_Z |s|^2 e^{-\kappa} < +\infty \quad \textit{and} \quad s \wedge dT \in \mathscr{I}_{(\varphi_Z+\kappa)|_Z},$$

there exists a section $S \in H^0(Y, K_Y + Z + H)$ such that

$$S|_Z = s \wedge dT \quad \textit{and} \quad \int_Y |S|^2 e^{-(\varphi_Z+\kappa)} \leq 8\pi(\mu+2) \int_Z |s|^2 e^{-\kappa}.$$

REMARK. In (i) and (ii) we are identifying the hypersurface Z with the line bundle associated to it.

REMARK. A crucial aspect of the L^2 Extension Theorem is the very loose dependence of the constant $8\pi(\mu+2)$ on the data; this number is not only independent of the section s, but it is also independent of the choice of line bundle H, divisor Z and hypersurface V, so long as the curvature hypotheses are satisfied. This universality of the constants will play a crucial role in our main application of the L^2 Extension Theorem, namely, the proof of Siu's Theorem on the deformation invariance of plurigenera.

1.3. Some corollaries

1.3.1. Bounded domains in $\mathbb{C}^n$. The simplest example comes from the following local and thus very useful setting. Let Y be a bounded pseudoconvex domain in $\mathbb{C}^n$, and let Z be the intersection of Y with an affine hyperplane in $\mathbb{C}^n$. By a linear change of coordinates, we can assume that $0 \in Y$ and that the hyperplane is cut out by the equation $z^n = 0$. The line bundle associated to the hyperplane is of course trivial, since it is cut out by a global holomorphic function, namely z^n. Moreover,

$$C := \sup_Y |z^n| < +\infty,$$

and clearly C is controlled by the diameter of Y. Take $T = z^n/C$, $H \to Y$ the trivial bundle, and κ a plurisubharmonic function. Since the line bundle associated to Z is trivial, we can choose $\varphi_Z \equiv 0$. In this case, we can take $\mu = 1$. The hypotheses of Theorem 1.2 are thus satisfied, and moreover the condition that $s \wedge dT \in \mathscr{I}_\kappa$ (since $\varphi_Z = 0$) is implied by the finiteness of the L^2-norm of s on Z. Thus we obtain the following result.

COROLLARY 1.3. *Let $\Omega \subset \mathbb{C}^n$ be a bounded, pseudoconvex domain containing the origin and φ a plurisubharmonic function. Let $\mathcal{H} \subset \mathbb{C}^n$ be the affine hyperplane defined by the equation $z^n = 0$. Then there is a constant K depending only on the diameter of Ω such that the following holds. For any holomorphic function f such that*

$$\int_{\mathcal{H}\cap\Omega} |f|^2 e^{-\varphi} dV_{n-1} < +\infty,$$

there is a holomorphic function F on Ω such that

$$F|_{\mathcal{H}\cap\Omega} = f \quad \textit{and} \quad \int_\Omega |F|^2 e^{-\varphi} dV_n \le K \int_{\mathcal{H}\cap\Omega} |f|^2 e^{-\varphi} dV_{n-1}.$$

1.3.2. Families of projective manifolds. Consider the following setting. Let X be an essentially Stein manifold of complex dimension $n \ge 2$ and $w : X \to \mathbb{D}$ a surjective holomorphic map to the unit disk such that (i) dw is nowhere zero, and (ii) every fiber $X_t := w^{-1}(t)$ is a compact manifold. (For example, the fibers of w can be projective manifolds.) Let $Z = w^{-1}(0)$ be the central fiber. The line bundle associated to Z is again trivial, because it is cut out by a holomorphic function, namely w. Again we take the metric $e^{-\varphi_Z} \equiv 1$ and so we can take $\mu = 1$ and also ignore the multiplier ideal membership hypothesis. We thus obtain the following result.

THEOREM 1.4 (Siu). *Let $w : X \to \mathbb{D}$ and X_0 be as above. Let $H \to X$ be a holomorphic line bundle and $e^{-\kappa}$ a singular Hermitian metric whose curvature current $\sqrt{-1}\partial\bar\partial\kappa$ is non-negative. Then for every H-valued holomorphic $(n-1)$-form s on X_0 such that*

$$\int_{X_0} |s|^2 e^{-\kappa} < +\infty$$

there exists an H-valued holomorphic n-form S such that

$$S|_{X_0} = s \wedge dw \quad \textit{and} \quad \int_X |S|^2 e^{-\kappa} \le 24\pi \int_{X_0} |s|^2 e^{-\kappa}.$$

1.4. The twisted basic estimate

Let Ω be a smoothly bounded pseudoconvex domain in our essentially Stein manifold Y. Let τ and A be positive smooth functions on Ω, and let $e^{-\psi}$ be a singular metric for $H + Z$ over Ω.

LEMMA 1.5. *For any $(n,1)$-form u in the domain of $\bar\partial$ and of the adjoint $\bar\partial^*_\psi$ of $\bar\partial$, the following inequality holds.*

$$\int_\Omega (\tau + A)\left|\bar\partial^*_\psi u\right|^2 e^{-\psi} + \int_\Omega \tau\left|\bar\partial u\right|^2 e^{-\psi} \geq \int_\Omega \left(\tau\sqrt{-1}\partial\bar\partial\psi - \sqrt{-1}\partial\bar\partial\tau - \frac{1}{A}\sqrt{-1}\partial\tau\wedge\bar\partial\tau\right)(u,u)e^{-\psi}. \tag{2.1}$$

PROOF. We begin with a special case of the usual Bochner-Kodaira Identity derived in lecture 1: for any H-valued $(n,1)$-form in the domain of $\bar\partial^*$ and the domain of $\bar\partial$,

$$\int_\Omega \left|\bar\partial^*_\varphi u\right|^2 e^{-\varphi} + \int_\Omega \left|\bar\partial u\right|^2 e^{-\varphi} = \int_\Omega \sqrt{-1}\partial\bar\partial\varphi(u,u)e^{-\varphi} + \int_\Omega \left|\overline{\nabla} u\right|^2 e^{-\varphi} + \int_{\partial\Omega}\sqrt{-1}\partial\bar\partial\rho(u,u)e^{-\varphi} \tag{2.2}$$

(In this setting of $H+Z$-valued canonical forms, the Kähler curvature cancels out the negative of the Ricci curvature and we have this very manageable formula.) Note that in the basic estimate, the last two terms of the right hand side are non-negative.

Next we pass the the twisted estimates. The idea is to consider a twist of the original metric for H. That is to say, we consider a metric $e^{-\psi}$ for H. For any such metric, there exists a positive function τ such that

$$e^{-\varphi} = \tau e^{-\psi}.$$

It is then an easy calculation that

$$\partial\bar\partial\varphi = \partial\bar\partial\psi - \tau^{-1}\partial\bar\partial\tau + \tau^{-2}\partial\tau\wedge\bar\partial\tau.$$

Also, using the formula

$$\bar\partial^*_\varphi u = -\sum_j e^{\varphi}\frac{\partial}{\partial z^j}(e^{-\varphi}u_{\bar j}),$$

where locally $u = \sum u_{\bar j} d\bar z^j$ with $u_{\bar j}$ canonical sections, we have the formula

$$\bar\partial^*_\varphi u = -\tau^{-1}\partial\tau(u) + \bar\partial^*_\psi u.$$

Substitution of these identities into the basic estimate (2.2) then gives the twisted Bochner-Kodaira Identity:

$$\begin{aligned}\int_\Omega (\tau + A)\left|\bar\partial^*_\psi u\right|^2 e^{-\psi} &+ \int_\Omega \tau\left|\bar\partial u\right|^2 e^{-\psi} \\ = &- \int_\Omega \sqrt{-1}\partial\bar\partial\tau(u,u)e^{-\psi} \\ &+ \int_\Omega \sqrt{-1}\tau\partial\bar\partial\psi(u,u)e^{-\psi} + \int_\Omega A\left|\bar\partial^*_\psi u\right|^2 e^{-\psi} + 2\mathrm{Re}\int_\Omega \partial\tau(u)\overline{\bar\partial^*_\psi u}e^{-\psi} \\ &+ \int_\Omega \tau\left|\overline{\nabla} u\right|^2 e^{-\psi} + \int_{\partial\Omega}\tau\sqrt{-1}\partial\bar\partial\rho(u,u)e^{-\psi}\end{aligned} \tag{2.3}$$

By the Cauchy-Schwarz Inequality,

$$2\mathrm{Re}\int_\Omega \partial\tau(u)\overline{\bar\partial^*_\psi u}e^{-\psi} \geq -\int_\Omega A\left|\bar\partial^*_\psi u\right|^2 e^{-\psi} - \int_\Omega \frac{1}{A}\sqrt{-1}\partial\tau\wedge\bar\partial\tau(u,u)e^{-\psi}.$$

Applying this estimate to the last term on the third line of (2.3) completes the proof. □

1.5. Proof of Theorem 1.2

1.5.1. Choices for τ, A and ψ, and an a priori estimate. First, we set

$$\tau = a + h(a) \quad \text{and} \quad A = \frac{(1+h'(a))^2}{-h''(a)},$$

where

$$h(x) = 1 - x + \frac{1}{\mu}\log(2e^{\mu(x-1)} - 1)$$

and $a : \Omega \to [1, \infty)$ is a function to be chosen shortly. Observe that for $x \geq 1$,

$$h'(x) = \frac{1}{2e^{\mu(x-1)} - 1} \in (0,1) \quad \text{and} \quad h''(x) = \frac{-2\mu e^{\mu(x-1)}}{(2e^{\mu(x-1)} - 1)^2} < 0,$$

and thus since $1 + \log r \geq \frac{1}{r}$ when $r \geq 1$,

$$\tau \geq 1 + h'(a).$$

Moreover, $A > 0$, which in necessary in our choice of A. We also take this opportunity to note that

$$A = \tfrac{1}{2\mu} e^{\mu(1-a)}.$$

With these choices of τ and A, we have

$$(2.4) \quad -\partial_\alpha \partial_{\bar\beta}\tau - \frac{\partial_\alpha \tau \overline{\partial_\beta \tau}}{A} = -\partial_\alpha\left((1+h'(a))\partial_{\bar\beta} a\right) - \frac{(1+h'(a))^2 \partial_\alpha a \overline{\partial_\beta a}}{A} = (1+h'(a))\left(-\partial_\alpha \partial_{\bar\beta} a\right)$$

Our next task is to construct the function a. To this end, define

$$v = \log|T|^2 - \varphi_Z = \log(|T|^2 e^{-\varphi_Z}).$$

We note that $v \leq 0$. Fix a constant $\gamma > 1$. We define the function a to be

$$a = a_\varepsilon := \gamma - \frac{1}{\mu}\log\left(e^v + \varepsilon^2\right),$$

where $\varepsilon > 0$ is chosen so small that $a \geq 1$. Later we will let $\varepsilon \to 0$ and $\gamma \to 1$.

We calculate that

$$\begin{aligned} -\sqrt{-1}\partial\bar\partial a &= \frac{\sqrt{-1}}{\mu}\partial\bar\partial \log(e^v + \varepsilon^2) \\ &= \frac{e^v}{\mu(e^v+\varepsilon^2)}\sqrt{-1}\partial\bar\partial v + \frac{4\varepsilon^2\sqrt{-1}\partial(e^{v/2}) \wedge \bar\partial(e^{v/2})}{\mu((e^{v/2})^2 + \varepsilon^2)^2} \\ &= -\frac{1}{\mu}\frac{e^v}{(e^v+\varepsilon^2)}\sqrt{-1}\partial\bar\partial\varphi_Z + \frac{4\varepsilon^2\sqrt{-1}\partial(e^{v/2}) \wedge \bar\partial(e^{v/2})}{\mu((e^{v/2})^2 + \varepsilon^2)^2}. \end{aligned}$$

In the last equality we have used the fact that

$$\sqrt{-1}\partial\bar\partial v = \pi[Z] - \sqrt{-1}\partial\bar\partial\varphi_Z,$$

where $[Z]$ is the current of integration over Z. The term involving the current of integration vanishes because $e^v|_Z \equiv 0$.

It remains to choose the metric $e^{-\psi}$. We take

$$\psi = \kappa + \log|T|^2.$$

Then

$$\begin{aligned}
& \tau\sqrt{-1}\partial\bar\partial\psi - \sqrt{-1}\partial\bar\partial\tau - \frac{\sqrt{-1}\partial\tau\wedge\bar\partial\tau}{A} \\
= \ & \tau\sqrt{-1}\partial\bar\partial\kappa + \tau\pi[Z] \\
& +(1+h'(a))\left(-\frac{e^v}{\mu(e^v+\varepsilon^2)}\sqrt{-1}\partial\bar\partial\varphi_Z + \frac{4\varepsilon^2\sqrt{-1}\partial(e^{v/2})\wedge\bar\partial(e^{v/2})}{\mu((e^{v/2})^2+\varepsilon^2)^2}\right) \\
\geq \ & \frac{4\varepsilon^2\sqrt{-1}\partial(e^{v/2})\wedge\bar\partial(e^{v/2})}{\mu((e^{v/2})^2+\varepsilon^2)^2}.
\end{aligned}$$

The inequality follows from assumption (i) and the fact that $\tau \geq 1 + h'(a) \geq 1$. Combining with (2.1), we obtain the following lemma.

LEMMA 1.6. *Let $T = \bar\partial \circ \sqrt{\tau + A}$ and $S = \sqrt{\tau}\bar\partial$. Then for any $(n,1)$-form u in the domain of S and of the adjoint T^* of T, the following inequality holds.*

$$\int_\Omega \left|\left\langle u, \bar\partial(e^{v/2})\right\rangle\right|^2 \frac{4\varepsilon^2}{\mu(e^v+\varepsilon^2)^2} e^{-\psi} \leq \left(||T^*u||^2_\psi + ||Su||^2_\psi\right)$$

1.5.2. A smooth extension and its holomorphic correction. Since Ω is Stein, we can extend $s \wedge dT$ to an $(H+Z)$-valued holomorphic n-form $\tilde s$ on Ω. By extending to a Stein neighborhood of Ω (which exists by hypothesis) we may also assume that

$$\int_\Omega |\tilde s|^2 e^{-(\varphi_Z+\kappa)} < +\infty.$$

(Here we have used the local integrability of $|s\wedge dT|^2 e^{-(\varphi_Z+\kappa)}$ on Z.) Note that from this construction we can deduce no better estimate on this $\tilde s$. In particular, the estimate could degenerate as Ω grows.

In order to tame the growth of this extension $\tilde s$, we first modify it to a smooth extension. To this end, let $\delta > 0$ and let $\chi \in \mathcal{C}^\infty_0([0,1))$ be a cutoff function with values in $[0,1]$ such that $\chi \equiv 1$ on $[0,\delta]$ and $|\chi'| \leq 1+\delta$. We write

$$\chi_\varepsilon := \chi\left(\frac{e^v}{\varepsilon^2}\right).$$

We distinguish the smooth $(n,1)$-form with values in $Z+H$

$$\alpha_\varepsilon := \bar\partial\chi_\varepsilon\tilde s.$$

Then one has the estimate

$$\begin{aligned}
|(u,\alpha_\varepsilon)_\psi|^2 &\leq \left(\int_\Omega |\left\langle u,\alpha_\varepsilon\right\rangle| e^{-\psi}\right)^2 \\
&= \left(\int_\Omega \left|\left\langle u, \chi'\left(\frac{e^v}{\varepsilon^2}\right)\tilde s \wedge \frac{2e^{v/2}\bar\partial(e^{v/2})}{\varepsilon^2}\right\rangle\right| e^{-\psi}\right)^2 \\
&\leq \mu\int_\Omega \left|\frac{\tilde s}{\varepsilon^2}\chi'\left(\frac{e^v}{\varepsilon^2}\right)\right|^2 \frac{(e^v+\varepsilon^2)^2}{\varepsilon^2} e^{-(\psi-v)} \\
&\quad \times \int_\Omega \left|\left\langle u, \bar\partial(e^{v/2})\right\rangle\right|^2 \frac{4\varepsilon^2}{\mu(e^v+\varepsilon^2)^2} e^{-\psi} \\
&\leq C_\varepsilon\left(||T^*u||^2_\psi + ||Su||^2_\psi\right)
\end{aligned}$$

where the last inequality follows from Lemma 1.6, and

$$C_\varepsilon := \frac{4\mu(1+\delta)^2}{\varepsilon^2}\int_{e^v\le\varepsilon^2}|\tilde{s}|^2\, e^{-(\varphi_Z+\kappa)} \overset{\varepsilon\to 0}{\longrightarrow} 8\pi\mu(1+\delta)^2\int_Z |s|^2 e^{-\kappa}.$$

As a result of this estimate, we obtain the following theorem.

THEOREM 1.7. *There exists a smooth n-form β_ε such that*

$$T\beta_\varepsilon = \alpha_\varepsilon \quad \text{and} \quad \int_\Omega |\beta_\varepsilon|^2 e^{-\psi} \le C_\varepsilon.$$

In particular,

$$\beta_\varepsilon|_Z \equiv 0.$$

PROOF. In view of the inequality

$$|(u,\alpha_\varepsilon)| \le C_\varepsilon\left(||T^*u||^2_\psi + ||Su||^2_\psi\right),$$

the existence of β_ε in L^2 follows from the Functional Analysis Lemma 1.4. Next, β_ε is smooth because

$$\beta_\varepsilon = (\beta_\varepsilon - \sqrt{\tau+A}\chi_\varepsilon\tilde{s}) + \sqrt{\tau+A}\chi_\varepsilon\tilde{s}$$

is the sum of a holomorphic section and a smooth section. It remains only to show that $\beta_\varepsilon|_Z \equiv 0$. But one notices that ψ is at least as singular as $\log|T|^2$, and thus $e^{-\psi}$ is not locally integrable at any point of Z. The desired vanishing of β_ε follows. □

CONCLUSION OF THE PROOF OF THEOREM 1.2. Note first that, since $\mu \ge 1$,

$$\begin{aligned}\limsup_{\varepsilon\to 0}\sup_\Omega(e^v(\tau+A)) &= \limsup_{\varepsilon\to 0}\sup_\Omega e^{\mu(\gamma-a)}(1+\tfrac{1}{\mu}\log(2e^{\mu(a-1)}-1)+\tfrac{1}{2\mu}e^{\mu(1-a)})\\ &\le e^{\mu(\gamma-1)}\sup_{x\ge 1}\tfrac{1}{x}(1+\tfrac{1}{\mu}\log 2x+\tfrac{1}{2\mu x})\\ &\le (1+\tfrac{2}{\mu})e^{\mu(\gamma-1)}.\end{aligned}$$

Next let

$$S_\varepsilon := \chi_\varepsilon\tilde{s} - \sqrt{\tau+A}\beta_\varepsilon.$$

Then S_ε is a holomorphic section, $S_\varepsilon|_Z = \tilde{s}|_Z = s\wedge dT$, and we have the estimate

$$\int_\Omega |S_\varepsilon|^2 e^{-(\varphi_Z+\kappa)} \le (1+\tfrac{2}{\mu})e^{\mu(\gamma-1)}(1+o(1))8\pi\mu(1+\delta)^2\int_Z|s|^2e^{-\kappa} \quad \text{as } \varepsilon\to 0.$$

Here we have used the estimate obtained in Theorem 1.7. The term $o(1)$ comes in because $\chi_\varepsilon\tilde{s}$ is a smooth, uniformly bounded section whose support approaches a set of measure zero, and thus the integral of the square norm of $\chi_\varepsilon\tilde{s}$ converges to zero.

Now, by the sub-mean value property, uniform L^2-estimates with plurisubharmonic (hence locally bounded above) weights implies locally uniform sup-norm estimates. It follows from Montel's Theorem that S_ε converges to a holomorphic section S. Evidently

$$S|_{Z\cap\Omega} = s\wedge dT$$

and

$$\int_\Omega |S|^2 e^{-(\varphi_Z+\kappa)} \le e^{\mu(\gamma-1)}\times 8\pi(\mu+2)(1+\delta)^2\int_Z|s|^2e^{-\kappa}.$$

Finally, by the uniformity of all estimates, we may let $\delta\to 0$, $\gamma\to 1$ and then $\Omega\to Y$. The proof is complete. □

1.6. What happened?

The proof of the L^2 Extension Theorem may seem rather mystifying on a number of levels. Let us try to summarize what has happened.

The basic scheme is very much analogous to the way in which we have used Hörmander's Theorem to extend sections from points in our proof of Kodaira's Embedding Theorem; we construct a smooth extension of our section, holomorphic in a neighborhood of the submanifold, and then try to correct it. However, there are two problem that arise in the present context.

The first problem is that we cannot get an L^2-estimate for an extension to a uniform-sized neighborhood. We can employ an exhaustion argument, but as our compact sets grow, we will have to shrink the neighborhood onto which we have an extension. The shrinking of this neighborhood will significantly worsen the $\bar{\partial}$ of the smooth extension, and therefore give much worse estimates for the correction. The second problem is that the weight we have is not sufficiently positive to solve $\bar{\partial}$ in order to make the holomorphic correction.

At this point, the magic trick is introduced which seems to handle both problems. We exchange our $\bar{\partial}$ operator for the operator $T = \bar{\partial} \circ \sqrt{\tau + A}$. This new operator is of the form $N\bar{\partial} + M$ where N and M are 0th order operators.

If we add into the picture the operator $S = \sqrt{\tau}\bar{\partial}$, then $ST = 0$, and moreover we have a lower bound for $||T^*\beta||^2 + ||S\beta||^2 = ((TT^* + S^*S)\beta, \beta)$. Since the operator T is somewhat like $\bar{\partial}+$ a lower order term, we cannot expect an identity like the Bochner-Kodaira identity. Nevertheless, an optimal lower bound for the eigenvalues of $TT^* + S^*S$ could be thought of as a "sub-curvature" for such operators. The point of the Ohsawa-Takegoshi technique is that while $\Box$ is not strictly positive, the twisted operator $TT^* + S^*S$ is! Moreover, the amount of positivity we can introduce significantly improves the estimates for the correction to the smooth extension.

Note that a function τ for which $-\sqrt{-1}\partial\bar{\partial}\tau - \frac{\sqrt{-1}}{A}\partial\tau \wedge \bar{\partial}\tau$ is strictly positive for some positive function A, cannot exist on a compact complex manifold, since then $-\tau$ is strictly plurisubharmonic. Thus we are forced to use the exhaustion technique, and this is why we need some kind of convexity hypothesis on $Y - V$. We use the hypothesis that $Y - V$ is Stein because we want to use singular Hermitian metrics. (Recall that we encountered the same issue in the proof of Hörmander's Theorem for singular Hermitian metrics.) The method can work in the more general setting of weakly pseudoconvex domains if the metrics $e^{-\kappa}$ and $e^{-(\kappa+\varphi_Z)}$ can be approximated from below by an increasing sequence of metrics with the appropriate curvature bounds.

Finally, at the end of the day we want results about holomorphic extensions. Since we are making our corrections with the perturbed equation $T\beta = \alpha$, the search for a holomorphic correction will require L^2-estimates for $\sqrt{\tau + A}\beta$. One way to obtain such estimates is by estimating the supremum of $e^v(\tau + A)$, which explains why we made the choices of τ and A: we want to (i) have the positivity of $-\sqrt{-1}\partial\bar{\partial}\tau - \frac{\sqrt{-1}}{A}\partial\tau \wedge \bar{\partial}\tau$, (ii) the sup-norm estimates for $e^v(\tau + A)$, and (iii) uniform estimates so that we can pass to the limit and obtain an extension on all of $Y - V$. (L^2-estimates will then give us the extension across V.)

In fact the choice of τ and A, while delicate, does not need to be sharp; one could make different choices and get different bounds for the extension produced by the L^2 Extension Theorem.

2. The deformation invariance of plurigenera

In this section we are going to prove Y.-T. Siu's theorem on the deformation invariance of plurigenera.

In the original proof of Siu, the deformation invariance of plurigenera is proved using two key tools. The first is the L^2 Extension Theorem (Theorem 1.2), and the second is the effective global generation of multiplier ideal sheaves. (We present a non-effective version, also due to Siu, at the end of the next lecture, namely Theorem 3.2. The effective result can be found in [**Siu-2002**].)

Recently Păun [**Păun-2007**] discovered a very short proof of Siu's Theorem on the deformation invariance of plurigenera. We present this proof here. The novelty of Păun's proof is that the effective global generation theorem is no longer required, and one can make do with only the L^2 extension theorem.

2.1. Statement of Siu's theorem

2.1.1. Holomorphic families. Let $\pi : X \to \mathbb{D}$ be a holomorphic map from a Kähler manifold X to the unit disk. Assume that the following properties hold.

(i) The differential $d\pi$ is nowhere zero on X.
(ii) Each fiber $X_t := \pi^{-1}(t)$ is a compact complex manifold.
(iii) X has a holomorphic line bundle $A \to X$ with a smooth metric $e^{-\psi}$ whose curvature is strictly positive. (In the presence of A, each fiber is a complex projective manifold, as seen by applying Kodaira's Embedding Theorem to $(X_t, A|_{X_t})$).

Such a fibration $\pi : X \to \mathbb{D}$ satisfying properties (i) and (ii) is called a *holomorphic family.* If property (iii) also holds, then $\pi : X \to \mathbb{D}$ is called an *algebraic family.*

PROPOSITION 2.1. *For any $s, t \in \mathbb{D}$, the fibers X_s and X_t of a holomorphic family are diffeomorphic.*

PROOF. The tangent spaces of the fibers of π define a complex codimension 1 subbundle V of T_X. Fix a Riemannian metric on T^*_X such that $|d\pi| = 1$, and let ξ be the Riemannian dual to $d\pi$. Since $d\pi$ is never zero, neither is ξ. Moreover, $|\xi|^2 = d\pi(\xi) = |d\pi|^2 = 1$ and for any $v \in V$, $\langle \xi, v\rangle = d\pi(v) = 0$.

Let φ^λ_ξ be the flow of ξ. This flow is not defined for all $\lambda \in \mathbb{R}$; the set of λ depends on the initial condition $x \in X$. But wherever is is defined, it satisfies

$$\pi(\varphi^\lambda_\xi x) = \lambda + \pi(x).$$

Indeed,

$$\frac{d}{d\lambda}\pi(\varphi^\lambda_\xi x) = d\pi_{\varphi^\lambda_\xi x}(\xi_{\varphi^\lambda_\xi x}) \equiv 1.$$

We claim that if $\lambda \mapsto \varphi^\lambda_\xi(x)$ exists for all $\lambda \in U(x) \subset \mathbb{R}$, then $\lambda \mapsto \varphi^\lambda_\xi(y)$ is defined for all $\lambda \in U(x)$ whenever $\pi(y) = \pi(x)$. This follows from the Existence and Uniqueness Theorem for ODE, together with the compactness of the fibers. Indeed, the only way that solutions can cease to exist if if they have escaped to infinity, but $\pi(\varphi^\lambda_\xi(y))$ has not escaped to the boundary of the disc, and $\varphi^\lambda_\xi(y)$ cannot escape to infinity along the fibers. This means that φ^λ_ξ is a diffeomorphism of fibers whenever it is defined. Using different initial conditions, we can identify any two fibers of π. □

2.1.2. Plurigenera, the invariance problem and Siu's extension theorem.

DEFINITION 2.2. *Let Y be a complex manifold. For each $m \in \mathbb{N}_{\geq 1}$ the number*

$$P_m(Y) := dim_{\mathbb{C}} H^0(Y, mK_Y)$$

is called the mth plurigenus of Y.

For a holomorphic family, we have a function

$$\mu : \mathbb{D} \ni t \mapsto P_m(X_t) \in \mathbb{N}.$$

The deformation invariance of plurigenera states that for each m the function μ is constant.

It is a known fact, proved for example in Hartshorne's Book (starting on page 281) that the function μ is upper semi-continuous. It follows the dimension of the space of pluricanonical sections on some X_t could be more than that of its immediate neighbors. Thus to prove the constancy of μ, it suffices to prove that every pluricanonical section on the central fiber X_0 extends to a pluricanonical section on the family X. The extension is in a sense that will be made more precise presently.

THEOREM 2.3 (Siu, Păun). *Let $\pi : X \to \mathbb{D}$ be an algebraic family of n-dimensional complex manifolds over the unit disk. Assume there is a Kähler form ω on X such that*

$$\int_X \omega^{n+1} < +\infty.$$

Suppose we have a line bundle $L \to X$ with singular Hermitian metric $e^{-\kappa}$, and a smooth Hermitian metric $e^{-\gamma}$ such that $e^{-\kappa}|_{X_0}$ is a singular Hermitian metric on X_0 and

$$\sup_X e^{\kappa-\gamma} < +\infty.$$

Let $s \in H^0(X_0, mK_{X_0} + L)$ such that

$$\int_{X_0} |s|^2 \omega^{-n(m-1)} e^{-\kappa} < +\infty.$$

Then there is a section $S \in H^0(X, mK_X + L)$ such that

$$S|_{X_0} = s \wedge (d\pi)^{\otimes m} \quad \text{and} \quad \int_X |S|^2 \omega^{-(n+1)(m-1)} e^{-\kappa} < +\infty.$$

REMARK. The finiteness of the volume of X can be achieved by shrinking X slightly, i.e., replacing X with $\pi^{-1}(\mathbb{D}(0,r)$ for some $r < 1$. Similarly, a metric $e^{-\gamma}$ with the desired properties can be obtained by shrinking X. In the latter case, we are using the fact that a plurisubharmonic function is locally bounded above.

REMARK. The constancy of the plurigenera of X follows by taking $L = 0$. One can also define twisted plurigenera, and these too will be invariant in families.

REMARK. The result as stated is due to Păun, and was conjectured by Siu, who proved it in the case of locally integrable $e^{-\kappa}$.

2.2. Strategy of the proof

Let s be the section to be extended. By the L^2 Extension Theorem 1.4, it suffices to find a metric $e^{-\mu}$ for $mK_X + L$ such that

$$\int_{X_0} |s|^2 e^{-\frac{(m-1)\mu+\kappa}{m}} < +\infty.$$

Such an estimate would hold if we could find a metric $e^{-\mu}$ such that

$$e^{-\mu}|_{X_0} = \frac{1}{|s|^2},$$

for then

$$\begin{aligned}\int_{X_0} |s|^2 e^{-\frac{(m-1)\mu+\kappa}{m}} &= \int_{X_0} |s|^{2/m} e^{-\frac{\kappa}{m}} \\ &\leq \left(\int_{X_0} \omega^n\right)^{\frac{m-1}{m}} \left(\int_{X_0} |s|^2 \omega^{-n(m-1)} e^{-\kappa}\right)^{\frac{1}{m}} \\ &< +\infty,\end{aligned}$$

where the inequality is a consequence of Hölder's Inequality.

In effect, we have substituted the problem of extending the section s for the problem of extending the metric $|s|^{-2}$. In general, extending metrics is harder than extending sections; at present there is no general method for extending non-negatively curved metrics. But fortunately, as it turns out, we can extend such an 'algebraic' metric by using extension theorems for sections.

It may seem like we have made no progress, but in fact, the situation for extending algebraic metrics is much better. The main reason for the improvement is that we can take roots of metrics. Thus we can try to extend the metric

$$|s^{\otimes k}|^{-2} e^{-\varphi}$$

for some sufficiently positive (but fixed!) metric $e^{-\varphi}$ of some line bundle A, which we can assume has a section, say a. Thus we try to extend the sections $s^{\otimes k} \otimes a$. This almost works; the problem is that a may have zeros, and these zeros make it harder to use the extension theorem. So instead, we assume A is sufficiently positive that there are sections $a_1, ..., a_N$ with no common zeros. (In fact, we will assume slightly more from A, for technical reasons that will be clear below.) It is then possible to extend the sections $s^{\otimes k} \otimes a_j$ to $\tilde{s}_j^{(k)}$, $1 \leq j \leq N$, and form the metric

$$e^{-\mu_k} := e^{-\frac{1}{k}\log(\sum_j |\tilde{s}^{(k)}|^2)}$$

for the $\mathbb{Q}$-line bundle $mK_X + \frac{1}{k}A$. Such extension is achieved using a slightly delicate but nevertheless straightforward induction argument. The uniformity of the bounds in the Extension Theorem 1.2 allows us to pass to the limit $k \to \infty$ and obtain a metric of non-negative curvature for mK_X.

We now make this discussion more precise.

2.3. Positively twisted canonical sections

Let A be a holomorphic line bundle with the following property:

(GG) For each $0 \leq p \leq m-1$ the global sections $H^0(X, pK_X + A)$ generate the sheaf of germs of holomorphic sections of $pK_X + A$.

Let us fix bases

$$\{\tilde{\sigma}_j^{(p)} \quad 1 \le j \le N_p\}$$

of $H^0(X, pK_X + A)$. We let $\sigma_j^{(p)} \in H^0(X_0, pK_{X_0} + A|_{X_0})$ be such that

$$\tilde{\sigma}_j^{(p)}|_{X_0} = \sigma_j^{(p)} \wedge (d\pi)^{\otimes p}.$$

Such $\sigma_j^{(p)}$ exist by adjunction.

PROPOSITION 2.4. *There exist a constant $C < +\infty$ and sections*

$$\{\tilde{\sigma}_j^{(km+p)} \in H^0(X, (km+p)K_X + kL + A) \ ; \ 1 \le j \le N_p\}_{0 \le p \le m-1, k=0,1,2,\ldots}$$

with the following properties.

(a) $\tilde{\sigma}_j^{(mk+p)}|_{X_0} = s^{\otimes k} \otimes \sigma_j^{(p)} \wedge (d\pi)^{km+p}$

(b) *If $k \ge 1$,*

$$\int_X \frac{\sum_{j=1}^{N_0} |\tilde{\sigma}_j^{(mk)}|^2 \omega^{-(n+1)mk} e^{-(k\gamma+\psi)}}{\sum_{j=1}^{N_{m-1}} |\tilde{\sigma}_j^{(mk-1)}|^2 \omega^{-(n+1)(mk-1)} e^{-((k-1)\gamma+\psi)}} \omega^{n+1} \le C.$$

(c) *For $1 \le p \le m-1$,*

$$\int_X \frac{\sum_{j=1}^{N_p} |\tilde{\sigma}_j^{(mk+p)}|^2 \omega^{-(n+1)(mk+p)} e^{-(k\gamma+\psi)}}{\sum_{j=1}^{N_{p-1}} |\tilde{\sigma}_j^{(mk+p-1)}|^2 \omega^{-(n+1)(mk+p-1)} e^{-(k\gamma+\psi)}} \omega^{n+1} \le C.$$

REMARK. We may take $e^{-\psi}$ to be any metric for $A \to X$. The choice of ψ does not affect the integrals.

PROOF. We induct on k and p.

Let $\widehat{C}$ denote the maximum of

$$\max_X \left\{ \frac{\sum_{j=1}^{N_0} |\tilde{\sigma}_j^{(0)}|^2 e^{-\psi}}{\sum_{j=1}^{N_p} |\tilde{\sigma}_j^{(m-1)}|^2 e^{-\psi} \omega^{-(n+1)(m-1)}} \right\},$$

$$\max_{0 \le p \le m-2} \max_X \left\{ \frac{\sum_{j=1}^{N_{p+1}} |\tilde{\sigma}_j^{(p+1)}|^2 e^{-\psi} \omega^{-(p+1)(n+1)}}{\sum_{j=1}^{N_p} |\tilde{\sigma}_j^{(p)}|^2 e^{-\psi} \omega^{-p(n+1)}} \right\}.$$

($k = 0$) As far as extension there is nothing to prove. Note that

$$\int_X \frac{\sum_{j=1}^{N_p} |\tilde{\sigma}_j^{(p)}|^2 \omega^{-(n+1)p} e^{-\psi}}{\sum_{j=1}^{N_{p-1}} |\tilde{\sigma}_j^{(p-1)}|^2 \omega^{-(n+1)(p-1)} e^{-\psi}} \omega^{n+1} \le \widehat{C} \int_X \omega^{n+1}.$$

($k \ge 1$) Assume the result has been proved for $k-1$.

<u>(($p = 0$))</u>: Consider the sections $s^{\otimes k} \otimes \sigma_j^{(0)}$, and define the semi-positively curved metric

$$\psi_{k,0} := \log \sum_{j=1}^{N_{m-1}} |\tilde{\sigma}_j^{(km-1)}|^2.$$

Observe that

$$\int_{X_0} |s^k \otimes \sigma_j^{(0)}|^2 e^{-(\psi_{k,0}+\kappa)} \le \int_{X_0} |s|^2 \frac{|\sigma_j^{(0)}|^2}{\sum_{j=1}^{N_{m-1}} |\sigma_j^{(m-1)}|^2} e^{-\kappa} < +\infty.$$

By the Extension Theorem 1.4 there exist sections

$$\tilde{\sigma}_j^{(km)} \in H^0(X, mkK_X + kL + A), \quad 1 \le j \le N_0$$

such that

$$\tilde{\sigma}_j^{(km)}|_{X_0} = s^{\otimes k} \otimes \sigma_j^{(0)} \wedge (d\pi)^{\otimes km}, \quad 1 \le j \le N_0,$$

and

$$\int_X |\tilde{\sigma}_j^{(km)}|^2 e^{-(\psi_{k,0}+\kappa)} \le 24\pi \int_{X_0} |s|^2 \frac{|\sigma_j^{(0)}|^2}{\sum_{j=1}^{N_{m-1}} |\sigma_j^{(m-1)}|^2} e^{-\kappa}.$$

Summing, we obtain

$$\begin{aligned}
\int_X \frac{\sum_{j=1}^{N_o} |\tilde{\sigma}_j^{(km)}|^2 e^{-\gamma}}{\sum_{j=1}^{N_{m-1}} |\tilde{\sigma}_j^{(km-1)}|^2} &\le \sup_X e^{\kappa-\gamma} \int_X \frac{\sum_{j=1}^{N_o} |\tilde{\sigma}_j^{(km)}|^2 e^{-\kappa}}{\sum_{j=1}^{N_{m-1}} |\tilde{\sigma}_j^{(km-1)}|^2} \\
&\le 24\pi \sup_X e^{\kappa-\gamma} \int_{X_0} |s|^2 \frac{\sum_{j=1}^{N_0} |\sigma_j^{(0)}|^2}{\sum_{j=1}^{N_{m-1}} |\sigma_j^{(m-1)}|^2} e^{-\kappa} \\
&\le 24\pi \widehat{C} \sup_X e^{\kappa-\gamma} \int_{X_0} |s|^2 \omega^{-n(m-1)} e^{-\kappa}
\end{aligned}$$

<u>$((1 \le p \le m-1))$</u>: Assume that we have obtained the sections $\tilde{\sigma}_j^{(km+p-1)}$, $1 \le j \le N_{p-1}$. Consider the non-negatively curved singular metric

$$\psi_{k,p} := \log \sum_{j=1}^{N_{p-1}} |\tilde{\sigma}_j^{(mk+p-1)}|^2$$

for $(km+p+1)K_X + kL + A$. We have

$$\int_{X_0} |s^k \otimes \sigma_j^{(p)}|^2 e^{-\psi_{k,p}} \le \int_{X_0} \frac{|\sigma_j^{(p)}|^2}{\sum_{j=1}^{N_{p-1}} |\sigma_j^{(p-1)}|^2} < +\infty.$$

By the L^2 Extension Theorem 1.4 there exist sections

$$\tilde{\sigma}_j^{(km+p)} \in H^0(X, (km+p+1)K_X + kL + A), \quad 1 \le j \le N_0$$

such that

$$\tilde{\sigma}_j^{(km+p)}|_{X_0} = s^{\otimes k} \otimes \sigma_j^{(p)} \wedge (d\pi)^{\otimes km+p}, \quad 1 \le j \le N_p,$$

and

$$\int_X |\tilde{\sigma}_j^{(km+p)}|^2 e^{-\psi_{k,p}} \le 24\pi \int_{X_0} \frac{|\sigma_j^{(p)}|^2}{\sum_{j=1}^{N_{p-1}} |\sigma_j^{(p-1)}|^2}.$$

Summing, we obtain

$$\int_X \frac{\sum_{j=1}^{N_o} |\tilde{\sigma}_j^{(km)}|^2}{\sum_{j=1}^{N_{m-1}} |\tilde{\sigma}_j^{(km-1)}|^2} \le 32\pi\widehat{C} \int_{X_0} \omega^n.$$

Let

$$C = \widehat{C} \times \max\left\{ \int_X \omega^{n+1}, 24\pi \int_{X_0} \omega^n, 24\pi \int_{X_0} |s|^2 \omega^{-n(m-1)} e^{-\kappa} \right\}.$$

The proof is finished. □

2.4. Construction of the metric

Fix a smooth metric $e^{-\psi}$ for $A \to X$. Consider the functions

$$\lambda_{km+p} := \log \sum_{j=1}^{N_p} |\tilde{\sigma}_j^{(km+p)}|^2 \omega^{-(n+1)(mk+p)} e^{-(k\gamma+\psi)}.$$

Observe that by Proposition 2.4 and the concavity of the logarithm, we have the bound

$$\frac{1}{\int_X \omega^{n+1}} \int_X (\lambda_N - \lambda_{N-1}) \omega^{n+1} \le \log\left(\frac{C}{\int_X \omega^{n+1}}\right).$$

(This inequality if the reverse of Jensen's inequality; it can be obtained from Jensen's Inequality by considering the convex function $-\log$.) It follows that the function

$$\Lambda_k = \frac{1}{k}\lambda_{mk}$$

satisfies the integral bound

$$\int_X \Lambda_k \omega^{n+1} \le mC'$$

for some uniform constant C'.

Now, locally, the functions λ_{km+p} are a sum of a smooth function and a subharmonic function, i.e., they are quasi-plurisubharmonic. Using the sub-mean value property for plurisubharmonic functions, we find that for any $r < 1$

$$\Lambda_k(x) \le C_r \int_X \Lambda_k \le mC_rC', \quad x \in \pi^{-1}(\mathbb{D}_0(r)).$$

It follows that the function

$$\Lambda(x) := \limsup_{y \to x} \limsup_{k \to \infty} \Lambda_k(y)$$

exists, and is locally the sum of a plurisubharmonic function and a continuous function.

Consider the singular Hermitian metric $e^{-\mu}$ for $mK_X + L$ defined by

$$e^{-\mu} = e^{-\Lambda} \omega^{-(n+1)m} e^{-\gamma}.$$

This singular metric is given by the formula

$$e^{-\mu(x)} = \exp\left(-\left(\limsup_{y \to x} \limsup_{k \to \infty} \mu_k(y)\right)\right),$$

where

$$e^{-\mu_k} = e^{-\Lambda_k} \omega^{-(n+1)m} e^{-\gamma}.$$

The curvature of $e^{-\mu_k}$ is thus

$$\begin{aligned} \sqrt{-1}\partial\bar{\partial}\mu_k &= \frac{1}{k}\sqrt{-1}\partial\bar{\partial}\log\sum_{j=1}^{N_0} |\tilde{\sigma}_j^{(mk)}|^2 - \frac{1}{k}\sqrt{-1}\partial\bar{\partial}\psi \\ &\ge -\frac{1}{k}\sqrt{-1}\partial\bar{\partial}\psi \end{aligned}$$

PROPOSITION 2.5. *The curvature of $e^{-\mu}$ is non-negative.*

PROOF. It suffices to work locally. Then we have that the function

$$\mu_k + \frac{1}{k}\psi$$

is plurisubharmonic. But

$$\limsup_{y\to x}\limsup_{k\to\infty}\mu_k + \frac{1}{k}\psi = \limsup_{y\to x}\limsup_{k\to\infty}\mu_k = \mu.$$

It follows that μ is plurisubharmonic, as desired. □

2.5. Examination of $e^{-\mu}$ on the central fiber: conclusion of the proof

Notice that

$$\Lambda_k|_{X_0} = \log(|s|^2\omega^{-(nm)}e^{-\gamma}) + \frac{1}{k}\left(\log\sum_{j=1}^{N_0}|\sigma_j^{(0)}|^2e^{-\psi} + \log\frac{\sqrt{-1}d\pi\wedge d\bar{\pi}}{\omega}\right),$$

and so, passing to the limit as $k\to\infty$, we have

$$e^{-\mu}|_{X_0} = \frac{1}{|s|^2}.$$

Thus by the discussion in (the beginning of) Paragraph 2.2, we are done. □

3. Pluricanonical extension on projective manifolds

Our next goal is to establish sufficient conditions for extending sections of the pluri-adjoint bundles $m(K_Z+E|_Z)$ from a smooth hypersurface Z to a projective manifold X. The main difference between this setting and the setting of the previous section is that while the normal bundle of $X_0 = \pi^{-1}(0)$ is automatically trivial, the normal bundle of Z in X need not be trivial. However, the modification needed to handle the possible positivity of the normal bundle of Z is already built into the Extension Theorem 1.2, and the way to proceed is directly analogous to what has been done in the previous section.

3.1. Statement of the theorem

THEOREM 3.1. *Let X be a projective algebraic manifold, and $Z\subset X$ a smooth complex submanifold of codimension 1. Denote by $T\in H^0(X,Z)$ the canonical holomorphic section whose zero divisor is Z. Let $E, B\to X$ holomorphic line bundles and assume there exist singular Hermitian metrics $e^{-\varphi_Z}$, $e^{-\varphi_E}$ and $e^{-\varphi_B}$ for the line bundle associated to Z, for E and for B respectively, with the following properties:*

(R) *The metrics $e^{-\varphi_Z}$, $e^{-\varphi_E}$ and $e^{-\varphi_B}$ restrict to singular Hermitian metrics on Z.*

(B) *The metric $e^{-\varphi_Z}$ satisfies the uniform bound*

$$\sup_X |T|^2e^{-\varphi_Z} < +\infty.$$

(P) *There is an integer $\mu > 0$ such that*

$$\sqrt{-1}\partial\bar{\partial}(\varphi_E+\varphi_B)\geq 0 \quad \text{and} \quad \mu\sqrt{-1}\partial\bar{\partial}(\varphi_E+\varphi_B)\geq\sqrt{-1}\partial\bar{\partial}\varphi_Z,$$
$$\sqrt{-1}\partial\bar{\partial}(m\varphi_E+\varphi_B)\geq 0 \quad \text{and} \quad \mu\sqrt{-1}\partial\bar{\partial}(m\varphi_E+\varphi_B)\geq\sqrt{-1}\partial\bar{\partial}\varphi_Z.$$

(T) *The multiplier ideal of $(\varphi_Z+\varphi_E)|_Z$ is trivial: $\mathscr{I}_{(\varphi_Z+\varphi_E)|_Z} = \mathcal{O}_Z$.*

Then every element of $H^0(Z,\mathcal{O}_Z(m(K_Z+E|_Z)+B|_Z)\otimes\mathscr{I}_{(\varphi_Z+\varphi_E+\varphi_B)|_Z})$ extends to a global holomorphic section in $H^0(X, m(K_X+Z+E)+B)$.

REMARK. When we say that a section $s \in H^0(Z, m(K_Z+E|_Z))$ extends to a section S in $H^0(X, m(K_X+Z+E))$, we mean that

$$S|_Z = s \wedge dT^{\otimes m}.$$

REMARK. Recall that a singular Hermitian metric is a Hermitian metric for a holomorphic line bundle such that, if the metric is represented locally by $e^{-\varphi}$, then φ is $L^1_{\ell oc}$. In most situations, people deal with such metrics only when φ is plurisubharmonic. Such φ are locally bounded above. Here we have a metric in the picture that need not be plurisubharmonic, namely $e^{-\varphi_Z}$. Thus we add to our definition of singular Hermitian metrics the additional requirement that the local potentials φ be uniformly bounded above on their domain of definition.

With this convention, a singular metric for a holomorphic line bundle does not induce a singular metric for the dual bundle. However, this asymmetry will not pose a problem for us.

3.2. Some corollaries

An immediate corollary of Theorem 3.1 is the following result.

COROLLARY 3.2. *Let the notation and hypotheses of Theorem 3.1 hold. In addition, assume that the metric $e^{-(\varphi_Z+\varphi_E+\varphi_B)}|_Z$ is locally integrable on Z. Then the natural restriction map*

$$H^0(X, m(K_X+Z+E)+B) \to H^0(Z, m(K_Z+E|_Z)+B|_Z)$$

is surjective. In particular, if $B = \mathcal{O}_X$ and we take $\varphi_B \equiv 0$, then condition (P) of Theorem 3.1 simplifies to $\sqrt{-1}\partial\bar{\partial}\varphi_E \geq 0$ and $\mu\sqrt{-1}\partial\bar{\partial}\varphi_E \geq \varphi_Z$, and

$$H^0(X, m(K_X+Z+E)) \to H^0(Z, m(K_Z+E|_Z))$$

is surjective.

Next we derive corollaries of Theorem 3.1 that can be phrased in terms of more algebro-geometric properties of E and Z. The first of these is the following theorem of Takayama

THEOREM 3.3. [**Takayama-2006**, Theorem 4.1] *Let X be a complex projective manifold, $Z \subset X$ a complex submanifold of codimension 1, and E an integral divisor on X. Assume that $E \sim_{\mathbb{Q}} A+D$ for some big and nef $\mathbb{Q}$-divisor A and some effective $\mathbb{Q}$-divisor D such that Z is in A-general position and $Z \not\subset$ Support(D), and that the pair $(Z, D|_Z)$ is klt. Then the natural restriction map*

$$H^0(X, m(K_X+Z+E)) \to H^0(Z, m(K_Z+E|_Z))$$

is surjective.

REMARK. Recall that if D is a $\mathbb{Q}$-divisor on Z, then
(i) the multiplier ideal $\mathscr{I}(D)$ is the multiplier ideal for the metric $e^{-\frac{1}{m}\log|s_{mD}|^2}$, where m is a positive integer such that mD is an integral divisor and s_{mD} is the section with divisor mD, and
(ii) one says that the pair (X, D) is *klt* (Kawamata Log Terminal) if $\mathscr{I}(D) = \mathcal{O}_X$.

Theorem 3.3 is a corollary of Theorem 3.1. To see this, we argue as follows. It is not hard to see (cf. Section 4 in [**Takayama-2006**]) that we may assume without loss of generality that A is an ample $\mathbb{Q}$-divisor. Thus we can construct a singular metric $e^{-\varphi_E}$ of positive curvature for E as follows: take a multiple mA that is very ample, and let s_D be the canonical multi-section of D whose $\mathbb{Q}$-divisor is D. Then we set

$$\varphi_E = \log |s_D|^2 + \log \left(\sum_{j=1}^{N} |s_j|^{2/m} \right),$$

where $s_1, ..., s_N$ is a basis for $H^0(X, mA)$. Since Z is not contained in the support of D, $e^{-\varphi_E}$ restricts to Z as a well-defined singular metric. Fix any smooth metric $e^{-\varphi_Z}$ for Z, and let $B = \mathcal{O}_X$ and $\varphi_B \equiv 0$. Then evidently hypotheses (P) and (T) of Theorem 3.1 are satisfied. Moreover, the multiplier ideal $\mathscr{I}(e^{-(\varphi_E+\varphi_Z)}|_Z)$ is supported away from Z because $(Z, D|_Z)$ is klt. Thus theorem 3.3 follows.

In Theorem 3.3, we would like to remove the hypothesis that E is big. In some sense, this is achieved in Theorem 3.1. (Indeed, the desire to handle the case where E is not necessarily big forms the initial impetus for the present article.) However, as we mentioned, we would like to state a result that uses more intrinsic properties of the divisors, rather than a result that includes a choice of metrics.

DEFINITION 3.4. *Let L be an integral divisor on X. For each integer $k > 0$, fix bases*

$$s_1^{(k)}, ..., s_{N_k^L}^{(k)} \in H^0(X, kL).$$

Then define

$$\psi_L = \log \sum_{k=1}^{\infty} \varepsilon_k \left(\sum_{j=1}^{N_k^L} |s_j^{(k)}|^2 \right)^{2/k}.$$

where $\varepsilon_k > 0$ are small enough to make the sum converge. We extend the definition to $\mathbb{Q}$-divisors L by setting

$$\psi_L = \frac{1}{m} \psi_{mL},$$

where $m > 0$ is the smallest integer such that mL is an integral divisor.

REMARK. Metrics of the form $e^{-\psi_L}$ have the following property: for every integer $m > 0$ such that mL is integral, the natural inclusion

$$H^0(X, mL \otimes \mathscr{I}(e^{-m\psi_L})) \to H^0(X, mL)$$

is an isomorphism. Indeed, if for all $m > 0$, $H^0(X, mL) = \{0\}$, there is nothing to prove. On the other hand, if $\sigma = \sum c^j s_j^{(m)} \in H^0(X, mL)$ then

$$|\sigma|^2 e^{-m\psi_L} \lesssim \frac{|\sigma|^2}{|s_1^{(m)}|^2 + ... + |s_{N_m}^{(m)}|^2} \leq \sum |c^j|^2$$

is bounded and thus integrable.

Recall that the set theoretic base locus $\mathbb{B}\mathrm{s}(|L|)$ of an integral divisor L is the common zero locus of all holomorphic sections of the line bundle associated to L.

THEOREM 3.5. *Let X be a projective algebraic manifold, $Z \subset X$ a smooth divisor, and E an integral divisor on X. Assume that one can write $E \sim_{\mathbb{Q}} E_1 + E_2$ such that the following properties hold.*

(P_a) *For some $\mu \in \mathbb{N}$ such that μE_1 is integral,*

$$\mathbb{B}\mathrm{s}(|\mu E_1|) \cup \mathbb{B}\mathrm{s}(|\mu E_1 - Z|) = \emptyset.$$

(T_a) *The singular metric* $e^{-\psi_{E_2}}$ *restricts to a singular metric on* Z, *and*

$$\mathscr{I}(e^{-\psi_{E_2}}|_Z) = \mathscr{O}_Z.$$

Then the restriction map

$$H^0(X, m(K_X + Z + E)) \to H^0(Z, m(K_Z + E|_Z))$$

is surjective.

PROOF OF THEOREM 3.5 FROM THEOREM 3.1. Let

$$\zeta = \psi_{\mu E_1} \quad \text{and} \quad \eta = \psi_{\mu E_1 - Z}.$$

By hypothesis (P_a), the curvature currents of the singular metrics $e^{-\zeta}$ and $e^{-\eta}$ for μE_1 and $\mu E_1 - Z$ respectively are non-negative and smooth. Let

$$\varphi_Z := \zeta - \eta, \quad \varphi_{E_1} = \frac{1}{\mu}\zeta \quad \text{and} \quad \varphi_E := \varphi_{E_1} + \psi_{E_2}.$$

Note the following.

(i) The metrics

$$\varphi_{E_1}, \quad \varphi_E \quad \text{and} \quad \mu\varphi_E - \varphi_Z = \mu\psi_{E_2} + \eta$$

have non-negative curvature currents.

(ii) φ_Z and ψ_{E_1} are smooth, and thus

$$\mathscr{I}(e^{-(\varphi_Z+\varphi_E)}|_Z) = \mathscr{I}(e^{-\psi_{E_2}}|_Z) = \mathscr{O}_Z$$

by hypothesis (T_a).

Now take $B = \mathscr{O}_X$, $\varphi_B \equiv 0$. The metrics $e^{-\varphi_Z}$, $e^{-\varphi_E}$ and $e^{-\varphi_B}$ satisfy the hypotheses of Theorem 3.1, and moreover

$$\mathscr{I}(e^{-(\varphi_Z+\varphi_E+\varphi_B)}|_Z) = \mathscr{O}_Z.$$

We thus obtain Theorem 3.5. □

REMARK. If one can write $E \sim_{\mathbb{Q}} E_1 + E_2$ where E_1 is an ample $\mathbb{Q}$-divisor and $(Z, E_2|_Z)$ is klt, then certainly properties (P_a) and (T_a) hold. We thus recover Theorem 3.3.

REMARK. A natural way in which the hypotheses of Theorem 3.5 might arise is the following. Suppose X and Y are projective manifolds, $\dim_{\mathbb{C}} Y > \dim_{\mathbb{C}} X$, and $\pi : Y \to X$ is a holomorphic map whose fibers have constant dimension. Let Z and E be divisors on X such that (X, Z, E) satisfy the hypotheses of Takayama's Theorem 3.3. Then the hypotheses of Theorem 3.5 hold for π^*E and π^*Z on Y.

3.3. Proof of Theorem 3.1

For the rest of the paper, we normalize our canonical section T of the line bundle associated to Z, so that

$$\sup_X |T|^2 e^{-\varphi_Z} = 1.$$

REMARK. Let $s \in H^0(Z, \mathscr{O}_Z(m(K_Z + E|_Z) + B|_Z) \otimes \mathscr{I}(e^{-(\varphi_B+\varphi_B+\varphi_Z)}|_Z))$ be the section to be extended. We note that in fact,

$$\int_Z |s|^2 \omega^{-(n-1)(m-1)} e^{-(m-1)\gamma_E} e^{-(\varphi_E+\varphi_B)} < +\infty,$$

since

$$e^{-(\varphi_E+\varphi_B)} \le \left(\sup_Z e^{\varphi_Z-\gamma_Z}\right) e^{\gamma_Z} e^{-(\varphi_E+\varphi_B+\varphi_Z)} \le C e^{\gamma_Z} e^{-(\varphi_E+\varphi_B+\varphi_Z)}.$$

The last inequality follows since, by our convention, the local potentials of singular metrics are locally bounded above.

3.4. The inductive construction

Fix a holomorphic line bundle $A \to X$ sufficiently positive as to have the following property:

(GG) For each $0 \le p \le m-1$ the global sections $H^0(X, p(K_X+Z+E)+A)$ generate the sheaf $\mathcal{O}_X(p(K_X+Z+E)+A)$.

Let us fix bases

$$\{\tilde{\sigma}_j^{(p)} \ ; \ 1 \le j \le N_p\}$$

of $H^0(X, p(K_X+Z+E)+A)$. We let $\sigma_j^{(p)} \in H^0(Z, p(K_Z+E|_Z)+A|_Z)$ be such that

$$\tilde{\sigma}_j^{(p)}|_Z = \sigma_j^{(p)} \wedge (dT)^{\otimes p}.$$

We also fix smooth metrics

$$e^{-\gamma_Z}, \ e^{-\gamma_E} \text{ and } e^{-\gamma_B} \text{ for } Z \to X, \ E \to X \text{ and } B \to X$$

respectively.

PROPOSITION 3.6. *There exist a constant $C < +\infty$ and sections*

$$\{\tilde{\sigma}_j^{(km+p)} \in H^0(X, (km+p)(K_X+Z+E)+kB+A) \ ; 1 \le j \le N_p\}_{0 \le p \le m-1, k=0,1,2,\dots}$$

with the following properties.

(a) $\tilde{\sigma}_j^{(mk+p)}|_Z = s^{\otimes k} \otimes \sigma_j^{(p)} \wedge (dT)^{(km+p)}$

(b) *If $k \ge 1$,*

$$\int_X \frac{\sum_{j=1}^{N_0} |\tilde{\sigma}_j^{(mk)}|^2 e^{-(\gamma_Z+\gamma_E+\gamma_B)}}{\sum_{j=1}^{N_{m-1}} |\tilde{\sigma}_j^{(mk-1)}|^2} \le C.$$

(c) *For $1 \le p \le m-1$,*

$$\int_X \frac{\sum_{j=1}^{N_p} |\tilde{\sigma}_j^{(mk+p)}|^2 e^{-(\gamma_Z+\gamma_E)}}{\sum_{j=1}^{N_{p-1}} |\tilde{\sigma}_j^{(mk+p-1)}|^2} \le C.$$

PROOF. Fix a constant $\widehat{C}$ such that the

$$\sup_X \frac{\sum_{j=1}^{N_0} |\tilde{\sigma}_j^{(0)}|^2 \omega^{n(m-1)} e^{(m-1)(\gamma_Z+\gamma_E)}}{\sum_{j=1}^{N_{m-1}} |\tilde{\sigma}_j^{(m-1)}|^2} \le \widehat{C}$$

and

$$\sup_Z \frac{\sum_{j=1}^{N_0} |\sigma_j^{(0)}|^2 \omega^{(n-1)(m-1)} e^{(m-1)\gamma_E}}{\sum_{j=1}^{N_{m-1}} |\sigma_j^{(m-1)}|^2} \le \widehat{C},$$

and for all $0 \le p \le m-2$,

$$\sup_X \frac{\sum_{j=1}^{N_{p+1}} |\tilde{\sigma}_j^{(p+1)}|^2 \omega^{-n} e^{-(\gamma_Z+\gamma_E)}}{\sum_{j=1}^{N_p} |\tilde{\sigma}_j^{(p)}|^2} \le \widehat{C},$$

and

$$\sup_Z \frac{\sum_{j=1}^{N_{p+1}} |\sigma_j^{(p+1)}|^2 \omega^{-(n-1)} e^{-\gamma_E}}{\sum_{j=1}^{N_p} |\sigma_j^{(p)}|^2} \le \widehat{C}.$$

($k = 0$) As far as extension there is nothing to prove. Note that

$$\int_X \frac{\sum_{j=1}^{N_p} |\tilde{\sigma}_j^{(p)}|^2 e^{-(\gamma_Z+\gamma_E)}}{\sum_{j=1}^{N_{p-1}} |\tilde{\sigma}_j^{(p-1)}|^2} \le \widehat{C} \int_X \omega^n.$$

($k \ge 1$) Assume the result has been proved for $k-1$.
<u>$((p=0))$</u>: Consider the sections $s^{\otimes k} \otimes \sigma_j^{(0)}$, and define the semi-positively curved metric

$$\psi_{k,0} := \log \sum_{j=1}^{N_{m-1}} |\tilde{\sigma}_j^{(km-1)}|^2$$

for the line bundle $(mk-1)(K_X + Z + E) + (k-1)B + A$. Observe that locally,

$$\begin{aligned} |(s \wedge dT^m)^k \otimes \sigma_j^{(0)}|^2 e^{-(\varphi_Z+\psi_{k,0}+\varphi_E+\varphi_B)} &= |s \wedge dT^m|^2 \frac{|\sigma_j^{(0)}|^2 e^{-(\varphi_Z+\varphi_E+\varphi_B)}}{\sum_{j=1}^{N_{m-1}} |\sigma_j^{(m-1)}|^2} \\ &\lesssim |s|^2 e^{-(\varphi_Z+\varphi_E+\varphi_B)}. \end{aligned}$$

Moreover, we have

$$\mu\sqrt{-1}\partial\bar{\partial}(\psi_{k,0} + \varphi_E + \varphi_B) \ge \max(\sqrt{-1}\partial\bar{\partial}\varphi_Z, 0).$$

Finally,

$$\begin{aligned} &\int_Z |s^k \otimes \sigma_j^{(0)}|^2 e^{-(\psi_{k,0}+\varphi_E+\varphi_B)} \\ = &\int_Z |s|^2 \frac{|\sigma_j^{(0)}|^2 e^{(m-1)\gamma_E} e^{-((m-1)\gamma_E+\varphi_E+\varphi_B)}}{\sum_{j=1}^{N_{m-1}} |\sigma_j^{(m-1)}|^2} < +\infty. \end{aligned}$$

We may thus apply Theorem 1.2 to obtain sections

$$\tilde{\sigma}_j^{(km)} \in H^0(X, mk(K_X + Z + E) + kB + A), \quad 1 \le j \le N_0$$

such that

$$\tilde{\sigma}_j^{(km)}|_Z = s^{\otimes k} \otimes \sigma_j^{(0)} \wedge (dT)^{\otimes km}, \quad 1 \le j \le N_0,$$

and

$$\int_X |\tilde{\sigma}_j^{(km)}|^2 e^{-(\psi_{k,0}+\varphi_Z+\varphi_E+\varphi_B)} \le 24\pi\mu \int_Z |s|^2 \frac{|\sigma_j^{(0)}|^2 e^{-(\varphi_E+\varphi_B)}}{\sum_{j=1}^{N_{m-1}} |\sigma_j^{(m-1)}|^2}.$$

Summing over j, we obtain

$$\begin{aligned}
&\int_X \frac{\sum_{j=1}^{N_o} |\tilde{\sigma}_j^{(km)}|^2 e^{-(\gamma_Z+\gamma_E+\gamma_B)}}{\sum_{j=1}^{N_{m-1}} |\tilde{\sigma}_j^{(km-1)}|^2} \\
\leq\ & \sup_X e^{\varphi_Z+\varphi_E+\varphi_B-\gamma_Z-\gamma_E-\gamma_B} \int_X \frac{\sum_{j=1}^{N_o} |\tilde{\sigma}_j^{(km)}|^2 e^{-(\varphi_Z+\varphi_E+\varphi_B)}}{\sum_{j=1}^{N_{m-1}} |\tilde{\sigma}_j^{(km-1)}|^2} \\
\leq\ & 24\pi\mu \sup_X e^{\varphi_Z+\varphi_E+\varphi_B-\gamma_Z-\gamma_E-\gamma_B} \int_Z |s|^2 \frac{\sum_{j=1}^{N_0} |\sigma_j^{(0)}|^2 e^{-(\varphi_E+\varphi_B)}}{\sum_{j=1}^{N_{m-1}} |\sigma_j^{(m-1)}|^2} e^{-\kappa} \\
\leq\ & 24\pi\mu\widehat{C} \sup_X e^{\varphi_Z+\varphi_E+\varphi_B-\gamma_Z-\gamma_E-\gamma_B} \int_Z |s|^2 \omega^{-(n-1)(m-1)} e^{-((m-1)\gamma_E+\varphi_E+\varphi_B)}
\end{aligned}$$

$((1 \leq p \leq m-1))$: Assume we have the sections $\tilde{\sigma}_j^{(km+p-1)}$, $1 \leq j \leq N_{p-1}$. Consider the non-negatively curved singular metric

$$\psi_{k,p} := \log \sum_{j=1}^{N_{p-1}} |\tilde{\sigma}_j^{(mk+p-1)}|^2$$

for $(km+p-1)(K_X+Z+E)+kB+A$. We have

$$|s^k \otimes \sigma_j^{(p)}|^2 e^{-(\varphi_Z+\psi_{k,p}+\varphi_E)} = \frac{|\sigma_j^{(p)}|^2 e^{-(\varphi_Z+\varphi_E)}}{\sum_{j=1}^{N_{p-1}} |\sigma_j^{(p-1)}|^2} \lesssim e^{-(\varphi_Z+\varphi_E)},$$

which is locally integrable by the hypothesis (T). Next,

$$\begin{aligned}
\int_Z |s^k \otimes \sigma_j^{(p)}|^2 e^{-(\psi_{k,p}+\varphi_E)} &= \int_Z \frac{|\sigma_j^{(p)}|^2 e^{-\varphi_E}}{\sum_{j=1}^{N_{p-1}} |\sigma_j^{(p-1)}|^2} \\
&\leq C^\star \int_Z e^{\gamma_Z} \frac{|\sigma_j^{(p)}|^2 e^{-(\varphi_Z+\varphi_E)}}{\sum_{j=1}^{N_{p-1}} |\sigma_j^{(p-1)}|^2} < +\infty,
\end{aligned}$$

where

$$C^\star := \sup_Z e^{\varphi_Z-\gamma_Z}.$$

By Theorem 1.2 there exist sections

$$\tilde{\sigma}_j^{(km+p)} \in H^0(X, (mk+p)(K_X+Z+E)+kB+A), \quad 1 \leq j \leq N_0$$

such that

$$\tilde{\sigma}_j^{(km+p)}|_Z = s^{\otimes k} \otimes \sigma_j^{(p)} \wedge (dT)^{\otimes km+p}, \quad 1 \leq j \leq N_p,$$

and

$$\int_X |\tilde{\sigma}_j^{(km+p)}|^2 e^{-(\psi_{k,p}+\varphi_Z+\varphi_E)} \leq 40\pi\mu \int_Z \frac{|\sigma_j^{(p)}|^2 e^{-\varphi_E}}{\sum_{j=1}^{N_{p-1}} |\sigma_j^{(p-1)}|^2}.$$

Summing over j, we obtain

$$\int_X \frac{\sum_{j=1}^{N_p} |\tilde{\sigma}_j^{(km+p)}|^2 e^{-(\gamma_Z+\gamma_E)}}{\sum_{j=1}^{N_{p-1}} |\tilde{\sigma}_j^{(km+p-1)}|^2} \leq 24\pi\mu \sup_X e^{\varphi_Z+\varphi_E-\gamma_Z-\gamma_E} \widehat{C} \int_Z e^{-\varphi_E}\omega^{n-1}.$$

Letting C be the maximum of the numbers

$$\widehat{C}\int_X \omega^n,$$

$$24\pi\mu\widehat{C}\sup_X e^{\varphi_Z+\varphi_E+\varphi_B-\gamma_Z-\gamma_E-\gamma_B}\int_Z |s|^2\omega^{-n-1(m-1)}e^{-((m-1)\gamma_E+\varphi_E+\varphi_B)}$$

and
$$24\pi\mu\sup_X e^{\varphi_Z+\varphi_E-\gamma_Z-\gamma_E}\widehat{C}\int_Z e^{-\varphi_E}\omega^{n-1}$$

completes the proof. □

3.5. Construction of the metric

Fix a smooth metric $e^{-\psi}$ for $A \to X$. Consider the functions

$$\lambda_N := \log\sum_{j=1}^{N_p} |\tilde{\sigma}_j^{(km+p)}|^2\omega^{-n(mk+p)}e^{-(km(\gamma_Z+\gamma_E)+k\gamma_B+\psi)},$$

where $N = km + p$. We then have the following lemma.

LEMMA 3.7. *For any non-empty open subset $V \subset X$ and any smooth function $f : \overline{V} \to \mathbb{R}_+$,*

$$\frac{1}{\int_V f\omega^n}\int_V (\lambda_N - \lambda_{N-1})f\omega^n \le \log\left(\frac{C\sup_V f}{\int_V f\omega^n}\right).$$

PROOF. Observe that by Proposition 3.6, there exists a constant C such that for any open subset $V \subset X$,

$$\int_V (e^{\lambda_N-\lambda_{N-1}})f\omega^n \le C\sup_V f.$$

The lemma follows from an application of (the concave version of) Jensen's inequality to the concave function log. □

Consider the function

$$\Lambda_k = \frac{1}{k}\lambda_{mk}.$$

Note that Λ_k is locally the sum of a plurisubharmonic function and a smooth function. By applying Lemma 3.7 and using the telescoping property, we see that for any open set $V \subset X$ and any smooth function $f : \overline{V} \to \mathbb{R}_+$,

$$\frac{1}{\int_V f\omega^n}\int_V \Lambda_k f\omega^n \le m\log\left(\frac{C\sup_V f}{\int_V f\omega^n}\right). \tag{2.5}$$

PROPOSITION 3.8. *There exists a constant C_o such that*

$$\Lambda_k(x) \le C_o, \quad x \in X.$$

PROOF. Let us cover X by coordinate charts $V_1, ..., V_N$ such that for each j there is a biholomorphic map F_j from V_j to the ball $B(0, 2)$ of radius 2 centered at the origin in $\mathbb{C}^n$, and such that if $U_j = F_j^{-1}(B(0, 1))$, then $U_1, ..., U_N$ is also an open cover. Let $W_j = V_j \setminus F_j^{-1}(B(0, 3/2))$.

Now, on each V_j, Λ_k is the sum of a plurisubharmonic function and a smooth function. Say $\Lambda_k = h + g$ on V_j, where h is plurisubharmonic and g is smooth.

Then for constant A_j we have

$$\begin{aligned}\sup_{U_j} \Lambda_k &\leq \sup_{U_j} g + \sup_{U_j} h \\ &\leq \sup_{U_j} g + A_j \int_{W_j} h \cdot F_{j*} dV \\ &\leq \sup_{U_j} g - A_j \int_{W_j} g \cdot F_{j*} dV + A_j \int_{W_j} \Lambda_k \cdot F_{j*} dV\end{aligned}$$

Let

$$C_j := \sup_{U_j} g - A_j \int_{W_j} g \cdot F_{j*} dV$$

and define the smooth function f_j by

$$f_j \omega^n = F_{j*} dV.$$

Then by (2.5) applied with $V = W_j$ and $f = f_j$, we have

$$\sup_{U_j} \Lambda_k \leq C_j + mA_j \log \left(\frac{C \sup_{W_j} f_j}{\int_{W_j} f_j \omega^n} \right) \int_{W_j} f_j \omega^n.$$

Letting

$$C_o := \max_{1 \leq j \leq N} \left\{ C_j + mA_j \log \left(\frac{C \sup_{W_j} f_j}{\int_{W_j} f_j \omega^n} \right) \int_{W_j} f_j \omega^n \right\}$$

completes the proof. □

Since the upper regularization of the lim sup of a uniformly bounded sequence of plurisubharmonic functions is plurisubharmonic (see, e.g., [**Hormander-1990**, Theorem 1.6.2]), we essentially have the following corollary.

COROLLARY 3.9. *The function*

$$\Lambda(x) := \limsup_{y \to x} \limsup_{k \to \infty} \Lambda_k(y)$$

is locally the sum of a plurisubharmonic function and a smooth function.

PROOF. One need only observe that the function Λ_k is obtained from a singular metric on the line bundle $m(K_X + Z + E) + B$ (this singular metric $e^{-\kappa_k}$ will be described shortly) by multiplying by a fixed smooth metric of the dual line bundle. □

Consider the singular Hermitian metric $e^{-\kappa}$ for $m(K_X + Z + E) + B$ defined by

$$e^{-\kappa} = e^{-\Lambda} \omega^{-nm} e^{-(m(\gamma_Z + \gamma_E) + \gamma_B)}.$$

This singular metric is given by the formula

$$e^{-\kappa(x)} = \exp \left(- \limsup_{y \to x} \limsup_{k \to \infty} \kappa_k(y) \right),$$

where

$$e^{-\kappa_k} = e^{-\Lambda_k} \omega^{-nm} e^{-(m(\gamma_Z + \gamma_E) + \gamma_B)}.$$

The curvature of $e^{-\kappa_k}$ is thus

$$\begin{aligned}\sqrt{-1}\partial\bar{\partial}\kappa_k &= \frac{\sqrt{-1}}{k}\partial\bar{\partial}\log\sum_{j=1}^{N_0}|\tilde{\sigma}_j^{(mk)}|^2 - \frac{1}{k}\sqrt{-1}\partial\bar{\partial}\psi \\ &\geq -\frac{1}{k}\sqrt{-1}\partial\bar{\partial}\psi\end{aligned}$$

We claim next that the curvature of $e^{-\kappa}$ is non-negative. To see this, it suffices to work locally. Then we have that the functions

$$\kappa_k + \frac{1}{k}\psi$$

are plurisubharmonic. But

$$\limsup_{y\to x}\limsup_{k\to\infty}\kappa_k + \frac{1}{k}\psi = \limsup_{y\to x}\limsup_{k\to\infty}\kappa_k = \kappa.$$

It follows that κ is plurisubharmonic, as desired.

3.6. Conclusion of the proof

Notice that, after identifying K_Z with $(K_X + Z)|_Z$ by dividing by dT,

$$\kappa_k|Z = \log|s|^2 + \frac{1}{k}\log\sum_{j=1}^{N_0}|\sigma_j^{(0)}|^2.$$

Thus we obtain

$$e^{-\kappa}|_Z = \frac{1}{|s|^2}.$$

It follows that

$$\begin{aligned}&\int_Z |s|^2 e^{-\frac{(m-1)\kappa+m\varphi_E+\varphi_B}{m}} \\ &= \int_Z |s|^{2/m} e^{-\frac{((m-1)+1)\varphi_E+\varphi_B}{m}} \\ &= \int_Z |s|^{2/m} e^{-\frac{(m-1)(\varphi_E-\gamma_E)+(m-1)\gamma_E+\varphi_E+\varphi_B}{m}} \\ &\leq \left(\int_Z e^{\gamma_E-\varphi_E}\omega^{n-1}\right)^{\frac{m-1}{m}}\left(\int_Z |s|^2\omega^{-(n-1)(m-1)}e^{-(\varphi_E+(m-1)\gamma_E+\varphi_B)}\right)^{\frac{1}{m}} \\ &< +\infty,\end{aligned}$$

where the first inequality is a consequence of Hölder's Inequality. Next, working locally on Z and identifying sections and metric with functions, we have

$$|s\wedge dT|^2 e^{-(\varphi_Z+\frac{(m-1)\kappa+m\varphi_E+\varphi_B}{m})} \sim |s|^{2/m}e^{-\frac{\varphi_Z+\varphi_E+\varphi_B}{m}}e^{-\frac{m-1}{m}(\varphi_Z+\varphi_E)}.$$

Now, by another application of Hölder's Inequality, we have (locally on Z) that

$$\begin{aligned}&\int |s|^{2/m}e^{-\frac{\varphi_Z+\varphi_E+\varphi_B}{m}}e^{-\frac{m-1}{m}(\varphi_Z+\varphi_E)} \\ &\leq \left(\int |s|^2 e^{-(\varphi_Z+\varphi_E+\varphi_B)}\right)^{1/m}\times\left(\int e^{-(\varphi_Z+\varphi_E)}\right)^{(m-1)/m},\end{aligned}$$

and thus we obtain the local integrability of

$$|s|^{2/m}e^{-\frac{\varphi_Z+\varphi_E+\varphi_B}{m}}e^{-\frac{m-1}{m}(\varphi_Z+\varphi_E)}.$$

Finally,

$$\mu m\sqrt{-1}\partial\bar{\partial}\left(\tfrac{m-1}{m}\kappa + \tfrac{m\varphi_E+\varphi_B}{m}\right) \geq \mu\sqrt{-1}\partial\bar{\partial}(m\varphi_E + \varphi_B) \geq \max(\sqrt{-1}\partial\bar{\partial}\varphi_Z, 0).$$

An application of Theorem 1.2 completes the proof of Theorem 3.1. □

4. Exercises

4.1. Demailly's approximation method

In this exercise, we explore Demailly's method for approximating plurisubharmonic functions by multiples of logarithms of holomorphic functions. The exercise culminates in a simple proof, due to Demailly [**Demailly-1992**], of a deep and powerful theorem of Siu [**Siu-1974**].

Fix a plurisubharmonic function φ on the unit ball $B \subset \mathbb{C}^n$. We recall that the *Lelong number* of φ at $x \in B$ is the number

$$\nu(\varphi, x) := \liminf_{z \to x} \frac{\varphi}{\log|z-x|^2} = \lim_{r\to 0} \frac{\sup_{B(x,r)}\varphi}{\log r^2}.$$

(1) Prove that there is a constant $C > 0$ with the following property: given any $x \in B$ and any plurisubharmonic function φ with $\varphi(x) \neq -\infty$ and any $a \in \mathbb{C}$ such that there is a holomorphic function $f \in \mathcal{O}(B)$ such that

$$f(x) = a \quad \text{and} \quad \int_B |f|^2 e^{-\varphi} \leq C|a|^2 e^{-\varphi(x)}. \tag{2.6}$$

(2) Fix a plurisubharmonic function φ on the unit ball in $\mathbb{C}^n$. Consider the space $\mathscr{H}(m\varphi)$ of all holomorphic functions f on B such that

$$||f||_m := \left(\int_B |f|^2 e^{-m\varphi}\right)^{1/2} < +\infty.$$

Fix an orthonormal basis $\{g_j\}$ of $\mathscr{H}(m\varphi)$ and define

$$\varphi_m := \frac{1}{m}\log\sum_{j\geq 1}|g_j|^2.$$

(a) Show that φ_m is independent of the choice of orthonormal basis.

(b) For $x \in B$, show that the subspace $V_x \subset \mathscr{H}(m\varphi)$ consisting of all $f \in \mathscr{H}(m\varphi)$ satisfying $f(x) = 0$ has codimension 1.

(c) By completing an orthonormal basis of V_x to an orthonormal basis of $\mathscr{H}(m\varphi)$, show that

$$\varphi_m(x) = \sup_{||f||_m=1} \tfrac{1}{m}\log|f(x)|^2.$$

Deduce in particular that φ_m is convergent.

(d) Use the sub-mean value property for subharmonic functions together with (c) to deduce that

$$\varphi_m(x) \leq \sup_{|\zeta-x|<r} \varphi(\zeta) - \frac{K}{m}\log r$$

for some absolute constant K depending only on n.

(e) Use Exercise (1) to prove that

$$\varphi_m(x) \geq \varphi(x) - \frac{\log C}{m}.$$

Hint: Choose a so that the right hand side of (2.6) is 1.

(3) Deduce that $\varphi_m \to \varphi$ pointwise and also that

$$\nu(\varphi, x) - \frac{n}{m} \le \nu(\varphi_m, x) \le \nu(\varphi, x).$$

(4) In this final exercise, we will prove Siu's Theorem on the Lelong super-level sets of plurisubharmonic functions.

Let X be a complex manifold, φ a plurisubharmonic function on X and $c > 0$. Recall that the Lelong super-level c set of φ is

$$E_c(\varphi) = \{x \in X \ ; \ \nu(\varphi, x) \ge c\}.$$

THEOREM 4.1 (Siu). *For each $c > 0$, $E_c(\varphi)$ is an at most countable intersection of analytic sets.*

(a) Show that the Lelong number $\nu(\log |f|^2, x)$ of the log-square-modulus of a holomorphic function is

$$\mathrm{Ord}_x(f) = \sup\{k \in \mathbb{N} \ ; \ D^\alpha f(x) = 0 \text{ for all } |\alpha| < k\}.$$

Conclude that, with $\{g_j\}$ as in Exercise (2),

$$E_b(\varphi_m) = \bigcap_{|\alpha|<mb, j\ge 1} \{z \ ; \ D^\alpha g_j(z) = 0\}.$$

(b) Show that

$$E_c(\varphi) = \bigcap_{m>>0} E_{c-n/m}(\varphi_m),$$

and thus deduce Siu's Theorem.

LECTURE 3

The Skoda division theorem

1. Statement of the division theorem

A long standing problem in several complex variables is the so-called corona problem: if $g^1, ..., g^p$ is a collection of bounded holomorphic functions on the unit ball in $\mathbb{C}^n$ satisfying

$$\sum_{j=1}^{p} |g^j|^2 \geq c > 0,$$

are there bounded holomorphic functions $h_1, ..., h_p$ on the unit ball such that

$$\sum_{j=1}^{p} g^j h_j = 1?.$$

There has been a lot of work on the problem, and we shall not go into the history here, except to say that for $n = 1$, the affirmative answer to the question is a famous theorem of Carleson [**Carleson-1962**], while if the unit ball is replaced by a bounded pseudoconvex domain, a result of Sibony [**Sibony-1987**] shows that in general the answer is no. (Sibony's example has a smooth boundary that is strictly pseudoconvex at all points except one.)

In his paper [**Skoda-1972**], Skoda formulated and solved an L^2 version of the problem. Since Skoda's work, there has been work of Demailly [**Demailly-1982**], Ohsawa [**Ohsawa-1999,2002**],McNeal [**McNeal-1996b**], Siu [**Siu-2005**], myself [**Varolin-2008**] and probably many others whose work I am not aware of. Recently there have been interesting new division results using integral kernel methods. (Some references are [**Andersson-Samuelsson-Sznajdman-2008**], [**Andersson-Götmark-2008**] and [**Götmark-2008**].)

In keeping with the geometric flavor of the notes, we will state and prove a version of Skoda's Theorem that is tailored to the setting of holomorphic sections of Hermitian holomorphic line bundles. (Such a version was deduced by Siu in [**Siu-2005**] for projective manifolds as a corollary of Skoda's original result, but we give a slightly more general version using Skoda's proof, naturally adapted to the geometric setting of essentially Stein manifolds.) We begin by stating the general division problem, and then turning to the statement and proof of Skoda's Division Theorem. The lecture concludes with two recent applications of Skoda's Theorem, due to Siu, and two long exercises that establish some useful results.

1.1. The division problem

Let X be an essentially Stein manifold. We fix on X two holomorphic line bundles $E \to X$ and $F \to X$, and a collection of sections

$$g^1, ..., g^p \in H^0(X, E)$$

where $p \geq 1$ is some integer. We seek to determine which sections $f \in H^0(X, F + K_X)$ can be divided by $g = (g^1, ..., g^p)$, in the sense that there exist sections

$$h_1, ..., h_p \in H^0(X, F - E + K_X)$$

satisfying the equality

$$f = \sum_{i=1}^{p} h_i g^i.$$

Moreover, if f satisfies some sort of L^2 estimate, what can we say about estimates for $h_1, ..., h_p$? We shall refer to this question as the *division problem.*

1.2. Skoda's theorem

The next theorem, which is the main result of this lecture, constitutes one of Skoda's solutions to the division problem, and the only one that we will discuss. For other results and applications, see [**Skoda-1972**] as well as [**Demailly-2001**].

THEOREM 1.1. *Let X be an essentially Stein manifold of complex dimension n, $F, E \to X$ holomorphic line bundles with singular Hermitian metrics $e^{-\psi}$ and $e^{-\eta}$ respectively, and $g^1, ..., g^p \in H^0(X, E)$. Fix $\alpha > 1$ and let $q := \min(p-1, n)$. Assume that*

$$\sqrt{-1}\partial\bar{\partial}\psi \geq \alpha q \sqrt{-1}\partial\bar{\partial}\eta.$$

Then for any $f \in H^0(X, K_X + F)$ such that

$$\int_X \frac{|f|^2 e^{-\psi}}{(|g|^2 e^{-\eta})^{\alpha q + 1}} < +\infty$$

there are p sections $h_1, ..., h_p \in H^0(X, K_X + F - E)$ such that

$$\sum_{k=1}^{p} h_k g^k = f \quad \textit{and} \quad \int_X \frac{|h|^2 e^{-(\psi - \eta)}}{(|g|^2 e^{-\eta})^{\alpha q}} \leq \frac{\alpha}{\alpha - 1} \int_X \frac{|f|^2 e^{-\psi}}{(|g|^2 e^{-\eta})^{\alpha q + 1}}.$$

2. Proof of the division theorem

2.1. Hilbert space theory

The Hilbert space theory needed to prove Skoda's Theorem is similar to but a little different from that needed in the proof of Hörmander's Theorem and the Ohsawa-Takegoshi Extension Theorem.

Summation convention. Though we did not do so in the statement of Skoda's Theorem, we return now to the use the the complex version of Einstein's summation convention. In addition, we introduce into our order of operations the rules

$$|a_i b^i|^2 = |a_1 b^1 + ...|^2, \quad \text{while} \quad |a_i|^2 |b^i|^2 = |a_1|^2 |b^1|^2 + ...$$

The functional analysis. Let $\mathscr{H}_0$, $\mathscr{H}_1$, $\mathscr{H}_2$ and $\mathscr{F}_1$ be Hilbert spaces with inner products $(\ ,\ )_0$, $(\ ,\ )_1$, $(\ ,\ )_2$ and $(\ ,\ )_*$ respectively. Suppose we have a bounded linear operator $T_2 : \mathscr{H}_0 \to \mathscr{H}_2$ and closed, densely defined operators $T_1 : \mathscr{H}_0 \to \mathscr{H}_1$ and $S_1 : \mathscr{H}_1 \to \mathscr{F}_1$ satisfying

$$S_1 T_1 = 0.$$

Let $\mathscr{K} = \text{Kernel}(T_1)$. We consider the following problem.

PROBLEM 2.1. Given $\eta \in \mathscr{H}_2$, is there an element $\xi \in \mathscr{K}$ such that $T_2 \xi = \eta$? If so, what can we say about $|\xi|_0$?

Problem 2.1 was solved by Skoda in [**Skoda-1972**]. The difference between Skoda's solution and the one we present here is that Skoda identified the subspace of all η for which the problem can be solved, whereas we aim to solve the problems one η at a time. This is a difference in presentation only; the two approaches are equivalent.

PROPOSITION 2.1. *Let $\eta \in \mathscr{H}_2$. Suppose there exists a constant $C > 0$ such that for all $u \in T_2(\mathscr{K})$ and all $\beta \in$ Domain(T_1^*),*

$$|(\eta, u)_2|^2 \le C\left(|T_2^* u + T_1^* \beta|_0^2 + |S_1 \beta|_*^2\right). \tag{3.1}$$

Then there exists $\xi \in \mathscr{K}$ such that $T_2 \xi = \eta$ and $|\xi|_0^2 \le C$.

PROOF. We first restrict (3.1) to $\beta \in$ Domain$(T_1^*) \cap$ Kernel(S_1). Note that

(i) since $S_1 T_1 = 0$, the image of T_1^* agrees with the image of the restriction of T_1^* to Kernel(S_1) (since in general, Kernel$(S_1) \perp$ Image(S_1^*)), and
(ii) the image of T_1^* is dense in $\mathscr{K}^\perp$.

Thus the estimate (3.1) may be rewritten

$$|(\eta, u)|^2 \le C|[T_2^* u]|^2, \tag{3.2}$$

where we denote by [ ] the projection to the quotient space $\mathscr{H}_0/\mathscr{K}^\perp$ and the norm on the right hand side is the norm induced on $\mathscr{H}_0/\mathscr{K}^\perp$ in the usual way. Observe that $\mathscr{K}$ is a closed subspace, since T_1 has closed graph and $\mathscr{K}$ is the intersection of the graph of T_1 with $\mathscr{H}_0 \times \{0\}$ in $\mathscr{H}_0 \times \mathscr{H}_1$. As is well known, $\mathscr{H}_0/\mathscr{K}^\perp$ with its induced norm is isomorphic to the subspace $\mathscr{K}$. (The isomorphism sends any, and thus every, member of $u + \mathscr{K}^\perp$ to its orthogonal projection onto $\mathscr{K}$.) We define a linear functional $\ell : [\text{Image}(T_2^*)] \to \mathbb{C}$ by

$$\ell([T_2^* u]) = (\eta, u)_2.$$

Then by (3.2) ℓ is continuous with norm $\le \sqrt{C}$. By defining ℓ to be zero on $[\text{Image}(T_2^*)]^\perp$, we may assume that ℓ is defined on all of $\mathscr{H}_0/\mathscr{K}^\perp$ with norm still bounded by $\sqrt{C}$. The Riesz Representation Theorem then tells us that ℓ is represented by inner product with respect to some element ξ of $\mathscr{H}_0/\mathscr{K}^\perp$ which we can identify with $\mathscr{K}$ at this point. Evidently we have $|\xi|_0^2 \le C$ and

$$(T_2 \xi, u)_2 = (\xi, T_2^* u + \mathscr{K}^\perp) = \ell([T_2^* u]) = (\eta, u)_2.$$

The proof is complete. □

Hilbert spaces of sections. Let Y be a Kähler manifold of complex dimension n and $H \to Y$ a holomorphic line bundle equipped with a singular Hermitian metric $e^{-\varphi}$. Given a smooth section f of $H + K_Y \to Y$, we can define its L^2-norm

$$||f||_\varphi^2 := \int_Y |f|^2 e^{-\varphi}.$$

As we saw on a number of previous occasions, this norm does not depend on the Kähler metric for Y. Indeed, we think of a section of $H + K_Y$ as an H-valued $(n,0)$-form. Then the functions $|f|^2 e^{-\varphi}$ transforms like the local representatives of a measure on Y, and may thus be integrated.

We define

$$L^2(Y, H + K_Y, e^{-\varphi})$$

to be the Hilbert space completion of the space of smooth sections f of $H + K_Y \to Y$ whose norm $||f||_\varphi$ is finite.

More generally, we have Hilbert spaces of $(0,q)$-forms with values in $H+K_Y$. Given such a $(0,q)$-form β, defined locally by

$$\beta = \beta_{\bar{J}} d\bar{z}^J,$$

where $J=(j_1,...,j_q)\in\{1,...,n\}^q$ is a multiindex and $d\bar{z}^J = d\bar{z}^{j_1}\wedge ...\wedge d\bar{z}^{j_q}$ and the $\beta_{\bar{J}}$ are skew-symmetric in J, we have a globally defined measure $|\beta|^2 e^{-\varphi}$ defined on any coordinate chart where H is locally trivial by

$$|\beta|^2 e^{-\varphi} := \bar{\beta}^I \beta_I e^{-\varphi}\left(\frac{\sqrt{-1}}{2}\right)^n dz^1\wedge d\bar{z}^1\wedge ...dz^n\wedge d\bar{z}^n.$$

We then define

$$||\beta||^2_{\varphi} = \int_Y |\beta|^2 e^{-\varphi}$$

and let

$$L^2_{0,q}(Y, H+K_Y, e^{-\varphi})$$

denote the Hilbert space closure of the space of smooth $(0,q)$-forms β with values in $H+K_Y$ such that $||\beta||^2_{\varphi}<+\infty$. Of course, these norms depend on the Kähler metric g as soon as $q\ge 1$.

We are only going to be (explicitly) interested in the cases $q=0$ and $q=1$, although $q=2$ will enter in an auxiliary way.

Choices. In employing Proposition 2.1, we shall consider the following spaces.

$$\begin{aligned}
\mathscr{H}_0 &:= (L^2(\Omega, K_X+F-E, e^{-\varphi_1}))^p\\
\mathscr{H}_1 &:= (L^2_{(0,1)}(\Omega, K_X+F-E, e^{-\varphi_1}))^p\\
\mathscr{H}_2 &:= L^2(\Omega, K_X+F, e^{-\varphi_2})\\
\mathscr{F}_1 &:= (L^2_{(0,2)}(\Omega, K_X+F-E, e^{-\varphi_1}))^p
\end{aligned}$$

Next we define the operators T_1 and T_2. Let

$$T: L^2(\Omega, K_X+F-E, e^{-\varphi_1})\to L^2_{0,1}(\Omega, K_X+F-E, e^{-\varphi_1})$$

be the densely defined linear operator whose action on smooth forms with compact support is

$$Tu = \bar{\partial} u$$

and whose domain, as usual, consists of those $u\in L^2(\Omega, K_X+F-E, e^{-\varphi_1})$ such that $\bar{\partial}u$, defined in the sense of currents, is represented by an element of $L^2_{0,1}(\Omega, K_X+F-E, e^{-\varphi_1})$. We let

$$T_1 : \mathscr{H}_0 \to \mathscr{H}_1$$

be defined by

$$T_1(u_1,...,u_p) = (Tu_1,...,Tu_p).$$

We remind the reader that T has formal adjoint $T^* = T^*_{\varphi_1}$ given by the formula

$$T^*\beta = -e^{\varphi_1}\partial_\nu(e^{-\varphi_1}\beta^\nu).$$

It follows that

$$T_1^*(\beta_1,...,\beta_p) = \left(-e^{\varphi_1}\partial_\nu\left(e^{-\varphi_1}\beta_1^\nu\right),...,-e^{\varphi_1}\partial_\nu\left(e^{-\varphi_1}\beta_p^\nu\right)\right).$$

We will also use the densely defined operator

$$S: L^2_{0,1}(\Omega, K_X+F-E, e^{-\varphi_1})\to L^2_{0,2}(\Omega, K_X+F-E, e^{-\varphi_1})$$

defined as $\bar\partial$ in the sense of currents, and with domain consisting of all elements $\beta \in L^2_{0,1}(\Omega, K_X + F - E, e^{-\varphi_1})$ such that the current $\bar\partial\beta$ is represented by an element of $L^2_{0,2}(\Omega, K_X + F - E, e^{-\varphi_1})$. We then define the associated densely defined operator

$$S_1(\beta_1, ..., \beta_p) = (S\beta_1, ..., S\beta_p).$$

We will not need the formal adjoint of S.

Next we let

$$T_2 : \mathscr{H}_0 \to \mathscr{H}_2$$

be defined by

$$T_2(h_1, ..., h_p) = h_i g^i.$$

We have

$$(T_2^* u, h)_0 = (u, T_2 h)_2 = \int_\Omega u\overline{h_i g^i} e^{-\varphi_2} = \int_\Omega e^{-(\varphi_2-\varphi_1)}\bar{g}^i u \bar{h}_i e^{-\varphi_1},$$

And thus

$$T_2^* u \quad = \quad \left(e^{-(\varphi_2-\varphi_1)}\bar{g}^1 u, ..., e^{-(\varphi_2-\varphi_1)}\bar{g}^p u\right).$$

Remark. Let us comment on the meaning of this *a priori* local formula. The sections g^j, $1 \le j \le p$, are sections of E, and $e^{-(\varphi_2-\varphi_1)}$ is a metric for $F - (F - E) = E$. Thus for each j, $|g^j|^2 e^{-(\varphi_2-\varphi_1)}$ is a globally defined function, and so the expressions

$$e^{-(\varphi_2-\varphi_1)}\bar{g}^j = e^{-(\varphi_2-\varphi_1)}|g^j|^2/g^j$$

transform like sections of $-E$. Since u takes values in $K_X + F$, the expressions

$$e^{-(\varphi_2-\varphi_1)}\bar{g}^j u$$

transform like sections of $K_X + F - E$, which is what we expect.

2.2. Classical L^2 identities and estimates

In this section we collect some L^2 identities.

The Bochner-Kodaira Identity. Let Ω be a domain in a complex manifold with smooth, $\mathbb{R}$-codimension-1 boundary $\partial\Omega$. Fix a proper smooth function ρ on a neighborhood of Ω such that

$$\Omega = \{\rho < 0\} \quad \text{and} \quad |\partial\rho| \equiv 1 \text{ on } \partial\Omega.$$

Let $H \to \Omega$ be a holomorphic line bundle with singular Hermitian metric $e^{-\varphi}$. The following identity was established in Lecture 1 and used again in Lecture 2.

Bochner-Kodaira Identity:
For any smooth $(0,1)$-form β with values in $K_X + H$ and lying in the domain of $\bar\partial^*$,

$$\int_\Omega |-e^{\varphi}\partial_\nu(\beta^\nu e^{-\varphi})|^2 e^{-\varphi} + \int_\Omega |\bar\partial\beta|^2 e^{-\varphi} = \int_\Omega \beta^\nu\overline{\beta^\mu}(\partial_\nu\partial_{\bar\mu}\varphi)e^{-\varphi}$$
$$+ \int_\Omega |\bar\nabla\beta|^2 e^{-\varphi} + \int_{\partial\Omega} \beta^\nu\overline{\beta^\mu}(\partial_\nu\partial_{\bar\mu}\rho)e^{-\varphi}.$$

Skoda's Identity. For $u \in T_2(\text{Kernel } T_1)$ and $\beta = (\beta_1, ..., \beta_p) \in \text{Domain}(T_1^*)$ we have

$$\begin{aligned} & (T_2^* u, T_1^* \beta)_0 \\ = & (T_1(T_2^*)u, \beta)_1 \\ = & \int_\Omega u \left\{ \overline{\beta_k^\nu \partial_\nu \left(g^k e^{-(\varphi_2 - \varphi_1)} \right)} \right\} e^{-\varphi_1}. \end{aligned}$$

It follows that if β is also in Domain(S_1) then

$$\begin{aligned} & ||T_1^* \beta + T_2^* u||_0^2 + ||S_1 \beta||_*^2 \\ = & ||T_1^* \beta||_0^2 + ||S_1 \beta||_*^2 + ||T_2^* u||_0^2 + 2\text{Re } (T_2^* u, T_1^* \beta)_0 \\ = & \sum_{k=1}^p \left(||T^* \beta_k||_{\varphi_1}^2 + ||S \beta_k||_{\varphi_1}^2 \right) + \int_\Omega e^{-2(\varphi_2 - \varphi_1)} |g|^2 |u|^2 e^{-\varphi_1} \\ & +2\text{Re} \int_\Omega u \left\{ \overline{\beta_k^\nu \partial_\nu \left(g^k e^{-(\varphi_2 - \varphi_1)} \right)} \right\} e^{-\varphi_1}. \end{aligned}$$

By applying the Bochner-Kodaira identity, we obtain the identity we have called SKODA'S IDENTITY:

$$\begin{aligned} (3.3) \qquad & ||T_1^* \beta + T_2^* u||_0^2 + ||S_1 \beta||_*^2 = \int_\Omega e^{-(\varphi_2 - \varphi_1)} |g|^2 |u|^2 e^{-\varphi_2} \\ & +2\text{Re} \int_\Omega u \left\{ \overline{\beta_k^\nu \partial_\nu \left(g^k e^{-(\varphi_2 - \varphi_1)} \right)} \right\} e^{-\varphi_1} + \int_\Omega (\beta_k^\nu \overline{\beta_k^\mu} \partial_\nu \partial_{\bar{\mu}} \varphi_1) e^{-\varphi_1} \\ & + ||\bar{\nabla} \beta||_*^2 + \int_{\partial\Omega} (\beta_k^\nu \overline{\beta_k^\mu} \partial_\nu \partial_{\bar{\mu}} \rho) e^{-\varphi_1}. \end{aligned}$$

Here $||\bar{\nabla} \beta||_*^2 = ||\bar{\nabla} \beta_1||_{\varphi_1}^2 + ... + ||\bar{\nabla} \beta_p||_{\varphi_1}^2$.

Skoda's inequality. To obtain an estimate from Skoda's identity, one makes use of the following inequality of Skoda.

THEOREM 2.2 (Skoda's Inequality). [**Skoda-1972**] *Let $g = (g^1, ..., g^p)$ be holomorphic functions on a domain $U \subset \mathbb{C}^n$, and let $q = \min(n, p-1)$. Then*

$$q(\beta_k^\nu \overline{\beta_k^\mu} \partial_\nu \partial_{\bar{\mu}} \log |g|^2) \geq |g|^2 \left| \beta_k^\nu \partial_\nu (g^k |g|^{-2}) \right|^2$$

REMARK. When passing to a global setting, it is helpful to keep in mind that $g^k |g|^{-2}$ transforms like a section of the anti-holomorphic line bundle $-\overline{E}$, and thus $\partial_\nu (g^k |g|^{-2})) dz^\nu$ transforms like a $(-\overline{E})$-valued $(1,0)$-form. In particular, both sides of Skoda's inequality consist of globally defined functions.

The main step in the proof of Skoda's inequality is the following lemma.

LEMMA 2.3. *Let n and p be integers, and let a_k, b_k^ν, c_ν^k ; $1 \leq k \leq p, 1 \leq \nu \leq n$ be complex numbers. Then, with $q = \min(n, p-1)$ and $|a|^2 = \sum_\ell |a_\ell|^2$, we have*

$$(3.4) \qquad \left| \sum_{j,k} \bar{a}^j (a_j b_k^\nu - a_k b_j^\nu) c_\nu^k \right|^2 \leq q |a|^2 \sum_k \sum_{j<m} \left| (a_j b_m^\nu - a_m b_j^\nu) c_\nu^k \right|^2$$

PROOF. The case $q = p-1$ is elementary. Indeed, by the Cauchy-Schwarz inequality we have

$$\begin{aligned}\Big|\sum_{j,k} \bar{a}^j(a_j b^\nu_k - a_k b^\nu_j)c^k_\nu\Big|^2 &\le |a|^2 \sum_j \Big|\sum_k (a_j b^\nu_k - a_k b^\nu_j)c^k_\nu\Big|^2 \\ &\le (p-1)|a|^2 \sum_{j,k} \Big|(a_j b^\nu_k - a_k b^\nu_j)c^k_\nu\Big|^2,\end{aligned}$$

and from this the desired inequality (3.4) follows easily (see below).

The case $q = n$ requires a little more work.

Fix a single $\nu_o \in \{1, ..., n\}$. (Thus, even though we are using summation convention, we suspend that convention here.) Let

$$A_{\nu_o} := \Big|\sum_{j,m} \bar{a}^m (a_m b^{\nu_o}_j - a_j b^{\nu_o}_m)c^j_{\nu_o}\Big|^2.$$

By permuting the roles of m and j, we have

$$\begin{aligned}A_{\nu_o} &= \Big|\tfrac{1}{2}\sum_{j,m}(a_m b^{\nu_o}_j - a_j b^{\nu_o}_m)(\bar{a}^m c^j_{\nu_o} - \bar{a}^j c^m_{\nu_o})\Big|^2 \\ &= \Big|\sum_{j<m}(a_m b^{\nu_o}_j - a_j b^{\nu_o}_m)(\bar{a}^m c^j_{\nu_o} - \bar{a}^j c^m_{\nu_o})\Big|^2 \\ &\le \Big(\sum_{j<m} |a_m b^{\nu_o}_j - a_j b^{\nu_o}_m|^2\Big)\Big(\sum_{j<m} |\bar{a}^m c^j_{\nu_o} - \bar{a}^j c^m_{\nu_o}|^2\Big) \\ &= \Big(\sum_{j<m} |a_m b^{\nu_o}_j - a_j b^{\nu_o}_m|^2\Big)\Big(|a|^2|c_{\nu_o}|^2 - \Big|\sum_j \bar{a}^j c^j_{\nu_o}\Big|^2\Big) \\ &\le \Big(\sum_{j<m} |a_m b^{\nu_o}_j - a_j b^{\nu_o}_m|^2\Big)|a|^2|c_{\nu_o}|^2\end{aligned}$$

where in the equality we have used Lagrange's identity

$$\sum_{i<j} |a^i b^j - a^j b^i|^2 = |a|^2|b|^2 - |a \cdot \bar{b}|^2.$$

We then calculate as follows.

$$\begin{aligned}\Big|\sum_{\nu_o=1}^{n}\sum_{j,k} \bar{a}^j(a_j b^{\nu_o}_k - a_k b^{\nu_o}_j)c^k_{\nu_o}\Big|^2 &\le n\sum_{\nu_o=1}^{n}\Big|\sum_{j,k} \bar{a}^j(a_j b^{\nu_o}_k - a_k b^{\nu_o}_j)c^k_{\nu_o}\Big|^2 \\ &\le n|a|^2 \sum_k \sum_{j<m} |a_j b^\nu_m - a_m b^\nu_j|^2 |c^k_\nu|^2.\end{aligned}$$

(The reader should recall our summation convention discussed at the beginning of the lecture.) To complete the proof, we must show that

$$\sum_k \sum_{j<m} |a_j b^\nu_m - a_m b^\nu_j|^2 |c^k_\nu|^2 = \sum_k \sum_{j<m} |(a_j b^\nu_m - a_m b^\nu_j)c^k_\nu|^2. \tag{3.5}$$

To see this, consider the Hermitian form on $\mathbb{C}^p$ given by

$$H(X,Y) = \sum_{m<j}(a_m x_j - a_j x_m)\overline{(a_m y_j - a_j y_m)}.$$

Write $B^{\nu} = (b^{\nu}_1, ..., b^{\nu}_p)$, $1 \le \nu \le n$, and let $V = \text{span}\{B^{\nu}\ ;\ 1 \le \nu \le n\}$ be the subspace of $\mathbb{C}^p$ spanned by the B^{ν}. The dimension of V is at most n. Let $\tilde{B}^1, ..., \tilde{B}^n$ be n vectors that are pairwise orthogonal for H and span V. (If the dimension n' of V is less than n, then $\tilde{B}^{n'+1} = \cdots = \tilde{B}^n = 0$.) Then there is a linear transformation $L = (L^{\mu}_{\nu})$ of V such that

$$B^{\nu} = L^{\nu}_{\mu}\tilde{B}^{\mu}.$$

Let

$$\tilde{c}^k_{\nu} = c^k_{\mu}L^{\mu}_{\nu}.$$

Then

$$\tilde{B}^{\nu}\tilde{c}^k_{\nu} = \tilde{B}^{\nu}L^{\mu}_{\nu}c^k_{\mu} = B^{\mu}c^k_{\mu},$$

and thus for all $1 \le j, k \le p$,

$$\tilde{b}^{\nu}_j\tilde{c}^k_{\nu} = b^{\mu}_j c^k_{\mu}.$$

Since only the expressions $\tilde{b}^{\nu}_j\tilde{c}^k_{\nu}$ appear in our inequalities, we may assume that the vectors $B^1, ..., B^n$ are pairwise orthogonal for H.

We have

$$\begin{aligned} & \sum_{j<m} |(a_j b^{\nu}_m - a_m b^{\nu}_j)c^k_{\nu}|^2 \\ = \ & H(B^{\nu}, B^{\mu})c^k_{\nu}\bar{c}^k_{\mu} \\ = \ & H(B^{\nu}, B^{\nu})|c^k_{\nu}|^2 \text{ (summed once over } \nu) \\ = \ & \sum_{j<m} |a_j b^{\nu}_m - a_m b^{\nu}_j|^2 |c^k_{\nu}|^2. \end{aligned}$$

This completes the proof. □

Proof of Skoda's Inequality. Let $\varphi = \log|g|^2$. Observe that

$$\partial_{\nu}\varphi = \frac{\bar{g}_k(\partial_{\nu}g^k)}{|g|^2}.$$

Thus

$$\begin{aligned} e^{\varphi}\beta^{\nu}_k\partial_{\nu}(g^k e^{-\varphi}) &= \beta^{\nu}_k(\partial_{\nu}g^k) - g^k\beta^{\nu}_k(\partial_{\nu}\varphi) \\ &= |g|^{-2}\left(|g|^2(\partial_{\nu}g^k)\beta^{\nu}_k - g^k\beta^{\nu}_k\bar{g}_i(\partial_{\nu}g^i)\right) \\ &= |g|^{-2}\left(\bar{g}_i(g^i\partial_{\nu}g^k - g^k\partial_{\nu}g^i)\beta^{\nu}_k\right) \end{aligned}$$

Next,

$$\begin{aligned} \beta^{\nu}_k\overline{\beta^{\mu}_k}\partial_{\nu}\partial_{\bar{\mu}}\varphi &= \beta^{\nu}_k\overline{\beta^{\mu}_k}\partial_{\nu}\left(\frac{g^i\overline{\partial_{\mu}g^i}}{|g|^2}\right) \\ &= \frac{|g|^2|\beta^{\nu}_k\partial_{\nu}g^i|^2 - |\overline{g^i}\beta^{\nu}_k\partial_{\nu}g^i|^2}{|g|^4} \\ &= |g|^{-4}\sum_{i<j}|\overline{g^i}\beta^{\nu}_k\partial_{\nu}g^j - \overline{g^j}\beta^{\nu}_k\partial_{\nu}g^i|^2 \end{aligned}$$

where in the last equality we have used Lagrange's Identity. The proof is completed by an application of Lemma 2.3. □

Skoda's Basic Estimate. From Theorem 2.2 and Skoda's Identity (3.3), we immediately obtain the following theorem of Skoda.

THEOREM 2.4 (Skoda's Basic Estimate). *Let X be an essentially Stein manifold, $E, F \to X$ holomorphic line bundle with singular metrics $e^{-\eta}$ and $e^{-\psi}$ respectively, $\Omega \subset X$ a pseudoconvex domain and $g^1, ..., g^p \in H^0(X, E)$ holomorphic sections. Let $\alpha > 1$. Set*

$$q = \min(n, p-1), \quad \varphi_1 = \psi + \alpha q \log(|g|^2 e^{-\eta}) \quad \textit{and} \quad \varphi_2 = \varphi_1 + \log|g|^2.$$

For any p-tuple of $F - E$-valued $(0,1)$-forms $\beta = (\beta_1, ..., \beta_p) \in$ Domain$(T_1^*) \cap$ Domain(S_1) and any $u \in T_2$(Kernel(T_1)) we have the estimate

$$\begin{aligned} &||T_1^*\beta + T_2^* u||_0^2 + ||S_1\beta||_*^2 \\ &\geq \frac{\alpha}{\alpha - 1}\int_\Omega |u|^2 e^{-\varphi_2} + \int_\Omega \beta_k^\nu \overline{\beta_k^\mu}\left(\partial_\nu \partial_{\bar\mu}\psi - \alpha q \partial_\nu \partial_{\bar\mu}\eta\right) e^{-\varphi_1}. \end{aligned} \tag{3.6}$$

PROOF. We are going to use Skoda's Identity (3.3). First note that, by the Cauchy-Schwartz Inequality, for any open set U we have

$$\begin{aligned} & 2\mathrm{Re}\int_U u\left\{\overline{\beta_k^\nu \partial_\nu\left(g^k e^{-(\varphi_2 - \varphi_1)}\right)}\right\} e^{-\varphi_1} \\ = \; & 2\mathrm{Re}\int_U u|g|^{-1}\left\{\overline{|g|\beta_k^\nu \partial_\nu\left(g^k|g|^{-2}\right)}\right\} e^{-\varphi_1} \\ \geq \; & -\int_U \frac{1}{\alpha}\frac{|u|^2}{|g|^2} e^{-\varphi_1} - \int_U \alpha|g|^2\left|\beta_k\partial_\nu(g^k|g|^{-2})\right|^2 e^{-\varphi_1} \\ \geq \; & -\int_U \frac{1}{\alpha}|u|^2 e^{-\varphi_2} - \int_U \alpha q(\beta_k^\nu \overline{\beta_k^\mu}\partial_\nu\partial_{\bar\mu}\log|g|^2) e^{-\varphi_1}, \end{aligned} \tag{3.7}$$

where the second inequality follows from Skoda's Inequality (Theorem 2.2). Since the integrands are globally defined, we may replace U by Ω. Substituting $\varphi_1 = \psi + \alpha q \log(|g|^2 e^{-\eta})$, combining the inequality (3.7) with Skoda's Identity (3.3) and dropping the positive terms

$$||\bar\nabla\beta||_*^2 \quad \text{and} \quad \int_{\partial\Omega}\beta_k^\nu\overline{\beta_k^\mu}\partial_\nu\partial_{\bar\mu}\rho e^{-\varphi_1}$$

finishes the proof. □

2.3. Proof of Theorem 1.1

Combining Skoda's Basic Estimate (3.6) with the hypotheses of Theorem 1.1 gives us the inequality

$$||T_1^*\beta + T_2^* u||_0^2 + ||S_1\beta||_*^2 \geq \frac{\alpha - 1}{\alpha}\int_\Omega |u|^2 e^{-\varphi_2}.$$

Let $f \in H^0(X, K_X + F)$ such that

$$\int_X \frac{|f|^2 e^{-\psi}}{(|g|^2 e^{-\eta})^{\alpha q + 1}} < +\infty.$$

Then for all $u \in T_2(\mathscr{K})$ and all $\beta \in$ Domain(T_1^*) we have

$$\begin{aligned} \left|\int_X \frac{f\bar u e^{-\psi}}{(|g|^2 e^{-\eta})^{\alpha q+1}}\right|^2 &\leq \left(\frac{\alpha}{\alpha-1}\int_X \frac{|f|^2 e^{-\psi}}{(|g|^2 e^{-\eta})^{\alpha q+1}}\right)\left(\frac{\alpha-1}{\alpha}\int_\Omega |u|^2 e^{-\varphi_2}\right) \\ &\leq \left(\frac{\alpha}{\alpha-1}\int_X \frac{|f|^2 e^{-\psi}}{(|g|^2 e^{-\eta})^{\alpha q+1}}\right)\left(||T_1^*\beta + T_2^* u||_0^2 + ||S_1\beta||_*^2\right). \end{aligned}$$

By Proposition 2.1 there exist sections $h_1, ..., h_p \in H^0(X, K_X + F - E)$ such that

$$h_k g^k = f \quad \text{and} \quad \int_X \frac{|h|^2 e^{-\psi}}{(|g|^2 e^{-\eta})^{\alpha q}} \le \frac{\alpha}{\alpha - 1} \int_X \frac{|f|^2 e^{-\psi}}{(|g|^2 e^{-\eta})^{\alpha q + 1}}.$$

The proof of Theorem 1.1 is complete. □

3. Global generation of multiplier ideal sheaves

In this section, we establish a Theorem of Siu [**Siu-1998**] on the global generation of multiplier ideal sheaves. The result was proved independently by Esnault and Viehweg using different methods; see [**Lazarsfeld-2005**] for the reference of the latter.

3.1. A preparation lemma

LEMMA 3.1. *Let $A \to Z$ be a very ample line bundle on a compact complex manifold of dimension n, with smooth metric $h = e^{-\varphi}$. For every point $x \in Z$, fix a coordinate neighborhood W_x with coordinates z_x such that*

$$W_x = \{|z_x| < 2\} \quad \textit{and} \quad U_x := \{|z_x| < 1\} \subset\subset W_x.$$

Then there exists a constant $C > 0$ with the following property. For every holomorphic line bundle $L \to Z$ with singular Hermitian metric $e^{-\kappa}$ having semi-positive curvature current, and global holomorphic section $s \in H^0(W_x, L + (n+1)A + K_Z)$ such that

$$C_s := \int_{U_x} |s|^2 e^{-(\kappa + (n+1)\varphi)} < +\infty,$$

there exist sections $\sigma \in H^0(Z, L + (n+1)A + K_Z)$ and $v_1, ..., v_n \in H^0(U_x, L + (n+1)A + K_Z)$ such that

$$s = \sigma + z_x^j v_j \quad \textit{on} \quad U_x, \qquad \int_Z |\sigma|^2 e^{-(\kappa + (n+1)\varphi)} \le C C_s$$

and

$$\int_{U_x} \frac{|v_j|^2}{|z_x|^{2(n-(n+1)\beta)}} e^{-(\kappa + (n+1)\varphi)} \le C C_s \quad \textit{for} \quad j = 1, ..., n.$$

PROOF. Fix sections $u_1, ..., u_N \in H^0(Z, (n+1)A)$ whose common zero locus is precisely $\{x\}$, and such that each u_j vanishes to order $n+1$ at x. Choose $\beta \in (0, \frac{1}{n+1})$, so that

$$\frac{e^{-\beta(n+1)\varphi}}{(\sum_{j=1}^N |u_j|^2)^{1-\beta}}$$

is a metric for $(n+1)A$ that is not locally integrable at x, but is locally integrable at every point of Z other than x.

Let $V_x := \{|z_x| < 1/2\} \subset\subset U_x$. Take $\chi \in \mathscr{C}_0^\infty(V_x)$ be such that $0 \le \chi \le 1$ and $\chi|_{V_x} \equiv 1$. Consider the form

$$\theta = \bar{\partial} \chi s.$$

Since $\theta \equiv 0$ on V_x, we have

$$\int_Z \frac{|\theta|^2 e^{-(\kappa + \beta(n+1)\varphi)}}{\left(\sum_{j=1}^N |u_j|^2\right)^{1-\beta}} \le C_0 \int_{U_x} |s|^2 e^{-(\kappa + (n+1)\varphi)} = C_0 C_s < +\infty.$$

Notice that the metric

$$\frac{e^{-(\kappa+\beta(n+1)\varphi)}}{\left(\sum_{j=1}^{N}|u_j|^2\right)^{1-\beta}}$$

has curvature at least $\beta(n+1)\sqrt{-1}\partial\bar{\partial}\varphi$, and also dominates some multiple of the metric $e^{-\kappa+(n+1)\varphi)}$. By Hörmander's Theorem, $L+(n+1)A+K_Z$ has a section u such that

$$\bar{\partial}u=\theta \quad \text{and} \quad \int_Z \frac{|u|^2e^{-(\kappa+\beta(n+1)\varphi)}}{\left(\sum_{j=1}^{N}|u_j|^2\right)^{1-\beta}} \le C_1C_s.$$

It follows that

$$\sigma=\chi s-u$$

is holomorphic on Z and satisfies

$$\begin{aligned}&\int_Z|\sigma|^2e^{-(\kappa+(n+1)\varphi)}\\ \le\ & 2\int_{U_x}|s|^2e^{-(\kappa+(n+1)\varphi)}+2C_2\int_Z\frac{|u|^2e^{-(\kappa+\beta(n+1)\varphi)}}{(\sum|u_j|^2)^{1-\beta}}\\ \le\ & C_3C_s.\end{aligned}$$

Next, let $\tau=s-\sigma=(1-\chi)s-u$. Then

$$\begin{aligned}\int_{U_x}\frac{|\tau|^2e^{-(\kappa+\beta(n+1)\varphi)}}{|z_x|^{2(n+1)(1-\beta)}} &\le 2\int_{U_x}\frac{|(1-\chi)s|^2e^{-(\kappa+\beta(n+1)\varphi)}}{|z_x|^{2(n+1)(1-\beta)}}+2\int_{U_x}\frac{|u|^2e^{-(\kappa+\beta(n+1)\varphi)}}{|z_x|^{2(n+1)(1-\beta)}}\\ &\le 2\int_{U_x}\frac{|(1-\chi)s|^2e^{-(\kappa+\beta(n+1)\varphi)}}{|z_x|^{2(n+1)(1-\beta)}}+2C'\int_Z\frac{|u|^2e^{-(\kappa+\beta(n+1)\varphi)}}{\left(\sum_j|u_j|^2\right)^{1-\beta}}\\ &\le C_4C_s<+\infty.\end{aligned}$$

We now apply Skoda's Theorem with $g=z_x=(z_x^1,\ldots,z_x^n)$. Thus $p=n$, so $q=n-1$. Choose α so that $\alpha q+1=(n+1)(1-\beta)$. Then

$$\alpha=\frac{(n+1)(1-\beta)-1}{q}=\frac{n-(n+1)\beta}{n-1}>1.$$

Thus by Skoda's Theorem there exist sections $v_1,\ldots,v_n\in H^0(U_x,L\otimes A^{\otimes(n+1)}\otimes K_Z)$ such that

$$\tau=z_x^jv_j \quad \text{and} \quad \int_{U_x}\frac{\sum|v_j|^2e^{-(\kappa+\beta(n+1)\varphi)}}{|z_x|^{2(n-(n+1)\beta)}}\le\frac{n-(n+1)\beta}{1-(n+1)\beta}C_4C_s=C_5C_s.$$

Following the proof, one can check that all the constants are independent of L and $e^{-\kappa}$, and the dependence on s is only through the constant C_s. The proof is complete. □

REMARK. In the above lemma and also in what follows, the line bundle $(n+1)A$ may be replaced by any holomorphic line bundle B such that for any point p of Z, there are global holomorphic sections of B all of which vanish to order at least $n+1$ at p and whose common zero locus consists of p alone.

3.2. Siu's theorem on global generation

THEOREM 3.2 (Siu). *Let $A \to Z$ be a very ample line bundle on a compact complex manifold of dimension n. Then for every holomorphic line bundle $L \to Z$ with Hermitian metric $e^{-\kappa}$ having semi-positive curvature current, the space of global sections $H^0(Z, \mathscr{I}_\kappa(L+(n+1)A+K_Z))$ generates $\mathscr{I}_\kappa(L+(n+1)A+K_Z)$.*

PROOF. Let $s \in \mathscr{I}_{\kappa,x}(L+(n+1)A+K_Z)$. Then s is defined on some neighborhood of x. By Lemma 3.1 we can write

$$s = \sigma + z_x^j v_j \tag{3.8}$$

for some global section $\sigma \in H^0(Z, \mathscr{I}_{\kappa,x}(L+(n+1)A+K_Z))$ and germs of sections $v_1, ..., v_n \in \mathscr{I}_\kappa(L+(n+1)A+K_Z)$. Let $\mathscr{J}$ be the ideal of holomorphic germs in $\mathscr{I}_\kappa(L+(n+1)A+K_Z)$ generated by $H^0(Z, \mathscr{I}_\kappa(L+(n+1)A+K_Z))$. By (3.8), we have

$$\mathscr{I}_\kappa(L+(n+1)A+K_Z) \subset \mathscr{J} + \mathfrak{m}_x \mathscr{I}_\kappa(L+(n+1)A+K_Z).$$

Modding out by $\mathscr{J}$, we have

$$\frac{\mathscr{I}_\kappa(L+(n+1)A+K_Z)}{\mathscr{J}} \subset \mathfrak{m}_x \frac{\mathscr{I}_\kappa(L+(n+1)A+K_Z)}{\mathscr{J}}.$$

By Nakayama's Lemma, $\mathscr{I}_\kappa(L+(n+1)A+K_Z) = \mathscr{J}$, and the proof is complete. □

REMARK. In his work [**Siu-2002**] on the deformation invariance of plurigenera, Siu also established what he called an effective version of global generation of multiplier ideals. Lemma 3.1, which is more precise than what is necessary for the proof of Theorem 3.2, is useful establishing the effective version of Siu's Theorem.

3.3. Siu's theorem on effective finite generation

In his paper [**Siu-2005**], Siu derived from Skoda's Theorem the following result in the case of algebraic manifolds. We give a slightly different proof of Siu's result, in the more general setting of essentially Stein manifolds.

THEOREM 3.3. *Let X be an essentially Stein manifold of complex dimension n, $L \to X$ a holomorphic line bundle, and $H \to X$ a holomorphic line bundle with non-negatively curved singular Hermitian metric $e^{-\varphi}$. Let $k \geq 1$ be an integer and fix sections $G^1, ..., G^p \in H^0(X, L)$. Define the multiplier ideals*

$$\mathscr{J}_{k+1} = \mathscr{I}(e^{-\varphi}|G|^{-2(n+k+1)}) \quad \text{and} \quad \mathscr{J}_k = \mathscr{I}(e^{-\varphi}|G|^{-2(n+k)}).$$

Then

$$H^0(X, \mathscr{J}_{k+1}((n+k+1)L+H+K_X) = \bigoplus_{j=1}^{p} G_j H^0(X, \mathscr{J}_k((n+k)L+H+K_X)).$$

PROOF. By taking $G^p = ...G^n = 0$, we may assume that $q = n$. Take $\alpha = (n+k)/n$, so that $\alpha q = n+k$. We are going to use Theorem 1.1 with $F = (n+k+1)L+H$, $E = L$ and $g^i = G^i$. Fix a metric $e^{-\eta}$ for L having non-negative curvature current, (for example, one could take $\eta = \log|G^1|^2$) and let $\psi = \varphi + (n+k+1)\eta$. Then

$$\sqrt{-1}\partial\bar\partial\psi - \alpha q\sqrt{-1}\partial\bar\partial\eta = \sqrt{-1}\partial\bar\partial\varphi + \sqrt{-1}\partial\bar\partial\eta \geq 0.$$

Suppose $f \in H^0(X, \mathscr{J}_{k+1}((n+k+1)L+H+K_X))$. By Skoda's Theorem 1.1 there exist sections $h_1, ..., h_p \in H^0(X, \mathcal{O}_X((n+k)L+H+K_X))$ such that $h_k G^k = f$. Moreover,

$$\begin{aligned}\int_X \frac{|h|^2 e^{-\varphi}}{|G|^{2(n+k)}} &= \int_X \frac{|h|^2 e^{-(\psi-\eta)}}{(|g|^2 e^{-\eta})^{n\alpha}} \\ &\leq \frac{\alpha}{\alpha-1}\int_X \frac{|f|^2 e^{-\psi}}{(|g|^2 e^{-\eta})^{n\alpha+1}} \\ &= \frac{n+k}{n+k-1}\int_X \frac{|f|^2 e^{-\varphi}}{|G|^{2(n+k+1)}} < +\infty.\end{aligned}$$

Thus in particular $h_i \in \mathscr{J}_k$ locally. The proof is complete. □

Using Theorem 3.3, we can now state and prove another result of Siu that reduces the problem of finite generation of the canonical ring to an albeit difficult estimate.

THEOREM 3.4 (Pluricanonical version). *Let X be a complex projective manifold of complex dimension n and $E \to X$ holomorphic line bundles. Assume we have non-negatively curved singular Hermitian metrics $e^{-\eta}$ and $e^{-\psi_E}$ for K_X and E respectively. Let $\mu \geq 0$, $m \geq 1$ be integers and fix section $G^1, ..., G^p \in H^0(X, mK_X)$. Define the multiplier ideals*

$$\mathscr{J}_{\mu+1} = \mathscr{I}(e^{-\psi_E} e^{-(\mu+(n+2)m)\eta} |G|^{-2(n+2)})$$

and

$$\mathscr{J}_{\mu} = \mathscr{I}(e^{-\psi_E} e^{-(\mu+(n+1)m)\eta} |G|^{-2(n+1)}).$$

Then

$$\begin{aligned}&H^0(X, \mathscr{J}_{\mu+1}((\mu+(n+2)m+1)K_X+E)) \\ &\quad= \bigoplus_{j=1}^{p} G_j H^0(X, \mathscr{J}_{\mu}((\mu+m(n+1)+1)K_X+E)).\end{aligned}$$

PROOF. Apply Theorem 3.3 with $k = 1$, $L = mK_X$, $H = \mu K_X + E$ and $\varphi = \mu\eta + \psi_E$. □

As a result of the latter division theorem, we obtain the following consequence.

COROLLARY 3.5 (Template for Finite Generation). *Let X be a complex projective manifold of complex dimension n and $E \to X$ holomorphic line bundles. Assume we have non-negatively curved singular Hermitian metrics $e^{-\varphi}$ and $e^{-\psi_E}$ for K_X and E respectively. Let $k > m \geq 1$ be integers. Fix a basis $\{G^1, ..., G^p\} \subset H^0(X, mK_X + E)$. Assume moreover that the following hold.*

(1) *The Singular metric $e^{-\psi_E}$ is locally integrable.*
(2) *For all $\mu \in \mathbb{N}$ and each $s \in H^0(X, \mu K_X + E)$,*

$$\sup_X |s|^2 e^{-(\mu\varphi+\xi_E)} < +\infty$$

for some (and thus any) smooth metric $e^{-\xi_E}$ for E.

(3) *There is a constant $C > 0$ such that*

$$|G|^2 e^{-m\varphi} \geq C.$$

Then for any $k > (n+2)m$,

$$H^0(X, \mathscr{I}_{\psi_E}(kK_X + E)) = H^0(X, mK_X) \otimes H^0(X, \mathscr{I}_{\psi_E}((k-m)K_X + E)).$$

In particular, taking $E = \mathcal{O}_X$ *and* $\psi_E \equiv 0$, *we find that if* (2) *and* (3) *hold for all* m *then the canonical ring*

$$R_X := \bigoplus_{\ell=1}^{\infty} H^0(X, \ell K_X)$$

is finitely generated, and in fact it is generated by

$$\bigoplus_{\ell=1}^{m(n+2)} H^0(X, \ell K_X).$$

PROOF. In this setting, the multiplier ideals in Theorem 3.3 (or more precisely in Theorem 3.4) are trivial, and thus the first result follows. The finite generation statement is obvious. □

REMARK. In Siu's approach to finite generation of the canonical ring, one takes the metric

$$\varphi := \log \left(\sum_{\mu=1}^{\infty} \varepsilon_\mu \left(\sum_{j=1}^{h^0(X,\mu K_X)} |s_j^{(\mu)}|^2 \right)^{1/\mu} \right)$$

where for each $\mu \geq 1$, $\{s_j^{(\mu)}\}$ is a basis for $H^0(X, \mu K_X)$, and the positive numbers ε_μ tend to zero as $\mu \to \infty$ so rapidly that the metric φ is smooth. It is obvious that for this φ (2) holds in Corollary 3.5, and the whole point is to establish the estimate (3).

4. Exercises

In the exercises we will use the version of Theorem 1.1 where X is a pseudoconvex domain in $\mathbb{C}^n$ and F and E are both the trivial bundle. We will take $\eta \equiv 0$, but keep φ as a general plurisubharmonic function. As before, $\alpha > 1$ and $g = (g^1, ..., g^p)$ are given, and $q = \min(p-1, n)$. Skoda's Theorem takes the following form.

THEOREM 4.1. *Let* $X \subset \mathbb{C}^n$ *be a pseudoconvex domain and* $\varphi : X \to [-\infty, \infty)$ *a plurisubharmonic function. Fix holomorphic functions* $g^1, ..., g^p$ *and a number* $\alpha > 1$, *and set* $q = \min(p-1, n)$. *Then for any* $f \in \mathcal{O}(X)$ *such that*

$$\int_X \frac{|f|^2 e^{-\varphi}}{|g|^{2\alpha q+2}} < +\infty$$

there are holomorphic functions $h_1, ..., h_p$ *such that*

$$\sum_{k=1}^{p} g^k h_k = f \quad \textit{and} \quad \int_X \frac{|h|^2 e^{-\varphi}}{|g|^{2\alpha q}} \leq \frac{\alpha}{\alpha - 1} \int_X \frac{|f|^2 e^{-\varphi}}{|g|^{2\alpha q+2}}.$$

4.1. Skoda's theorem and the Nullstellensatz

Hilbert's Nullstellensatz is the statement that the ideal of the variety of an ideal J is the radical of J. For the ring of polynomials on $\mathbb{C}^n$, it is equivalent to the following result.

THEOREM 4.2. *If $g^1, ..., g^p$ are p polynomials with no common zeros, then there are polynomials $h_1, ..., h_p$ such that*

$$\sum_{k=1}^{p} g^k h_k = 1.$$

In this exercise, we will prove Theorem 4.2 and more.

(1) Show that if $g^1, ..., g^p$ are p polynomials of degree $\le d$ and with no common zeros, then there are constants $C > C' > 0$ and an integer N such that

$$C'(1+|z|^2)^{-N} \le |g(z)|^2 \le C(1+|z|^2)^d.$$

Hint: It may be useful to consider the situation in $\mathbb{P}_n$ rather than $\mathbb{C}^n$.

(2) Show that there are coefficient polynomials $h_1, ..., h_p$ such that $h_k g^k = 1$ and $\deg(h_j) \le N + n + 1$. (You will need an appropriate weight φ.)

(3) Estimate

$$\sum_{|\alpha| \le d} |a_{j,\alpha}|^2,$$

where

$$h_j(z) = \sum_{|\alpha| \le d} a_{j,\alpha} z^\alpha.$$

4.2. The Briançon-Skoda theorem

In this exercise we will use Theorem 4.1 to establish the Briançon-Skoda Theorem [**Skoda-Briançon-1974**]. We begin with some definitions.

DEFINITION 4.3. *Let $\mathscr{I} = (g^1, ..., g^p) \subset \mathcal{O}_{\mathbb{C}^n}$ be an ideal.*

- *The ideal $\bar{\mathscr{I}}^{(k)}$ consists of germs $u \in \mathcal{O}_{\mathbb{C}^n}$ such $|u| \le C|g|^k$ for some $C > 0$.*
- *The ideal $\hat{\mathscr{I}}^{(k)}$ consists of germs $u \in \mathcal{O}_{\mathbb{C}^n}$ such that for all sufficiently small $\varepsilon > 0$ and some sufficiently small ball Ω,*

$$\int_\Omega \frac{|u|^2}{|g|^{2k(1+\varepsilon)}} dV < +\infty.$$

(1) Show that
  (a) $\bar{\mathscr{I}}^{(k)} \subset \hat{\mathscr{I}}^{(k)}$.
  (b) $\mathscr{I}^{(k)} \subset \bar{\mathscr{I}}^{(k)}$.
  (c) $\bar{\mathscr{I}}^{(k)} \bar{\mathscr{I}}^{(\ell)} \subset \bar{\mathscr{I}}^{(k+\ell)}$.
  (d) $\bar{\mathscr{I}}^{(k)} \hat{\mathscr{I}}^{(\ell)} \subset \hat{\mathscr{I}}^{(k+\ell)}$.

(2) Show that if $p > n$ then there are elements $\tilde{g}^1, ..., \tilde{g}^n \in \mathscr{I}$ such that for some $C > 1$,

$$\frac{1}{C}|g|^2 \le |\tilde{g}|^2 \le C|g|^2. \tag{3.9}$$

In fact, show that generic linear combinations

$$\tilde{g}^j = \sum_{k=1}^{p} a^j_k g^k, \quad 1 \le j \le n$$

will do, by justifying the following steps.

(a) Fix a small ball Ω such that $g \in \mathcal{O}(\Omega)^p$ and define

$$A := \left\{(z, [w^1], ..., [w^n]) \in \Omega \times (\mathbb{P}_{p-1})^n \ ; \ w^i_k g^k(z) = 0, \ 1 \le i \le n\right\}$$

and

$$A^* := \bigcup \text{irreducible components of } A \text{ not contained in } g^{-1}(0) \times (\mathbb{P}_{p-1})^n.$$

Show that $\dim_{\mathbb{C}} A^* = n(r-1)$. Conclude that

$$A^*_0 := A^* \cap (\{0\} \times (\mathbb{P}_{p-1})^n)$$

is a proper subvariety of $\{0\} \times (\mathbb{P}_{p-1})^n$.

(b) Show that for any $a^i \in \mathbb{C}^p - \{0\}$, $1 \le i \le n$ such that $(0, [a^1], ..., [a^n]) \notin A^*_0$, there is an $\varepsilon > 0$ such that

$$\left(A^* \cap (B(0, 2\varepsilon) \times (\mathbb{P}_{p-1})^n)\right) \cap \left(B(0, \varepsilon) \times \prod_{i=1}^{n} [B(a_j, \varepsilon)]\right) = \emptyset.$$

(c) Show that there is some $j \in \{1, ..., n\}$ such that for all $z \in B(0, \varepsilon)$,

$$\varepsilon |g(z)| \le |a^j_k g^k(z)|.$$

(d) Complete the proof of (3.9).

(3) Show that $\hat{\mathscr{I}}^{(k+1)} \subset \mathscr{I} \hat{\mathscr{I}}^{(k)}$ for $k \ge \min(n-1, p-1)$ as follows.

(a) First assume that $p \le n$. For $f \in \hat{\mathscr{I}}^{(k+1)}$, use Skoda's Theorem 4.1 to write $f = \sum h_j g^j$ for some $h_j \in \mathscr{I}^{(k)}$, and deduce that $f \in \mathscr{I} \hat{\mathscr{I}}^{(k)}$.

(b) If $p > n$, show that there is an ideal $\mathscr{J} \subset \mathscr{I}$ with n generators such that $\hat{\mathscr{J}}^{(k)} = \hat{\mathscr{I}}^{(k)}$.

(c) Use step (b) $\hat{\mathscr{I}}^{(k+1)} \subset \mathscr{I} \hat{\mathscr{I}}^{(k)}$ for $k \ge n-1$.

(4) Prove the Briançon-Skoda Theorem: let $r = \min(n-1, p-1)$. Then

(a) $\hat{\mathscr{I}}^{(k+1)} = \mathscr{I} \hat{\mathscr{I}}^{(k)} = \bar{\mathscr{I}} \hat{\mathscr{I}}^{(k)}$ for $k \ge r$, and

(b) $\bar{\mathscr{I}}^{(k+r)} \subset \hat{\mathscr{I}}^{(k+r)} \subset \mathscr{I}^k$ for all $k \ge 1$.

Bibliography

[Andersson-Götmark-2008] *Explicit representation of membership of polynomial ideals.* arXiv:0806.2592

[Andersson-Samuelsson-Sznajdman-2008] *On the Briancon-Skoda theorem on a singular variety* arXiv:0806.3700

[Andreotti-Vesentini-1965] Andreotti, A., Vesentini, E., *Carleman estimates for the Laplace-Beltrami equation on complex manifolds.* Publ. IHES 25 (1965) 81-130.

[Berndtsson-1996] Berndtsson, B., *The extension theorem of Ohsawa-Takegoshi and the theorem of Donnelly-Fefferman.* Ann. Inst. Fourier (Grenoble) 46 (1996), no. 4, 1083–1094.

[Berndtsson-Ortega Cerdà-1995] Berndtsson, B.; Ortega Cerdà, J., *On interpolation and sampling in Hilbert spaces of analytic functions.* J. Reine Angew. Math. 464 (1995), 109–128.

[Carleson-1962] Carleson, L., *Interpolations by bounded analytic functions and the corona problem.* Ann. of Math. (2) 76 1962 547–559.

[Catlin-1983] Catlin, D. *Necessary conditions for subellipticity of the $\bar\partial$-Neumann problem.* Ann. of Math. (2) 117 (1983), no. 1, 147–171.

[Catlin-1987] Catlin, D. *Subelliptic estimates for the $\overline{\partial}$-Neumann problem on pseudoconvex domains.* Ann. of Math. (2) 126 (1987), no. 1, 131–191.

[D'Angelo-1979] D'Angelo, J., *Finite type conditions for real hypersurfaces.* J. Differential Geom. 14 (1979), no. 1, 59–66 (1980).

[D'Angelo-1982] D'Angelo, J., *Real hypersurfaces, orders of contact, and applications.* Ann. of Math. (2) 115 (1982), no. 3, 615–637.

[Demailly-1982] Demailly, J.-P. *Estimations L^2 pour l'opérateur $\bar\partial$ d'un bré vectoriel holomorphe semi-positif au dessus d'une variété kählérienne complète.* Ann. Sci. Ec. Norm. Sup., 15 (1982), 457511.

[Demailly-1992] Demailly, J.-P., *Regularization of closed positive currents and Intersection Theory.* J. Alg. Geom., 1 (1992), 361409.

[Demailly-2000] Demailly, J.-P., *On the Ohsawa-Takegoshi-Manivel L^2 extension theorem.* Complex analysis and geometry (Paris, 1997), 47–82, Progr. Math., 188, Birkhuser, Basel, 2000.

[Demailly-2001] Demailly, J.-P. *Multiplier ideal sheaves and analytic methods in algebraic geometry.* School on Vanishing Theorems and Effective Results in Algebraic Geometry (Trieste, 2000), 1–148, ICTP Lect. Notes, 6, Abdus Salam Int. Cent. Theoret. Phys., Trieste, 2001. See also `http://www-fourier.ujf-grenoble.fr/~demailly/books.html`

[Demailly-Book] Demailly, J.-P. *Complex analytic and algebraic geometry.* Available at `http://www-fourier.ujf-grenoble.fr/~demailly/books.html`

[Diederich-Fornæss-1978] Diederich, K.; Fornaess, J., *Pseudoconvex domains with real-analytic boundary.* Ann. Math. (2) 107 (1978), no. 2, 371–384.

[Fornæss-Sibony-1993] Fornæss, J. E.; Sibony, N., *Smooth pseudoconvex domains in $\mathbb{C}^2$ for which the corona theorem and L^p estimates for $\overline{\partial}$ fail.* Complex analysis and geometry, 209–222, Univ. Ser. Math., Plenum, New York, 1993.

[Griffiths-Harris-1978] Griffiths, P., Harris, J. *Principles of Algebraic Geometry.* Wiley Interscience (1978).

[Götmark-2008] Götmark, E., *Explicit solutions of division problems for matrices of polynomials* arXiv:0806.2592

[Hörmander-1965] Hörmander, L., *L^2 estimates and existence theorems for the $\bar\partial$ operator.* Acta Math. 113 1965 89–152.

[Hormander-1990] Hörmander, L., *An introduction to complex analysis in several variables.* Third edition. North-Holland Mathematical Library, 7. North-Holland Publishing Co., Amsterdam, 1990.

[Kohn-1963] Kohn, J. J. *Harmonic integrals on strongly pseudo-convex manifolds. I.* Ann. of Math. (2) 78 1963 112–148.

[Kohn-1964] Kohn, J.J. *Kohn, J. J. Harmonic integrals on strongly pseudo-convex manifolds. II.* Ann. of Math. (2) 79 1964 450–472.

[Kohn-1979] Kohn, J. J. *Subellipticity of the $\bar\partial$-Neumann problem on pseudo-convex domains: sufficient conditions.* Acta Math. 142 (1979), no. 1-2, 79–122.

[Lazarsfeld-2005] Lazarsfeld, R., *Positivity in algebraic geometry. I & II.* Springer 2005

[Manivel-1996] Manivel, L. *Un théorème de prolongement L^2 de sections holomorphes d'un fibré hermitien. (French) [An L^2 extension theorem for the holomorphic sections of a Hermitian bundle]* Math. Z. 212 (1993), no. 1, 107–122.

[McNeal-1996a] McNeal, J., *On large values of L^2 holomorphic functions* Math. Res. Let. 3 (1996), 247-259.

[McNeal-1996b] McNeal, J., *A proof of a Skoda-type Theorem from L^2 extension.* Private communication; unpublished.

[McNeal-Varolin-2007] McNeal, J.; Varolin, D., *Analytic inversion of adjunction: L^2 extension theorems with gain.* Ann. Inst. Fourier (Grenoble) 57 (2007), no. 3, 703–718.

[Ohsawa-1999] Ohsawa, T., *Extension and division on complex manifolds. Analytic extension formulas and their applications.* (Fukuoka, 1999/Kyoto, 2000), 189–206, Int. Soc. Anal. Appl. Comput., 9, Kluwer Acad. Publ., Dordrecht, 2001.

[Ohsawa-2002] Ohsawa, T., *A precise L^2 division theorem.* Complex geometry (Gttingen, 2000), 185–191, Springer, Berlin, 2002.

[Ohsawa-Takegoshi-1987] Ohsawa, T., Takegoshi, K., *On the extension of L^2 holomorphic functions.* Math. Z. 195 (1987), no. 2, 197–204.

[Păun-2007] Păun, M., *Siu's invariance of plurigenera: a one-tower proof.* J. Differential Geom. 76 (2007), no. 3, 485–493.

[Seip-1992] Seip, K., *Density theorems for sampling and interpolation in the Bargmann-Fock space. I.* J. Reine Angew. Math. 429 (1992), 91–106.

[Sibony-1987] Sibony, N., *Problème de la couronne pour des domaines pseudoconvexes àbord lisse.* Ann. of Math. (2) 126 (1987), no. 3, 675–682.

[Siu-1974] Siu, Y.-T., *Analyticity of sets associated to Lelong numbers and the extension of closed positive currents.* Invent. Math. 27 (1974), 53–156.

[Siu-1996] Siu, Y.-T., *The Fujita conjecture and the extension theorem of Ohsawa-Takegoshi* Geometric Complex Analysis, Hayama. World Scientific (1996), 577-592.

[Siu-1998] Siu, Y.-T., *Invariance of plurigenera.* Invent. Math. 134 (1998), no. 3, 661–673.

[Siu-2002] Siu, Y.-T., *Extension of twisted pluricanonical sections with plurisubharmonic weight and invariance of semipositively twisted plurigenera for manifolds not necessarily of general type.* Complex geometry. Collection of papers dedicated to Hans Grauert. Springer-Verlag, Berlin, 2002. (223–277)

[Siu-2005] Siu, Y.-T., *Multiplier Ideal Sheaves in Complex and Algebraic Geometry.* Sci. China Ser. A 48 (2005), suppl., 1–31.

[Siu-2006] Siu, Y.-T., *A General Non-Vanishing Theorem and an Analytic Proof of the Finite Generation of the Canonical Ring.* arXiv:math/0610740

[Siu-2008] Siu, Y.-T., *Finite generation of canonical ring by analytic method.* Sci. China Ser. A 51 (2008), no. 4, 481–502.

[Skoda-1972] Skoda, H., *Application des techniques L^2 la théorie des idéaux d'une algèbre de fonctions holomorphes avec poids.* Ann. Sci. École Norm. Sup. (4) 5 (1972), 545–579.

[Skoda-Briançon-1974] Skoda, H.; Briançon, J., *Sur la clôture intgrale d'un idal de germes de fonctions holomorphes en un point de $\mathbb{C}^n$.* C. R. Acad. Sci. Paris Sér. A 278 (1974), 949–951.

[Takayama-2006] Takayama, S, Pluricanonical systems on algebraic varieties of general type. Invent. Math. 165 (2006), no. 3, 551–587.

[Varolin-2007] Varolin, D., *A Takayama-type extension theorem.* Compos. Math. 144 (2008), no. 2, 522–540.

[Varolin-2008] Varolin, D., *Division theorems and twisted complexes.* Math. Z. 259 (2008), no. 1, 1–20.

Structure Theorems for Projective and Kähler Varieties

Jean-Pierre Demailly

IAS/Park City Mathematics Series
Volume 17, 2008

Structure Theorems for Projective and Kähler Varieties

Jean-Pierre Demailly

0. Introduction

The main purpose of these notes is to describe some basic structure theorems for projective or compact Kähler varieties and their cohomology, using recent techniques of complex analysis and potential theory. One central unifying concept is the concept of positivity, which can ve viewed either in algebraic terms (positivity of divisors and algebraic cycles), or in more analytic terms (plurisubharmonicity, psoitive currents, hermitian connections with positive curvature). In the course of the 20th century, following work by early precursors such as B. Riemann and H. Poincaré, powerful L^2 techniques have emerged in the hands of W.V.D. Hodge, S. Bochner, K. Kodaira, J.J. Kohn, L. Hörmander and many others. We refer to Dror Varolin's notes [Var09] in the present volume for an exposition of the Bochner-Kodaira technique and its main consequences, namely the fundamental L^2 estimates of Hörmander [Hör65] (resolution of $\overline{\partial}$ equations), Skoda [Sko72b, 78] (surjectivity theorem for morphisms of holomorphic vector bundles) and Ohsawa-Takegoshi [OT87] (L^2 extension theorem). All these theorems have an incredible amount of geometric consequences: among those of the Ohsawa-Takegoshi theorem, for instance, let us mention Siu's theorem [Siu74] on the analyticity of Lelong numbers, a more recent approximation theorem of closed positive $(1,1)$-currents by divisors due to [Dem92], the subadditivity property $\mathcal{I}(\varphi+\psi) \subset \mathcal{I}(\varphi)\mathcal{I}(\psi)$ of multiplier ideals [DEL00], the restriction formula $\mathcal{I}(\varphi_{|Y}) \subset \mathcal{I}(\varphi)_{|Y}, \ldots$. A suitable combination of these results can be used to reprove Fujita's result [Fuj94] on approximate Zariski decomposition, as detailed in section 4. Sections 5 and 6 are devoted to the study of positive cones in Kähler or projective geometry. Recent "algebro-analytic" characterizations of the Kähler cone ([DP04]) and the pseudo-effective cone of divisors ([BDPP04]) are explained in detail. This leads to a discussion of the important concepts of volume and mobile intersections, following S.Boucksom's PhD work [Bou02]. As a consequence, we show that a projective algebraic manifold has a pseudo-effective canonical line bundle if and only if it is not uniruled. The section 7 presents some important ideas of H. Tsuji, later refined by Berndtsson and Păun, concerning the so-called "super-canonical metrics", and their interpretation in terms of the invariance of plurigenera and of the abundance conjecture. In the final section, we state Păun's version of the Shokurov-Hacon-McKernan-Siu non vanishing theorem and give an account of the very recent approach of the proof of

Université de Grenoble I, Institut Fourier
E-mail address: demailly@fourier.ujf-grenoble.fr

the finiteness of the canonical ring by Birkar-Păun [BiP09], based on the ideas of Hacon-McKernan and Siu.

I would like to thank the organizers of the Graduate Summer School on Analytic and Algebraic Geometry held at the Park City Mathematical Institute in July 2008 for their invitation to give a series of lectures, and thus for the opportunity of publishing these notes.

1. Numerically effective and pseudo-effective (1,1) classes

1.A. Pseudo-effective line bundles and metrics with minimal singularities

The concept of pseudo-effectivity is quite general and makes sense on an arbitrary compact complex manifold X (no projective or Kähler assumption is needed). Let us recall that all usual concepts of cohomology (de Rham, resp. Dolbeault, ...) can be calculated with complexes of smooth forms or with complexes of currents, alternatively. This comes from the fact that both types of complexes provide resolutions of the same sheaves, namely germs of locally constant functions, resp. germs of holomorphic p-forms.

In the general context of compact complex manifolds, it is in fact more appropriate to work with Bott-Chern cohomology, instead of de Rham of Dolbeault cohomology classes: we define

$$(1.1)\qquad H^{p,q}_{\mathrm{BC}}(X) = \{d\text{-closed } (p,q)\text{-forms}\}/\{\partial\overline{\partial}\text{-exact } (p,q)\text{-forms}\}.$$

By means of the Frölicher spectral sequence, it is easily shown that these cohomology groups are finite dimensional and can be computed either with spaces of smooth forms or with currents. In both cases, the quotient topology of $H^{p,q}_{\mathrm{BC}}(X)$ induced by the Fréchet topology of smooth forms or by the weak topology of currents is Hausdorff. Clearly $H^{\bullet}_{\mathrm{BC}}(X)$ is a bigraded algebra. This algebra can be shown to be isomorphic to the usual De Rham cohomology algebra $H^{\bullet}(X,\mathbb{C})$ if X is Kähler or more generally if X is in the Fujiki class $\mathcal{C}$ of manifolds bimeromorphic to Kähler manifolds.

(1.2) Definition. *Let L we a holomorphic line bundle on a compact complex manifold X. we say that L pseudo-effective if $c_1(L) \in H^{1,1}_{\mathrm{BC}}(X)$ is the cohomology class of some closed positive current T, i.e. if L can be equipped with a singular hermitian metric h with $T = \frac{\mathrm{i}}{2\pi}\Theta_{L,h} \geq 0$ as a current.*

Recall that for any hermitian metric h on a holomorphic vector bundle E, one can define a unique compatible Chern connection $D_{E,h}$ such that $D^{0,1}_{E,h} = \overline{\partial}$. We then denote $\Theta_{E,h} = D^2_{E,h} \in C^\infty(X, \Lambda^{1,1}T^*_X \otimes \mathrm{Hom}(E,E))$ the associated curvature tensor. If $E = L$ is a holomorphic line bundle with a singular metric h such that $\log h \in L^1_{\mathrm{loc}}$, the curvature can be computed in the sense of distributions. The locus where h has singularities turns out to be extremely important. The following definition was introduced in [DPS00].

(1.3) Definition. *Let L be a pseudo-effective line bundle on a compact complex manifold X. Consider two hermitian metrics h_1, h_2 on L with curvature $i\Theta_{L,h_j} \geq 0$ in the sense of currents.*

(i) *We will write $h_1 \preccurlyeq h_2$, and say that h_1 is less singular than h_2, if there exists a constant $C > 0$ such that $h_1 \leq C h_2$.*

(ii) *We will write* $h_1 \sim h_2$, *and say that* h_1, h_2 *are equivalent with respect to singularities, if there exists a constant* $C > 0$ *such that* $C^{-1}h_2 \le h_1 \le Ch_2$.

Of course $h_1 \preccurlyeq h_2$ if and only if the associated weights in suitable trivializations locally satisfy $\varphi_2 \le \varphi_1 + C$. This implies in particular $\nu(\varphi_1, x) \le \nu(\varphi_2, x)$ at each point. The above definition is motivated by the following observation.

(1.4) Theorem. *For every pseudo-effective line bundle* L *over a compact complex manifold* X, *there exists up to equivalence of singularities a unique class of hermitian metrics* h *with minimal singularities such that* $i\Theta_{L,h} \ge 0$.

Proof. The proof is almost trivial. We fix once for all a smooth metric h_∞ (whose curvature is of random sign and signature), and we write singular metrics of L under the form $h = h_\infty e^{-\psi}$. The condition $i\Theta_{L,h} \ge 0$ is equivalent to $\frac{i}{2\pi}\partial\overline{\partial}\psi \ge -u$ where $u = \frac{i}{2\pi}\Theta_{L,h_\infty}$. This condition implies that ψ is plurisubharmonic up to the addition of the weight φ_∞ of h_∞, and therefore locally bounded from above. Since we are concerned with metrics only up to equivalence of singularities, it is always possible to adjust ψ by a constant in such a way that $\sup_X \psi = 0$. We now set

$$h_{\min} = h_\infty e^{-\psi_{\min}}, \qquad \psi_{\min}(x) = \sup_\psi \psi(x)$$

where the supremum is extended to all functions ψ such that $\sup_X \psi = 0$ and $\frac{i}{2\pi}\partial\overline{\partial}\psi \ge -u$. By standard results on plurisubharmonic functions (see Lelong [Lel69]), $\psi_{\min}$ still satisfies $\frac{i}{2\pi}\partial\overline{\partial}\psi_{\min} \ge -u$ (i.e. the weight $\varphi_\infty + \psi_{\min}$ of $h_{\min}$ is plurisubharmonic), and $h_{\min}$ is obviously the metric with minimal singularities that we were looking for. [In principle one should take the upper semicontinuous regularization $\psi^*_{\min}$ of $\psi_{\min}$ to really get a plurisubharmonic weight, but since $\psi^*_{\min}$ also participates to the upper envelope, we obtain here $\psi_{\min} = \psi^*_{\min}$ automatically]. □

(1.5) Remark. In general, the supremum $\psi = \sup_{j\in I} \psi_j$ of a locally dominated family of plurisubharmonic functions ψ_j is not plurisubharmonic strictly speaking, but its "upper semi-continuous regularization" $\psi^*(z) = \limsup_{\zeta\to z} \psi(\zeta)$ is plurisubharmonic and coincides almost everywhere with ψ, with $\psi^* \ge \psi$. However, in the context of (1.5), ψ^* still satisfies $\psi^* \le 0$ and $\frac{i}{2\pi}\partial\overline{\partial}\psi \ge -u$, hence ψ^* participates to the upper envelope. As a consequence, we have $\psi^* \le \psi$ and thus $\psi = \psi^*$ is indeed plurisubharmonic. Under a strict positivity assumption, namely if L is a big line bundle (i.e. the curvature can be taken to be strictly positive in the sense of currents, see 1.12) and (1.17 ii) below), then $h_{\min}$ can be shown to possess some regularity properties. The reader may consult [BmD09] for a rather general (but certainly non trivial) proof that $\psi_{\min}$ possesses locally bounded second derivatives $\partial^2\psi_{\min}/\partial z_j\partial\overline{z}_k$ outside an analytic set $Z \subset X$; in other words, $i\Theta_{L,h_{\min}}$ has locally bounded coefficients on $X \smallsetminus Z$.

Following Nadel [Nad89], one defines the *multiplier ideal sheaf* $\mathcal{I}(\varphi)$ of a psh function φ to be the sheaf of germs of holomorphic functions f such that $|f|^2e^{-2\varphi}$ is locally summable (see also [Var09] in this volume; some authors simply write φ where we have put 2φ, but this is just a matter of notation). When φ is plurisubharmonic, $\mathcal{I}(\varphi)$ is a coherent ideal sheaf which is generated over any bounded Stein coordinate open set $U \subset\subset X$ by a Hilbert basis (g_j) of the L^2 space

$$\mathcal{H}_U(\varphi) = \{f \in \mathcal{O}_X(U)\,;\ \int_U |f(z)|^2 e^{-2\varphi(z)} dV(z) < +\infty\}.$$

Similarly, if (L, h) is a singular hermitian line bundle and h is trivialized locally as $h = e^{-2\varphi}$, we simply put $\mathcal{I}(h) = \mathcal{I}(\varphi)$.

(1.6) Definition. *Let L be a pseudo-effective line bundle. If h is a singular hermitian metric such that $i\Theta_{L,h} \geq 0$ and*

$$H^0(X, mL \otimes \mathcal{I}(h^{\otimes m})) \simeq H^0(X, mL) \qquad \textit{for all } m \geq 0,$$

we say that h is an analytic Zariski decomposition of L.

In other words, we require that h has singularities so mild that the vanishing conditions prescribed by the multiplier ideal sheaves $\mathcal{I}(h^{\otimes m})$ do not kill any sections of L and its multiples.

(1.7) Exercise. A special case is when there is an isomorphism $pL = A+E$ where A and E are effective divisors such that $H^0(X, mpL) = H^0(X, mA)$ for all m and $\mathcal{O}(A)$ is generated by sections. Then A possesses a smooth hermitian metric h_A, and this metric defines a singular hermitian metric h on L with poles $\frac{1}{p}E$ and curvature $\frac{1}{p}\Theta_{A,h_A} + \frac{1}{p}[E]$. Show that this metric h is an analytic Zariski decomposition.
Note: when X projective and there is a decomposition $pL = A + E$ with A nef (see (1.9) below), E effective and $H^0(X, mpL) = H^0(X, mA)$ for all m, one says that this is an *algebraic Zariski decomposition* of L. It can be shown that Zariski decompositions exist in dimension 2, but in higher dimension one can see that they do not exist.

(1.8) Theorem. *The metric $h_{\min}$ with minimal singularities provides an analytic Zariski decomposition.*

It follows that an analytic Zariski decomposition always exists (while algebraic decompositions do not exist in general, especially in dimension 3 and more.)

Proof. Let $\sigma \in H^0(X, mL)$ be any section. Then we get a singular metric h on L by putting $|\xi|_h = |\xi/\sigma(x)^{1/m}|$ for $\xi \in L_x$, and it is clear that $|\sigma|_{h^m} = 1$ for this metric. Hence $\sigma \in H^0(X, mL \otimes \mathcal{I}(h^{\otimes m}))$, and a fortiori $\sigma \in H^0(X, mL \otimes \mathcal{I}(h_{\min}^{\otimes m}))$ since $h_{\min}$ is less singular than h. □

1.B. Nef line bundles

Many problems of algebraic geometry (e.g. problems of classification of algebraic surfaces or higher dimensional varieties) lead in a natural way to the study of line bundles satisfying semipositivity conditions. It turns out that semipositivity in the sense of curvature (at least, as far as smooth metrics are considered) is not a very satisfactory notion. A more flexible notion perfectly suitable for algebraic purposes is the notion of *numerical effectivity*. The goal of this section is to give a few fundamental algebraic definitions and to discuss their differential geometric counterparts. We first suppose that X is a projective algebraic manifold, $\dim X = n$.

(1.9) Definition. *A holomorphic line bundle L over a projective manifold X is said to be numerically effective, nef for short, if $L \cdot C = \int_C c_1(L) \geq 0$ for every curve $C \subset X$.*

If L is nef, it can be shown that $L^p \cdot Y = \int_Y c_1(L)^p \geq 0$ for any p-dimensional subvariety $Y \subset X$ (see e.g. [Har70]). In relation to this, let us recall the Nakai-Moishezon ampleness criterion: a line bundle L is ample if and only if $L^p \cdot Y > 0$ for every p-dimensional subvariety Y. From this, we easily infer

(1.10) Proposition. *Let L be a line bundle on a projective algebraic manifold X, on which an ample line bundle A and a hermitian metric ω are given. The following properties are equivalent:*

(a) *L is nef;*

(b) *for any integer $k \geq 1$, the line bundle $kL + A$ is ample;*

(c) *for every $\varepsilon > 0$, there is a smooth metric h_ε on L such that $\mathrm{i}\Theta_{L,h_\varepsilon} \geq -\varepsilon\omega$.*

Proof. (a) $\Rightarrow$ (b). If L is nef and A is ample then clearly $kL + A$ satisfies the Nakai-Moishezon criterion, hence $kL + A$ is ample.

(b) $\Rightarrow$ (c). Condition (c) is independent of the choice of the hermitian metric, so we may select a metric h_A on A with positive curvature and set $\omega = \mathrm{i}\Theta_{A,h_A}$. If $kL + A$ is ample, this bundle has a metric h_{kL+A} of positive curvature. Then the metric $h_L = (h_{kL+A} \otimes h_A^{-1})^{1/k}$ has curvature

$$\mathrm{i}\Theta_{L,h_L} = \frac{1}{k}\big(\mathrm{i}\Theta(kL + A) - \mathrm{i}\Theta_A\big) \geq -\frac{1}{k}\mathrm{i}\Theta_{A,h_A}\,;$$

in this way the negative part can be made smaller than $\varepsilon\,\omega$ by taking k large enough.

(c) $\Rightarrow$ (a). Under hypothesis (c), we get $L \cdot C = \int_C \frac{\mathrm{i}}{2\pi}\Theta_{L,h_\varepsilon} \geq -\frac{\varepsilon}{2\pi}\int_C \omega$ for every curve C and every $\varepsilon > 0$, hence $L \cdot C \geq 0$ and L is nef. □

Let now X be an arbitrary compact complex manifold. Since there need not exist any curve in X, Property 1.10 c) is simply taken as a definition of nefness ([DPS94]):

(1.11) Definition. *A line bundle L on a compact complex manifold X is said to be nef if for every $\varepsilon > 0$, there is a smooth hermitian metric h_ε on L such that $\mathrm{i}\Theta_{L,h_\varepsilon} \geq -\varepsilon\omega$.*

In general, it is not possible to extract a smooth limit h_0 such that $\mathrm{i}\Theta_{L,h_0} \geq 0$. The following simple example is given in [DPS94] (Example 1.7). Let E be a non trivial extension $0 \to \mathcal{O} \to E \to \mathcal{O} \to 0$ over an elliptic curve C and let $X = \mathbb{P}(E)$ (with notation as in (4.12)) be the corresponding ruled surface over C. Then $L = \mathcal{O}_{\mathbb{P}(E)}(1)$ is nef but does not admit any smooth metric of nonnegative curvature. This example answers negatively a question raised by Fujita [Fuj83].

Let us now introduce the important concept of *Kodaira-Iitaka dimension* of a line bundle.

(1.12) Definition. *If L is a line bundle, the Kodaira-Iitaka dimension $\kappa(L)$ is the supremum of the rank of the canonical maps*

$$\Phi_m : X \smallsetminus B_m \longrightarrow \mathbb{P}(V_m),\ x \longmapsto H_x = \{\sigma \in V_m\,;\, \sigma(x) = 0\},\ m \geq 1$$

with $V_m = H^0(X, mL)$ *and* $B_m = \bigcap_{\sigma \in V_m} \sigma^{-1}(0) =$ *base locus of* V_m. *In case* $V_m = \{0\}$ *for all* $m \geq 1$, *we set* $\kappa(L) = -\infty$.

A line bundle is said to be big *if* $\kappa(L) = \dim X$.

The following lemma is well-known (the proof is a rather elementary consequence of the Schwarz lemma).

(1.13) Serre-Siegel lemma ([Ser54], [Sie55]). *Let* L *be any line bundle on a compact complex manifold. Then we have*

$$h^0(X, mL) \leq O(m^{\kappa(L)}) \qquad \text{for } m \geq 1,$$

and $\kappa(L)$ *is the smallest constant for which this estimate holds.* □

1.C. Description of the positive cones

Let us recall that an integral cohomology class in $H^2(X, \mathbb{Z})$ is the first Chern class of a holomorphic (or algebraic) line bundle if and only if it lies in the *Neron-Severi* group

$$\text{(1.14)} \qquad \mathrm{NS}(X) = \mathrm{Ker}\left(H^2(X, \mathbb{Z}) \to H^2(X, \mathcal{O}_X)\right)$$

(this fact is just an elementary consequence of the exponential exact sequence $0 \to \mathbb{Z} \to \mathcal{O} \to \mathcal{O}^\star \to 0$). If X is compact Kähler, as we will suppose from now on in this section, this is the same as saying that the class is of type $(1,1)$ with respect to Hodge decomposition.

Let us consider the real vector space $\mathrm{NS}_{\mathbb{R}}(X) = \mathrm{NS}(X) \otimes_{\mathbb{Z}} \mathbb{R}$, which can be viewed as a subspace of the space $H^{1,1}(X, \mathbb{R})$ of real $(1,1)$ cohomology classes. Its dimension is by definition the Picard number

$$\text{(1.15)} \qquad \rho(X) = \mathrm{rank}_{\mathbb{Z}} \mathrm{NS}(X) = \dim_{\mathbb{R}} \mathrm{NS}_{\mathbb{R}}(X).$$

We thus have $0 \leq \rho(X) \leq h^{1,1}(X)$, and the example of complex tori shows that all intermediate values can occur when $n = \dim X \geq 2$.

The positivity concepts for line bundles considered in sections 1.A and 1.B possess in fact natural generalizations to $(1,1)$ classes which are not necessarily integral or rational – and this works at least in the category of compact Kähler manifolds (in fact, by using Bott-Chern cohomology, one could even extend these concepts to arbitrary compact complex manifolds).

(1.16) Definition. *Let* (X, ω) *be a compact Kähler manifold.*

(i) *The Kähler cone is the set* $\mathcal{K} \subset H^{1,1}(X, \mathbb{R})$ *of cohomology classes* $\{\omega\}$ *of Kähler forms. This is an* open convex cone.

(ii) *The closure* $\overline{\mathcal{K}}$ *of the Kähler cone consists of classes* $\{\alpha\} \in H^{1,1}(X, \mathbb{R})$ *such that for every* $\varepsilon > 0$ *the sum* $\{\alpha + \varepsilon\omega\}$ *is Kähler, or equivalently, for every* $\varepsilon > 0$, *there exists a smooth function* φ_ε *on* X *such that* $\alpha + i\partial\overline{\partial}\varphi_\varepsilon \geq -\varepsilon\omega$. *We say that* $\overline{\mathcal{K}}$ *is the cone of* nef $(1,1)$*-classes.*

(iii) *The pseudo-effective cone is the set* $\mathcal{E} \subset H^{1,1}(X, \mathbb{R})$ *of cohomology classes* $\{T\}$ *of closed positive currents of type* $(1,1)$. *This is a* closed convex cone.

(iv) *The interior* $\mathcal{E}^\circ$ *of* $\mathcal{E}$ *consists of classes which still contain a closed positive current after one subtracts* $\varepsilon\{\omega\}$ *for* $\varepsilon > 0$ *small, in other words, they are classes of closed* $(1,1)$*-currents* T *such that* $T \geq \varepsilon\omega$. *Such a current will be called a* Kähler current, *and we say that* $\{T\} \in H^{1,1}(X, \mathbb{R})$ *is a* big $(1,1)$-class.

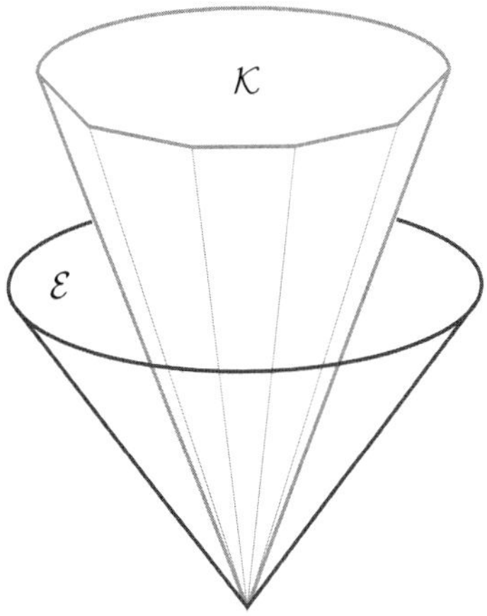

$\mathcal{K}$ = Kähler cone in $H^{1,1}(X,\mathbb{R})$ [open]

$\overline{\mathcal{K}}$ = nef cone in $H^{1,1}(X,\mathbb{R})$ [closure of $\mathcal{K}$]

$\mathcal{E}$ = pseudo-effective cone in $H^{1,1}(X,\mathbb{R})$ [closed]

$\mathcal{E}^\circ$ = big cone in $H^{1,1}(X,\mathbb{R})$ [interior of $\mathcal{E}$]

The openness of $\mathcal{K}$ is clear by definition, and the closedness of $\mathcal{E}$ is a consequence of the fact that bounded sets of currents are weakly compact (as follows from the similar weak compacteness property for bounded sets of positive measures). It is then clear that $\overline{\mathcal{K}} \subset \mathcal{E}$.

In spite of the fact that cohomology groups can be defined either in terms of forms or currents, it turns out that the cones $\overline{\mathcal{K}}$ and $\mathcal{E}$ are in general different. To see this, it is enough to observe that a Kähler class $\{\alpha\}$ satisfies $\int_Y \alpha^p > 0$ for every p-dimensional analytic set. On the other hand, if X is the surface obtained by blowing-up $\mathbb{P}^2$ in one point, then the exceptional divisopr $E \simeq \mathbb{P}^1$ has a cohomology class $\{\alpha\}$ such that $\int_E \alpha = E^2 = -1$, hence $\{\alpha\} \notin \overline{\mathcal{K}}$, although $\{\alpha\} = \{[E]\} \in \mathcal{E}$.

In case X is projective, all Chern classes $c_1(L)$ of line bundles lie by definition in $\mathrm{NS}(X)$, and likewise, all classes of real divisors $D = \sum c_j D_j$, $c_j \in \mathbb{R}$, lie in $\mathrm{NS}_\mathbb{R}(X)$. In order to deal with such *algebraic classes*, we therefore introduce the intersections

$$\mathcal{K}_{\mathrm{NS}} = \mathcal{K} \cap \mathrm{NS}_\mathbb{R}(X), \qquad \mathcal{E}_{\mathrm{NS}} = \mathcal{E} \cap \mathrm{NS}_\mathbb{R}(X),$$

and refer to classes of $H^{1,1}(X,\mathbb{R})$ not contained in $\mathrm{NS}_\mathbb{R}(X)$ as *transcendental classes.*

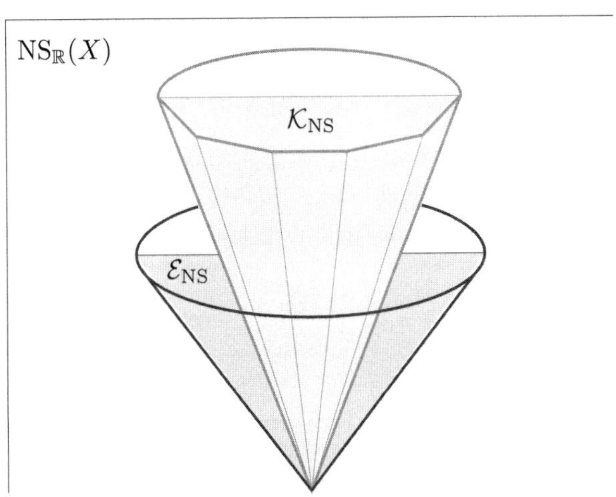

A very important fact is that all four cones $\mathcal{K}_{\mathrm{NS}}$, $\mathcal{E}_{\mathrm{NS}}$, $\overline{\mathcal{K}}_{\mathrm{NS}}$, $\mathcal{E}^{\circ}_{\mathrm{NS}}$ have simple algebraic interpretations.

(1.17) Theorem. *Let X be a projective manifold. Then*

(i) *$\mathcal{K}_{\mathrm{NS}}$ is the open cone generated by classes of* ample *(or very ample) divisors A (Recall that a divisor A is said to be very ample if the linear system $H^0(X,\mathcal{O}(A))$ provides an embedding of X in projective space).*

(ii) *The interior $\mathcal{E}^{\circ}_{\mathrm{NS}}$ is the cone generated by classes of* big *divisors, namely divisors D such that $h^0(X,\mathcal{O}(kD)) \geq c\,k^{\dim X}$ for k large.*

(iii) *$\mathcal{E}_{\mathrm{NS}}$ is the closure of the cone generated by classes of* effective *divisors, i.e. divisors $D = \sum c_j D_j$, $c_j \in \mathbb{R}_+$.*

(iv) *The closed cone $\overline{\mathcal{K}}_{\mathrm{NS}}$ consists of the closure of the cone generated by* nef *divisors D (or nef line bundles L), namely effective integral divisors D such that $D\cdot C \geq 0$ for every curve C.*

In other words, the terminology "nef", "big", "pseudo-effective" used for classes of the full transcendental cones appear to be a natural extrapolation of the algebraic case.

Proof. (i) is just Kodaira's embedding theorem, and properties (iii) and (iv) are obtained easily by passing to the closure of the open cones. We will thus give details of the proof only for (ii) which is possibly not as well known.

By looking at points of $\mathrm{NS}_{\mathbb{Q}}(X) = \mathrm{NS}(X)\otimes_{\mathbb{Z}}\mathbb{Q}$ and multiplying by a denominator, it is enough to check that a line bundle L such that $c_1(L) \in \mathcal{E}^{\circ}$ is big. However, this means that L possesses a singular hermitian metric h_L such that $\Theta_{L,h_L} \geq \varepsilon\omega$ for some Kähler metric ω. For some integer $p_0 > 0$, we can then produce a singular hermitian metric with positive curvature and with a given logarithmic pole $h_L^{p_0} e^{-\theta(z)\log|z-x_0|^2}$ in a neighborhood of every point $x_0 \in X$ (here θ is a smooth cut-off function supported on a neighborhood of x_0). Then Hörmander's L^2 existence theorem [Hör65] can be used to produce sections of L^k which generate all jets of order $(k/p_0) - n$ at points x_0, so that L is big.

Conversely, if L is big and A is a (smooth) very ample divisor, the exact sequence $0 \to \mathcal{O}_X(kL-A) \to \mathcal{O}_X(kL) \to \mathcal{O}_A(kL_{\restriction A}) \to 0$ and the estimates $h^0(X,\mathcal{O}_X(kL)) \geq ck^n$, $h^0(A,\mathcal{O}_A(kL_{\restriction A})) = O(k^{n-1})$ imply that $\mathcal{O}_X(kL-A)$ has a section for k large, thus $kL - A \equiv D$ for some effective divisor D. This means that there exists a singular metric h_L on L such that

$$\frac{\mathrm{i}}{2\pi}\Theta_{L,h_L} = \frac{1}{k}\Big(\frac{\mathrm{i}}{2\pi}\Theta_{A,h_A} + [D]\Big) \geq \frac{1}{k}\omega$$

where $\omega = \frac{\mathrm{i}}{2\pi}\Theta_{A,h_A}$, hence $c_1(L) \in \mathcal{E}^{\circ}$. □

Before going further, we need a lemma.

(1.18) Lemma. *Let X be a compact Kähler n-dimensional manifold, let L be a nef line bundle on X, and let E be an arbitrary holomorphic vector bundle. Then $h^q(X,\mathcal{O}(E)\otimes\mathcal{O}(kL)) = o(k^n)$ as $k \to +\infty$, for every $q \geq 1$. If X is projective algebraic, the following more precise bound holds:*

$$h^q(X,\mathcal{O}(E)\otimes\mathcal{O}(kL)) = O(k^{n-q}), \qquad \forall q \geq 0.$$

Proof. The Kähler case will be proved in Section 12, as a consequence of the holomorphic Morse inequalities. In the projective algebraic case, we proceed by induction on $n = \dim X$. If $n = 1$ the result is clear, as well as if $q = 0$. Now let A be a nonsingular ample divisor such that $E \otimes \mathcal{O}(A - K_X)$ is Nakano positive. Then the Nakano vanishing theorem applied to the vector bundle $F = E \otimes \mathcal{O}(kL + A - K_X)$ shows that $H^q(X, \mathcal{O}(E) \otimes \mathcal{O}(kL + A)) = 0$ for all $q \geq 1$. The exact sequence

$$0 \to \mathcal{O}(kL) \to \mathcal{O}(kL + A) \to \mathcal{O}(kL + A)_{\restriction A} \to 0$$

twisted by E implies

$$H^q(X, \mathcal{O}(E) \otimes \mathcal{O}(kL)) \simeq H^{q-1}(A, \mathcal{O}(E_{\restriction A} \otimes \mathcal{O}(kL + A)_{\restriction A}),$$

and we easily conclude by induction since $\dim A = n-1$. Observe that the argument does not work any more if X is not algebraic. It seems to be unknown whether the $O(k^{n-q})$ bound still holds in that case. □

(1.19) Corollary. *If L is nef, then L is big (i.e. $\kappa(L) = n$) if and only if $L^n > 0$. Moreover, if L is nef and big, then for every $\delta > 0$, L has a singular metric $h = e^{-2\varphi}$ such that $\max_{x \in X} \nu(\varphi, x) \leq \delta$ and $\mathrm{i}\Theta_{L,h} \geq \varepsilon\,\omega$ for some $\varepsilon > 0$. The metric h can be chosen to be smooth on the complement of a fixed divisor D, with logarithmic poles along D.*

Proof. By Lemma 1.18 and the Riemann-Roch formula, we have $h^0(X, kL) = \chi(X, kL) + o(k^n) = k^n L^n / n! + o(k^n)$, whence the first statement. By the proof of (1.17 ii), there exists a singular metric h_1 on L such that

$$\frac{\mathrm{i}}{2\pi}\Theta_{L,h_1} = \frac{1}{k}\Big(\frac{\mathrm{i}}{2\pi}\Theta_{A,h_A} + [D]\Big) \geq \frac{1}{k}\omega, \qquad \omega = \frac{\mathrm{i}}{2\pi}\Theta_{A,h_A}.$$

Now, for every $\varepsilon > 0$, there is a smooth metric h_ε on L such that $\frac{\mathrm{i}}{2\pi}\Theta_{L,h_\varepsilon} \geq -\varepsilon\omega$. The convex combination of metrics $h'_\varepsilon = h_1^{k\varepsilon} h_\varepsilon^{1-k\varepsilon}$ is a singular metric with poles along D which satisfies

$$\frac{\mathrm{i}}{2\pi}\Theta_{L,h'_\varepsilon} \geq \varepsilon(\omega + [D]) - (1 - k\varepsilon)\varepsilon\omega \geq k\varepsilon^2\omega.$$

Its Lelong numbers are $\varepsilon\nu(D, x)$ and they can be made smaller than δ by choosing $\varepsilon > 0$ small. □

We still need a few elementary facts about the numerical dimension of nef line bundles.

(1.20) Definition. *Let L be a nef line bundle on a compact Kähler manifold X. One defines the numerical dimension of L to be*

$$\mathrm{nd}(L) = \max\big\{k = 0, \ldots, n\,;\ c_1(L)^k \neq 0 \text{ in } H^{2k}(X, \mathbb{R})\big\}.$$

By Corollary 1.19, we have $\kappa(L) = n$ if and only if $\mathrm{nd}(L) = n$. In general, we merely have an inequality.

(1.21) Proposition. *If L is a nef line bundle on a compact Kähler manifold, then $\kappa(L) \leq \mathrm{nd}(L)$.*

Proof. By induction on $n = \dim X$. If $\mathrm{nd}(L) = n$ or $\kappa(L) = n$ the result is true, so we may assume $r := \kappa(L) \leq n - 1$ and $k := \mathrm{nd}(L) \leq n - 1$. Fix $m > 0$ so that

$\Phi = \Phi_{|mL|}$ has generic rank r. Select a nonsingular ample divisor A in X such that the restriction of $\Phi_{|mL|}$ to A still has rank r (for this, just take A passing through a point $x \notin B_{|mL|}$ at which $\mathrm{rank}(d\Phi_x) = r < n$, in such a way that the tangent linear map $d\Phi_{x\restriction T_{A,x}}$ still has rank r). Then $\kappa(L_{\restriction A}) \geq r = \kappa(L)$ (we just have an equality because there might exist sections in $H^0(A, mL_{\restriction A})$ which do not extend to X). On the other hand, we claim that $\mathrm{nd}(L_{\restriction A}) = k = \mathrm{nd}(L)$. The inequality $\mathrm{nd}(L_{\restriction A}) \geq \mathrm{nd}(L)$ is clear. Conversely, if we set $\omega = \frac{\mathrm{i}}{2\pi}\Theta_{A,h_A} > 0$, the cohomology class $c_1(L)^k$ can be represented by a closed positive current of bidegree (k,k)

$$T = \lim_{\varepsilon \to 0} \left(\frac{\mathrm{i}}{2\pi}\Theta_{L,h_\varepsilon} + \varepsilon\omega\right)^k$$

after passing to some subsequence (there is a uniform bound for the mass thanks to the Kähler assumption, taking wedge products with ω^{n-k}). The current T must be non zero since $c_1(L)^k \neq 0$ by definition of $k = \mathrm{nd}(L)$. Then $\{[A]\} = \{\omega\}$ as cohomology classes, and

$$\int_A c_1(L_{\restriction A})^k \wedge \omega^{n-1-k} = \int_X c_1(L)^k \wedge [A] \wedge \omega^{n-1-k} = \int_X T \wedge \omega^{n-k} > 0.$$

This implies $\mathrm{nd}(L_{\restriction A}) \geq k$, as desired. The induction hypothesis with X replaced by A yields

$$\kappa(L) \leq \kappa(L_{\restriction A}) \leq \mathrm{nd}(L_{\restriction A}) \leq \mathrm{nd}(L). \qquad \square$$

(1.22) Remark. It may happen that $\kappa(L) < \mathrm{nd}(L)$: take e.g.

$$L \to X = X_1 \times X_2$$

equal to the total tensor product of an ample line bundle L_1 on a projective manifold X_1 and of a unitary flat line bundle L_2 on an elliptic curve X_2 given by a representation $\pi_1(X_2) \to U(1)$ such that no multiple kL_2 with $k \neq 0$ is trivial. Then $H^0(X, kL) = H^0(X_1, kL_1) \otimes H^0(X_2, kL_2) = 0$ for $k > 0$, and thus $\kappa(L) = -\infty$. However $c_1(L) = \mathrm{pr}_1^\star c_1(L_1)$ has numerical dimension equal to $\dim X_1$. The same example shows that the Kodaira dimension may increase by restriction to a subvariety (if $Y = X_1 \times \{\mathrm{point}\}$, then $\kappa(L_{\restriction Y}) = \dim Y$). $\square$

1.D. The Kawamata-Viehweg vanishing theorem

We derive here an algebraic version of the Nadel vanishing theorem in the context of nef line bundles. This algebraic vanishing theorem has been obtained independently by Kawamata [Kaw82] and Viehweg [Vie82], who both reduced it to the Akizuki-Kodaira-Nakano vanishing theorem [AN54] by cyclic covering constructions. Since then, a number of other proofs have been given, one based on connections with logarithmic singularities [EV86], another on Hodge theory for twisted coefficient systems [Kol85], a third one on the Bochner technique [Dem89] (see also [EV92] for a general survey). Since the result is best expressed in terms of multiplier ideal sheaves (avoiding then any unnecessary desingularization in the statement), we feel that the direct approach via Nadel's vanishing theorem is extremely natural.

If $D = \sum \alpha_j D_j \geq 0$ is an effective $\mathbb{Q}$-divisor, we define the *multiplier ideal sheaf* $\mathcal{I}(D)$ to be equal to $\mathcal{I}(\varphi)$ where $\varphi = \sum \alpha_j |g_j|$ is the corresponding psh function defined by generators g_j of $\mathcal{O}(-D_j)$. If D is a divisor with normal crossings, it is an easy exercise to show that

$$\mathcal{I}(D) = \mathcal{O}(-\lfloor D \rfloor), \qquad \text{where} \quad \lfloor D \rfloor = \sum \lfloor \alpha_j \rfloor D_j \tag{1.23}$$

is the integer part of D. In general, the computation of $\mathcal{I}(D)$ can be made algebraically thanks to Hironaka [Hir64], by using desingularizations $\mu : \widetilde{X} \to X$ such that $\mu^\star D$ becomes a divisor with normal crossings on $\widetilde{X}$. In fact, if $\mu : \widetilde{X} \to X$ is a proper modification and φ is a plurisubharmonic function on X, one can easily prove the direct image formula

$$\mathcal{O}_X(K_X) \otimes \mathcal{I}(\varphi) = \mu_*\big(\mathcal{O}_{\widetilde{X}}(K_{\widetilde{X}}) \otimes \mathcal{I}(\varphi \circ \mu)\big) \tag{1.24}$$

$$\mathcal{I}(\varphi) = \mu_*\big(\mathcal{O}_{\widetilde{X}}(K_{\widetilde{X}/X}) \otimes \mathcal{I}(\varphi \circ \mu)\big) \tag{1.24'}$$

in terms of the relative canonical sheaf $K_{\widetilde{X}/X} = K_{\widetilde{X}} \otimes \mu^*(K_X^{-1})$; the proof is immediate from the L^2 definition.

(1.25) Kawamata-Viehweg vanishing theorem. *Let X be a projective algebraic manifold and let F be a line bundle over X such that some positive multiple mF can be written $mF = L + D$ where L is a nef line bundle and D an effective divisor. Then*

$$H^q\big(X, \mathcal{O}(K_X + F) \otimes \mathcal{I}(m^{-1}D)\big) = 0 \text{ for } q > n - \mathrm{nd}(L).$$

(1.26) Special case. *If F is a nef line bundle, then*

$$H^q\big(X, \mathcal{O}(K_X + F)\big) = 0 \text{ for } q > n - \mathrm{nd}(F).$$

Proof of Theorem 1.25. First suppose that $\mathrm{nd}(L) = n$, i.e. that L is big. By the proof of 1.16 ii), there is a singular hermitian metric h_0 on L such that the corresponding weight φ_0 has algebraic singularities and

$$\mathrm{i}\Theta_{L,h_0} = 2\mathrm{i}d'd''\varphi_0 \geq \varepsilon_0\omega$$

for some $\varepsilon_0 > 0$. On the other hand, since L is nef, there are metrics given by weights φ_ε such that $\frac{\mathrm{i}}{2\pi}\Theta_{L,h_\varepsilon} \geq -\varepsilon\omega$ for every $\varepsilon > 0$, ω being a Kähler metric. Let $\varphi_D = \sum \alpha_j \log |g_j|$ be the weight of the singular metric on $\mathcal{O}(D)$. We define a singular metric on F by

$$\varphi_F = \frac{1}{m}\big((1-\delta)\varphi_{L,\varepsilon} + \delta\varphi_{L,0} + \varphi_D\big)$$

with $\varepsilon \ll \delta \ll 1$, δ rational. Then φ_F has algebraic singularities, and by taking δ small enough we find $\mathcal{I}(\varphi_F) = \mathcal{I}(\frac{1}{m}\varphi_D) = \mathcal{I}(\frac{1}{m}D)$. In fact, $\mathcal{I}(\varphi_F)$ can be computed by taking integer parts of certain $\mathbb{Q}$-divisors, and adding $\delta\varphi_{L,0}$ does not change the integer part of the rational numbers involved when δ is small. Now

$$\begin{aligned} dd^c\varphi_F &= \frac{1}{m}\big((1-\delta)dd^c\varphi_{L,\varepsilon} + \delta dd^c\varphi_{L,0} + dd^c\varphi_D\big) \\ &\geq \frac{1}{m}\big(-(1-\delta)\varepsilon\omega + \delta\varepsilon_0\omega + [D] \geq \frac{\delta\varepsilon}{m}\omega, \end{aligned}$$

if we choose $\varepsilon \leq \delta\varepsilon_0$. Nadel's theorem [Var09, sect. 8.4.2] thus implies the desired vanishing result for all $q \geq 1$.

Now, if $\mathrm{nd}(L) < n$, we use hyperplane sections and argue by induction on $n = \dim X$. Since the sheaf $\mathcal{O}(K_X) \otimes \mathcal{I}(m^{-1}D)$ behaves functorially with respect to modifications (and since the L^2 cohomology complex is "the same" upstairs and downstairs), we may assume after blowing-up that D is a divisor with normal crossings. Then the multiplier ideal sheaf $\mathcal{I}(m^{-1}D) = \mathcal{O}(-\lfloor m^{-1}D \rfloor)$ is locally free. By Serre duality, the expected vanishing is equivalent to

$$H^q(X, \mathcal{O}(-F) \otimes \mathcal{O}(\lfloor m^{-1}D \rfloor)) = 0 \qquad \text{for } q < \mathrm{nd}(L).$$

Select a nonsingular ample divisor A such that A meets all components D_j transversally, and take A positive enough so that $\mathcal{O}(A + F - \lfloor m^{-1}D \rfloor)$ is ample. Then $H^q(X, \mathcal{O}(-A-F) \otimes \mathcal{O}(\lfloor m^{-1}D \rfloor)) = 0$ for $q < n$ by Kodaira vanishing, and the exact sequence $0 \to \mathcal{O}_X(-A) \to \mathcal{O}_X \to (i_A)_\star \mathcal{O}_A \to 0$ twisted by $\mathcal{O}(-F) \otimes \mathcal{O}(\lfloor m^{-1}D \rfloor)$ yields an isomorphism

$$H^q(X, \mathcal{O}(-F) \otimes \mathcal{O}(\lfloor m^{-1}D \rfloor)) \simeq H^q(A, \mathcal{O}(-F_{\restriction A}) \otimes \mathcal{O}(\lfloor m^{-1}D_{\restriction A} \rfloor)).$$

The proof of 1.21 showed that $\mathrm{nd}(L_{\restriction A}) = \mathrm{nd}(L)$, hence the induction hypothesis implies that the cohomology group on A on the right hand side is zero for $q < \mathrm{nd}(L)$. □

1.E. A uniform global generation property due to Y.T. Siu

Let X be a projective manifold, and (L, h) a pseudo-effective line bundle. The "uniform global generation property" states in some sense that the tensor product sheaf $L \otimes \mathcal{I}(h)$ has a uniform positivity property, for any singular hermitian metric h with nonnegative curvature on L.

(1.27) Theorem Y.T. Siu, ([Siu98]). *Let X be a projective manifold. There exists an ample line bundle G on X such that for every pseudo-effective line bundle (L, h), the sheaf $\mathcal{O}(G+L) \otimes \mathcal{I}(h)$ is generated by its global sections. In fact, G can be chosen as follows: pick any very ample line bundle A, and take G such that $G - (K_X + nA)$ is ample, e.g. $G = K_X + (n+1)A$.*

Proof. Let φ be the weight of the metric h on a small neighborhood of a point $z_0 \in X$. Assume that we have a local section u of $\mathcal{O}(G+L) \otimes \mathcal{I}(h)$ on a coordinate open ball $B = B(z_0, \delta)$, such that

$$\int_B |u(z)|^2 e^{-2\varphi(z)} |z - z_0|^{-2(n+\varepsilon)} dV(z) < +\infty.$$

Then Skoda's division theorem [Sko72b] (see also Corollary 8.21 below) implies $u(z) = \sum (z_j - z_{j,0}) v_j(z)$ with

$$\int_B |v_j(z)|^2 e^{-2\varphi(z)} |z - z_0|^{-2(n-1+\varepsilon)} dV(z) < +\infty,$$

in particular $u_{z_0} \in \mathcal{O}(G+L) \otimes \mathcal{I}(h) \otimes \mathfrak{m}_{X,z_0}$. Select a very ample line bundle A on X. We take a basis $\sigma = (\sigma_j)$ of sections of $H^0(X, G \otimes \mathfrak{m}_{X,z_0})$ and multiply the metric h of G by the factor $|\sigma|^{-2(n+\varepsilon)}$. The weight of the above metric has singularity $(n+\varepsilon)\log|z - z_0|^2$ at z_0, and its curvature is

$$i\Theta_G + (n+\varepsilon) i\partial\overline{\partial} \log |\sigma|^2 \geq i\Theta_G - (n+\varepsilon)\Theta_A. \tag{1.28}$$

Now, let f be a local section in $H^0(B, \mathcal{O}(G+L) \otimes \mathcal{I}(h))$ on $B = B(z_0, \delta)$, δ small. We solve the global $\overline{\partial}$ equation

$$\overline{\partial} u = \overline{\partial}(\theta f) \qquad \text{on } X$$

with a cut-off function θ supported near z_0 and with the weight associated with our above choice of metric on $G + L$. Thanks to Nadel's Theorem, the solution exists if the metric of $G + L - K_X$ has positive curvature. As $i\Theta_{L,h} \geq 0$ in the sense of currents, (1.28) shows that a sufficient condition is $G - K_X - nA > 0$ (provided that ε is small enough). We then find a smooth solution u such that $u_{z_0} \in \mathcal{O}(G + L) \otimes \mathcal{I}(h) \otimes \mathfrak{m}_{X,z_0}$, hence

$$F := \theta f - u \in H^0(X, \mathcal{O}(G + L) \otimes \mathcal{I}(h))$$

is a global section differing from f by a germ in $\mathcal{O}(G+L)\otimes\mathcal{I}(h)\otimes\mathfrak{m}_{X,z_0}$. Nakayama's lemma implies that $H^0(X, \mathcal{O}(G+L)\otimes\mathcal{I}(h))$ generates the stalks of $\mathcal{O}(G+L)\otimes\mathcal{I}(h)$.

1.F. Hard Lefschetz theorem with multiplier ideal sheaves

We state here without proof a basic surjectivity theorem, first appeared in [DPS00], which extends the classical "Hard Lefschetz Theorem".

(1.29) Theorem. *Let (L, h) be a pseudo-effective line bundle on a compact Kähler manifold (X, ω), of dimension n, let $\Theta_{L,h} \geq 0$ be its curvature current and $\mathcal{I}(h)$ the associated multiplier ideal sheaf. Then, the wedge multiplication operator $\omega^q \wedge \bullet$ induces a surjective morphism*

$$\Phi^q_{\omega,h} : H^0(X, \Omega^{n-q}_X \otimes L \otimes \mathcal{I}(h)) \longrightarrow H^q(X, \Omega^n_X \otimes L \otimes \mathcal{I}(h)).$$

The special case when L is nef is due to Takegoshi [Tak97]. An even more special case is when L is semi-positive, i.e. possesses a smooth metric with semi-positive curvature. In that case the multiple ideal sheaf $\mathcal{I}(h)$ coincides with $\mathcal{O}_X$ and we get the following consequence already observed by Mourougane [Mou99].

(1.30) Corollary. *Let (L, h) be a semi-positive line bundle on a compact Kähler manifold (X, ω) of dimension n. Then, the wedge multiplication operator $\omega^q \wedge \bullet$ induces a surjective morphism*

$$\Phi^q_\omega : H^0(X, \Omega^{n-q}_X \otimes L) \longrightarrow H^q(X, \Omega^n_X \otimes L).$$

It should be observed that although all objects involved in Theorem (1.29) are algebraic when X is a projective manifold, there are no known algebraic proof of the statement; it is not even clear how to define algebraically $\mathcal{I}(h)$ for the case when $h = h_{min}$ is a metric with minimal singularities. However, even in the special circumstance when L is nef, the multiplier ideal sheaf is crucially needed.

In fact, let B be an elliptic curve and let V be the rank 2 vector bundle over B which is defined as the (unique) non split extension

$$0 \to \mathcal{O}_B \to V \to \mathcal{O}_B \to 0. \tag{1.31}$$

In particular, the bundle V is numerically flat, i.e. $c_1(V) = 0$, $c_2(V) = 0$. We consider the ruled surface $X = \mathbb{P}(V)$. On that surface there is a unique section $C = \mathbb{P}(\mathcal{O}_B) \subset X$ with $C^2 = 0$ and

$$\mathcal{O}_X(C) = \mathcal{O}_{\mathbb{P}(V)}(1)$$

is thus a nef line bundle. It is easy to see that

$$h^0(X, \mathcal{O}_X(mC)) = h^0(X, \mathcal{O}_{\mathbb{P}(V)}(m)) = h^0(B, S^m V) = 1 \tag{1.32}$$

for all $m \in \mathbb{N}$ (otherwise we would have $mC = aC + M$ where aC is the fixed part of the linear system $|mC|$ and $M \neq 0$ the moving part, thus $M^2 \geq 0$ and $C \cdot M > 0$, contradiction). On the other hand, rather simple calculations show that

$$h^0(X, \Omega^1_X(mC)) = 2 \tag{1.33}$$

for all $k \geq 2$. This shows that the generalized Hard Lefschetz Theorem cannot hold true, even for nef line bundles, when the multiplier ideal sheaf $\mathcal{I}(h)$ is omitted in the statement.

2. Holomorphic Morse inequalities

Let X be a compact Kähler manifold, E a holomorphic vector bundle of rank r and L a line bundle over X. If L is equipped with a smooth metric h of curvature form $\Theta_{L,h}$, we define the q-index set of L to be the open subset

$$X(q, L) = \left\{ x \in X\ ;\ \mathrm{i}\Theta_{L,h}(x) \text{ has } \begin{matrix} q & \text{negative eigenvalues} \\ n-q & \text{positive eigenvalues} \end{matrix} \right\} \tag{2.1}$$

for $0 \leq q \leq n$. Hence X admits a partition $X = \Delta \cup \bigcup_q X(q, L)$ where $\Delta = \{x \in X\,;\, \det(\Theta_{L,h}(x)) = 0\}$ is the degeneracy set. We also introduce

$$X(\leq q, L) = \bigcup_{0 \leq j \leq q} X(j, L). \tag{2.1$'$}$$

(2.2) Morse inequalities ([Dem85]). *For any hermitian holomorphic line bundle $L, h)$ and any holomorphic vector bundle E over a compact complex manifold X, the cohomology groups $H^q\big(X, E \otimes \mathcal{O}(kL)\big)$ satisfy the following asymptotic inequalities as $k \to +\infty$:*

(a) *Weak Morse inequalities*

$$h^q\big(X, E \otimes \mathcal{O}(kL)\big) \leq r\frac{k^n}{n!} \int_{X(q,L)} (-1)^q \Big(\frac{\mathrm{i}}{2\pi}\Theta_{L,h}\Big)^n + o(k^n).$$

(b) *Strong Morse inequalities*

$$\sum_{0 \leq j \leq q} (-1)^{q-j} h^j\big(X, E \otimes \mathcal{O}(kL)\big) \leq r\frac{k^n}{n!} \int_{X(\leq q,L)} (-1)^q \Big(\frac{\mathrm{i}}{2\pi}\Theta_{L,h}\Big)^n + o(k^n).$$

The proof is based on the spectral theory of the complex Laplace operator, using either a localization procedure or, alternatively, a heat kernel technique. These inequalities are a useful complement to the Riemann-Roch formula when information is needed about individual cohomology groups, and not just about the Euler-Poincaré characteristic.

One difficulty in the application of these inequalities is that the curvature integral is in general quite uneasy to compute, since it is neither a topological nor an algebraic invariant. However, the Morse inequalities can be reformulated in a more algebraic setting in which only algebraic invariants are involved. We give here two such reformulations.

(2.3) Theorem. *Let $L = F - G$ be a holomorphic line bundle over a compact Kähler manifold X, where F and G are numerically effective line bundles. Then* *for every* $q = 0, 1, \ldots, n = \dim X$, *there is an asymptotic strong Morse inequality*

$$\sum_{0\le j\le q} (-1)^{q-j} h^j(X, kL) \le \frac{k^n}{n!} \sum_{0\le j\le q} (-1)^{q-j} \binom{n}{j} F^{n-j}\cdot G^j + o(k^n).$$

Proof. By adding ε times a Kähler metric ω to the curvature forms of F and G, $\varepsilon > 0$ one can write $\frac{i}{2\pi}\Theta_L = \theta_{F,\varepsilon} - \theta_{G,\varepsilon}$ where $\theta_{F,\varepsilon} = \frac{i}{2\pi}\Theta_F + \varepsilon\omega$ and $\theta_{G,\varepsilon} = \frac{i}{2\pi}\Theta_G + \varepsilon\omega$ are positive definite. Let $\lambda_1 \ge \ldots \ge \lambda_n > 0$ be the eigenvalues of $\theta_{G,\varepsilon}$ with respect to $\theta_{F,\varepsilon}$. Then the eigenvalues of $\frac{i}{2\pi}\Theta_L$ with respect to $\theta_{F,\varepsilon}$ are the real numbers $1-\lambda_j$ and the set $X(\le q, L)$ is the set $\{\lambda_{q+1} < 1\}$ of points $x \in X$ such that $\lambda_{q+1}(x) < 1$. The strong Morse inequalities yield

$$\sum_{0\le j\le q} (-1)^{q-j} h^j(X, kL) \le \frac{k^n}{n!} \int_{\{\lambda_{q+1}<1\}} (-1)^q \prod_{1\le j\le n} (1-\lambda_j)\theta_{F,\varepsilon}^n + o(k^n).$$

On the other hand we have

$$\binom{n}{j} \theta_{F,\varepsilon}^{n-j} \wedge \theta_{G,\varepsilon}^j = \sigma_n^j(\lambda)\, \theta_{F,\varepsilon}^n,$$

where $\sigma_n^j(\lambda)$ is the j-th elementary symmetric function in $\lambda_1, \ldots, \lambda_n$, hence

$$\sum_{0\le j\le q} (-1)^{q-j} \binom{n}{j} F^{n-j}\cdot G^j = \lim_{\varepsilon\to 0} \int_X \sum_{0\le j\le q} (-1)^{q-j} \sigma_n^j(\lambda)\, \theta_{F,\varepsilon}^n.$$

Thus, to prove the Lemma, we only have to check that

$$\sum_{0\le j\le n} (-1)^{q-j} \sigma_n^j(\lambda) - \mathbf{1}_{\{\lambda_{q+1}<1\}}(-1)^q \prod_{1\le j\le n} (1-\lambda_j) \ge 0$$

for all $\lambda_1 \ge \ldots \ge \lambda_n \ge 0$, where $\mathbf{1}_{\{\ldots\}}$ denotes the characteristic function of a set. This is easily done by induction on n (just split apart the parameter λ_n and write $\sigma_n^j(\lambda) = \sigma_{n-1}^j(\lambda) + \sigma_{n-1}^{j-1}(\lambda)\,\lambda_n$). □

In the case $q = 1$, we get an especially interesting lower bound (this bound has been observed and used by S. Trapani [Tra95] in a similar context).

(2.4) Consequence. $h^0(X, kL) - h^1(X, kL) \ge \frac{k^n}{n!}(F^n - nF^{n-1}\cdot G) - o(k^n)$. *Therefore some multiple* kL *has a section as soon as* $F^n - nF^{n-1}\cdot G > 0$.

(2.5) Remark. The weaker inequality

$$h^0(X, kL) \ge \frac{k^n}{n!}(F^n - nF^{n-1}\cdot G) - o(k^n)$$

is easy to prove if X is projective algebraic. Indeed, by adding a small ample $\mathbb{Q}$-divisor to F and G, we may assume that F, G are ample. Let m_0G be very ample and let k' be the smallest integer $\ge k/m_0$. Then $h^0(X, kL) \ge h^0(X, kF - k'm_0G)$. We select k' smooth members G_j, $1 \le j \le k'$ in the linear system $|m_0G|$ and use the exact sequence

$$0 \to H^0(X, kF - \sum G_j) \to H^0(X, kF) \to \bigoplus H^0(G_j, kF_{|G_j}).$$

Kodaira's vanishing theorem yields $H^q(X,kF)=0$ and $H^q(G_j,kF_{|G_j})=0$ for $q\ge 1$ and $k\ge k_0$. By the exact sequence combined with Riemann-Roch, we get

$$\begin{aligned}h^0(X,kL)&\ge h^0(X,kF-\textstyle\sum G_j)\\&\ge \frac{k^n}{n!}F^n-O(k^{n-1})-\sum\Big(\frac{k^{n-1}}{(n-1)!}F^{n-1}\cdot G_j-O(k^{n-2})\Big)\\&\ge\frac{k^n}{n!}\Big(F^n-n\frac{k'm_0}{k}F^{n-1}\cdot G\Big)-O(k^{n-1})\\&\ge\frac{k^n}{n!}\Big(F^n-n\,F^{n-1}\cdot G\Big)-O(k^{n-1}).\end{aligned}$$

(This simple proof is due to F. Catanese.) □

(2.6) Corollary. *Suppose that F and G are nef and that F is big. Some multiple of $mF-G$ has a section as soon as*

$$m>n\,\frac{F^{n-1}\cdot G}{F^n}.$$

In the last condition, the factor n is sharp: this is easily seen by taking $X=\mathbb{P}_1^n$ and $F=\mathcal{O}(a,\ldots,a)$ and $G=\mathcal{O}(b_1,\ldots,b_n)$ over $\mathbb{P}_1^n$; the condition of the Corollary is then $m>\sum b_j/a$, whereas $k(mF-G)$ has a section if and only if $m\ge\sup b_j/a$; this shows that one cannot replace n by $n(1-\varepsilon)$. □

3. Approximation of closed positive (1,1)-currents by divisors

3.A. Local approximation theorem through Bergman kernels

We prove here, as an application of the Ohsawa-Takegoshi extension theorem, that every psh function on a pseudoconvex open set $\Omega\subset\mathbb{C}^n$ can be approximated very accurately by functions of the form $c\log|f|$, where $c>0$ and f is a holomorphic function. The main idea is taken from [Dem92]. For other applications to algebraic geometry, see [Dem93] and Demailly-Kollár [DK01]. Recall that the Lelong number of a function $\varphi\in\mathrm{Psh}(\Omega)$ at a point x_0 is defined to be

$$\nu(\varphi,x_0)=\liminf_{z\to x_0}\frac{\varphi(z)}{\log|z-x_0|}=\lim_{r\to 0_+}\frac{\sup_{B(x_0,r)}\varphi}{\log r}. \tag{3.1}$$

In particular, if $\varphi=\log|f|$ with $f\in\mathcal{O}(\Omega)$, then $\nu(\varphi,x_0)$ is equal to the vanishing order

$$\mathrm{ord}_{x_0}(f)=\sup\{k\in\mathbb{N}\,;D^\alpha f(x_0)=0,\ \forall|\alpha|<k\}.$$

(3.2) Theorem. *Let φ be a plurisubharmonic function on a bounded pseudoconvex open set $\Omega\subset\mathbb{C}^n$. For every $m>0$, let $\mathcal{H}_\Omega(m\varphi)$ be the Hilbert space of holomorphic functions f on Ω such that $\int_\Omega|f|^2e^{-2m\varphi}d\lambda<+\infty$ and let $\varphi_m=\frac{1}{2m}\log\sum|\sigma_\ell|^2$ where (σ_ℓ) is an orthonormal basis of $\mathcal{H}_\Omega(m\varphi)$. Then there are constants $C_1,C_2>0$ independent of m such that*

(a) $\varphi(z)-\dfrac{C_1}{m}\le\varphi_m(z)\le\sup_{|\zeta-z|<r}\varphi(\zeta)+\dfrac{1}{m}\log\dfrac{C_2}{r^n}$

for every $z\in\Omega$ and $r<d(z,\partial\Omega)$. In particular, φ_m converges to φ pointwise and in L^1_{loc} topology on Ω when $m\to+\infty$ and

(b) $\nu(\varphi,z)-\dfrac{n}{m}\le\nu(\varphi_m,z)\le\nu(\varphi,z)$ *for every $z\in\Omega$.*

Proof. (a) Note that $\sum|\sigma_\ell(z)|^2$ is the square of the norm of the evaluation linear form $\mathrm{ev}_z : f \mapsto f(z)$ on $\mathcal{H}_\Omega(m\varphi)$, since $\sigma_\ell(z) = \mathrm{ev}_z(\sigma_\ell)$ is the ℓ-th coordinate of ev_z in the orthonormal basis (σ_ℓ). In other words, we have

$$\sum|\sigma_\ell(z)|^2 = \sup_{f\in B(1)} |f(z)|^2$$

where $B(1)$ is the unit ball of $\mathcal{H}_\Omega(m\varphi)$ (The sum is called the *Bergman kernel* associated with $\mathcal{H}_\Omega(m\varphi)$.) As φ is locally bounded from above, the L^2 topology is actually stronger than the topology of uniform convergence on compact subsets of Ω. It follows that the series $\sum|\sigma_\ell|^2$ converges uniformly on Ω and that its sum is real analytic. Moreover, by what we just explained, we have

$$\varphi_m(z) = \sup_{f\in B(1)} \frac{1}{m}\log|f(z)|.$$

For $z_0 \in \Omega$ and $r < d(z_0, \partial\Omega)$, the mean value inequality applied to the psh function $|f|^2$ implies

$$\begin{aligned}|f(z_0)|^2 &\leq \frac{1}{\pi^n r^{2n}/n!}\int_{|z-z-0|<r} |f(z)|^2 d\lambda(z)\\ &\leq \frac{1}{\pi^n r^{2n}/n!}\exp\Big(2m\sup_{|z-z_0|<r}\varphi(z)\Big)\int_\Omega |f|^2 e^{-2m\varphi}d\lambda.\end{aligned}$$

If we take the supremum over all $f \in B(1)$ we get

$$\varphi_m(z_0) \leq \sup_{|z-z_0|<r}\varphi(z) + \frac{1}{2m}\log\frac{1}{\pi^n r^{2n}/n!}$$

and the second inequality in (a) is proved – as we see, this is an easy consequence of the mean value inequality. Conversely, the Ohsawa-Takegoshi extension theorem (Corollary 8.9) applied to the 0-dimensional subvariety $\{z_0\} \subset \Omega$ shows that for any $a \in \mathbb{C}$ there is a holomorphic function f on Ω such that $f(z_0) = a$ and

$$\int_\Omega |f|^2 e^{-2m\varphi}d\lambda \leq C_3|a|^2 e^{-2m\varphi(z_0)},$$

where C_3 only depends on n and $\operatorname{diam}\Omega$. We fix a such that the right hand side is 1. Then $\|f\| \leq 1$ and so we get

$$\varphi_m(z_0) \geq \frac{1}{m}\log|f(z_0)| = \frac{1}{m}\log|a| = \varphi(z) - \frac{\log C_3}{2m}.$$

The inequalities given in (a) are thus proved. Taking $r = 1/m$, we find that $\lim_{m\to+\infty}\sup_{|\zeta-z|<1/m}\varphi(\zeta) = \varphi(z)$ by the upper semicontinuity of φ, and therefore $\lim\varphi_m(z) = \varphi(z)$, since $\lim\frac{1}{m}\log(C_2 m^n) = 0$.

(b) The above estimates imply

$$\sup_{|z-z_0|<r}\varphi(z) - \frac{C_1}{m} \leq \sup_{|z-z_0|<r}\varphi_m(z) \leq \sup_{|z-z_0|<2r}\varphi(z) + \frac{1}{m}\log\frac{C_2}{r^n}.$$

After dividing by $\log r < 0$ when $r \to 0$, we infer

$$\frac{\sup_{|z-z_0|<2r}\varphi(z) + \frac{1}{m}\log\frac{C_2}{r^n}}{\log r} \leq \frac{\sup_{|z-z_0|<r}\varphi_m(z)}{\log r} \leq \frac{\sup_{|z-z_0|<r}\varphi(z) - \frac{C_1}{m}}{\log r},$$

and from this and definition (3.1), it follows immediately that

$$\nu(\varphi, x) - \frac{n}{m} \le \nu(\varphi_m, z) \le \nu(\varphi, z). \qquad \square$$

Theorem 3.2 implies in a straightforward manner the deep result of [Siu74] on the analyticity of the Lelong number upperlevel sets.

(3.3) Corollary [Siu74]). *Let φ be a plurisubharmonic function on a complex manifold X. Then, for every $c > 0$, the Lelong number upperlevel set*

$$E_c(\varphi) = \big\{z \in X\,;\ \nu(\varphi, z) \ge c\big\}$$

is an analytic subset of X.

Proof. Since analyticity is a local property, it is enough to consider the case of a psh function φ on a pseudoconvex open set $\Omega \subset \mathbb{C}^n$. The inequalities obtained in 3.2 b) imply that

$$E_c(\varphi) = \bigcap_{m \ge m_0} E_{c-n/m}(\varphi_m).$$

Now, it is clear that $E_c(\varphi_m)$ is the analytic set defined by the equations $\sigma_\ell^{(\alpha)}(z) = 0$ for all multi-indices α such that $|\alpha| < mc$. Thus $E_c(\varphi)$ is analytic as a (countable) intersection of analytic sets. $\square$

(3.4) Remark. It can be easily shown that the Lelong numbers of any closed positive (p, p)-current coincide (at least locally) with the Lelong numbers of a suitable plurisubharmonic potential φ (see Skoda [Sko72a]). Hence Siu's theorem also holds true for the Lelong number upperlevel sets $E_c(T)$ of any closed positive (p, p)-current T.

3.B. Global approximation of closed (1,1)-currents on a compact complex manifold

We take here X to be an arbitrary compact complex manifold (no Kähler assumption is needed). Now, let T be a closed $(1,1)$-current on X. We assume that T is *almost positive*, i.e. that there exists a $(1,1)$-form γ with continuous coefficients such that $T \ge \gamma$; the case of positive currents ($\gamma = 0$) is of course the most important.

(3.5) Lemma. *There exists a smooth closed $(1,1)$-form α representing the same $\partial\overline{\partial}$- cohomology class as T and an* almost psh *function φ on X such that $T = \alpha + \frac{i}{\pi}\partial\overline{\partial}\varphi$. (We say that a function φ is almost psh if its complex Hessian is bounded below by a $(1,1)$-form with locally bounded coefficients, that is, if $i\partial\overline{\partial}\varphi$ is almost positive).*

Proof. Select an open covering (U_j) of X by coordinate balls such that $T = \frac{i}{\pi}\partial\overline{\partial}\varphi_j$ over U_j, and construct a global function $\varphi = \sum \theta_j\varphi_j$ by means of a partition of unity (θ_j) subordinate to U_j. Now, we observe that $\varphi - \varphi_k$ is smooth on U_k because all differences $\varphi_j - \varphi_k$ are smooth in the intersections $U_j \cap U_k$ and $\varphi - \varphi_k = \sum \theta_j(\varphi_j - \varphi_k))$. Therefore $\alpha := T - \frac{i}{\pi}\partial\overline{\partial}\varphi$ is smoth. $\square$

By replacing T with $T-\alpha$ and γ with $\gamma-\alpha$, we can assume without loss of generality that $\{T\}=0$, i.e. that $T=\frac{i}{\pi}\partial\overline{\partial}\varphi$ with an almost psh function φ on X such that $\frac{i}{\pi}\partial\overline{\partial}\varphi \geq \gamma$.

Our goal is to approximate T in the weak topology by currents $T_m=\frac{i}{\pi}\partial\overline{\partial}\varphi_m$ such their potentials φ_m have analytic singularities in the sense of Definition 1.10, more precisely, defined on a neighborhood V_{x_0} of any point $x_0 \in X$ in the form $\varphi_m(z)=c_m \log\sum_j |\sigma_{j,m}|^2+O(1)$, where $c_m>0$ and the $\sigma_{j,m}$ are holomorphic functions on V_{x_0}.

We select a finite covering (W_ν) of X with open coordinate charts. Given $\delta>0$, we take in each W_ν a maximal family of points with (coordinate) distance to the boundary $\geq 3\delta$ and mutual distance $\geq \delta/2$. In this way, we get for $\delta>0$ small a finite covering of X by open balls U'_j of radius δ (actually every point is even at distance $\leq \delta/2$ of one of the centers, otherwise the family of points would not be maximal), such that the concentric ball U_j of radius 2δ is relatively compact in the corresponding chart W_ν. Let $\tau_j : U_j \longrightarrow B(a_j,2\delta)$ be the isomorphism given by the coordinates of W_ν. Let $\varepsilon(\delta)$ be a modulus of continuity for γ on the sets U_j, such that $\lim_{\delta\to 0}\varepsilon(\delta)=0$ and $\gamma_x-\gamma_{x'}\leq \frac{1}{2}\varepsilon(\delta)\,\omega_x$ for all $x,x'\in U_j$. We denote by γ_j the $(1,1)$-form with constant coefficients on $B(a_j,2\delta)$ such that $\tau_j^\star\gamma_j$ coincides with $\gamma-\varepsilon(\delta)\,\omega$ at $\tau_j^{-1}(a_j)$. Then we have

$$0\leq \gamma-\tau_j^\star\gamma_j\leq 2\varepsilon(\delta)\,\omega \text{ on } U'_j \tag{3.6}$$

for $\delta>0$ small. We set $\varphi_j=\varphi\circ\tau_j^{-1}$ on $B(a_j,2\delta)$ and let q_j be the homogeneous quadratic function in $z-a_j$ such that $\frac{i}{\pi}\partial\overline{\partial}q_j=\gamma_j$ on $B(a_j,2\delta)$. Finally, we set

$$\psi_j(z)=\varphi_j(z)-q_j(z) \text{ on } B(a_j,2\delta). \tag{3.7}$$

Then ψ_j is plurisubharmonic, since

$$\frac{i}{\pi}\partial\overline{\partial}(\psi_j\circ\tau_j)=T-\tau_j^\star\gamma_j\geq\gamma-\tau_j^\star\gamma_j\geq 0.$$

We let $U'_j\subset\subset U''_j\subset\subset U_j$ be concentric balls of radii δ, $1.5\,\delta$, 2δ respectively. On each open set U_j the function $\psi_j:=\varphi-q_j\circ\tau_j$ defined in (3.7) is plurisubharmonic, so Theorem (3.2) applied with $\Omega=U_j$ produces functions

$$\psi_{j,m}=\frac{1}{2m}\log\sum_\ell|\sigma_{j,\ell}|^2,\ (\sigma_{j,\ell})=\text{basis of }\mathcal{H}_{U_j}(m\psi_j). \tag{3.8}$$

These functions approximate ψ_j as m tends to $+\infty$ and satisfy the inequalities

$$\psi_j(x)-\frac{C_1}{m}\leq\psi_{j,m}(x)\leq\sup_{|\zeta-x|<r}\psi_j(\zeta)+\frac{1}{m}\log\frac{C_2}{r^n}. \tag{3.9}$$

The functions $\psi_{j,m}+q_j\circ\tau_j$ on U_j then have to be glued together by a partition of unity technique. For this, we rely on the following "discrepancy" lemma, estimating the variation of the approximating functions on overlapping balls.

(3.10) Lemma. *There are constants $C_{j,k}$ independent of m and δ such that the almost psh functions $w_{j,m}=2m(\psi_{j,m}+q_j\circ\tau_j)$, i.e.*

$$w_{j,m}(x)=2m\,q_j\circ\tau_j(x)+\log\sum_\ell\big|\sigma_{j,\ell}(x)\big|^2,\ x\in U''_j,$$

satisfy

$$|w_{j,m}-w_{k,m}|\leq C_{j,k}\big(\log\delta^{-1}+m\varepsilon(\delta)\delta^2\big)\text{ on } U''_j\cap U''_k.$$

Proof. The details will be left as an exercise to the reader. The main idea is the following: for any holomorphic function $f_j \in \mathcal{H}_{U_j}(m\psi_j)$, a $\overline{\partial}$ equation $\overline{\partial}u = \overline{\partial}(\theta f_j)$ can be solved on U_k, where θ is a cut-off function with support in $U_j'' \cap U_k''$, on a ball of radius $< \delta/4$, equal to 1 on the ball of radius $\delta/8$ centered at a given point $x_0 \in U_j'' \cap U_k''$. We apply the L^2 estimate with respect to the weight $(n+1)\log|x-x_0|^2 + 2m\psi_k$, where the first term is picked up so as to force the solution u to vanish at x_0, in such a way that $F_k = u - \theta f_j$ is holomorphic and $F_k(x_0) = f_j(x_0)$. The discrepancy between the weights on U_j'' and U_k'' is

$$\psi_j(x) - \psi_k(x) = -\big(q_j \circ \tau_j(x) - q_k \circ \tau_k(x)\big)$$

and the $\partial\overline{\partial}$ of this difference is $O(\varepsilon(\delta))$, so it is easy to correct the discrepancy up to a $O(\varepsilon(\delta)\delta^2)$ term by multiplying our functions by an invertible holomorphic function G_{jk}. In this way, we get a uniform L^2 control on the L^2 norm of the solution $f_k = G_{jk}F_k = G_{jk}(u - \theta f_j)$ of the form

$$\int_{U_k} |f_k|^2 e^{-2m\psi_k} \leq C_{j,k}\delta^{-2n-4} e^{mO(\varepsilon(\delta)\delta^2)} \int_{U_j} |f_j|^2 e^{-2m\psi_j}.$$

The required estimate follows, using the fact that

$$e^{2m\psi_{j,m}(x)} = \sum_\ell |\sigma_{j,\ell}(x)|^2 = \sup_{f \in \mathcal{H}_{U_j}(m\psi_j),\, \|f\| \leq 1} |f(x)|^2 \text{ on } U_j,$$

and the analogous equality on U_k. □

Now, the actual glueing of our almost psh functions is performed using the following elementary partition of unity argument, which follows from a brute force calculation (see e.g. [Dem82]).

(3.11) Lemma. *Let $U_j' \subset\subset U_j''$ be locally finite open coverings of a complex manifold X by relatively compact open sets, and let θ_j be smooth nonnegative functions with support in U_j'', such that $\theta_j \leq 1$ on U_j'' and $\theta_j = 1$ on U_j'. Let $A_j \geq 0$ be such that*

$$i(\theta_j \partial\overline{\partial}\theta_j - \partial\theta_j \wedge \overline{\partial}\theta_j) \geq -A_j\omega \text{ on } U_j'' \smallsetminus U_j'$$

for some positive $(1,1)$-form ω. Finally, let w_j be almost psh functions on U_j with the property that $i\partial\overline{\partial}w_j \geq \gamma$ for some real $(1,1)$-form γ on M, and let C_j be constants such that

$$w_j(x) \leq C_j + \sup_{k \neq j,\, U_k' \ni x} w_k(x) \text{ on } U_j'' \smallsetminus U_j'.$$

Then the function $w = \log\big(\sum \theta_j^2 e^{w_j}\big)$ is almost psh and satisfies

$$i\partial\overline{\partial}w \geq \gamma - 2\Big(\sum_j \mathbf{1}_{U_j'' \smallsetminus U_j'} A_j e^{C_j}\Big)\omega.$$

We apply Lemma (3.11) to functions $\widetilde{w}_{j,m}$ which are just slight modifications of the functions $w_{j,m} = 2m(\psi_{j,m} + q_j \circ \tau_j)$ occurring in (3.10) :

$$\begin{aligned}\widetilde{w}_{j,m}(x) &= w_{j,m}(x) + 2m\Big(\frac{C_1}{m} + C_3\varepsilon(\delta)(\delta^2/2 - |\tau_j(x)|^2)\Big)\\ &= 2m\Big(\psi_{j,m}(x) + q_j \circ \tau_j(x) + \frac{C_1}{m} + C_3\varepsilon(\delta)(\delta^2/2 - |\tau_j(x)|^2)\Big)\end{aligned}$$

where $x \mapsto z = \tau_j(x)$ is a local coordinate identifying U_j to $B(0, 2\delta)$, C_1 is the constant occurring in (3.9) and C_3 is a sufficiently large constant. It is easy to see that we can take $A_j = C_4\delta^{-2}$ in Lemma (3.11). We have

$$\widetilde{w}_{j,m} \geq w_{j,m} + 2C_1 + m\frac{C_3}{2}\varepsilon(\delta)\delta^2 \text{ on } B(x_j, \delta/2) \subset U'_j,$$

since $|\tau_j(x)| \leq \delta/2$ on $B(x_j, \delta/2)$, while

$$\widetilde{w}_{j,m} \leq w_{j,m} + 2C_1 - mC_3\varepsilon(\delta)\delta^2 \text{ on } U''_j \smallsetminus U'_j.$$

For $m \geq m_0(\delta) = (\log \delta^{-1}/(\varepsilon(\delta)\delta^2)$, Lemma (3.10) implies

$$|w_{j,m} - w_{k,m}| \leq C_5 m\varepsilon(\delta)\delta^2$$

on $U''_j \cap U''_k$. Hence, for C_3 large enough, we get

$$\widetilde{w}_{j,m}(x) \leq \sup_{k\neq j,\, B(x_k,\delta/2)\ni x} w_{k,m}(x) \leq \sup_{k\neq j,\, U'_k\ni x} w_{k,m}(x) \text{ on } U''_j \smallsetminus U'_j,$$

and we can take $C_j = 0$ in the hypotheses of Lemma (3.11). The associated function $w = \log\big(\sum \theta_j^2 e^{\widetilde{w}_{j,m}}\big)$ is given by

$$w = \log \sum_j \theta_j^2 \exp\Big(2m\big(\psi_{j,m} + q_j \circ \tau_j + \frac{C_1}{m} + C_3\varepsilon(\delta)(\delta^2/2 - |\tau_j|^2)\big)\Big).$$

If we define $\varphi_m = \frac{1}{2m}w$, we get

$$\varphi_m(x) := \frac{1}{2m}w(x) \geq \psi_{j,m}(x) + q_j \circ \tau_j(x) + \frac{C_1}{m} + \frac{C_3}{4}\varepsilon(\delta)\delta^2 > \varphi(x)$$

in view of (3.9), by picking an index j such that $x \in B(x_j, \delta/2)$. In the opposite direction, the maximum number N of overlapping balls U_j does not depend on δ, and we thus get

$$w \leq \log N + 2m\Big(\max_j \{\psi_{j,m}(x) + q_j \circ \tau_j(x)\} + \frac{C_1}{m} + \frac{C_3}{2}\varepsilon(\delta)\delta^2\Big).$$

By definition of ψ_j we have $\sup_{|\zeta - x|<r} \psi_j(\zeta) \leq \sup_{|\zeta - x|<r} \varphi(\zeta) - q_j \circ \tau_j(x) + C_5 r$ thanks to the uniform Lipschitz continuity of $q_j \circ \tau_j$, thus by (3.9) again we find

$$\varphi_m(x) \leq \frac{\log N}{2m} + \sup_{|\zeta - x|<r} \varphi(\zeta) + \frac{C_1}{m} + \frac{1}{m}\log\frac{C_2}{r^n} + \frac{C_3}{2}\varepsilon(\delta)\delta^2 + C_5 r$$

By taking for instance $r = 1/m$ and $\delta = \delta_m \to 0$, we see that φ_m converges to φ. On the other hand (3.6) implies $\frac{i}{\pi}\partial\overline{\partial} q_j \circ \tau_j(x) = \tau_j^\star \gamma_j \geq \gamma - 2\varepsilon(\delta)\omega$, thus

$$\frac{i}{\pi}\partial\overline{\partial}\widetilde{w}_{j,m} \geq 2m\big(\gamma - C_6\varepsilon(\delta)\omega\big).$$

Lemma (3.11) then produces the lower bound

$$\frac{i}{\pi}\partial\overline{\partial} w \geq 2m\big(\gamma - C_6\varepsilon(\delta)\omega\big) - C_7\delta^{-2}\omega,$$

whence

$$\frac{i}{\pi}\partial\overline{\partial}\varphi_m \geq \gamma - C_8\varepsilon(\delta)\omega$$

for $m \geq m_0(\delta) = (\log \delta^{-1})/(\varepsilon(\delta)\delta^2)$. We can fix $\delta = \delta_m$ to be the smallest value of $\delta > 0$ such that $m_0(\delta) \leq m$, then $\delta_m \to 0$ and we have obtained a sequence of quasi psh functions φ_m satisfying the following properties.

(3.12) Theorem. *Let φ be an almost psh function on a compact complex manifold X such that $\frac{i}{\pi}\partial\overline{\partial}\varphi \geq \gamma$ for some continuous $(1,1)$-form γ. Then there is a sequence of almost psh functions φ_m such that φ_m has the same singularities as a logarithm of a sum of squares of holomorphic functions and a decreasing sequence $\varepsilon_m > 0$ converging to 0 such that*

(i) $\varphi(x) < \varphi_m(x) \leq \sup_{|\zeta - x| < r} \varphi(\zeta) + C\Big(\frac{|\log r|}{m} + r + \varepsilon_m\Big)$
with respect to coordinate open sets covering X. In particular, φ_m converges to φ pointwise and in $L^1(X)$ and

(ii) $\nu(\varphi, x) - \frac{n}{m} \leq \nu(\varphi_m, x) \leq \nu(\varphi, x)$ *for every $x \in X$;*

(iii) $\frac{i}{\pi}\partial\overline{\partial}\varphi_m \geq \gamma - \varepsilon_m\omega$.

In particular, we can apply this to an arbitrary positive or almost positive closed $(1,1)$-current $T = \alpha + \frac{i}{\pi}\partial\overline{\partial}\varphi$.

(3.13) Corollary. *Let T be an almost positive closed $(1,1)$-current on a compact complex manifold X such that $T \geq \gamma$ for some continuous $(1,1)$-form γ. Then there is a sequence of currents T_m whose local potentials have the same singularities as $1/m$ times a logarithm of a sum of squares of holomorphic functions and a decreasing sequence $\varepsilon_m > 0$ converging to 0 such that*

(i) *T_m converges weakly to T,*

(ii) $\nu(T, x) - \frac{n}{m} \leq \nu(T_m, x) \leq \nu(T, x)$ *for every $x \in X$;*

(iii) $T_m \geq \gamma - \varepsilon_m\omega$.

We say that our currents T_m are approximations of T possessing logarithmic poles.

By using blow-ups of X, the structure of the currents T_m can be better understood. In fact, consider the coherent ideals $\mathcal{J}_m$ generated locally by the holomorphic functions $(\sigma_{j,m}^{(k)})$ on U_k in the local approximations

$$\varphi_{k,m} = \frac{1}{2m}\log\sum_j |\sigma_{j,m}^{(k)}|^2 + O(1)$$

of the potential φ of T on U_k. These ideals are in fact globally defined, because the local ideals $\mathcal{J}_m^{(k)} = (\sigma_{j,m}^{(k)})$ are integrally closed, and they coincide on the intersections $U_k \cap U_\ell$ as they have the same order of vanishing by the proof of Lemma (13,10). By Hironaka [Hir64], we can find a composition of blow-ups with smooth centers $\mu_m : \widetilde{X}_m \to X$ such that $\mu_m^*\mathcal{J}_m$ is an invertible ideal sheaf associated with a normal crossing divisor D_m. Now, we can write

$$\mu_m^*\varphi_{k,m} = \varphi_{k,m} \circ \mu_m = \frac{1}{m}\log|s_{D_m}| + \widetilde{\varphi}_{k,m}$$

where s_{D_m} is the canonical section of $\mathcal{O}(-D_m)$ and $\widetilde{\varphi}_{k,m}$ is a smooth potential. This implies

$$\mu_m^*T_m = \frac{1}{m}[D_m] + \beta_m \qquad (3.14)$$

where $[D_m]$ is the current of integration over D_m and β_m is a smooth closed $(1,1)$-form which satisfies the lower bound $\beta_m \geq \mu_m^*(\gamma - \varepsilon_m\omega)$. (Recall that the pull-back of a closed $(1,1)$-current by a holomorphic map f is always well-defined, by taking a local plurisubharmonic potential φ such that $T = i\partial\overline{\partial}\varphi$ and writing

$f^*T = i\partial\overline{\partial}(\varphi \circ f)$). In the remainder of this section, we derive from this a rather important geometric consequence, first appeared in [DP04]). We need two related definitions.

(3.15) Definition. *A Kähler current on a compact complex space X is a closed positive current T of bidegree $(1,1)$ which satisfies $T \geq \varepsilon\omega$ for some $\varepsilon > 0$ and some smooth positive hermitian form ω on X.*

(3.16) Definition. *A compact complex manifold is said to be in the* Fujiki class $\mathcal{C}$) *if it is bimeromorphic to a Kähler manifold (or equivalently, using Hironaka's desingularization theorem, if it admits a proper Kähler modification).*

(3.17) Theorem. *A compact complex manifold X is bimeromorphic to a Kähler manifold (i.e. $X \in \mathcal{C}$) if and only if it admits a Kähler current.*

Proof. If X is bimeromorphic to a Kähler manifold Y, Hironaka's desingularization theorem implies that there exists a blow-up $\widetilde{Y}$ of Y (obtained by a sequence of blow-ups with smooth centers) such that the bimeromorphic map from Y to X can be resolved into a modification $\mu : \widetilde{Y} \to X$. Then $\widetilde{Y}$ is Kähler and the push-forward $T = \mu_*\widetilde{\omega}$ of a Kähler form $\widetilde{\omega}$ on $\widetilde{Y}$ provides a Kähler current on X. In fact, if ω is a smooth hermitian form on X, there is a constant C such that $\mu^*\omega \leq C\widetilde{\omega}$ (by compactness of $\widetilde{Y}$), hence

$$T = \mu_*\widetilde{\omega} \geq \mu_*(C^{-1}\mu^*\omega) = C^{-1}\omega.$$

Conversely, assume that X admits a Kähler current $T \geq \varepsilon\omega$. By Theorem 3.13 (iii), there exists a Kähler current $\widetilde{T} = T_m \geq \frac{\varepsilon}{2}\omega$ (with $m \gg 1$ so large that $\varepsilon_m \leq \varepsilon/2$) in the same $\partial\overline{\partial}$-cohomology class as T, possessing logarithmic poles. Observation (3.14) implies the existence of a composition of blow-ups $\mu : \widetilde{X} \to X$ such that

$$\mu^*\widetilde{T} = [\widetilde{D}] + \widetilde{\beta} \qquad \text{on } \widetilde{X},$$

where $\widetilde{D}$ is a $\mathbb{Q}$-divisor with normal crossings and $\widetilde{\beta}$ a smooth closed $(1,1)$-form such that $\widetilde{\beta} \geq \frac{\varepsilon}{2}\mu^*\omega$. In particular $\widetilde{\beta}$ is positive outside the exceptional locus of μ. This is not enough yet to produce a Kähler form on $\widetilde{X}$, but we are not very far. Suppose that $\widetilde{X}$ is obtained as a tower of blow-ups

$$\widetilde{X} = X_N \to X_{N-1} \to \cdots \to X_1 \to X_0 = X,$$

where X_{j+1} is the blow-up of X_j along a smooth center $Y_j \subset X_j$. Denote by $E_{j+1} \subset X_{j+1}$ the exceptional divisor, and let $\mu_j : X_{j+1} \to X_j$ be the blow-up map. Now, we use the following simple

(3.18) Lemma. *For every Kähler current T_j on X_j, there exists $\varepsilon_{j+1} > 0$ and a smooth form u_{j+1} in the $\partial\overline{\partial}$-cohomology class of $[E_{j+1}]$ such that*

$$T_{j+1} = \mu_j^\star T_j - \varepsilon_{j+1}u_{j+1}$$

is a Kähler current on X_{j+1}.

Proof. The line bundle $\mathcal{O}(-E_{j+1})|E_{j+1}$ is equal to $\mathcal{O}_{P(N_j)}(1)$ where N_j is the normal bundle to Y_j in X_j. Pick an arbitrary smooth hermitian metric on N_j, use this metric to get an induced Fubini-Study metric on $\mathcal{O}_{P(N_j)}(1)$, and finally extend this metric as a smooth hermitian metric on the line bundle $\mathcal{O}(-E_{j+1})$. Such a metric has positive curvature along tangent vectors of X_{j+1} which are tangent to the fibers of $E_{j+1} = P(N_j) \to Y_j$. Assume furthermore that $T_j \geq \delta_j \omega_j$ for some hermitian form ω_j on X_j and a suitable $0 < \delta_j \ll 1$. Then

$$\mu_j^\star T_j - \varepsilon_{j+1} u_{j+1} \geq \delta_j \mu_j^\star \omega_j - \varepsilon_{j+1} u_{j+1}$$

where $\mu_j^* \omega_j$ is semi-positive on X_{j+1}, positive definite on $X_{j+1} \smallsetminus E_{j+1}$, and also positive definite on tangent vectors of $T_{X_{j+1}|E_{j+1}}$ which are not tangent to the fibers of $E_{j+1} \to Y_j$. The statement is then easily proved by taking $\varepsilon_{j+1} \ll \delta_j$ and by using an elementary compactness argument on the unit sphere bundle of $T_{X_{j+1}}$ associated with any given hermitian metric. □

End of proof of Theorem 3.17. If $\widetilde{u}_j$ is the pull-back of u_j to the final blow-up $\widetilde{X}$, we conclude inductively that $\mu^\star \widetilde{T} - \sum \varepsilon_j \widetilde{u}_j$ is a Kähler current. Therefore the smooth form

$$\widetilde{\omega} := \widetilde{\beta} - \sum \varepsilon_j \widetilde{u}_j = \mu^\star \widetilde{T} - \sum \varepsilon_j \widetilde{u}_j - [\widetilde{D}]$$

is Kähler and we see that $\widetilde{X}$ is a Kähler manifold. □

(3.19) Remark. A special case of Theorem (3.16) is the following characterization of Moishezon varieties (i.e. manifolds which are bimeromorphic to projective algebraic varieties or, equivalently, whose algebraic dimension is equal to their complex dimension):

A compact complex manifold X is Moishezon if and only if X possesses a Kähler current T such that the De Rham cohomology class $\{T\}$ is rational, i.e. $\{T\} \in H^2(X, \mathbb{Q})$.

In fact, in the above proof, we get an integral current T if we take the push forward $T = \mu_* \widetilde{\omega}$ of an integral ample class $\{\widetilde{\omega}\}$ on Y, where $\mu : Y \to X$ is a projective model of Y. Conversely, if $\{T\}$ is rational, we can take the $\varepsilon_j' s$ to be rational in Lemma 3.5. This produces at the end a Kähler metric $\widetilde{\omega}$ with rational De Rham cohomology class on $\widetilde{X}$. Therefore $\widetilde{X}$ is projective by the Kodaira embedding theorem. This result was already observed in [JS93] (see also [Bon93, Bon98] for a more general perspective based on a singular version of holomorphic Morse inequalities).

3.C. Global approximation by divisors

We now translate our previous approximation theorems into a more algebro-geometric setting. Namely, we assume that T is a closed positive $(1,1)$-current which belongs to the first Chern class $c_1(L)$ of a holomorphic line bundle L, and we assume here X to be algebraic (i.e. projective or at the very least Moishezon).

Our goal is to show that T can be approximated by divisors which have roughly the same Lelong numbers as T. The existence of weak approximations by divisors has already been proved in [Lel72] for currents defined on a pseudoconvex open set $\Omega \subset \mathbb{C}^n$ with $H^2(\Omega, \mathbb{R}) = 0$, and in [Dem92, 93b] in the situation considered here (cf. also [Dem82], although the argument given there is somewhat incorrect). We take the opportunity to present here a slightly simpler derivation.

Let X be a projective manifold and L a line bundle over X. A singular hermitian metric h on L is a metric such that the weight function φ of h is L^1_{loc} in any local trivialization (such that $L_{|U} \simeq U \times \mathbb{C}$ and $\|\xi\|_h = |\xi| e^{-\varphi(x)}$, $\xi \in L_x \simeq \mathbb{C}$). The curvature of L can then be computed in the sense of distributions

$$T = \frac{\mathrm{i}}{2\pi}\Theta_{L,h} = \frac{\mathrm{i}}{\pi}\partial\overline{\partial}\varphi,$$

and L is said to be pseudo-effective if L admits a singular hermitian metric h such that the curvature current $T = \frac{\mathrm{i}}{2\pi}\Theta_{L,h}$ is semi-positive [The weight functions φ of L are thus plurisubharmonic]. In what follows, we sometimes use an additive notation for $\mathrm{Pic}(X)$, i.e. kL is meant for the line bundle $L^{\otimes k}$.

We will also make use of the concept of complex singularity exponent, following e.g. [Var82, 83], [ArGV85] and [DK01]. A quasi-plurisubharmonic (quasi-psh) function is by definition a function φ which is locally equal to the sum of a psh function and of a smooth function, or equivalently, a locally integrable function φ such that $i\partial\overline{\partial}\varphi$ is locally bounded below by $-C\omega$ where ω is a hermitian metric and C a constant.

(3.20) Definition. *If K is a compact subset of X and φ is a quasi-psh function defined near K, we define*

(a) *the complex singularity exponent $c_K(\varphi)$ to be the supremum of all positive numbers c such that $e^{-2c\varphi}$ is integrable in a neighborhood of every point $z_0 \in K$, with respect to the Lebesgue measure in holomorphic coordinates centered at z_0. In particular $c_K(\varphi) = \inf_{z_0 \in K}(\varphi)$.*

(b) *The concept is easily extended to hermitian metrics $h = e^{-2\varphi}$ by putting $c_K(h) = c_K(\varphi)$, to holomorphic functions f by $c_K(f) = c_K(\log|f|)$, to coherent ideals $\mathcal{J} = (g_1, \ldots, g_N)$ by $c_K(\mathcal{J}) = c_K(\varphi)$ where $\varphi = \frac{1}{2}\log\sum|g_j|^2$. Also for an effective $\mathbb{R}$-divisor D, we put $c_K(D) = c_K(\log|\sigma_D|)$ where σ_D is the canonical section.*

The main technical result of this section can be stated as follows, in the case of big line bundles (cf. Proposition (1.17 ii)).

(3.21) Theorem. *Let L be a line bundle on a compact complex manifold X possessing a singular hermitian metric h with $\Theta_{L,h} \geq \varepsilon\omega$ for some $\varepsilon > 0$ and some smooth positive definite hermitian $(1,1)$-form ω on X. For every real number $m > 0$, consider the space $\mathcal{H}_m = H^0(X, L^{\otimes m} \otimes \mathcal{I}(h^m))$ of holomorphic sections σ of $L^{\otimes m}$ on X such that*

$$\int_X |\sigma|^2_{h^m} dV_\omega = \int_X |\sigma|^2 e^{-2m\varphi} dV_\omega < +\infty,$$

where $dV_\omega = \frac{1}{m!}\omega^m$ is the hermitian volume form. Then for $m \gg 1$, $\mathcal{H}_m$ is a non zero finite dimensional Hilbert space and we consider the closed positive $(1,1)$-current

$$T_m = \frac{i}{\pi}\partial\overline{\partial}\Big(\frac{1}{2m}\log\sum_k |g_{m,k}|^2\Big) = \frac{i}{\pi}\partial\overline{\partial}\Big(\frac{1}{2m}\log\sum_k |g_{m,k}|^2_h\Big) + \Theta_{L,h}$$

where $(g_{m,k})_{1 \leq k \leq N(m)}$ is an orthonormal basis of $\mathcal{H}_m$. Then :

(i) *For every trivialization $L_{|U} \simeq U \times \mathbb{C}$ on a cordinate open set U of X and every compact set $K \subset U$, there are constants $C_1, C_2 > 0$ independent of m and φ such that*

$$\varphi(z) - \frac{C_1}{m} \leq \psi_m(z) := \frac{1}{2m}\log\sum_k |g_{m,k}(z)|^2 \leq \sup_{|x-z|<r}\varphi(x) + \frac{1}{m}\log\frac{C_2}{r^n}$$

for every $z \in K$ *and* $r \le \frac{1}{2}d(K, \partial U)$. *In particular,* ψ_m *converges to* φ *pointwise and in* L^1_{loc} *topology on* Ω *when* $m \to +\infty$, *hence* T_m *converges weakly to* $T = \Theta_{L,h}$.

(ii) *The Lelong numbers* $\nu(T, z) = \nu(\varphi, z)$ *and* $\nu(T_m, z) = \nu(\psi_m, z)$ *are related by*

$$\nu(T, z) - \frac{n}{m} \le \nu(T_m, z) \le \nu(T, z) \quad \text{for every } z \in X.$$

(iii) *For every compact set* $K \subset X$, *the complex singularity exponents of the metrics given locally by* $h = e^{-2\varphi}$ *and* $h_m = e^{-2\psi_m}$ *satisfy*

$$c_K(h)^{-1} - \frac{1}{m} \le c_K(h_m)^{-1} \le c_K(h)^{-1}.$$

Proof. The major part of the proof is a variation of the arguments already explained in section 3.A.

(i) We note that $\sum |g_{m,k}(z)|^2$ is the square of the norm of the evaluation linear form $\sigma \mapsto \sigma(z)$ on $\mathcal{H}_m$, hence

$$\psi_m(z) = \sup_{\sigma \in B(1)} \frac{1}{m} \log |\sigma(z)|$$

where $B(1)$ is the unit ball of $\mathcal{H}_m$. For $r \le \frac{1}{2}d(K, \partial\Omega)$, the mean value inequality applied to the plurisubharmonic function $|\sigma|^2$ implies

$$\begin{aligned} |\sigma(z)|^2 &\le \frac{1}{\pi^n r^{2n}/n!} \int_{|x-z|<r} |\sigma(x)|^2 d\lambda(x) \\ &\le \frac{1}{\pi^n r^{2n}/n!} \exp\Big(2m \sup_{|x-z|<r} \varphi(x)\Big) \int_\Omega |\sigma|^2 e^{-2m\varphi} d\lambda. \end{aligned}$$

If we take the supremum over all $\sigma \in B(1)$ we get

$$\psi_m(z) \le \sup_{|x-z|<r} \varphi(x) + \frac{1}{2m} \log \frac{1}{\pi^n r^{2n}/n!}$$

and the right hand inequality in (i) is proved. Conversely, the Ohsawa-Takegoshi extension theorem [OhT87], [Ohs88] applied to the 0-dimensional subvariety $\{z\} \subset U$ shows that for any $a \in \mathbb{C}$ there is a holomorphic function f on U such that $f(z) = a$ and

$$\int_U |f|^2 e^{-2m\varphi} d\lambda \le C_3 |a|^2 e^{-2m\varphi(z)},$$

where C_3 only depends on n and $\operatorname{diam} U$. Now, provided a remains in a compact set $K \subset U$, we can use a cut-off function θ with support in U and equal to 1 in a neighborhood of a, and solve the $\overline{\partial}$-equation $\overline{\partial} g = \overline{\partial}(\theta f)$ in the L^2 space associated with the weight $2m\varphi + 2(n+1)|\log|z-a|$, that is, the singular hermitian metric $h(z)^m |z-a|^{-2(n+1)}$ on $L^{\otimes m}$. For this, we apply the standard Andreotti-Vesentini-Hörmander L^2 estimates (see e.g. [Dem82] for the required version). This is possible for $m \ge m_0$ thanks to the hypothesis that $\Theta_{L,h} \ge \varepsilon\omega > 0$, even if X is non Kähler (X is in any event a Moishezon variety from our assumptions). The bound m_0 depends only on ε and the geometry of a finite covering of X by compact sets $K_j \subset U_j$, where U_j are coordinate balls (say); it is independent of the point a

and even of the metric h. It follows that $g(a) = 0$ and therefore $\sigma = \theta f - g$ is a holomorphic section of $L^{\otimes m}$ such that

$$\int_X |\sigma|^2_{h^m} dV_\omega = \int_X |\sigma|^2 e^{-2m\varphi} dV_\omega \leq C_4 \int_U |f|^2 e^{-2m\varphi} dV_\omega \leq C_5 |a|^2 e^{-2m\varphi(z)},$$

in particular $\sigma \in \mathcal{H}_m = H^0(X, L^{\otimes m} \otimes \mathcal{I}(h^m))$. We fix a such that the right hand side is 1. This gives the inequality

$$\psi_m(z) \geq \frac{1}{m} \log |a| = \varphi(z) - \frac{\log C_5}{2m}$$

which is the left hand part of statement (i).

(ii) The first inequality in (i) implies $\nu(\psi_m, z) \leq \nu(\varphi, z)$. In the opposite direction, we find

$$\sup_{|x-z|<r} \psi_m(x) \leq \sup_{|x-z|<2r} \varphi(x) + \frac{1}{m} \log \frac{C_2}{r^n}.$$

Divide by $\log r < 0$ and take the limit as r tends to 0. The quotient by $\log r$ of the supremum of a psh function over $B(x,r)$ tends to the Lelong number at x. Thus we obtain

$$\nu(\psi_m, x) \geq \nu(\varphi, x) - \frac{n}{m}.$$

(iii) Again, the first inequality (in (i) immediately yields $h_m \leq C_6 h$, hence $c_K(h_m) \geq c_K(h)$. For the converse inequality, since we have $c_{\cup K_j}(h) = \min_j c_{K_j}(h)$, we can assume without loss of generality that K is contained in a trivializing open patch U of L. Let us take $c < c_K(\psi_m)$. Then, by definition, if $V \subset X$ is a sufficiently small open neighborhood of K, the Hölder inequality for the conjugate exponents $p = 1 + mc^{-1}$ and $q = 1 + m^{-1}c$ implies, thanks to equality $\frac{1}{p} = \frac{c}{mq}$,

$$\begin{aligned}\int_V e^{-2(m/p)\varphi} dV_\omega &= \int_V \Big(\sum_{1\leq k\leq N(m)} |g_{m,k}|^2 e^{-2m\varphi} \Big)^{1/p} \Big(\sum_{1\leq k\leq N(m)} |g_{m,k}|^2 \Big)^{-c/mq} dV_\omega \\ &\leq \bigg(\int_X \sum_{1\leq k\leq N(m)} |g_{m,k}|^2 e^{-2m\varphi} dV_\omega \bigg)^{1/p} \bigg(\int_V \Big(\sum_{1\leq k\leq N(m)} |g_{m,k}|^2 \Big)^{-c/m} dV_\omega \bigg)^{1/q} \\ &= N(m)^{1/p} \bigg(\int_V \Big(\sum_{1\leq k\leq N(m)} |g_{m,k}|^2 \Big)^{-c/m} dV_\omega \bigg)^{1/q} < +\infty.\end{aligned}$$

From this we infer $c_K(h) \geq m/p$, i.e., $c_K(h)^{-1} \leq p/m = 1/m + c^{-1}$. As $c < c_K(\psi_m)$ was arbitrary, we get $c_K(h)^{-1} \leq 1/m + c_K(h_m)^{-1}$ and the inequalities of (iii) are proved. □

(3.22) Remark. The proof would also work, with a few modifications, when X is a Stein manifold and L is an arbitrary holomorphic line bundle.

(3.23) Corollary. *Let $L \to X$ be a holomorphic line bundle and $T = \frac{i}{2\pi}\Theta_{L,h}$ the curvature current of some singular hermitian metric h on L.*

(i) *If L is big and $\Theta_{L,h} \geq \varepsilon\omega > 0$, there exists a sequence of holomorphic sections $h_s \in H^0(X, q_s L)$ with $\lim q_s = +\infty$ such that the $\mathbb{Q}$-divisors $D_s = \frac{1}{q_s}\,\mathrm{div}(h_s)$*

satisfy $T = \lim D_s$ *in the weak topology and* $\sup_{x\in X} |\nu(D_s,x) - \nu(T,x)| \to 0$ *as* $s \to +\infty$.

(ii) *If* L *is just pseudo-effective and* $\Theta_{L,h} \geq 0$, *for any ample line bundle* A, *there exists a sequence of non zero sections* $h_s \in H^0(X, p_sA + q_sL)$ *with* $p_s, q_s > 0$, $\lim q_s = +\infty$ *and* $\lim p_s/q_s = 0$, *such that the divisors* $D_s = \frac{1}{q_s}\operatorname{div}(h_s)$ *satisfy* $T = \lim D_s$ *in the weak topology and* $\sup_{x\in X} |\nu(D_s,x) - \nu(T,x)| \to 0$ *as* $s \to +\infty$.

Proof. Part (ii) is a rather straightforward consequence of part (i) applied to $mL + A$ and $T_m = \frac{1}{m}\Theta_{mL+A,h^m h_A} = T + \frac{1}{m}\Theta_{A,h_A} \to T$ when m tends to infinity. Therefore, it suffices to prove (i).

(i) By Theorem (3.21), we can find sections $g_1, \ldots, g_N \in H^0(X, mL)$ (omitting the index m for simplicity of notation), such that

$$T_m = \frac{i}{\pi}\partial\overline{\partial}\Big(\frac{1}{2m}\log\sum_{1\leq j\leq N}|g_j|_h^2\Big) + \Theta_{L,h} = \frac{i}{\pi}\partial\overline{\partial}\Big(\frac{1}{2m}\log\sum_{1\leq j\leq N}|g_j|^2\Big)$$

converges weakly to T and satisfies $\nu(T,x) - n/m \leq \nu(T_m,x) \leq \nu(T,x)$. In fact, since the number N of sections grows at most as $O(m^n)$, we can replace $\sum_{1\leq j\leq N}|g_j|^2$ by $\max_{1\leq j\leq N}|g_j|^2$, as the difference of the potentials tends uniformly to 0 with the help of the renormalizing constant $\frac{1}{2m}$. Hence, we can use instead the approximating currents

$$\widetilde{T}_m = \frac{i}{\pi}\partial\overline{\partial}u_m, \qquad u_m = \frac{1}{m}\log\max_{1\leq j\leq N}|g_j|.$$

Now, as L is big, we can write $k_0L = A + D$ where A is an ample divisor and D is an effective divisor, for some $k_0 > 0$. By enlarging k_0, we can even assume that A is very ample. Let σ_D be the canonical section of D and let $h_1, \ldots, h_N$ be sections of $H^0(X, A)$. We get a section of $H^0(X, (m\ell + k_0)L)$ by considering

$$u_{\ell,m} = (g_1^\ell h_1 + \ldots + g_N^\ell h_N)\sigma_D$$

By enlarging N if necessary and putting e.g. $g_j = g_N$ for $j > N$, we can assume that the sections h_j generate all 1-jets of sections of A at every point (actually, one can always achieve this with $n+1$ sections only, so this is not really a big demand). Then, for almost every N-tuple $(h_1, \ldots, h_N)$, Lemma 3.24 below and the weak continuity of $\partial\overline{\partial}$ imply that

$$\Delta_{\ell,m} = \frac{1}{\ell m}\frac{i}{\pi}\partial\overline{\partial}\log|u_{\ell,m}| = \frac{1}{\ell m}\operatorname{div}(u_{\ell,m})$$

converges weakly to $\widetilde{T}_m = \frac{i}{\pi}\partial\overline{\partial}u_m$ as ℓ tends to $+\infty$, and that

$$\nu(T_m, x) \leq \nu\Big(\frac{1}{\ell m}\Delta_{\ell,m}, x\Big) \leq \nu(T,x) + \frac{\mu+1}{\ell m},$$

where $\mu = \max_{x\in X}\operatorname{ord}_x(\sigma_D)$. This, together with the first step, implies the proposition for some subsequence $D_s = \Delta_{\ell(s),s}$, $\ell(s) \gg s \gg 1$. We even obtain the more explicit inequality

$$\nu(T,x) - \frac{n}{m} \leq \nu\Big(\frac{1}{\ell m}\Delta_{\ell,m}, x\Big) \leq \nu(T,x) + \frac{\mu+1}{\ell m}. \qquad \square$$

(3.24) Lemma. *Let Ω be an open subset in $\mathbb{C}^n$ and let $g_1, \ldots, g_N \in H^0(\Omega, \mathcal{O}_\Omega)$ be non zero functions. Let $S \subset H^0(\Omega, \mathcal{O}_\Omega)$ be a finite dimensional subspace whose elements generate all 1-jets at any point of Ω. Finally, set $u = \log \max_j |g_j|$ and*

$$u_\ell = g_1^\ell h_1 + \ldots + g_N^\ell h_N, \ h_j \in S \smallsetminus \{0\}.$$

Then for all $(h_1, \ldots, h_N)$ in $(S \smallsetminus \{0\})^N$ except a set of measure 0, the sequence $\frac{1}{\ell}\log|u_\ell|$ converges to u in $L^1_{\mathrm{loc}}(\Omega)$ and

$$\nu(u,x) \le \nu\Big(\frac{1}{\ell}\log|u_\ell|\Big) \le \nu(u,x) + \frac{1}{\ell}, \ \forall x \in X, \ \forall \ell \ge 1.$$

Proof. The argument is a fairly standard Whitney type argument based on a count of parameters and their dimensions. The details will be omitted. □

(3.25) Exercise. When L is ample and h is a smooth metric with $T = \frac{i}{2\pi}\Theta_{L,h} > 0$, show that the approximating divisors can be taken smooth (and thus irreducible if X is connected).
Hint. In the above proof of Corollary (3.23), the sections g_j have no common zeroes and one can take $\sigma_D = 1$. Moreover, a smooth divisor Δ in an ample linear system is always connected, otherwise two disjoint parts Δ', Δ'' would be big and nef and $\Delta' \cdot \Delta'' = 0$ would contradict the Hovanskii-Teissier inequality when X is connected.

(3.26) Corollary. *On a projective manifold X, effective $\mathbb{Q}$-divisors are dense in the weak topology in the cone $P^{1,1}_{\mathrm{NS}}(X)$ of closed positive $(1,1)$-currents T whose cohomology class $\{T\}$ belongs to the Neron-Severi space $\mathrm{NS}_{\mathbb{R}}(X)$.*

Proof. We may add ε times a Kähler metric ω so as to get $T + \varepsilon\omega > 0$, and then perturb by a small combination $\sum \delta_j \alpha_j$ of classes α_j in a $\mathbb{Z}$-basis of $\mathrm{NS}(X)$ so that $\Theta = T + \varepsilon\omega + \sum \delta_j\alpha_j \ge \frac{\varepsilon}{2}\omega$ and $\{\Theta\} \in H^2(X, \mathbb{Q})$. Then Θ can be approximated by $\mathbb{Q}$-divisors by Corollary (3.23), and the conclusion follows. □

(3.27) Comments. We can rephrase the above results by saying that the cone of closed positive currents $P^{1,1}_{\mathrm{NS}}(X)$ is a completion of the cone of effective $\mathbb{Q}$-divisors. A considerable advantage of using currents is that the cone of currents is locally compact in the weak topology, namely the section of the cone consisting of currents T of mass $\int_X T \wedge \omega^{n-1} = 1$ is compact. This provides a very strong tool for the study of the asymptotic behaviour of linear systems, as required for instance in the Minimal Model Program of Kawamata-Mori-Shokurov. One should be aware, however, that the cone of currents is really huge and contains objects which are very far from being algebraic in any reasonable sense. This occurs very frequently in the realm of complex dynamics. For instance, if $P_m(z,c)$ denotes the m-th iterate of the quadratic polynomial $z \mapsto z^2 + c$, then $P_m(z,c)$ defines a polynomial of degree 2^m on $\mathbb{C}^2$, and the sequence of $\mathbb{Q}$-divisors $D_m = \frac{1}{m}\frac{i}{\pi}\partial\overline{\partial}\log|P_m(z,c)|$ which have mass 1 on $\mathbb{C}^2 \subset \mathbb{P}^2_{\mathbb{C}}$ can be shown to converge to a closed positive current T of mass 1 on $\mathbb{P}^2_{\mathbb{C}}$. The support of this current T is extremely complicated : its slices $c = c_0$ are the Julia sets J_c of the quadratic polynomial $z \mapsto z^2 + c$, and the slice $z = 0$ is the famous Mandelbrot set M. Therefore, in general, limits of divisors in asymptotic linear systems may exhibit a fractal behavior.

3.D. Singularity exponents and log canonical thresholds

The goal of this section to relate "log canonical thresholds" with the α invariant introduced by G. Tian [Tia87] for the study of the existence of Kähler-Einstein metrics. The approximation technique of closed positive $(1,1)$-currents introduced above can be used to show that the α invariant actually coincides with the log canonical threshold (see also [DK01], [JK01], [BGK05], [Dem08]).

Usually, in these applications, only the case of the anticanonical line bundle $L = -K_X$ is considered. Here we will consider more generally the case of an arbitrary line bundle L (or $\mathbb{Q}$-line bundle L) on a complex manifold X, with some additional restrictions which will be stated later. We introduce a generalized version of Tian's invariant α, as defined in [Tia87] (see also [Siu88]).

(3.28) Definition. *Assume that X is a compact manifold and that L is a pseudo-effective line bundle, i.e. L admits a singular hermitian metric h_0 with $\Theta_{L,h_0} \geq 0$. If K is a compact subset of X, we put*

$$\alpha_K(L) = \inf_{\{h,\,\Theta_{L,h}\geq 0\}} c_K(h)$$

where h runs over all singular hermitian metrics on L such that $\Theta_{L,h} \geq 0$.

In algebraic geometry, it is more usual to look instead at linear systems defined by a family of linearly independent sections $\sigma_0, \sigma_1, \ldots \sigma_N \in H^0(X, L^{\otimes m})$. We denote by Σ the vector subspace generated by these sections and by

$$|\Sigma| := P(\Sigma) \subset |mL| := P(H^0(X, L^{\otimes m}))$$

the corresponding linear system. Such an $(N+1)$-tuple of sections $\sigma = (\sigma_j)_{0\leq j\leq N}$ defines a singular hermitian metric h on L by putting in any trivialization

$$|\xi|_h^2 = \frac{|\xi|^2}{\left(\sum_j |\sigma_j(z)|^2\right)^{1/m}} = \frac{|\xi|^2}{|\sigma(z)|^{2/m}}, \qquad \xi \in L_z,$$

hence $h(z) = |\sigma(z)|^{-2/m}$ with $\varphi(z) = \frac{1}{m}\log|\sigma(z)| = \frac{1}{2m}\log\sum_j |\sigma_j(z)|^2$ as the associated weight function. Therefore, we are interested in the number $c_K(|\sigma|^{-2/m})$. In the case of a single section σ_0 (corresponding to a one-point linear system), this is the same as the log canonical threshold $\mathrm{lct}_K(X, \frac{1}{m}D) = c_K(\frac{1}{m}D)$ of the associated divisor D, in the notation of Section 1 of [CS08]. We will also use the formal notation $c_K(\frac{1}{m}|\Sigma|)$ in the case of a higher dimensional linear system $|\Sigma| \subset |mL|$. The main result of this section is

(3.29) Theorem. *Let L be a big line bundle on a compact complex manifold X. Then for every compact set K in X we have*

$$\alpha_K(L) = \inf_{\{h,\,\Theta_{L,h}\geq 0\}} c_K(h) = \inf_{m\in\mathbb{Z}_{>0}} \inf_{D\in|mL|} c_K\Big(\frac{1}{m}D\Big).$$

Proof. Observe that the inequality

$$\inf_{m\in\mathbb{Z}_{>0}} \inf_{D\in|mL|} c_K\Big(\frac{1}{m}D\Big) \geq \inf_{\{h,\,\Theta_{L,h}\geq 0\}} c_K(h)$$

is trivial, since any divisor $D \in |mL|$ gives rise to a singular hermitian metric h.

The converse inequality will follow from the approximation techniques discussed above. Given a big line bundle L on X, there exists a modification $\mu : \widetilde{X} \to X$

of X such that $\widetilde{X}$ is projective and $\mu^*L = \mathcal{O}(A+E)$ where A is an ample divisor and E an effective divisor with rational coefficients. By pushing forward by μ a smooth metric h_A with positive curvature on A, we get a singular hermitian metric h_1 on L such that $\Theta_{L,h_1} \geq \mu_*\Theta_{A,h_A} \geq \varepsilon\omega$ on X. Then for any $\delta > 0$ and any singular hermitian metric h on L with $\Theta_{L,h} \geq 0$, the interpolated metric $h_\delta = h_1^\delta h^{1-\delta}$ satisfies $\Theta_{L,h_\delta} \geq \delta\varepsilon\omega$. Since h_1 is bounded away from 0, it follows that $c_K(h) \geq (1-\delta)c_K(h_\delta)$ by monotonicity. By theorem (3.21, iii) applied to h_δ, we infer

$$c_K(h_\delta) = \lim_{m\to+\infty} c_K(h_{\delta,m}),$$

and we also have

$$c_K(h_{\delta,m}) \geq c_K\Big(\frac{1}{m}D_{\delta,m}\Big)$$

for any divisor $D_{\delta,m}$ associated with a section $\sigma \in H^0(X, L^{\otimes m}\otimes\mathcal{I}(h_\delta^m))$, since the metric $h_{\delta,m}$ is given by $h_{\delta,m} = (\sum_k |g_{m,k}|^2)^{-1/m}$ for an orthornormal basis of such sections. This clearly implies

$$c_K(h) \geq \liminf_{\delta\to 0}\ \liminf_{m\to+\infty} c_K\Big(\frac{1}{m}D_{\delta,m}\Big) \geq \inf_{m\in\mathbb{Z}_{>0}}\ \inf_{D\in|mL|} c_K\Big(\frac{1}{m}D\Big). \qquad \square$$

In the applications, it is frequent to have a finite or compact group G of automorphisms of X and to look at G-invariant objects, namely G-equivariant metrics on G-equivariant line bundles L ; in the case of a reductive algebraic group G we simply consider a compact real form $G^{\mathbb{R}}$ instead of G itself.

One then gets an α invariant $\alpha_{K,G}(L)$ by looking only at G-equivariant metrics in Definition (3.28). All contructions made are then G-equivariant, especially $\mathcal{H}_m \subset |mL|$ is a G-invariant linear system. For every G-invariant compact set K in X, we thus infer

$$\text{(3.30)}\quad \alpha_{K,G}(L) := \inf_{\{h\ G-\text{equiv.},\,\Theta_{L,h}\geq 0\}} c_K(h) = \inf_{m\in\mathbb{Z}_{>0}}\ \inf_{|\Sigma|\subset|mL|,\ \Sigma^G=\Sigma} c_K\Big(\frac{1}{m}|\Sigma|\Big).$$

When G is a finite group, one can pick for m large enough a G-invariant divisor $D_{\delta,m}$ associated with a G-invariant section σ, possibly after multiplying m by the order of G. One then gets the slightly simpler equality

$$\text{(3.31)}\qquad \alpha_{K,G}(L) := \inf_{\{h\ G-\text{equiv.},\,\Theta_{L,h}\geq 0\}} c_K(h) = \inf_{m\in\mathbb{Z}_{>0}}\ \inf_{D\in|mL|^G} c_K\Big(\frac{1}{m}D\Big).$$

In a similar manner, one can work on an orbifold X rather than on a non singular variety. The L^2 techniques work in this setting with almost no change (L^2 estimates are essentially insensitive to singularities, since one can just use an orbifold metric on the open set of regular points).

The main interest of Tian's invariant $\alpha_{X,G}$ (and of the related concept of log canonical threshold) is that it provides a neat criterion for the existence of Kähler-Einstein metrics for Fano manifolds (see [Tia87], [Siu88], [Nad89], [DK01]).

(3.32) Theorem. *Let X be a Fano manifold, i.e. a projective manifold with $-K_X$ ample. Assume that X admits a compact group of automorphisms G such that $\alpha_{X,G}(K_X) > n/(n+1)$. Then X possesses a G-invariant Kähler-Einstein metric.*

We will not give here the details of the proof, which rely on very delicate C^k-estimates (successively for $k = 0, 1, 2, \dots$) for the Monge-Ampère operator. In fine, the required estimates can be shown to depend only on the boundedness of the integral $\int_X e^{-2\gamma\varphi}$ for a suitable constant $\gamma \in]\frac{n}{n+1}, 1]$, where φ is the potential of the Kähler metric $\omega \in c_1(X)$ (also viewed as the weight of a hermitian metric on K_X). Now, one can restrict the estimate to G-invariant weights φ, and this translates into the sufficient condition (3.32). The approach explained in [DK01] simplifies the analysis developped in earlier works by proving first a general semi-continuity theorem which implies the desired a priori bound under the assumption of Theorem 3.32. The semi-continuity theorem states as

(3.33) Theorem ([DK01]). *Let K be a compact set in a complex manifold X. Then the map $\varphi \mapsto c_K(\varphi)^{-1}$ is upper semi-continuous with respect to the weak ($= L^1_{\rm loc}$) topology on the space of plurisubharmonic functions. Moreover, if $\gamma < c_K(\varphi)$, then $\int_K |e^{-2\gamma\psi} - e^{-2\gamma\varphi}|$ converges to 0 when ψ converges to φ in the weak topology.*

Sketch of proof. We will content ourselves by explaining the main points. It is convenient to observe (by a quite easy integration argument suggested to us by J. McNeal) that $c_K(\varphi)$ can be calculated by estimating the Lebesgue volume $\mu_U(\{\varphi < \log r\}$ of tubular neighborhoods as $r \to 0$:

(3.34)

$$c_K(\varphi) = \sup\big\{c \geq 0\,;\, r^{-2c}\mu_U(\{\varphi < \log r\}) \text{ is bounded as } r \to 0, \text{ for some } U \supset K\big\}.$$

The first step is the following important monotonicity result, which is a straightforward consequence of the L^2 extension theorem.

(3.35) Proposition. *Let φ be a quasi-psh function on a complex manifold X, and let $Y \subset X$ be a complex submanifold such that $\varphi_{|Y} \not\equiv -\infty$ on every connected component of Y. Then, if K is a compact subset of Y, we have*

$$c_K(\varphi_{|Y}) \leq c_K(\varphi).$$

(Here, of course, $c_K(\varphi)$ is computed on X, i.e., by means of neighborhoods of K in X).

We need only proving monotonicity for $c_{z_0}(\varphi_{|Y})$ when z_0 is a point of Y. This is done by just extending the holomorphic function $f(z) = 1$ on $B(z_0, r) \cap Y$ with respect to the weight $e^{-2\gamma\varphi}$ whenever $\gamma < c_{z_0}(\varphi_{|Y})$.

(3.36) Proposition. *Let X, Y be complex manifolds of respective dimensions n, m, let $\mathcal{I} \subset \mathcal{O}_X$, $\mathcal{J} \subset \mathcal{O}_Y$ be coherent ideals, and let $K \subset X$, $L \subset Y$ be compact sets. Put $\mathcal{I} \oplus \mathcal{J} := \mathrm{pr}_1^\star \mathcal{I} + \mathrm{pr}_2^\star \mathcal{J} \subset \mathcal{O}_{X\times Y}$. Then*

$$c_{K\times L}(\mathcal{I} \oplus \mathcal{J}) = c_K(\mathcal{I}) + c_L(\mathcal{J}).$$

Proof. It is enough to show that $c_{(x,y)}(\mathcal{I} \oplus \mathcal{J}) = c_x(\mathcal{I}) + c_y(\mathcal{J})$ at every point $(x, y) \in X \times Y$. Without loss of generality, we may assume that $X \subset \mathbb{C}^n$, $Y \subset \mathbb{C}^m$ are open sets and $(x, y) = (0, 0)$. Let $g = (g_1, \ldots, g_p)$, resp. $h = (h_1, \ldots, h_q)$, be systems of generators of $\mathcal{I}$ (resp. $\mathcal{J}$) on a neighborhood of 0. Set

$$\varphi = \log \sum |g_j|^2, \qquad \psi = \log \sum |h_k|^2.$$

Then $\mathcal{I} \oplus \mathcal{J}$ is generated by the $p + q$-tuple of functions

$$g \oplus h = (g_1(x), \ldots g_p(x), h_1(y), \ldots, h_q(y))$$

and the corresponding psh function $\Phi(x,y) = \log\big(\sum |g_j(x)|^2 + \sum |h_k(y)|^2\big)$ has the same behavior along the poles as $\Phi'(x,y) = \max(\varphi(x),\psi(y))$ (up to a term $O(1) \le \log 2$). Now, for sufficiently small neighborhoods U, V of 0, we have

$$\mu_{U\times V}\big(\{\max(\varphi(x),\psi(y)) < \log r\}\big) = \mu_U\big(\{\varphi < \log r\} \times \mu_V(\{\psi < \log r\}\big),$$

and one can derive from this that

$$C_1 r^{2(c+c')} \le \mu_{U\times V}\big(\{\max(\varphi(x),\psi(y)) < \log r\}\big) \le C_2 r^{2(c+c')}\,|\log r|^{n-1+m-1}$$

with $c = c_0(\varphi) = c_0(\mathcal{I})$ and $c' = c_0(\psi) = c_0(\mathcal{J})$. We infer

$$c_{(0,0)}(\mathcal{I}\oplus\mathcal{J}) = c + c' = c_0(\mathcal{I}) + c_0(\mathcal{J}). \qquad \square$$

(3.37) Proposition. *Let f, g be holomorphic on a complex manifold X. Then, for every $x \in X$,*

$$c_x(f+g) \le c_x(f) + c_x(g).$$

More generally, if $\mathcal{I}$ and $\mathcal{J}$ are coherent ideals, then

$$c_x(\mathcal{I}+\mathcal{J}) \le c_x(\mathcal{I}) + c_x(\mathcal{J}).$$

Proof. Let Δ be the diagonal in $X \times X$. Then $\mathcal{I}+\mathcal{J}$ can be seen as the restriction of $\mathcal{I}\oplus\mathcal{J}$ to Δ. Hence Prop. 3.35 and 3.36 combined imply

$$c_x(\mathcal{I}+\mathcal{J}) = c_{(x,x)}((\mathcal{I}\oplus\mathcal{J})_{|\Delta}) \le c_{(x,x)}(\mathcal{I}\oplus\mathcal{J}) = c_x(\mathcal{I}) + c_x(\mathcal{J}).$$

Since $(f+g) \subset (f) + (g)$, we get

$$c_x(f+g) \le c_x((f)+(g)) \le c_x(f) + c_x(g). \qquad \square$$

Now we can explain in rough terms the strategy of proof of Theorem 3.33. We start by approximating psh singularities with analytic singularities, using theorem 3.21. By the argument of Corollary 3.23, we can even reduce ourselves to the case of invertible ideals (f) near $z_0 = 0$, and look at what happens when we have a uniformly convergent sequence $f_\nu \to f$. In this case, we use the Taylor expansion of f at 0 to write $f = p_N + s_N$ where p_N is a polynomial of degree N and $s_N(z) = O(|z|^{N+1})$. Clearly $c_0(s_N) \le n/(N+1)$, and from this we infer $|c_0(f) - c_0(P_N)| \le n/(N+1)$ by 3.37. Similarly, we get the uniform estimate $|c_0(f_\nu) - c_0(P_{\nu,N})| \le n/(N+1)$ for all indices ν. This means that the proof of the semi-continuity theorem is reduced to handling the situation of a finite dimensional space of polynomials. This case is well-known – one can apply Hironaka's desingularization theorem, in a relative version involving the coefficients of our polynomials as parameters. The conclusion is obtained by putting together carefully all required uniform estimates (which involve a lot of L^2 estimates). $\square$

4. Subadditivity of multiplier ideals and Fujita's approximate Zariski decomposition theorem

We first notice the following basic restriction formula for multiplier ideals, which is just a rephrasing of the Ohsawa-Takegoshi extension theorem.

(4.1) Restriction formula. *Let φ be a plurisubharmonic function on a complex manifold X, and let $Y \subset X$ be a submanifold. Then*

$$\mathcal{I}(\varphi_{|Y}) \subset \mathcal{I}(\varphi)_{|Y}.$$

Thus, in some sense, the singularities of φ can only get worse if we restrict to a submanifold (if the restriction of φ to some connected component of Y is identically $-\infty$, we agree that the corresponding multiplier ideal sheaf is zero). The proof is straightforward and just amounts to extending locally a germ of function f on Y near a point $y_0 \in Y$ to a function $\tilde{f}$ on a small Stein neighborhood of y_0 in X, which is possible by the Ohsawa-Takegoshi extension theorem. As a direct consequence, we get:

(4.2) Subadditivity Theorem.

(i) *Let X_1, X_2 be complex manifolds, $\pi_i : X_1 \times X_2 \to X_i$, $i = 1, 2$ the projections, and let φ_i be a plurisubharmonic function on X_i. Then*

$$\mathcal{I}(\varphi_1 \circ \pi_1 + \varphi_2 \circ \pi_2) = \pi_1^\star \mathcal{I}(\varphi_1) \cdot \pi_2^\star \mathcal{I}(\varphi_2).$$

(ii) *Let X be a complex manifold and let φ, ψ be plurisubharmonic functions on X. Then*

$$\mathcal{I}(\varphi + \psi) \subset \mathcal{I}(\varphi) \cdot \mathcal{I}(\psi)$$

Proof. (i) Let us fix two relatively compact Stein open subsets $U_1 \subset X_1$, $U_2 \subset X_2$. Then $\mathcal{H}^2(U_1 \times U_2, \varphi_1 \circ \pi_1 + \varphi_2 \circ \pi_2, \pi_1^\star dV_1 \otimes \pi_2^\star dV_2)$ is the Hilbert tensor product of $\mathcal{H}^2(U_1, \varphi_1, dV_1)$ and $\mathcal{H}^2(U_2, \varphi_2, dV_2)$, and admits $(f'_k \boxtimes f''_l)$ as a Hilbert basis, where (f'_k) and (f''_l) are respective Hilbert bases. Since $\mathcal{I}(\varphi_1 \circ \pi_1 + \varphi_2 \circ \pi_2)_{|U_1 \times U_2}$ is generated as an $\mathcal{O}_{U_1 \times U_2}$ module by the $(f'_k \boxtimes f''_l)$ (from the proof of the coherence of multiplier ideals, see [Var09, section 8.4.1]), we conclude that (i) holds true.

(ii) We apply (i) to $X_1 = X_2 = X$ and the restriction formula to $Y =$ diagonal of $X \times X$. Then

$$\begin{aligned}\mathcal{I}(\varphi + \psi) &= \mathcal{I}\big((\varphi \circ \pi_1 + \psi \circ \pi_2)_{|Y}\big) \subset \mathcal{I}\big(\varphi \circ \pi_1 + \psi \circ \pi_2\big)_{|Y} \\ &= \Big(\pi_1^\star \mathcal{I}(\varphi) \otimes \pi_2^\star \mathcal{I}(\psi)\Big)_{|Y} = \mathcal{I}(\varphi) \cdot \mathcal{I}(\psi).\end{aligned}$$

(4.3) Proposition. *Let $f : X \to Y$ be an arbirary holomorphic map, and let φ be a plurisubharmonic function on Y. Then $\mathcal{I}(\varphi \circ f) \subset f^\star \mathcal{I}(\varphi)$.*

Proof. Let

$$\Gamma_f = \{(x, f(x)\,;\, x \in X\} \subset X \times Y$$

be the graph of f, and let $\pi_X : X \times Y \to X$, $\pi_Y : X \times Y \to Y$ be the natural projections. Then we can view $\varphi \circ f$ as the restriction of $\varphi \circ \pi_Y$ to Γ_f, as π_X is a biholomorphism from Γ_f to X. Hence the restriction formula implies

$$\mathcal{I}(\varphi \circ f) = \mathcal{I}\big((\varphi \circ \pi_Y)_{|\Gamma_f}\big) \subset \mathcal{I}(\varphi \circ \pi_Y)_{|\Gamma_f} = \big(\pi_Y^\star \mathcal{I}(\varphi)\big)_{|\Gamma_f} = f^\star \mathcal{I}(\varphi). \qquad \square$$

As an application of subadditivity, we now reprove a result of Fujita [Fuj94], relating the growth of sections of multiples of a line bundle to the Chern numbers of its "largest nef part". Fujita's original proof is by contradiction, using the Hodge index theorem and intersection inequalities.

Let X be a projective n-dimensional algebraic variety and L a line bundle over X. We define the *volume* of L to be

$$\mathrm{Vol}(L) = \limsup_{k\to+\infty} \frac{n!}{k^n} h^0(X, kL) \in [0, +\infty[.$$

In view of Definition 1.12 and of the Serre-Siegel Lemma 1.13, the line bundle is *big* if and only if $\mathrm{Vol}(L) > 0$. If L is ample, we have $h^q(X, kL) = 0$ for $q \geq 1$ and $k \gg 1$ by the Kodaira-Serre vanishing theorem, hence

$$h^0(X, kL) \sim \chi(X, kL) \sim \frac{L^n}{n!} k^n$$

by the Riemann-Roch formula. Thus $\mathrm{Vol}(L) = L^n$ $(= c_1(L)^n)$ if L is ample. This is still true if L is nef (numerically effective), i.e. if $L \cdot C \geq 0$ for every effective curve C. In fact, one can show that $h^q(X, kL) = O(k^{n-q})$ in that case. The following well-known proposition characterizes big line bundles.

(4.4) Proposition. *The line bundle L is big if and only if there is a multiple $m_0 L$ such that $m_0 L = E + A$, where E is an effective divisor and A an ample divisor.*

Proof. If the condition is satisfied, the decomposition $km_0L = kE + kA$ gives rise to an injection $H^0(X, kA) \hookrightarrow H^0(X, km_0L)$, thus $\mathrm{Vol}(L) \geq m_0^{-n} \mathrm{Vol}(A) > 0$. Conversely, assume that L is big, and take A to be a very ample nonsingular divisor in X. The exact sequence

$$0 \longrightarrow \mathcal{O}_X(kL - A) \longrightarrow \mathcal{O}_X(kL) \longrightarrow \mathcal{O}_A(kL_{|A}) \longrightarrow 0$$

gives rise to a cohomology exact sequence

$$0 \to H^0(X, kL - A) \longrightarrow H^0(X, kL) \longrightarrow H^0(A, kL_{|A}),$$

and $h^0(A, kL_{|A}) = O(k^{n-1})$ since $\dim A = n - 1$. Now, the assumption that L is big implies that $h^0(X, kL) > ck^n$ for infinitely many k, hence $H^0(X, m_0L - A) \neq 0$ for some large integer m_0. If E is the divisor of a section in $H^0(X, m_0L - A)$, we find $m_0L - A = E$, as required. □

(4.5) Lemma. *Let G be an arbitrary line bundle. For every $\varepsilon > 0$, there exists a positive integer m and a sequence $\ell_\nu \uparrow +\infty$ such that*

$$h^0\big(X, \ell_\nu(mL - G)\big) \geq \frac{\ell_\nu^m m^n}{n!}\big(\mathrm{Vol}(L) - \varepsilon\big),$$

in other words, $\mathrm{Vol}(mL - G) \geq m^n(\mathrm{Vol}(L) - \varepsilon)$ *for m large enough.*

Proof. Clearly, $\mathrm{Vol}(mL - G) \geq \mathrm{Vol}(mL - (G + E))$ for every effective divisor E. We can take E so large that $G + E$ is very ample, and we are thus reduced to the case where G is very ample by replacing G with $G + E$. By definition of $\mathrm{Vol}(L)$, there exists a sequence $k_\nu \uparrow +\infty$ such that

$$h^0(X, k_\nu L) \geq \frac{k_\nu^n}{n!}\Big(\mathrm{Vol}(L) - \frac{\varepsilon}{2}\Big).$$

We take $m \gg 1$ (to be precisely chosen later), and $\ell_\nu = \big[\frac{k_\nu}{m}\big]$, so that $k_\nu = \ell_\nu m + r_\nu$, $0 \leq r_\nu < m$. Then

$$\ell_\nu(mL - G) = k_\nu L - (r_\nu L + \ell_\nu G).$$

Fix a constant $a \in \mathbb{N}$ such that $aG - L$ is an effective divisor. Then $r_\nu L \le maG$ (with respect to the cone of effective divisors), hence

$$h^0\big(X, \ell_\nu(mL - G)\big) \ge h^0\big(X, k_\nu L - (\ell_\nu + am)G\big).$$

We select a smooth divisor D in the very ample linear system $|G|$. By looking at global sections associated with the exact sequences of sheaves

$$0 \to \mathcal{O}(-(j+1)D) \otimes \mathcal{O}(k_\nu L) \to \mathcal{O}(-jD) \otimes \mathcal{O}(k_\nu L) \to \mathcal{O}_D(k_\nu L - jD) \to 0,$$

$0 \le j < s$, we infer inductively that

$$\begin{aligned} h^0(X, k_\nu L - sD) &\ge h^0(X, k_\nu L) - \sum_{0 \le j < s} h^0\big(D, \mathcal{O}_D(k_\nu L - jD)\big) \\ &\ge h^0(X, k_\nu L) - s\, h^0\big(D, k_\nu L_{|D}\big) \\ &\ge \frac{k_\nu^n}{n!}\Big(\operatorname{Vol}(L) - \frac{\varepsilon}{2}\Big) - s\, C k_\nu^{n-1} \end{aligned}$$

where C depends only on L and G. Hence, by putting $s = \ell_\nu + am$, we get

$$\begin{aligned} h^0\big(X, \ell_\nu(mL - G)\big) &\ge \frac{k_\nu^n}{n!}\Big(\operatorname{Vol}(L) - \frac{\varepsilon}{2}\Big) - C(\ell_\nu + am)k_\nu^{n-1} \\ &\ge \frac{\ell_\nu^n m^n}{n!}\Big(\operatorname{Vol}(L) - \frac{\varepsilon}{2}\Big) - C(\ell_\nu + am)(\ell_\nu + 1)^{n-1} m^{n-1} \end{aligned}$$

and the desired conclusion follows by taking $\ell_\nu \gg m \gg 1$. □

We are now ready to prove Fujita's decomposition theorem, as reproved in [DEL00].

(4.6) Theorem (Fujita). *Let L be a big line bundle. Then for every $\varepsilon > 0$, there exists a modification $\mu : \widetilde{X} \to X$ and a decomposition $\mu^\star L = E + A$, where E is an effective $\mathbb{Q}$-divisor and A an ample $\mathbb{Q}$-divisor, such that $A^n > \operatorname{Vol}(L) - \varepsilon$.*

(4.7) Remark. Of course, if $\mu^\star L = E + A$ with E effective and A nef, we get an injection

$$H^0(\widetilde{X}, kA) \hookrightarrow H^0(\widetilde{X}, kE + kA) = H^0(\widetilde{X}, k\mu^\star L) = H^0(X, kL)$$

for every integer k which is a multiple of the denominator of E, hence $A^n \le \operatorname{Vol}(L)$.

(4.8) Remark. Once Theorem 4.6 is proved, the same kind of argument easily shows that

$$\operatorname{Vol}(L) = \lim_{k \to +\infty} \frac{n!}{k^n} h^0(X, kL),$$

because the formula is true for every ample line bundle A.

Proof of Theorem 4.6. It is enough to prove the theorem with A being a big and nef divisor. In fact, Proposition 4.4 then shows that we can write $A = E' + A'$ where E' is an effective $\mathbb{Q}$-divisor and A' an ample $\mathbb{Q}$-divisor, hence

$$E + A = E + \varepsilon E' + (1 - \varepsilon)A + \varepsilon A'$$

where $A'' = (1 - \varepsilon)A + \varepsilon A'$ is ample and the intersection number A''^n approaches A^n as closely as we want. Let G be as in Theorem (1.25) (Siu's theorem on uniform global generation). Lemma 4.5 implies that $\operatorname{Vol}(mL - G) > m^n(\operatorname{Vol}(L) - \varepsilon)$ for m

large. By Theorem (1.8) on the existence of analytic Zariski decomposition, there exists a hermitian metric h_m of weight φ_m on $mL - G$ such that

$$H^0\big(X, \ell(mL-G)\big) = H^0\big(X, \ell(mL-G) \otimes \mathcal{I}(\ell\varphi_m)\big)$$

for every $\ell \geq 0$. We take a smooth modification $\mu : \widetilde{X} \to X$ such that

$$\mu^\star \mathcal{I}(\varphi_m) = \mathcal{O}_{\widetilde{X}}(-E)$$

is an invertible ideal sheaf in $\mathcal{O}_{\widetilde{X}}$. This is possible by taking the blow-up of X with respect to the ideal $\mathcal{I}(\varphi_m)$ and by resolving singularities (Hironaka [Hir64]). Theorem 1.25 applied to $L' = mL - G$ implies that $\mathcal{O}(mL) \otimes \mathcal{I}(\varphi_m)$ is generated by its global sections, hence its pull-back $\mathcal{O}(m\,\mu^\star L - E)$ is also generated. This implies

$$m\,\mu^\star L = E + A$$

where E is an effective divisor and A is a nef (semi-ample) divisor in $\widetilde{X}$. We find

$$\begin{aligned}
H^0(\widetilde{X}, \ell A) &= H^0\big(\widetilde{X}, \ell(m\,\mu^\star L - E)\big) \\
&\supset H^0\big(\widetilde{X}, \mu^\star\big(\mathcal{O}(\ell m L) \otimes \mathcal{I}(\varphi_m)^\ell\big)\big) \\
&\supset H^0\big(\widetilde{X}, \mu^\star\big(\mathcal{O}(\ell m L) \otimes \mathcal{I}(\ell\varphi_m)\big)\big),
\end{aligned}$$

thanks to the subadditivity property of multiplier ideals. Moreover, the direct image $\mu_\star\mu^\star\mathcal{I}(\ell\varphi_m)$ coincides with the integral closure of $\mathcal{I}(\ell\varphi_m)$, hence with $\mathcal{I}(\ell\varphi_m)$, because a multiplier ideal sheaf is always integrally closed. From this we infer

$$\begin{aligned}
H^0(\widetilde{X}, \ell A) &\supset H^0\big(X, \mathcal{O}(\ell m L) \otimes \mathcal{I}(\ell\varphi_m)\big) \\
&\supset H^0\big(X, \mathcal{O}(\ell(mL-G)) \otimes \mathcal{I}(\ell\varphi_m)\big) \\
&= H^0\big(X, \mathcal{O}(\ell(mL-G))\big).
\end{aligned}$$

By Lemma 4.5, we find

$$h^0(\widetilde{X}, \ell A) \geq \frac{\ell^n}{n!} m^n \big(\mathrm{Vol}(L) - \varepsilon\big)$$

for infinitely many ℓ, therefore $\mathrm{Vol}(A) = A^n \geq m^n(\mathrm{Vol}(L) - \varepsilon)$. Theorem 4.6 is proved, up to a minor change of notation $E \mapsto \frac{1}{m}E$, $A \mapsto \frac{1}{m}A$. □

We conclude by using Fujita's theorem to establish a geometric interpretation of the volume $\mathrm{Vol}(L)$. Suppose as above that X is a smooth projective variety of dimension n, and that L is a big line bundle on X. Given a large integer $k \gg 0$, denote by $B_k \subset X$ the base-locus of the linear system $|kL|$. The *moving self-intersection number* $(kL)^{[n]}$ of $|kL|$ is defined by choosing n general divisors $D_1, \ldots, D_n \in |kL|$ and putting

$$(kL)^{[n]} = \#\Big(D_1 \cap \ldots \cap D_n \cap (X - B_k)\Big).$$

In other words, we simply count the number of intersection points away from the base locus of n general divisors in the linear system $|kL|$. This notion arises for example in Matsusaka's proof of his "big theorem". We show that the volume $\mathrm{Vol}(L)$ of L measures the rate of growth with respect to k of these moving self-intersection numbers:

(4.9) Proposition. *One has*

$$\mathrm{Vol}(L) = \limsup_{k \to \infty} \frac{(kL)^{[n]}}{k^n}.$$

Proof. We start by interpreting $(kL)^{[n]}$ geometrically. Let $\mu_k : X_k \longrightarrow X$ be a modification of $|kL|$ such that $\mu_k^\star|kL| = |V_k| + F_k$, where

$$P_k := \mu_k^\star(kL) - F_k$$

is generated by sections, and $H^0(X, \mathcal{O}_X(kL)) = V_k = H^0(X_k, \mathcal{O}_{X_k}(P_k))$, so that $B_k = \mu_k(F_k)$. Then evidently $(kL)^{[n]}$ counts the number of intersection points of n general divisors in P_k, and consequently

$$(kL)^{[n]} = (P_k)^n.$$

Since then P_k is big (and nef) for $k \gg 0$, we have $\mathrm{Vol}(P_k) = (P_k)^n$. Also, $\mathrm{Vol}(kL) \geq \mathrm{Vol}(P_k)$ since P_k embeds in $\mu_k^\star(kL)$. Hence

$$\mathrm{Vol}(kL) \geq (kL)^{[n]} \qquad \forall k \gg 0.$$

On the other hand, an easy argument in the spirit of Lemma (4.5) shows that $\mathrm{Vol}(kL) = k^n \cdot \mathrm{Vol}(L)$ (cf. also [ELN96], Lemma 3.4), and so we conclude that

$$\mathrm{Vol}(L) \geq \frac{(kL)^{[n]}}{k^n}. \tag{4.10}$$

for every $k \gg 0$.

For the reverse inequality we use Fujita's theorem. Fix $\varepsilon > 0$, and consider the decomposition $\mu^\star L = A + E$ on $\mu : \widetilde{X} \longrightarrow X$ constructed in Fujita's theorem. Let k be any positive integer such that kA is integral and globally generated. By taking a common resolution we can assume that X_k dominates $\widetilde{X}$, and hence we can write

$$\mu_k^\star kL \sim A_k + E_k$$

with A_k globally generated and

$$(A_k)^n \geq k^n \cdot (\mathrm{Vol}(L) - \varepsilon).$$

But then A_k embeds in P_k and both $\mathcal{O}(A_k)$ and $\mathcal{O}(P_k)$ are globally generated, consequently

$$(A_k)^n \leq (P_k)^n = (kL)^{[n]}.$$

Therefore

$$\frac{(kL)^{[n]}}{k^n} \geq \mathrm{Vol}(L) - \varepsilon. \tag{4.11}$$

But (4.11) holds for any sufficiently large and divisible k, and in view of (4.10) the Proposition follows. □

5. Numerical characterization of the Kähler cone

5.A. Positive classes in intermediate (p,p) bidegrees

We first discuss some general positivity concepts for cohomology classes of type (p,p), although we will not be able to say much about these. Recall that we have a Serre duality pairing

$$H^{p,q}(X, \mathbb{C}) \times H^{n-p,n-q}(X, \mathbb{C}) \longrightarrow \mathbb{C}, \qquad (\alpha, \beta) \longmapsto \int_X \alpha \wedge \beta \in \mathbb{C}. \tag{5.1}$$

In particular, if we restrict to real classes, this yields a duality pairing

$$H^{p,p}(X, \mathbb{R}) \times H^{n-p,n-p}(X, \mathbb{R}) \longrightarrow \mathbb{R}, \qquad (\alpha, \beta) \longmapsto \int_X \alpha \wedge \beta \in \mathbb{R}. \tag{5.2}$$

Now, one can define $H^{p,p}_{\mathrm{SP}}(X,\mathbb{R})$ to be the closure of the cone of classes of d-closed strongly positive smooth (p,p)-forms (a (p,p)-form in $\Lambda^{p,p}T^*_X$ is by definition strongly positive if it is in the convex cone generated by decomposable (p,p) forms $iu_1\wedge\overline{u}_1\wedge\ldots\wedge iu_p\wedge\overline{u}_p$ where the u_j are $(1,0)$-forms). Clearly, $H^{1,1}_{\mathrm{SP}}(X,\mathbb{R})=\overline{\mathcal{K}}$ and the cup product defines a multilinear map

$$\overline{\mathcal{K}}\times\ldots\times\overline{\mathcal{K}}\longrightarrow H^{p,p}_{\mathrm{SP}}(X,\mathbb{R}) \tag{5.3}$$

on the p-fold product of the Kähler cone and its closure. We also have $H^{p,p}_{\mathrm{SP}}(X,\mathbb{R})\subset H^{p,p}_{\geq 0}(X,\mathbb{R})$ where $H^{p,p}_{\geq 0}(X,\mathbb{R})$ is the cone of classes of d-closed weakly positive currents of type (p,p), and the Serre duality pairing induces a positive intersection product

$$H^{p,p}_{\mathrm{SP}}(X,\mathbb{R})\times H^{n-p,n-p}_{\geq 0}(X,\mathbb{R})\longrightarrow\mathbb{R}_+,\qquad(\alpha,T)\longmapsto\int_X\alpha\wedge T\in\mathbb{R}_+ \tag{5.4}$$

(notice that if α is strongly positive and $T\geq 0$, then $\alpha\wedge T$ is a positive measure).

If $\mathcal{C}$ is a convex cone in a finite dimensional vector space E, we denote by $\mathcal{C}^\vee$ the dual cone, i.e. the set of linear forms $u\in E^\star$ which take nonnegative values on all elements of $\mathcal{C}$. By the Hahn-Banach theorem, we always have $\mathcal{C}^{\vee\vee}=\overline{\mathcal{C}}$. A basic problem would be to investigate whether $H^{p,p}_{\mathrm{SP}}(X,\mathbb{R})$ and $H^{n-p,n-p}_{\geq 0}(X,\mathbb{R})$ are always dual cones, and another even harder question, which somehow encompasses the Hodge conjecture, would be to relate these cones to the cones generated by cohomology classes of effective analytic cycles. We are essentially unable to address these extremely difficult questions, except in the special cases $p=1$ or $p=n-1$ which are much better understood and are the main target of the following sections.

5.B. Numerically positive classes of type (1,1)

We describe here the main results obtained in [DP04]. The upshot is that the Kähler cone depends only on the intersection product of the cohomology ring, the Hodge structure and the homology classes of analytic cycles. More precisely, we have:

(5.5) Theorem. *Let X be a compact Kähler manifold. Let $\mathcal{P}$ be the set of real $(1,1)$ cohomology classes $\{\alpha\}$ which are* numerically positive *on analytic cycles, i.e. such that $\int_Y\alpha^p>0$ for every irreducible analytic set Y in X, $p=\dim Y$. Then the Kähler cone $\mathcal{K}$ of X is one of the connected components of $\mathcal{P}$.*

(5.6) Special case. *If X is projective algebraic, then $\mathcal{K}=\mathcal{P}$.*

These results (which are new even in the projective case) can be seen as a generalization of the well-known Nakai-Moishezon criterion. Recall that the Nakai-Moishezon criterion provides a necessary and sufficient criterion for a line bundle to be ample: *a line bundle $L\to X$ on a projective algebraic manifold X is ample if and only if*

$$L^p\cdot Y=\int_Y c_1(L)^p>0,$$

for every algebraic subset $Y\subset X$, $p=\dim Y$.

It turns out that the numerical conditions $\int_Y\alpha^p>0$ also characterize arbitrary transcendental Kähler classes when X is projective: this is precisely the meaning of the special case 5.6.

(5.7) Example. The following example shows that the cone $\mathcal{P}$ need not be connected (and also that the components of $\mathcal{P}$ need not be convex, either). Let us consider for instance a complex torus $X = \mathbb{C}^n/\Lambda$. It is well-known that a generic torus X does not possess any analytic subset except finite subsets and X itself. In that case, the numerical positivity is expressed by the single condition $\int_X \alpha^n > 0$. However, on a torus, $(1,1)$-classes are in one-to-one correspondence with constant hermitian forms α on $\mathbb{C}^n$. Thus, for X generic, $\mathcal{P}$ is the set of hermitian forms on $\mathbb{C}^n$ such that $\det(\alpha) > 0$, and Theorem 5.5 just expresses the elementary result of linear algebra saying that the set $\mathcal{K}$ of positive definite forms is one of the connected components of the open set $\mathcal{P} = \{\det(\alpha) > 0\}$ of hermitian forms of positive determinant (the other components, of course, are the sets of forms of signature (p,q), $p+q=n$, q even. They are not convex when $p>0$ and $q>0$).

Sketch of proof of Theorems 5.5 and 5.6. By definition (1.16) (iv), a *Kähler current* is a closed positive current T of type $(1,1)$ such that $T \geq \varepsilon\omega$ for some smooth Kähler metric ω and $\varepsilon > 0$ small enough. The crucial steps of the proof of Theorem 5.5 are contained in the following statements.

(5.8) Proposition (Păun [Pau98a, 98b]). *Let X be a compact complex manifold (or more generally a compact complex space). Then*

(a) *The cohomology class of a closed positive $(1,1)$-current $\{T\}$ is nef if and only if the restriction $\{T\}_{|Z}$ is nef for every irreducible component Z in any of the Lelong sublevel sets $E_c(T)$.*

(b) *The cohomology class of a Kähler current $\{T\}$ is a Kähler class (i.e. the class of a* smooth *Kähler form) if and only if the restriction $\{T\}_{|Z}$ is a Kähler class for every irreducible component Z in any of the Lelong sublevel sets $E_c(T)$.*

The proof of Proposition 5.8 is not extremely hard if we take for granted the fact that Kähler currents can be approximated by Kähler currents with logarithmic poles, a fact which was first proved in section 3.B (see also [Dem92]). Thus in (b), we may assume that $T = \alpha + i\partial\overline{\partial}\varphi$ is a current with analytic singularities, where φ is a quasi-psh function with logarithmic poles on some analytic set Z, and φ smooth on $X \smallsetminus Z$. Now, we proceed by an induction on dimension (to do this, we have to consider analytic spaces rather than with complex manifolds, but it turns out that this makes no difference for the proof). Hence, by the induction hypothesis, there exists a smooth potential ψ on Z such that $\alpha_{|Z} + i\partial\overline{\partial}\psi > 0$ along Z. It is well known that one can then find a potential $\widetilde{\psi}$ on X such that $\alpha + i\partial\overline{\partial}\widetilde{\psi} > 0$ in a neighborhood V of Z (but possibly non positive elsewhere). Essentially, it is enough to take an arbitrary extension of ψ to X and to add a large multiple of the square of the distance to Z, at least near smooth points; otherwise, we stratify Z by its successive singularity loci, and proceed again by induction on the dimension of these loci. Finally, we use a a standard gluing procedure : the current $T = \alpha + i\max_\varepsilon(\varphi, \widetilde{\psi} - C)$, $C \gg 1$, will be equal to $\alpha + i\partial\overline{\partial}\varphi > 0$ on $X \smallsetminus V$, and to a smooth Kähler form on V. □

The next (and more substantial step) consists of the following result which is reminiscent of the Grauert-Riemenschneider conjecture ([Siu84], [Dem85]).

(5.9) Theorem ([DP04]). *Let X be a compact Kähler manifold and let $\{\alpha\}$ be a nef class (i.e. $\{\alpha\} \in \overline{\mathcal{K}}$). Assume that $\int_X \alpha^n > 0$. Then $\{\alpha\}$ contains a Kähler current T, in other words $\{\alpha\} \in \mathcal{E}^\circ$.*

Step 1. The basic argument is to prove that for every irreducible analytic set $Y \subset X$ of codimension p, the class $\{\alpha\}^p$ contains a closed positive (p,p)-current Θ such that $\Theta \geq \delta[Y]$ for some $\delta > 0$. For this, we use in an essentail way the Calabi-Yau theorem [Yau78] on solutions of Monge-Ampère equations, which yields the following result as a special case:

(5.10) Lemma ([Yau78]). *Let (X,ω) be a compact Kähler manifold and $n = \dim X$. Then for any smooth volume form $f > 0$ such that $\int_X f = \int_X \omega^n$, there exist a Kähler metric $\widetilde{\omega} = \omega + i\partial\overline{\partial}\varphi$ in the same Kähler class as ω, such that $\widetilde{\omega}^n = f$.* □

We exploit this by observing that $\alpha + \varepsilon\omega$ is a Kähler class. Hence we can solve the Monge-Ampère equation

(5.10 a) $$(\alpha + \varepsilon\omega + i\partial\overline{\partial}\varphi_\varepsilon)^n = C_\varepsilon \omega_\varepsilon^n$$

where (ω_ε) is the family of Kähler metrics on X produced by Lemma 3.4 (iii), such that their volume is concentrated in an ε-tubular neighborhood of Y.

$$C_\varepsilon = \frac{\int_X \alpha_\varepsilon^n}{\int_X \omega_\varepsilon^n} = \frac{\int_X (\alpha + \varepsilon\omega)^n}{\int_X \omega^n} \geq C_0 = \frac{\int_X \alpha^n}{\int_X \omega^n} > 0.$$

Let us denote by

$$\lambda_1(z) \leq \ldots \leq \lambda_n(z)$$

the eigenvalues of $\alpha_\varepsilon(z)$ with respect to $\omega_\varepsilon(z)$, at every point $z \in X$ (these functions are continuous with respect to z, and of course depend also on ε). The equation (5.10a) is equivalent to the fact that

(5.10 b) $$\lambda_1(z) \ldots \lambda_n(z) = C_\varepsilon$$

is constant, and the most important observation for us is that the constant C_ε is bounded away from 0, thanks to our assumption $\int_X \alpha^n > 0$.

Fix a regular point $x_0 \in Y$ and a small neighborhood U (meeting only the irreducible component of x_0 in Y). By Lemma 3.4, we have a uniform lower bound

(5.10 c) $$\int_{U \cap V_\varepsilon} \omega_\varepsilon^p \wedge \omega^{n-p} \geq \delta_p(U) > 0.$$

Now, by looking at the p smallest (resp. $(n-p)$ largest) eigenvalues λ_j of α_ε with respect to ω_ε, we find

(5.10 d) $$\alpha_\varepsilon^p \geq \lambda_1 \ldots \lambda_p \, \omega_\varepsilon^p,$$

(5.10 e) $$\alpha_\varepsilon^{n-p} \wedge \omega_\varepsilon^p \geq \frac{1}{n!}\lambda_{p+1} \ldots \lambda_n \, \omega_\varepsilon^n,$$

The last inequality (5.10 e) implies

$$\int_X \lambda_{p+1} \ldots \lambda_n \, \omega_\varepsilon^n \leq n! \int_X \alpha_\varepsilon^{n-p} \wedge \omega_\varepsilon^p = n! \int_X (\alpha + \varepsilon\omega)^{n-p} \wedge \omega^p \leq M$$

for some constant $M > 0$ (we assume $\varepsilon \leq 1$, say). In particular, for every $\delta > 0$, the subset $E_\delta \subset X$ of points z such that $\lambda_{p+1}(z) \ldots \lambda_n(z) > M/\delta$ satisfies $\int_{E_\delta} \omega_\varepsilon^n \leq \delta$, hence

$$\int_{E_\delta} \omega_\varepsilon^p \wedge \omega^{n-p} \le 2^{n-p} \int_{E_\delta} \omega_\varepsilon^n \le 2^{n-p}\delta. \tag{5.10 f}$$

The combination of (5.10 c) and (5.10 f) yields

$$\int_{(U\cap V_\varepsilon)\smallsetminus E_\delta} \omega_\varepsilon^p \wedge \omega^{n-p} \ge \delta_p(U) - 2^{n-p}\delta.$$

On the other hand (5.10 b) and (5.10 d) imply

$$\alpha_\varepsilon^p \ge \frac{C_\varepsilon}{\lambda_{p+1}\dots\lambda_n}\omega_\varepsilon^p \ge \frac{C_\varepsilon}{M/\delta}\omega_\varepsilon^p \qquad \text{on } (U\cap V_\varepsilon)\smallsetminus E_\delta.$$

From this we infer

$$\int_{U\cap V_\varepsilon} \alpha_\varepsilon^p\wedge\omega^{n-p} \ge \frac{C_\varepsilon}{M/\delta}\int_{(U\cap V_\varepsilon)\smallsetminus E_\delta} \omega_\varepsilon^p\wedge\omega^{n-p} \ge \frac{C_\varepsilon}{M/\delta}(\delta_p(U)-2^{n-p}\delta) > 0 \tag{5.10 g}$$

provided that δ is taken small enough, e.g. $\delta = 2^{-(n-p+1)}\delta_p(U)$. The family of (p,p)-forms α_ε^p is uniformly bounded in mass since

$$\int_X \alpha_\varepsilon^p \wedge \omega^{n-p} = \int_X (\alpha+\varepsilon\omega)^p \wedge \omega^{n-p} \le \text{Const}.$$

Inequality (5.10 g) implies that any weak limit Θ of (α_ε^p) carries a positive mass on $U\cap Y$. By Skoda's extension theorem [Sko82], $\mathbf{1}_Y\Theta$ is a closed positive current with support in Y, hence $\mathbf{1}_Y\Theta = \sum c_j[Y_j]$ is a combination of the various components Y_j of Y with coefficients $c_j > 0$. Our construction shows that Θ belongs to the cohomology class $\{\alpha\}^p$. Step 1 of Theorem 5.9 is proved.

Step 2. The second and final step consists in using a "diagonal trick": for this, we apply Step 1 to

$$\widetilde{X} = X\times X, \qquad \widetilde{Y} = \text{diagonal}\,\Delta \subset \widetilde{X}, \qquad \widetilde{\alpha} = \text{pr}_1^*\,\alpha + \text{pr}_2^*\,\alpha.$$

It is then clear that $\widetilde{\alpha}$ is nef on $\widetilde{X}$ and that

$$\int_{\widetilde{X}} (\widetilde{\alpha})^{2n} = \binom{2n}{n}\Big(\int_X \alpha^n\Big)^2 > 0.$$

It follows by Step 1 that the class $\{\widetilde{\alpha}\}^n$ contains a Kähler current Θ of bidegree (n,n) such that $\Theta \ge \delta[\Delta]$ for some $\delta > 0$. Therefore the push-forward

$$T := (\text{pr}_1)_*(\Theta\wedge \text{pr}_2^*\,\omega)$$

is a positive $(1,1)$-current such that

$$T \ge \delta(\text{pr}_1)_*([\Delta]\wedge\text{pr}_2^*\,\omega) = \delta\omega.$$

It follows that T is a Kähler current. On the other hand, T is numerically equivalent to $(\text{pr}_1)_*(\widetilde{\alpha}^n\wedge\text{pr}_2^*\,\omega)$, which is the form given in coordinates by

$$x \mapsto \int_{y\in X} \big(\alpha(x)+\alpha(y)\big)^n \wedge \omega(y) = C\alpha(x)$$

where $C = n\int_X \alpha(y)^{n-1}\wedge\omega(y)$. Hence $T \equiv C\alpha$, which implies that $\{\alpha\}$ contains a Kähler current. Theorem 5.9 is proved. □

End of Proof of Theorems 5.5 and 5.6. Clearly the open cone $\mathcal{K}$ is contained in $\mathcal{P}$, hence in order to show that $\mathcal{K}$ is one of the connected components of $\mathcal{P}$, we need only show that $\mathcal{K}$ is closed in $\mathcal{P}$, i.e. that $\overline{\mathcal{K}}\cap\mathcal{P}\subset\mathcal{K}$. Pick a class $\{\alpha\}\in\overline{\mathcal{K}}\cap\mathcal{P}$.

In particular $\{\alpha\}$ is nef and satisfies $\int_X \alpha^n > 0$. By Theorem 5.9 we conclude that $\{\alpha\}$ contains a Kähler current T. However, an induction on dimension using the assumption $\int_Y \alpha^p$ for all analytic subsets Y (we also use resolution of singularities for Y at this step) shows that the restriction $\{\alpha\}_{|Y}$ is the class of a Kähler current on Y. We conclude that $\{\alpha\}$ is a Kähler class by 5.8 (b), therefore $\{\alpha\} \in \mathcal{K}$, as desired. □

The projective case 5.6 is a consequence of the following variant of Theorem 5.5.

(5.11) Corollary. *Let X be a compact Kähler manifold. A $(1,1)$ cohomology class $\{\alpha\}$ on X is Kähler if and only if there exists a Kähler metric ω on X such that $\int_Y \alpha^k \wedge \omega^{p-k} > 0$ for all irreducible analytic sets Y and all $k = 1, 2, \ldots, p = \dim Y$.*

Proof. The assumption clearly implies that

$$\int_Y (\alpha + t\omega)^p > 0$$

for all $t \in \mathbb{R}_+$, hence the half-line $\alpha + (\mathbb{R}_+)\omega$ is entirely contained in the cone $\mathcal{P}$ of numerically positive classes. Since $\alpha + t_0\omega$ is Kähler for t_0 large, we conclude that the half-line in entirely contained in the connected component $\mathcal{K}$, and therefore $\alpha \in \mathcal{K}$. □

In the projective case, we can take $\omega = c_1(H)$ for a given very ample divisor H, and the condition $\int_Y \alpha^k \wedge \omega^{p-k} > 0$ is equivalent to

$$\int_{Y\cap H_1 \cap \ldots \cap H_{p-k}} \alpha^k > 0$$

for a suitable complete intersection $Y \cap H_1 \cap \ldots \cap H_{p-k}$, $H_j \in |H|$. This shows that algebraic cycles are sufficient to test the Kähler property, and the special case 5.6 follows. On the other hand, we can pass to the limit in 5.11 by replacing α by $\alpha + \varepsilon\omega$, and in this way we get also a characterization of nef classes.

(5.12) Corollary. *Let X be a compact Kähler manifold. A $(1,1)$ cohomology class $\{\alpha\}$ on X is nef if and only if there exists a Kähler metric ω on X such that $\int_Y \alpha^k \wedge \omega^{p-k} \geq 0$ for all irreducible analytic sets Y and all $k = 1, 2, \ldots, p = \dim Y$.*

By a formal convexity argument, one can derive from 5.11 or 5.12 the following interesting consequence about the dual of the cone $\mathcal{K}$.

(5.13) Theorem. *Let X be a compact Kähler manifold.*

(a) *A $(1,1)$ cohomology class $\{\alpha\}$ on X is nef if and only for every irreducible analytic set Y in X, $p = \dim X$ and every Kähler metric ω on X we have $\int_Y \alpha \wedge \omega^{p-1} \geq 0$. (Actually this numerical condition is needed only for Kähler classes $\{\omega\}$ which belong to a 2-dimensional space $\mathbb{R}\{\alpha\} + \mathbb{R}\{\omega_0\}$, where $\{\omega_0\}$ is a given Kähler class).*

(b) *The dual of the nef cone $\overline{\mathcal{K}}$ is the closed convex cone in $H^{n-1,n-1}(X,\mathbb{R})$ generated by cohomology classes of currents of the form $[Y]\wedge\omega^{p-1}$ in $H^{n-1,n-1}(X,\mathbb{R})$, where Y runs over the collection of irreducible analytic subsets of X and $\{\omega\}$ over the set of Kähler classes of X. This dual cone coincides with $H^{n-1,n-1}_{\geq 0}(X,\mathbb{R})$.*

Proof. (a) Clearly a nef class $\{\alpha\}$ satisfies the given numerical condition. The proof of the converse is more tricky. First, observe that for every integer $p \geq 1$, there exists a polynomial identity of the form

$$(5.14) \qquad (y-\delta x)^p - (1-\delta)^p x^p = (y-x)\int_0^1 A_p(t,\delta)\big((1-t)x+ty\big)^{p-1}\,dt$$

where $A_p(t,\delta) = \sum_{0\leq m\leq p} a_m(t)\delta^m \in \mathbb{Q}[t,\delta]$ is a polynomial of degree $\leq p-1$ in t (moreover, the polynomial A_p is unique under this limitation for the degree). To see this, we observe that $(y-\delta x)^p - (1-\delta)^p x^p$ vanishes identically for $x = y$, so it is divisible by $y - x$. By homogeneity in (x,y), we have an expansion of the form

$$(y-\delta x)^p - (1-\delta)^p x^p = (y-x)\sum_{0\leq \ell\leq p-1,\,0\leq m\leq p} b_{\ell,m}x^\ell y^{p-1-\ell}\delta^m$$

in the ring $\mathbb{Z}[x,y,\delta]$. Formula (5.14) is then equivalent to

$$(5.14') \qquad b_{\ell,m} = \int_0^1 a_m(t)\binom{p-1}{\ell}(1-t)^\ell t^{p-1-\ell}\,dt.$$

Since $(U,V)\mapsto \int_0^1 U(t)V(t)dt$ is a non degenerate linear pairing on the space of polynomials of degree $\leq p-1$ and since $(\binom{p-1}{\ell}(1-t)^\ell t^{p-1-\ell})_{0\leq \ell\leq p-1}$ is a basis of this space, (5.14′) can be achieved for a unique choice of the polynomials $a_m(t)$. A straightforward calculation shows that $A_p(t,0) = p$ identically. We can therefore choose $\delta_0 \in [0,1[$ so small that $A_p(t,\delta) > 0$ for all $t \in [0,1]$, $\delta \in [0,\delta_0]$ and $p = 1,2,\ldots,n$.

Now, fix a Kähler metric ω such that $\omega' = \alpha+\omega$ yields a Kähler class $\{\omega'\}$ (just take a large multiple $\omega = k\omega_0$, $k \gg 1$, of the given Kähler metric ω_0 to initialize the process). A substitution $x = \omega$ and $y = \omega'$ in our polynomial identity yields

$$(\alpha+(1-\delta)\omega)^p - (1-\delta)^p\omega^p = \int_0^1 A_p(t,\delta)\,\alpha\wedge\big((1-t)\omega+t\omega'\big)^{p-1}dt.$$

For every irreducible analytic subset $Y \subset X$ of dimension p we find

$$\int_Y(\alpha+(1-\delta)\omega)^p - (1-\delta)^p\int_Y\omega^p = \int_0^1 A_p(t,\delta)dt\Big(\int_Y \alpha\wedge\big((1-t)\omega+t\omega'\big)^{p-1}\Big).$$

However, $(1-t)\omega+t\omega'$ is a Kähler class (contained in $\mathbb{R}\{\alpha\}+\mathbb{R}\{\omega_0\}$) and therefore $\int_Y \alpha\wedge\big((1-t)\omega+t\omega'\big)^{p-1} \geq 0$ by the numerical condition. This implies $\int_Y(\alpha+(1-\delta)\omega)^p > 0$ for all $\delta \in [0,\delta_0]$. We have produced a segment entirely contained in $\mathcal{P}$ such that one extremity $\{\alpha+\omega\}$ is in $\mathcal{K}$, so the other extremity $\{\alpha+(1-\delta_0)\omega\}$ is also in $\mathcal{K}$. By repeating the argument inductively after replacing ω with $(1-\delta_0)\omega$, we see that $\{\alpha+(1-\delta_0)^\nu\omega\} \in \mathcal{K}$ for every integer $\nu \geq 0$. From this we infer that $\{\alpha\}$ is nef, as desired. (b) Part (a) can be reformulated by saying that the dual cone $\overline{\mathcal{K}}^\vee$ is the closure of the convex cone generated by $(n-1,n-1)$ cohomology classes of the form $[Y]\wedge\omega^{p-1}$. Since these classes are contained in $H^{n-1,n-1}_{\geq 0}(X,\mathbb{R})$ which is also contained in $\overline{\mathcal{K}}^\vee$ by (5.6), we infer that

$$(5.15) \qquad \overline{\mathcal{K}}^\vee = H^{n-1,n-1}_{\geq 0}(X,\mathbb{R}) = \overline{\mathrm{Cone}(\{[Y]\wedge\omega^{p-1}\})}. \qquad \square$$

5.C. Deformations of compact Kähler manifolds

Our main Theorem 5.5 also has an important application to the deformation theory of compact Kähler manifolds.

(5.16) Theorem. *Let $\pi : \mathcal{X} \to S$ be a deformation of compact Kähler manifolds over an irreducible base S. Then there exists a countable union $S' = \bigcup S_\nu$ of analytic subsets $S_\nu \subsetneq S$, such that the Kähler cones $\mathcal{K}_t \subset H^{1,1}(X_t, \mathbb{C})$ of the fibers $X_t = \pi^{-1}(t)$ are invariant over $S \smallsetminus S'$ under parallel transport with respect to the $(1,1)$-projection $\nabla^{1,1}$ of the Gauss-Manin connection ∇ in the decomposition of*

$$\nabla = \begin{pmatrix} \nabla^{2,0} & * & 0 \\ * & \nabla^{1,1} & * \\ 0 & * & \nabla^{0,2} \end{pmatrix}$$

on the Hodge bundle $H^2 = H^{2,0} \oplus H^{1,1} \oplus H^{0,2}$.

We moreover conjecture that for an arbitrary deformation $\mathcal{X} \to S$ of compact complex manifolds, the Kähler property is open with respect to the countable Zariski topology on the base S of the deformation.

Let us recall the general fact that all fibers X_t of a deformation over a connected base S are diffeomorphic, since $\mathcal{X} \to S$ is a locally trivial differentiable bundle. This implies that the cohomology bundle

$$S \ni t \mapsto H^k(X_t, \mathbb{C})$$

is locally constant over the base S. The corresponding (flat) connection of this bundle is called the Gauss-Manin connection, and will be denoted here by ∇. As is well known, the Hodge filtration

$$F^p(H^k(X_t, \mathbb{C})) = \bigoplus_{r+s=k, r \geq p} H^{r,s}(X_t, \mathbb{C})$$

defines a *holomorphic* subbundle of $H^k(X_t, \mathbb{C})$ (with respect to its locally constant structure). On the other hand, the Dolbeault groups are given by

$$H^{p,q}(X_t, \mathbb{C}) = F^p(H^k(X_t, \mathbb{C})) \cap \overline{F^{k-p}(H^k(X_t, \mathbb{C}))}, \qquad k = p + q,$$

and they form *real analytic* subbundles of $H^k(X_t, \mathbb{C})$. We are interested especially in the decomposition

$$H^2(X_t, \mathbb{C}) = H^{2,0}(X_t, \mathbb{C}) \oplus H^{1,1}(X_t, \mathbb{C}) \oplus H^{0,2}(X_t, \mathbb{C})$$

and the induced decomposition of the Gauss-Manin connection acting on H^2

$$\nabla = \begin{pmatrix} \nabla^{2,0} & * & * \\ * & \nabla^{1,1} & * \\ * & * & \nabla^{0,2} \end{pmatrix}.$$

Here the stars indicate suitable bundle morphisms – actually with the lower left and upper right stars being zero by Griffiths' transversality property, but we do not really care here. The notation $\nabla^{p,q}$ stands for the induced (real analytic, not necessarily flat) connection on the subbundle $t \mapsto H^{p,q}(X_t, \mathbb{C})$.

Sketch of Proof of Theorem 5.16. The result is local on the base, hence we may assume that S is contractible. Then the family is differentiably trivial, the Hodge bundle $t \mapsto H^2(X_t, \mathbb{C})$ is the trivial bundle and $t \mapsto H^2(X_t, \mathbb{Z})$ is a trivial lattice. We use the existence of a relative cycle space $C^p(\mathcal{X}/S) \subset C^p(\mathcal{X})$ which consists of all cycles contained in the fibres of $\pi : X \to S$. It is equipped with a canonical holomorphic projection

$$\pi_p : C^p(\mathcal{X}/S) \to S.$$

We then define the S_ν's to be the images in S of those connected components of $C^p(\mathcal{X}/S)$ which do not project onto S. By the fact that the projection is proper on each component, we infer that S_ν is an analytic subset of S. The definition of the S_ν's imply that the cohomology classes induced by the analytic cycles $\{[Z]\}$, $Z \subset X_t$, remain exactly the same for all $t \in S \smallsetminus S'$. This result implies in its turn that the conditions defining the numerically positive cones $\mathcal{P}_t$ remain the same, except for the fact that the spaces $H^{1,1}(X_t, \mathbb{R}) \subset H^2(X_t, \mathbb{R})$ vary along with the Hodge decomposition. At this point, a standard calculation implies that the $\mathcal{P}_t$ are invariant by parallel transport under $\nabla^{1,1}$. This is done as follows.

Since S is irreducible and S' is a countable union of analytic sets, it follows that $S \smallsetminus S'$ is arcwise connected by piecewise smooth analytic arcs. Let

$$\gamma : [0,1] \to S \smallsetminus S', \qquad u \mapsto t = \gamma(u)$$

be such a smooth arc, and let $\alpha(u) \in H^{1,1}(X_{\gamma(u)}, \mathbb{R})$ be a family of real $(1,1)$-cohomology classes which are constant by parallel transport under $\nabla^{1,1}$. This is equivalent to assuming that

$$\nabla(\alpha(u)) \in H^{2,0}(X_{\gamma(u)}, \mathbb{C}) \oplus H^{0,2}(X_{\gamma(u)}, \mathbb{C})$$

for all u. Suppose that $\alpha(0)$ is a numerically positive class in $X_{\gamma(0)}$. We then have

$$\alpha(0)^p \cdot \{[Z]\} = \int_Z \alpha(0)^p > 0$$

for all p-dimensional analytic cycles Z in $X_{\gamma(0)}$. Let us denote by

$$\zeta_Z(t) \in H^{2q}(X_t, \mathbb{Z}), \qquad q = \dim X_t - p,$$

the family of cohomology classes equal to $\{[Z]\}$ at $t = \gamma(0)$, such that $\nabla\zeta_Z(t) = 0$ (i.e. constant with respect to the Gauss-Manin connection). By the above discussion, $\zeta_Z(t)$ is of type (q,q) for all $t \in S$, and when $Z \subset X_{\gamma(0)}$ varies, $\zeta_Z(t)$ generates all classes of analytic cycles in X_t if $t \in S \smallsetminus S'$. Since ζ_Z is ∇-parallel and $\nabla\alpha(u)$ has no component of type $(1,1)$, we find

$$\frac{d}{du}(\alpha(u)^p \cdot \zeta_Z(\gamma(u)) = p\alpha(u)^{p-1} \cdot \nabla\alpha(u) \cdot \zeta_Z(\gamma(u)) = 0.$$

We infer from this that $\alpha(u)$ is a numerically positive class for all $u \in [0,1]$. This argument shows that the set $\mathcal{P}_t$ of numerically positive classes in $H^{1,1}(X_t, \mathbb{R})$ is invariant by parallel transport under $\nabla^{1,1}$ over $S \smallsetminus S'$.

By a standard result of Kodaira-Spencer [KS60] relying on elliptic PDE theory, every Kähler class in X_{t_0} can be deformed to a nearby Kähler class in nearby fibres X_t. This implies that the connected component of $\mathcal{P}_t$ which corresponds to the Kähler cone $\mathcal{K}_t$ must remain the same. The theorem is proved. □

As a by-product of our techniques, especially the regularization theorem for currents, we also get the following result for which we refer to [DP04].

(5.17) Theorem. *A compact complex manifold carries a Kähler current if and only if it is bimeromorphic to a Kähler manifold (or equivalently, dominated by a Kähler manifold).*

This class of manifolds is called the *Fujiki class* $\mathcal{C}$. If we compare this result with the solution of the Grauert-Riemenschneider conjecture, it is tempting to make the following conjecture which would somehow encompass both results.

(5.18) Conjecture. *Let X be a compact complex manifold of dimension n. Assume that X possesses a nef cohomology class $\{\alpha\}$ of type $(1,1)$ such that $\int_X \alpha^n > 0$. Then X is in the Fujiki class $\mathcal{C}$. [Also, $\{\alpha\}$ would contain a Kähler current, as it follows from Theorem 5.9 if Conjecture 5.18 is proved].*

We want to mention here that most of the above results were already known in the cases of complex surfaces (i.e. in dimension 2), thanks to the work of N. Buchdahl [Buc99, 00] and A. Lamari [Lam99a, 99b].

Shortly after the original [DP04] manuscript appeared in April 2001, Daniel Huybrechts [Huy01] informed us that Theorem 5.5 can be used to calculate the Kähler cone of a very general hyperkähler manifold: the Kähler cone is then equal to a suitable connected component of the positive cone defined by the Beauville-Bogomolov quadratic form. In the case of an arbitrary hyperkähler manifold, S.Boucksom [Bou02] later showed that a $(1,1)$ class $\{\alpha\}$ is Kähler if and only if it lies in the positive part of the Beauville-Bogomolov quadratic cone and moreover $\int_C \alpha > 0$ for all *rational curves* $C \subset X$ (see also [Huy99]).

6. Structure of the pseudo-effective cone and mobile intersection theory

6.A. Classes of mobile curves and of mobile $(n-1,n-1)$-currents

We introduce various positive cones in $H^{n-1,n-1}(X,\mathbb{R})$, some of which exhibit certain "mobility" properties, in the sense that they can be more or less freely deformed. Ampleness is clearly such a property, since a very ample divisor A can be moved in its linear system $|A|$ so as to cover the whole ambient variety. By extension, a Kähler class $\{\omega\} \in H^{1,1}(X,\mathbb{R})$ is also considered to be mobile, as illustrated alternatively by the fact that the Monge-Ampère volume form $(\omega + i\partial\overline{\partial}\varphi)^n$ of a Kähler metric in the same cohomology class can be taken to be equal to an arbitrary volume form $f > 0$ with $\int_X f = \int_X \omega^n$ (thanks to Yau's theorem [Yau78]).

(6.1) Definition. *Let X be a smooth projective variety.*

(i) *One defines* $\mathrm{NE}(X)$ *to be the convex cone generated by cohomology classes of all effective curves in* $H^{n-1,n-1}(X,\mathbb{R})$

(ii) *We say that C is a* mobile curve *if $C = C_{t_0}$ is a member of an analytic family $(C_t)_{t\in S}$ such that $\bigcup_{t\in S} C_t = X$ and, as such, is a reduced irreducible 1-cycle. We define the mobile cone* $\mathrm{ME}(X)$, *to be the convex cone generated by all mobile curves.*

(iii) *If X is projective, we say that an effective 1-cycle C is a* strongly mobile *if we have*

$$C = \mu_\star(\widetilde{A}_1 \cap \ldots \cap \widetilde{A}_{n-1})$$

for suitable very ample divisors $\widetilde{A}_j$ on $\widetilde{X}$, where $\mu : \widetilde{X} \to X$ is a modification. We let $\mathrm{ME}^s(X)$ *be the convex cone generated by all strongly mobile effective 1-cycles (notice that by taking $\widetilde{A}_j$ general enough these classes can be represented by reduced irreducible curves; also, by Hironaka, one could just restrict oneself to compositions of blow-ups with smooth centers).*

Clearly, we have $\mathrm{ME}^s(X) \subset \mathrm{ME}(X) \subset \mathrm{NE}(X)$. The cone $\mathrm{NE}(X)$ is contained in the analogue of the Neron-Severi group for $(n-1,n-1)$-classes, namely

$$\mathrm{NS}^{n-1}_{\mathbb{R}}(X) := (H^{n-1,n-1}(X,\mathbb{R}) \cap H^{2n-2}(X,\mathbb{Z})/\mathrm{tors}) \otimes_{\mathbb{Z}} \mathbb{R}$$

(sometimes also denoted $N_1(X)$ in the litterature). We wish to introduce similar concepts for cones of non necessarily integral classes, on arbitrary compact Kähler manifolds. The relevant definition is as follows.

(6.2) Definition. *Let X be a compact Kähler manifold.*

(i) *We define the set $\mathcal{N} = H^{n-1,n-1}_{\geq 0}(X,\mathbb{R})$ to be the (closed) convex cone in $H^{n-1,n-1}(X,\mathbb{R})$ generated by classes of positive currents T of type $(n-1,n-1)$, i.e., of bidimension $(1,1)$.*

(ii) *We define the cone $\mathcal{M}^s \subset H^{n-1,n-1}(X,\mathbb{R})$ of* strongly mobile classes *to be the closure of the convex cone generated by classes of currents of the form*

$$\mu_\star(\widetilde{\omega}_1 \wedge \ldots \wedge \widetilde{\omega}_{n-1})$$

where $\mu : \widetilde{X} \to X$ is an arbitrary modification, and the $\widetilde{\omega}_j$ are Kähler forms on $\widetilde{X}$.

(iii) *We define the cone $\mathcal{M} \subset H^{n-1,n-1}(X,\mathbb{R})$ of* mobile classes *to be the closure of the convex cone generated by classes of currents of the form*

$$\mu_\star([\widetilde{Y}_{t_0}] \wedge \widetilde{\omega}_1 \wedge \ldots \wedge \widetilde{\omega}_{p-1})$$

where $\mu : \widetilde{X} \to X$ is an arbitrary modification, the $\widetilde{\omega}_j$ are Kähler forms on $\widetilde{X}$ and $(\widetilde{Y}_t)_{t\in S}$ is an analytic family of effective p-dimensional analytic cycles covering $\widetilde{X}$ such that $\widetilde{Y}_{t_0}$ is reduced and irreducible, with p running over all $\{1,2,\ldots,n\}$.

Clearly, we have

$$\mathcal{M}^s \subset \mathcal{M} \subset \mathcal{N}.$$

For X projective, it is also immediately clear from the definitions that

$$\tag{6.3} \begin{cases} \overline{\mathrm{NE}(X)} \subset \mathcal{N}_{\mathrm{NS}} := \mathcal{N} \cap \mathrm{NS}^{n-1}_{\mathbb{R}}(X), \\ \overline{\mathrm{ME}(X)} \subset \mathcal{M}_{\mathrm{NS}} := \mathcal{M} \cap \mathrm{NS}^{n-1}_{\mathbb{R}}(X), \\ \overline{\mathrm{ME}^s(X)} \subset \mathcal{M}^s_{\mathrm{NS}} := \mathcal{M}^s \cap \mathrm{NS}^{n-1}_{\mathbb{R}}(X). \end{cases}$$

The upshot of these definitions lie in the following easy observation.

(6.4) Proposition. *Let X be a compact Kähler manifold. The Serre duality pairing*

$$H^{1,1}(X,\mathbb{R}) \times H^{n-1,n-1}(X,\mathbb{R}) \longrightarrow \mathbb{R}, \qquad (\alpha,\beta) \longmapsto \int_X \alpha \wedge \beta$$

takes nonnegative values

(a) *for all pairs $(\alpha,\beta) \in \overline{\mathcal{K}} \times \mathcal{N}$;*

(b) *for all pairs $(\alpha,\beta) \in \mathcal{E} \times \mathcal{M}$.*

Proof. (a) is obvious. In order to prove (b), we may assume that

$$\beta = \mu_\star([Y_{t_0}] \wedge \widetilde{\omega}_1 \wedge \ldots \wedge \widetilde{\omega}_{p-1})$$

for some modification $\mu : \widetilde{X} \to X$, where $\{\alpha\} = \{T\}$ is the class of a positive $(1,1)$-current on X and $\widetilde{\omega}_j$ are Kähler forms on $\widetilde{X}$. Then for $t \in S$ generic

$$\begin{aligned}\int_X \alpha \wedge \beta &= \int_X T \wedge \mu_\star([\widetilde{Y}_t] \wedge \widetilde{\omega}_1 \wedge \ldots \wedge \widetilde{\omega}_{p-1})\\ &= \int_X \mu^* T \wedge [\widetilde{Y}_t] \wedge \widetilde{\omega}_1 \wedge \ldots \wedge \widetilde{\omega}_{p-1}\\ &= \int_{\widetilde{Y}_t} (\mu^* T)_{\restriction \widetilde{Y}_t} \wedge \widetilde{\omega}_1 \wedge \ldots \wedge \widetilde{\omega}_{p-1} \geq 0 \end{aligned} \tag{6.5}$$

provided that we show that the final integral is well defined and that the formal calulations involved in (6.5) are correct. Here, we have used the fact that a closed positive $(1,1)$-current T always has a pull-back $\mu^\star T$, which follows from the observation that if $T = \alpha + i\partial\overline{\partial}\varphi$ with α smooth and φ quasi-psh, we may always set $\mu^\star T = \mu^*\alpha + i\partial\overline{\partial}(\varphi \circ \mu)$, with $\varphi \circ \mu$ quasi-psh and not identically $-\infty$ on $\widetilde{X}$. Similarly, we see that the restriction $(\mu^* T)_{\restriction \widetilde{Y}_t}$ is a well defined positive $(1,1)$-current for t generic, by putting

$$(\mu^* T)_{\restriction \widetilde{Y}_t} = (\mu^*\alpha)_{\restriction \widetilde{Y}_t} + i\partial\overline{\partial}\big((\varphi \circ \mu)_{\restriction \widetilde{Y}_t}\big)$$

and choosing t such that $\widetilde{Y}_t$ is not contained in the pluripolar set of $-\infty$ poles of $\varphi \circ \mu$ (this is possible thanks to the assumption that $\widetilde{Y}_t$ covers $\widetilde{X}$; locally near any given point we can modify α so that $\alpha = 0$ on a small neighborhood V, and then φ is psh on V). Finally, in order to justify the formal calculations we can use a regularization argument for T, writing $T = \lim T_k$ with $T_k = \alpha + i\partial\overline{\partial}\varphi_k$ and a decreasing sequence of smooth almost plurisubharmonic potentials $\varphi_k \downarrow \varphi$ such that the Levi forms have a uniform lower bound $i\partial\overline{\partial}\varphi_k \geq -C\omega$ (such a sequence exists by [Dem92]). Then $(\mu^* T_k)_{\restriction \widetilde{Y}_t} \to (\mu^* T)_{\restriction \widetilde{Y}_t}$ in the weak topology of currents. □

Proposition 6.4 leads to the natural question whether the cones $(\overline{\mathcal{K}}, \mathcal{N})$ and $(\mathcal{E}, \mathcal{M})$ are dual under Serre duality, The second part of the question is addressed in the next section. The results proved in § 5 yield a complete answer to the first part – even in the general Kähler setting.

(6.6) Theorem. *Let X be a compact Kähler manifold. Then*

(i) *$\overline{\mathcal{K}}$ and $\mathcal{N}$ are dual cones.*

(ii) *If X is projective algebraic, then $\overline{\mathcal{K}}_{\mathrm{NS}} = \mathrm{Nef}(X)$ and $\mathcal{N}_{\mathrm{NS}} = \overline{\mathrm{NE}(X)}$ and these cones are dual.*

Proof. (i) is a weaker version of (5.13 b).

(ii) The equality $\overline{\mathcal{K}}_{\mathrm{NS}} = \mathrm{Nef}(X)$ has already been discussed and is a consequence of the Kodaira embedding theorem. Now, we know that

$$\overline{\mathrm{NE}(X)} \subset \mathcal{N}_{\mathrm{NS}} \subset \overline{\mathcal{K}}_{\mathrm{NS}}^{\vee} = \mathrm{Nef}(X)^{\vee},$$

where the second inclusion is a consequence of (5.8 a). However, it is already well-known that $\overline{\mathrm{NE}(X)}$ and $\overline{\mathrm{NE}(X)}$ are dual cones (see [Har70]), hence the inclusions are equalities (we could also obtain a self-contained proof by reconsidering the arguments used for (5.13 a) when α and ω_0 are rational classes; one sees by the density of the rationals that the numerical condition for α is needed only for elements of the form $[Y]\wedge\omega^{p-1}$ with $\omega\in\mathbb{Q}\{\alpha\}+\mathbb{Q}\{\omega_0\}$ a rational class, so $[Y]\wedge\omega^{p-1}$ is then a $\mathbb{Q}$-effective curve). □

6.B. Zariski decomposition and mobile intersections

Let X be compact Kähler and let $\alpha\in\mathcal{E}^\circ$ be in the *interior* of the pseudo–effective cone. In analogy with the algebraic context such a class α is called "big", and it can then be represented by a *Kähler current* T, i.e. a closed positive $(1,1)$-current T such that $T\geq\delta\omega$ for some smooth hermitian metric ω and a constant $\delta\ll 1$. We first need a variant of the regularization theorem proved in section 3.B.

(6.7) Regularization theorem for currents. *Let X be a compact complex manifold equipped with a hermitian metric ω. Let $T=\alpha+i\partial\overline{\partial}\varphi$ be a closed $(1,1)$-current on X, where α is smooth and φ is a quasi-plurisubharmonic function. Assume that $T\geq\gamma$ for some real $(1,1)$-form γ on X with real coefficients. Then there exists a sequence $T_m=\alpha+i\partial\overline{\partial}\varphi_m$ of closed $(1,1)$-currents such that*

(i) *φ_m (and thus T_m) is smooth on the complement $X\smallsetminus Z_m$ of an analytic set Z_m, and the Z_m's form an increasing sequence*

$$Z_0\subset Z_1\subset\ldots\subset Z_m\subset\ldots\subset X.$$

(ii) *There is a uniform estimate $T_m\geq\gamma-\delta_m\omega$ with $\lim\downarrow\delta_m=0$ as m tends to $+\infty$.*

(iii) *The sequence (φ_m) is non increasing, and we have $\lim\downarrow\varphi_m=\varphi$. As a consequence, T_m converges weakly to T as m tends to $+\infty$.*

(iv) *Near Z_m, the potential φ_m has logarithmic poles, namely, for every $x_0\in Z_m$, there is a neighborhood U of x_0 such that $\varphi_m(z)=\lambda_m\log\sum_\ell|g_{m,\ell}|^2+O(1)$ for suitable holomorphic functions $(g_{m,\ell})$ on U and $\lambda_m>0$. Moreover, there is a (global) proper modification $\mu_m:\widetilde{X}_m\to X$ of X, obtained as a sequence of blow-ups with smooth centers, such that $\varphi_m\circ\mu_m$ can be written locally on $\widetilde{X}_m$ as*

$$\varphi_m\circ\mu_m(w)=\lambda_m\Big(\sum n_\ell\log|\widetilde{g}_\ell|^2+f(w)\Big)$$

where $(\widetilde{g}_\ell=0)$ are local generators of suitable (global) divisors D_ℓ on $\widetilde{X}_m$ such that $\sum D_\ell$ has normal crossings, n_ℓ are positive integers, and the f's are smooth functions on $\widetilde{X}_m$.

Sketch of proof. We essentially repeat the proofs of Theorems (3.2) and (3.12) with additional considerations. One fact that does not follow readily from these proofs is the monotonicity of the sequence φ_m (which we will not really need anyway). For this, we can take $m=2^\nu$ and use the subadditivity technique already explained in Step 3 of the proof of Theorem (11.3 b). The map μ_m is obtained by blowing-up the (global) ideals $\mathcal{J}_m$ defined by the holomorphic functions $(g_{j,m})$ in the local approximations $\varphi_m\sim\frac{1}{2m}\log\sum_j|g_{j,m}|^2$. By Hironaka [Hir64], we can achieve that $\mu_m^*\mathcal{J}_m$ is an invertible ideal sheaf associated with a normal crossing divisor. □

(6.8) Corollary. *If T is a Kähler current, then one can write $T=\lim T_m$ for a sequence of Kähler currents T_m which have logarithmic poles with coefficients in $\frac{1}{m}\mathbb{Z}$, i.e. there are modifications $\mu_m:X_m\to X$ such that*

$$\mu_m^\star T_m=[E_m]+\beta_m$$

where E_m is an effective $\mathbb{Q}$-divisor on X_m with coefficients in $\frac{1}{m}\mathbb{Z}$ (the "fixed part") and β_m is a closed semi-positive form (the "mobile part").

Proof. We apply Theorem (6.7) with $\gamma = \varepsilon\omega$ and m so large that $\delta_m \le \varepsilon/2$. Then T_m has analytic singularities and $T_m \ge \frac{\varepsilon}{2}\omega$, so we get a composition of blow-ups $\mu_m : X_m \to X$ such

$$\mu_m^* T_m = [E_m] + \beta_m,$$

where E_m is an effective $\mathbb{Q}$-divisor and $\beta_m \ge \frac{\varepsilon}{2}\mu_m^*\omega$. In particular, β_m is strictly positive outside the exceptional divisors, by playing with the multiplicities of the components of the exceptional divisors in E_m, we could even achieve that β_m is a Kähler class on X_m. Notice also that by construction, μ_m is obtained by blowing-up the multiplier ideal sheaves $\mathcal{I}(mT) = \mathcal{I}(m\varphi)$ associated to a potential φ of T. □

The more familiar algebraic analogue would be to take $\alpha = c_1(L)$ with a big line bundle L and to blow-up the base locus of $|mL|$, $m \gg 1$, to get a $\mathbb{Q}$-divisor decomposition

$$\mu_m^\star L \sim E_m + D_m, \qquad E_m \text{ effective, } D_m \text{ free.}$$

Such a blow-up is usually referred to as a "log resolution" of the linear system $|mL|$, and we say that $E_m + D_m$ is an approximate Zariski decomposition of L. We will also use this terminology for Kähler currents with logarithmic poles.

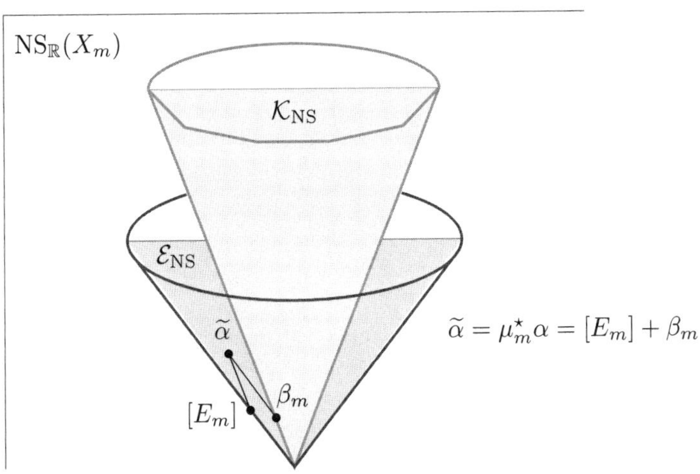

(6.9) Definition. *We define the* volume, *or* mobile self-intersection *of a big class $\alpha \in \mathcal{E}^\circ$ to be*

$$\mathrm{Vol}(\alpha) = \sup_{T\in\alpha} \int_{\widetilde{X}} \beta^n > 0$$

where the supremum is taken over all Kähler currents $T \in \alpha$ with logarithmic poles, and $\mu^\star T = [E] + \beta$ with respect to some modification $\mu : \widetilde{X} \to X$.

By Fujita [Fuj94] and Demailly-Ein-Lazarsfeld [DEL00], if L is a big line bundle, we have

$$\mathrm{Vol}(c_1(L)) = \lim_{m\to+\infty} D_m^n = \lim_{m\to+\infty} \frac{n!}{m^n} h^0(X, mL),$$

and in these terms, we get the following statement.

(6.10) Proposition. *Let L be a big line bundle on the projective manifold X. Let $\epsilon > 0$. Then there exists a modification $\mu : X_\epsilon \to X$ and a decomposition $\mu^*(L) = E + \beta$ with E an effective $\mathbb{Q}$-divisor and β a big and nef $\mathbb{Q}$-divisor such that*

$$\mathrm{Vol}(L) - \varepsilon \le \mathrm{Vol}(\beta) \le \mathrm{Vol}(L).$$

It is very useful to observe that the supremum in Definition 6.9 is actually achieved by a collection of currents whose singularities satisfy a filtering property. Namely, if $T_1 = \alpha + i\partial\overline{\partial}\varphi_1$ and $T_2 = \alpha + i\partial\overline{\partial}\varphi_2$ are two Kähler currents with logarithmic poles in the class of α, then

$$T = \alpha + i\partial\overline{\partial}\varphi, \qquad \varphi = \max(\varphi_1, \varphi_2) \tag{6.11}$$

is again a Kähler current with weaker singularities than T_1 and T_2. One could define as well

$$T = \alpha + i\partial\overline{\partial}\varphi, \qquad \varphi = \frac{1}{2m}\log(e^{2m\varphi_1} + e^{2m\varphi_2}), \tag{6.11'}$$

where $m = \mathrm{lcm}(m_1, m_2)$ is the lowest common multiple of the denominators occuring in T_1, T_2. Now, take a simultaneous log-resolution $\mu_m : X_m \to X$ for which the singularities of T_1 and T_2 are resolved as $\mathbb{Q}$-divisors E_1 and E_2. Then clearly the associated divisor in the decomposition $\mu_m^\star T = [E] + \beta$ is given by $E = \min(E_1, E_2)$. By doing so, the volume $\int_{X_m} \beta^n$ gets increased, as we shall see in the proof of Theorem 6.12 below.

(6.12) Theorem (Boucksom [Bou02]). *Let X be a compact Kähler manifold. We denote here by $H^{k,k}_{\ge 0}(X)$ the cone of cohomology classes of type (k,k) which have non-negative intersection with all closed semi-positive smooth forms of bidegree $(n-k, n-k)$.*

(i) *For each integer $k = 1, 2, \ldots, n$, there exists a canonical "mobile intersection product"*

$$\mathcal{E} \times \cdots \times \mathcal{E} \to H^{k,k}_{\ge 0}(X), \quad (\alpha_1, \ldots, \alpha_k) \mapsto \langle \alpha_1 \cdot \alpha_2 \cdots \alpha_{k-1} \cdot \alpha_k \rangle$$

such that $\mathrm{Vol}(\alpha) = \langle \alpha^n \rangle$ *whenever α is a big class.*

(ii) *The product is increasing, homogeneous of degree 1 and superadditive in each argument, i.e.*

$$\langle \alpha_1 \cdots (\alpha_j' + \alpha_j'') \cdots \alpha_k \rangle \ge \langle \alpha_1 \cdots \alpha_j' \cdots \alpha_k \rangle + \langle \alpha_1 \cdots \alpha_j'' \cdots \alpha_k \rangle.$$

It coincides with the ordinary intersection product when the $\alpha_j \in \overline{\mathcal{K}}$ are nef classes.

(iii) *The mobile intersection product satisfies the Teissier-Hovanskii inequalities ([Hov79], [Tei79, 82])*

$$\langle \alpha_1 \cdot \alpha_2 \cdots \alpha_n \rangle \ge (\langle \alpha_1^n \rangle)^{1/n} \ldots (\langle \alpha_n^n \rangle)^{1/n} \qquad (\textit{with } \langle \alpha_j^n \rangle = \mathrm{Vol}(\alpha_j)).$$

(iv) *For $k = 1$, the above "product" reduces to a (non linear) projection operator*

$$\mathcal{E} \to \mathcal{E}_1, \qquad \alpha \to \langle \alpha \rangle$$

onto a certain convex subcone $\mathcal{E}_1$ *of* $\mathcal{E}$ *such that* $\overline{\mathcal{K}} \subset \mathcal{E}_1 \subset \mathcal{E}$. *Moreover, there is a "divisorial Zariski decomposition"*

$$\alpha = \{N(\alpha)\} + \langle\alpha\rangle$$

where $N(\alpha)$ *is a uniquely defined effective divisor which is called the "negative divisorial part" of* α. *The map* $\alpha \mapsto N(\alpha)$ *is homogeneous and subadditive, and* $N(\alpha) = 0$ *if and only if* $\alpha \in \mathcal{E}_1$.

(v) *The components of* $N(\alpha)$ *always consist of divisors whose cohomology classes are linearly independent, especially* $N(\alpha)$ *has at most* $\rho = \operatorname{rank}_{\mathbb{Z}} \operatorname{NS}(X)$ *components.*

Proof. We essentially repeat the arguments developped in [Bou02], with some simplifications arising from the fact that X is supposed to be Kähler from the start.

(i) First assume that all classes α_j are big, i.e. $\alpha_j \in \mathcal{E}^\circ$. Fix a smooth closed $(n-k, n-k)$ *semi-positive* form u on X. We select Kähler currents $T_j \in \alpha_j$ with logarithmic poles, and a simultaneous log-resolution $\mu : \widetilde{X} \to X$ such that

$$\mu^\star T_j = [E_j] + \beta_j.$$

We consider the direct image current $\mu_\star(\beta_1 \wedge \ldots \wedge \beta_k)$ (which is a closed positive current of bidegree (k,k) on X) and the corresponding integrals

$$\int_{\widetilde{X}} \beta_1 \wedge \ldots \wedge \beta_k \wedge \mu^\star u \geq 0.$$

If we change the representative T_j with another current T'_j, we may always take a simultaneous log-resolution such that $\mu^\star T'_j = [E'_j] + \beta'_j$, and by using (6.11′) we can always assume that $E'_j \leq E_j$. Then $D_j = E_j - E'_j$ is an effective divisor and we find $[E_j] + \beta_j \equiv [E'_j] + \beta'_j$, hence $\beta'_j \equiv \beta_j + [D_j]$. A substitution in the integral implies

$$\begin{aligned}\int_{\widetilde{X}} &\beta'_1 \wedge \beta_2 \wedge \ldots \wedge \beta_k \wedge \mu^\star u \\ &= \int_{\widetilde{X}} \beta_1 \wedge \beta_2 \wedge \ldots \wedge \beta_k \wedge \mu^\star u + \int_{\widetilde{X}} [D_1] \wedge \beta_2 \wedge \ldots \wedge \beta_k \wedge \mu^\star u \\ &\geq \int_{\widetilde{X}} \beta_1 \wedge \beta_2 \wedge \ldots \wedge \beta_k \wedge \mu^\star u.\end{aligned}$$

Similarly, we can replace successively all forms β_j by the β'_j, and by doing so, we find

$$\int_{\widetilde{X}} \beta'_1 \wedge \beta'_2 \wedge \ldots \wedge \beta'_k \wedge \mu^\star u \geq \int_{\widetilde{X}} \beta_1 \wedge \beta_2 \wedge \ldots \wedge \beta_k \wedge \mu^\star u.$$

We claim that the closed positive currents $\mu_\star(\beta_1 \wedge \ldots \wedge \beta_k)$ are uniformly bounded in mass. In fact, if ω is a Kähler metric in X, there exists a constant $C_j \geq 0$ such that $C_j\{\omega\} - \alpha_j$ is a Kähler class. Hence $C_j\omega - T_j \equiv \gamma_j$ for some Kähler form γ_j on X. By pulling back with μ, we find $C_j\mu^\star\omega - ([E_j] + \beta_j) \equiv \mu^\star\gamma_j$, hence

$$\beta_j \equiv C_j\mu^\star\omega - ([E_j] + \mu^\star\gamma_j).$$

By performing again a substitution in the integrals, we find

$$\int_{\widetilde{X}} \beta_1 \wedge \ldots \wedge \beta_k \wedge \mu^\star u \leq C_1 \ldots C_k \int_{\widetilde{X}} \mu^\star\omega^k \wedge \mu^\star u = C_1 \ldots C_k \int_X \omega^k \wedge u$$

and this is true especially for $u = \omega^{n-k}$. We can now arrange that for each of the integrals associated with a countable dense family of forms u, the supremum is achieved by a sequence of currents $(\mu_m)_\star(\beta_{1,m} \wedge \ldots \wedge \beta_{k,m})$ obtained as direct images by a suitable sequence of modifications $\mu_m : \widetilde{X}_m \to X$. By extracting a subsequence, we can achieve that this sequence is weakly convergent and we set

$$\langle \alpha_1 \cdot \alpha_2 \cdots \alpha_k \rangle = \lim_{m \to +\infty} \uparrow \{(\mu_m)_\star(\beta_{1,m} \wedge \beta_{2,m} \wedge \ldots \wedge \beta_{k,m})\}$$

(the monotonicity is not in terms of the currents themselves, but in terms of the integrals obtained when we evaluate against a smooth closed semi-positive form u). By evaluating against a basis of positive classes $\{u\} \in H^{n-k,n-k}(X)$, we infer by Serre duality that the class of $\langle \alpha_1 \cdot \alpha_2 \cdots \alpha_k \rangle$ is uniquely defined (although, in general, the representing current is not unique).

(ii) It is indeed clear from the definition that the mobile intersection product is homogeneous, increasing and superadditive in each argument, at least when the α_j's are in $\mathcal{E}^\circ$. However, we can extend the product to the closed cone $\mathcal{E}$ by monotonicity, by setting

$$\langle \alpha_1 \cdot \alpha_2 \cdots \alpha_k \rangle = \lim_{\delta \downarrow 0} \downarrow \langle (\alpha_1 + \delta\omega) \cdot (\alpha_2 + \delta\omega) \cdots (\alpha_k + \delta\omega) \rangle$$

for arbitrary classes $\alpha_j \in \mathcal{E}$ (again, monotonicity occurs only where we evaluate against closed semi-positive forms u). By weak compactness, the mobile intersection product can always be represented by a closed positive current of bidegree (k, k).

(iii) The Teissier-Hovanskii inequalities are a direct consequence of the fact that they hold true for nef classes, so we just have to apply them to the classes $\beta_{j,m}$ on $\widetilde{X}_m$ and pass to the limit.

(iv) When $k = 1$ and $\alpha \in \mathcal{E}^0$, we have

$$\alpha = \lim_{m \to +\infty} \{(\mu_m)_\star T_m\} = \lim_{m \to +\infty} (\mu_m)_\star[E_m] + \{(\mu_m)_\star \beta_m\}$$

and $\langle \alpha \rangle = \lim_{m \to +\infty}\{(\mu_m)_\star \beta_m\}$ by definition. However, the images $F_m = (\mu_m)_\star F_m$ are effective $\mathbb{Q}$-divisors in X, and the filtering property implies that F_m is a decreasing sequence. It must therefore converge to a (uniquely defined) limit $F = \lim F_m := N(\alpha)$ which is an effective $\mathbb{R}$-divisor, and we get the asserted decomposition in the limit.

Since $N(\alpha) = \alpha - \langle \alpha \rangle$ we easily see that $N(\alpha)$ is subadditive and that $N(\alpha) = 0$ if α is the class of a smooth semi-positive form. When α is no longer a big class, we define

$$\langle \alpha \rangle = \lim_{\delta \downarrow 0} \downarrow \langle \alpha + \delta\omega \rangle, \qquad N(\alpha) = \lim_{\delta \downarrow 0} \uparrow N(\alpha + \delta\omega)$$

(the subadditivity of N implies $N(\alpha + (\delta + \varepsilon)\omega) \leq N(\alpha + \delta\omega)$). The divisorial Zariski decomposition follows except maybe for the fact that $N(\alpha)$ might be a convergent countable sum of divisors. However, this will be ruled out when (v) is proved. As $N(\bullet)$ is subadditive and homogeneous, the set $\mathcal{E}_1 = \{\alpha \in \mathcal{E} \,;\, N(\alpha) = 0\}$ is a closed convex conne, and we find that $\alpha \mapsto \langle \alpha \rangle$ is a projection of $\mathcal{E}$ onto $\mathcal{E}_1$ (according to [Bou02], $\mathcal{E}_1$ consists of those pseudo-effective classes which are "nef in codimension 1").

(v) Let $\alpha \in \mathcal{E}^\circ$, and assume that $N(\alpha)$ contains linearly dependent components F_j. Then already all currents $T \in \alpha$ should be such that $\mu^\star T = [E] + \beta$ where $F = \mu_\star E$

contains those linearly dependent components. Write $F = \sum \lambda_j F_j$, $\lambda_j > 0$ and assume that

$$\sum_{j\in J} c_j F_j \equiv 0$$

for a certain non trivial linear combination. Then some of the coefficients c_j must be negative (and some other positive). Then E is numerically equivalent to

$$E' \equiv E + t\mu^\star\Big(\sum \lambda_j F_j\Big),$$

and by choosing $t > 0$ appropriate, we obtain an effective divisor E' which has a zero coefficient on one of the components $\mu^\star F_{j_0}$. By replacing E with $\min(E, E')$ via (6.11′), we eliminate the component $\mu^\star F_{j_0}$. This is a contradiction since $N(\alpha)$ was supposed to contain F_{j_0}. □

(6.13) Definition. *For a class $\alpha \in H^{1,1}(X, \mathbb{R})$, we define the* numerical dimension $\mathrm{nd}(\alpha)$ *to be* $\mathrm{nd}(\alpha) = -\infty$ *if α is not pseudo-effective, and*

$$\mathrm{nd}(\alpha) = \max\{p \in \mathbb{N}\,;\ \langle\alpha^p\rangle \neq 0\}, \qquad \mathrm{nd}(\alpha) \in \{0, 1, \ldots, n\}$$

if α is pseudo-effective.

By the results of [DP04], a class is big ($\alpha \in \mathcal{E}^\circ$) if and only if $\mathrm{nd}(\alpha) = n$. Classes of numerical dimension 0 can be described much more precisely, again following Boucksom [Bou02].

(6.14) Theorem. *Let X be a compact Kähler manifold. Then the subset $\mathcal{D}_0$ of irreducible divisors D in X such that $\mathrm{nd}(D) = 0$ is countable, and these divisors are rigid as well as their multiples. If $\alpha \in \mathcal{E}$ is a pseudo-effective class of numerical dimension 0, then α is numerically equivalent to an effective $\mathbb{R}$-divisor $D = \sum_{j\in J} \lambda_j D_j$, for some finite subset $(D_j)_{j\in J} \subset \mathcal{D}_0$ such that the cohomology classes $\{D_j\}$ are linearly independent and some $\lambda_j > 0$. If such a linear combination is of numerical dimension 0, then so is any other linear combination of the same divisors.*

Proof. It is immediate from the definition that a pseudo-effective class is of numerical dimension 0 if and only if $\langle\alpha\rangle = 0$, in other words if $\alpha = N(\alpha)$. Thus $\alpha \equiv \sum \lambda_j D_j$ as described in 6.14, and since $\lambda_j\langle D_j\rangle \leq \langle\alpha\rangle$, the divisors D_j must themselves have numerical dimension 0. There is at most one such divisor D in any given cohomology class in $NS(X) \cap \mathcal{E} \subset H^2(X, \mathbb{Z})$, otherwise two such divisors $D \equiv D'$ would yield a blow-up $\mu : \widetilde{X} \to X$ resolving the intersection, and by taking $\min(\mu^\star D, \mu^\star D')$ via (6.11′), we would find $\mu^\star D \equiv E + \beta$, $\beta \neq 0$, so that $\{D\}$ would not be of numerical dimension 0. This implies that there are at most countably many divisors of numerical dimension 0, and that these divisors are rigid as well as their multiples. □

(6.15) Remark. If L is an arbitrary holomorphic line bundle, we define its numerical dimension to be $\mathrm{nd}(L) = \mathrm{nd}(c_1(L))$. Using the cananical maps $\Phi_{|mL|}$ and pulling-back the Fubini-Study metric it is immediate to see that $\mathrm{nd}(L) \geq \kappa(L)$ (which generalizes the analogue inequality already seen for nef line bundles, see (1.21)).

The above general concept of numerical dimension leads to a very natural formulation of the abundance conjecture for Kähler varieties.

(6.16) Generalized abundance conjecture. *Let X be an arbitrary compact Kähler manifold X.*

(a) *The Kodaira dimension of X should be equal to its numerical dimension : $\kappa(K_X) = \mathrm{nd}(K_X)$.*

(b) *More generally, let Δ be a $\mathbb{Q}$-divisor which is klt (Kawamata log terminal, i.e. such that $c_X(\Delta) > 1$). Then $\kappa(K_X + \Delta) = \mathrm{nd}(K_X + \Delta)$.*

This appears to be a fairly strong statement. In fact, already in the case $\Delta = 0$, it is not difficult to show that the generalized abundance conjecture would contain the $C_{n,m}$ conjectures.

(6.17) Remark. It is obvious that abundance holds in the case $\mathrm{nd}(K_X) = -\infty$ (if L is not pseudo-effective, no multiple of L can have sections), or in the case $\mathrm{nd}(K_X) = n$ which implies K_X big (the latter property follows e.g. from the solution of the Grauert-Riemenschneider conjecture in the form proven in [Dem85], see also [DP04]).

In the remaining cases, the most tractable situation is the case when $\mathrm{nd}(K_X) = 0$. In fact Theorem 6.14 then gives $K_X \equiv \sum \lambda_j D_j$ for some effective divisor with numerically independent components, $\mathrm{nd}(D_j) = 0$. It follows that the λ_j are rational and therefore

$$(*) \qquad K_X \sim \sum \lambda_j D_j + F \qquad \text{where } \lambda_j \in \mathbb{Q}^+,\ \mathrm{nd}(D_j) = 0 \text{ and } F \in \mathrm{Pic}^0(X).$$

If we assume additionally that $q(X) = h^{0,1}(X)$ is zero, then mK_X is linearly equivalent to an integral divisor for some multiple m, and it follows immediately that $\kappa(X) = 0$. The case of a general projective manifold with $\mathrm{nd}(K_X) = 0$ and positive irregularity $q(X) > 0$ has been solved by Campana-Peternell [CP04], Corollary 3.7. It would be interesting to understand the Kähler case as well.

6.C. The orthogonality estimate

The goal of this section is to show that, in an appropriate sense, approximate Zariski decompositions are almost orthogonal.

(6.18) Theorem. *Let X be a projective manifold, and let $\alpha = \{T\} \in \mathcal{E}^{\circ}_{\mathrm{NS}}$ be a big class represented by a Kähler current T. Consider an approximate Zariski decomposition*

$$\mu_m^\star T_m = [E_m] + [D_m]$$

Then

$$(D_m^{n-1} \cdot E_m)^2 \le 20\,(C\omega)^n \big(\mathrm{Vol}(\alpha) - D_m^n\big)$$

where $\omega = c_1(H)$ is a Kähler form and $C \ge 0$ is a constant such that $\pm\alpha$ is dominated by $C\omega$ (i.e., $C\omega \pm \alpha$ is nef).

Proof. For every $t \in [0,1]$, we have

$$\mathrm{Vol}(\alpha) = \mathrm{Vol}(E_m + D_m) \ge \mathrm{Vol}(tE_m + D_m).$$

Now, by our choice of C, we can write E_m as a difference of two nef divisors

$$E_m = \mu^\star\alpha - D_m = \mu_m^\star(\alpha + C\omega) - (D_m + C\mu_m^\star\omega).$$

(6.19) Lemma. *For all nef $\mathbb{R}$-divisors A, B we have*

$$\mathrm{Vol}(A-B) \geq A^n - nA^{n-1}\cdot B$$

as soon as the right hand side is positive.

Proof. In case A and B are integral (Cartier) divisors, this is a consequence of the holomorphic Morse inequalities 2.4 (see [Dem01]); one can also argue by an elementary estimate of to $H^0(X, mA - B_1 - \ldots - B_m)$ via the Riemann-Roch formula (assuming A and B very ample, $B_1, \ldots, B_m \in |B|$ generic). If A and B are $\mathbb{Q}$-Cartier, we conclude by the homogeneity of the volume. The general case of $\mathbb{R}$-divisors follows by approximation using the upper semi-continuity of the volume [Bou02, 3.1.26]. □

(6.20) Remark. We hope that Lemma 6.19 also holds true on an arbitrary Kähler manifold for arbitrary nef (non necessarily integral) classes. This would follow from a generalization of holomorphic Morse inequalities to non integral classes. However the proof of such a result seems technically much more involved than in the case of integral classes.

(6.21) Lemma. *Let $\beta_1, \ldots, \beta_n$ and $\beta'_1, \ldots, \beta'_n$ be nef classes on a compact Kähler manifold $\widetilde{X}$ such that each difference $\beta'_j - \beta_j$ is pseudo-effective. Then the n-th intersection products satisfy*

$$\beta_1 \cdots \beta_n \leq \beta'_1 \cdots \beta'_n.$$

Proof. We can proceed step by step and replace just one β_j by $\beta' j \equiv \beta_j + T_j$ where T_j is a closed positive $(1,1)$-current and the other classes $\beta'_k = \beta_k$, $k \neq j$ are limits of Kähler forms. The inequality is then obvious. □

End of proof of Theorem 6.18. In order to exploit the lower bound of the volume, we write

$$tE_m + D_m = A - B, \qquad A = D_m + t\mu_m^\star(\alpha + C\omega), \quad B = t(D_m + C\mu_m^\star\omega).$$

By our choice of the constant C, both A and B are nef. Lemma 6.19 and the binomial formula imply

$$\begin{aligned}\mathrm{Vol}(tE_m + D_m) &\geq A^n - nA^{n-1}\cdot B \\ &= D_m^n + nt\, D_m^{n-1}\cdot \mu_m^\star(\alpha + C\omega) + \sum_{k=2}^n t^k \binom{n}{k} D_m^{n-k}\cdot \mu_m^\star(\alpha + C\omega)^k \\ &\quad - nt\, D_m^{n-1}\cdot (D_m + C\mu_m^\star\omega) \\ &\quad - nt^2 \sum_{k=1}^{n-1} t^{k-1}\binom{n-1}{k} D_m^{n-1-k}\cdot \mu_m^\star(\alpha + C\omega)^k \cdot (D_m + C\mu_m^\star\omega).\end{aligned}$$

Now, we use the obvious inequalities

$$D_m \leq \mu_m^\star(C\omega), \qquad \mu_m^\star(\alpha + C\omega) \leq 2\mu_m^\star(C\omega), \qquad D_m + C\mu_m^\star\omega \leq 2\mu_m^\star(C\omega)$$

in which all members are nef (and where the inequality $\le$ means that the difference of classes is pseudo-effective). We use Lemma 6.21 to bound the last summation in the estimate of the volume, and in this way we get

$$\operatorname{Vol}(tE_m + D_m) \ge D_m^n + ntD_m^{n-1} \cdot E_m - nt^2 \sum_{k=1}^{n-1} 2^{k+1} t^{k-1} \binom{n-1}{k} (C\omega)^n.$$

We will always take t smaller than $1/10n$ so that the last summation is bounded by $4(n-1)(1+1/5n)^{n-2} < 4ne^{1/5} < 5n$. This implies

$$\operatorname{Vol}(tE_m + D_m) \ge D_m^n + nt\, D_m^{n-1} \cdot E_m - 5n^2t^2(C\omega)^n.$$

Now, the choice $t = \frac{1}{10n}(D_m^{n-1} \cdot E_m)((C\omega)^n)^{-1}$ gives by substituting

$$\frac{1}{20} \frac{(D_m^{n-1} \cdot E_m)^2}{(C\omega)^n} \le \operatorname{Vol}(E_m + D_m) - D_m^n \le \operatorname{Vol}(\alpha) - D_m^n$$

(and we have indeed $t \le \frac{1}{10n}$ by Lemma 6.21), whence Theorem 6.18. Of course, the constant 20 is certainly not optimal. □

(6.22) Corollary. *If $\alpha \in \mathcal{E}_{\mathrm{NS}}$, then the divisorial Zariski decomposition $\alpha = N(\alpha) + \langle \alpha \rangle$ is such that*

$$\langle \alpha^{n-1} \rangle \cdot N(\alpha) = 0.$$

Proof. By replacing α with $\alpha + \delta c_1(H)$, one sees that it is sufficient to consider the case where α is big. Then the orthogonality estimate implies

$$\begin{aligned}(\mu_m)_\star(D_m^{n-1}) \cdot (\mu_m)_\star E_m &= D_m^{n-1} \cdot (\mu_m)^\star(\mu_m)_\star E_m \\ &\le D_m^{n-1} \cdot E_m \\ &\le C(\operatorname{Vol}(\alpha) - D_m^n)^{1/2}.\end{aligned}$$

Since $\langle \alpha^{n-1} \rangle = \lim(\mu_m)_\star(D_m^{n-1})$, $N(\alpha) = \lim(\mu_m)_\star E_m$ and $\lim D_m^n = \operatorname{Vol}(\alpha)$, we get the desired conclusion in the limit. □

6.D. Dual of the pseudo-effective cone

The following statement was first proved in [BDPP04].

(6.23) Theorem. *If X is projective, the cones $\mathcal{E}_{\mathrm{NS}} = \overline{\operatorname{Eff}(X)}$ and $\overline{\operatorname{ME}^s(X)}$ are dual.*

In other words, a line bundle L is pseudo-effective if (and only if) $L \cdot C \ge 0$ for all *mobile curves*, i.e., $L \cdot C \ge 0$ for every very generic curve C (not contained in a countable union of algebraic subvarieties). In fact, by definition of $\operatorname{ME}^s(X)$, it is enough to consider only those curves C which are images of generic complete intersection of very ample divisors on some variety $\widetilde{X}$, under a modification $\mu : \widetilde{X} \to X$. By a standard blowing-up argument, it also follows that a line bundle L on a normal Moishezon variety is pseudo-effective if and only if $L \cdot C \ge 0$ for every mobile curve C.

Proof. By (6.4 b) we have in any case

$$\mathcal{E}_{\mathrm{NS}} \subset (\operatorname{ME}^s(X))^\vee.$$

If the inclusion is strict, there is an element $\alpha \in \partial\mathcal{E}_{\mathrm{NS}}$ on the boundary of $\mathcal{E}_{\mathrm{NS}}$ which is in the interior of $\operatorname{ME}^s(X)^\vee$.

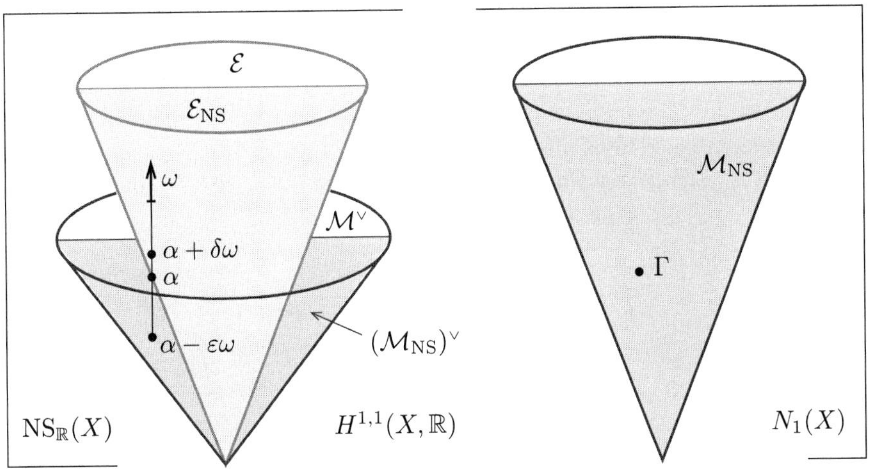

Let $\omega = c_1(H)$ be an ample class. Since $\alpha \in \partial\mathcal{E}_{\rm NS}$, the class $\alpha + \delta\omega$ is big for every $\delta > 0$, and since $\alpha \in ((\mathrm{ME}^s(X))^\vee)^\circ$ we still have $\alpha - \varepsilon\omega \in (\mathrm{ME}^s(X))^\vee$ for $\varepsilon > 0$ small. Therefore

$$\alpha \cdot \Gamma \geq \varepsilon\omega \cdot \Gamma \tag{6.24}$$

for every strongly mobile curve Γ, and therefore for every $\Gamma \in \overline{\mathrm{ME}^s(X)}$. We are going to contradict (6.24). Since $\alpha + \delta\omega$ is big, we have an approximate Zariski decomposition

$$\mu_\delta^\star(\alpha + \delta\omega) = E_\delta + D_\delta.$$

We pick $\Gamma = (\mu_\delta)_\star(D_\delta^{n-1}) \in \overline{\mathrm{ME}^s(X)}$. By the Hovanskii-Teissier concavity inequality

$$\omega \cdot \Gamma \geq (\omega^n)^{1/n}(D_\delta^n)^{(n-1)/n}.$$

On the other hand

$$\begin{aligned}\alpha \cdot \Gamma &= \alpha \cdot (\mu_\delta)_\star(D_\delta^{n-1})\\ &= \mu_\delta^\star\alpha \cdot D_\delta^{n-1} \leq \mu_\delta^\star(\alpha + \delta\omega) \cdot D_\delta^{n-1}\\ &= (E_\delta + D_\delta) \cdot D_\delta^{n-1} = D_\delta^n + D_\delta^{n-1} \cdot E_\delta.\end{aligned}$$

By the orthogonality estimate, we find

$$\begin{aligned}\frac{\alpha \cdot \Gamma}{\omega \cdot \Gamma} &\leq \frac{D_\delta^n + \big(20(C\omega)^n(\mathrm{Vol}(\alpha + \delta\omega) - D_\delta^n)\big)^{1/2}}{(\omega^n)^{1/n}(D_\delta^n)^{(n-1)/n}}\\ &\leq C'(D_\delta^n)^{1/n} + C''\frac{(\mathrm{Vol}(\alpha + \delta\omega) - D_\delta^n)^{1/2}}{(D_\delta^n)^{(n-1)/n}}.\end{aligned}$$

However, since $\alpha \in \partial\mathcal{E}_{\rm NS}$, the class α cannot be big so

$$\lim_{\delta\to 0} D_\delta^n = \mathrm{Vol}(\alpha) = 0.$$

We can also take D_δ to approximate $\mathrm{Vol}(\alpha + \delta\omega)$ in such a way that the term $(\mathrm{Vol}(\alpha + \delta\omega) - D_\delta^n)^{1/2}$ tends to 0 much faster than D_δ^n. Notice that $D_\delta^n \geq \delta^n\omega^n$, so in fact it is enough to take

$$\mathrm{Vol}(\alpha + \delta\omega) - D_\delta^n \le \delta^{2n},$$

which gives $(\alpha \cdot \Gamma)/(\omega \cdot \Gamma) \le (C' + C'')\delta$. This contradicts (6.24) for δ small. □

(6.25) Conjecture. The Kähler analogue should be :
For an arbitrary compact Kähler manifold X, the cones $\mathcal{E}$ and $\mathcal{M}$ are dual.

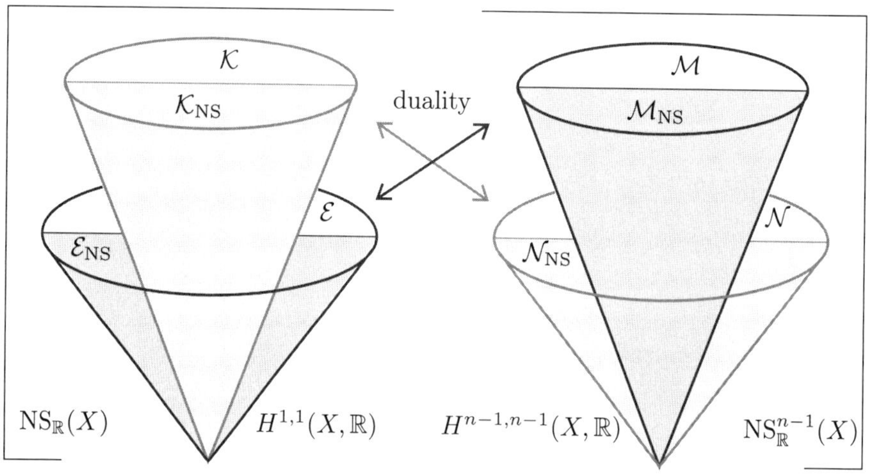

If holomorphic Morse inequalities were known also in the Kähler case, we would infer by the same proof that "α not pseudo-effective" implies the existence of a blow-up $\mu : \widetilde{X} \to X$ and a Kähler metric $\widetilde{\omega}$ on $\widetilde{X}$ such that $\alpha \cdot \mu_\star(\widetilde{\omega})^{n-1} < 0$. In the special case when $\alpha = K_X$ is not pseudo-effective, we would expect the Kähler manifold X to be covered by rational curves. The main trouble is that characteristic p techniques are no longer available. On the other hand it is tempting to approach the question via techniques of symplectic geometry :

(6.26) Question. *Let (M,ω) be a compact real symplectic manifold. Fix an almost complex structure J compatible with ω, and for this structure, assume that $c_1(M) \cdot \omega^{n-1} > 0$. Does it follow that M is covered by rational J-pseudoholomorphic curves ?*

The relation between the various cones of mobile curves and currents in (6.1) and (6.2) is now a rather direct consequence of Theorem 6.23. In fact, using ideas hinted in [DPS96], we can say a little bit more. Given an irreducible curve $C \subset X$, we consider its normal "bundle" $N_C = \mathrm{Hom}(\mathcal{I}/\mathcal{I}^2, \mathcal{O}_C)$, where $\mathcal{I}$ is the ideal sheaf of C. If C is a general member of a covering family (C_t), then N_C is nef. Now [DPS96] says that the dual cone of the pseudo-effective cone of X contains the closed cone spanned by curves with nef normal bundle, which in turn contains the cone of mobile curves. In this way we get :

(6.27) Theorem. *Let X be a projective manifold. Then the following cones coincide.*

(i) *the cone* $\mathcal{M}_{\mathrm{NS}} = \mathcal{M} \cap \mathrm{NS}^{n-1}_{\mathbb{R}}(X)$;

(ii) *the cone* $\mathcal{M}^s_{\mathrm{NS}} = \mathcal{M}^s \cap \mathrm{NS}^{n-1}_{\mathbb{R}}(X)$;

(iii) *the closed cone* $\overline{\mathrm{ME}^s(X)}$ *of strongly mobile curves* ;

(iv) *the closed cone* $\overline{\mathrm{ME}(X)}$ *of mobile curves* ;

(v) *the closed cone* $\overline{\mathrm{ME}_{\mathrm{nef}}(X)}$ *of curves with nef normal bundle.*

Proof. We have already seen that

$$\mathrm{ME}^s(X) \subset \mathrm{ME}(X) \subset \mathrm{ME}_{\mathrm{nef}}(X) \subset (\mathcal{E}_{\mathrm{NS}})^\vee$$

and

$$\mathrm{ME}^s(X) \subset \mathcal{M}^s_{\mathrm{NS}}(X) \subset \mathcal{M}_{\mathrm{NS}} \subset (\mathcal{E}_{\mathrm{NS}})^\vee$$

by 6.4 (iii). Now Theorem 6.23 implies $(\mathcal{M}_{\mathrm{NS}})^\vee = \overline{\mathrm{ME}^s(X)}$, and 6.27 follows. □

(6.28) Corollary. *Let X be a projective manifold and L a line bundle on X.*

(i) *L is pseudo-effective if and only if $L \cdot C \geq 0$ for all curves C with nef normal sheaf N_C.*

(ii) *If L is big, then $L \cdot C > 0$ for all curves C with nef normal sheaf N_C.*

Corollary 6.28 (i) strenghtens results from [PSS99]. It is however not yet clear whether $\mathcal{M}_{\mathrm{NS}}$ is equal to the closed cone of curves with *ample* normal bundle (although we certainly expect this to be true). The most important special case of Theorem 6.23 is

(6.29) Theorem. *If X is a projective manifold, then K_X is pseudo-effective (i.e. $K_X \in \mathcal{E}_{\mathrm{NS}}$), if and only if X is not uniruled (i.e. not covered by rational curves).*

Proof. If X is covered by rational curves C_t, then it is well-known that the normal bundle N_{C_t} is nef for a general member C_t, thus

$$K_X \cdot C_t = K_{C_t} \cdot C_t - N_{C_t} \cdot C_t \leq -2,$$

and K_X cannot be pseudo-effective. Conversely, if $K_X \notin \mathcal{E}_{\mathrm{NS}}$, Theorem 6.23 shows that there is a mobile curve C_t such that $K_X \cdot C_t < 0$. The standard "bend-and-break" lemma of Mori theory then produces a covering family Γ_t of rational curves with $K_X \cdot \Gamma_t < 0$, so X is uniruled. □

Notice that the generalized abundance conjecture 6.16 would then yield the stronger result :

(6.30) Conjecture. *Let X be a projective manifold. If X is not uniruled, then K_X is a $\mathbb{Q}$-effective divisor and $\kappa(X) = \mathrm{nd}(K_X) \geq 0$.*

7. Super-canonical metrics and abundance

7.A. Construction of super-canonical metrics

Let X be a compact complex manifold and $(L, h_{L,\gamma})$ a holomorphic line bundle over X equipped with a singular hermitian metric $h_{L,\gamma} = e^{-\gamma} h_L$ with satisfies $\int e^{-\gamma} < +\infty$ locally on X, where h_L is a smooth metric on L. In fact, we can more generally consider the case where $(L, h_{L,\gamma})$ is a "hermitian $\mathbb{R}$-line bundle";

by this we mean that we have chosen a smooth real d-closed $(1,1)$ form α_L on X (whose dd^c cohomology class is equal to $c_1(L)$), and a specific current $T_{L,\gamma}$ representing it, namely $T_{L,\gamma} = \alpha_L + dd^c\gamma$, such that γ is a locally integrable function satisfying $\int e^{-\gamma} < +\infty$. An important special case is obtained by considering a klt (Kawamata log terminal) effective divisor Δ. In this situation $\Delta = \sum c_j \Delta_j$ with $c_j \in \mathbb{R}$, and if g_j is a local generator of the ideal sheaf $\mathcal{O}(-\Delta_j)$ identifying it to the trivial invertible sheaf $g_j\mathcal{O}$, we take $\gamma = \sum c_j \log|g_j|^2$, $T_{L,\gamma} = \sum c_j[\Delta_j]$ (current of integration on Δ) and α_L given by any smooth representative of the same dd^c-cohomology class; the klt condition precisely means that

$$\int_V e^{-\gamma} = \int_V \prod |g_j|^{-2c_j} < +\infty \tag{7.1}$$

on a small neighborhood V of any point in the support $|\Delta| = \bigcup \Delta_j$ (condition (7.1) implies $c_j < 1$ for every j, and this in turn is sufficient to imply Δ klt if Δ is a normal crossing divisor; the line bundle L is then the real line bundle $\mathcal{O}(\Delta)$, which makes sens as a genuine line bundle only if $c_j \in \mathbb{Z}$). For each klt pair (X, Δ) such that $K_X + \Delta$ is pseudo-effective, H. Tsuji [Tsu07a, Tsu07b] has introduced a "super-canonical metric" which generalizes the metric introduced by Narasimhan and Simha [NS68] for projective algebraic varieties with ample canonical divisor. We take the opportunity to present here a simpler, more direct and more general approach.

We assume from now on that $K_X + L$ is *pseudo-effective*, i.e. that the class $c_1(K_X) + \{\alpha_L\}$ is pseudo-effective, and under this condition, we are going to define a "super-canonical metric" on K_X+L. Select an arbitrary smooth hermitian metric ω on X. We then find induced hermitian metrics h_{K_X} on K_X and $h_{K_X+L} = h_{K_X}h_L$ on $K_X + L$, whose curvature is the smooth real $(1,1)$-form

$$\alpha = \Theta_{K_X+L,h_{K_X+L}} = \Theta_{K_X,\omega} + \alpha_L.$$

A singular hermitian metric on $K_X + L$ is a metric of the form $h_{K_X+L,\varphi} = e^{-\varphi}h_{K_X+L}$ where φ is locally integrable, and by the pseudo-effectivity assumption, we can find quasi-psh functions φ such that $\alpha + dd^c\varphi \geq 0$. The metrics on L and $K_X + L$ can now be "subtracted" to give rise to a metric

$$h_{L,\gamma}h^{-1}_{K_X+L,\varphi} = e^{\varphi-\gamma}h_L h^{-1}_{K_X+L} = e^{\varphi-\gamma}h^{-1}_{K_X} = e^{\varphi-\gamma}dV_\omega$$

on $K_X^{-1} = \Lambda^n T_X$, since $h^{-1}_{K_X} = dV_\omega$ is just the hermitian (n,n) volume form on X. Therefore the integral $\int_X h_{L,\gamma}h^{-1}_{K_X+L,\varphi}$ has an intrinsic meaning, and it makes sense to require that

$$\int_X h_{L,\gamma}h^{-1}_{K_X+L,\varphi} = \int_X e^{\varphi-\gamma}dV_\omega \leq 1 \tag{7.2}$$

in view of the fact that φ is locally bounded from above and of the assumption $\int e^{-\gamma} < +\infty$. Observe that condition (7.2) can always be achieved by subtracting a constant to φ. Now, we can generalize Tsuji's super-canonical metrics on klt pairs (cf. [Tsu07b]) as follows.

(7.3) Definition. *Let X be a compact complex manifold and let (L, h_L) be a hermitian $\mathbb{R}$-line bundle on X associated with a smooth real closed $(1,1)$ form α_L. Assume that $K_X + L$ is pseudo-effective and that L is equipped with a singular hermitian metric $h_{L,\gamma} = e^{-\gamma}h_L$ such that $\int e^{-\gamma} < +\infty$ locally on X. Take a hermitian* *metric ω on X and define $\alpha = \Theta_{K_X+L,h_{K_X+L}} = \Theta_{K_X,\omega} + \alpha_L$. Then we define the super-canonical metric $h_{\rm can}$ of $K_X + L$ to be*

$$h_{K_X+L,{\rm can}} = \inf_{\varphi} h_{K_X+L,\varphi} \quad \textit{i.e.} \quad h_{K_X+L,{\rm can}} = e^{-\varphi_{\rm can}} h_{K_X+L}, \ \textit{where}$$

$$\varphi_{\rm can}(x) = \sup_{\varphi} \varphi(x) \ \textit{for all } \varphi \textit{ with } \alpha + dd^c\varphi \geq 0, \ \int_X e^{\varphi-\gamma} dV_\omega \leq 1.$$

In particular, this gives a definition of the super-canonical metric on $K_X + \Delta$ for every klt pair (X, Δ) such that $K_X + \Delta$ is pseudo-effective, and as an even more special case, a super-canonical metric on K_X when K_X is pseudo-effective.

In the sequel, we assume that γ has analytic singularities, otherwise not much can be said. The mean value inequality then immediately shows that the quasi-psh functions φ involved in definition (7.3) are globally uniformly bounded outside of the poles of γ, and therefore everywhere on X, hence the envelopes $\varphi_{\rm can} = \sup_\varphi \varphi$ are indeed well defined and bounded above. As a consequence, we get a "super-canonical" current $T_{\rm can} = \alpha + dd^c\varphi_{\rm can} \geq 0$ and $h_{K_X+L,{\rm can}}$ satisfies

$$\int_X h_{L,\gamma} h^{-1}_{K_X+L,{\rm can}} = \int_X e^{\varphi_{\rm can}-\gamma} dV_\omega < +\infty. \tag{7.4}$$

It is easy to see that in Definition (7.3) the supremum is a maximum and that $\varphi_{\rm can} = (\varphi_{\rm can})^*$ everywhere, so that taking the upper semicontinuous regularization is not needed. In fact if $x_0 \in X$ is given and we write

$$(\varphi_{\rm can})^*(x_0) = \limsup_{x\to x_0} \varphi_{\rm can}(x) = \lim_{\nu\to+\infty} \varphi_{\rm can}(x_\nu) = \lim_{\nu\to+\infty} \varphi_\nu(x_\nu)$$

with suitable sequences $x_\nu \to x_0$ and (φ_ν) such that $\int_X e^{\varphi_\nu-\gamma} dV_\omega \leq 1$, the well-known weak compactness properties of quasi-psh functions in L^1 topology imply the existence of a subsequence of (φ_ν) converging in L^1 and almost everywhere to a quasi-psh limit φ. Since $\int_X e^{\varphi_\nu-\gamma} dV_\omega \leq 1$ holds true for every ν, Fatou's lemma implies that we have $\int_X e^{\varphi-\gamma} dV_\omega \leq 1$ in the limit. By taking a subsequence, we can assume that $\varphi_\nu \to \varphi$ in $L^1(X)$. Then for every $\varepsilon > 0$ the mean value $⨍_{B(x_\nu,\varepsilon)} \varphi_\nu$ satisfies

$$⨍_{B(x_0,\varepsilon)} \varphi = \lim_{\nu\to+\infty} ⨍_{B(x_\nu,\varepsilon)} \varphi_\nu \geq \lim_{\nu\to+\infty} \varphi_\nu(x_\nu) = (\varphi_{\rm can})^*(x_0),$$

hence we get $\varphi(x_0) = \lim_{\varepsilon\to 0} ⨍_{B(x_0,\varepsilon)} \varphi \geq (\varphi_{\rm can})^*(x_0) \geq \varphi_{\rm can}(x_0)$, and therefore the sup is a maximum and $\varphi_{\rm can} = \varphi^*_{\rm can}$. By elaborating on this argument, one can infer certain regularity properties of the envelope.

(7.5) Theorem ([BmD09]). *Let X be a compact complex manifold and (L, h_L) a holomorphic $\mathbb{R}$-line bundle such that $K_X + L$ is big. Assume that L is equipped with a singular hermitian metric $h_{L,\gamma} = e^{-\gamma} h_L$ with analytic singularities such that $\int e^{-\gamma} < +\infty$ (klt condition). Denote by Z_0 the set of poles of a singular metric $h_0 = e^{-\psi_0} h_{K_X+L}$ with analytic singularities on $K_X + L$ and by Z_γ the poles of γ (assumed analytic). Then the associated super-canonical metric $h_{\rm can}$ is continuous on $X \smallsetminus (Z_0 \cup Z_\gamma)$.*

In fact, using the regularization techniques of [Dem94], it is shown in [BmD09] that $h_{\rm can}$ possesses some computable logarithmic modulus of continuity. In order to shorten the exposition, we will only give a proof of the continuity in the algebraic case, using approximation by pluri-canonical sections.

(7.6) Algebraic version of the super-canonical metric. Since the klt condition is open and $K_X + L$ is assumed to be big, we can always perturb L a little bit, and after blowing-up X, assume that X is projective and that $(L, h_{L,\gamma})$ is obtained as a sum of $\mathbb{Q}$-divisors

$$L = G + \Delta$$

where Δ is klt and G is equipped with a smooth metric h_G (from which $h_{L,\gamma}$ is inferred, with Δ as its poles, so that $\Theta_{L,h_{L,\gamma}} = \Theta_{G,L_G} + [\Delta]$). Clearly this situation is "dense" in what we have been considering before, just as $\mathbb{Q}$ is dense in $\mathbb{R}$. In this case, it is possible to give a more algebraic definition of the super-canonical metric φ_{can}, following the original idea of Narasimhan-Simha [NS68] (see also H. Tsuji [Tsu07a]) – the case considered by these authors is the special situation where $G = 0$, $h_G = 1$ (and moreover $\Delta = 0$ and K_X ample, for [NS68]). In fact, if m is a large integer which is a multiple of the denominators involved in G and Δ, we can consider sections

$$\sigma \in H^0(X, m(K_X + G + \Delta)).$$

We view them rather as sections of $m(K_X + G)$ with poles along the support $|\Delta|$ of our divisor. Then $(\sigma \wedge \overline{\sigma})^{1/m} h_G$ is a volume form with integrable poles along $|\Delta|$ (this is the klt condition for Δ). Therefore one can normalize σ by requiring that

$$\int_X (\sigma \wedge \overline{\sigma})^{1/m} h_G = 1.$$

Each of these sections defines a singular hermitian metric on $K_X + L = K_X + G + \Delta$, and we can take the regularized upper envelope

$$\varphi_{\mathrm{can}}^{\mathrm{alg}} = \left(\sup_{m,\sigma} \frac{1}{m} \log |\sigma|^2_{h^m_{K_X+L}} \right)^* \tag{7.7}$$

of the weights associated with a smooth metric h_{K_X+L}. It is clear that $\varphi_{\mathrm{can}}^{\mathrm{alg}} \le \varphi_{\mathrm{can}}$ since the supremum is taken on the smaller set of weights $\varphi = \frac{1}{m} \log |\sigma|^2_{h^m_{K_X+L}}$, and the equalities

$$e^{\varphi-\gamma} dV_\omega = |\sigma|^{2/m}_{h^m_{K_X+L}} e^{-\gamma} dV_\omega = (\sigma \wedge \overline{\sigma})^{1/m} e^{-\gamma} h_L = (\sigma \wedge \overline{\sigma})^{1/m} h_{L,\gamma} = (\sigma \wedge \overline{\sigma})^{1/m} h_G$$

imply $\int_X e^{\varphi-\gamma} dV_\omega \le 1$. We claim that the inequality $\varphi_{\mathrm{can}}^{\mathrm{alg}} \le \varphi_{\mathrm{can}}$ is an equality. The proof is an immediate consequence of the following statement based in turn on the Ohsawa-Takegoshi theorem and the approximation technique of [Dem92].

(7.8) Proposition. *With $L = G + \Delta$, ω, $\alpha = \Theta_{K_X+L, h_{K_X+L}}$, γ as above and $K_X + L$ assumed to be big, fix a singular hermitian metric $e^{-\varphi} h_{K_X+L}$ of curvature $\alpha + dd^c\varphi \ge 0$, such that $\int_X e^{\varphi-\gamma} dV_\omega \le 1$. Then φ is equal to a regularized limit*

$$\varphi = \left(\limsup_{m \to +\infty} \frac{1}{m} \log |\sigma_m|^2_{h^m_{K_X+L}} \right)^*$$

for a suitable sequence $\sigma_m \in H^0(X, m(K_X+G+\Delta))$ with $\int_X (\sigma_m \wedge \overline{\sigma}_m)^{1/m} h_G \le 1$.

Proof. By our assumption, there exists a quasi-psh function ψ_0 with analytic singularity set Z_0 such that

$$\alpha + dd^c\psi_0 \ge \varepsilon_0 \omega > 0$$

and we can assume $\int_C e^{\psi_0-\gamma}dV_\omega < 1$ (the strict inequality will be useful later). For $m \geq p \geq 1$, this defines a singular metric $\exp(-(m-p)\varphi - p\psi_0)h^m_{K_X+L}$ on $m(K_X+L)$ with curvature $\geq p\varepsilon_0\omega$, and therefore a singular metric

$$h_{L'} = \exp(-(m-p)\varphi - p\psi_0)h^m_{K_X+L}h^{-1}_{K_X}$$

on $L' = (m-1)K_X + mL$, whose curvature $\Theta_{L',h_{L'}} \geq (p\varepsilon_0 - C_0)\omega$ is arbitrary large if p is large enough. Let us fix a finite covering of X by coordinate balls. Pick a point x_0 and one of the coordinate balls B containing x_0. By the Ohsawa-Takegoshi extension theorem applied on the ball B, we can find a section σ_B of $K_X + L' = m(K_X+L)$ which has norm 1 at x_0 with respect to the metric $h_{K_X+L'}$ and $\int_B |\sigma_B|^2_{h_{K_X+L'}} dV_\omega \leq C_1$ for some uniform constant C_1 depending on the finite covering, but independent of m, p, x_0 . Now, we use a cut-off function $\theta(x)$ with $\theta(x) = 1$ near x_0 to truncate σ_B and solve a $\overline{\partial}$-equation for $(n,1)$-forms with values in L to get a global section σ on X with $|\sigma(x_0)|_{h_{K_X+L'}} = 1$. For this we need to multiply our metric by a truncated factor $\exp(-2n\theta(x)\log|x-x_0|)$ so as to get solutions of $\overline{\partial}$ vanishing at x_0. However, this perturbs the curvature by bounded terms and we can absorb them again by taking p larger. In this way we obtain

$$\int_X |\sigma|^2_{h_{K_X+L'}} dV_\omega = \int_X |\sigma|^2_{h^m_{K_X+L}} e^{-(m-p)\varphi - p\psi_0} dV_\omega \leq C_2. \tag{7.9}$$

Taking $p > 1$, the Hölder inequality for congugate exponents m, $\frac{m}{m-1}$ implies

$$\begin{aligned}
\int_X (\sigma\wedge\overline{\sigma})^{\frac{1}{m}} h_G &= \int_X |\sigma|^{2/m}_{h^m_{K_X+L}} e^{-\gamma} dV_\omega \\
&= \int_X \Big(|\sigma|^2_{h^m_{K_X+L}} e^{-(m-p)\varphi-p\psi_0}\Big)^{\frac{1}{m}} \Big(e^{(1-\frac{p}{m})\varphi + \frac{p}{m}\psi_0 - \gamma}\Big) dV_\omega \\
&\leq C_2^{\frac{1}{m}} \bigg(\int_X \Big(e^{(1-\frac{p}{m})\varphi+\frac{p}{m}\psi_0-\gamma}\Big)^{\frac{m}{m-1}} dV_\omega\bigg)^{\frac{m-1}{m}} \\
&\leq C_2^{\frac{1}{m}} \bigg(\int_X (e^{\varphi-\gamma})^{\frac{m-p}{m-1}} \Big(e^{\frac{p}{p-1}(\psi_0-\gamma)}\Big)^{\frac{p-1}{m-1}} dV_\omega\bigg)^{\frac{m-1}{m}} \\
&\leq C_2^{\frac{1}{m}} \bigg(\int_X e^{\frac{p}{p-1}(\psi_0-\gamma)} dV_\omega\bigg)^{\frac{p-1}{m}}
\end{aligned}$$

using the hypothesis $\int_X e^{\varphi-\gamma}dV_\omega \leq 1$ and another application of Hölder's inequality. Since klt is an open condition and $\lim_{p\to+\infty}\int_X e^{\frac{p}{p-1}(\psi_0-\gamma)}dV_\omega = \int_X e^{\psi_0-\gamma}dV_\omega < 1$, we can take p large enough to ensure that

$$\int_X e^{\frac{p}{p-1}(\psi_0-\gamma)} dV_\omega \leq C_3 < 1.$$

Therefore, we see that

$$\int_X (\sigma\wedge\overline{\sigma})^{\frac{1}{m}} h_G \leq C_2^{\frac{1}{m}} C_3^{\frac{p-1}{m}} \leq 1$$

for p large enough. On the other hand

$$|\sigma(x_0)|^2_{h_{K_X+L'}} = |\sigma(x_0)|^2_{h^m_{K_X+L}} e^{-(m-p)\varphi(x_0)-p\psi_0(x_0)} = 1,$$

thus

$$\frac{1}{m}\log|\sigma(x_0)|^2_{h^m_{K_X+L}} = \Big(1-\frac{p}{m}\Big)\varphi(x_0) + \frac{p}{m}\psi_0(x_0) \tag{7.10}$$

and, as a consequence

$$\frac{1}{m}\log|\sigma(x_0)|^2_{h^m_{K_X+L}} \longrightarrow \varphi(x_0)$$

whenever $m \to +\infty$, $\frac{p}{m} \to 0$, as long as $\psi_0(x_0) > -\infty$. In the above argument, we can in fact interpolate in finitely many points $x_1, x_2, \ldots, x_q$ provided that $p \geq C_4 q$. Therefore if we take a suitable dense subset $\{x_q\}$ and a "diagonal" sequence associated with sections $\sigma_m \in H^0(X, m(K_X+L))$ with $m \gg p = p_m \gg q = q_m \to +\infty$, we infer that

$$\left(\limsup_{m\to+\infty}\frac{1}{m}\log|\sigma_m(x)|^2_{h^m_{K_X+L}}\right)^* \geq \limsup_{x_q\to x}\varphi(x_q) = \varphi(x) \tag{7.11}$$

(the latter equality occurring if $\{x_q\}$ is suitably chosen with respect to φ). In the other direction, (7.9) implies a mean value estimate

$$\frac{1}{\pi^n r^{2n}/n!}\int_{B(x,r)}|\sigma(z)|^2_{h^m_{K_X+L}}\,dz \leq \frac{C_5}{r^{2n}}\sup_{B(x,r)} e^{(m-p)\varphi+p\psi_0}$$

on every coordinate ball $B(x,r) \subset X$. The function $|\sigma_m|^2_{h^m_{K_X+L}}$ is plurisubharmonic after we correct the non necessarily positively curved smooth metric h_{K_X+L} by a factor of the form $\exp(C_6|z-x|^2)$, hence the mean value inequality shows that

$$\frac{1}{m}\log|\sigma_m(x)|^2_{h^m_{K_X+L}} \leq \frac{1}{m}\log\frac{C_5}{r^{2n}} + C_6 r^2 + \sup_{B(x,r)}\Big(1-\frac{p_m}{m}\Big)\varphi + \frac{p_m}{m}\psi_0.$$

By taking in particular $r = 1/m$ and letting $m \to +\infty$, $p_m/m \to 0$, we see that the opposite of inequality (7.9) also holds. □

(7.12) Remark. We can rephrase our results in slightly different terms. In fact, let us put

$$\varphi_m^{\mathrm{alg}} = \sup_\sigma \frac{1}{m}\log|\sigma|^2_{h^m_{K_X+L}},\qquad \sigma \in H^0(X, m(K_X+G+\Delta)),$$

with normalized sections σ such that $\int_X(\sigma\wedge\overline{\sigma})^{1/m}h_G = 1$. Then φ_m^{alg} is quasi-psh (the supremum is taken over a compact set in a finite dimensional vector space) and by passing to the regularized supremum over all σ and all φ in (7.10) we get

$$\varphi_{\mathrm{can}} \geq \varphi_m^{\mathrm{alg}} \geq \Big(1-\frac{p}{m}\Big)\varphi_{\mathrm{can}}(x) + \frac{p}{m}\psi_0(x).$$

As φ_{can} is bounded from above, we find in particular

$$0 \leq \varphi_{\mathrm{can}} - \varphi_m^{\mathrm{alg}} \leq \frac{C}{m}(|\psi_0(x)|+1).$$

This implies that $(\varphi_m^{\mathrm{alg}})$ converges uniformly to φ_{can} on every compact subset of $X \subset Z_0$, and in this way we infer again (in a purely qualitative manner) that φ_{can} is continuous on $X \smallsetminus Z_0$. Moreover, we also see that in (7.7) the upper semicontinuous regularization is not needed on $X \smallsetminus Z_0$; in case K_X+L is ample, it is not needed at all and we have uniform convergence of $(\varphi_m^{\mathrm{alg}})$ towards φ_{can} on the whole of X. Obtaining such a uniform convergence when K_X+L is just big looks like a more delicate question, related e.g. to abundance of K_X+L on those subvarieties Y where the restriction $(K_X+L)_{|Y}$ would be e.g. nef but not big.

(7.13) Generalization. In the general case where L is a $\mathbb{R}$-line bundle and $K_X + L$ is merely pseudo-effective, a similar algebraic approximation can be obtained. We take instead sections

$$\sigma \in H^0(X, mK_X + \lfloor mG \rfloor + \lfloor m\Delta \rfloor + p_m A)$$

where (A, h_A) is a positive line bundle, $\Theta_{A,h_A} \geq \varepsilon_0 \omega$, and replace the definition of $\varphi_{\rm can}^{\rm alg}$ by

$$\varphi_{\rm can}^{\rm alg} = \Big(\limsup_{m\to+\infty}\ \sup_{\sigma}\ \frac{1}{m}\log|\sigma|^2_{h_{mK_X+\lfloor mG\rfloor+p_m A}}\Big)^*, \tag{7.14}$$

$$\int_X (\sigma\wedge\overline{\sigma})^{\frac{2}{m}} h^{\frac{1}{m}}_{\lfloor mG\rfloor+p_m A} \leq 1, \tag{7.15}$$

where $m \gg p_m \gg 1$ and $h^{1/m}_{\lfloor mG\rfloor}$ is chosen to converge uniformly to h_G.

We then find again $\varphi_{\rm can} = \varphi_{\rm can}^{\rm alg}$, with an almost identical proof – though we no longer have a sup in the envelope, but just a lim sup. The analogue of Proposition (7.8) also holds true in this context, with an appropriate sequence of sections $\sigma_m \in H^0(X, mK_X + \lfloor mG \rfloor + \lfloor m\Delta \rfloor + p_m A)$.

(7.16) Remark. It would be nice to have a better understanding of the super-canonical metrics. In case X is a curve, this should be easier. In fact X then has a hermitian metric ω with constant curvature, which we normalize by requiring that $\int_X \omega = 1$, and we can also suppose $\int_X e^{-\gamma}\omega = 1$. The class $\lambda = c_1(K_X + L) \geq 0$ is a number and we take $\alpha = \lambda\omega$. Our envelope is $\varphi_{\rm can} = \sup\varphi$ where $\lambda\omega + dd^c\varphi \geq 0$ and $\int_X e^{\varphi-\gamma}\omega \leq 1$. If $\lambda = 0$ then φ must be constant and clearly $\varphi_{\rm can} = 0$. Otherwise, if $G(z,a)$ denotes the Green function such that $\int_X G(z,a)\,\omega(z) = 0$ and $dd^c G(z,a) = \delta_a - \omega(z)$, we find

$$\varphi_{\rm can}(z) \geq \sup_{a\in X}\Big(\lambda G(z,a) - \log\int_{z\in X} e^{\lambda G(z,a)-\gamma(z)}\omega(z)\Big)$$

by taking already the envelope over $\varphi(z) = \lambda G(z,a) - \mathrm{Const}$. It is natural to ask whether this is always an equality, i.e. whether the extremal functions are always given by one of the Green functions, especially when $\gamma = 0$.

7.B. Invariance of plurigenera and positivity of curvature of super-canonical metrics

The concept of super-canonical metric can be used to give a very interesting result on the positivity of relative pluricanonical divisors, which itself can be seen to imply the invariance of plurigenera. The main idea is due to H. Tsuji [Tsu07a], and some important details were fixed by Berndtsson and Păun [BnP09], using techniques inspired from their results on positivity of direct images [Bnd06], [BnP08].

(7.17) Theorem. *Let $\pi : \mathcal{X} \to S$ be a deformation of projective algebraic manifolds over some irreducible complex space S (π being assumed locally projective over S). Let $\mathcal{L} \to \mathcal{X}$ be a holomorphic line bundle equipped with a hermitian metric $h_{\mathcal{L},\gamma}$ of weight γ such that $i\Theta_{\mathcal{L},h_{\mathcal{L},\gamma}} \geq 0$ (i.e. γ is plurisubharmonic), and $\int_{X_t} e^{-\gamma} < +\infty$, i.e. we assume the metric to be klt over all fibers $X_t = \pi^{-1}(t)$. Then the metric defined on $K_{\mathcal{X}} + \mathcal{L}$ as the fiberwise super-canonical metric has semi-positive* *curvature over $\mathcal{X}$. In particular, $t \mapsto h^0(X_t, m(K_{X_t} + \mathcal{L}_{\restriction X_t}))$ is constant for all $m > 0$.*

Once the metric is known to have a plurisuharmonic weight on the total space of $\mathcal{X}$, the Ohsawa-Takegoshi theorem can be used exactly as at the end of the proof of lemma (12.3). Therefore the final statement is just an easy consequence. The cases when $\mathcal{L} = \mathcal{O}_{\mathcal{X}}$ is trivial or when $\mathcal{L}_{\restriction X_t} = \mathcal{O}(\Delta_t)$ for a family of klt $\mathbb{Q}$-divisors are especially interesting.

Proof (Sketch). By our assumptions, there exists (at least locally over S) a relatively ample line bundle $\mathcal{A}$ over $\mathcal{X}$. We have to show that the weight of the global super-canonical metric is plurisubharmonic, and for this, it is enough to look at analytic disks $\Delta \to S$. We may thus as well assume that $S = \Delta$ is the unit disk. Consider the super-canonical metric $h_{\mathrm{can},0}$ over the fiber X_0. The approximation argument seen above (see (7.9) and remark (7.13)) show that $h_{\mathrm{can},0}$ has a weight $\varphi_{\mathrm{can},0}$ which is a regularized upper limit

$$\varphi^{\mathrm{alg}}_{\mathrm{can},0} = \left(\limsup_{m\to+\infty} \frac{1}{m} \log|\sigma_m|^2\right)^*$$

defined by sections $\sigma_m \in H^0(X_0, m(K_{X_0} + \mathcal{L}_{\restriction X_0}) + p_m\mathcal{A}_{\restriction X_0})$ such that

$$\int_{X_0} |\sigma|^2 e^{-(m-p_m)\varphi_{\mathrm{can},0} - p_m\psi_0} dV_\omega \le C_2.$$

with the suitable weights. Now, by section 12, these sections extend to sections $\widetilde{\sigma}_m$ defined on the whole family $\mathcal{X}$, satisfying a similar L^2 estimate (possibly with a slightly larger constant C_2' under control). If we set

$$\Phi = \left(\limsup_{m\to+\infty} \frac{1}{m} \log|\widetilde{\sigma}_m|^2\right)^*,$$

then Φ is plurisubharmonic by construction, and $\varphi_{\mathrm{can}} \ge \Phi$ by the defining property of the super-canonical metric. Finally, we also have $\varphi_{\mathrm{can},0} = \Phi_{\restriction X_0}$ from the approximation technique. It follows easily that φ_{can} satisfies the mean value inequality with respect to any disk centered on the central fiber X_0. Since we can consider arbitrary analytic disks $\Delta \to S$, the plurisubharmonicity of φ_{can} follows. □

7.C. Tsuji's strategy for studying abundance

H. Tsuji [Tsu07c] has recently proposed the following interesting prospective approach of the abundance conjecture.

(7.18) Conjecture/question. *Let (X, Δ) be a klt pair such that $K_X + \Delta$ is pseudoeffective and has numerical dimension $\mathrm{nd}(K_X + \Delta) > 0$. Then for every point $x \in X$ there exists a closed positive current $T_x \in c_1(K_X + \Delta)$ such that the Lelong number at x satisfies $\nu(T_x, x) > 0$.*

It would be quite tempting to try to produce such currents e.g. by a suitable modification of the construction of super-canonical metrics, trying to enforce singularities of the metric at any prescribed point $x \in X$. A related procedure would be to enforce enough vanishing of sections of $A + m(K_X + \Delta)$ at point x, where A is a sufficiently ample line bundle. The number of these sections grows as cm^p where $p = \mathrm{nd}(K_X + \Delta)$. Hence, by an easy linear algebra argument, one can prescribe a vanishing order $s \sim c'm^{p/n}$ of such a section σ, whence a Lelong number $\sim c'm^{\frac{p}{n}-1}$

for the corresponding rescaled current of integration $T = \frac{1}{m}[Z_\sigma]$ on the zero divisor. Unfortunately, this tends to 0 as $m \to +\infty$ whenever $p < n$. Therefore, one should use a more clever argument which takes into account the fact that, most probably, all directions do not behave in an "isotropic way", and vanishing should be prescribed only in certain directions.

Assuming that (7.18) holds true, a simple semi-continuity argument would imply that there exists a small number $c > 0$ such that the analytic set $Z_x = E_c(T_x)$ contains x, and one would expect conjecturally that these sets can be reorganized as the generic fibers of a reduction map $f : X \dashrightarrow Y$, together with a klt divisor Δ' on Y such that (in first approximation, and maybe only after replacing X, Y by suitable blow-ups), one has $K_X + \Delta = f^*(K_Y + \Delta' + R_f) + \beta$ where R_f is a suitable orbifold divisor (in the sense of Campana [Cam04]) and β a suitable pseudo-effective class. The expectation is that $\dim Y = p = \mathrm{nd}(K_X + \Delta)$ and that (Y, Δ') is of general type, i.e. $\mathrm{nd}(K_Y + \Delta') = p$.

8. Siu's analytic approach and Păun's non vanishing theorem

We describe here briefly some recent developments without giving much detail about proofs. Recall that given a pair (X, Δ) where X is a normal projective variety and Δ an effective $\mathbb{R}$-divisor, the transform of (X, Δ) by a birational morphism $\mu : \widetilde{X} \to X$ of normal varieties is the unique pair $(\widetilde{X}, \widetilde{\Delta})$ such that $K_{\widetilde{X}} + \widetilde{\Delta} = \mu^*(K_X + \Delta) + E$ where E is an effective μ-exceptional divisor (we assume here that $K_X + \Delta$ and $K_{\widetilde{X}} + \widetilde{\Delta}$ are $\mathbb{R}$-Cartier divisors).

In [BCHM06], Birkar, Cascini, Hacon and McKernan proved old-standing conjectures concerning the existence of minimal models and finiteness of the canonical ring for arbitrary projective varieties. The latter result was also announced independently by Siu in [Siu06]. The main results can be summarized in the following statement.

(8.1) Theorem. *Let (X, Δ) be a klt pair where Δ is big.*

(i) *If $K_X + \Delta$ is pseudo-effective, (X, Δ) has a log-minimal model, i.e. there is a birational transformation $(\widetilde{X}, \widetilde{\Delta})$ with $\widetilde{X}$ $\mathbb{Q}$-factorial, such that $K_{\widetilde{X}} + \widetilde{\Delta}$ is nef and satisfies additionally strict inequalities for the discrepancies of μ-exceptional divisors.*

(ii) *If $K_X + \Delta$ is not pseudo-effective, then (X, Δ) has a Mori fiber space, i.e. there exists a birational transformation $(\widetilde{X}, \widetilde{\Delta})$ and a morphism $\varphi : \widetilde{X} \to Y$ such that $-(K_{\widetilde{X}} + \widetilde{\Delta})$ is φ-ample.*

(iii) *If moreover Δ is a $\mathbb{Q}$-divisor, the log-canonical ring $\bigoplus_{m \geq 0} H^0(X, m(K_X + \Delta))$ is finitely generated.*

The proof, for which we can only refer to [BCHM06], is an extremely subtle induction on dimension involving finiteness of flips (a certain class of birational transforms improving positivity of $K_X + \Delta$ step by step), and a generalization of Shokurov's non vanishing theorem [Sho85]. The original proof of this non vanishing result was itself based on an induction on dimension, using the existence of minimal models in dimension $n-1$. Independently, Y.T. Siu [Siu06] announced an analytic proof of the finiteness of canonical rings $\bigoplus_{m \geq 0} H^0(X, mK_X)$, along with an analytic variant of Shokurov's non vanishing theorem; in his approach, multiplier ideals and Skoda's division theorem are used in crucial ways. Let us mention a basic statement in this direction which illustrates the connection with Skoda's result, and is interesting for two reasons : i) it does not require any strict positivity assumption, ii) it shows that it is enough to have a sufficiently good approximation of the minimal singularity metric $h_{\min}$ by sections of sufficiently large linear systems $|pK_X|$.

(8.2) Proposition. *Let X be a projective n-dimensional manifold with K_X pseudo-effective. Let $h_{\min} = e^{-\varphi_{\min}}$ be a metric with minimal singularity on K_X (e.g. the super-canonical metric), and let $c_0 > 0$ be the log canonical threshold of $\varphi_{\min}$, i.e. $h_{\min}^{c_0-\delta} = e^{-(c_0-\delta)\varphi_{\min}} \in L^1$ for $\delta > 0$ small. Assume that there exists an integer $p > 0$ so that the linear system $|pK_X|$ provides a weight $\psi_p = \frac{1}{p}\log\sum|\sigma_j|^2$ whose singularity approximates $\varphi_{\min}$ sufficiently well, namely*

$$\psi_p \geq \Big(1 + \frac{1+c_0-\delta}{pn}\Big)\varphi_{\min} + O(1) \qquad \text{for some } \delta > 0.$$

Then $\bigoplus_{m\geq 0} H^0(X, mK_X)$ is finitely generated, and a set of generators is actually provided by a basis of sections of $\bigoplus_{0\leq m\leq np+1} H^0(X, mK_X)$.

Proof. A simple argument based on the curve selection lemma (see e.g. [Dem01], Lemma 11.16) shows that one can extract a system $g = (g_1, \ldots g_n)$ of at most n sections from (σ_j) in such a way that the singularities are unchanged, i.e. $C_1 \log|\sigma| \leq \log|g| \leq C_2 \log|\sigma|$. We apply Skoda's division (7.20) with $E = \mathcal{O}_X^{\oplus n}$, $Q = \mathcal{O}(pK_X)$ and $L = \mathcal{O}((m-p-1)K_X)$ [so that $K_X \otimes Q \otimes L = \mathcal{O}_X(mK_X)$], and with the metric induced by $h_{\min}$ on K_X. By definition of a metric with minimal singularities, every section f in $H^0(X, mK_X) = H^0(X, K_X \otimes Q \otimes L)$ is such that $|f|^2 \leq Ce^{m\varphi_{\min}}$. The weight of the metric on $Q \otimes L$ is $(m-1)\varphi_{\min}$. Accordingly, we find

$$\begin{aligned}|f|^2|g|_{h_{min}}^{-2n-2\varepsilon} e^{-(m-1)\varphi_{\min}} &\leq C\exp\Big(m\varphi_{\min} - p(n+\varepsilon)(\psi_p - \varphi_{\min}) - (m-1)\varphi_{\min}\Big)\\ &\leq C'\exp\big(-(c_0 - \delta/2)\varphi_{\min}\big)\end{aligned}$$

for $\varepsilon > 0$ small, thus the left hand side is in L^1. Skoda's theorem implies that we can write $f = g \cdot h = \sum g_j h_j$ with $h_j \in H^0(X, K_X \otimes L) = H^0(X, (m-p)K_X)$. The argument holds as soon as the curvature condition $m - p - 1 \geq (n - 1 + \varepsilon)p$ is satisfied, i.e. $m \geq np + 2$. Therefore all multiples $m \geq np + 2$ are generated by sections of lower degree $m - p$, and the result follows. □

Recently, Păun [Pau08] has been able to provide a very strong Shokurov-type analytic non vanishing statement, and in the vein of Siu's approach [Siu06], he gave a very detailed independent proof which does not require any intricate induction on dimension (i.e. not involving the existence of minimal models).

(8.3) Theorem (Păun [Pau08]). *Let X be a projective manifold, and let $\alpha_L \in \mathrm{NS}_{\mathbb{R}}(X)$ be a cohomology class in the real Neron-Severi space of X, such that :*

(a) *The adjoint class $c_1(K_X) + \alpha_L$ is pseudoeffective, i.e. there exist a closed positive current*

$$\Theta_{K_X+L} \in c_1(K_X) + \alpha_L;$$

(b) *The class α_L contains a Kähler current Θ_L (so that α_L is big), such that the respective potentials φ_L of Θ_L and φ_{K_X+L} of Θ_{K_X+L} satisfy*

$$e^{(1+\varepsilon)(\varphi_{K_X+L}-\varphi_L)} \in L^1_{\mathrm{loc}}$$

where ε is a positive real number.

Then the adjoint class $c_1(K_X) + \alpha_L$ *contains an effective* $\mathbb{R}$*-divisor.*

The proof is a clever application of the Kawamata-Viehweg-Nadel vanishing theorem, combined with a perturbation trick of Shokurov [Sho85] and with diophantine approximation to reduce the situation to the case of $\mathbb{Q}$-divisors. Shokurov's trick allows to single out components of the divisors involved, so as to be able to take restrictions and apply induction on dimension. One should notice that the poles of φ_L may help in achieving condition (8.3 b), so one obtains a stronger condition by requiring (b′) $\exp((1+\varepsilon)\varphi_{K_X+L}) \in L^1_{\mathrm{loc}}$ for $\varepsilon > 0$ small, namely that $c_1(K_X) + \alpha_L$ is klt. The resulting weaker statement then makes sense in a pure algebraic setting. In [BrP09], Birkar and Păun announced a relative version of (8.3), and they have shown that this can be used to reprove a relative version of Theorem (8.1); the notes from Mihai Păun [Pau09] in the present volume give a fairly precise account of these ideas. A similar purely algebraic approach has been described by C. Hacon last year in his Oberwolfach notes [Hac08], as well as in the present volume.

Bibliography

[AN54] Y. Akizuki and S. Nakano. – *Note on Kodaira-Spencer's proof of Lefschetz theorems*, Proc. Jap. Acad., **30** (1954), 266–272.

[BmD09] R. Berman and J.-P. Demailly. – *Regularity of plurisubharmonic upper envelopes in big cohomology classes*, arXiv:math.CV/0905.1246v1.

[Bnd06] B. Berndtsson. – *Subharmonicity properties of the Bergman kernel and some other functions associated to pseudoconvex domains*, Ann. Inst. Fourier (Grenoble), **56** (2006), 1633–1662.

[BnP08] B. Berndtsson and M. Păun. – *Bergman kernels and the pseudoeffectivity of relative canonical bundles*, Duke Math. J, **145** (2008), 341–378.

[BnP09] B. Berndtsson and M. Păun. – *Canonical metrics for projective families*, Privately communicated manuscript, February 2009.

[BCHM06] C. Birkar, P. Cascini, C.D. Hacon, J. McKernan. – *Existence of minimal models for varieties of log general type*, arXiv:math.AG/0610203v2.

[BrP09] C. Birkar, M. Păun. – *Minimal models, flips and finite generation : a tribute to V.V. Shokurov and Y.-T. Siu*, arXiv:math.AG/0904.2936.

[BGK05] Ch. P. Boyer, K. Galicki, J. Kollár. – *Einstein Metrics on Spheres*, Annals of Math, **162** (2005), 557–580.

[Bon93] L. Bonavero. – *Inégalités de Morse holomorphes singulières*, Prépublication Université de Grenoble I, Institut Fourier, n° 259, 1993.

[Bou02] S. Boucksom. – *Cônes positifs des variétés complexes compactes*, Thesis, Grenoble 2002.

[BDPP04] S. Boucksom, J.-P. Demailly, M. Păun, Th. Peternell. – *The pseudo-effective cone of a compact Kähler manifold and varieties of negative Kodaira dimension*, manuscript May 2004, math.AG/0405285.

[Buc99] N. Buchdahl. – *On compact Kähler surfaces*, Ann. Inst. Fourier, **49** (1999), 287–302.

[Buc00] N. Buchdahl. – *A Nakai-Moishezon criterion for non-Kähler surfaces*, Ann. Inst. Fourier, **50** (2000), 1533–1538.

[Cam04] F. Campana. – *Orbifolds, special varieties and classification theory*, Ann. Inst. Fourier (Grenoble), **54** (2004), 499–630.

[CP04] F. Campana, Th. Peternell. – *Geometric stability of the cotangent bundle and the universal cover of a projective manifold*, arXiv:math.AG/0405093.

[Dem82] J.-P. Demailly. – *Estimations L^2 pour l'opérateur $\overline{\partial}$ d'un fibré vectoriel holomorphe semi-positif au dessus d'une variété kählérienne complète*, Ann. Sci. Ec. Norm. Sup., **15** (1982), 457–511.

[Dem85] J.-P. Demailly. – *Champs magnétiques et inégalités de Morse pour la d''-cohomologie*, Ann. Inst. Fourier (Grenoble), **35** (1985), 189–229.

[Dem92] J.-P. Demailly. – *Regularization of closed positive currents and Intersection Theory*, J. Alg. Geom., **1** (1992), 361–409.

[Dem93] J.-P. Demailly. – *A numerical criterion for very ample line bundles*, J. Differential Geom., **37** (1993), 323–374.

[Dem94] J.-P. Demailly. – *Regularization of closed positive currents of type* (1, 1) *by the flow of a Chern connection*, Actes du Colloque en l'honneur de P. Dolbeault (Juin 1992), édité par H. Skoda et J.-M. Trépreau, Aspects of Mathematics, Vol. E 26, Vieweg, 1994, 105–126.

[Dem01] J.-P. Demailly. – *Multiplier ideal sheaves and analytic methods in algebraic geometry*, Lecture Notes, School on "Vanishing theorems and effective results in Algebraic Geometry, ICTP Trieste, April 2000, Publications of ICTP, 2001.

[Dem08] J.-P. Demailly. – *Appendix to I. Cheltsov and C. Shramov's article "Log canonical thresholds of smooth Fano threefolds": On Tian's invariant and log canonical thresholds*, arXiv: math.AG 0806.2107.

[DEL00] J.-P. Demailly, L. Ein and R. Lazarsfeld. – *A subadditivity property of multiplier ideals*, Michigan Math. J., 2000, to appear.

[DK01]] J.-P. Demailly, J. Kollár. – *Semicontinuity of complex singularity exponents and Kähler-Einstein metrics on Fano orbifolds*, math.AG/9910118; Ann. Ec. Norm. Sup, **34** (2001), 525–556.

[DP04] J.-P. Demailly, M. Păun. – *Numerical characterization of the Kähler cone of a compact Kähler manifold*, arXiv: math.AG/0105176, Annals of Math, **159** (2004), 1247–1274.

[DPS94] J.-P. Demailly, Th. Peternell, M. Schneider. – *Compact complex manifolds with numerically effective tangent bundles*, J. Algebraic Geometry, **3** (1994), 295–345.

[DPS96] J.-P. Demailly, Th. Peternell, M. Schneider. – *Holomorphic line bundles with partially vanishing cohomology*, Conf. in honor of F. Hirzebruch, Israel Mathematical Conference Proceedings Vol. **9** (1996), 165–198.

[DPS00] J.-P. Demailly, Th. Peternell, M. Schneider. – *Pseudo-effective line bundles on compact Kähler manifolds*, June 2000, Algebraic Geometry e-prints math.AG/0006205.

[DSk79] J.-P. Demailly, H. Skoda. – *Relations entre les notions de positivité de P.A. Griffiths et de S. Nakano*, Séminaire P. Lelong-H. Skoda (Analyse), année 1978/79, Lecture notes in Math. no **822**, Springer-Verlag, Berlin (1980) 304–309.

[ELN96] L. Ein, R. Lazarsfeld and M. Nakamaye. – *Zero estimates, intersection theory, and a theorem of Demailly*, in Higher Dimensional Complex Varieties (Trento, 1994), de Gruyter (1996), 183–207.

[Fuj83] T. Fujita. – *Semipositive line bundles*, J. Fac. Sci. Univ. of Tokyo, **30** (1983), 353–378.

[Fuj94] T. Fujita. – *Approximating Zariski decomposition of big line bundles*, Kodai Math. J., **17** (1994), 1–3.

[Hac08] C. Hacon. – *Extension theorems and the existence of flips*, Series of 8 lectures given at Oberwolfach Math. Institute, Oct. 12-18, 2008.

[HaMK07] C. Hacon, J. McKernan. – *Extension theorems and the existence of flips*, In "flips for 3-folds and 4-folds" (ed. A. Corti), Oxford lecture series in mathematics and its applications **35**, Oxford University Press, 2007.

[HaMK08] C. Hacon, J. McKernan. – *Existence of minimal models for varieties of general type, II*, (Existence of pl-flips), arXiv:math.AG/0808.1929v1.

[Har70] R. Hartshorne. – *Ample subvarieties of algebraic varieties*, Lecture Notes in Math., Vol. 156, Springer-Verlag, Berlin (1970).

[Hir64] H. Hironaka. – *Resolution of singularities of an algebraic variety over a field of characteristic zero*, Ann. of Math., **79** (1964), 109–326.

[Hör65] L. Hörmander. – *L^2 estimates and existence theorems for the $\overline{\partial}$ operator*, Acta Math., **113** (1965), 89–152.

[Hov79] A.G. Hovanski. – *Geometry of convex bodies and algebraic geometry*, Uspehi Mat. Nauk, **34** (4) (1979), 160–161.

[Huy99] D. Huybrechts. – *Compact Hyperkähler Manifolds: Basic Results*, Invent. Math., **135** (1999), 63–113.

[Huy01] D. Huybrechts. – *The Projectivity Criterion for Hyperkähler manifolds as a Consequence of the Demailly-Păun Theorem*, personal communication, manuscript Köln University, May 2001, 3 p, to appear in Invent. Math.

[JK01] J.M. Johnson, J. Kollár. – *Kähler-Einstein metrics on log del Pezzo surfaces in weighted projective 3-spaces*, Ann. Inst. Fourier, **51** (2001), 69–79.

[Kaw82] Y. Kawamata. – *A generalization of Kodaira-Ramanujam's vanishing theorem*, Math. Ann., **261** (1982), 43–46.

[Kis78] C.O. Kiselman. – *The partial Legendre transformation for plurisubharmonic functions*, Invent. Math., **49** (1978), 137–148.

[Kis79] C.O. Kiselman. – *Densité des fonctions plurisousharmoniques*, Bull. Soc. Math. France, **107** (1979), 295–304.

[Kis84] C.O. Kiselman. – *Sur la définition de l'opérateur de Monge-Ampère complexe*, Analyse Complexe, Proceedings of the Journées Fermat (SMF), Toulouse, 1983, Lecture Notes in Math., Vol. 1094, Springer-Verlag, Berlin (1984), 139–150.

[Kod53] K. Kodaira. – *On a differential geometric method in the theory of analytic stacks*, Proc. Nat. Acad. Sci. USA, **39** (1953), 1268–1273.

[Kod54] K. Kodaira. – *On Kähler varieties of restricted type*, Ann. of Math., **60** (1954), 28–48.

[Koh63] J.J. Kohn. – *Harmonic integrals on strongly pseudo-convex manifolds I*, Ann. Math., **78** (1963), 206–213.

[Koh64] J.J. Kohn. – *Harmonic integrals on strongly pseudo-convex manifolds II*, Ann. Math., **79** (1964), 450–472.

[Kol85] J. Kollár. – *Vanishing theorems for cohomology groups*, Algebraic Geometry, Bowdoin 1985, Proc. Symp. Pure Math., Vol. 46, 1987, 233–243.

[Lam99a] A. Lamari. – *Courants kählériens et surfaces compactes*, Ann. Inst. Fourier, **49** (1999), 263–285.

[Lam99b] A. Lamari. – *Le cone Kählérien d'une surface*, J. Math. Pures Appl., **78** (1999), 249–263.

[Lel69] P. Lelong. – *Plurisubharmonic functions and positive differential forms*, Gordon and Breach, New-York, and Dunod, Paris, 1969.

[Mou99] Ch. Mourougane. – *Théorèmes d'annulation génériques pour les fibrés vectoriels semi-négatifs*, Bull. Soc. Math. Fr., **127** (1999), 115–133.

[Nad89] A.M. Nadel. – *Multiplier ideal sheaves and Kähler-Einstein metrics of positive scalar curvature*, Proc. Nat. Acad. Sci. U.S.A., **86** (1989), 7299–7300 and Annals of Math., **132** (1990), 549–596.

[Ohs88] T. Ohsawa. – *On the extension of L^2 holomorphic functions, II*, Publ. RIMS, Kyoto Univ., **24** (1988), 265–275.

[OT87] T. Ohsawa and K. Takegoshi. – *On the extension of L^2 holomorphic functions*, Math. Zeitschrift, **195** (1987), 197–204.

[Pau98a] M. Păun. – *Sur l'effectivité numérique des images inverses de fibrés en droites*, Math. Ann, **310** (1998), 411–421.

[Pau98b] M. Păun. – *Fibrés en droites numériquement effectifs et variétés Kählériennes compactes à courbure de Ricci nef*, Thèse, Université de Grenoble 1, (1998), 80 p.

[Pau00] M. Păun. – *Semipositive $(1,1)$-cohomology classes on projective manifolds*, Preprint de l'I.R.M.A., Strasbourg.

[Pau07] M. Păun. – *Siu's invariance of plurigenera: a one-tower proof*, J. Differential Geom, **76** (2007), 485-493.

[Pau08] M. Păun. – *Relative critical exponents, non-vanishing and metrics with minimal singularities*, arXiv:math.CV/0807.3109.

[Ser54] J.-P. Serre. – *Fonctions automorphes: quelques majorations dans le cas où X/G est compact*, Sém. Cartan (1953-54) 2–1 à 2–9.

[Sho85] V.V. Shokurov. – *A nonvanishing theorem*, Izv. Akad. Nauk SSSR Ser. Mat, **49** (1985), 635-651.

[Sie55] C.L. Siegel. – *Meromorphic Funktionen auf kompakten Mannigfaltigkeiten* , Nachrichten der Akademie der Wissenschaften in Göttingen, Math.-Phys. Klasse, **4** (1955), 71–77.

[Siu74] Y.T. Siu. – *Analyticity of sets associated to Lelong numbers and the extension of closed positive currents*, Invent. Math., **27** (1974), 53–156.

[Siu88] Y.T. Siu. – *The existence of Kähler-Einstein metrics on manifolds with positive anticanonical line bundle and a suitable finite symmetry group*, Ann. of Math, **127** (1988), 585–627.

[Siu98] Y.T. Siu. – *Invariance of Plurigenera*, Invent. Math., **134** (1998), 661–673.

[Siu06] Y.-T. Siu. – *A General Non-Vanishing Theorem and an Analytic Proof of the Finite Generation of the Canonical Ring*, arXiv:math.AG/0610740.

[Siu08] Y.-T. Siu. – *Finite Generation of Canonical Ring by Analytic Method*, arXiv: math.CV/0803.2454.

[Sko72a] H. Skoda. – *Sous-ensembles analytiques d'ordre fini ou infini dans $\mathbb{C}^n$*, Bull. Soc. Math. France, **100** (1972), 353–408.

[Sko72b] H. Skoda. – *Applications des techniques L^2 à la théorie des idéaux d'une algèbre de fonctions holomorphes avec poids*, Ann. Scient. Ec. Norm. Sup. 4e Série, **5** (1972), 545–579.

[Sko78] H. Skoda. – *Morphismes surjectifs de fibrés vectoriels semi-positifs*, Ann. Sci. Ecole Norm. Sup., **11** (1978), 577–611.

[Sko82] H. Skoda. – *Prolongement des courants positifs fermés de masse finie*, Invent. Math., **66** (1982), 361–376.

[Tak97] K. Takegoshi. – *On cohomology groups of nef line bundles tensorized with multiplier ideal sheaves on compact Kähler manifolds*, Osaka J. Math., **34** (1997), 783–802.

[Tei79] B. Teissier. – *Du théorème de l'index de Hodge aux inégalités isopérimétriques*, C. R. Acad. Sc. Paris, sér. A, **288** (29 Janvier 1979), 287–289.

[Tei82] B. Teissier. – *Bonnesen-type inequalities in algebraic geometry*, Sem. on Diff. Geom. edited by S.T. Yau, 1982, Princeton Univ. Press, 85–105.

[Tia87] G. Tian. – *On Kähler-Einstein metrics on certain Kähler manifolds with* $c_1(M) > 0$, Invent. Math, **89** (1987), 225–246.

[Tra95] S. Trapani. – *Numerical criteria for the positivity of the difference of ample divisors*, Math. Zeitschrift, **219** (1995), 387–401.

[Tsu07a] H. Tsuji. – *Canonical singular hermitian metrics on relative canonical bundles*, prepublication Sophia University, Japan, arXiv:math.AG/0704.0566.

[Tsu07b] H. Tsuji. – *Canonical volume forms on compact Kähler manifolds*, prepublication Sophia University, Japan, arXiv:math.AG/0707.0111.

[Tsu07c] H. Tsuji. – *Talk and personal communication*, MSRI Hot-Topics Workshop "Minimal and canonical models in algebraic geometry" help at Berkeley in April 2007.

[Var09] D. Varolin. – *Three variations on a theme in complex analytic geometry*, IAS/Park City Summer Institute, 18th annual session, July 6 -26, 2008, this volume.

[Vie82] E. Viehweg. – *Vanishing theorems*, J. Reine Angew. Math., **335** (1982), 1–8.

[Yau78] S.T. Yau. – *On the Ricci curvature of a complex Kähler manifold and the complex Monge-Ampère equation I*, Comm. Pure and Appl. Math., **31** (1978), 339–411.

Lecture Notes on Rational Polytopes and Finite Generation

Mihai Păun

IAS/Park City Mathematics Series
Volume 17, 2008

Lecture Notes on Rational Polytopes and Finite Generation

Mihai Păun

0. Introduction

Let X be a non-singular, projective manifold, and let $(Y_j)_{j=1,\dots,N}$ be a set of non-singular hypersurfaces of X, having strictly normal crossings. We denote by $A \to X$ an ample $\mathbb{Q}$-bundle on X. Following V. Shokurov, (cf. [26]) we define the set

$$\mathcal{E}_{Y,A} := \{\tau \in [0,1]^N \ : K_X + Y_\tau + A \in \mathrm{Psef}(X)\}$$

where $Y_\tau := \sum_{j=1}^N \tau^j Y_j$.

Let $\mathcal{L} \subset \mathbb{R}^r$ be an r-dimensional polytope, whose vertices have rational coordinates (i.e. a *rational polytope*). For $j = 1, .., N$, let $l^j \ : \mathbb{R}^r \to \mathbb{R}$ be a set of affine forms defined over $\mathbb{Q}$, such that

$$0 \le l^j(\theta) \le 1 - \varepsilon_0$$

for all $\theta \in \mathcal{L}$, where ε_0 is a positive real number.

We denote by $d := (d^0, ..., d^r)$ an element of $\mathbb{Z}^{r+1}$, and we introduce the set

$$\Gamma_d := \{(m, \theta) \in \mathbb{Z}_+ \times \mathcal{L} \ : \forall j = 0, ..., r, \quad m\theta^j \in d^j\mathbb{Z}, \quad l(\theta) \in \mathcal{E}_{Y,A}\} \tag{1}$$

where by convention we put $\theta^0 := 1$ in the relations above.

Our focus in these notes will be on the following ring of holomorphic sections

$$\mathcal{A}_r(X) := \bigoplus_{(m,\theta)\in\Gamma_d} H^0\Big(X, m(K_X + \sum_{j=1}^N l^j(\theta)Y_j + A)\Big),$$

where we assume that the coordinates of d are divisible enough.

In the framework of the *minimal models program*, the following result was established by C. Birkar, P. Cascini, C. Hacon and J. McKernan in [2] (one can profitably consult the enlightening presentations in [10], [15] of this article).

0.1. Theorem [2]. *The following assertions are true.*

(i) *The set $\mathcal{E}_{Y,A}$ is a rational polytope ;*

(ii) *The ring $\mathcal{A}_r(X)$ defined above is finitely generated.* □

A first remark about 0.1 is that the properties of the set $\mathcal{E}_{Y,A}$ are crucial for the original proof given in [2]. Concerning the finite generation part of theorem 0.1, we mention the very elegant approach due to Y.-T. Siu (see [30], [31], [32]), which

Institut Élie Cartan, Nancy, and Korea Institute for Advanced Study, Seoul
E-mail address: paun@iecn.u-nancy.fr

does not use V. Shokurov polytopes technique. Instead, he introduces new invariants associated to linear series (the "minimal discrepancy subspace"), which most certainly will be useful in other contexts.

An *intuitive* reason for the formulation of the point ii) above is that in this way we take into account a consistent part of statements of the inductive process in the original article [2] ; a more "down to Earth" explanation is provided by the approach used in the section 3.

In this survey we propose a proof of theorem 0.1 guided by the same principles as in [2], [21] and [30]. *In spirit,* we follow to a large extent the article [21] ; however, as one can see in sections 2.2, 2.3 and 3 the main *technical aspects* of our approach are quite different. Actually, the core of our arguments relies *exclusively* on the extension theorems established in [24]. The theory of *closed positive currents* and its relationship with the algebraic geometry as developed by J.-P. Demailly and Y.-T. Siu over many years plays an important rôle in this text. Needless to say that the original proof [2] does not use such tools, and we do not claim at all that they are indispensable ; we just hope that the flexibility of the techniques employed here may be suitable for forthcoming research around these topics.

Finally, we would like to mention that part of the arguments for 0.1 presented here are only sketched and/or commented ; still we hope that the main ideas and techniques of the proof are quite transparent. □

1. Basic definitions and notations

In this paragraph we fix some notations and we collect a few results which will be needed in the proof of 0.1.

Let $\{\alpha\} \in \mathrm{Psef}(X)$ be a pseudoeffective cohomology class of (1,1) type, where α is a non-singular and closed differential form. The important notion of *current with minimal singularities in the sense of pluripotential theory* associated to the class $\{\alpha\}$ was introduced in [7]. The corresponding object will be denoted by $\Theta_{\min}$, and will be refered here as *current with minimal singularities in the sense of Demailly.* We recall next its definition ; for a more ample presentation of its properties, see [4], [5], [7], [8], [9].

We consider the family of potentials

$$\mathcal{P} := \{f \in L^1(X) \ : f \leq 0 \text{ and } \alpha + \sqrt{-1}\partial\bar{\partial} f \geq 0\}$$

where the above inequality is assumed to hold in the sense of currents on X. Then we define $f_{\min}$ to be the *regularized upper envelope* of the above family, and then

$$\Theta_{\min} := \alpha + \sqrt{-1}\partial\bar{\partial} f_{\min}.$$

In the geometric context $\{\alpha\} = c_1(L)$, where $L \to X$ is a $\mathbb{Q}$-bundle, one can restrict the family of functions $\mathcal{P}$ above, and only take into account the potentials f induced by holomorphic sections of L and its multiples (suitably normalized). Let $f_{\min,\sigma}$ be the regularized upper envelope of this smaller family (see [28]) ; the corresponding current is denoted by $\Theta_{\min,\sigma}$ and we call it *current with minimal singularities in the sense of Siu.* The inequality

$$f_{\min} \geq f_{\min,\sigma}$$

is obvious from the definitions. If the bundle L is big, then the above currents have equivalent singularities if and only if the algebra of sections defined by L is of finite type, as remarked in [5]. The relationship between these two currents play an important rôle in the proof of 0.1.

Let $W \subset X$ be an irreducible hypersurface ; following [5], [22], we denote by $\nu_W(\{\alpha\})$ the *minimal multiplicity of the class* $\{\alpha\}$ along W. We recall that the minimal multiplicity $\nu_W(\{\alpha\})$ is in general *strictly smaller* than the Lelong number of the current with minimal singularities of $\{\alpha\}$, cf [5].

Next, we have the notion of *numerical dimension* of a pseudoeffective class $\{\alpha\}$ generalizing the corresponding definition in the setting of the nef line bundles. In the present text we will denote this invariant by $\mathrm{nd}(\{\alpha\})$.
We will not reproduce the precise definition of the preceding invariants ; their relevance for our article is explained in the next statement.

1.1. Theorem [5], [22]. *Let X be a compact complex manifold, and let $\{\alpha\} \in \mathrm{Psef}(X)$ be a pseudoeffective class.*
(a) *The function $\{\alpha\} \to \nu_W(\{\alpha\})$ is convex and lower semi-continuous on the closed set* $\mathrm{Psef}(X)$.
(b) *We assume that* $\mathrm{nd}(\{\alpha\}) = 0$ *; then there exists an unique closed positive current*

$$T = \sum_{i=1}^{N} a_T^j [W_j] \in \{\alpha\}$$

and moreover we have $a_T^j = \nu_{W_j}(\{\alpha\})$. □

We recall next the notion of *asymptotic vanishing order* of a divisor along an irreducible hypersurface $W \subset X$.

Let D be an effective $\mathbb{Q}$-divisor ; following [11], we define

$$\mathrm{ord}_W\left(\|D\|\right) := \lim_{p\to\infty} \frac{\mathrm{ord}_W(|pD|)}{p}$$

where $\mathrm{ord}_W(|pD|)$ is the vanishing order along W of a generic representative of the linear system $|pD|$. The order function can be extended by continuity to the cone of effective $\mathbb{R}$-divisors ; we refer to [11] for the proof of this assertion.
The following important result was equally established in [11], section 4. Let $D_1, ..., D_r$ be effective $\mathbb{Q}$-divisors, such that the Cox ring

$$\mathcal{R} := \bigoplus_{(m^1,...,m^r)\in d\mathbb{Z}_+} H^0(X, m^1 D_1 + ... + m^r D_r)$$

is finitely generated, where d is a divisible enough integer.

1.2. Theorem [11]. *Under the above assumption, there exists a smooth fan Ξ refining $\mathbb{N}^r$ such that on each cone of Ξ, the function*

$$(m^1, ...m^r) \to \mathrm{ord}_W\left(\|m^1 D_1 + ... + m^r D_r\|\right)$$

is linear. □

In particular, there exists a refinement (Δ_j) of the standard polytope

$$\Delta := \{\tau = (\tau^1, ..., \tau^r) : \tau^j \geq 0, j = 1, ..., r \text{ and } \tau^1 + ... + \tau^r = 1\}$$

such that each Δ_j is generated by finite set of *vertices with rational coordinates*, and such that if we denote by $D_\tau := \sum \tau^j D_j$, then we have

$$\mathrm{ord}_W\left(\|(1-\eta)D_{\tau_0} + \eta D_{\tau_1}\|\right) = (1-\eta)\,\mathrm{ord}_W\left(\|D_{\tau_0}\|\right) + \eta\,\mathrm{ord}_W\left(\|D_{\tau_1}\|\right) \tag{2}$$

for any τ_0, τ_1 within the same Δ_j, and any $0 \leq \eta \leq 1$. The equality (2) follows directly from theorem 1.2 if τ_1, τ_2 and η have rational coordinates ; in general, we use a continuity argument.

We note that if D_1 and D_2 are *numerically equivalent* big divisors, then we have $\mathrm{ord}_W\left(\|D_1\|\right) = \mathrm{ord}_W\left(\|D_2\|\right)$, as stated in theorem A of [11]. If $D_1 = K_X + L_1$ where L_1 is klt and big, then the same equality holds true, as soon as D_1 is (pseudo)effective (i.e. in this case we do not need the "big" hypothesis), see [24]. □

To conclude this paragraph, we recall an elementary result concerning the set $\mathcal{E}_{Y,A}$ defined in the introduction by using the hypersurfaces $(Y_j)_{j=1,\ldots,N}$.

In order to study its "shape" near an arbitrary point τ_0, it would be convenient to have $0 \leq \tau_0^j \leq 1-\varepsilon_0$, for some strictly positive ε_0 and all the indexes j. We show that we can assume that such inequalities holds, provided that we modify slightly the ample part A.

We use the following *translation* technique which goes back at least to [2], as follows. Let $0 < \varepsilon_0 < 1/2$ be a rational number, such that $A + \varepsilon_0 \sum_{j\in J} Y_j$ is still ample, for any subset $J \subset \{1,..., N\}$. We define the set

$$J_0 := \{1 \leq j \leq N \ : \tau_0^j > 1 - \varepsilon_0\}$$

and then for any $\tau \in [0,1]^N$ we have

$$K_X + \sum_{j\in\{1,\ldots,N\}\setminus J_0} \tau^j Y_j + \sum_{j\in J_0} \tau^j Y_j + A \equiv K_X + \sum_{j\in\{1,\ldots,N\}\setminus J_0} \tau^j Y_j + \sum_{j\in J_0} (\tau^j - \varepsilon_0) Y_j + A_0$$

where A_0 is the ample bundle $A + \varepsilon_0 \sum_{j\in J_0} Y_j$. We remark that $0 \leq \tau_0^j - \varepsilon_0 \leq 1-\varepsilon_0$ for any $j \in J_0$, and therefore there exists a small open set $\Omega \subset \mathcal{E}_{Y,A}$ containing the point τ_0, such that the map $\Phi \ : \Omega \to \mathcal{E}_{Y,A_0}$ given by the relation

$$\tau \to (\tau', \tau'' - \varepsilon_0)$$

is well defined, where τ' (respectively τ'') corresponds to the components of τ whose indexes belongs to $\{1,..., N\} \setminus J_0$ (respectively J_0).

In conclusion, while analyzing the set $\mathcal{E}_{Y,A}$ near one of its points τ_0, we can assume that $0 \leq \tau_0^j \leq 1 - \varepsilon_0$, for all $1 \leq j \leq N$. □

2. Proof of (i)

We start with a few preliminary remarks ; in the first place, the set $\mathcal{E}_{Y,A}$ is convex ; it is equally closed by the usual properties of the pseudoeffective cone.

Also, we note that a point $\tau_0 \in \mathcal{E}_{Y,A}$ having at least one non-rational coordinate cannot be *extremal.* The argument is an immediate consequence of the non-vanishing theorem (see [2], [24]), as follows.

Let $\sum_{j\in I} \rho_0^j [W_j]$ be an effective $\mathbb{R}$-divisor, numerically equivalent with $K_X + Y_{\tau_0} + A$.

As in [24], paragraph 1.I, we consider the set

$$\mathcal{J} := \left\{(x,\tau) \in \mathbb{R}^{|I|} \times \mathbb{R}^N \ : \sum_{j\in I} x^j [W_j] \equiv K_X + Y_\tau + A\right\}$$

and we note that $\mathcal{J}$ is an affine subspace of $\mathbb{R}^{|I|} \times \mathbb{R}^N$, which is defined over the rational numbers. Our given data (ρ_0, τ_0) corresponds to a point in $\mathcal{J}$; let $\eta > 0$ be a positive number. We can construct the rational approximations $(\rho_{\eta s}, \tau_{\eta s}) \in \mathcal{J}$ of (ρ_0, τ_0) such that :

(i) There exists $q_{\eta s} \in \mathbb{Z}_+$ such that $q_{\eta s}(\rho_{\eta s}, \tau_{\eta s})$ has integer coordinates, and such that the next Dirichlet inequality is satisfied $q_{\eta s}\|\rho_0 - \rho_{\eta s}\| < \eta$ (and a similar relation for the τ_0).

(ii) The point (ρ_0, τ_0) belongs to the convex hull of $(\rho_{\eta s}, \tau_{\eta s})$.

We remark that the coordinates of $\rho_{\eta s}$ are positive rational numbers, and we clearly have $\tau_{\eta s} \in \big([0,1] \cap \mathbb{Q}\big)^N$ as soon as $\eta \ll 1$; this is a consequence of (i) above. In conclusion, if at least one component of τ_0 is not rational, then this point cannot be extremal. □

By the classical theory of convex sets, the first part of 0.1 can be reformulated as follows.

2.0. Claim. *The set of extremal points of the set $\mathcal{E}_{Y,A}$ is isolated.* □

Our proof of claim 2.0 is in some sense a generalization of the arguments in [24] : there are two main cases we are forced to consider.

First we analyze an extremal point $\tau_0 \in \mathcal{E}_{Y,A}$ for which the numerical dimension of the corresponding $\mathbb{Q}$-bundle $K_X + Y_{\tau_0} + A$ is equal to zero. In order to establish the claim under this hypothesis, the main results we invoke here are due to S. Boucksom (see [5]).

If the numerical dimension of the bundle $K_X + Y_{\tau_0} + A$ is at least one, then given any sequence $\tau_k \to \tau_0$ as $k \to \infty$ we will create a center S adapted to (τ_k) on a modification of X, and we will use the full force of 0.1 in lower dimensions for the proof of 2.0. □

2.1. The case $\mathrm{nd}(\{K_X + Y_{\tau_0} + A\}) = 0$

Let $\tau_0 \in \mathcal{E}_{Y,A}$ be a point such that $\mathrm{nd}(\{K_X + Y_{\tau_0} + A\}) = 0$; we do not assume it rational or extremal, for the moment. By the result 1.1 recalled in the preceding paragraph, there exists an unique closed positive current

$$\Theta_0 := \sum_{j \in I} a^j_{\min}(\tau_0)[W_j]$$

in the class $K_X + Y_{\tau_0} + A$; we denote by I above a finite set.

Let $\tau_k \in \mathcal{E}_{Y,A}$ be a sequence converging to τ_0 as $k \to \infty$, and let

$$\Theta_k := \sum_{j \in I} a^j_{\min}(\tau_k)[W_j] + \Lambda_k$$

be a current with minimal singularities in the class $\{K_X + Y_{\tau_k} + A\}$; along the next few lines, we reproduce the arguments in [5] to show that we have $\lim_k a^j_{\min}(\tau_k) = a^j_{\min}(\tau_0)$.

In the first place, the sequence of currents Θ_k (can be assumed to) converge to Θ_0. Indeed this is clear, by the uniqueness part of the statement 1.1 above, combined with standard properties of closed positive currents, see [27]. By the semi-continuity of the Lelong numbers of closed positive currents, we have

$$a^j_{\min}(\tau_0) \geq \lim_k a^j_{\min}(\tau_k). \tag{3}$$

On the other hand, for each $k \geq 1$ and each index j we have

$$a^j_{\min}(\tau_k) \geq \nu_{W_j}(\{K_X + Y_{\tau_k} + A\}). \tag{4}$$

This inequality combined with the lower-semicontinuity result 1.1 (a) yields

$$a^j_{\min}(\tau_0) \leq \lim_k a^j_{\min}(\tau_k) \tag{5}$$

and thus we obtain

$$a^j_{\min}(\tau_0) = \lim_k a^j_{\min}(\tau_k). \tag{6}$$

For any positive real $\eta > 0$, we consider the current

$$T_{k,\eta} := \Theta_k + \eta(\Theta_k - \Theta_0). \tag{7}$$

Then we have $T_{k,\eta} \in \{K_X + Y_{\tau_{k,\eta}} + A\}$, where $\tau_{k,\eta} := \tau_k + \eta(\tau_k - \tau_0)$. The next simple considerations enable us to conclude :

- the components of $\tau_{k,\eta}$ belong to the interval $[0, 1]$, as soon as $k \gg 0$ (we recall that the components of τ_0 are assumed to be positive and smaller than $1 - \varepsilon_0$) ;
- the current $T_{k,\eta}$ is positive – it is at this point that we are using the equality (6) in an essential manner.

In conclusion, the quantity τ_k can be written as a convex combination of τ_0 and $\tau_{k,\eta}$; each of these two points belong to $\mathcal{E}_{Y,A}$ and neither of them is equal to τ_k. Therefore the set of extremal points of $\mathcal{E}_{Y,A}$ is isolated at any τ_0 for which the associated class $\{K_X + Y_{\tau_0} + A\}$ has numerical dimension zero. □

2.2. The "x method" for sequences

In the course of paragraphs 2.2 and 2.3, our main goal will be to analyse the set of *extremal points* near $\tau_0 \in \mathcal{E}_{Y,A} \cap \mathbb{Q}^N$, such that

$$\mathrm{nd}(\{K_X + Y_{\tau_0} + A\}) \geq 1.$$

Let $\tau_k \in \mathcal{E}_{Y,A}$ be a sequence of points with rational coordinates converging to τ_0 as $k \to \infty$; for the moment, we do not assume that (τ_k) are extremal. In this subsection we will show that we can apply the so-called *x-method* in the version explained in [2] and [24], simultaneously for (a subsequence of) $(K_X + Y_{\tau_k} + A)_{k\in\mathbb{N}}$. Since there are practically no serious difficulties while transposing the arguments from the case of a single point to our set-up, we will not provide here a full treatment of the lemma which will soon appear, but rather try to make it "believable". □

We can assume that the coordinates of the vector $\tau_0 \in \mathcal{E}_{Y,A}$ belong to the interval $[0, 1 - \varepsilon_0]$, for some $\varepsilon_0 > 0$, thanks to the discussion at the end of the paragraph 1. Along the next few lines, we construct the analogue of an *incomplete linear system* in the usual x-method corresponding to each τ_k, whose singularities do not change as $k \to \infty$.

The class $\{K_X + Y_{\tau_k} + A\}$ is pseudoeffective, and we denote by Θ_k its current with minimal singularities in the sense of Demailly. There exists $x_0 \in X$ a very general point, such that $\nu\big(\Theta_k, x_0\big) = 0$, for any $k \geq 0$. By the *folklore* results recalled e.g. in [20], there exists a Kähler current $T_0 \in \{m_0(K_X + Y_{\tau_0} + A) + A\}$, with logarithmic poles and such that

$$\nu(T_0, x_0) \geq \dim(X) + 1 \tag{8}$$

where m_0 is a large enough integer. We can also assume that the singularities of T_0 are rational numbers, in the sense that there exists a birational map $f : \widehat{X} \to X$

such that $f^\star(T_0)$ can be written as an effective divisor with rational coefficients plus a non-singular $(1,1)$-form. Let $\eta > 0$ be a rational number ; we define

$$T_k \ := T_0 - \eta \sum_{j=1}^{N} \theta_j + m_0 \sum_{j=1}^{N} (\tau_k^j - \tau_0^j + \eta)[Y_j]$$

where θ_j is a non-singular representative of $m_0 c_1(Y_j)$. We remark that the current

$$\Xi \ := T_0 - \eta \sum_{j=1}^{N} \theta_j$$

is a Kähler current, provided that $\eta \ll 1$; we fix such a quantity, and then there exists $k_\eta \gg 0$, such that $\tau_k^j - \tau_0^j + \eta > 0$, for all $j = 1, ..., N$ and for all $k \geq k_\eta$.

Hence the current T_k can be written as follows

$$T_k = \Xi + \sum_{j=0}^{N} f^j(\tau_k)[Y_j], \tag{9}$$

where $f^j(\tau) \ := m_0(\tau^j - \tau_0^j + \eta)$. Moreover, it belongs to the class $m_0(K_X + Y_{\tau_k} + A) + A$. We consider a log-resolution $\widetilde{\mu} \ : \widetilde{X} \to X$ of $T_0 + \sum_{j=1}^{N} Y_j$. The $\widetilde{\mu}$–inverse image of T_k reads as follows

$$\widetilde{\mu}^\star(T_k) \equiv \sum_{i\in I} a_\Xi^i(\tau_k)[\widetilde{Y}_i] + \Lambda_\Xi \tag{10}$$

where Λ_Ξ is a Kähler form whose cohomology class is rational, and a_Ξ^i are affine forms defined over $\mathbb{Q}$ (we use here the fact that Ξ is a Kähler current). Moreover, there exists a $\delta_0 > 0$ such that $a_\Xi^i(\tau) > 0$ for any $\|\tau - \tau_0\| \leq \delta_0$ (given the expression of the forms f^j).

Similarly, we write

$$\widetilde{\mu}^\star(Y_{\tau_k}) = \sum_{i\in I} l^i(\tau_k)\widetilde{Y}_i$$

where $l^i \ : \mathbb{R}^N \to \mathbb{R}$ are affine functions with positive and rational coefficients. The relative canonical bundle of $\widetilde{\mu}$ is written as follows

$$K_{\widetilde{X}/X} = \sum_{i\in I} a^i_{\widetilde{X}/X}[\widetilde{Y}_i].$$

As in the previous section, we denote by Θ_k a current with minimal singularities in the sense of Demailly of the class $\{K_X + Y_{\tau_k} + A\}$; its $\widetilde{\mu}$-inverse image can be decomposed according to the set of $(\widetilde{Y}_i)$, as follows :

$$\widetilde{\mu}^\star(\Theta_k) = \sum_{i\in I} a^i_{\min}(\tau_k)[\widetilde{Y}_i] + \Lambda_k \tag{11}$$

where $a^i_{\min}(\tau_k)$ are positive real numbers, and where Λ_k is a closed positive current, whose generic Lelong number along each of $(\widetilde{Y}_i)$ is equal to zero.

A slight difficulty in what will follow is the fact that *a-priori*, we ignore the variation of the quantity $a^i_{\min}(\tau_k)$ with respect to k, but still it is a bounded sequence, and for our purposes we can assume that $a^i_{\min}(\tau_k) \to a^i_\infty$ as $k \to \infty$. We remark that at this point there is no connection between a^i_∞ and the *expected* singularity $a^i_{\min}(\tau_0)$ of the minimal current of the class $\{K_X + Y_{\tau_0} + A\}$.

We have the next statement, which is a first step towards the proof of claim 2.0.

2.2.1. Lemma. *Under the assumptions and notations above, there exists* $t^0, q^j \in \mathbb{Q}_+$, *a family of affine forms* $r^i : \mathbb{R}^{1+N} \to \mathbb{R}$ *with rational coefficients, and a set of strictly normal crossings hypersurfaces* $(\widetilde{Y_j})_{j\in\widetilde{I}} \subset \widetilde{X}$ *such that*

$$(1+m_0t^0)\Lambda_k + \sum_{i\in I_n\cup I_p} \widetilde{l^i_k}(\tau_k)[\widetilde{Y_i}] \equiv K_{\widetilde{X}} + \widetilde{S} + \sum_{i\in I_p} r^i\big(\tau_k, a^0_{\min}(\tau_k)\big)\widetilde{Y_i} + \widetilde{A}$$

where $\widetilde{I} = I_p \cup I_n \cup \{0\}$ *is a partition of* $\widetilde{I}$, *and where we use the next notations*

$$\widetilde{l^i_k}(\tau) := (1+m_0t^0)a^j_{\min}(\tau_k) + a^j_{\widetilde{X}/X} - t^0 a^j_{\Xi} - l^j(\tau) \tag{12}$$

for $i \in I_n$, *as well as*

$$\widetilde{l^i_k}(\tau) := (1+m_0t^0)\big(a^j_{\min}(\tau_k) - q^j\big) \tag{13}$$

for $i \in I_p$ *; we also denote by* $\widetilde{S} := \widetilde{Y_0}$. *In addition, the following relations holds true provided that* k *belongs to a well chosen subsequence of natural numbers.*

(a) *For each* $j \in I_n \cup I_p$ *we have* $0 \leq \widetilde{l^i_k}(\tau_k) \leq (1+m_0t^0)a^j_{\min}(\tau_k) + a^j_{\widetilde{X}/X}$.

(b) *For each* $i \in I_p$ *we have* $0 \leq r^i\big(\tau_k, a^i_{\min}(\tau_k)\big) \leq 1-\varepsilon_0$, *where* ε_0 *is a positive real number.*

(c) *The bundle* $\widetilde{A}$ *is ample on* $\widetilde{X}$. □

As in [24], the starting point of the proof of 2.2.1 is the following identity

$$\begin{aligned}(1+m_0t)\Lambda_k \equiv& K_{\widetilde{X}} + \sum_{i\in I}(ta^i_{\Xi}(\tau_k) + l^i(\tau_k) - (1+m_0t)a^i_{\min}(\tau_k) - a^i_{\widetilde{X}/X})\widetilde{Y_i} + \\ &+ t\Lambda_{\Xi} + (1-t)\mu^\star(A).\end{aligned} \tag{14}$$

We remark that for $t = 0$ the coefficients of $\widetilde{Y_i}$ above are strictly smaller than 1, whereas for $t = 1$, at least one of these numbers is strictly greater than 1 (given the singularities of Ξ).

The approach one has to follow for the proof is quite clear : first we determine the center $\widetilde{S}$ by using $k \to \infty$ in the formula above, and then thanks to the fact that the family of "incomplete linear systems" T_k we have just constructed is explicit enough, we can describe in a very precise manner the variation of the coefficients when we replace τ_0 by τ_k. We will not provide further details here, since they are completely standard (and rather tedious). □

Let $\tau \in [0,1]^N$ and $a \in \mathbb{R}_+$; we define the $\mathbb{R}$-divisor

$$\begin{aligned}E(\tau, a) :=& \big(a^0_{\widetilde{X}/X} + (1+m_0t^0)a\big)\widetilde{S} + \\ &+ \sum_{i\in I_n}\big(t^0 a^j_{\Xi}(\tau) + l^j(\tau)\big)\widetilde{Y_j} + \sum_{i\in I_p}\big((1+m_0t^0)q^j + a^j_{\widetilde{X}/X}\big)\widetilde{Y_j}\end{aligned} \tag{15}$$

As one can easily see, the support of the effective divisor $E(\tau, a)$ is not necessarily contractible with respect to $\widetilde{\mu}$, and we have

$$E\big(\tau_k, a^0_{\min}(\tau_k)\big) \leq (1+m_0t^0)\widetilde{\mu}^\star(\Theta_k) + \sum_{i\in I} a^i_{\widetilde{X}/X}[\widetilde{Y_i}]$$

in the sense of currents on $\widetilde{X}$.

From the definition (15), together with the formulas (12), (13) and (14) we infer the next numerical equivalence relation

$$E\big(\tau_k, a^0_{\min}(\tau_k)\big)+K_{\widetilde{X}}+\widetilde{S}+\sum_{i\in I_p} r^i\big(\tau_k, a^0_{\min}(\tau_k)\big)\widetilde{Y}_i+\widetilde{A}_1\equiv \equiv(1+m_0t^0)\mu^\star(K_X+Y_{\tau_k}+A)+K_{\widetilde{X}/X} \tag{16}$$

which will be very useful in a moment. □

We recall now the following result : *there exists a birational map $\widehat{\mu}:\widehat{X}\to\widetilde{X}$ such that the inverse image of $\widetilde{S}$ is equal to its proper transform, and such that for any set of real numbers $0\le x^i\le 1-\varepsilon_0$ we have*

$$\widehat{\mu}^\star\big(K_{\widetilde{X}}+\widetilde{S}+\sum_{i\in I}x^i\widetilde{Y}_i\big)-\sum_{j\in J_1}\widehat{r}^i(x)E_i\equiv K_{\widehat{X}}+\widehat{S}+\sum_{j\in J_2}\widehat{r}^i(x)\widehat{Y}_i.$$

where E_i are $\widehat{\mu}$-exceptional, the $\widehat{Y}_i$ are mutually disjoint, and $\widehat{r}^i$ are affine forms, such that $\widehat{r}^i(x)\le 1-\varepsilon_0$ and such that $\widehat{r}^i(x)\ge 0$ if and only if $i\in J_2$. We stress on the fact that in the formula above, only the decomposition $J_1\cup J_2$ depends on the particular vector x we are dealing with, provided that ε_0 is fixed ; this is just a small variation on the very classical result stating the same thing without pointing out the (in)dependence of the data $(\widehat{\mu},\widehat{r}^i)$ on the particular sequence (x^i). Therefore we will refer to the arguments provided in [16], from which the previous statement follows.

In our particular case, this translates as follows. There exists a map $\widehat{\mu}:\widehat{X}\to\widetilde{X}$, together with a set of non-singular and mutually disjoint hypersurfaces $(\widehat{Y}_j)$ such that

$$\widehat{\mu}^\star(\widetilde{\Theta}_k)+\sum_{j\in J_n}\widehat{l}^j\big(\tau_k, a^0_{\min}(\tau_k)[E_i]\equiv K_{\widehat{X}}+S+\sum_{j\in J_p}\widehat{r}^j\big(\tau_k, a^0_{\min}(\tau_k)\widehat{Y}_i+\mu^\star(\widetilde{A}), \tag{17}$$

where $(\widehat{r}^j,\widehat{l}^j)$ are affine forms defined over $\mathbb{Q}$, and (E_j) are contracted by $\widehat{\mu}$. In particular, the class $\alpha\big(\tau_k, a^0_{\min}(\tau_k)\big):=\{K_{\widehat{X}}+S+\sum_{j\in J_p}\widehat{r}^i\big(\tau_k, a^0_{\min}(\tau_k)\big)\widehat{Y}_i+\mu^\star(\widetilde{A})\}$ is pseudoeffective, and the Lelong number along S of its minimal singularity current is equal to zero.

The analytic methods are now particularly useful, in the following context. By the non-vanishing theorem in [24] and its consequences (cf. sections 2 and 3 of this article), the cohomology class $\alpha\big(\tau_k, a^0_{\min}(\tau_k)\big)$ contains an effective $\mathbb{R}$-divisor, whose support does not contains S. In particular, if we denote by

$$\widehat{\Theta}_k\in\alpha\big(\tau_k, a^0_{\min}(\tau_k)\big) \tag{18}$$

a current with minimal singularities in the above class, then its restriction to S denoted by $\widehat{\Theta}_{k|S}$ is a well-defined closed positive current in the restriction of the class ; we have

$$\widehat{\Theta}_{k|S}:=\sum_{j\in J_p}\rho^j_{\min}(\tau_k)[\widehat{Y}_{j|S}]+\Lambda_{k,S} \tag{19}$$

so the class $\alpha\big(\tau_k, a^0_{\min}(\tau_k)\big)_{|S}-\Big\{\sum_{j\in J_p}\min\Big(\widehat{r}^i\big(\tau_k, a^0_{\min}(\tau_k)\big),\rho^j_{\min}(\tau_k\Big)[\widehat{Y}_{j|S}]\Big\}$ is pseudoeffective. □

2.3. The induced polytope and its properties

We recall that our primary goal is to show that the elements of the sequence (τ_k) cannot be *extremal* points of $\mathcal{E}_{Y,A}$, as soon as k is large enough. For example, it would be enough to determine $\tau_{k0} \in \mathcal{E}_{Y,A}$ such that τ_k belongs to the interior of the segment $[\tau_0, \tau_{k0}]$.

To this end, we will use the *pseudoeffectivity criteria* established in [24], in the next framework. Let $\overline{X}$ be a projective manifold, and let $\overline{S}, \overline{Y}_j$ be a set of strictly normal crossing hypersurfaces, such that $\overline{Y}_j \cap \overline{Y}_k = \emptyset$ if $j \neq k$. We fix a $\mathbb{Q}$-bundle $\overline{A}$ on $\overline{X}$, such that for every $\delta > 0$, there exists a set $0 < \delta^j < \delta$ of positive real numbers, such that $A - \sum_j \delta^j Y_j$ is ample. Then we have the next statements, which are implicit in [24].

2.3.1. Theorem [24]. *Let $0 \leq \nu^j < 1$; we assume that for all $\varepsilon \ll 1$, there exists a current*

$$T_\varepsilon \in \{K_{\overline{X}} + \overline{S} + \sum_j \nu^j \overline{Y}_j + \overline{A}\}$$

whose Lelong number along $\overline{S}$ is equal to zero, and such that $T_\varepsilon \geq -\varepsilon\omega$, where ω is a Kähler form on X. Then the class $\{K_X + \overline{S} + \sum_j \nu^j \widehat{Y}_j + \widehat{A}\}$ contains an effective $\mathbb{R}$-divisor whose support does not include $\overline{S}$.

One of the important tools in the proof of the above theorem is the following statement (see [24], paragraph 1.H).

2.3.2. Theorem [24]. *We assume that the numbers ν^j above are rational ; there exists a positive real $\varepsilon \ll 1$ such that the following property holds true.*

Any section $u \in H^0\big(\overline{S}, q(K_{\overline{X}} + \overline{S} + \sum_j \nu^j \overline{Y}_j + \overline{A})_{|S}\big)$ extends to $\overline{X}$, provided that there exists $T \in \{K_{\overline{X}} + \overline{S} + \sum_j \nu^j \overline{Y}_j + \overline{A}\}$ a closed current whose restriction to $\overline{S}$ is well-defined, such that $T \geq -\varepsilon/q\omega$ and such that

$$\mathrm{ord}_{\overline{Y}_{j|\overline{S}}}(u) \geq q \min\Big(\nu(T_{|\overline{S}}, Y_{j|\overline{S}}), \nu^j\Big) - \varepsilon$$

for all j. In particular, the bundle $K_{\overline{X}} + \overline{S} + \sum_j \nu^j \overline{Y}_j + \overline{A}$ is (pseudo)effective, if a couple (u, T) as above exists.

We refer to [3], [6], [8], [9], [12], [15], [16], [17], [23], [28], [29], [33], [34], [35], [36] (to quote only a few...) and the references therein for similar results/ideas. Coming back to the discussion at the begining of this paragraph, we intend to use the statement above in order to determine the point τ_{k0}. The couple (T, u) with the properties required by 2.3.2 will be obtained via theorem 0.1 in lower dimensions ; it is precisely at this moment that the polytopes of Shokurov will play a crucial role. □

The definition of the following set is modeled after the properties (16)-(19) of the sequence (τ_k), by *decoupling the variables.*

Let $C > 0$ be an upper bound for the sequences $\big(a^0_{\min}(\tau_k)\big)$ and $\big(\rho^j_{\min}(\tau_k)\big)$; a *finite* upper bound indeed exists, thanks to the fact that the cohomology classes that contains the currents having the above singularities belong to a bounded set in $\mathrm{Psef}(X)$, respectively $\mathrm{Psef}(S)$. We introduce the following set

$$\mathcal{E}_{|S} := \{(\tau, a, \rho) \in [0,1]^N \times [0, C] \times [0, C]^{|J_p|} : \mathbf{C.1} - \mathbf{C.4} \text{ below are satisfied }\}$$

C.1. We have

$$1-\varepsilon_0 \geq \widehat{r}^i(\tau,a) \geq 0, \quad \widehat{l}^i(\tau,a) \geq 0, \quad a^i_{\Xi}(\tau) \geq 0\ ;$$

C.2. The cohomology class of the $\mathbb{R}$-bundle

$$D(\tau,a,\rho) := K_S + \sum_{j\in J_p} \left(\widehat{r}^i(\tau,a)-\rho^i\right)_+ \widehat{Y}_i + \mu^\star(\widetilde{A})$$

is pseudoeffective on S, where we denote by $(x)_+ := \max(x,0)$;

C.3. We have the next numerical equivalence

$$K_{\widehat{X}} + S + \sum_{j\in J_p} \widehat{r}^i(\tau,a)\widehat{Y}_i + \widehat{\mu}^\star(\widetilde{A}) \equiv$$

$$\widehat{\mu}^\star\Big((1+t^0m_0)\widetilde{\mu}^\star(K_X+Y_\tau+A\}) - E(\tau,a) + K_{\widetilde{X}/X}\Big)\Big) + \sum_{j\in J_n} \widehat{l}^i(\tau,a)E_i$$

(we use the notation in (34)) ;

C.4. We have $\mathrm{ord}_{\widehat{Y_j}\cap S}\left(\|D(\tau,a,\rho)\|\right) = \left(\rho^j - \widehat{r}^j(\tau,a)\right)_+$ for each $j\in J_p$. □

Concerning the conditions above, our first claim is that *the set $\mathcal{E}_1$ defined just by the conditions* **C.1** – **C.3** *is a rational polytope.* Indeed, the requirements **C.1** and **C.3** are affine (in)equations, and the linear forms defining them have rational coefficients. As for condition **C.2**, we consider the set

$$\mathcal{E}_{\widehat{W},\widehat{\mu}^\star(\widetilde{A})} := \big\{\eta = (\eta_j) \in [0,1-\varepsilon_0]^{|J_p|}\ : K_S + \sum_{j\in J_p} \eta^j\widehat{Y}_i + \mu^\star(\widetilde{A}) \in \mathrm{Psef}(S)\big\}.$$

The induction hypothesis shows that $\mathcal{E}_{\widehat{W},\widehat{\mu}^\star(\widetilde{A})}$ is indeed a rational polytope, and then condition **C.3** reads as

$$\left(\widehat{r}^i(\tau,a)-\rho^i\right)_+ \in \mathcal{E}_{\widehat{W},\widehat{\mu}^\star(\widetilde{A})},$$

and thus the set $\mathcal{E}_1$ is a rational polytope ; we fix its vertices $(\tau_j,a_j,\rho_j)_{j=1,\dots,G}$, and let d be its dimension.

For each $k=1,\dots,G$, we define $D_k := D(\tau_k,a_k,\rho_k)$. It is a pseudoeffective $\mathbb{Q}$-line bundle, and therefore it is effective, by the non-vanishing theorem [2], [24]. The Cox ring associated to (D_k) is finitely generated, by induction, and therefore for each $\widehat{Y}_{j|S}$ the associated *asymptotic vanishing order* function is piecewise affine, by the result [11] recalled in the section 1.

We consider a decomposition of $\mathcal{E}_1 = \bigcup_{k=1}^{M} \mathcal{C}_k$ into standard simplexes, such that the functions below

$$(\tau,a,\rho) \to \left(\widehat{r}^i(\tau,a)-\rho^i\right)_+ \tag{20}$$

and

$$(\tau,a,\rho) \to \mathrm{ord}_{\widehat{Y_j}\cap S}\left(\|D(\tau,a,\rho)\|\right) \tag{21}$$

becomes *affine* when restricted to any $\mathcal{C}_k$; we remark that the existence of such a decomposition is slightly different from the assertion in the statement 1.2, but it can be seen to follow, by a quick linear algebra argument. Thus, the equation **C.4** imposes *affine requirements* on the parameters (τ,a,ρ), and in conclusion, we have just proved the next statement.

2.3.3. Lemma. *The set $\mathcal{E}_{|S}$ defined by the relation* **C.1** – **C.4** *above is a rational polytope.* □

We stress the fact that the extension of the function "ord" to the $\mathbb{R}$-divisors we are using in **C.4** depends on the polytope $\mathcal{E}_1$; hopefully, this will not cause too much confusion.

We denote by $(\tau_j, a_j, \rho_j)_{j\in K}$ the vertices of $\mathcal{E}_{|S}$; they have rational coordinates, and for the purposes of the next corollary we can assume that they are independent in the affine sense- this can always be achieved modulo a subdivision.

Given a point $(\tau, a, \rho) \in \mathcal{E}_{|S}$, we can write it as follows

$$(\tau, a, \rho) = \sum_{j\in K} \lambda^j(\tau_j, a_j, \rho_j), \tag{22}$$

where $\sum_j \lambda_j = 1$; moreover, if the coordinates of the vector above are rational, then its affine coordinates (λ_j) are rational as well.

The main use of the linear structure of the set $\mathcal{E}_{|S}$ is revealed by the next statement, which can be seen as a *uniform non-vanishing*, see also [2], [3], and especially [18].

2.3.4. Proposition. *There exists an integer $q_0 \in \mathbb{Z}_+$ with the following property : let*

$$(\tau, a, \rho) \in \mathcal{E}_{|S},$$

be a point with rational coordinates, and let $q \in \mathbb{Z}_+$ be a common denominator of its affine coordinates (λ_j) in the equality (44). Then the bundle

$$qq_0\big(K_S + \sum_{j\in J_p} \widehat{r}^j(\tau, a)\widehat{Y}_{i|S} + \mu^\star(\widetilde{A})\big)$$

has a section whose vanishing order along the set $\widehat{Y}_{i|S}$ is exactly $qq_0\rho^i$.

Proof. We consider the $\mathbb{Q}$-bundle $D(\tau_j, a_j, \rho_j)$ corresponding to a vertex of $\mathcal{E}_{|S}$. Its asymptotic vanishing order at the generic point of $\widehat{Y}_{i|S}$ is given by condition **C.4**, that is to say $\mathrm{ord}_{\widehat{Y}_i\cap S}(\|D(\tau_j, a_j, \rho_j\|) = (\rho^i - \widehat{r}^j(\tau_j, a_j))_+$. Since by induction we already know that the ring associated to $D(\tau_j, a_j, \rho_j)$ is finitely generated, we obtain a section u_j of the bundle $q_jD(\tau_j, a_j, \rho_j)$ such whose vanishing order on $\widehat{Y}_{i|S}$ is equal to $q_j(\rho^i - \widehat{r}^i(\tau_j, a_j))_+$, for each $i \in J_p$. Then we obtain -by twisting with an appropriate divisor- a section

$$v_j \in H^0\Big(S, q_j(K_S + \sum_{k\in J_p} \widehat{r}^k(\tau_j, a_j)\widehat{Y}_{k|S} + \mu^\star(\widetilde{A}))\Big)$$

whose vanishing order at the generic point of $\widehat{Y}_{i|S}$ is equal to $q_j\rho^i$, for each $i \in J_p$. The section we seek is obtained by convex combination of the v_j, so the proposition is proved.

The same argument shows that for any point $(\tau, a, \rho) \in \mathcal{E}_{|S}$, we can construct an effective $\mathbb{R}$-divisor $\Xi \in \{K_S + \sum_{k\in J_p} \widehat{r}^k(\tau, a)\widehat{Y}_{k|S} + \mu^\star(\widetilde{A})\}$ whose vanishing order along $\widehat{Y}_{m|S}$ is equal to ρ^m, for every $m \in J_n$. □

Our next goal is to show that we have $(\tau_k, a^0_{\min}(\tau_k), \rho_{\min}(\tau_k)) \in \mathcal{E}_{|S}$; this will be the consequence of a more general result. Before stating it, we introduce one more notation.

Let

$$\mathcal{G} \subset [0,1]^N \times [0, C] \tag{23}$$

(where C is the constant fixed at the begining of this paragraph) be the set described by the following three conditions.

C.5. We have $0 \leq \widehat{r}^i(\tau, a) \leq 1 - \varepsilon_0$ for all $i \in J_p$;

C.6. The class $\{K_{\widehat{X}} + S + \sum_{i \in J_p} \widehat{r}^i(\tau, a)\widehat{Y}_i + \widehat{\mu}^\star(\widetilde{A})\}$ is pseudoeffective.

C.7. We denote by $\widehat{\Theta}(\tau, a) \in \{K_{\widehat{X}} + S + \sum_{j \in J_p} \widehat{r}^i(\tau, a)\widehat{Y}_i + \widehat{\mu}^\star(\widetilde{A})\}$ the current with minimal singularities ; then we have $\nu\big(\Theta_{\min}(\tau, a), S\big) = 0$.

An important remark is that the set $\mathcal{G}$ is *compact and convex* : the fact that it is convex is immediate from its definition, whereas its closeness is a *direct* consequence of theorem 2.3.1.

The restriction $\widehat{\Theta}(\tau, a)_{|S}$ is well defined, and we decompose the restriction current as follows

$$\widehat{\Theta}(\tau, a)_{|S} = \sum_{j \in J_p} \rho^i_{\min}(\tau, a)\widehat{Y}_{i|S} + \Lambda_S(\tau, a), \tag{24}$$

where the Lelong number of $\Lambda_S(\tau, a)$ on each $Y_{j|S}$ is equal to zero. Therefore, by the previous equality we define the function

$$\rho_{\min} : \mathcal{G} \to \mathbb{R}_+^{|J_p|} ;$$

if $\tau \in \mathcal{E}_{Y,A}$ is an element such that $\big(\tau, a^0_{\min}(\tau)\big) \in \mathcal{G}$, then we use the notation

$$\rho^i_{\min}(\tau) := \rho^i_{\min}(\tau, a^0_{\min}(\tau)) \tag{25}$$

and we remark that this is consistent with the notations in (19) for $\tau = \tau_k$, an element of our initial sequence.

If the point $(\tau, a) \in \mathcal{G}$ has rational coordinates, then by the results in [24] we know that $\rho^j_{\min}(\tau, a) \in \mathbb{Q}$, for any $j \in J_p$. An important result of this subsection is the following statement.

2.3.5. Proposition. *We assume that the requirements* **C.1** *and* **C.3** *are fulfilled by some element* $(\tau, a) \in \mathcal{G} \cap \mathbb{Q}^{1+N}$. *Then the point* $\big(\tau, a, \rho_{\min}(\tau, a)\big)$ *belongs to the polytope* $\mathcal{E}_{|S}$. *In particular, we have* $\big(\tau_k, a^0_{\min}(\tau_k), \rho_{\min}(\tau_k)\big) \in \mathcal{E}_{|S}$ *for all* $k \in \mathbb{Z}_+$.

PROOF. By the relation (46), we obtain

$$\sum_{j \in J_p} \rho^i_{\min}(\tau, a)\widehat{Y}_{i|S} + \Lambda_S(\tau, a) \equiv K_S + \sum_{j \in J_p} \widehat{r}^i(\tau, a)\widehat{Y}_{i|S} + \widehat{\mu}^\star(\widetilde{A})$$

and therefore we have

$$\sum_{j \in J_p} \big(\rho^i_{\min}(\tau, a) - \widehat{r}^i(\tau, a)\big)_+ \widehat{Y}_{i|S} + \Lambda_S(\tau, a)$$
$$\equiv K_S + \sum_{j \in J_p} \big(\widehat{r}^i(\tau, a) - \rho^i_{\min}(\tau, a)\big)_+ \widehat{Y}_{i|S} + +\widehat{\mu}^\star(\widetilde{A}).$$

In conclusion, the $\mathbb{Q}$-bundle $D\big(\tau, a, \rho_{\min}(\tau, a)\big)$ is pseudoeffective, and this proves that the point $\big(\tau, a, \rho_{\min}(\tau, a)\big)$ satisfies condition **C.2**.

Let u_0 be a section of the bundle

$$q\big(K_S + \sum_{i\in J_p} \big(\widehat{r}^i(\tau,a) - \rho^i_{\min}(\tau,a)\big)_+ \widehat{Y}_{i|S} + \mu^\star(\widetilde{A})\big) ;$$

such an object defines canonically a section u of the bundle

$$q\big(K_S + \sum_{i\in J_p} \widehat{r}^i(\tau,a)\widehat{Y}_{i|S} + \mu^\star(\widetilde{A})\big)$$

whose zero set contains the divisor $\sum_{i\in J_p} \min\{\rho^i_{\min}(\tau,a), \widehat{r}^i(\tau,a))\}\widehat{Y}_{i|S}$. But then the section u admits an extension to $\widehat{X}$, by 2.3.2 (see also [1], [12], [14]), so in particular we have

$$\operatorname{ord}_{\widehat{Y_j}\cap S}\Big(\|D\big(\tau,a,\rho_{\min}(\tau,a)\big)\|\Big) \geq \big(\rho^j_{\min}(\tau,a) - \widehat{r}^j(\tau,a)\big)_+ \tag{26}$$

for all $j\in J$, because the metric induced by the extension of u is more singular than the metric with minimal singularities of the corresponding bundle.

The ring of sections associated to $D\big(\tau,a,\rho_{\min}(\tau,a)\big)$ is finitely generated by induction, therefore we have

$$\operatorname{ord}_{\widehat{Y_j}\cap S}\Big(\|D\big(\tau,a,\rho_{\min}(\tau,a)\big)\|\Big) = \nu\big(\Theta_\sigma(\tau,a), \widehat{Y_j}\cap S\big)$$

where we denote by $\Theta_\sigma(\tau,a)$ the current associated to the algebra of sections of the bundle $D\big(\tau,a,\rho_{\min}(\tau,a)\big)$, i.e. the current with minimal singularities in the sense of Siu, whose definition was recalled in the paragraph 1.

Next, the closed positive current

$$\widehat{\Theta}(\tau,a)_{|S} - \sum_{i\in J_p} \min\{\rho^i_{\min}(\tau,a), \widehat{r}^i(\tau,a)\}[\widehat{Y}_{i|S}] \in \{D\big(\tau,a,\rho_{\min}(\tau,a)\big)\}$$

is certainly *more singular* than the current with minimal singularities of the class above.

Summing up the previous considerations, we have shown that the quantity we are interested in $\big(\rho^j_{\min}(\tau,a) - \widehat{r}^j(\tau,a\big)_+$ is *smaller* than the generic Lelong number of the current $\Theta_\sigma(\tau,a)$ on $\widehat{Y_j}\cap S$, and *greater* than the generic Lelong number of the current with minimal singularities of the class $\{D\big(\tau,a,\rho_{\min}(\tau,a)\big)\}$ along $\widehat{Y_j}\cap S$.

Since the bundle $D\big(\tau,a,\rho_{\min}(\tau,a)\big)$ can be written as K_S+L, where L is big and klt, the current with minimal singularities in the sense of Siu *coincides* with the current with minimal singularities in the sense of Demailly, as it was established in [24] ; in conclusion, the relation (48) becomes an equality, and 2.3.5 is proved. □

The following corollary is similar to the results established in [21], and can be seen as a consequence of the preceding arguments, together with elementary convex geometry considerations. The main idea of the proof goes back at least to [14] (see the paragraph concerning the rationality of the restricted algebras ; see equally [24] for a use of this idea in analytic setting).

2.3.6. Corollary. *Up to the choice of a subsequence, for all $k\gg 0$, there exists an element $\tau_{k0}\in\mathcal{E}_{Y,A}$ such that :*

(i) *The point τ_k belongs to the interior of the segment $[\tau_0,\tau_{k0}]$;*

(ii) *The restriction of the function $\tau\to a^0_{\min}(\tau)$ to the segment $[\tau_0,\tau_{k0}]$ is affine.*

(iii) *For each* $\tau \in [\tau_0, \tau_{k0}]$, *we have* $(\tau, a^0_{\min}(\tau)) \in \mathcal{G}$, *and the functions* $\tau \to \rho^j_{\min}(\tau)$ *defined in (47) are affine on* $[\tau_0, \tau_{k0}]$.

PROOF. By proposition 2.3.5 we have $(\tau_k, a^0_{\min}(\tau_k), \rho_{\min}(\tau_k)) \in \mathcal{E}_{|S}$, for any $k \gg 0$; again, we can assume that the vertices $(\tau_j, a_j, \rho_j)_{j=1,\dots,d+1}$ of $\mathcal{E}_{|S}$ are affinely independent (by passing to a subdivision of the polytope if necessary).

The affine coordinates of the point $(\tau_k, a^0_{\min}(\tau_k), \rho_{\min}(\tau_k))$ are written as follows

$$(\tau_k, a^0_{\min}(\tau_k), \rho_{\min}(\tau_k)) = \sum_{j=1}^{d+1} \lambda^j_k(\tau_j, a_j, \rho_j),$$

where λ^j_k are positive rational numbers, such that $\sum_j \lambda^j_k = 1$. For each index j, we can assume that $\lim_{k\to\infty} \lambda^j_k := \lambda^j_0$, and we consider the corresponding point

$$(\tau_0, a_0, \rho_0) = \sum_{j=1}^{d+1} \lambda^j_0(\tau_j, a_j, \rho_j) \in \mathcal{E}_{|S},$$

where the first component is indeed τ_0, since $\tau_k \to \tau_0$.

The function $\tau \to a^0_{\min}(\tau)$ defined on $\mathcal{E}_{Y,A}$ is *convex and bounded*, therefore we have

$$a_0 \le a^0_{\min}(\tau_0). \tag{27}$$

We will assume for the rest of the proof that the coordinates of the vector $V_0 := (\lambda^j_0)$ *are rational numbers, since this particular case already illustrates the main ideas of the proof.*

Let q be the common denominator of the coefficients λ^j_0 of the vector V_0. By proposition 2.3.4, there exists a section u of the bundle

$$q_0 q(K_S + \sum_{j \in J_p} (\widehat{r}^i(\tau_0, a_0)\widehat{Y}_{i|S} + \mu^\star(\widetilde{A}))$$

whose vanishing order on $\widehat{Y}_{i|S}$ is precisely $q_0 q \rho^i_0$.

The theorem 2.3.2 implies that the section u admits an extension to $\widehat{X}$. Indeed, the current T in 2.3.2 is obtained by moving the current $\widehat{\Theta}_k$ (see (18)) into the class $\{K_X + S + \sum_{i\in J_p} (\widehat{r}^i(\tau_0, a_0)\widehat{Y}_i + \mu^\star(\widetilde{A})\}$; the loss of positivity induced by this operation is of order $\mathcal{O}(\|\tau_k - \tau_0\| + |a^0_{\min}(\tau_k) - a_0|)$. We have

$$(\tau_k, a^0_{\min}(\tau_k), \rho_{\min}(\tau_k)) \to (\tau_0, a_0, \rho_0) = \sum_{j=1}^{d+1} \lambda^j_0(\tau_j, a_j, \rho_j)$$

and therefore the hypothesis of 2.3.2 are satisfied as soon as k is large enough.

As a consequence, we have $(\tau_0, a_0) \in \mathcal{G}$; it follows that the divisor of zeroes of u is greater than

$$q_0 q \sum_{j \in J_p} \rho^j_{\min}(\tau_0, a_0)[\widehat{Y}_{j|S}],$$

and this translates as

$$\rho^i_0 \ge \rho^i_{\min}(\tau_0, a_0) \tag{28}$$

for every $i \in J_p$.

Furthermore, we have $(\tau_0, a_0, \rho_0) \in \mathcal{E}_{|S}$ and thus the *numerical* relation **C.3** shows that the extension of the section u to $\widehat{X}$ induces a closed positive current in the class $\{K_X + Y_{\tau_0} + A\}$, whose Lelong number along S is equal to a_0 ; in particular, we have

$$a_0 \geq a^0_{\min}(\tau_0). \tag{29}$$

By combining (27) and (29) we obtain

$$a_0 = a^0_{\min}(\tau_0), \tag{30}$$

and we equally infer that

$$\rho^i_0 \geq \rho^i_{\min}(\tau_0) \tag{31}$$

as a consequence of (28).

Next, we introduce the notation $V_k := (\lambda^j_k)$, and we consider the vector $q(V_0 - V_k)$. Its norm tends to zero as $k \to \infty$ and the coordinates of qV_0 are integers. Under these circumstances, there exists a positive integer q'_k and a vector V'_{k0} such that the coordinates of $q'_k V'_{k0}$ are integers, and such that we have

$$q(V_0 - V_k) = q'_k(V_k - V'_{k0}) \tag{32}$$

(see e.g. [2], or "any good book on diophantine approximation"). The relation (32) implies in particular that the coordinates of V'_{k0} are positive rational numbers, and that their sum is equal to 1.

We consider the vector

$$(\tau'_{k0}, a'_{k0}, \rho'_{k0}) := \sum_{j \in I_1} \lambda'^j_{k0}(\tau_j, a_j, \rho_j), \tag{33}$$

where (λ'^j_{k0}) are the coordinates of V'_{k0}. The same extension arguments which were used a few lines above for (τ_0, a_0, ρ_0) will show in the first place that $(\tau'_{k0}, a'_{k0}) \in \mathcal{G}$, provided that $k \gg 0$. We also obtain that $K_X + Y_{\tau'_{k0}} + A \in \mathrm{Psef}(X)$ together with

$$a'_{k0} \geq a^0_{\min}(\tau'_{k0}) \tag{34}$$

and

$$\rho'^j_{k0} \geq \rho^j_{\min}(\tau'_{k0}, a'_{k0}). \tag{35}$$

The relation (32) implies that

$$V_k = \frac{q}{q + q'_{k0}} V_0 + \frac{q'_{k0}}{q + q'_{k0}} V'_{k0}$$

and therefore we obtain

$$(\tau_k, a^0_{\min}(\tau_k), \rho_{\min}(\tau_k)) = \frac{q}{q + q'_{k0}}(\tau_0, a_0, \rho_0) + \frac{q'_{k0}}{q + q'_{k0}}(\tau'_{k0}, a'_{k0}, \rho'_{k0}). \tag{36}$$

The equality above shows that τ_k is the non-trivial convex combination of two points of $\mathcal{E}_{Y,A}$, and therefore the first part of corollary 2.3.7 is proved, under the additional rationality assumption at the beginning of the proof.

Using the same circle of ideas we prove the point ii) and iii) of 2.3.6, as follows. The convexity of the function $a^0_{\min}$, together with the relations (52) and (56) shows that this function is in fact affine on the segment $[\tau_0, \tau'_{k0}]$, and that we have $a'_{k0} = a^0_{\min}(\tau'_{k0})$. Furthermore, this implies in the first place that the functions

$$\tau \to \rho^j_{\min}(\tau)$$

are *convex* on the segment $[\tau_0, \tau'_{k0}]$. When combined with the inequalities (31), (35) and with the relation (36), we obtain that they are affine on the segment $[\tau_0, \tau'_{k0}]$, for every $k \gg 0$.

The complete proof of the corollary involves an additional diophantine approximation process, but the basic ideas are the same. $\square$

We consider the following data (similar to the set-up in the introduction). Let $\mathcal{L} \subset \mathbb{R}^r$ be a rational polytope, and for let $l^j : \mathbb{R}^r \to \mathbb{R}$ be a set of affine forms defined over $\mathbb{Q}$, where $j = 2, ...N$. We assume that $0 \leq l^j(\theta) \leq 1 - \varepsilon_0$, for all $\theta \in \mathcal{L}$, where ε_0 is a positive real number. If the point $\tau := (1, l^2(\theta), ..., l^N(\theta))$ belongs to $\mathcal{E}_{Y,A}$, then we denote by $T(\theta)$ the current with minimal singularities in the class $K_X + Y_1 + \sum_{j=2}^{N} l^j(\theta) Y_j + A$. We have the next statement (cf. [21], and implicitly in [2]), which will be useful in the next paragraph.

2.3.7. Corollary.
a) *We consider the set*

$$\mathcal{E}^1 := \{\theta \in \mathcal{L} : (1, l^2(\theta), ..., l^N(\theta)) \in \mathcal{E}_{Y,A}, \quad \nu(T(\theta), Y_1) = 0\}.$$

Then $\mathcal{E}^1$ is a rational polytope.

b) *We assume furthermore that the hypersurfaces $(Y_j)_{j \neq 1}$ are mutually disjoint, and that A is not necessarily ample, but it has the next property : for each $\delta > 0$, there exists $\delta > \varepsilon^j > 0$ such that $A - \sum_{i \neq 1} \varepsilon^i Y_i$ is ample. As in (46), we define the functions $\theta \to \rho^j_{\min}(\theta)$ by the restriction of $T(\theta)$ to Y_1 ; then $\rho^j_{\min}$ are piecewise affine on $\mathcal{E}^1$.*

Proof. Part a) of the above statement is a direct consequence of the proof of i). Indeed, $\mathcal{E}^1$ above is a renamed version of the set $\mathcal{G}$ which was considered before 2.3.5, so in the first place it is *closed*. The analysis of its extremal points is carried out as above, without any modification ; in fact, it is much simpler, since we already have the (analogue of the) set S above. The substitute for the polytope $\mathcal{E}_{|S}$ -previously constructed by using of the sequence (τ_k)- is defined as follows. We denote by $\mathcal{E}^1_{|Y_1}$ the set of elements $(\theta, \rho) \in \mathcal{E}^1 \times [0, C]^{N-1}$ such that :

- The class $\alpha(\theta, \rho) := \{K_{Y_1} + \sum_{j=2}^{N} (l^j(\theta) - \rho^j)_+ Y_{j|Y_1} + A\}$ is pseudoeffective ;
- We have $\mathrm{ord}_{Y_j \cap Y_1}(\|\alpha(\theta, \rho)\|) = (\rho^j - l^j(\theta))_+$.

As in 2.3.3 we verify that $\mathcal{E}^1_{|Y_1}$ is a rational polytope, which contains the points $(q, \rho_{\min}(q))$ for any $q \in \mathcal{E}^1 \cap \mathbb{Q}^r$, see 2.3.4. We conclude by the (analogue of the) corollary 2.3.6. $\square$

Let $\theta_0 \in \mathcal{E}^1$ be an arbitrary point. We consider the intersection of $\mathcal{E}^1$ with the *minimal affine subspace* in $\mathbb{R}^r$ defined over $\mathbb{Q}$, which contains θ_0 ; this rational polytope is denoted by $A(\theta) \subset \mathcal{E}^1$. Since $(q, \rho_{\min}(q)) \in \mathcal{E}^1_{|Y_1}$ for any rational point $q \in A(\theta)$ we deduce that $(\theta_0, \rho^j_{\min}(\theta_0)) \in \mathcal{E}^1_{|Y_1}$, by a limit argument.

Next, we remark that given an arbitrary sequence $(\theta_k) \subset \mathcal{E}^1$ converging to θ_0 such that $\theta_k \neq \theta_0$ there exists a sequence $\theta_{0k} \subset \mathcal{E}^1$ such that θ_k belongs to the interior of the segment $[\theta_0, \theta_{0k}]$ and moreover the restriction of the function $\rho^j_{\min}$ to the above segment is affine (again, modulo the choice of a subsequence). This claim was already verified at the end of the proof of corollary 2.3.6 under some additional rationality assumptions ; we remark that the *main* use of the rationality of θ_k was to infer that

$$(\theta_k, \rho_{\min}(\theta_k)) \in \mathcal{E}^1_{|Y_1}. \tag{37}$$

The relation (37) was shown to hold a few lines ago regardless to the rationality of the components of θ_k, thanks to the polytope structure of $\mathcal{E}^1$.

The fact that the restriction of $\rho^j_{\min}$ to a segment $[\theta_0, \theta_{0k}]$ is affine follows as in the proof of 2.3.6 above, up to simple modifications which will only be sketched along the next bullets :

- We can assume that $\rho^j_{\min}(\theta_k) \to \rho^j_0$ as $k \to \infty$; if some of the coordinates of the limit of (37) are not rational, then for any $\eta > 0$, we obtain a finite family $(\theta_{\eta s}, \rho_{\eta s}) \in \mathcal{E}^1_{|Y_1}$ (by diophantine approximation of $(\theta_0, \rho_0) \in \mathcal{E}^1_{|Y_1}$), such that $q_{\eta s}\|(\theta_0, \rho_0) - (\theta_{\eta s}, \rho_{\eta s})\| \leq \eta$ where $q_{\eta s}$ are integers and $q_{\eta s}(\theta_{\eta s}, \rho_{\eta s})$ has integer coefficients. Moreover, we can insure that the $(\theta_{\eta s})$ is a set of affine generators of $A(\theta)$, and that (θ_0, ρ_0) belongs to the convex hull of $(\theta_{\eta s}, \rho_{\eta s})$.
- By using proposition 2.3.4, together with the extension argument used in 2.3.6, we obtain $\rho^j_{\eta s} \geq \rho^j_{\min}(\theta_{\eta s})$ for all j. The convexity shows that in fact each function $\rho^j_{\min}$ is affine when restricted to the convex hull of $(\theta_{\eta s})$;
- We construct $V'_{\eta s}$, the analogue of the vector V'_{k0} as in (32), and the relation (36) will show the fact that the restriction of $\rho^j_{\min}$ to some segment $[\theta_{\eta s}, \theta'_{\eta s}]$ containing θ_k in its interior is affine ; the conclusion follows.

Another important consequence of these arguments is the *continuity* of the functions $\rho_{\min}$ up to the boundary of $\mathcal{E}^1$.

Let $f : \mathcal{E}^1 \to [0, C]$ be one of the functions $\rho^j_{\min}$ we are interested in ; it is continuous and convex, by the preceding considerations. We consider the *epigraph* of f, defined as

$$\mathrm{Epi}(f) := \{(v, y) \in \mathcal{E}^1 \times [0, C] : y \geq f(v)\}.$$

We have just proved that for any sequence of points $(\theta_k) \subset \mathcal{E}^1$ converging to θ_0, there exists a $k_0 \gg 0$ such that for any $k \geq k_0$, the restriction of f to a segment $[\theta_0, \theta_{k0}]$ containing θ_k in its interior is affine. This implies that $\mathrm{Epi}(f)$ can only have finitely many extremal points on the graph of f, so it follows that $\mathrm{Epi}(f)$ is a polytope (the set $\mathrm{Epi}(f)$ is closed, thanks to the continuity of f). This in turn is equivalent to the fact that the function f is polyhedral (or piecewise affine), which is what we wanted to prove. □

Finally, we note that there exists a subdivision of $\mathcal{E}^1$ into *rational polytopes* say $\mathcal{C}_k$, such that the restriction of $\rho_{\min}$ to each $\mathcal{C}_k$ is affine. The rationality assertion is a consequence of the fact that the function $\rho_{\min}$ is locally affine when restricted to the minimal rational affine space containing an arbitrary point θ_0. □

3. Proof of (ii)

We use the notations in the introduction ; our first remark is that a decomposition of the polytope $\mathcal{L}$ into simplexes induce a decomposition of the algebra $\mathcal{A}_r(X)$ as a direct sum, so we can assume that $\mathcal{L}$ is a r-dimensional simplex, generated by the vertices $\xi_1, ..., \xi_{r+1}$; by hypothesis, the coordinates $(\xi_j^i)$of each ξ_j is a rational number.

For each point $\theta \in \mathcal{L}$ we can define its affine coordinates

$$\theta = \sum_{j=1}^{r+1} \eta^j \xi_j \tag{38}$$

where $\eta^j \geq 0$ and $\sum_j \eta^j = 1$. The relation (38) above implies that each η^j is an affine function of the coordinates of θ (with respect to the canonical basis of $\mathbb{R}^r$).

The $\mathbb{Q}$-bundle $K_X + \sum_{j=1}^N l^j(\xi_k)Y_j + A$ is pseudoeffective for each $k = 1, ..., r+1$, thanks to the property (ii). By the non-vanishing theorem [2], [24] there exists an effective $\mathbb{Q}$-divisor

$$\Xi_k := \sum_{j=1}^N a_k^j Y_j \tag{39}$$

linearly equivalent with $K_X + \sum_{j=1}^N l^j(\xi_k)Y_j + A$ (strictly speaking, this is not entirely correct, since the support of the zero set of the section Ξ_k is not obliged to be contained in (Y_j), but it becomes so after a blow-up ; we do not detail this step here, since it is completely standard). Furthermore, *we can assume that* $\Xi_k \neq 0$ *for all* $k = 1, ..., r+1$. Indeed, assume that say $\Xi_{r+1} = 0$; then we can reduce the study of the algebra $\mathcal{A}_r(X)$ to a smaller rank algebra, as follows (see [2]).

Let $\alpha \in \mathcal{L}$; we can write it as a convex combination

$$\alpha = (1 - \eta^{r+1})\alpha_1 + \eta^{r+1}\xi_{r+1}$$

where θ_1 belongs to the polytope whose vertices are $\xi_1, ..., \xi_r$. Since $\Xi_{r+1} = 0$, we infer that the bundle $K_X + \sum_{j=1}^N l^j(\xi_{r+1})Y_j + A$ is linearly equivalent to zero, and then the current

$$T := \frac{1}{1 - \eta^{r+1}}\Theta_{\min}(\alpha)$$

belongs to the class $\{K_X + \sum_{j=1}^N l^j(\alpha_1)Y_j + A\}$ (we denote by $\Theta_{\min}(\alpha)$ the current with minimal singularities of the class $\{K_X + \sum_{j=1}^N l^j(\alpha)Y_j + A\}$). The equality above implies that $\Theta_{\min}(\alpha) = (1 - \eta^{r+1})\Theta_{\min}(\alpha_1)$, so the finite generation of the algebra $\mathcal{A}_r(X)$ is obtained by induction on r, under the assumption above (we do no treat the case $r = 1$ separately, since it is completely similar to the general case). □

Hence we will assume from now on that $\Xi_k \neq 0$ for all $k = 1, ..., r+1$, and let m_0 be a positive integer, multiple of the denominators of a_k^j, ξ_j^i and of the coefficients of (l^j). We obtain a non-zero section

$$s_k \in H^0\Big(X, m_0\big(K_X + \sum_{j=1}^N l^j(\xi_k)Y_j + A\big)\Big) \tag{40}$$

whose zeroes divisor is $m_0\Xi_k$.

Our approach for the finite generation statement 0.1 (ii) is based upon the next simple result.

3.1. Proposition. *Let $\delta > 0$ be a positive real number. We assume that there exists an integer $d_\infty \in \mathbb{Z}_+$ divisible enough, and a finite set of sections*

$$\sigma_k \in H^0\big(X, m_k(K_X + \sum_{j=1}^{N} l^j(\theta_k)Y_j + A)\big)$$

where $k = 1, ..., K$ and $(m_k, \theta_k) \in \Gamma_d$ such that the next requirements are satisfied.

(R.1) *There exists a finite set of points $\alpha_q \in \mathcal{L}$, together with the positive reals $\delta > \delta_q > 0$ such that $\mathcal{L} \subset \cup_q B(\alpha_q, \delta_q)$, where $B(\alpha_q, \delta_q)$ is the ball in $\mathbb{R}^r$ centered at α_q with radius δ_s with respect to the norm $\|\theta\| \;:= \sup_j |\theta^j|$.*

(R.2) *Let u be a section of the bundle $m\big(K_X + \sum_{j=1,..,N} l^j(\theta)Y_j + A\big)$ such that $(m, \theta) \in \Gamma_d$ and moreover $m\theta^j \in d_\infty d^j\mathbb{Z}_+$ for all $j = 0, ..., r$; we assume that $\theta \in B(\alpha_s, \delta_s)$. Then we have*

$$u = u_1 + \sum_{p \in I(m,\theta)} \lambda_d \sigma_1^{\otimes p_1} \otimes ... \otimes \sigma_K^{\otimes p_k}$$

where $I(m, \theta)$ is the set of multi-indexes p such that

$$\sum m_k p_k = m, \quad \sum p_k m_k \theta_k = m\theta,$$

and the section u_1 satisfies the vanishing conditions $\mathrm{ord}_{Y_q}(u_1) \geq m\delta_s$, for any integer $1 \leq q \leq N$ such that $\sum_{k=1}^{r+1} \eta^k(\alpha_s)a_k^q \neq 0$.

Then the algebra $\mathcal{A}_r(X)$ is finitely generated. □

PROOF. We denote by $\mathcal{A}_r^{(d_\infty)}(X)$ the truncation of $\mathcal{A}_r(X)$ corresponding to the degrees m which satisfy the arithmetic conditions in $\mathbf{R_2}$. A first observation is that the integer m_0 in (61) can be assumed to be a multiple of d_∞. We will prove next that $\mathcal{A}_r(X)$ is finitely generated by the elements $(\sigma_j), (s_k)$, together with a finite number of pluricanonical sections of small degree.

To this end, let $u \in \mathcal{A}_r^{(d_\infty)}(X)$ be a section of $m\big(K_X + \sum_{j=1}^{N} l^j(\theta)Y_j + A\big)$, such that $\|\theta - \alpha_1\| \leq \delta_1$. By hypothesis we have $u = P(\sigma_1, ..., \sigma_N) + u_1$ where u_1 has the vanishing properties stated in $\mathbf{R}.2$. We can assume that

$$m\delta_1 \geq \max_{j,p}\{m_0 a_p^j\} \tag{41}$$

since otherwise u belongs to a fixed, finite subset of $\mathcal{A}_r^{(d_\infty)}(X)$.

Let k be an integer such that $\eta^k(\alpha_1) \neq 0$. According to the hypothesis of the proposition, we have $\mathrm{ord}_{Y_q}(u_1) \geq \delta_1 m$ in particular for all the indexes q such that Y_q belongs to the support of Ξ_k. Then the section u_1 can be written as

$$u_1 = s_k v_k$$

and we verify next that there exists *some* k for which the section v_k above belongs to the truncation of the algebra $\mathcal{A}_r(X)$ as fixed in the preceding proposition-because we are willing to use the induction on the degree of u. In any case we have

$$v_k \in H^0\big(X, (m-m_0)(K_X + \sum_{j=1,\dots,N} l^j(\theta')Y_j + A)\big)$$

where θ' is defined by the relation

$$m_0\xi_k + (m-m_0)\theta' = m\theta. \tag{42}$$

Our goal is to show that for some index k the point θ' belongs to the polytope $\mathcal{L}$. We remark that the affine coordinates of θ' are given by

$$\eta^j(\theta') = \frac{m}{m-m_0}\eta^j(\theta)$$

if $j \neq k$ and

$$\eta^k(\theta') = \frac{m\eta^k(\theta) - m_0}{m-m_0},$$

so we will have $\theta' \in \mathcal{L}$ if

$$m\eta^k(\theta) \geq m_0. \tag{43}$$

If the inequality (43) fails to hold for all k such that $\eta^k(\alpha_1) \neq 0$, then we obtain an upper bound for m, because $\|\theta - \alpha_1\| \leq \delta_1$, so we will have

$$m \leq C\delta_1 m + dm_0$$

and of course we can assume that $C\delta_1 < 1$. Therefore, unless m is smaller than a fixed constant, there exists an index k for which the inequality (43) is satisfied ; moreover, the relations $(m-m_0)\theta'^j \in d_\infty d^j\mathbb{Z}$ follows from (42).

In conclusion, the section v_k belongs to $\mathcal{A}_r^{(d_\infty)}(X)$, and the finite generation of this algebra is deduced by induction on the degree m. Thus, the proposition is completely proved, since the d_∞ power of any section of $\mathcal{A}_r(X)$ belongs to $\mathcal{A}_r^{(d_\infty)}(X)$. □

3.1. The first step

The construction of the generators (σ_j) will be performed by induction on $\dim(X)$, by using the compactness of $\mathcal{L}$; we begin by fixing an arbitrary point $\alpha_0 \in \mathcal{L}$.

By a suitable convex combination of the set of effective $\mathbb{Q}$-divisors Ξ_k corresponding to the vertices of $\mathcal{L}$ we obtain

$$K_X + \sum_{j=1}^N l^j(\alpha_0)Y_j + A \equiv \sum_{i=1}^N r^i(\alpha_0)Y_i \tag{44}$$

where $r^i(\theta) := \sum_{p=1}^{r+1} \eta^p(\theta)a_p^i \geq 0$ is an affine form with rational coefficients, for each i, cf. (38), (39).

Let t be a real number ; we consider the following identity, taken from [21] (which is nothing but the threshold of the $\mathbb{Q}$-divisor constructed in (44) with respect to the measure $\dfrac{d\lambda}{\prod |\sigma_{Y_j}|^{l^j(\alpha_0)}}$ in the language used in [24]) :

$$(1+t)\big(K_X + \sum_{j=1}^N l^j(\alpha_0)Y_j + A\big) \equiv K_X + \sum_{j=1}^N (l^j(\alpha_0) + tr^j(\alpha_0)Y_j + A.$$

The procedure we will describe in this section can be used under more general hypothesis on the linear forms (r^j, l^j) ; the precise requirements are the following.

Initial data. *There exists* $t^0, \delta^0 \in \mathbb{R}_+$ *such that we have*

$$r^j(\theta) \geq 0, \quad l^j(\theta) + tr^j(\theta) \geq 0$$

if $\|\theta - \alpha_0\| \leq \delta^0$ *and* $t \geq t^0$. *Moreover we have* $l^j(\alpha_0) + t^0 r^j(\alpha_0) < 1$ *for all* $j = 1, ..., N$, *and* $r^j(\alpha_0) > 0$ *for each* $j \in \Lambda(\alpha_0)$. □

We consider the set $\Lambda(\alpha_0) := \{j \in J \; : r^j(\alpha_0) \neq 0\}$; thanks to the ampleness of A we can assume that the following quantities

$$\widetilde{t^j} := \frac{1 - l^j(\alpha_0)}{r^j(\alpha_0)} \tag{45}$$

are distinct, for $j \in \Lambda(\alpha_0)$.

The *smallest* among the real numbers above is assumed to be $\widetilde{t^1}$; therefore there exists $\varepsilon_1 > 0$ such that

$$l^j(\alpha_0) + \widetilde{t^1} r^j(\alpha_0) \leq 1 - \varepsilon_1 \tag{46}$$

for all $j \neq 1$; we also have

$$l^1(\alpha_0) + \widetilde{t^1} r^1(\alpha_0) = 1, \tag{47}$$

so we can write

$$(1 + \widetilde{t^1})\big(K_X + \sum_{j=1}^{N} l^j(\alpha_0) Y_j + A\big) \equiv K_X + Y_1 + \sum_{j=2}^{N} (l^j(\alpha_0) + \widetilde{t^1} r^j(\alpha_0) Y_j + A.$$

We have $t^0 < \widetilde{t^1}$, and it may happen that $\widetilde{t^1}$ is not rational, so during the next lines we will construct a rational approximation t^1 of $\widetilde{t^1}$, and we will determine a real $\delta > 0$ such that

$$(1 + t^1)\big(K_X + \sum_{j=1}^{N} l^j(\theta) Y_j + A\big) \equiv K_X + Y_1 + \sum_{j=2}^{N_1} \widetilde{l^j}(\theta) Y_j + A_1,$$

where $\widetilde{l^j}$ are linear forms with rational coefficients, such that $0 \leq \widetilde{l^j}(\theta) \leq 1 - \dfrac{\varepsilon_1}{2}$ if $\|\theta - \alpha_0\| \leq \delta$.

In order to achieve this, we will use the same arguments as in paragraph 2.2. Let γ, δ be positive real numbers and let $t^1 \in [(1-\gamma)\widetilde{t^1}, \widetilde{t^1}] \cap \mathbb{Q}$. For any $\theta \in \mathcal{L}$ such that $\|\theta - \alpha_0\| \leq \delta$, a quick computation shows that we have

$$\big|l^j(\alpha_0) + \widetilde{t^1} r^j(\alpha_0) - l^j(\theta) - t^1 r^j(\theta)\big| \leq C(\widetilde{t^1})(\gamma + \delta) \tag{48}$$

where $C(\widetilde{t^1}) \geq 1$ is a constant depending on $\widetilde{t^1}$. The first requirement we impose to δ and γ is

$$C(\widetilde{t^1})(\gamma + \delta) \leq \frac{\varepsilon_1}{2} \tag{49}$$

and then we will have

$$0 \leq l^j(\theta) + t^1 r^j(\theta) \leq 1 - \frac{\varepsilon_1}{2}$$

if $\|\theta - \alpha_0\| \leq \delta$ and $j \neq 1$.

The next conditions on δ, γ are needed in order to transform the expression $\big(l^1(\theta) + t^1 r^1(\theta) - 1\big) Y_1 + A$. Let

$$l_0 := \max_j l^j(\alpha_0) < 1$$

and let C_A be a positive and divisible enough integer, such that

$$C_A(1-l_0) \geq 2 \tag{50}$$

and such that C_AA+Y_j, C_AA-Y_j and C_AA are very ample line bundles. Let $q \geq C(\widetilde{t^1})(\gamma+\delta)$ be a rational number ; we have

$$\big(l^1(\theta)+t^1r^1(\theta)-1\big)Y_1+A \equiv \sum_{j=1}^{3} \widetilde{l}^{N+j}(\theta)H_j + 1/2A \tag{51}$$

where $H_{N+1}, H_{N+2}, H_{N+3}$ are respectively generic hyperplane sections of the linear systems $|Y_1+C_AA|, |C_AA-Y_1|$ and $|C_AA|$, and where we define

$$\widetilde{l}^{N+1}(\theta) := l^1(\theta)+t^1r^1(\theta)-1+q, \quad \widetilde{l}^{N+2}(\theta) := q,$$

and

$$\widetilde{l}^{N+3}(\theta) := \big(1+\frac{1}{2C_A}-l^1(\theta)-t^1r^1(\theta)-2q\big)$$

We choose now the parameters as follows

$$\gamma^1 := \min\Big(\frac{\varepsilon_1}{4C(\widetilde{t^1})}, \frac{1}{12C_AC(\widetilde{t^1})}, \frac{2C_A-1}{2C_A(1-\min_i l^i(\alpha_0))}\Big) \leq \frac{1}{2}, \tag{52}$$

$q := \varepsilon_1$ and let t^1 be any rational number contained in the interval $[(1-\gamma^1)\widetilde{t^1}, \widetilde{t^1}]$; we do not yet fix δ, but instead impose the condition $\delta \leq \gamma^1$. The first two quantities in (52) are dictated by (49) and by the bounds we wish to impose to the affine forms $\widetilde{l}^{N+j}$.

In conclusion, we have the linear equivalence relation

$$(1+t^1)\big(K_X+\sum_{j=1}^{N} l^j(\theta)Y_j+A\big) \equiv K_X+Y_1+\sum_{j=2}^{N+3}\widetilde{l^j}(\theta)Y_j+A_1, \tag{53}$$

where $A_1 = 1/2A$, and $\widetilde{l^j}(\theta) := l^j(\theta)+t^1r^j(\theta)$ for $j=2,...,N$.
As a motivation for what will follow, we recall that in order to be able to apply proposition 3.1 above, we have to show in particular that the algebra $\mathcal{A}_r(X)$ is finitely generated modulo the sections whose normalized vanishing order along Y_1 is greater than a fixed constant. Thus, it is natural to consider the expression

$$(1+t^1)\big(K_X+\sum_{j=1}^{N} l^j(\theta)Y_j+A\big)-\theta^{r+1}Y_1 \equiv K_X+Y_1+\sum_{j=2}^{N+3}\widetilde{l^j}(\theta)Y_j+$$
$$+A_1-\theta^{r+1}Y_1$$

where the additional parameter θ^{r+1} corresponds to the normalized vanishing order along Y_1. The last term in the preceding formula can be written as follows

$$A_1-\theta^{r+1}Y_1 = \theta^{r+1}\big(2C_AA_1-Y_1\big)+\big(1/2-2C_A\theta^{r+1}\big)A_1+1/2A_1. \tag{54}$$

Next we define ε^1 by the next equality

$$(1+t^1)\varepsilon^1 := \frac{1}{4C_A} \tag{55}$$

and then we have

$$(1+t^1)\big(K_X+\sum_{j=1}^{N} l^j(\theta)Y_j+A\big)-\theta^{r+1}Y_1 \equiv K_X+Y_1+\sum_{j=2}^{N+5}\widetilde{l^j}(\theta,\theta^{r+1})Y_j+A_2 \tag{56}$$

where the forms/hypersurfaces corresponding to $j = N+4, N+5$ are obtained via (54), and $A_2 = 1/4A$. The choice of ε^1 as in (55), together with (52) and the fact that $t^1 \geq (1-\gamma^1)\widetilde{t}^1$ implies

$$r^1(\alpha_0) \geq 2\varepsilon^1, \quad l^1(\alpha_0) + t^1 r^1(\alpha_0) \geq 2(1+t^1)\varepsilon^1. \tag{57}$$

Now we fix $\delta^1 \leq \gamma^1$ small enough in order to insure that

$$r^1(\theta) \geq \varepsilon^1, \quad l^1(\theta) + t^1 r^1(\theta) \geq (1+t^1)\varepsilon^1. \tag{58}$$

for any $\theta \in \mathcal{L}$ such that $\|\theta - \alpha_0\| \leq \delta^1$. This implies that

$$l^1(\theta) + tr^1(\theta) \geq (1+t)\varepsilon^1. \tag{59}$$

for any $t \geq t^1$. If $0 \leq \theta^{r+1} \leq (1+t^1)\varepsilon^1$ and $\|\theta - \alpha_0\| \leq \delta^1$ we notice that $0 \leq \widetilde{l^j}(\theta, \theta^{r+1}) \leq 1 - \varepsilon_1/2$. □

After all this preliminaries, we introduce the following algebra. We first define the polytope $\mathcal{L}' \subset \mathcal{L} \times \left[0, \varepsilon^1(1+t^1)\right]$ given by the couples (θ, θ^{r+1}) satisfying the following conditions :

(a) We have $\|\theta - \alpha_0\| \leq \delta^1$;

(b) The bundle $K_X + Y_1 + \sum_{j=2}^{N+5} \widetilde{l^j}(\theta, \theta^{r+1}) Y_j + A_2 \in \mathrm{Psef}(X)$, and its generic Lelong number across Y_1 is equal to zero.

The fact that indeed $\mathcal{L}'$ is a polytope is a consequence of corollary 2.3.7.

Now we recall that in the definition of the algebra $\mathcal{A}_r(X)$ we use the element $d = (d^0, ..., d^r) \in \mathbb{Z}^{r+1}$ in order to define the set Γ_d. Let n_1 be an integer such that $\dfrac{n_1}{1+t^1} \in \mathbb{Z}$; we introduce the following set

$$\Gamma'_d := \{(m', \theta') : \theta' = (\theta, \theta^{r+1}) \in \mathcal{L}' \cap \mathbb{Q}^{r+1}, \ m'\theta^j \in \frac{n_1}{1+t^1} d^j \mathbb{Z}_+, \ m'\theta^{r+1} \in \mathbb{Z}\}$$

and the associated algebra

$$\mathcal{A}_{r+1}(X, \alpha_0) := \bigoplus_{(m', \theta') \in \Gamma'_d} H^0\big(X, m'(K_X + Y_1 + \sum_{j=1}^{N+5} \widetilde{l^j}(\theta') Y_j + A_2)\big). \tag{60}$$

We remark that the previous definition is meaningful, as soon as n_1 is divisible enough. We also remark the asymmetry of the arithmetic conditions imposed to θ and θ^{r+1} in the definition of Γ'_d ; the reason for this will appear in a moment.

Let $\mathcal{A}_{r+1}(X, \alpha_0)_{|Y_1}$ be the restricted algebra associated to $\mathcal{A}_{r+1}(X, \alpha_0)$; we have the next simple consequence of 2.3.7.

3.1.1. Lemma. *The restricted algebra $\mathcal{A}_{r+1}(X, \alpha_0)_{|Y_1}$ is finitely generated.* □

PROOF. Our first observation is that the coefficients of Y_j above satisfy the inequality $0 \leq \widetilde{l^j}(\theta') \leq 1 - \epsilon_1/2$, for each $\theta' \in \mathcal{L}'$, as a consequence of the definitions/choices previously made. In this context, there exists a birational map $\mu : \widehat{X} \to X$ such that -up to a subdivision of $\mathcal{L}'$- we have

$$\mu^\star\Big(K_X + Y_1 + \sum_{j=1}^{N+5} \widetilde{l^j}(\theta') Y_j + A_2\Big) + \sum_{i \in I_-} \widehat{l^i}(\theta') E_i = K_{\widehat{X}} + \widehat{Y_1} + \sum_{i \in I_+} \widehat{l^j}(\theta') \widehat{Y_j} + \mu^\star(A_2)$$

where the hypersurfaces $\widehat{Y_j}$ corresponding to I_+ are mutually disjoint, and the μ-proper transform of Y_1 is equal to its total inverse image (again, we refer to the arguments in [14] for the existence of μ).
As it was already discussed in the first paragraph of the present article (and also [1], [14]), via the map μ we can easily identify a truncation of the restricted algebra $\mathcal{A}_{r+1}(X, \alpha_0)_{|Y_1}$, as follows.

For each $\theta' \in \mathcal{L}'$, the current with minimal singularities corresponding to the bundle $K_X + Y_1 + \sum_{j=2}^{N+5} \widetilde{l^j}(\theta')Y_j + A_2$ admits a well-defined restriction to Y_1, and let $\rho^j_{\min}(\theta')$ be the Lelong number of this restriction at the generic point of $Y_{j|Y_1}$. By corollary 2.3.7, the functions $\rho^j_{\min}$ are piecewise affine, defined over $\mathbb{Q}$, and let p_0 be an integer, which is a multiple of all the denominators of their coefficients. Then the p_0-truncation of the restricted algebra $\mathcal{A}_{r+1}(X, \alpha_0)_{|Y_1}$ is isomorphic to

$$(61)\ \mathcal{A} := \bigoplus_{(m',\theta')\in\Gamma'_{p_0 d}} H^0\Big(\widehat{Y_1}, m'\big(K_{\widehat{Y_1}} + \sum_{j\in I_+} \big(\widetilde{l^j}(\theta') - \rho^j_{\min}(\theta')\big)_+ \widehat{Y}_{j|\widehat{Y}_1} + \mu^\star(A_2)\big)\Big).$$

where the set $\Gamma'_{p_0 d}$ in the expression above is defined as $(m', \theta') \in \Gamma'_d$ such that

$$m'\theta^k \in p_0 d^k \frac{n_1}{1+t^1}\mathbb{Z}_+,$$

if $0 \leq k \leq r$ and $m'\theta^{r+1} \in p_0\mathbb{Z}$. By induction, the algebra $\mathcal{A}$ is finitely generated, and then the restricted algebra has the same property ; thus lemma 3.1.1 is completely proved. □

3.1.2. Remark. By the construction of the algebra $\mathcal{A}_{r+1}(X, \alpha_0)$ we infer the following important fact. Let $v \in H^0\big(X, m'(K_X + Y_1 + \sum_{j=1}^{N+5} \widetilde{l^j}(\theta')Y_j + A_2)\big)$ be an element of this algebra. We have $\theta' = (\theta, \theta^{r+1})$, and we consider the section

$$u := v \otimes \sigma_{Y_1}^{\otimes m'\theta^{r+1}} \ ;$$

we denote by σ_{Y_1} the canonical section associated to Y_1. The above definition is legitimate, since we have $m'\theta^{r+1} \in \mathbb{Z}_+$.

The claim is that the section u belongs to $A_r(X)$ and indeed this is obvious, since by the relation (56) we have

$$u \in H^0\big(X, m'(1+t^1)(K_X + \sum_{j=1}^{N} l^j(\theta)Y_j + A)\big)$$

as well as $\theta \in \mathcal{L}$ and $m'(1+t^1)\theta^j \in n_1 d^j \mathbb{Z}_+$, for $j = 0, ..., r$. □

Before explaining the end of our proof, we will discuss next the impact of the finite generation of the restricted algebra $\mathcal{A}_{r+1}(X, \alpha_0)_{|Y_1}$ on the properties of our initial object $\mathcal{A}_r(X)$.

Let $\mathcal{A}_r^{(n_1)}(X)$ be the truncation of $\mathcal{A}_r(X)$, corresponding to the set $(m, \theta) \in \Gamma_d$ such that $m\theta^k \in n_1 d^k \mathbb{Z}$, for $k = 0, ..., r$ (where the integer n_1 is *the same* as the one appearing in the construction of the algebra $\mathcal{A}_{r+1}(X, \alpha_0)$ a few lines above). Let

$$u \in H^0\big(X, m(K_X + \sum_{j=1}^{N} l^j(\theta)Y_j + A)\big)$$

be an element of $\mathcal{A}_r^{(n_1)}(X)$, such that $\|\theta - \alpha_0\| \leq \delta$. We write $m = (1+t^1)m'$ where $m' \in \mathbb{Z}$ thanks to the fact that $m \in (1+t^1)\mathbb{Z}$. By the formula (56), the section u becomes an element of the group

$$H^0\big(X, m'(K_X + Y_1 + \sum_{j=2}^{N+3} \widetilde{l^j}(\theta)Y_j + A_1)\big) \tag{62}$$

(because we tacitly assume that n_1 is divisible enough). Let k_1 be the vanishing order of u along the hypersurface Y_1 ; we have

$$u = \sigma_{Y_1}^{\otimes k_1} u_1.$$

If $\dfrac{k_1}{m} \geq \varepsilon^1$, then we stop ; if not, we observe that u_1 is a section of the bundle

$$m'\big(K_X + Y_1 + \sum_{j=1}^{N+5} \widetilde{l^j}(\theta')Y_j + A_2\big),$$

where $\theta^{r+1} = k_1/m'$. In other words, u_1 is an element of $\mathcal{A}_{r+1}(X, \alpha_0)$, as one can deduce from the relations (a), (b) and the definition of Γ'_d.
We denote by $\beta_1, ..., \beta_K$ the generators of the restricted algebra $\mathcal{A}_{r+1}(X, \alpha_0)_{|Y_1}$, and then we have

$$u_1 = \sum_p \lambda_p \beta_1^{\otimes p_1} \otimes ... \otimes \beta_K^{\otimes p_K} + \sigma_{Y_1}^{k_2} u_2$$

where $k_2 \geq 1$ and where the restriction of u_2 to Y_1 is non-identically zero.

If $\beta_k \in H^0\Big(X, m'_k\big(K_X + Y_1 + \sum_{j=1}^{N+5} \widetilde{l^j}(\theta'_k)Y_j + A_2\big)\Big)$, then the equality above implies that

$$m'(\theta, \theta^{r+1}) = \sum_{k=1}^{K} p_k m'_k(\theta_k, \theta_k^{r+1})$$

so we have

$$m'\theta^{r+1} = \sum_{k=1}^{K} p_k m'_k \theta_k^{r+1}.$$

For $k = 1, ..., K$, we define

$$\omega_k := \beta_k \otimes \sigma_{Y_1}^{m'_k \theta_k^{r+1}} \in H^0\Big(X, (1+t^1)m'_k\big(K_X + \sum_{j=1}^{N} l^j(\theta_k)Y_j + A\big)\Big).$$

Thanks to the remark 3.1.2 above, we have $\omega_k \in \mathcal{A}_r^{(n_1)}(X)$ and we get

$$u = \sum_s \lambda_p \omega_1^{\otimes p_1} \otimes ... \otimes \omega_K^{\otimes p_K} + \sigma_{Y_1}^{k_1+k_2} u_2.$$

Again, if $\dfrac{k_1 + k_2}{m} \geq \varepsilon^1$, then we stop ; if not, we repeat the same procedure with u_2.
In conclusion, after a finite number of steps, we can write

$$u = \sum_s \lambda_s \omega_1^{\otimes s_1} \otimes ... \otimes \omega_K^{\otimes s_K} + \sigma_{Y_1}^{\otimes m\varepsilon^1} u_{[2]} \; ; \tag{$\star$}$$

where $u_{[2]}$ is a section of the bundle $m(K_X + \sum_{j=1}^{N} l^j(\theta)Y_j + A) - m\varepsilon^1 Y_1$.

Therefore, after this first step we have shown that our algebra satisfies a consistent part of the requirements of proposition 3.1. □

3.2. Iteration scheme

We turn now our attention to the sections of the algebra $\mathcal{A}_r(X)$, whose normalized vanishing order along Y_1 is greater than ε^1.

Let $t \geq t^1$; we have the identity

$$(1+t)\big(K_X + \sum_{j=1}^{N} l^j(\theta)Y_j + A - \varepsilon^1 Y_1\big) \equiv K_X + \big(l^1(\theta) + tr^1(\theta) - (1+t)\varepsilon^1\big)Y_1 +$$

$$+ \sum_{j\in J, j\neq 1} \big(l^j(\theta) + tr^j(\theta)Y_j + A \tag{63}$$

In order to make the exposition a bit cleaner, we introduce further notations :

- $l_2^j := l^j$ and $r_2^j := r^j$, if $j \neq 1$, and
- $l_2^1 := l^1 - \varepsilon^1$ and $r_2^1 := r^1 - \varepsilon^1$ for $j = 1$.

Then we observe that we have

$$r_2^j(\theta) \geq 0, \quad l_2^j(\theta) + tr_2^j(\theta) \geq 0 \tag{64}$$

for any $\theta \in \mathcal{L}$ such that $\|\theta - \alpha_0\| \leq \delta^1$, and for any $t \geq t^1$. For $j = 1$, we remark that this is a consequence of the inequality (59), whose importance appear clearly at this moment.

We also note that we have

$$l_2^j(\alpha_0) + t^1 r_2^j(\alpha_0) < 1, \quad r_2^p(\alpha_0) > 0 \tag{65}$$

for all $j = 1, ..., N$ and $p \in \Lambda(\alpha_0)$. Therefore, after the first step, the new affine forms (l_2^j, r_2^j) are satisfying the assumptions in the "initial data" of the preceding section.

We consider the quantities

$$\widetilde{t_2^j} := \frac{1 - l_2^j(\alpha_0)}{r_2^j(\alpha_0)} \tag{66}$$

for $j \in \Lambda(\alpha_0)$; we remark that the smallest of them is strictly greater than t^1, by the inequality (65).

The procedure described in 3.1 applied in the current setting will give the numbers $\gamma^2, \delta^2, \varepsilon^2$ and $t^2 \geq t^1$ such that the next relations are satisfied

$$(1+t^2)\varepsilon^2 := \frac{1}{4C_A} \tag{67}$$

and

$$r_2^{j_2}(\theta) \geq \varepsilon^2, \quad l_2^{j_2}(\theta) + tr_2^{j_2}(\theta) \geq (1+t)\varepsilon^2. \tag{68}$$

for any $\theta \in \mathcal{L}$ such that $\|\theta - \alpha_0\| \leq \delta^2$ and for any $t \geq t^2$. The integer j_2 is the index for which the minimum of the quantities (66) is achieved ; we note that we have $j_2 \in \Lambda(\alpha_0)$.

We observe that *we are not forced to increase the constant* C_A, because the condition $C_A\big(1 - \max_j l_2^j(\alpha_0)\big) \geq 2$ is automatically satisfied, since $l_2^j \leq l^j$. Also, we will get the analogue of the relation (56) as follows

$$(1+t^2)\big(K_X + \sum_{j=1}^{N} l_2^j(\theta)Y_j + A\big) - \theta^{r+1}Y_1 \equiv K_X + Y_{j_2} + \sum_{j=1,..,N+5, j\neq j_2} \widetilde{l_2^j}(\theta, \theta^{r+1})Y_j + A_2.$$

With this data, we introduce the analogue of the algebra $\mathcal{A}_{r+1}(X, \alpha_0)$, and at the end of the second step, we will obtain the following conclusion. If $j_2 = 1$, then we get the finite generation of the algebra $\mathcal{A}_r(X)$ locally near the point (a-priori chosen) α_0, modulo sections having the normalized vanishing order along Y_1 greater than $\varepsilon_1 + \varepsilon_2$. If not, the same conclusion holds, modulo sections with normalized vanishing order greater than ε_1 along Y_1 and greater than ε^2 along Y_{j_2}.

The rest is clear : we keep on iterating the procedure above, so assume that we have performed this say k times. This means that for each $1 \leq p \leq k$ we have constructed the following objects.

(a) We have linear forms l^j_p and r^j_p, such that

$$l^j_p := l^j - \sum_{q\in\Lambda_j, q\leq p-1} \varepsilon^q, \quad r^j_p := r^j - \sum_{q\in\Lambda_j, q\leq p-1} \varepsilon^q \tag{69}$$

where we have $q \in \Lambda_j$ if and only if the index j achieve the minimum of the expressions (66) at the qth step.

(b) We obtain a set of real numbers $(\gamma^p, \delta^p, \varepsilon^p, t^p)$ such that

$$\min_{j=1,\dots,N} \frac{1 - l^j_p(\alpha_0)}{r^j_p(\alpha_0)} \geq t^p \geq t^{p-1}, \quad (1 + t^p)\varepsilon^p := \frac{1}{4C_A}$$

and such that

$$r^j_p(\theta) \geq 0, \quad l^j_p(\theta) + t r^j_p(\theta) \geq 0 \tag{70}$$

together with

$$r^q_p(\alpha_0) > 0, \quad r^{j_p}_p(\theta) \geq \varepsilon^p, \quad l^{j_p}_p(\theta) + t r^{j_p}_p(\theta) \geq (1+t)\varepsilon^p$$

if $\|\theta - \alpha_0\| \leq \delta^p$, $t \geq t^p$ and $q \in \Lambda(\alpha_0)$; we denote by j_p the index for which the minimum of the quantities (66) is achieved at the begining of the step p.

(c) We have the identity

$$\begin{aligned}(1 + t^p)\big(K_X + \sum_{j=1}^{N} l^j_p(\theta)Y_j + A\big) - \theta^{r+1}Y_{j_p} \equiv \\ \equiv K_X + Y_{j_p} + \sum_{j=1,..,N+5, j\neq j_p} \widetilde{l^j_p}(\theta, \theta^{r+1})Y_j + A_2.\end{aligned} \tag{71}$$

(d) We define the polytope $\mathcal{L}'_p \subset \mathcal{L} \times \big[0, \varepsilon^p(1+t^p)\big]$ given by the couples (θ, θ^{r+1}) satisfying the following conditions :

(d') We have $\|\theta - \alpha_0\| \leq \delta^p$;

(d'') The bundle $K_X + Y_{j_p} + \sum_{j\neq j_p} \widetilde{l^j_p}(\theta, \theta^{r+1})Y_j + A_2 \in \text{Psef}(X)$, and its generic Lelong number across Y_{j_p} is equal to zero.

(e) Let n_p be an integer which is divisible enough, such that $\dfrac{n_p}{1+t^p} \in \mathbb{Z}$; we introduce the set

$$\begin{aligned}\Gamma'_d(p) := \big\{(m', \theta') \ : \theta' = (\theta, \theta^{r+1}) \in \mathcal{L}'_p \cap \mathbb{Q}^{r+1}, \\ m'\theta^j \in \frac{n_p}{1+t^p} d^j \mathbb{Z}_+, \ m'\theta^{r+1} \in \mathbb{Z}\big\}\end{aligned}$$

and the associated algebra

$$\mathcal{A}_{r+1}(X,\alpha_0\ ;p)\ :=\ \bigoplus_{(m',\theta')\in\Gamma'_d(p)} H^0\big(X, m'(K_X+Y_{j_p}+\sum_{j\neq j_p}\widetilde{l^j_p}(\theta')Y_j+A_2)\big).$$

We note that the divisibility constraints we have to impose to n_p also depends on the previous steps, as $m'\theta^j$ must clear the denominators of the coefficients of of $\widetilde{l^q_p}$ for $q=1,\dots,N+5$; in particular, we require $n_p\varepsilon^q\in\mathbb{Z}$ for each $q\le p$. However, the total number of such constraints is *finite*. □

As we did after the proof of lemma 3.1.1, we explain next the progression we have achieved in the direction of lemma 3.1 after $k\ge 2$ iterations.

In the first place, the analogue of the remark 3.1.2 reads in the *iterated context* as follows. Let

$$v\in H^0\big(X, m'(K_X+Y_{j_k}+\sum_{j\neq j_k}\widetilde{l^j_k}(\theta')Y_j+A_2)\big)$$

be an element of $\mathcal{A}_{r+1}(X,\alpha_0\ ;k)$. Then we define

$$u\ :=v\otimes\sigma_{Y_{j_k}}^{\otimes m'\theta^{r+1}}\otimes\prod_{p=1}^{k-1}\sigma_{Y_{j_p}}^{\otimes m\varepsilon^p}$$

where $m\ :=(1+t^k)m'$; we remark that the above expressions are meaningful, by the divisibility properties of the (n_p) in the point (e). The observation is that the section u above belongs to the algebra $\mathcal{A}_r(X)$; the verification is immediate.

Next, we define $n^k\ :=\prod_{p=1}^k n_p$, and let $\mathcal{A}_r^{(n^k)}(X)$ be the truncation of $\mathcal{A}_r(X)$, corresponding to the set $(m,\theta)\in\Gamma_d$ such that $m\theta^s\in n^kd^s\mathbb{Z}$, for $s=0,\dots,r$. Let

$$\sigma\in H^0\big(X, m(K_X+\sum_{j=1}^N l^j(\theta)Y_j+A)\big)$$

be an element of $\mathcal{A}_r^{(n^k)}(X)$; we assume that $\|\theta-\alpha_0\|\le\min_j\delta^j$. By the relation $(\star)$, the section σ can be written as a polynomial of a fixed number of generators, plus $\sigma_{Y_{j_1}}^{\otimes m\varepsilon^1}\sigma_{[2]}$, where

$$\sigma_{[2]}\in H^0\big(X, m(K_X+\sum_{j=1}^N l^j_2(\theta)Y_j+A)\big).$$

Let ν be the vanishing order of $\sigma_{[2]}$ along the hypersurface Y_{j_2}. If $\nu/m\ge\varepsilon^2$, then we stop ; if not, there exists a section

$$v_{[2]}\in H^0\big(X, m^{(2)}(K_X+Y_{j_2}+\sum_{j\neq j_2}\widetilde{l^j_2}(\theta')Y_j+A_2)\big)$$

such that

$$\sigma_{[2]}=\sigma_{Y_{j_2}}^{\otimes\nu}v_{[2]}$$

where $m=(1+t^2)m^{(2)}$ (thanks to the relations (93) above). In other words, the section $v_{[2]}$ becomes an element of the algebra $\mathcal{A}_{r+1}(X,\alpha_0\ ;2)$.

The procedure described at the end of 3.1 will show that the section u is equivalent with $\sigma_{Y_{j_1}}^{\otimes m\varepsilon^1}\sigma_{Y_{j_2}}^{\otimes m\varepsilon^2}\sigma_{[3]}$, modulo a polynomial of a finite number of generators (corresponding to the finite generation of the restricted algebras of $\mathcal{A}_{r+1}(X, \alpha_0)$ and $\mathcal{A}_{r+1}(X, \alpha_0\ ; 2)$).

In conclusion, after k steps we obtain

$$u = \sum_s \lambda_s \omega_1^{\otimes s_1} \otimes \ldots \otimes \omega_F^{\otimes s_F} + \sigma_{Y_{j_1}}^{\otimes m\varepsilon^1} \otimes \ldots \otimes \sigma_{Y_{j_k}}^{\otimes m\varepsilon^k} \otimes u_{[k+1]}. \tag{$\dagger$}$$

The following lemma shows that if $k \gg 0$, then the hypothesis of 3.1 will be fulfilled.

3.2.1. Lemma. *After a finite number of iterations of the procedure above, each $j \in \Lambda(\alpha_0)$ will be the index for which the minimum of the quantities (66) is obtained.*

PROOF. We argue by contradiction : assume that there exists an index $j_0 \in \Lambda(\alpha_0)$ such that the minimum of (66) is not obtained for $j = j_0$, for any number of iterations. Then we infer that the sequence (t^k) is bounded, because at each step t^k is a minimum, in particular smaller than

$$\frac{1 - l^{j_0}(\alpha_0)}{r^{j_0}(\alpha_0)} \tag{72}$$

and this quantity will be unchanged during the whole iteration process. Then the relation (55) gives a lower bound for ε^k, and this contradicts the positivity of r_j^p for some index j, for which the cardinal of the set Λ_j is large enough. Thus the lemma is proved, and so is theorem 0.1. □

3.2.2. Remark. The extension theorems for pluricanonical forms are crucial for all the main steps of the arguments presented in this survey. The structure of the proof presented in this text shows clearly that in order to obtain new non-vanishing and/or finite generation results (e.g. using as little as possible of the positivity of the $\mathbb{Q}$-bundle A), this is the technique to be further investigated and refined. □

References

[1] Ambro, F. – *Restrictions of log canonical algebras of general type*, arXiv :math/0510212.

[2] Birkar, C., Cascini, P., Hacon, C., McKernan, J. – *Existence of minimal models for varieties of log general type*, on the web.

[3] Birkar, C., Păun, M. – *Minimal models, flips and finite generation : a tribute to V.V. SHOKUROV and Y.-T. SIU*, arXiv :0904.2936.

[4] Boucksom, S.. – *Divisorial Zariski decompositions on compact complex manifolds*, Ann. Sci. Ecole Norm. Sup. (4) 37 (2004), no. 1, 45–76.

[5] Boucksom, S., Eyssidieux, Ph., Guedj, V., Zeriahi, A.. – *Monge-Ampre equations in big cohomology classes*, arXiv :0812.3674.

[6] Claudon, B. – *Invariance for multiples of the twisted canonical bundle*, math.AG/0511736, Ann. Inst. Fourier (Grenoble) 57 (2007), no. 1, 289–300.

[7] Demailly, J.-P., Peternell, Th., Schneider, M. :. – *Pseudoeffective line bundles on compact Kähler manifolds ;* , Internat. J. Math. 12 (2001), no. 6.

[8] Demailly, J.-P. – *Kähler manifolds and transcendental techniques in algebraic geometry*, Plenary talk and Proceedings of the Internat. Congress of Math., Madrid (2006), 34p, volume I.

[9] Demailly, J.-P. – *Analytic approach of the minimal model program and of the abundance conjectures, ;* , Park City Math Institute Lecture Notes, July 2008.

[10] Druel, S. – *Existence de modèles minimaux pour les variétés de type général*, Exposé 982, Séminaire Bourbaki, 2007/08.

[11] Ein, L., Lazarsfeld, R., Mustaţă, M., Nakamaye, M., Popa, M. :. – *Asymptotic invariants of base loci* , Ann. Inst. Fourier (Grenoble) **56** (2006), no. 6, 1701–1734.

[12] Ein, L., Popa, M. – *Adjoint ideals and extension theorems*, preprint in preparation ; june 2007.

[13] Hacon, C., McKernan, J. – *Boundedness of pluricanonical maps of varieties of general type*, Invent. Math. Volume **166**, Number 1 / October, 2006, 1-25.
[14] Hacon, C., McKernan, J. – *On the existence of flips ;* , math.AG/0507597.
[15] Kawamata, Y. – *Finite generation of a canonical ring*, arXiv :0804.3151.
[16] Kawamata, Y., Matsuda, K., Matsuki, K. – *Introduction to the minimal model program*, Algebraic Geometry, Sendai (T. Oda, ed.), Kinokuniya-North- Holland, 1987, Adv. Stud. Pure Math.
[17] Kim, D. – *L^2 extension of adjoint line bundle sections*, arXiv :0802.3189.
[18] Kollár, J. – *Effective base point freeness*, Math. Ann. 296 (1993), no. 4.
[19] Kollár, J., Mori, S. – *Birational geometry of algebraic varieties.*, Cambridge University Press, Cambridge, 1998.
[20] Lazarsfeld, R. – *Positivity in Algebraic Geometry*, Springer, Ergebnisse der Mathematik und ihrer Grenzgebiete.
[21] Lazić, V. – *Adjoint rings are finitely generated*, arXiv :0905.2707.
[22] Nakayama, N. – *Zariski decomposition and abundance*, MSJ Memoirs **14**, Tokyo (2004).
[23] Păun, M. – *Siu's Invariance of Plurigenera : a One-Tower Proof*, preprint IECN (2005), J. Differential Geom. 76 (2007), no. 3, 485493.
[24] Păun, M. – *Relative critical exponents, non-vanishing and metrics with minimal singularities*, arXiv :0807.3109.
[25] Shokurov, V. – *A non-vanishing theorem*, Izv. Akad. Nauk SSSR (49) 1985.
[26] Shokurov, V. – *3-fold log models.*, Algebraic geometry 4, J. Math. Sci. 81 (1996), no. 3, 26672699.
[27] Siu, Y.-T. – *Analyticity of sets associated to Lelong numbers and the extension of closed positive currents*, Invent. Math. 27 (1974), 53–156.
[28] Siu, Y.-T. – *Invariance of Plurigenera*, Inv. Math., **134** (1998), 661-673.
[29] Siu, Y.-T. – *Extension of twisted pluricanonical sections with plurisubharmonic weight and invariance of semipositively twisted plurigenera for manifolds not necessarily of general type*, Complex geometry (Göttingen, **2000**), 223–277, Springer, Berlin, 2002.
[30] Siu, Y.-T. – *A General Non-Vanishing Theorem and an Analytic Proof of the Finite Generation of the Canonical Ring*, arXiv :math/0610740.
[31] Siu, Y.-T. – *Finite Generation of Canonical Ring by Analytic Method*, arXiv :0803.2454.
[32] Siu, Y.-T. – *Techniques for the Analytic Proof of the Finite Generation of the Canonical Ring*, arXiv :0811.1211.
[33] Takayama, S :. – *On the Invariance and Lower Semi–Continuity of Plurigenera of Algebraic Varieties*, J. Algebraic Geom. **16** (2007), no. 1, 1–18.
[34] Takayama, S. – *Pluricanonical systems on algebraic varieties of general type*, Invent. Math. Volume **165**, Number 3 / September, 2005, 551-587.
[35] Tsuji, H. – *Extension of log pluricanonical forms from subvarieties*, math.CV/0511342 .
[36] Varolin, D. – *A Takayama-type extension theorem*, math.CV/0607323, to appear in Comp. Math.

Introduction to Resolution of Singularities

Mircea Mustaţă

IAS/Park City Mathematics Series
Volume 17, 2008

Introduction to Resolution of Singularities

Mircea Mustaţă

These notes have been prepared for my graduate lectures in the 2008 PCMI program "Analytic and algebraic geometry: common problems, different methods". Their goal is to give an introductory account of resolution of singularities in arbitrary dimension. For a long time, this has been an esoteric topic. While the results of the fundamental work of Hironaka [**Hir**] have been widely quoted, the proof of the main theorem had been understood by only a few experts.

Due to a string of simplifications worked out by many people, it is the case that nowadays the proof can be part of the background of the working algebraic geometer. The fact that it is entirely elementary, makes it accessible to a student with a one semester course in algebraic geometry. While the beautiful ideas and constructions involved in the proof did not seem to find so far applications outside resolution of singularities, one can hope that this might change once the proof becomes part of the mainstream.

The goal of this five lecture introductory course was to present the complete proof of resolution of singularities, in a way accessible to beginner algebraic geometry students. These notes claim no originality: they are based on Włodarczyk's work [**Wło**], and on further simplifications due to Kollár. In particular, for the reader who would like to go deeper into the subject, we highly recommend Kollár's monograph [**Kol**].

Acknowledgment. I am indebted to all the participants in the PCMI program for the comments and suggestions I received during the course. I am also grateful to János Kollár for pointing out several errors in an earlier version of the notes, and to Jarek Włodarczyk for patiently answering my questions. Last but not least, I would like to thank the PCMI staff, and especially to Catherine Giesbrecht, for their help and support.

Department of Mathematics, University of Michigan, Ann Arbor, MI 48109, USA
E-mail address: mmustata@umich.edu
During the preparation of these lecture notes, I was partially supported by NSF grant DMS-0758454, and by a Packard Fellowship

LECTURE 1

Resolutions and principalizations

In everything that follows, we work over an algebraically closed field k of characteristic zero. The main results extend easily to arbitrary fields, in a way compatible with field extensions (see [**Kol**]), but we will not worry about this aspect. A variety is a separated, reduced scheme of finite type over k, not necessarily irreducible. All topological notions refer to the Zariski topology. Given an ideal I on a variety X, with a slight abuse we denote by $\mathrm{Supp}(I)$ the subset of X defined by I.

1.1. The main theorems

Theorem 1.1. *Given an irreducible variety X, there is a projective, birational morphism $f\colon Y \to X$, such that Y is a smooth variety.*

A morphism f as in the theorem is called a *resolution of singularities* of X. One also says that Y is a *desingularization* of X.

One typically imposes more conditions on f. For example, one can require that f be an isomorphism over the smooth locus of X (see Exercise 1.11 below), or one can impose good properties of the exceptional locus (see Exercise 1.12 below).

The next theorem deals with the case when instead of desingularizing a variety, one considers an ambient smooth variety X, and one wants to "resolve" an ideal on X. The idea is that the simplest ideal on a smooth variety is a principal ideal whose support, in suitable coordinates, looks like a union of coordinate hyperplanes in $\mathbf{A}^n$.

To make this precise, we introduce some terminology. Let X be a smooth variety. A system of coordinates on an open subset $U \subseteq X$ consists of $x_1, \ldots, x_n \in \mathcal{O}_X(U)$ such that for every point $p \in U$, the ideal of p is generated by $x_1 - x_1(p), \ldots, x_n - x_n(p)$. Equivalently, the differential forms $dx_1, \ldots, dx_n$ trivialize the cotangent bundle of X over U. The dual basis of the tangent bundle consists of the vector fields $\frac{\partial}{\partial x_1}, \ldots, \frac{\partial}{\partial x_n}$.

Since X is smooth, around every point in X there is a system of coordinates. We say that $x_1, \ldots, x_n$ are coordinates *at* p if, in addition, $x_i(p) = 0$ for all i. In this case, every element f in the completion $\widehat{\mathcal{O}_{X,p}}$ can be uniquely written as a formal power series in $x_1, \ldots, x_n$, and $\frac{\partial f}{\partial x_i}$ has the obvious meaning.

An effective divisor D on the smooth variety X has *simple normal crossings* (SNC, for short) if for every $p \in X$ there is an open neighborhood U of p, and coordinates $x_1, \ldots, x_n$ on U such that $D|_U$ is defined by $x_1^{a_1} \cdots x_n^{a_n}$ for nonnegative integers $a_1, \ldots, a_n$. More generally, if $Z \subseteq X$ is a closed subvariety and $D = \sum_{i=1}^{r} a_i E_i$, then *$D$ has simple normal crossings with Z* if for every $p \in X$ there are local coordinates $x_1, \ldots, x_n$ in some neighborhood U of p, such that each of $Z, E_1, \ldots, E_r$ that contains p is defined in U by a subset of the x_j. In particular, in this case every intersection $Z \cap E_{i_1} \cap \cdots E_{i_\ell}$ is smooth. Note that if D has SNC with Z, then given any irreducible component Z_0 of Z, the divisor $\sum_{Z_0 \not\subseteq E_i} a_i E_i|_{Z_0}$ has SNC in Z_0.

Exercise 1.2. Let X be a smooth variety, and D an effective divisor on X having SNC with the subvariety Z. Show that if $h\colon \mathrm{Bl}_Z(X) \to X$ is the blow-up along Z, with exceptional divisor E, then $h^*(D) + E$ has SNC.

All our ideal sheaves are assumed to be coherent. If I is an ideal sheaf on a variety X, we will say that I is everywhere nonzero if the restriction of I to every irreducible component of X is nonzero.

Theorem 1.3. *Given a smooth variety X, and an everywhere nonzero ideal sheaf I on X, there is a projective, birational morphism $f\colon Y \to X$ such that Y is smooth, and $I \cdot \mathcal{O}_Y$ is the ideal of a SNC divisor on Y.*

Such f is called a *principalization* of I (one also says that f is a *log resolution* of the pair (X, I)). One can combine Theorems 1.1 and 1.4 to get

Theorem 1.4. *Given an irreducible variety X, and a nonzero ideal sheaf I on X, there is a projective, birational morphism $f\colon Y \to X$ such that Y is smooth, and $I \cdot \mathcal{O}_Y$ is the ideal of a SNC divisor on Y.*

It is clear that Theorem 1.4 is implied by Theorems 1.1 and 1.3. Indeed, given an arbitrary X and I, we first apply Theorem 1.1 to get $g\colon Z \to X$ that is projective and birational, and with Z smooth, and we then apply Theorem 1.3 for Z and $I \cdot \mathcal{O}_Z$ to get $h\colon Y \to Z$. If we take $f = g \circ h$, then f has the properties in the statement of Theorem 1.4.

More surprisingly, as we will see in the next section, a more precise version of Theorem 1.3 implies Theorem 1.1. The advantage is that when proving Theorem 1.3 instead of Theorem 1.1, we switch from a "geometric" setting, in which we need to keep track of the structure of the varieties involved, to an "algebraic" setting, in which we work with ideals on a smooth variety.

1.2. Strengthenings of Theorem 1.3

There are three extra requirements on the morphism in Theorem 1.3, that are very important both for the proof of the theorem, and for its applications.

1.2.1. The first improvement: functoriality

A fundamental property is that principalization can be done in a functorial way with respect to smooth morphisms. In order to formalize this, consider the category $\mathcal{C}$ whose objects are pairs (X, I), with X a smooth variety, and I an everywhere nonzero ideal sheaf on X. Morphisms in $\mathcal{C}$ from (X', I') to (X, I) are smooth morphisms $\phi\colon X' \to X$ such that $I \cdot \mathcal{O}_{X'} = I'$. On the other hand, let $\mathcal{B}ir$ be the category whose objects are projective birational morphisms between smooth varieties, and whose maps are given by Cartesian diagrams.

It is part of Theorem 1.3 that principalization can be extended to a functor from $\mathcal{C}$ to $\mathcal{B}ir$. In other words, one can associate to every pair (X, I) a principalization $f\colon Y \to X$, such that if $\phi\colon X' \to X$ is a smooth morphism, and $I' = I \cdot \mathcal{O}_{X'}$, then the principalizations $f\colon Y \to X$ and $f'\colon Y' \to X'$ of I and I', respectively, fit in a

Cartesian diagram

$$\begin{array}{ccc} Y' & \xrightarrow{\phi'} & Y \\ {\scriptstyle f'}\downarrow & & \downarrow{\scriptstyle f} \\ X' & \xrightarrow{\phi} & X \end{array}$$

Remark 1.5. Note that since f and f' are birational, if there is ϕ' making the above diagram commutative, then such ϕ' is unique.

Remark 1.6. It is clear that if we have a principalization $Y \to X$ of (X, I), and if $\phi\colon X' \to X$ is smooth, then $Y \times_X X' \to X'$ is a principalization of $I \cdot \mathcal{O}_{X'}$. On the other hand, it is not clear that starting with a principalization of $(X', I \cdot \mathcal{O}_{X'})$ we would be able to "descend" it to a principalization of (X, I).

Note that this functoriality implies, in particular, that our construction of principalizations commutes with open embeddings. It also commutes with étale morphisms, and this will be a key fact for the proof of the theorem.

Functoriality also gives interesting information about principalizations. A first consequence is that the principalization $f\colon Y \to X$ given by the theorem for an ideal I on X is an isomorphism over the complement of $\mathrm{Supp}(I)$. Indeed, by commutativity with open immersions, it is enough to prove the assertion when $\mathrm{Supp}(I)$ is empty, that is, when $I = \mathcal{O}_X$. In this case $\phi\colon X \to \mathrm{Spec}(k)$ is smooth, and I is the pull-back of $\mathcal{O}_{\mathrm{Spec}(k)}$, hence by functoriality it is enough to prove the assertion for $\mathrm{Spec}(k)$, when it is clear.

Another consequence of functoriality is that principalization commutes with pulling-back via projections. More precisely, suppose that I is an everywhere nonzero ideal on the smooth variety X, and that Z is another smooth variety. Since the projection $X \times Z \to X$ is smooth, functoriality implies that if $f\colon Y \to X$ is the principalization of I, then the principalization of $I \cdot \mathcal{O}_{X \times Z}$ is given by $f \times \mathrm{id}_Z\colon Y \times Z \to X \times Z$.

A more interesting fact is that functoriality gives for free equivariant principalization. Indeed, suppose that X is a smooth variety, I is an everywhere nonzero ideal on X, and we have an algebraic group G acting on X, such that it induces an action of G on the closed subscheme of X defined by I.

Exercise 1.7. Let $\mu\colon G \times X \to X$ be the morphism giving the action of G on X.

i) Show that if $p\colon G \times X \to X$ is the projection, then the pull-backs of I via μ and p are equal.
ii) Show that μ is a smooth morphism.
iii) Deduce from functoriality with respect to smooth morphisms that if $f\colon Y \to X$ is the principalization of I given by Theorem 1.3, then we have an induced action of G on Y such that f is G-equivariant.

1.2.2. The second improvement: the principalization as a composition of blow-ups with smooth centers

The morphism f in Theorem 1.3 can be obtained as a composition of blow-ups with smooth centers (we emphasize that in order to guarantee functoriality, one has to allow disconnected centers). In fact, as we will see, more is true: f will be given as a composition

$$Y = X_r \xrightarrow{f_r} X_{r-1} \to \cdots \to X_1 \xrightarrow{f_1} X_0 = X,$$

where $X_{i+1} = \mathrm{Bl}_{C_i}(X_i)$, with exceptional divisor E_{i+1}, and each C_i has SNC with E_i, and with the proper transforms of $E_1, \ldots, E_{i-1}$. Note that one advantage of working on a smooth variety is the fact that the description of such blow-ups is particularly easy.

Remark 1.8. In the above setting, we see that the hypothesis on C_i implies (by induction on i) that E_{i+1} has simple normal crossings with the proper transforms of $E_1, \ldots, E_i$.

A key fact about the above stronger version of Theorem 1.3 is that it implies Theorem 1.1, at least when X can be embedded as a closed subvariety of a smooth variety W (for the general case, see Remark 1.10 below). After replacing W by the product with an affine space, we may assume $\mathrm{codim}(X, W) \geq 2$. Let $h\colon Z \to W$ be the principalization given by Theorem 1.3 for the ideal $I_{X/W}$ defining X in W.

We know that h is a composition of blow-ups with smooth centers. Moreover, we have seen that f is an isomorphism over $W \smallsetminus X$. Therefore each blow-up center of codimension ≥ 2 (the other ones give isomorphisms, and at this point we can ignore them) is contained in the inverse image of X. As long as we only blow-up centers whose image in W is strictly contained in X, on the resulting variety W_i over W we have the proper transform of X. This is an irreducible component of $\mathrm{Supp}(I_{X/W} \cdot \mathcal{O}_{W_i})$ of codimension ≥ 2. In order to get a principalization, it follows that at some point one of the blow-up centers has to be the proper transform of X. If Y is this proper transform, we see that Y has to be smooth. The induced morphism $Y \to X$ gives a resolution of singularities of X.

Remark 1.9. It is clear from the above construction that the resolution of an irreducible variety X is obtained as a composition of blow-ups, the centers being given by intersecting the centers on the ambient varieties with the proper transforms of X. However, these centers on the blow-ups of X are not necessarily smooth. We will see in Example 5.12 below that for the algorithm we will present in these notes this can indeed be the case.

1.2.3. The third improvement: compatibility with closed embeddings

Another useful property of the construction of principalizations in Theorem 1.3 is that it commutes with closed embeddings, in the following sense. Suppose that X is a variety, and that I is an everywhere nonzero ideal on X. If $X \hookrightarrow Z$ is a closed embedding, with both X and Z smooth and of pure dimension, and if $J \subseteq \mathcal{O}_Z$ is the inverse image of I via $\mathcal{O}_Z \to \mathcal{O}_X$, then the principalization of J is defined by the same sequence of blow-up centers that gives the principalization of I. More precisely, suppose that the principalization of I is given by the composition

$$Y = X_r \xrightarrow{f_r} X_{r-1} \to \cdots \to X_1 \xrightarrow{f_1} X_0 = X,$$

with $X_{i+1} = \mathrm{Bl}_{C_i}(X_i)$. In this case, X_1 is contained in $Z_1 := \mathrm{Bl}_{C_0}(Z)$, hence we can consider $Z_2 := \mathrm{Bl}_{C_1}(Z_1)$, and continue this way. The assertion is that the principalization of J given by Theorem 1.3 is the composition $Z_r \to \cdots \to Z_2 \to Z_1 \to Z$.

As an application, we now combine this fact with functoriality to show that the resolution of singularities constructed in §1.2.2 does not depend on the choice of embedding in a smooth variety. Suppose that $i\colon X \hookrightarrow W$ and $i'\colon X \hookrightarrow W'$ are two closed embeddings in smooth varieties W and W', both of codimension at least

two, and we need to show the resolutions f and f' of X induced by the respective principalizations of $I_{X/W}$ and $I_{X/W'}$ are canonically isomorphic.
Step 1. Suppose first that both W and W' (hence also X) are affine. After embedding W and W' in suitable affine spaces, it follows from the compatibility of principalizations with closed embeddings that we may assume $W = \mathbf{A}^n$ and $W' = \mathbf{A}^m$.
Step 2. Let $\phi\colon \mathbf{A}^n \to \mathbf{A}^m$ be a morphism such that $\phi \circ i = i'$. We have a closed embedding $\Phi\colon \mathbf{A}^n \hookrightarrow \mathbf{A}^{m+n}$ given by $\Phi(x) = (x, \phi(x))$. Since $\Phi \circ i = (i, i')$, compatibility with respect to closed embeddings shows that the resolution of X corresponding to i is given by the same sequence of centers as the resolution corresponding to (i, i'). A similar argument compares the resolutions corresponding to i' and (i, i'). It follows that f and f' are canonically isomorphic.
Step 3. Suppose now that W and W' are arbitrary. Note first that if U is an open subset of W, then $f^{-1}(U \cap X) \to U \cap X$ is the resolution of $U \cap X$ induced by the embedding $U \cap X \hookrightarrow U$ (we use here functoriality of principalizations with respect to open immersions). The same thing applies to the open subsets of W'. Suppose now that $W = \bigcup_i U_i$ are $W' = \bigcup_j U'_j$ are affine open covers. For every i and j, we can cover $U_i \cap U'_j \cap X$ by open subsets V_{ijk} that are principal affine with respect to both $U_i \cap X$ and $U'_j \cap X$. It follows that we can write $V_{ijk} = A_{ijk} \cap X = A'_{ijk} \cap X$, for affine open subsets $A_{ijk} \subseteq U_i$, $A'_{ijk} \subseteq U'_j$. It follows from Step 2 that we have canonical isomorphisms $f^{-1}(V_{ijk}) \simeq (f')^{-1}(V_{ijk})$ over V_{ijk}. Each intersection of two subsets V_{ijk} can be covered by principal affine open subsets with respect to the two V_{ijk}. By canonicity of the isomorphisms in Step 2, we see that we can glue the isomorphisms over each V_{ijk} to an isomorphism of f and f'.

Remark 1.10. One can use the above invariance to give a proof of Theorem 1.1 in the case when X can not be embedded in a smooth variety. The key point is to first extend the construction we gave to disjoint unions of irreducible varieties. Once this is done, the invariance with respect to embedding can be used to patch local resolutions. While we do not go into details here, we will explain an analogous argument in the context of the proof of Theorem 1.3. For more details about the proof of Theorem 1.1, we refer to [**Kol**]. Note however that the approach in *loc. cit.* is slightly different. It proceeds by first constructing resolutions on affine open subsets, and then by gluing these local resolutions.

Exercise 1.11. We will see in Remark 5.13 below that if X is a proper closed subvariety of Z, with both X and Z smooth and of pure dimension, then the principalization of $I_{X/Z}$ is given by $\mathrm{Bl}_X(Z) \to Z$. Use this to show that for every irreducible X (that can be embedded in a smooth variety), the resolution $Y \to X$ that we constructed above is an isomorphism over X_{sm}.

Exercise 1.12. Let X be a normal variety, and $I_1, \ldots, I_s$ nonzero ideals on X. Show that there is a projective, birational morphism $f\colon Y \to X$ such that

i) Y is smooth.
ii) The exceptional locus $\mathrm{Ex}(f)$ is a divisor, and $I_i \cdot \mathcal{O}_Y$ is the ideal of a divisor D_i, for every i.
iii) The divisor $\mathrm{Ex}(f) + \sum_{i=1}^{s} D_i$ has SNC.

1.3. Historical comments

Let us give a few references, though we make no attempt at a detailed historical account. The two main results in this section are due to Hironaka [**Hir**]. Various simplifications and improvements have been made over the years. The approach we follow is based on a combination of [**Wło**] and [**Kol**]. This in turn is based on previous work of Giraud [**Gir**] and Villamayor [**Vil1**] and [**Vil2**]. Among several of the other important contributions to resolution of singularities, we mention the work of Bierstone and Milman [**BM**] and Encinas and Hauser [**EH**].

The positive characteristic version of the main results in this lecture is still missing in arbitrary dimension (though a flurry of recent activity in this direction gives hope that such a result might be quite close). At this point, the best result in positive (and also mixed) characteristic is the result of de Jong [**dJ**]. Instead of providing a resolution of singularities of X, it provides an *alteration* of X, that is, a projective, generically finite morphism $f\colon Y \to X$, with Y nonsingular. It is worth pointing out that while the proof of the desingularization result in Theorem 1.3 is very much algebraic, de Jong's proof is quite geometric.

The situation is better in small dimensions: in dimension ≤ 3, resolution of singularities in the sense of Theorem 1.1 is known also in positive characteristic. This is due to work of Abhyankar, Cossart-Piltant, and Cutkosky (see [**CP1**], [**CP2**], and [**Cut**]).

LECTURE 2

Marked ideals

In this lecture we set up the basic definitions. Recall that our goal is to prove principalization (in its strengthened form). The only invariant of the singularities of an ideal that comes up in the proof is the order of vanishing. It turns out that things become conceptually simpler if one includes a reference number for the order: this leads to the notion of marked ideal. These are the objects that we will desingularize. In this lecture we mostly follow the terminology and the presentation from [**Wło**].

2.1. Marked ideals

2.1.1. The order of an ideal

We start by reviewing the notion of order of an ideal at a point on a smooth variety.

Definition 2.1. Let I be an ideal sheaf on the smooth variety X. If p is a point in X, then the *order* $\mathrm{ord}_p(I)$ of I at p is the largest integer r such that $I \cdot \mathcal{O}_{X,p} \subseteq \mathfrak{m}_{X,p}^r$, where $\mathfrak{m}_{X,p}$ is the maximal ideal in the local ring $\mathcal{O}_{X,p}$ (we make the convention that $\mathrm{ord}_p(I) = \infty$ if $I = (0)$ around p).

Note that if $x_1, \ldots, x_n$ are local coordinates around p, then $\mathrm{ord}_p(I) \geq m$ if and only if for every $f \in I$ (or in a system of generators of I), we have $\frac{\partial^{|\alpha|} f}{\partial x^\alpha}(p) = 0$ whenever the multi-index $\alpha = (\alpha_1, \ldots, \alpha_n)$ has length $|\alpha| := \sum_i \alpha_i$ less than m. Indeed, for this we may assume that $x_i(p) = 0$ for all i. Since computing the order can be done in $\widehat{\mathcal{O}_{X,p}} \simeq k[\![x_1, \ldots, x_n]\!]$, the assertion follows from the fact that the order of a formal power series f is the smallest degree of a monomial in f.

In particular, we see that the set $\{p \in X \mid \mathrm{ord}_p(I) \geq m\}$ is closed in X. Note also that if H is the hypersurface defined by f, then p is a smooth point of H if and only if $\mathrm{ord}_p(f) = 1$.

Definition 2.2. More generally, if I is an ideal sheaf on the smooth variety X, and if C is an irreducible closed subvariety of X, then the order $\mathrm{ord}_C(I)$ is the largest integer r such that $I \cdot \mathcal{O}_{X,C} \subseteq \mathfrak{m}_{X,C}^r$, where $\mathfrak{m}_{X,C}$ is the maximal ideal in the local ring $\mathcal{O}_{X,C}$.

Exercise 2.3. Let C be a smooth and irreducible closed subvariety of the smooth variety X, and let I be an ideal sheaf on X.

i) Suppose that $x_1, \ldots, x_n$ are coordinates at $p \in C$, such that C is defined in a neighborhood of p by $(x_1, \ldots, x_d)$. Show that $\mathrm{ord}_C(I) \geq m$ if and only if for every $f \in I$, if we write f as a power series in $k[\![x_1, \ldots, x_n]\!]$, then each term in f has degree $\geq m$ in $x_1, \ldots, x_d$.

ii) Show that if the regular function f contains the monomial $x_1^{a_1}\cdots x_n^{a_n}$ with $a_1+\cdots+a_d=m$ in its expression in $k[\![x_1,\ldots,x_n]\!]$, then $\frac{\partial^m f}{\partial x_1^{a_1}\cdots\partial x_r^{a_r}}$ does not vanish identically on C. Deduce that $\mathrm{ord}_C(I)=\min_{p\in C}\mathrm{ord}_p(I)$.

Suppose now that C is a smooth and irreducible closed subvariety of X. Let $\sigma\colon X_1=\mathrm{Bl}_C X\to X$ be the blow-up of X with center C. Suppose that $x_1,\ldots,x_n$ are local coordinates in some open subset U meeting C, such that C is defined in U by $(x_1,\ldots,x_d)$. A typical chart on X_1 has coordinates $x_1,y_2,\ldots,y_d,x_{d+1},\ldots,x_n$, where $x_i=x_1y_i$ for $2\leq i\leq d$, and the exceptional divisor E_1 on X_1 is defined in this chart by (x_1).

Exercise 2.4. Show that the largest r such that $I\cdot\mathcal{O}_{X_1}\subseteq\mathcal{O}_{X_1}(-rE_1)$ is $\mathrm{ord}_C(I)$.

When constructing a principalization of an ideal I, in the presence of a blow-up as above, one commonly replaces I by its *strict transform* $\sigma_*^{-1}(I)$, which is the ideal on X_1 such that $I\cdot\mathcal{O}_{X_1}=\sigma_*^{-1}(I)\cdot\mathcal{O}_{X_1}(-rE_1)$, with $r=\mathrm{ord}_C(I)$. This operation has some unpleasant features (for example, it can happen that we have $I\subseteq J$, but $\sigma_*^{-1}(I)\not\subseteq\sigma_*^{-1}(J)$). This drawback will disappear when working with marked ideals.

Note that we allow blow-ups whose centers contain codimension one components. In this case we distinguish between the proper transform and the strict transform of a divisor. If $\sigma\colon X_1=\mathrm{Bl}_C(X)\to X$ is the blow-up with center C, then the *strict transform* $\widetilde{F}$ of a reduced divisor F is the divisor defined by the strict transform of $\mathcal{O}_X(-F)$, that is, the closure of $\sigma^{-1}(F\smallsetminus C)$. On the other hand, the *proper transform* of F is the unique divisor F_1 on X_1 such that $\sigma_*(F_1)=F$. The difference is when there is a codimension one component C_0 of C that is also a component of F: $\sigma^{-1}(F)$ does not appear in $\widetilde{F}$, but it appears in F_1.

2.1.2. The definition of a marked ideal

A *marked ideal* (X,I,m) consists of the following data:

i) A smooth variety X.
ii) A (coherent) ideal sheaf I on X.
iii) A positive integer m, the *marking* of the marked ideal.

When the ambient variety X is understood, we drop it from the notation.

A *marked ideal with divisor* (X,I,m,E) consists of a marked ideal (X,I,m), and an ordered collection $E=\left(E^{(1)},\ldots,E^{(\ell)}\right)$, where each $E^{(i)}$ is a smooth (but not necessarily connected) divisor on X, such that $\sum_i E^{(i)}$ is a reduced divisor with simple normal crossings (in particular, the $E^{(i)}$ have no common components). Of course, a marked ideal can be identified with a marked ideal with divisor, for which the divisorial part is empty.

The marked ideals with divisors are the key players in constructing principalizations. It is natural to consider the divisor E as part of the data, since we need to keep track of the exceptional divisors that appear during the various steps of our resolution algorithm. However, for many constructions the divisor is irrelevant, and in those cases we will discuss the main concepts in the context of marked ideals.

The *support* of a marked ideal (X,I,m) is the set

$$\mathrm{Supp}(X,I,m):=\{p\in X\,|\,\mathrm{ord}_p(I)\geq m\}.$$

This is a closed subset of X. If I is everywhere nonzero, then each component of $\mathrm{Supp}(X,I,m)$ has codimension ≥ 1. Note also that $\mathrm{Supp}(X,I,1)=\mathrm{Supp}(I)$.

Given a marked ideal (X, I, m), we will deal with the following type of blow-up, called *admissible blow-up*: $\sigma\colon X_1 = \mathrm{Bl}_C X \to X$, where C is a nonempty smooth subvariety of X (not necessarily connected), such that $C \subseteq \mathrm{Supp}(I, m)$. If we deal with a marked ideal with divisor (X, I, m, E), then σ is admissible if, in addition, $\sum_j E^{(j)}$ has SNC with C. We denote by F the exceptional divisor $\sigma^{-1}(C)$ of σ.

Note that by definition, if σ is an admissible blow-up as above, then $\mathrm{ord}_C(I) \geq m$. Therefore there is an ideal I_1 on X_1 such that $I \cdot \mathcal{O}_{X_1} = I_1 \cdot \mathcal{O}_{X_1}(-mF)$. The *controlled transform* $\sigma^c(X, I, m)$ is the marked ideal (X_1, I_1, m) (we follow the terminology in [**Wło**]; this marked ideal is the *birational transform* in [**Kol**]).

If σ is admissible with respect to a marked ideal with divisor (X, I, m, E), where $E = \left(E^{(1)}, \ldots, E^{(\ell)}\right)$, then the controlled transform is again a marked ideal with divisor, namely (X, I_1, m, E_1), where $E_1 = \left(\widetilde{E^{(1)}}, \ldots, \widetilde{E^{(\ell)}}, F\right)$, each $\widetilde{E^{(i)}}$ being the strict transform of $E^{(i)}$. Note that because of our convention in the definition of strict transform, the divisors in E_1 have no common components.

We emphasize that we allow the blow-up of codimension one subvarieties. In this case σ is an isomorphism, but the ideal (and also the corresponding sequence of divisors, in the case of marked ideals with divisors) changes.

A *sequence of admissible blow-ups* with respect to a marked ideal with divisor (X, I, m, E) is a sequence $\sigma = (\sigma_1, \ldots, \sigma_r)$, where

$$X_r \overset{\sigma_r}{\to} \cdots \overset{\sigma_2}{\to} X_1 \overset{\sigma_1}{\to} X = X_0,$$

and for every i, with $1 \leq i \leq r$

i) $\sigma_i\colon X_i \to X_{i-1}$ is an admissible blow-up with respect to $(X_{i-1}, I_{i-1}, m, E_{i-1})$.
ii) We have $(X_i, I_i, m, E_i) = \sigma_i^c(X_{i-1}, I_{i-1}, m, E_{i-1})$ for all $i \geq 1$, and $(X_0, I_0, m, E_0) = (X, I, m, E)$.

The *controlled transform* of (X, I, m, E) with respect to σ is the marked ideal with divisor

$$\sigma^c(X, I, m, E) := (X_r, I_r, m, E_r).$$

Note that the sequence is determined not just by the composition of the maps, but by the whole sequence of blow-up centers. When we only deal with marked ideals, we get corresponding notions of sequence of admissible blow-ups, and controlled transform, by simply ignoring the divisorial part in the previous definition.

A *resolution* of the marked ideal with divisor (X, I, m, E) is a sequence of admissible blow-ups as above, such that the support of $\sigma^c(X, I, m, E)$ is empty. Note that if we start with an ideal I on the smooth variety X, then a resolution of $(X, I, 1, \emptyset)$ gives a principalization of I. Indeed, the support of the controlled transform $(X_r, I_r, 1, E_r)$ being empty, means that $I_r = \mathcal{O}_{X_r}$, and it follows from definition that $I \cdot \mathcal{O}_{X_r}$ is the ideal of a divisor supported on the union of the components of E_r (and which therefore has SNC). Hence in order to prove Theorem 1.3 we will concentrate on resolving marked ideals with divisors.

We have an obvious notion of pull-back of marked ideals (possibly, with divisors) under smooth morphisms. Indeed, for $f\colon Y \to X$ smooth, we put

$$f^*(X, I, m, E) = (Y, I \cdot \mathcal{O}_Y, m, f^*E).$$

Note that while pulling-back divisors does not preserve connectedness, it preserves the SNC condition.

If we have a sequence of admissible blow-ups σ with respect to a marked ideal (possibly, with divisor), then pulling back by a smooth morphism gives a sequence of admissible blow-ups with respect to the pull-back of the marked ideal. Note, however, that by pulling-back some blow-up centers may become empty (the corresponding blow-up will be called *empty blow-up*). In this case, we delete the trivial blow-ups and renumber the elements in the sequence accordingly.

2.1.3. Equivalence and order relations on marked ideals

We want to identify two marked ideals if they have the same resolutions. Therefore we say that two marked ideals $\mathcal{I} = (X, I, m)$ and $\mathcal{I}' = (X, I', m')$ are *equivalent*, and write $\mathcal{I} \sim \mathcal{I}'$ if the two marked ideals have the same sequences of admissible blow-ups, and for every such sequence σ, the two controlled transforms $\sigma^c(\mathcal{I})$ and $\sigma^c(\mathcal{I}')$ have the same support. Two marked ideals with divisors are equivalent if, in addition to the previous condition, the two sequences of divisors are the same (note that the condition is now slightly weaker, since we might have fewer sequences of blow-ups to consider).
It is clear that this is an equivalence relation on marked ideals (possibly, with divisors). We see that if two marked ideals with divisors are equivalent, then a resolution for one of them is automatically a resolution for the other.

Exercise 2.5. Show that if $\mathcal{I} = (X, I, m)$ is a marked ideal, then $\mathcal{I}$ is equivalent to (X, I^r, mr).

Exercise 2.6. Show that if $\mathcal{I} = (X, I, m)$ is a marked ideal, then $\mathcal{I} \sim (X, \overline{I}, m)$, where $\overline{I}$ is the integral closure of I.

It is convenient to also consider some order relations on marked ideals. The first such relation is defined as follows. If $\mathcal{I} = (X, I, m)$ and $\mathcal{I}' = (X, I', m')$ are marked ideals, then $\mathcal{I} \leq \mathcal{I}'$ if the following hold:

i) Every sequence of admissible blow-ups for $\mathcal{I}'$ is also an admissible sequence for $\mathcal{I}$.
ii) For every such sequence σ, we have $\mathrm{Supp}(\sigma^c(\mathcal{I}')) \subseteq \mathrm{Supp}(\sigma^c(\mathcal{I}))$.

It is clear that this gives an order relation on marked ideals. We have $\mathcal{I} \sim \mathcal{I}'$ if and only if $\mathcal{I} \leq \mathcal{I}'$ and $\mathcal{I}' \leq \mathcal{I}$.

Another natural order relation is induced by the inclusion of ideals. Given marked ideals $\mathcal{I} = (X, I, m)$ and $\mathcal{I}' = (X, I', m')$, we put $\mathcal{I} \subseteq \mathcal{I}'$ if $m = m'$ and $I \subseteq I'$.

Exercise 2.7. Show that if $\mathcal{I}$ and $\mathcal{I}'$ are marked ideals with $\mathcal{I} \subseteq \mathcal{I}'$, then every sequence of admissible blow-ups σ with respect to $\mathcal{I}'$ is also a sequence of admissible blow-ups with respect to $\mathcal{I}$, and we have $\sigma^c(\mathcal{I}) \subseteq \sigma^c(\mathcal{I}')$. In particular, $\mathcal{I} \leq \mathcal{I}'$.

Similar definitions can be made for marked ideals with divisors for which the two sequences of divisors coincide.

2.1.4. Operations on marked ideals

In this subsection we assume that all marked ideals are on the same variety. We first define the product of two such marked ideals (I_1, m_1) and (I_2, m_2) by

$$(I_1, m_1) \cdot (I_2, m_2) := (I_1 I_2, m_1 + m_2).$$

This operation is commutative and associative.

Exercise 2.8. Let $\mathcal{I}_1 = (I_1, m_1)$ and $\mathcal{I}_2 = (I_2, m_2)$ be two marked ideals.

i) Show that $\mathrm{Supp}(\mathcal{I}_1) \cap \mathrm{Supp}(\mathcal{I}_2) \subseteq \mathrm{Supp}(\mathcal{I}_1 \cdot \mathcal{I}_2)$.
ii) Show that a sequence σ of admissible blow-ups for both $\mathcal{I}_1$ and $\mathcal{I}_2$ is also a sequence of admissible blow-ups for $\mathcal{I}_1 \cdot \mathcal{I}_2$, and $\sigma^c(\mathcal{I}_1 \cdot \mathcal{I}_2) = \sigma^c(\mathcal{I}_1) \cdot \sigma^c(\mathcal{I}_2)$.

The sum of marked ideals is only slightly more subtle than that of product. We first define the sum in the case when all marked ideals have the same order. Given marked ideals $(I_1, m), \ldots, (I_r, m)$, their sum is defined by

$$(I_1, m) + \cdots + (I_r, m) := \left(\sum_{j=1}^{r} I_j, m \right).$$

Exercise 2.9. Let $\mathcal{I}_1 = (I_1, m), \ldots, \mathcal{I}_r = (I_r, m)$ be marked ideals.

i) Show that $\mathrm{Supp}(\mathcal{I}_1 + \cdots + \mathcal{I}_r) = \bigcap_i \mathrm{Supp}(\mathcal{I}_i)$.
ii) Show that a sequence of blow-ups σ is admissible with respect to the sum if and only if it is admissible with respect to each of the terms. Moreover, in this case we have

$$\sigma^c(\mathcal{I}_1 + \cdots + \mathcal{I}_r) = \sigma^c(\mathcal{I}_1) + \cdots + \sigma^c(\mathcal{I}_r).$$

Exercise 2.10. Show that the sum of marked ideals is compatible with the equivalence and the order relations on marked ideals.

Suppose now that we deal with marked ideals $\mathcal{I}_1 = (I_1, m_1), \ldots, \mathcal{I}_r = (I_r, m_r)$. If some of the m_i's are distinct, it is natural to consider

$$\mathcal{I}_1 +' \cdots +' \mathcal{I}_r := \sum_{j=1}^{r} \mathcal{I}_j^{m_1 \cdots \widehat{m_j} \cdots m_r}.$$

Exercise 2.11. Note that this version of sum of marked ideals is not associative. However, show that it is associative up to equivalence of marked ideals.

Exercise 2.12. Show that the analogues of the assertions in Exercises 2.9 and 2.10 hold for this version of sum.

Exercise 2.13. Show that if both versions of sum are defined, then the results are equivalent:

$$(I_1, m) +' \cdots +' (I_r, m) \sim (I_1, m) + \cdots + (I_r, m).$$

2.2. Derived ideals

We now come to one of the key concepts in this story, the ideal generated by all the (first) derivatives of the elements of a given ideal. Iterating this construction will lead us to equations for the hypersurfaces of maximal contact that will be discussed in the next lecture.

2.2.1. Properties of derived ideals

Suppose that I is an ideal sheaf on a smooth variety X. The *derived ideal* $D(I)$ of I is the image of

$$\mathrm{Der}_k(\mathcal{O}_X) \otimes_k I \to \mathcal{O}_X.$$

Note that this is again a coherent sheaf of ideals on X.

The ideal $D(I)$ contains I. To see this, choose local coordinates $x_1, \ldots, x_n$, and note that for every $f \in I$, we have

$$f = \frac{\partial(fx_i)}{\partial x_i} - x_i \frac{\partial f}{\partial x_i} \in D(I).$$

Exercise 2.14. Show that if $x_1, \ldots, x_n$ are local coordinates on an affine open subset U of X, and if I is generated on U by $f_1, \ldots, f_r$, then the ideal $D(I)$ is generated on U by

$$\left\{ f_i, \frac{\partial f_i}{\partial x_1}, \ldots, \frac{\partial f_i}{\partial x_n} | 1 \leq i \leq r \right\}.$$

Lemma 2.15. *If C is a smooth and irreducible closed subvariety of X, contained in* Supp(I), *then*

$$\mathrm{ord}_C(D(I)) = \mathrm{ord}_C(I) - 1.$$

PROOF. It follows from Exercise 2.3 that it is enough to prove the lemma when $C = \{p\}$ is a point. Fix a system of coordinates $x_1, \ldots, x_n$ at p. The assertion of the lemma follows from the fact that if $f \in k[\![x_1, \ldots, x_n]\!]$ has order $m \geq 1$, then all $\frac{\partial f}{\partial x_i}$ have order $\geq m - 1$, and there is at least one such partial derivative that has order exactly $m - 1$ (this is the only point where we use the fact that we are in characteristic zero, but it is a crucial point). □

In light of the lemma, given a marked ideal $\mathcal{I} = (X, I, m)$ with $m \geq 2$, we put $D(\mathcal{I}) = (X, D(I), m - 1)$. The lemma implies that $\mathrm{Supp}(D(\mathcal{I})) = \mathrm{Supp}(\mathcal{I})$.

One can iterate the definition of derived ideals, as follows. If $\mathcal{I}$ is a marked ideal with marking m, then for $i \leq m - 1$ we recursively define $D^i(\mathcal{I}) = D(D^{i-1}(\mathcal{I}))$. We see that $\mathrm{Supp}(D^i(\mathcal{I})) = \mathrm{Supp}(\mathcal{I})$ for every $i \leq m - 1$.

Exercise 2.16. Show that if $f\colon Y \to X$ is a smooth morphism, and if (I, m) is a marked ideal on X, then $D^i(f^*(I, m)) = f^*(D^i(I, m))$ for every $i \leq m - 1$.

2.2.2. Behavior with respect to sequences of admissible blow-ups

The following proposition is one of the key results for what follows. It describes the behavior of taking the derived ideal with respect to admissible blow-ups.

Proposition 2.17. *Let $\mathcal{I} = (X, I, m)$ be a marked ideal, and let $i \leq m - 1$.*

i) *Every sequence σ of admissible blow-ups for $\mathcal{I}$ is also a sequence of admissible blow-ups for $D^i(\mathcal{I})$.*
ii) *For every such σ, we have $\sigma^c(D^i(\mathcal{I})) \subseteq D^i(\sigma^c(\mathcal{I}))$.*

PROOF. It is clear that it is enough to prove the case $i = 1$, since the general case then follows by induction on i. Moreover, it is enough to prove that if σ is an admissible blow-up for $\mathcal{I}$, then it is an admissible blow-up for $D(\mathcal{I})$, and $\sigma^c(D(\mathcal{I})) \subseteq D(\sigma^c(\mathcal{I}))$. Indeed, then the assertion follows by induction on the length r of the sequence of blow-ups: given such a sequence σ

$$X_r \xrightarrow{\sigma_r} \cdots \xrightarrow{\sigma_1} X_0 = X, \tag{2.1}$$

applying the above inclusion for σ_1 gives $\sigma_1^c(D(\mathcal{I})) \subseteq D(\sigma_1^c(\mathcal{I}))$. If τ is the sequence given by $\sigma_2, \ldots, \sigma_r$, applying the inductive hypothesis for τ and $\sigma_1^c(\mathcal{I})$ implies that τ is a sequence of admissible blow-ups for $D(\sigma_1^c(\mathcal{I}))$, hence for $\sigma_1^c(D(\mathcal{I}))$ and

$$\sigma^c(D(\mathcal{I})) = \tau^c(\sigma_1^c(D(\mathcal{I}))) \subseteq \tau^c(D(\sigma_1^c(\mathcal{I})) \subseteq D(\sigma^c(\mathcal{I})).$$

Suppose now that σ is an admissible blow-up for $\mathcal{I}$. Since $\mathrm{Supp}(D(\mathcal{I})) = \mathrm{Supp}(\mathcal{I})$, this is also admissible for $D(\mathcal{I})$. In order to show that $\sigma^c(D(\mathcal{I})) \subseteq D(\sigma^c(\mathcal{I}))$, we choose local coordinates $x_1, \ldots, x_n$ on X, such that the blow-up center is defined by $(x_1, \ldots, x_d)$. On the blow-up, let us consider the chart with coordinates $x_1, y_2, \ldots, y_d, x_{d+1}, \ldots, x_n$, where $x_i = x_1 y_i$ for $2 \leq i \leq d$. Note that the exceptional divisor is defined by (x_1). The ideal in $\sigma^c(\mathcal{I})$ is generated by $\{\sigma^c(f) \mid f \in I\}$, where we write $\sigma^*(f) = x_1^m \cdot \sigma^c(f)$. Similarly, the ideal in $\sigma^c(D(\mathcal{I}))$ is generated by $x_1\sigma^c(f)$ and $\sigma^c(\partial f/\partial x_j)$, for $f \in I$ and $1 \leq j \leq n$, where $\sigma^*(\partial f/\partial x_j) = x_1^{m-1} \cdot \sigma^c(\partial f/\partial x_j)$.

Lemma 2.18. *With the above notation, we have*

$$\sigma^c\left(\frac{\partial f}{\partial x_j}\right) = x_1 \frac{\partial \sigma^c(f)}{\partial x_j} \text{ for } j > d, \tag{2.2}$$

$$\sigma^c\left(\frac{\partial f}{\partial x_j}\right) = \frac{\partial \sigma^c(f)}{\partial y_j} \text{ for } 2 \leq j \leq d, \tag{2.3}$$

$$\sigma^c\left(\frac{\partial f}{\partial x_1}\right) = m\sigma^c(f) + x_1 \frac{\partial \sigma^c(f)}{\partial x_1} - \sum_{j=2}^{d} y_j \frac{\partial \sigma^c(f)}{\partial y_j}. \tag{2.4}$$

PROOF. We prove these formulas by differentiating the equality

$$x_1^m \sigma^c(f) = f(x_1, x_1y_2, \ldots, x_1y_d, x_{d+1}, \ldots, x_n) \tag{2.5}$$

using the Chain Rule. Differentiating with respect to x_j, for $j > d$, gives

$$x_1^m \frac{\partial \sigma^c(f)}{\partial x_j} = \frac{\partial f}{\partial x_j}(x_1, x_1y_2, \ldots, x_1y_d, x_{d+1}, \ldots, x_n) = x_1^{m-1}\sigma^c\left(\frac{\partial f}{\partial x_j}\right).$$

If we differentiate (2.5) with respect to y_j, for $2 \leq j \leq d$ we get

$$x_1^m \frac{\partial \sigma^c(f)}{\partial y_j} = x_1 \cdot \frac{\partial f}{\partial x_j}(x_1, x_1y_2, \ldots, x_1y_d, x_{d+1}, \ldots, x_n) = x_1^m \sigma^c\left(\frac{\partial f}{\partial x_j}\right).$$

Differentiating (2.5) with respect to x_1 gives

$$m x_1^{m-1}\sigma^c(f) + x_1^m \frac{\partial \sigma^c(f)}{\partial x_1} = \frac{\partial f}{\partial x_1}(\ldots) + \sum_{j=2}^{d} y_j \cdot \frac{\partial f}{\partial x_j}(\ldots)$$

$$= x_1^{m-1}\sigma^c\left(\frac{\partial f}{\partial x_1}\right) + \sum_{j=2}^{d} y_j x_1^{m-1}\sigma^c\left(\frac{\partial f}{\partial x_j}\right).$$

We deduce the formulas in the lemma. □

We now return to the proof of Proposition 2.17. The formulas (2.2) and (2.3) show that $\sigma^c(\partial f/\partial x_j)$, for $j \geq 2$, lie in the ideal corresponding to $D(\sigma^c(\mathcal{I}))$. Moreover, since $\sigma^c(f)$ and $\frac{\partial \sigma^c(f)}{\partial x_1}$ lie in the ideal corresponding to $D(\sigma^c(\mathcal{I}))$, we see from (2.4) that also $\sigma^c(\partial f/\partial x_1)$ lies in this ideal. This proves $\sigma^c(D(\mathcal{I})) \subseteq D(\sigma^c(I))$, and completes the proof of the proposition. □

Remark 2.19. Note that the proof of the above proposition applies also when $i = m$. In other words, if $\sigma^c(\mathcal{I}) = (X_r, I_r, m)$, then $D^m(I) \cdot \mathcal{O}_{X_r} \subseteq D^m(I_r)$.

Corollary 2.20. *If $\mathcal{I} = (X, I, m)$ is a marked ideal, and σ is a sequence of admissible blow-ups for $\mathcal{I}$, then* $\mathrm{Supp}(\sigma^c(\mathcal{I})) \subseteq \mathrm{Supp}(\sigma^c(D^i(\mathcal{I}))$ *for every* $i \leq m-1$.

PROOF. The assertion follows from Proposition 2.17, since $\mathrm{Supp}(\sigma^c(\mathcal{I})) = \mathrm{Supp}(D(\sigma^c(\mathcal{I})) \subseteq \mathrm{Supp}(\sigma^c(D(\mathcal{I}))$. $\square$

The fact that in Proposition 2.17 ii) we do not have equality will cause some complications later. We will give in Proposition 3.13 below a formula that completely describes $D^i(\sigma^c(I))$, but this will also involve a log version of the derived ideals.

LECTURE 3

Hypersurfaces of maximal contact and coefficient ideals

The proof of principalization proceeds by induction on dimension of the ambient variety. In this lecture we introduce the fundamental concept of hypersurface of maximal contact. The idea is that resolving a given marked ideal on X is equivalent with resolving on the hypersurface of maximal contact the restriction of a related ideal on X. Describing this ideal, the coefficient ideal, is the content of the second part of the lecture. Our presentation follows [**Wło**], though the proof of a technical property of coefficient ideals is based on [**Kol**].

3.1. Hypersurfaces of maximal contact

In this section we consider a marked ideal $\mathcal{I} = (X, I, m)$ of *maximal order*. This means that $m \geq \max_{p \in X} \operatorname{ord}_p(I)$. Note that the equality being strict means precisely that $\operatorname{Supp}(\mathcal{I}) = \emptyset$. One can also rephrase the condition for $\mathcal{I}$ to be of maximal order by saying that $D^m(I) = \mathcal{O}_X$.

Given a marked ideal of maximal order $\mathcal{I} = (X, I, m)$, a *hypersurface of maximal contact* of $\mathcal{I}$ is a smooth hypersurface defined by a section of $T(\mathcal{I}) := D^{m-1}(I)$. Note that by our assumption on $\mathcal{I}$, for every $p \in X$ we have $\operatorname{ord}_p(D^{m-1}(I)) \leq 1$. It follows that for every $p \in X$, there is a hypersurface of maximal contact in some open neighborhood of p.

Example 3.1. Let (X, I, m) be a marked ideal with $\max_{p \in X} \operatorname{ord}_p(I) = m$. Suppose that $p \in X$, and f is a section of I with $\operatorname{ord}_p(f) = m$, such that in a system of coordinates $x_1, \ldots, x_n$ at p, f is a Weierstrass polynomial

$$f = x_n^m + \sum_{i=1}^{m} a_i(x_1, \ldots, x_{n-1}) x_n^{m-i}.$$

Since $\frac{\partial^{m-1} f}{\partial x_n^{m-1}} = m! \cdot x_n + (m-1)! \cdot a_1(x_1, \ldots, x_{n-1})$, we see that the hyperurface $H = \left(x_n = -\frac{1}{m} a_1(x_1, \ldots, x_{n-1})\right)$ is a hypersurface of maximal contact for (X, I, m) around p.

The fundamental idea in the resolution algorithm is to reduce resolving the ideal $\mathcal{I}$ on X to resolving another ideal on a hypersurface of maximal contact. We will discuss this ideal in the following section. However, since such a hypersurface might not exist globally, it follows that we need to argue locally. The fact that a hypersurface of maximal contact is not unique was the reason for much headache in the old versions of the algorithm. An idea from [**Wło**] greatly simplified the previous treatments, by replacing our given ideal with a "homogenized" version that is equivalent to the given ideal, but for whom "all choices of hypersurfaces of maximal contact are equivalent". We will explain this approach in the next lecture.

We now use Proposition 2.17 to give a first hint of the importance of hypersurfaces of maximal contact. Suppose that H is a hypersurface of maximal contact for $\mathcal{I} = (X, I, m)$. Since the equation defining H lies in $D^{m-1}(I)$, it follows that $H \supseteq \mathrm{Supp}(D^{m-1}(\mathcal{I})) = \mathrm{Supp}(\mathcal{I})$. Moreover, the following proposition shows that the analogous inclusion holds when considering a sequence of admissible blow-ups.

Proposition 3.2. *Let $\mathcal{I} = (X, I, m)$ be a marked ideal of maximal order, and suppose that H is a hypersurface of maximal contact for $\mathcal{I}$.*

i) *For every admissible blow-up $\sigma\colon X_1 = \mathrm{Bl}_C(X) \to X$, the center C is contained in H. Moreover, the marked ideal $\mathcal{I}_1 = \sigma^c(\mathcal{I})$ is of maximal order, and the proper transform H_1 of H is a hypersurface of maximal contact for $\mathcal{I}_1$.*
ii) *If σ is a sequence of admissible blow-ups for $\mathcal{I}$, then at every step the proper transform of H is a hypersurface of maximal contact for the corresponding controlled transform of $\mathcal{I}$, and it contains the blow-up center at that level.*

PROOF. It is clear that it is enough to prove i), since the assertion in ii) then follows by induction on the length of the sequence. In order to show that $\sigma^c(\mathcal{I})$ has maximal order, we need to show that $D^m(I_1) = \mathcal{O}_{X_1}$, where $\sigma^c(\mathcal{I}) = (I_1, m)$. Remark 2.19 gives

$$\mathcal{O}_{X_1} = D^m(I) \cdot \mathcal{O}_{X_1} \subseteq D^m(I_1),$$

which shows that (I_1, m) is of maximal order.

Suppose now that h is a local equation of H. The blow-up σ_1 is admissible, hence its center is contained in $\mathrm{Supp}(\mathcal{I}) \subseteq H$. Since H is smooth, we deduce that we can write $\sigma^*(h) = h_1 u$, where u is a local equation for the exceptional divisor, and h_1 is an equation for H_1. Therefore h_1 is a local section of $\sigma^c(D^{m-1}(I), 1)$, and this ideal is contained in $D^{m-1}(\sigma^c(\mathcal{I}))$ by Proposition 2.17. Since H_1 is smooth, as the proper transform of a smooth subvariety containing the blow-up center, it follows that H_1 is indeed a hypersurface of maximal contact for $\sigma^c(\mathcal{I})$. □

Exercise 3.3. If $f\colon Y \to X$ is a smooth morphism, and if H is a hypersurface of maximal contact for the marked ideal (X, I, m), show that $f^{-1}(H)$ is a hypersurface of maximal contact for $f^*(X, I, m)$.

3.2. The coefficient ideal

We now introduce a marked ideal equivalent with a given one, that has the property that its support restricts well to smooth hypersurfaces (and furthermore, the same property is propagated under sequences of admissible blow-ups). As a consequence, resolving the ideal we have started with is equivalent to resolving the restriction of this new ideal to a hypersurface of maximal contact.

Let us fix a marked ideal $\mathcal{I} = (I, m)$ on X. We are interested in restricting the marked ideal to smooth hypersurfaces in X. Once things are set up for marked ideals, dealing also with divisors causes no trouble.

Suppose that H is a smooth hypersurface in X, and consider the restriction $\mathcal{I}|_H := (I \cdot \mathcal{O}_H, m)$. It is clear that we have

$$\mathrm{Supp}(\mathcal{I}) \cap H \subseteq \mathrm{Supp}(\mathcal{I}|_H). \tag{3.1}$$

Indeed, let us choose local coordinates $x_1, \ldots, x_n$ at a point $p \in H$, such that $H = (x_1 = 0)$. If $f \in k[\![x_1, \ldots, x_n]\!]$ corresponds to a local section in I, and if all monomials in f have degree $\geq m$, then the same holds for the monomials in $f|_{x_1=0}$, which gives (3.1).

There is one case when we always have equality in (3.1): this happens when $m = 1$. For $m > 1$, in order to go in the direction of the converse, let us consider the Taylor expansion

$$f = \sum_{i \geq 0} \frac{1}{i!} \cdot (\partial^i f / \partial x_1^i)|_{x_1=0} x_1^i. \tag{3.2}$$

We see that $\mathrm{ord}_p(f) \geq m$ if and only if $\mathrm{ord}_p(\partial^i f/\partial x_1^i)|_H \geq m - i$ for $i \leq m-1$. This suggests the following definition.

3.2.1. Definition of coefficient ideal

The *coefficient ideal* $\mathcal{C}(I, m)$ of (I, m) is the sum of marked ideals

$$\mathcal{C}(I,m) := (I,m)^{m!/m} + D(I,m)^{m!/(m-1)} + \cdots + D^{m-1}(I,m)^{m!}.$$

Note that by Exercise 2.12 we have $\mathrm{Supp}(\mathcal{C}(I,m)) = \bigcap_{i=0}^{m-1} \mathrm{Supp}(D^i(I,m)) = \mathrm{Supp}(I,m)$. In fact, we will see in Proposition 3.5 below that $\mathcal{C}(I,m) \sim (I,m)$.

Proposition 3.4. *Let $\mathcal{I} = (I,m)$ be a marked ideal on X. For every smooth hypersurface H in X, we have*

$$\mathrm{Supp}(\mathcal{C}(\mathcal{I})) \cap H = \mathrm{Supp}(\mathcal{C}(\mathcal{I})|_H) = \mathrm{Supp}(\mathcal{I}) \cap H.$$

PROOF. The inclusion $\mathrm{Supp}(\mathcal{C}(\mathcal{I})) \cap H \subseteq \mathrm{Supp}(\mathcal{C}(\mathcal{I})|_H)$ follows by general considerations, see (3.1). On the other hand, the inclusion $\mathrm{Supp}(\mathcal{C}(\mathcal{I})|_H) \subseteq \mathrm{Supp}(\mathcal{I}) \cap H$ is a consequence of the Taylor expansion (3.2). Since we have already seen that $\mathcal{I}$ and $\mathcal{C}(\mathcal{I})$ have the same support, this proves the proposition. □

Proposition 3.5. *If $\mathcal{I} = (I,m)$ is a marked ideal on X, then $\mathcal{C}(\mathcal{I}) \sim \mathcal{I}$.*

PROOF. Since $\mathcal{C}(\mathcal{I}) \supseteq \mathcal{I}^{(m-1)!} \sim \mathcal{I}$, it follows that $\mathcal{I} \leq \mathcal{C}(\mathcal{I})$. If we show that $\mathcal{C}(\mathcal{I}) \leq \mathcal{I}$, then it follows that the two marked ideals are equivalent. Therefore it is enough to show that if σ is a sequence of $r \geq 0$ admissible blow-ups with respect to both $\mathcal{I}$ and $\mathcal{C}(\mathcal{I})$, then $\mathrm{Supp}(\sigma^c(I)) \subseteq \mathrm{Supp}(\sigma^c(\mathcal{C}(\mathcal{I})))$. When $r = 0$, this simply means that $\mathrm{Supp}(\mathcal{I}) \subseteq \mathrm{Supp}(\mathcal{C}(\mathcal{I}))$, which we have already noticed.

If $r > 0$, using the fact that the product and sum of marked ideals is compatible with controlled transform (see Exercises 2.8 and 2.9), we see that

$$\sigma^c(\mathcal{C}(\mathcal{I})) = \sum_{i=0}^{m-1} \sigma^c(D^i(\mathcal{I}))^{m!/i} \subseteq \sum_{i=0}^{m-1} D^i(\sigma^c(\mathcal{I}))^{m!/i} = \mathcal{C}(\sigma^c(\mathcal{I})),$$

where the inclusion follows from Proposition 2.17. Using the fact that $\sigma^c(\mathcal{I})$ and $\mathcal{C}(\sigma^c(\mathcal{I}))$ have the same support, we deduce that $\mathrm{Supp}(\sigma^c(\mathcal{I})) \subseteq \mathrm{Supp}(\sigma^c(\mathcal{C}(\mathcal{I})))$. □

For future reference, we record the assertion that we have seen in the above proof.

Corollary 3.6. *Let $\mathcal{I} = (I,m)$ be a marked ideal on X. If σ is a sequence of admissible blow-ups with respect to $\mathcal{I}$, then σ is also a sequence of admissible blow-ups with respect to $\mathcal{C}(\mathcal{I})$, and $\sigma^c(\mathcal{C}(\mathcal{I})) \subseteq \mathcal{C}(\sigma^c(\mathcal{I}))$.*

Remark 3.7. It follows from Exercise 2.16 and the definition that if $f\colon Y \to X$ is a smooth morphism, and (I,m) is a marked ideal on X, then $\mathcal{C}(f^*(I,m)) = f^*(\mathcal{C}(I,m))$.

3.2.2. Behavior with respect to sequences of admissible blow-ups
We now want to extend Proposition 3.4 to the case when we have in the picture a sequence of admissible blow-ups. Suppose that $\mathcal{I} = (I, m)$ is a marked ideal on X, and that H is a smooth hypersurface in X. Let τ be a sequence of blow-ups

$$H_r \stackrel{\tau_r}{\to} \cdots \stackrel{\tau_2}{\to} H_1 \stackrel{\tau_1}{\to} H = H_0,$$

where $H_{i+1} = \mathrm{Bl}_{C_i}(H_i)$, with each C_i smooth (in particular, each component of C_i is a proper subset of the corresponding component of H_i). Using the same centers, we get a sequence σ of blow-ups

$$X_r \stackrel{\sigma_r}{\to} \cdots \stackrel{\sigma_2}{\to} X_1 \stackrel{\sigma_1}{\to} X = X_0.$$

Proposition 3.8. *With the above notation, the following hold:*

i) *The sequence σ consists of admissible blow-ups with respect to $\mathcal{I}$ if and only if τ consists of admissible blow-ups with respect to $\mathcal{C}(\mathcal{I})|_H$.*
ii) *If the equivalent conditions in* i) *hold, then*

$$\mathrm{Supp}\,(\tau^c\,(\mathcal{C}(\mathcal{I})|_H)) = \mathrm{Supp}\,(\sigma^c\,(\mathcal{C}(\mathcal{I}))) \cap H_r = \mathrm{Supp}\,(\sigma^c(\mathcal{I})) \cap H_r.$$

Proof. It is enough to show that, assuming both conditions in i) hold, then the equalities in ii) hold as well. Indeed, then i) follows by induction on r, since ii) implies that for $C \subseteq H_r$ we have $C \subseteq \mathrm{Supp}\,(\tau^c(\mathcal{C}(\mathcal{I})|_H))$ if and only if $C \subseteq \mathrm{Supp}(\sigma^c(\mathcal{I}))$. Note also that the second equality in ii) follows from the fact that $\mathcal{I}$ and $\mathcal{C}(\mathcal{I})$ are equivalent.

The inclusion $\mathrm{Supp}\,(\tau^c\,(\mathcal{C}(\mathcal{I})|_H)) \supseteq \mathrm{Supp}\,(\sigma^c(\mathcal{C}(\mathcal{I}))) \cap H_r$ follows from the fact that $\tau^c(\mathcal{C}(\mathcal{I})|_H) = \sigma^c(\mathcal{C}(\mathcal{I}))|_{H_r}$ and (3.1). We only prove here the inclusion $\mathrm{Supp}(\sigma^c(\mathcal{C}(\mathcal{I}))|_{H_r}) \subseteq \mathrm{Supp}(\sigma^c(\mathcal{I}))$ in the special case $r = 1$. The general case is more subtle, and we will prove it in the next subsection, after some preparation.

Let $p \in \mathrm{Supp}(\sigma^c(\mathcal{C}(\mathcal{I}))|_{H_1})$. Since p lies on the proper transform of H, we may assume that we have coordinates $x_1, \ldots, x_n$ on X, such that C is defined by $(x_1, \ldots, x_d)$, H is defined by (x_d), and p lies in the chart with coordinates $x_1, y_2, \ldots, y_d, x_{d+1}, \ldots, x_n$, where $x_i = x_1 y_i$ for $2 \le i \le d$. Note that H_1 is defined by (y_d), and by assumption $d \ge 2$.

We need to show that for every $h \in I$, we have $\mathrm{ord}_p(\sigma^c(h)) \ge m$, where $\sigma^*(h) = x_1^m \sigma^c(h)$. Using the formula (3.2) for $f = \sigma^c(h)$ with respect to $H_1 \subset X_1$, we see that it is enough to show that $\mathrm{ord}_p\left(\frac{\partial^i f}{\partial y_d^i}|_{H_1}\right) \ge m - i$ for all $i \le m - 1$. Using the notation in Lemma 2.18, as well as formula (2.3), we see that

$$\frac{\partial^i f}{\partial y_d^i} = \sigma^c\left(\frac{\partial^i h}{\partial x_d^i}\right) \in \sigma^c(D^i(\mathcal{I})).$$

By assumption, $p \in \mathrm{Supp}(\sigma^c(\mathcal{C}(\mathcal{I}))|_{H_1})$, hence $p \in \mathrm{Supp}(\sigma^c(D^i(\mathcal{I}))|_{H_1})$ for $i \le m - 1$. We conclude that $\mathrm{ord}_p\left(\frac{\partial^i f}{\partial y_d^i}|_{H_1}\right) \ge m - i$ for $i \le m - 1$, and this completes the proof of the proposition in the case of one admissible blow-up. □

In practice we start with a marked ideal with divisor (X, I, m, E) of maximal order, and find (at least, locally) a hypersurface of maximal contact H for $\mathcal{I} = (I, m)$. Recall that $\mathcal{I}|_H = (I|_H, m)$. In order to restrict to H also the divisorial part, we need to assume that H has SNC with the sum of the components of E, and that none of the components of H appears among the components of E. In this case, if $E = \left(E^{(1)}, \ldots, E^{(\ell)}\right)$, we put $E|_H = \left(E^{(1)}|_H, \ldots, E^{(\ell)}|_H\right)$. We can not always choose directly a hypersurface of maximal contact with this property. However, we will see later how this can be achieved during the resolution algorithm.

For now, let us assume that H has this property. Furthermore, let us assume that no component of H is contained in $\mathrm{Supp}(\mathcal{I})$. We claim that under these assumptions we can identify the resolutions of $(X, \mathcal{I}, E)$ with the resolutions of $(H, \mathcal{C}(\mathcal{I})|_H, E|_H)$, such that the blow-up centers of the corresponding sequences are the same. Indeed, suppose first that σ is a sequence of admissible blow-ups $X_r \to \cdots \to X_1 \to X$ that resolves $(X, \mathcal{I}, E)$. Proposition 3.2 implies that each blow-up center $C_i \subseteq X_i$ is contained in the proper transform H_i of H, which is a hypersurface of maximal contact for (X_i, I_i, m, E_i). Furthermore, H_i again has simple normal crossings with the sum of the components of E_i, and no component of H_i is a component of E_i. Note also that no component of H_i is contained in $\mathrm{Supp}(I_i, m)$: otherwise the image of this component in X would give a component of H contained in $\mathrm{Supp}(I, m)$. Therefore each component of C_i is a proper subset of the corresponding component of H_i. We can therefore apply Proposition 3.8 to deduce that τ is a sequence of admissible blow-ups with respect to $\mathcal{C}(\mathcal{I})|_H$, and the support of $\tau^c(\mathcal{C}(\mathcal{I})|_H)$ is contained in the support of $\sigma^c(\mathcal{I})$, hence it is empty.

Conversely, suppose that τ is a sequence of admissible blow-ups $H_r \to \cdots \to H_1 \to H$ resolving $(H, \mathcal{C}(\mathcal{I})|_H, E|_H)$. Using the same sequence of centers we get a sequence σ of blow-ups $X_r \to \cdots \to X_1 \to X$. This is admissible with respect to $(X, \mathcal{I}, E)$ by Proposition 3.8. Since H is a hypersurface of maximal contact for $\mathcal{I}$, Proposition 3.2 implies that the support of $\sigma^c(\mathcal{I})$ is contained in H_r, hence it is equal to the support of $\tau^c(\mathcal{C}(\mathcal{I})|_H)$ by Proposition 3.8, and therefore it is empty.

This identification of resolutions is the main ingredient in the inductive approach to principalization. We still need to deal with the fact that hypersurfaces of maximal contact exist only locally. Therefore, in order to be able to patch together the local resolutions, we need to have independence on the choice of such hypersurface. This will be achieved in the next lecture by "homogenizing" (I, m), so that two choices of hypersurfaces of maximal contact differ only by a (formal, or étale) automorphism.

Remark 3.9. The key property of coefficient ideals has been formalized in [**Kol**], as follows. A marked ideal of maximal order (I, m) is *D-balanced* if $D^i(I)^m \subseteq I^{m-i}$ for every $i < m$. One can show that the coefficient ideal $\mathcal{C}(I, m)$ has this property. Furthermore, it is shown in *loc. cit.*, Theorem 3.84, that this property ensures good behavior with respect to restriction to a hypersurface (on the given variety, as well as after a sequence of admissible blow-ups).

3.2.3. Behavior of derived ideals under admissible blow-ups, revisited

Our goal now is to give a more precise formula for the derived ideals of the controlled transform of a marked ideal. We achieve this in terms of suitable log derived ideals, following [**Kol**].

As we have seen in the proof of the case $r = 1$ of Proposition 3.8, the key point was that certain derivatives behave well with respect to the controlled transform. In order to isolate these derivatives, one uses the notion of log derivations. Suppose that H is a smooth hypersurface on the smooth variety X. The sheaf of *log derivations along* H is the subsheaf $\mathrm{Der}(\mathcal{O}_X)(-\log H)$ of $\mathrm{Der}(\mathcal{O}_X)$ that consists of all derivations that preserve the ideal $\mathcal{O}_X(-H)$. If $x_1, \ldots, x_n$ are local coordinates on U such that H is defined by (x_1), then $\mathrm{Der}(\mathcal{O}_X)(-\log H)|_U$ is a free $\mathcal{O}_U$-module with basis $x_1\partial_1, \partial_2, \ldots \partial_n$, where we put $\partial_i := \frac{\partial}{\partial x_i}$.

If I is a (coherent) sheaf of ideals on X, then the *log derived* ideal $D(-\log H)(I)$ is the image of $\mathrm{Der}(\mathcal{O}_X)(-\log H)\otimes_k I \to \mathcal{O}_X$. We inductively define $D(-\log H)^i(I)$ in the obvious way. If $\mathcal{I} = (I, m)$ is a marked ideal, we put $D(-\log H)^i(\mathcal{I}) := (D(-\log H)^i(I), m-i)$ for $i \leq m-1$.

Exercise 3.10. Show that for every marked ideal $\mathcal{I} = (I, m)$ and every $i \leq m-1$, we have

$$D^i(\mathcal{I}|_H) = D(-\log H)^i(\mathcal{I})|_H.$$

Note that this is false if on the right-hand side we put $D^i(\mathcal{I})|_H$.

Exercise 3.11. Let I be an ideal on X, and $x_1, \ldots, x_n$ be coordinates on X such that H is defined by (x_1). For every $i \geq 0$, we denote by $\partial_1^i(I)$ the image of $\mathcal{O}_X\partial_1^i \otimes I \to \mathcal{O}_X$. If $\mathcal{I} = (I, m)$ is a marked ideal, and $i \leq m-1$, we put $\partial_1^i(I, m) := (\partial_1^i(I), m-i)$.

i) Show that $D(I) = D(-\log H)(I) + \partial_1(I)$.

ii) Deduce that for every marked ideal, and every $1 \leq i \leq m-1$, we have

$$D^i(I, m) = \sum_{j=0}^{i} \left(D(-\log H)^{i-j}(\partial_1^j(I, m))\right).$$

Exercise 3.12. Use the formulas in Lemma 2.18 to deduce the following log version of Proposition 2.17. Let $\mathcal{I} = (I, m)$ be a marked ideal, H a smooth hypersurface in X, and $i \leq m-1$. If $\sigma\colon X_1 = \mathrm{Bl}_C(X) \to X$ is an admissible blow-up with respect to $\mathcal{I}$, with $C \subseteq H$ and $\mathrm{codim}(C, X) \geq 2$, then

$$\sigma^c\left(D(-\log H)^i(\mathcal{I})\right) \subseteq D(-\log H_1)^i(\sigma^c(\mathcal{I})),$$

where H_1 is the proper transform of H.

Our main interest in log derived ideals comes from the following description of derived ideals of controlled transforms.

Proposition 3.13. *Let $\mathcal{I} = (X, I, m)$ be a marked ideal, and let $i \leq m-1$. If H is a smooth hypersurface in X, and if $\sigma = (\sigma_1, \ldots, \sigma_r)$ is a sequence of admissible blow-ups with respect to $\mathcal{I}$, such that the center of each σ_p has codimension ≥ 2, and is contained in the proper transform $H_{p-1} \subset X_{p-1}$ of H, then we have*

$$D^i(\sigma^c(\mathcal{I})) = \sum_{j=0}^{i} D(-\log H_r)^{i-j}\left(\sigma^c(D^j(\mathcal{I}))\right).$$

Proof. The inclusion "$\supseteq$" is easy: by Proposition 2.17 we have $\sigma^c(D^j(\mathcal{I})) \subseteq D^j(\sigma^c(\mathcal{I}))$, hence

$$D(-\log H_r)^{i-j}\left(\sigma^c(D^j(\mathcal{I}))\right) \subseteq D^{i-j}(D^j(\sigma^c(\mathcal{I})) = D^i(\sigma^c(\mathcal{I})).$$

We now prove the reverse inclusion by induction on r. If $r = 1$, then let $x_1, \ldots, x_n$ be local coordinates on X such that the center of σ_1 is defined by $(x_1, \ldots, x_d)$, and H is defined by (x_d). The inclusion to prove is trivial away from the proper transform H_1 (consider $j = 0$), hence we only need to consider charts that intersect H_1. A typical such chart has coordinates $x_1, y_2, \ldots, y_d, x_{d+1}, \ldots, x_n$, with $x_j = x_1y_j$ for $2 \leq j \leq d$, and in this chart H_1 is defined by (y_d). The formula in Exercise 3.11 gives

$$D^i(\sigma^c(\mathcal{I})) = \sum_{j=0}^{i} D(-\log H_1)^{i-j}\left(\partial_{y_d}^j(\sigma^c(\mathcal{I}))\right),$$

and the formula (2.3) in Lemma 2.18 implies $\partial^j_{y_d}(\sigma^c(\mathcal{I})) = \sigma^c(\partial^j_{x_d}(\mathcal{I})) \subseteq \sigma^c(D^j(\mathcal{I}))$. This gives the inclusion "$\subseteq$" in the proposition when $r = 1$.

Suppose now that $r > 1$, and let τ denote the sequence $(\sigma_1, \ldots, \sigma_{r-1})$. Applying what we have already proved for σ_r gives

$$D^i(\sigma^c(I)) \subseteq \sum_{\ell=0}^{i} D(-\log H_r)^{i-\ell}\left(\sigma_r^c(D^\ell(\tau^c(\mathcal{I})))\right),$$

while the inductive hypothesis for τ and Exercise 3.12 give

$$\begin{aligned}\sigma_r^c\left(D^\ell(\tau^c(\mathcal{I}))\right) &\subseteq \sum_{j=0}^{\ell} \sigma_r^c\left(D(-\log H_{r-1})^{\ell-j}\left(\tau^c(D^j(\mathcal{I}))\right)\right)\\ &\subseteq \sum_{j=0}^{\ell} D(-\log H_r)^{\ell-j}\left(\sigma^c(D^j(\mathcal{I}))\right).\end{aligned}$$

Combining these two formulas gives the inclusion "$\subseteq$" in the statement. □

Corollary 3.14. *Under the assumptions in the proposition, we have an induced sequence τ of admissible blow-ups $H_r \to \cdots \to H$ with respect to each $D^j(\mathcal{I}|_H)$, for $j \leq m-1$, and*

$$\text{Supp}\left(\sigma^c(\mathcal{I})\right) \cap H_r = \bigcap_{j=0}^{m-1} \text{Supp}\left(\tau^c(D^j(\mathcal{I})|_H)\right). \tag{3.3}$$

Proof. Note first that (3.3) holds when $r = 0$ by Proposition 3.4). In order to prove the corollary, it is enough to show that if the first assertion holds, then the second one does, as well (then the first assertion follows by induction on r). We deduce from Proposition 3.13 that

$$D^{m-1}(\sigma^c(\mathcal{I})) = \sum_{j=0}^{m-1} D(-\log H_r)^{m-1-j}\left(\sigma^c(D^j(\mathcal{I}))\right). \tag{3.4}$$

Since $D^{m-1}(\sigma^c(\mathcal{I}))$ has marking 1, it follows that taking its support commutes with restricting to H_r, hence

$$\text{Supp}(\sigma^c(I)) \cap H_r = \text{Supp}\left(D^{m-1}(\sigma^c(\mathcal{I}))\right) \cap H_r = \text{Supp}\left(D^{m-1}(\sigma^c(\mathcal{I}))|_{H_r}\right).$$

On the other hand, if we restrict the right-hand side of (3.4) to H_r, we see using Exercise 3.10 that we get

$$\sum_{j=0}^{m-1} D^{m-1-j}\left(\sigma^c(D^j(\mathcal{I}))|_{H_r}\right) = \sum_{j=0}^{m-1} D^{m-1-j}\left(\tau^c(D^j(\mathcal{I})|_H)\right).$$

The support of this sum is equal to $\bigcap_{j=0}^{m-1} \text{Supp}\left(\tau^c(D^j(\mathcal{I})|_H)\right)$, hence the assertion in the corollary. □

We can now prove in general the fact that taking the support of the controlled transform of the coefficient ideal commutes with restricting to the proper transform on a sequence of admissible blow-ups.

PROOF OF PROPOSITION 3.8, CONTINUED. In order to complete the proof, it is enough to show that if σ is a sequence of admissible blow-ups with respect to $\mathcal{I} = (I, m)$

$$X_r \overset{\sigma_r}{\to} \cdots \overset{\sigma_2}{\to} X_1 \overset{\sigma_1}{\to} X_0 = X,$$

such that for every i the center C_{i-1} of σ_i has codimension ≥ 2 in X, and is contained in the proper transform H_{i-1} of H, then

$$\operatorname{Supp}\left(\sigma^c(\mathcal{C}(\mathcal{I}))|_{H_r}\right) \subseteq \operatorname{Supp}(\sigma^c(\mathcal{I})) \cap H_r. \tag{3.5}$$

Note that the left-hand side of (3.5) is equal to

$$\operatorname{Supp}\left(\sum_{j=0}^{m-1} \left(\sigma^c(D^j(\mathcal{I}))|_{H_r}\right)^{m!/j}\right) = \bigcap_{j=0}^{m-1} \operatorname{Supp}\left(\sigma^c(D^j(\mathcal{I}))|_{H_r}\right)$$
$$= \bigcap_{j=0}^{m-1} \operatorname{Supp}\left(\tau^c(D^j(\mathcal{I})|_H)\right).$$

The assertion now follows from Corollary 3.14. □

LECTURE 4

Homogenized ideals

The construction we discuss in this lecture replaces a given marked ideal with a "homogenized" one, which has the property that "it looks the same" with respect to any choice of hypersurface of maximal contact (up to étale transformations). This allows us to deal easily with the non-uniqueness of hypersurfaces of maximal contact. In this presentation we follow [**Wło**].

In practice, we will replace a given marked ideal (I, m) with the coefficient ideal of the homogenization of (I, m). Instead of doing this in two steps, one can do this in one step, as in [**Kol**], by using a slightly different ideal, that enjoys the good properties of both the coefficient ideal, and of the homogenization. However, for pedagogical reasons, we have decided to use the two-step construction, in order to emphasize the two ideas involved.

4.1. Basics of homogenized ideals

4.1.1. The definition of the homogenized ideal

Let $\mathcal{I} = (I, m)$ be a marked ideal of maximal order on X. Recall that $T(\mathcal{I})$ denotes the ideal $D^{m-1}(I)$, such that hypersurfaces of maximal contact for $\mathcal{I}$ are smooth hypersurfaces defined by elements of $T(\mathcal{I})$. We would like to replace $\mathcal{I}$ by another marked ideal $\mathcal{H}(\mathcal{I})$ that is equivalent to $\mathcal{I}$, and which has the following property. Given two hypersurfaces of maximal contact H_1 and H_2 for $\mathcal{H}(\mathcal{I})$ in some open subset $U \subseteq X$, there is an automorphism of U that takes $\mathcal{H}(\mathcal{I})$ to itself, and H_1 to H_2. This might be impossible to achieve, since there might not be nontrivial such automorphisms. However, it turns out that for our purpose it will be enough to find automorphisms of the completion $\widehat{\mathcal{O}_{X,p}}$ with the analogous properties, for every $p \in \mathrm{Supp}(\mathcal{I})$. Indeed, we will show that such automorphisms extend to automorphisms of suitable étale neighborhoods of p. The fact that our resolution algorithm constructed in lower dimension is functorial with respect to étale morphisms will allow us to compare the resolutions of the restrictions to the two hypersurfaces of maximal contact.

In order to motivate the definition of $\mathcal{H}(\mathcal{I})$, note that we want to enlarge (I, m) to an ideal (J, m) such that if $p \in \mathrm{Supp}(\mathcal{I})$, and $h_1, h_2 \in T(\mathcal{I})_p$ are local equations of hypersurfaces of maximal contact, then there is an automorphism ϕ of $\widehat{\mathcal{O}_{X,p}}$ such that $\phi(J) = J$ and $\phi(h_1) = h_2$. Let us write $h_2 = h_1 + w$. Since hypersurfaces of maximal contact are smooth, it follows that we can find regular functions $x_1, \ldots, x_{n-1}$ around p, where $n = \dim(\widehat{\mathcal{O}_{X,p}})$, such that both $x_1, \ldots, x_{n-1}, h_1$, and $x_1, \ldots, x_{n-1}, h_2$ generate the maximal ideal in $\widehat{\mathcal{O}_{X,p}}$. It follows that we can define an automorphism ϕ of $\widehat{\mathcal{O}_{X,p}}$ such that $\phi(x_i) = x_i$ for all i, and $\phi(h_1) = h_2$.

Note that given $f \in I$, its image in $\widehat{\mathcal{O}_{X,p}}$ can be written as $F(x_1, \ldots, x_{n-1}, h_1)$ for some formal power series F. In this case $\phi(f)$ is equal to

$$F(x_1, \ldots, x_{n-1}, h_1 + w) = \sum_{i \in \mathbf{N}} \frac{1}{i!} \cdot \frac{\partial^i F}{\partial x_n^i}(x_1, \ldots, x_{n-1}, h_1) w^i.$$

Since we need $\phi(J) \subseteq J$, it follows that we need at least $\phi(f) \in J$ for every $f \in I$. The above Taylor expansion shows that we have this property if $D^i(I) \cdot T(\mathcal{I})^i \subseteq J$ for every i.

Remark 4.1. We need to put the above condition only for $i \leq m - 1$, since for $i \geq m$ we have $D^i(I) \cdot T(\mathcal{I})^i \subseteq T(\mathcal{I})^m = D^{m-1}(I) \cdot T(\mathcal{I})^{m-1}$.

Let $\mathcal{I} = (I, m)$ be a marked ideal of maximal order. The *homogenized marked ideal* $\mathcal{H}(\mathcal{I})$ is defined by

$$\mathcal{H}(\mathcal{I}) = \sum_{i=0}^{m-1} D^i(\mathcal{I}) \cdot (T(\mathcal{I}), 1)^i = \left(\sum_{i=0}^{m-1} D^i(I) \cdot T(\mathcal{I})^i, m \right).$$

Note that if $m = 1$, then $\mathcal{H}(\mathcal{I}) = (T(\mathcal{I}), 1) = \mathcal{I}$.

4.1.2. Properties of the homogenized ideal

Proposition 4.2. *The two marked ideals $\mathcal{I}$ and $\mathcal{H}(\mathcal{I})$ are equivalent.*

PROOF. Since the inclusion $\mathcal{I} \subseteq \mathcal{H}(\mathcal{I})$ is clear from definition, it is enough to show that $\mathcal{H}(\mathcal{I}) \leq \mathcal{I}$. We need to show that if σ is a sequence of admissible blow-ups with respect to both $\mathcal{I}$ and $\mathcal{H}(\mathcal{I})$ (possibly of length zero), then $\mathrm{Supp}(\sigma^c(\mathcal{I})) \subseteq \mathrm{Supp}(\sigma^c(\mathcal{H}(\mathcal{I})))$. It follows from the properties of sums and products of marked ideals that

$$\mathrm{Supp}(\sigma^c(\mathcal{H}(\mathcal{I}))) = \bigcap_{i=0}^{m-1} \mathrm{Supp}\left(\sigma^c(D^i(\mathcal{I})) \cdot \sigma^c(T(\mathcal{I})^i, i)\right)$$

$$\supseteq \bigcap_{i=0}^{m-1} \left(\mathrm{Supp}\left(\sigma^c(D^i(I))\right) \cap \mathrm{Supp}\left(\sigma^c(T(\mathcal{I}), 1)\right)\right).$$

On the other hand, it follows from Corollary 2.20 that

$$\mathrm{Supp}(\sigma^c(\mathcal{I})) \subseteq \mathrm{Supp}(\sigma^c(D^i(\mathcal{I}))) \quad \text{for } i \leq m - 1.$$

In particular, $\mathrm{Supp}(\sigma^c(\mathcal{I})) \subseteq \mathrm{Supp}(\sigma^c(T(\mathcal{I}), 1))$. We conclude that $\mathrm{Supp}(\sigma^c(\mathcal{I})) \subseteq \mathrm{Supp}(\sigma^c(\mathcal{H}(\mathcal{I})))$. □

Lemma 4.3. *If $\mathcal{I} = (I, m)$ is a marked ideal of maximal order, then $D^i(\mathcal{H}(\mathcal{I})) \subseteq \mathcal{H}(D^i(\mathcal{I}))$ for every $i \leq m - 1$.*

PROOF. Note first that for all such i, the marked ideal $D^i(\mathcal{I})$ is of maximal order, and $T(D^i(\mathcal{I})) = T(\mathcal{I})$. It is enough to prove the lemma when $i = 1$, as the general case then follows immediately by induction on i. In this case the assertion follows from the Product Formula:

$$D(\mathcal{H}(I, m)) = \left(\sum_{j=0}^{m-1} D(D^i(I) \cdot T(\mathcal{I})^i), m - 1 \right)$$

$$\subseteq \left(\sum_{i=0}^{m-1} D^{i+1}(I) \cdot T(\mathcal{I})^i, m - 1 \right) = \mathcal{H}(D(\mathcal{I})).$$

□

Corollary 4.4. *If $\mathcal{I}$ is a marked ideal of maximal order, then $T(\mathcal{H}(\mathcal{I})) = T(\mathcal{I})$.*

Proof. Since $\mathcal{I} \subseteq \mathcal{H}(\mathcal{I})$, it is clear that $T(\mathcal{I}) \subseteq T(\mathcal{H}(\mathcal{I}))$. On the other hand, if we apply the lemma for $i = m-1$, we get $D^{m-1}(\mathcal{H}(\mathcal{I})) \subseteq \mathcal{H}(T(\mathcal{I}), 1) = (T(\mathcal{I}), 1)$, hence $T(\mathcal{H}(\mathcal{I})) \subseteq T(\mathcal{I})$. □

Corollary 4.5. *If $\mathcal{I}$ is a marked ideal of maximal order, then $\mathcal{H}(\mathcal{H}(\mathcal{I})) = \mathcal{H}(\mathcal{I})$.*

Proof. We just need to show that $\mathcal{H}(\mathcal{H}(\mathcal{I})) \subseteq \mathcal{H}(\mathcal{I})$. Since $T(\mathcal{H}(\mathcal{I})) = T(\mathcal{I})$, we see that we need to show that for every $i \geq 1$, we have $D^i(\mathcal{H}(\mathcal{I})) \cdot T(\mathcal{I})^i \subseteq \sum_{j\geq 0} D^i(\mathcal{I}) \cdot T(\mathcal{I})^i$. As in the proof of Lemma 4.3, one sees using the Product Rule that $D^i(\mathcal{H}(\mathcal{I})) \subseteq \sum_{j\geq 0} D^{i+j}(I) \cdot T(\mathcal{I})^j$. We deduce that $D^i(\mathcal{H}(\mathcal{I})) \cdot T(\mathcal{I})^i \subseteq \sum_{j\geq i} D^j(\mathcal{I}) \cdot T(\mathcal{I})^j \subseteq \mathcal{H}(\mathcal{I})$. □

Remark 4.6. It follows from Exercise 2.16 and the definition that if $f\colon Y \to X$ is a smooth morphism, and if (I, m) is a marked ideal of maximal order on X, then $\mathcal{H}(f^*(I, m)) = f^*(\mathcal{H}(I, m))$.

Exercise 4.7. Let $\mathcal{I} = (I, 2)$ be a marked ideal.

i) Show that $\mathcal{C}(\mathcal{I}) = (I + D(I)^2, 2)$.
ii) Show that $\mathcal{H}(\mathcal{I}) = (I + D(I)^2, 2)$.
iii) Deduce that $\mathcal{C}(\mathcal{H}(\mathcal{I})) = (I + D(I)^2, 2)$.

4.2. Comparing hypersurfaces of maximal contact: formal equivalence

We can now prove the main result about homogenized ideals. With a slight abuse of notation, we sometimes use the same letter to denote both a marked ideal, and the corresponding ideal. We hope that the meaning will be clear from the context.

Proposition 4.8. *Let $\mathcal{I} = (I, m)$ be a marked ideal of maximal order on X. If $p \in \mathrm{Supp}(\mathcal{I})$, and if $h_1, h_2 \in T(\mathcal{I})_p \subseteq \mathcal{O}_{X,p}$ define hypersurfaces of maximal contact H_1, H_2 around p, then there is an automorphism ϕ of $\widehat{\mathcal{O}_{X,p}}$ such that*

i) *$\phi(\widehat{\mathcal{H}(\mathcal{I})_p}) = \widehat{\mathcal{H}(\mathcal{I})_p}$.*
ii) *$\phi(h_1) = h_2$.*
iii) *$\phi(w) - w \in \widehat{T(\mathcal{I})_p}$ for every $w \in \widehat{\mathcal{O}_{X,p}}$. In particular, if J is the image in $\widehat{\mathcal{O}_{X,p}}$ of the ideal defining $\mathrm{Supp}(I, m)$ (with the reduced structure), then ϕ induces the identity on $\widehat{\mathcal{O}_{X,p}}/J$.*

Furthermore, if we have a marked ideal with divisor $(\mathcal{I}, E)$, and if around p each H_i has SNC with the sum of the divisors in E, and is not contained in this sum, then we may assume that ϕ leaves invariant the images in $\widehat{\mathcal{O}_{X,p}}$ of the equations of the divisors in E.

Proof. We construct ϕ as described in §4.1.1: if $\dim(\widehat{\mathcal{O}_{X,p}}) = n$, we choose $x_1, \ldots, x_{n-1}$ such that the maximal ideal in $\widehat{\mathcal{O}_{X,p}}$ is generated by $x_1, \ldots, x_{n-1}, h_1$, and also by $x_1, \ldots, x_{n-1}, h_2$. The automorphism ϕ is defined by $\phi(x_i) = x_i$ for every i, and $\phi(h_1) = h_2$. We have seen that $\phi(\widehat{I_p}) \subseteq \widehat{\mathcal{H}(\mathcal{I})_p}$, and applying the same argument for $\mathcal{H}(\mathcal{I})$, we get that $\phi(\widehat{\mathcal{H}(\mathcal{I})_p})$ is contained in the completion of $\mathcal{H}(\mathcal{H}(\mathcal{I}))_p$, which is equal to the completion of $\mathcal{H}(\mathcal{I})_p$ by Corollary 4.5. This gives the inclusion "$\subseteq$" in i), and applying the same argument for ϕ^{-1} we get the reverse inclusion.

Note that ii) follows from the definition of ϕ. Since $\phi(h_1)-h_1 = h_2-h_1 \in \widehat{T(\mathcal{I})_p}$, and $\phi(x_i) - x_i = 0$ for every i, it follows that $\phi(w) - w \in \widehat{T(\mathcal{I})_p}$ for every $w \in \widehat{\mathcal{O}_{X,p}}$ (since $\widehat{\mathcal{O}_{X,p}}$ is a formal power series ring in $x_1, \ldots, x_{n-1}, h_1$, we just need to check the statement for these elements). For the last assertion in the proposition, it is enough to note that under those assumptions, we may choose the x_i such that any component of a divisor in E passing through p is defined by one of the x_i. □

4.3. Comparing hypersurfaces of maximal contact: étale equivalence

For applications, we will need a version of the above proposition in which we pass from the formal neighborhood to an étale neighborhood. Suppose that (X, I, m, E) is a marked ideal with divisor of maximal order, and suppose that H_1 and H_2 are hypersurfaces of maximal contact for (I, m) such that each H_i has SNC with the sum of the divisors in E, and no component of H_i is contained in this sum.

Proposition 4.9. *With the above notation, there is a marked ideal with divisor* $(\overline{X}, \overline{I}, m, \overline{E})$ *and two étale maps* $f, g \colon \overline{X} \to X$ *whose images contain an open neighborhood of* $\mathrm{Supp}(I, m)$ *such that*

i) $f^*(X, \mathcal{H}(I, m), E) = (\overline{X}, \overline{I}, m, \overline{E}) = g^*(X, \mathcal{H}(I, m), E)$.

ii) $f^{-1}(H_1) = g^{-1}(H_2) =: \overline{H}$ *is a hypersurface of maximal contact for* $(\overline{X}, \overline{I}, m, \overline{E})$. *It has SNC with the sum of the divisors in* $\overline{E}$, *and no component of* $\overline{H}$ *is contained in this sum.*

iii) *For every* $\overline{q} \in \mathrm{Supp}(\overline{I}, m)$, *we have* $f(\overline{q}) = g(\overline{q}) =: q$. *Moreover, if* $w \in \mathcal{O}_{X,q}$, *then* $f^*(w) - g^*(w) \in T(\overline{I}, m)$ *in a neighborhood of* $\overline{q}$.

PROOF. It is enough to show that for every $p \in \mathrm{Supp}(I, m)$, we can find $(\overline{X}, \overline{I}, m, \overline{E})$, and étale morphisms f and g satisfying i), ii), and iii), and such that p lies in the image of f and g. Indeed, then the disjoint union of finitely many such $\overline{X}$, with images covering $\mathrm{Supp}(I, m)$, satisfies the conditions in the proposition.

Let h_1 and h_2 be local equations for H_1 and H_2 at p (note that since p lies in the support of (I, m), it lies on every hypersurface of maximal contact). In order to construct $\overline{X}$, we start as in the proof of Proposition 4.8. Let $x_1, \ldots, x_{n-1}$ be regular functions on an affine open neighborhood U of p, vanishing at p, such that both $x_1, \ldots, x_{n-1}, h_1$, and $x_1 \ldots, x_{n-1}, h_2$ give systems of coordinates on U. We may further assume that for every component of E intersecting U, its equation in U is given by one of the x_i. We define morphisms $\alpha, \beta \colon U \to \mathbf{A}^n$ such that if $y_1, \ldots, y_n$ are the standard coordinates on $\mathbf{A}^n$, we have $\alpha^*(y_i) = x_i = \beta^*(y_i)$ for $i \leq n-1$, and $\alpha^*(y_n) = h_1$ and $\beta^*(y_n) = h_2$. Our assumption implies that both α and β are étale, hence if we consider the Cartesian diagram

$$\begin{array}{ccc} Y = U \times_{\mathbf{A}^n} U & \xrightarrow{g} & U \\ {\scriptstyle f}\downarrow & & \downarrow{\scriptstyle \beta} \\ U & \xrightarrow{\alpha} & \mathbf{A}^n \end{array}$$

both f and g are étale. In particular, Y is smooth. Let $\overline{p} \in Y$ be the point corresponding to (p, p). Since f and g are étale, we see that f and g induce isomorphisms $\phi_1, \phi_2 \colon \widehat{\mathcal{O}_{X,p}} \to \widehat{\mathcal{O}_{Y,\overline{p}}}$. Note that the automorphism ϕ we constructed in the proof of Proposition 4.8 is $\phi_2^{-1} \circ \phi_1$. If two ideals on Y have the same image in $\widehat{\mathcal{O}_{Y,\overline{p}}}$, then the two ideals are equal in some open neighborhood of $\overline{p}$. We deduce from Proposition 4.8 that if we replace Y by a suitable open neighborhood $\overline{X}$ of $\overline{p}$, then we have a marked ideal with divisor $(\overline{X}, \overline{I}, m, \overline{E})$ such that i) holds. Similarly, after possibly replacing $\overline{X}$ by a smaller open subset, we have $\overline{H} := f^{-1}(H_1) = g^{-1}(H_2)$. Moreover, $\overline{H}$ is smooth since H is smooth and f is étale; we may also assume that it is defined by a section of $T(\overline{I}, m)$, since ϕ_1 and ϕ_2 pull-back $\widehat{T(I,m)_p}$ to $\widehat{T(\overline{I},m)_{\overline{p}}}$. Therefore $\overline{H}$ is a hypersurface of maximal contact for $(\overline{X}, \overline{I}, m, \overline{E})$. The last assertion in ii) follows from the fact that f is étale, and H has the corresponding properties with respect to the sum of the divisors in E.

In order to prove iii), let us choose generators $w_1, \ldots, w_s$ of $\mathcal{O}_X(U)$ as a k-algebra. Using one more time Proposition 4.8 iii), after possibly replacing $\overline{X}$ by an open neighborhood of $\overline{p}$, we may assume that each $f^*(w_i) - g^*(w_i)$ is a global section of $T(\overline{I}, m)$. If $\overline{q} \in \operatorname{Supp}(\overline{I}, m)$, then

$$w_i(f(\overline{q})) = f^*(w_i)(\overline{q}) = g^*(w_i)(\overline{q}) = w_i(g(\overline{q}))$$

for all i, hence $f(\overline{q}) = g(\overline{q})$.

Furthermore, we see that $f^*(w) - g^*(w)$ is a global section of $T(\overline{I}, m)$ for every $w \in \mathcal{O}_X(U)$. For the second property in iii), it is enough to show that given $w, w' \in \mathcal{O}_X(U)$, the difference $f^*(w/w') - g^*(w/w')$ is a section of $T(\overline{I}, m)$ over the open subset $f^{-1}(w' \neq 0) \cap g^{-1}(w' \neq 0)$. If we write $f^*(w) = g^*(w) + b$ and $f^*(w') = g^*(w') + b'$, with b, b' global sections of $T(\overline{I}, m)$, then

$$f^*(w/w') - g^*(w/w') = \frac{g^*(w) + b}{g^*(w') + b'} - \frac{g^*(w)}{g^*(w')} = \frac{b}{f^*(w')} - \frac{b' g^*(w)}{f^*(w') g^*(w')}, \tag{4.1}$$

and this is a section of $T(\overline{I}, m)$ over $f^{-1}(w' \neq 0) \cap g^{-1}(w' \neq 0)$. □

Remark 4.10. In [**Kol**] one formalizes the main property of homogenized ideals as *MC-invariance*. A marked ideal of maximal order (I, m) is MC-invariant if $T(I, m) \cdot D(I) \subseteq I$. It is clear that $\mathcal{H}(I, m)$ is MC-invariant, and it is shown in *loc. cit.*, Theorem 3.92, that MC-invariant ideals satisfy the analogue of Proposition 4.8.

We will apply the above proposition as follows. Given two hypersurfaces of maximal contact for (X, I, m, E), we will use them to construct two resolutions for this marked ideal with divisor. Since the two hypersurfaces pull-back to the same hypersurface on $\overline{X}$, induction will allow us to conclude that the two sequences of blow-ups pull-back to the same sequence on $\overline{X}$. The following proposition will then allow us to conclude that they already agree on X.

Proposition 4.11. *Let (X, I, m) be a marked ideal of maximal order, and consider two sequences σ and σ' of admissible blow-ups. Suppose that $f, g\colon \overline{X} \to X$ are two étale morphisms, whose images contain some neighborhood of* $\operatorname{Supp}(I, m)$, *such that*

i) $f^*(I) = g^*(I) =: \overline{I}$.
ii) *For every $\overline{q} \in \operatorname{Supp}(\overline{I}, m)$, we have $f(\overline{q}) = g(\overline{q}) =: q$. Moreover, if $w \in \mathcal{O}_{X,q}$ then $f^*(w) - g^*(w) \in T(\overline{I}, m)$ in some neighborhood of $\overline{q}$.*
iii) $f^*(\sigma) = g^*(\sigma')$.

In this case $\sigma = \sigma'$.

PROOF. It is clear that the two sequences have the same length, and we do induction on this length. To simplify the notation, we only treat the first step in the proof, but the induction then follows in a straightforward manner.

Suppose that C and C' are the first centers of blow-up in σ and σ', respectively. By assumption, we have $C, C' \subseteq \mathrm{Supp}(I, m)$. Since $f = g$ on

$$\mathrm{Supp}(\overline{I}, m) = f^{-1}(\mathrm{Supp}(I, m)) = g^{-1}(\mathrm{Supp}(I, m)),$$

and $f^{-1}(C) = g^{-1}(C')$, we conclude that $C = C'$. If $X_1 = \mathrm{Bl}_C(X)$ and $\overline{X}_1 = \mathrm{Bl}_{f^{-1}(C)}(\overline{X})$, then we have a Cartesian square

$$\begin{array}{ccc} \overline{X}_1 & \xrightarrow{\overline{\sigma}_1} & \overline{X} \\ {\scriptstyle f_1}\big\downarrow & & \big\downarrow{\scriptstyle f} \\ X_1 & \xrightarrow{\sigma_1} & X \end{array}$$

and a similar one, with f and f_1 replaced by g and g_1, respectively. Note that f_1 and g_1 are étale, and their images contain an open neighborhood of $\sigma_1^{-1}(\mathrm{Supp}(I, m))$, hence of $\mathrm{Supp}(I_1, m)$, where I_1 is the controlled transform of I. Since the exceptional divisor of $\overline{\sigma}_1$ is the pull-back by f_1 (or g_1) of the exceptional divisor of σ_1, we deduce from i) that $f_1^*(I_1) = g_1^*(I_1) = \overline{I}_1$, where $(\overline{I}_1, m) = \overline{\sigma}_1^c(\overline{I}, m)$. In order to be able to continue by induction, it is enough to show that f_1 and g_1 also satisfy condition ii).

Given $\overline{q} \in \mathrm{Supp}(\overline{I}_1, m)$, let $\overline{p} = \overline{\sigma}_1(\overline{q}) \in \mathrm{Supp}(\overline{I}, m)$, and $p = f(\overline{p}) = g(\overline{p})$. Choose local coordinates $x_1, \ldots, x_n$ in some affine open neighborhood W of p, with all $x_i(p) = 0$, and such that C is defined by the ideal $(x_1, \ldots, x_d)$. By assumption, in some neighborhood of $\overline{p}$ we can write $f^*(x_i) - g^*(x_i) = b_i \in T(\overline{I}, m)$.

After relabeling the x_i, we may assume that both $f_1(\overline{q})$ and $g_1(\overline{q})$ lie in the chart W_1 on X_1 with coordinates $y_1, \ldots, y_{d-1}, x_d, \ldots, x_n$, with $x_i = x_d y_i$ for $1 \leq i \leq d-1$. If we show that for all $j \geq d$ and all $i \leq d-1$

$$f_1^*(x_j) - g_1^*(x_j), f_1^*(y_i) - g_1^*(y_i) \in T(\overline{I}_1, m) \tag{4.2}$$

in some neighborhood of $\overline{q}$, then we are done. Indeed, we first deduce that these differences vanish at $\overline{q}$, and this gives $f_1(\overline{q}) = g_1(\overline{q}) =: q$. Moreover, since $\mathcal{O}_{X_1}(W_1)$ is generated over $\mathcal{O}_X(W)$ by $y_1, \ldots, y_{d-1}, x_d, \ldots, x_n$, it follows that for every $w \in \mathcal{O}_{X_1}(W_1)$ we have $f_1^*(w) - g_1^*(w) \in T(\overline{I}_1, m)$ in some neighborhood of $\overline{q}$. The fact that we get the same assertion for every regular function defined in the neighborhood of q now follows as in the last part of the proof of Proposition 4.9.

Note that by Proposition 2.17 we have $\overline{\sigma}_1^c(T(\overline{I}, m)) \subseteq T(\overline{I}_1, m)$, hence we may write $\overline{\sigma}_1^*(b_i) = hb_i'$, with $b_i' \in T(\overline{I}_1, m)$ in a neighborhood of $\overline{q}$, and with h a local equation for the exceptional divisor. Note that $f_1^*(x_d)/h$ and $g_1^*(x_d)/h$ are invertible around $\overline{q}$. Since $f_1^*(x_j) - g_1^*(x_j) = \overline{\sigma}_1^*(b_j)$ for $j \geq d$, it is clear that this difference lies in $T(\overline{I}_1, m)$ in some neighborhood of $\overline{q}$. Consider now for $i \leq d-1$

$$f_1^*(y_i) - g_1^*(y_i) = \frac{f_1^*(x_i)}{f_1^*(x_d)} - \frac{g_1^*(x_i)}{g_1^*(x_d)} = \frac{g_1^*(x_i) + hb_i'}{g_1^*(x_d) + hb_d'} - \frac{g_1^*(x_i)}{g_1^*(x_d)}$$

$$= \frac{h}{f_1^*(x_d)} b_i' - \frac{h}{f_1^*(x_d)} g_1^*(y_i) b_d' \in T(\overline{I}_1, m)$$

in some neighborhood of $\overline{q}$. This proves (4.2). Therefore we can now iterate the argument to prove the proposition in the case of blow-up sequences of arbitrary length. □

LECTURE 5

Proof of principalization

In this lecture we put together the tools developed in the previous ones, and give the proof of a stronger version of Theorem 1.3. In doing this we follow closely Kollár's presentation in [**Kol**] (with the exception of §5.3.1 and §5.3.2, in which we modify slightly the construction in *loc. cit.*).

5.1. The statements

Our main goal is to prove the following stronger version of Theorem 1.3. In this statement we deal with triples (X, I, E), where X is a smooth variety over an algebraically closed field k of characteristic zero, $I \subseteq \mathcal{O}_X$ is an everywhere nonzero ideal sheaf on X, and E is a SNC divisor on X (note that in this case E does not stand for a sequence of divisors).

Theorem 5.1. *One can associate to every triple (X, I, E) as above a blow-up sequence*

$$\mathcal{P}(X,I,E): \quad X_r = \mathrm{Bl}_{C_{r-1}}(X_{r-1}) \xrightarrow{\sigma_r} \cdots \xrightarrow{\sigma_2} X_1 = \mathrm{Bl}_{C_0}(X_0) \xrightarrow{\sigma_1} X_0 = X$$

such that

i) *Each $C_i \subseteq X_i$ is smooth, and has SNC with the proper transform of E.*
ii) *If $E = \emptyset$, then each C_i has SNC with the sum D_i of the proper transforms of the exceptional divisors on $X_1, \ldots, X_i$.*
iii) *$I \cdot \mathcal{O}_{X_r}$ is the ideal of a divisor having SNC with the proper transform of E (or with D_r, if $E = \emptyset$).*
iv) *$\sigma_1 \circ \cdots \circ \sigma_r$ is an isomorphism over $X \smallsetminus \mathrm{Supp}(I)$.*
v) *$\mathcal{P}$ is functorial with respect to smooth morphisms.*
vi) *$\mathcal{P}$ commutes with field extensions.*
vii) *$\mathcal{P}$ commutes with closed embeddings on triples (X, I, E) with $E = \emptyset$.*

Property vii) means that if X is a closed subvariety of Z, with both X and Z smooth and of pure dimension, and if $J \subseteq \mathcal{O}_Z$ is the inverse image of the everywhere nonzero ideal $I \subseteq \mathcal{O}_X$ via the surjection $\mathcal{O}_Z \to \mathcal{O}_X$, then $\mathcal{P}(X, I, \emptyset)$ and $\mathcal{P}(Z, J, \emptyset)$ are defined by the same sequence of blow-up centers. The meaning of functoriality in v) is the following. Suppose that $f\colon X' \to X$ is a smooth morphism, and consider the sequence of morphisms obtained by pulling-back $\mathcal{P}(X, I, E)$:

$$X'_r = X_r \times_X X' \to \cdots \to X'_1 = X_1 \times_X X' \to X'. \tag{5.1}$$

Note that the morphism $X'_{i+1} \to X'_i$ is the blow-up along $C'_i = C_i \times_X X' \subseteq X'_i$, but this center may be empty (recall that $X'_{i+1} \to X'_i$ is then called *empty blow-up*). The assertion in v) is that $\mathcal{P}(X', f^*(I), f^*(E))$ is obtained from the sequence (5.1) by deleting the empty blow-ups, and by renumbering the remaining ones. Note that if f is surjective, then all C_i' are nonempty, and therefore there are no empty blow-ups to delete.

We will prove the above theorem by successively lowering the order of the ideal. This will be done with reference to a fixed marking. For technical reasons we will also need to order the components of E. In other words, we will deal with marked ideals with divisors. The following is the main result of this lecture.

Theorem 5.2. *One can associate to every marked ideal with divisor* (X, I, m, E), *with* I *everywhere nonzero, a sequence of admissible blow-ups*

$$\mathcal{PM}(X, I, m, E): \quad X_r = \mathrm{Bl}_{C_{r-1}}(X_{r-1}) \overset{\sigma_r}{\to} \cdots \overset{\sigma_2}{\to} X_1 = \mathrm{Bl}_{C_0}(X_0) \overset{\sigma_1}{\to} X_0 = X$$

such that

i) *The support of the controlled transform* (I_r, m) *on* X_r *is empty.*
ii) $\mathcal{PM}$ *commutes with smooth morphisms.*
iii) $\mathcal{PM}$ *commutes with field extensions.*
iv) *If* X *is a closed subvariety of* Z, *with both* X *and* Z *smooth and pure-dimensional, and if the ideal* J *on* Z *is the inverse image of* I *via the surjection* $\mathcal{O}_Z \to \mathcal{O}_X$, *then* $\mathcal{PM}(Z, J, 1, \emptyset)$ *is defined by the same sequence of centers that gives* $\mathcal{PM}(X, I, 1, \emptyset)$.

Note that under the assumptions of iv), we have $\max_{p \in Z} \mathrm{ord}_p(J) = 1$, unless $X = Z$. The meaning of functoriality with respect to smooth morphisms is the same as in Theorem 5.1.

The proof of Theorem 5.2 is by induction, and consists of two parts. The first step will be to show that assuming constructed $\mathcal{PM}(X, I, m, E)$ when $\dim(X) \leq n-1$, we can associate to every (X, I, m, E) of *maximal order*, with $\dim(X) = n$ a sequence of admissible blow-ups $\mathcal{R}(X, I, m, E)$ that satisfies i), ii), and iii) in the theorem. The second step will be to use $\mathcal{R}$ to construct $\mathcal{PM}$ in dimension n for an arbitrary (X, I, m, E). The starting point of the induction is when $\dim(X) = 0$. In this case, since I is everywhere nonzero it follows that $I = \mathcal{O}_X$, and we take $\mathcal{PM}(X, I, m, E)$ to be the empty sequence (that is, $r = 0$).

5.2. Part I: the maximal order case

We now assume that we have constructed $\mathcal{PM}(X, I, m, E)$ whenever $\dim(X) \leq n-1$, such that it commutes with field extensions, and is functorial with respect to smooth morphisms in dimension $\leq n-1$. Suppose that (X, I, m, E) is of maximal order, and $\dim(X) = n \geq 1$. The strategy is the following: we first construct $\mathcal{R}(X, I, m, E)$ when there is a hypersurface of maximal contact H. In this case, we do the construction by induction, using the results in the previous lectures. We will see that the construction is independent of the choice of hypersurface, and that it is functorial with respect to smooth morphisms. This will allow us to globalize the construction. We point out that in the local case some subtlety is due to the fact that in order to be able to restrict to the hypersurface of maximal contact, we first need to ensure that this hypersurface has SNC with the sum of the components of E.

Remark 5.3. Note that if (X, I, m, E) is of maximal order and has nonempty support, then for every admissible blow-up $X_1 \to X$, the strict transform of I is equal to the controlled transform I_1. Furthermore, we have seen in Proposition 3.2 that I_1 is again of maximal order.

5.2.1. Separating the support from the proper transform of a divisor

The following lemma will allow us to put the hypersurface of maximal contact in SNC with the sum of the components of E.

Lemma 5.4. *Assuming Theorem 5.2 in dimension $< n$, the following holds: given any $j \geq 1$, one can define for every marked ideal with divisor (X, I, m, E) of maximal order, with $\dim(X) = n$ and $E = \left(E^{(1)}, \ldots, E^{(s)}\right)$ a sequence ϕ of admissible blow-ups*

$$\mathcal{R}_j(X, I, m, E): \quad X_r \xrightarrow{\phi_r} \cdots \to X_1 \xrightarrow{\phi_1} X = X_0$$

such that

i) $\mathrm{Supp}(I_r, m)$ *is disjoint from the proper transform of* $E^{(j)}$.
ii) $\mathcal{R}_j$ *commutes with smooth morphisms in dimension* $\leq n$.
iii) $\mathcal{R}_j$ *commutes with field extensions.*

Proof. We first consider the components of $E^{(j)}$ that lie in $\mathrm{Supp}(I, m)$ (if any), and let $\phi_1 \colon X_1 \to X$ be the simultaneous blow-up of these components. Note that since (I, m) is of maximal order, a prime divisor F is contained in $\mathrm{Supp}(I, m)$ if and only if $I = \mathcal{O}_X(-mF)$ in a neighborhood of F, and in this case F is smooth. Note that none of the components of the proper transform of $E^{(j)}$ is contained in the support of (I_1, m); moreover, the exceptional divisor $\mathrm{Exc}(\phi_1)$ of ϕ_1 is disjoint from $\mathrm{Supp}(I_1, m)$.

We now have the marked ideal with divisor (X_1, I_1, m, E_1). Let $\overline{E_1}$ denote denote the same sequence of divisors as in E_1, but with $E_1^{(j)}$ removed. Since no component of $E_1^{(j)}$ is contained in $\mathrm{Supp}(\mathcal{C}(\mathcal{H}(I_1, m))) = \mathrm{Supp}(I_1, m)$, it follows from Propositions 3.4 and 4.2 that $\mathcal{C}(\mathcal{H}(I_1, m))|_{E_1^{(j)}}$ is everywhere nonzero. Therefore we may consider $\mathcal{PM}(E_1^{(j)}, \mathcal{C}(\mathcal{H}(I_1, m))|_{E_1^{(j)}}, \overline{E_1}|_{E_1^{(j)}})$. By Propositions 3.8 and 4.2, the same sequence of centers gives a sequence of admissible blow-ups with respect to (X_1, I_1, m, E_1). We get $\mathcal{R}_j(X, I, m, E)$ by concatenating this sequence of blow-ups and ϕ_1.

Let $(I_r, m) = \phi^c(I, m)$, and $(J_r, m) = \phi^c(\mathcal{C}(\mathcal{H}(I, m)))$. Note that the proper transform of $E^{(j)}$ is equal to the sum of $E_r^{(j)}$ and of the proper transform of $\mathrm{Exc}(\phi_1)$ (which does not meet $\mathrm{Supp}(I_r, m)$ since $\mathrm{Exc}(\phi_1) \cap \mathrm{Supp}(I_1, m) = \emptyset$). Another application of Propositions 3.8 and 4.2 gives that $\mathrm{Supp}(I_r, m) \cap E_r^{(j)} = \mathrm{Supp}(J_r, m) \cap E_r^{(j)}$ is equal to the support of the controlled transform of $\mathcal{C}(\mathcal{H}(I_1, m))|_{E_1^{(j)}}$, which is empty. This proves i), and ii) and iii) follow since $\mathcal{PM}$ has the corresponding properties in dimension $< n$. □

Remark 5.5. There is no need in using the homogenized ideal in the above proof. However, we will apply the lemma also in a context when the construction will involve the choice of a hypersurface of maximal contact. In that case, using the homogenized ideal makes the construction independent of this choice.

5.2.2. Construction in the presence of a hypersurface of maximal contact

We now assume Theorem 5.2 in dimension $< n$, and construct $\mathcal{R}(X, I, m, E)$ that satisfies i), ii) and iii) in Theorem 5.2 when (X, I, m, E) is a marked ideal with divisor of maximal order, having a hypersurface of maximal contact H, and $\dim(X) = n$.

Given such (X, I, m, E) and H, with $E = (E^{(1)}, \ldots, E^{(\ell)})$, we apply successively $\mathcal{R}_1, \ldots, \mathcal{R}_\ell$ to (X, I, m, E) to get $\pi \colon X' \to X$. Let $(X', I', m, E') = \pi^c(X, I, m, E)$. By construction, the proper transform of $\sum_{j=1}^{\ell} E^{(j)}$ is disjoint from the support of (I', m).

Proposition 3.2 implies that the centers in the blow-up sequence so far are contained in the corresponding proper transforms of H, and the proper transform $H' \subset X'$ is a hypersurface of maximal contact for (I', m). We deduce that the sum of the prime divisors in E' that are not proper transforms of divisors in E has SNC with H'. Therefore H' has SNC with E' is a neighborhood of $\mathrm{Supp}(I', m)$.

Furthermore, in some open neighborhood of $\mathrm{Supp}(I', m)$, no component of H' appears in E'. Indeed, such a component would be the proper transform on X' of a component of some exceptional divisor corresponding to the blow-up along a codimension one center. However, after such a blow-up, that exceptional divisor is disjoint from the support of the corresponding controlled transform of (I, m), hence any component of H' that appears in E' is disjoint from $\mathrm{Supp}(I', m)$. Since from now on all blow-ups will have centers over $\mathrm{Supp}(I', m)$, without any loss of generality we may assume that H' has SNC with the sum of the prime divisors that appear in E', and that no component of H' is contained in this sum.

We now apply $\mathcal{R}_1(X', I', m, (H', E'))$. By concatenating all these sequences, we get $\mathcal{R}(X, I, m, E)$. If (X_r, I_r, m, E_r) is the last marked ideal in this sequence, and if H_r is the proper transform of H on X_r, then Proposition 3.2 implies that H_r is a hypersurface of maximal contact of (I_r, m), hence it contains $\mathrm{Supp}(I_r, m)$. On the other hand, by construction H_r and $\mathrm{Supp}(I_r, m)$ are disjoint, hence $\mathrm{Supp}(I_r, m)$ is empty.

Note that while in the above process we have replaced several times a marked ideal (J, m) on one of the X_i by $\mathcal{C}(\mathcal{H}(J, m))$, we have chosen only one hypersurface of maximal contact, namely H on X.

5.2.3. Independence of the choice of hypersurface of maximal contact We now show that the above definition of $\mathcal{R}(X, I, m, E)$ is independent of H. Indeed, suppose that $\widetilde{H}$ is another hypersurface of maximal contact for (X, I, m, E). The first part in the construction is independent of H. This choice plays a role only in the blow-ups starting from X'. Recall the construction from this point on, following the proof of Lemma 5.4. We first simultaneously blow-up the components of H' that are contained in $\mathrm{Supp}(I', m)$ to get X_1'. These are precisely the prime divisors F on X' such that $I' = \mathcal{O}_{X'}(-mF)$ around F (such F is necessarily smooth, and is contained in every hypersurface of maximal contact of (I', m)). In particular this step is independent of H.

Let H_1' and $\widetilde{H}_1'$ be the proper transforms of H and $\widetilde{H}$, respectively, on X_1'. We denote by G the sequence of divisors that we have on X_1'. If (I_1', m) is the controlled transform of (I, m) on X_1', then we consider $\mathcal{C}(\mathcal{H}(I_1', m))$. We need to show that

$$\mathcal{PM}(H_1', \mathcal{C}(\mathcal{H}(I_1', m))|_{H_1'}, G|_{H_1'}) \text{ and } \mathcal{PM}(\widetilde{H}_1', \mathcal{C}(\mathcal{H}(I_1', m))|_{\widetilde{H}_1'}, G|_{\widetilde{H}_1'})$$

induce the same sequence of blow-ups on X_1' (recall that as we have seen in §5.2.2, we may assume that no component of H_1' and $\widetilde{H}_1'$ appears in G). We construct étale morphisms $f, g \colon \overline{X_1'} \to X_1'$ as in Proposition 4.9, such that in particular

$$f^*(\mathcal{H}(I_1', m), G) = g^*(\mathcal{H}(I_1', m), G) \text{ and } \overline{H} := f^{-1}(H_1') = g^{-1}(\widetilde{H}_1').$$

Note also that by Exercise 3.7, pulling-back by smooth morphisms commutes with taking the coefficient ideal, hence

$$f^*(\mathcal{C}(\mathcal{H}(I_1', m))) = \mathcal{C}(f^*(\mathcal{H}(I_1', m))) = \mathcal{C}(g^*(\mathcal{H}(I_1', m))) = g^*(\mathcal{C}(\mathcal{H}(I_1', m))).$$

Let $u\colon \overline{H} \to H_1'$ and $w\colon \overline{H} \to \widetilde{H}_1'$ denote the restrictions of f and g, respectively. It follows from the functoriality property of $\mathcal{PM}$ in lower dimension that

$$u^*\mathcal{PM}(H_1', \mathcal{C}(\mathcal{H}(I_1', m))|_{H_1'}, G|_{H_1'}) = \mathcal{PM}(\overline{H}, f^*(\mathcal{C}(\mathcal{H}(I_1', m)))|_{\overline{H}}, f^*(G)|_{\overline{H}})$$

$$= \mathcal{PM}(\overline{H}, g^*(\mathcal{C}(\mathcal{H}(I_1', m)))|_{\overline{H}}, g^*(G)|_{\overline{H}}) = w^*\mathcal{PM}(\widetilde{H}_1', \mathcal{C}(\mathcal{H}(I_1', m))|_{\widetilde{H}_1'}, G|_{\widetilde{H}_1'}).$$

If we consider the corresponding sequences of blow-ups over X_1', then we can apply Proposition 4.11 for f and g to see that the two sequences are the same.

Once we have proved independence, functoriality and compatibility with field extensions (when we have hypersurfaces of maximal contact) is easy. Suppose that $f\colon Y \to X$ is a smooth morphism between varieties of dimension $\leq n$, and let (X, I, m, E) be a marked ideal with divisor, of maximal order. If $H \subset X$ is a hypersurface of maximal contact for (I, m), then on Y we can use the hypersurface of maximal contact $f^{-1}(H)$ for $f^*(X, I, m, E)$. Since taking the coefficient and homogenized ideals commutes with smooth pull-back, functoriality follows from definition and functoriality in smaller dimension. Similarly, compatibility with field extensions follows from such compatibility in smaller dimensions, and the fact that taking coefficient and homogenized ideals is compatible with field extensions.

5.2.4. Globalization

Suppose now that (X, I, m, E) is an arbitrary marked ideal of maximal order, with $\dim(X) = n$. We can find a finite open cover $X = \bigcup_{i=1}^N U_i$, such that each $(I, m)|_{U_i}$ has a hypersurface of maximal contact $H_i \subset U_i$ (this is the case if X has no components of dimension zero; however, on such components $\mathrm{Supp}(I, m) = \emptyset$, hence we can simply ignore the zero-dimensional components).

Let us consider on the disjoint union $X' := \bigsqcup_i U_i$ the marked ideal with divisor (I', m, E') that on each U_i is given by $(I, m, E)|_{U_i}$. We similarly get a marked ideal with divisor on $X'' := \bigsqcup_{i\leq j}(U_i \cap U_j)$, denoted by (I'', m, E''). We have a morphism $f\colon X' \to X$ that on each U_i is given by inclusion. We similarly have $g, h\colon X'' \to X'$ defined such that g and h map $U_i \cap U_j$ by the inclusion map to U_i and, respectively, U_j. It is clear that f, g, and h are surjective and étale (in fact, they are open immersions locally on their domains). Moreover, we have

$$g^*(X', I', m, E') = (X'', I'', m, E'') = h^*(X', I', m, E').$$

Note that $H' := \bigsqcup_i H_i$ is a hypersurface of maximal contact for (X', I', m, E'), while $H'' := \bigsqcup_{i\leq j}(H_i \cap U_j)$ is a hypersurface of maximal contact for (X'', I'', m, E''). In particular, we may consider $\mathcal{R}(X', I', m, E')$ and $\mathcal{R}(X'', I'', m, E'')$, and we see that

$$g^*\mathcal{R}(X', I', m, E') = \mathcal{R}(X'', I'', m, E'') = h^*\mathcal{R}(X', I', m, E'). \tag{5.2}$$

This is a consequence of the independence of the choice of hypersurface of maximal contact in §5.2.3, and of functoriality with respect to smooth morphisms. Note that since g and h are surjective, there are no empty blow-ups to delete in pulling-back via g and h. Therefore the equality in (5.2) implies that the blow-up centers giving $\mathcal{R}(X', I', m, E')$ can be glued together to give a sequence of admissible blow-ups with respect to (X, I, m, E), whose pull-back via the smooth surjective morphism

f is $\mathcal{R}(X', I', m, E')$. This sequence defines $\mathcal{R}(X, I, m, E)$, and it is clear that it gives a resolution of (I, m).

Remark 5.6. We point out a subtlety in the above globalization of the definition of resolutions. Suppose that (X, I, m, E) is a marked ideal of maximal order, and $X = U \cup V$ is an open cover such that we know how to define $\mathcal{R}((I, m, E)|_U)$ and $\mathcal{R}((I, m, E)|_V)$. This does not determine in general $\mathcal{R}(X, I, m, E)$. Indeed, if the first blow-up centers C_U and C_V in the two resolutions are disjoint, then it is not clear whether in the first step of $\mathcal{R}(X, I, m, E)$ we should blow-up C_U, C_V, or their union. The trick of using $U \sqcup V$, and functoriality with respect to smooth surjective morphisms was introduced in [**Kol**], and greatly simplified previous approaches.

Exercise 5.7. Use functoriality with respect to smooth morphisms in the definition of $\mathcal{R}(X, I, m, E)$– in the context when there is a hypersurface of maximal contact–to show that the above definition is independent of the open cover $X = \bigcup_i U_i$.

Note also that the above definition of $\mathcal{R}(X, I, m, E)$ is functorial with respect to smooth morphisms. Indeed, given a smooth morphism $\phi \colon X' \to X$, and an open cover $X = \bigcup_i U_i$ as in the above definition, we can use the open cover $X' = \bigcup_i f^{-1}(U_i)$ to define $\mathcal{R}(X', I \cdot \mathcal{O}_{X'}, m, f^*(E))$. In this case, functoriality follows from the case when there is a hypersurface of maximal contact. Compatibility with field extensions for the above definition is also easy. Therefore we have defined $\mathcal{R}(X, I, m, E)$ for every marked ideal with divisor of maximal order, and with $\dim(X) = n$, such that the definition is functorial in dimension $\leq n$, and compatible with field extensions.

5.3. Part II: the general case

In order to complete the proof of Theorem 5.2, we need to show that if we can construct $\mathcal{R}(X, I, m, E)$ in dimension n, when (I, m) is of maximal order, then we can construct $\mathcal{PM}(X, I, m, E)$ for every (X, I, m, E), with $\dim(X) = n$. A natural approach is to start instead with (X, I, s, E), where $s = \max_{p \in X} \mathrm{ord}_p(I)$, and then repeatedly apply the maximal order case, to keep lowering the order until it drops below m. The subtlety comes from the fact that in this process we always take the strict transform, while we are interested in the controlled transform (with respect to m).

In order to keep track of this difference, it is more convenient to slightly change the procedure. Given a pair (X, I, m, E) with $E = \left(E^{(1)}, \ldots, E^{(\ell)}\right)$, we factor I as $\mathcal{M}(I, E) \cdot \mathcal{N}(I, E)$, where $\mathcal{M}(I, E) = \mathcal{O}(-\sum_i a_i F_i)$ for suitable a_i, where each F_i is an irreducible component of $\sum_{j=1}^{\ell} E^{(j)}$, and where no irreducible component of $\sum_{j=1}^{\ell} E^{(j)}$ is contained in the subscheme defined by $\mathcal{N}(I, E)$. We call $\mathcal{M}(I, E)$ and $\mathcal{N}(I, E)$ the *monomial* (respectively, *non-monomial*) *factor* of I. Our first goal is to make I monomial, that is, to reduce to the case when $I = \mathcal{M}(I, E)$. We do this by constructing a sequence π of admissible blow-ups with respect to (I, m, E) such that if $\pi^c(I, m, E) = (I', m, E')$, then $\mathcal{N}(I', E')$ is the structure sheaf in a neighborhood of $\mathrm{Supp}(I', m)$.

5.3.1. Reduction to the monomial case

Step 1. Let $\overline{m} := \max_{p \in X} \mathrm{ord}_p(\mathcal{N}(I, E))$. We first want to reduce to the case $\overline{m} < m$. If $\overline{m} \geq m$, we consider $\mathcal{R}(X, \mathcal{N}(I, E), \overline{m}, E)$ to get a sequence ϕ of admissible blow-ups with respect to $(X, \mathcal{N}(I, E), \overline{m}, E)$. Let $\phi^c(\mathcal{N}(I, E), \overline{m}, E) =$

$(\mathcal{N}(I,E)'',\overline{m},E'')$. We clearly have $(I,m) \leq (\mathcal{N}(I,E),\overline{m})$, hence ϕ is also admissible with respect to (I,m,E), and let $\phi^c(I,m,E) = (I'',m,E'')$. Note that $\phi^*(\mathcal{M}(I,E)) \subseteq \mathcal{M}(I'',E'')$, and furthermore, I'' is the product of $\mathcal{N}(I,E)''$ and of a monomial factor involving the (proper transforms of the) exceptional divisors over X. Therefore $\mathcal{N}(I'',E'') = \mathcal{N}(I,E)''$, whose order at every point is $< \overline{m}$. After repeating this step at most $(\overline{m}-m+1)$ times, we may replace (X,I,m,E) by its controlled transform, and hence assume that $\overline{m} < m$.

Step 2. From this point on we need to pay attention that we only blow-up loci inside $\mathrm{Supp}(I,m)$. Let $\widetilde{m} := \max_{p\in\mathrm{Supp}(I,m)} \mathrm{ord}_p(\mathcal{N}(I,E)) < m$. Since

$$\mathrm{Supp}(\mathcal{N}(I,E)^m + I^{\widetilde{m}}, m\cdot\widetilde{m}) = \mathrm{Supp}(\mathcal{N}(I,E),\widetilde{m}) \cap \mathrm{Supp}(I,m)$$

(and similar equalities for the controlled transforms), it follows that if ψ is the sequence of blow-ups given by $\mathcal{R}(X,\mathcal{N}(I,E)^m+I^{\widetilde{m}},m\cdot\widetilde{m},E)$ (note that the marked ideal with divisor to which we apply $\mathcal{R}$ is of maximal order), then ψ is admissible with respect to both (I,m,E) and $(\mathcal{N}(I,E),\widetilde{m},E)$, and therefore also with respect to $(\mathcal{M}(I,E),m-\widetilde{m},E)$. Let $\psi^c(I,m,E) = (I',m,E')$, and

$$\psi^c(\mathcal{N}(I,E),\widetilde{m}) = (\mathcal{N}(I,E)',\widetilde{m}),\ \psi^c(\mathcal{M}(I,E),m-\widetilde{m}) = (\mathcal{M}(I,E)',m-\widetilde{m}).$$

Note that we have $I' = \mathcal{N}(I,E)'\cdot\mathcal{M}(I,E)'$, and $\mathcal{M}(I,E)'$ is monomial with respect to E', hence $\mathcal{N}(I',E') = \mathcal{N}(\mathcal{N}(I,E)',E')$. Therefore

$$\max_{p\in\mathrm{Supp}(I',m))}\{\mathrm{ord}_p\mathcal{N}(I',E')\} < \widetilde{m}.$$

After at most $\widetilde{m}$ such steps, and after replacing (X,I,m,E) by its controlled transform, we may assume that $\mathrm{Supp}(I,m)\cap\mathrm{Supp}(\mathcal{N}(I,E)) = \emptyset$. From this point on, we will only consider sequences of blow-ups with centers over $\mathrm{Supp}(I,m)$. Therefore, after restricting to a suitable open neighborhood of $\mathrm{Supp}(I,m)$, we may assume that $\mathcal{N}(I,E) = \mathcal{O}_X$.

5.3.2. The monomial case

Suppose that we have the marked ideal with divisor (X,I,E,m), where $E = (E^{(1)},\ldots,E^{(\ell)})$, such that $I = \mathcal{O}(-\sum_F a_F F)$, where the sum is over the set Γ of irreducible components of $\sum_{j=1}^{\ell} E^{(j)}$. In this part of the algorithm we will use in an essential way the fact that the divisors in E are ordered. For every F in Γ, let $\lambda(F)$ be the (unique) j such that F appears in $E^{(j)}$.

Step 1. Suppose that there is F in Γ such that $a_F \geq m$. Let $b_1 := \max_F\{a_F\}$, and $j_{1,1} := \min\{\lambda(F) \mid a_F = b_1\}$. We consider the blow-up of X along $\sum_F F$, where the sum is over those F with $a_F = b_1$ that appear in $E^{(j_{1,1})}$. Note that for the controlled transform of (I,m,E) we have that either b_1 goes down, or b_1 is constant and $j_{1,1}$ goes up, but staying $\leq \ell$. It follows that after repeating this step finitely many times we may assume that $a_F < m$ for all F in Γ.

Step q. Here $2 \leq q \leq n$. Suppose that for all distinct $F_1,\ldots,F_{q-1}$ in Γ, with $F_1\cap\ldots\cap F_{q-1} \neq \emptyset$, we have $a_{F_1}+\cdots+a_{F_{q-1}} < m$. If there are distinct $F_1,\ldots,F_q$ in Γ with $F_1\cap\ldots\cap F_q \neq \emptyset$ and $a_{F_1}+\cdots+a_{F_q} \geq m$, let b_q denote the maximum $a_{F_1}+\cdots+a_{F_q}$ for all such $F_1,\ldots,F_q$. Let us consider all $(F_1,\ldots,F_q)$ with $F_1\cap\ldots\cap F_q \neq \emptyset$, $a_{F_1}+\cdots+a_{F_q} = b_q$, and such that $(\lambda(F_1),\ldots,\lambda(F_q)) = (j_{q,1},\ldots,j_{q,q})$ is minimal in the lexicographic order. Note that $j_{q,1} < \ldots < j_{q,q}$ (this follows from the minimality with respect to the lexicographic order, and the fact that each $E^{(j)}$ is smooth, hence any two distinct components of $E^{(j)}$ do not intersect).

Let Z be the union of all such $F_1 \cap \ldots \cap F_q$. This is a disjoint union of smooth irreducible codimension q subsets. The blow-up σ of X with center Z is admissible with respect to (I, m, E), and let us write $\sigma^c(I, m, E) = (I', m, E')$.

We first show that (I', m, E') also satisfies the hypothesis in Step q. It is clear that I' is again monomial with respect to E'. Fix an irreducible component G of the exceptional divisor corresponding to an irreducible component of $F_1 \cap \ldots \cap F_q$. The coefficient a_G of G in the divisor corresponding to I' is $a_{F_1} + \ldots + a_{F_q} - m$. Consider a nonempty intersection of G with $(q-2)$ of the (proper transforms of the) divisors in Γ, say $G_1, \ldots, G_{q-2}$. Note first that $F_1 \cap \ldots \cap F_q \cap G_1 \cap \ldots \cap G_{q-2} \neq \emptyset$. If i is such that F_i is not in $\{G_1, \ldots, G_{q-2}\}$, then

$$a_G + a_{G_1} + \cdots + a_{G_{q-2}} = \big((a_{F_1} + \cdots + a_{F_q}) - a_{F_i}\big) - m + (a_{F_i} + a_{G_1} + \cdots + a_{G_{q-2}}) < m,$$

by the hypothesis in Step q for (I, m, E). Therefore (I', m, E') satisfies the hypothesis in Step q (the other inequalities being clear).

We now consider a nonempty intersection $G \cap G_1 \cap \ldots \cap G_{q-1}$, where G is as before, and where the G_i are (proper transforms of) elements in Γ. We have

$$a_G + a_{G_1} + \cdots + a_{G_{q-1}} = a_{F_1} + \cdots + a_{F_q} + (a_{G_1} + \cdots + a_{G_{q-1}} - m) < a_{F_1} + \cdots + a_{F_q} = b_q.$$

This shows that after replacing (I, m, E) by (I', m, E') either b_q goes down, or the number of those $(F_1, \ldots, F_q)$ with $F_1 \cap \ldots \cap F_q \neq \emptyset$ and $a_{F_1} + \cdots + a_{F_q} = b_q$ goes down. It follows that after repeating this step finitely many times, we may assume that for all $F_1, \ldots, F_q$ in Γ with $F_1 \cap \ldots \cap F_q \neq \emptyset$ we have $a_{F_1} + \cdots + a_{F_q} < m$. In this case we go to Step $(q+1)$.

After being done with Step n, we achieve a resolution of (X, I, m, E) by a sequence of admissible blow-ups. Checking functoriality with respect to smooth morphisms, and compatibility with field extensions is straightforward. Both assertions follow since we have

i) Functoriality with respect to smooth morphisms and compatibility with field extensions in the maximal order case.
ii) The compatibility of the decomposition $I = \mathcal{M}(I, E) \cdot \mathcal{N}(I, E)$ with pull-back by smooth morphisms, and with field extensions.
iii) In the monomial case, the algorithm is compatible with these two operations.

The proof of Theorem 5.2 is complete, modulo the compatibility with closed immersions.

Remark 5.8. Note that for a marked ideal (X, I, m) of maximal order and with no divisorial part, our two definitions agree, that is

$$\mathcal{PM}(X, I, m, \emptyset) = \mathcal{R}(X, I, m, \emptyset).$$

Indeed, it is enough to note that $\mathcal{N}(I, \emptyset) = I$. Moreover, given a marked ideal I with $\max_{p \in X} \operatorname{ord}_p(I) = m' \geq m$, the first sequence of blow-ups in $\mathcal{PM}(X, I, m, \emptyset)$ is given by $\mathcal{R}(X, I, m, \emptyset)$.

Remark 5.9. In replacing a marked ideal (I, m) of maximal order by $\mathcal{C}(\mathcal{H}(I, m))$, we go from the marking m to the marking $m!$. Since this operation is repeated many times through the resolution process, it makes the whole algorithm very hard to apply in even simple-minded examples.

5.3.3. Compatibility with closed immersions

Suppose that X is a closed subvariety of Z, with both X and Z smooth and pure-dimensional, and that J is the inverse image of I via $\mathcal{O}_Z \to \mathcal{O}_X$. We want to show that the sequence of centers defining $\mathcal{PM}(X, I, 1, \emptyset)$ defines also $\mathcal{PM}(Z, J, 1, \emptyset)$. In order to show this, we may replace Z by any open neighborhood of X.

Of course, we may assume that $\operatorname{codim}(X, Z) = d \geq 1$. After replacing Z by the finite disjoint union of various open neighborhoods of points in X (see §5.2.4), we may assume that $X = H_1 \cap \ldots \cap H_d$, where each H_i is a smooth subvariety of Z of pure codimension one, with $\sum_i H_i$ having simple normal crossings. An obvious induction on d now reduces us to the case when X is a smooth subvariety of Z, of pure codimension one.

Note that $(J, 1)$ has maximal order, and X is a hypersurface of maximal contact for $(J, 1)$. Moreover, we have $\mathcal{C}(J, 1) = (J, 1) = (T(J, 1), 1) = \mathcal{H}(J, 1)$. Since our divisor is empty, it follows from the algorithm we gave in the case of maximal order (see Remark 5.8) that $\mathcal{PM}(Z, J, 1, \emptyset)$ is given by the same sequence of centers as

$$\mathcal{PM}(X, \mathcal{C}(\mathcal{H}(J, 1))|_X, 1, \emptyset) = \mathcal{PM}(X, J|_X, 1, \emptyset).$$

Since $J|_X = I$, this proves iv), and completes the proof of Theorem 5.2.

5.4. Proof of principalization

We now show that Theorem 5.2 implies all assertions in Theorem 5.1. We want to apply Theorem 5.2 to a marked ideal with divisor $(X, I, 1, F)$, for a suitable $F = (F^{(1)}, \ldots, F^{(\ell)})$. In order to simply get a principalization of (X, I, E), we could for example define F by arbitrarily ordering the irreducible components of E. However, in order to guarantee functoriality, we need to eliminate arbitrary choices. We will do this by taking some preliminary blow-ups.

If E is smooth (that is, if its components are pairwise disjoint), then we take $F = (E)$. Our goal is to take some blow-ups to reduce to this situation. Suppose that N is the largest number of components of E that intersect, and let C_0 be the union of all N-fold intersections of components of E. Since there are no $(N+1)$-fold intersections, it follows that C_0 is smooth, and C_0 has SNC with E. After replacing X by $X_1 := \mathrm{Bl}_{C_0}(X)$, I by $I \cdot \mathcal{O}_{X_1}$, and E by its proper transform on X_1, we may assume that there are at most $(N-1)$-fold intersections of the components of E. After at most N such steps, we arrive at a pair $(\widetilde{X}, I \cdot \mathcal{O}_{\widetilde{X}}, \widetilde{E})$ such that $\widetilde{E}$ is smooth. We now define $\mathcal{P}(X, I, E)$ to be the composition of the above sequence of blow-ups with $\mathcal{PM}(\widetilde{X}, I \cdot \mathcal{O}_{\widetilde{X}}, 1, (\widetilde{E}))$. If the final marked ideal with divisor in this sequence if $(X_r, I_r, 1, E_r)$, then $\mathrm{Supp}(I_r) = \emptyset$, hence $I_r = \mathcal{O}_{X_r}$. It follows that $I \cdot \mathcal{O}_{X_r}$ is the ideal of a divisor supported on the union of the components of E_r. By construction, this divisor has SNC with the proper transform of E, and with the sum of the proper transforms of the exceptional divisors in the sequence when $E = \emptyset$. The fact that $\mathcal{P}(X, I, E)$ is functorial with respect to smooth morphisms follows from the functoriallity of the above process, and of the algorithm in Theorem 5.2. The same argument gives compatibility with field extensions.

We have seen in Lecture 1 that functoriality implies the assertion in iv). Since vii) follows from Theorem 5.2 iv), the proof of Theorem 5.1 is complete.

Example 5.10. Let $I = (y^2 z, y^4) \subset k[y, z]$, and let us compute $\mathcal{PM}(\mathbf{A}^2, I, 2, \emptyset)$. The first sequence of blow-ups is given by $\mathcal{PM}(\mathbf{A}^2, I, 3, \emptyset)$, that we describe by restricting to a hypersurface of maximal contact Z.

We have $D(I) = (y^2, yz)$ and $D^2(I) = (y, z)$, hence we may take $Z = (y = 0)$. We compute $\mathcal{H}(I, 3) = \big((y, z)^3, 3\big)$, and $\mathcal{C}(\mathcal{H}(I, 3)) = \big((y, z)^6, 6\big)$. Therefore we have $(\mathcal{C}(\mathcal{H}(I, 3))|_Z = ((z^6), 6)$, hence $\mathcal{PM}(\mathbf{A}^2, I, 3, \emptyset)$ is given by $\sigma_1 \colon X_1 = \mathrm{Bl}_0(\mathbf{A}^2) \to \mathbf{A}^2$. Let $\sigma^c(I, 2) = (I_1, 2)$, and denote by E_1 the exceptional divisor. In the chart U_1 on X_1 with coordinates y and z/y, we have $I_1 = y(y, z/y)$ and $E_1 = (y = 0)$. In the chart V_1 on X_1 with coordinates y/z and z, we have $I_1 = ((y/z)^2 z)$, and $E_1 = (z = 0)$. Hence the factorization of I_1 into monomial and non-monomial parts is given by $I_1 = \mathcal{O}(-E_1) \cdot \mathcal{N}$, where $\mathcal{N}|_{U_1} = (y, z/y)$, while $\mathcal{N}|_{V_1} = (y/z)^2$. Note that $\mathrm{Supp}(\mathcal{N})$ is the union of the proper transform L of $(y = 0)$, with the point $Q \in U_1$ given by $(y = z/y = 0)$. We have $\max_{p \in X_1} \mathrm{ord}_p(\mathcal{N}) = 2$, and this maximum is attained precisely for $p \in L$. It is easy to deduce that $\mathcal{PM}(\mathbf{A}^2, I, 2, \emptyset)$ is given by $(\sigma_1, \sigma_2, \sigma_3)$, where $\sigma_2 \colon X_2 = \mathrm{Bl}_L(X_1) \to X_1$, and σ_3 is the blow-up of $X_2 = X_1$ at Q.

Example 5.11. Let $J = (x^2 + y^2 z) \subset k[x, y, z]$, and let us describe the beginning of $\mathcal{P}(\mathbf{A}^3, J, \emptyset) = \mathcal{PM}(\mathbf{A}^2, J, 1, \emptyset)$. Note that $\max_{p \in \mathbf{A}^3} \mathrm{ord}_p(J) = 2$, with $\mathrm{Supp}(J, 2) = (x = y = 0)$. Therefore the first sequence of blow-ups is given by $\mathcal{PM}(\mathbf{A}^3, J, 2, \emptyset)$.

Note that $D(J) = (x, y^2, yz)$, hence $H = (x = 0)$ is a hypersurface of maximal contact for $(J, 2)$. It follows from Exercise 4.7 that

$$\mathcal{C}(\mathcal{H}(J, 2))|_H = \big((J + D(J)^2)|_H, 2\big) = (I, 2),$$

where I is the ideal in the previous exercise. It follows that $\mathcal{PM}(\mathbf{A}^3, J, 2, \emptyset)$ is given by $\tau = (\tau_1, \tau_2, \tau_3)$, where $\tau_1 \colon X_1 = \mathrm{Bl}_0(\mathbf{A}^3) \to \mathbf{A}^3$, $\tau_2 \colon X_2 \to X_1$ is the blow-up of the proper transform on X_1 of the line $(x = y = 0)$ and $\tau_3 \colon X_3 \to X_2$ is the blow-up of a point in X_2. We leave as an exercise for the reader the task of completely describing the principalization of J.

Example 5.12. As we promised in Remark 1.9, we give an example to show that if we construct resolutions of singularities as in Lecture 1, following the above algorithm for principalization, then we might need to blow-up along singular subvarieties. This example is due to Bierstone and Milman (see [**Kol**], Example 3.106).

Let $W = \mathbf{A}^4$, and X defined by $I = (x^3 - y^2, x^4 + xz^2 - w^3)$. It is easy to see that X is irreducible, of dimension 2, hence it is a complete intersection. Using the Jacobian criterion, one sees that the singular locus of X is defined, set-theoretically, by $x = y = w = 0$. In particular, X is generically reduced, and since it is Cohen-Macaulay, it is reduced.

Recall that in order to construct the resolution of singularities of X, we consider the principalization of $(W, I, \emptyset)$, which is given by $\mathcal{PM}(W, I, 1, \emptyset)$. Note that $\max_{p \in X} \mathrm{ord}_p(I) = 2$, hence the first step consists of $\mathcal{PM}(W, I, 2, \emptyset)$. We first compute

$$D(I) = (y, x^2, 4x^3 + z^2, xz, w^2).$$

We see that a hypersurface of maximal contact is the hyperplane $H = (y = 0)$. By Exercise 4.7, we have $\mathcal{C}(\mathcal{H}(I, 2)) = (I + D(I)^2, 2)$. The above formula for $D(I)$ shows that the ideal in $\mathcal{C}(\mathcal{H}(I, 2))|_H$ is $(x^3, xz^2 - w^3, x^2 z^2, x^2 w^2, z^2 w^2, xz^3, xzw^2, w^4)$. This ideal is supported on the line $L = (x = w = 0)$ in H. The ideal has order 3 at the origin, and order 1 everywhere else on L. Hence in order to resolve this restriction, we need to blow-up the origin.

It follows that the first step in the principalization of I is given by the blowing-up W_1 of the origin. Consider the chart $U_1 \subset W_1$ with coordinates x_1, y_1, z_1, w_1, with $x = x_1$, $y = x_1y_1$, $z = x_1z_1$, and $w = x_1w_1$. The controlled transform of $(I, 2)$ has ideal $I_1 = (x_1 - y_1^2, x_1(x_1 + z_1^2 - w_1^3))$, and the exceptional divisor is $E_1 = (x_1 = 0)$. Since I_1 has order 1 at every point in its support, whatever happens in the following steps of $\mathcal{PM}(W, I, 2, \emptyset)$ is a trivial blow-up over U_1.

We are interested in the next sequence of blow-ups after $\mathcal{PM}(W, I, 2, \emptyset)$. Over U_1, this is the same as $\mathcal{PM}(U_1, I_1, 1, E_1)$. Since $\mathrm{Supp}(I_1, 1)$ meets E_1, we first need to take $\mathcal{PM}(E_1, I_1|_{E_1}, 1)$. We have $I_1|_{E_1} = (y_1^2)$, hence we first have to blow-up the subvariety Z of W_1 defined by $(x_1 = y_1 = 0)$.

On the other hand, the proper transform X_1 of X on W_1 is defined in U_1 by $(x_1 - y_1^2, x_1 + z_1^2 - w_1^3)$. Its intersection with Z is the cuspidal curve $(z_1^2 - w_1^3 = 0)$. It follows that in the resolution process of the surface X, we need to blow-up along a singular curve.

Remark 5.13. If X is a proper closed subvariety of Z, with both X and Z smooth and pure-dimensional, then $\mathcal{P}(Z, I_{X/Z}, \emptyset)$ is given by $\mathrm{Bl}_X(Z) \to Z$. Indeed, by compatibility with closed embeddings, and arguing as in §5.3.3, we see that it is enough to show this when X has pure codimension one in Z. In this case $(I_{X/Z}, 1)$ has maximal order, and X is a hypersurface of maximal contact for this marked ideal. It follows from the proof of Theorem 5.1 that $\mathcal{P}(Z, I_{X/Z}, \emptyset) = \mathcal{PM}(Z, I_{X/Z}, 1, \emptyset)$, and we see from the proof of Theorem 5.2 that this is simply given by the blowing-up along X (see the first step in the proof of Lemma 5.4).

Bibliography

[BM] E. Bierstone and P. Milman, Canonical desingularization in characteristic zero by blowing up the maximum strata of a local invariant, Invent. Math. **128** (1997), 207–302.

[CP1] V. Cossart and O. Piltant, Resolution of singularities of threefolds in positive characteristic II, J. Algebra **321** (2009), 1836–1976.

[CP2] V. Cossart and O. Piltant, Resolution of singularities of threefolds in positive characteristic I. Reduction to local uniformization on Artin-Schreier and purely inseparable coverings, J. Algebra **320** (2008), 1051–1082.

[Cut] S. D. Cutkosky, Resolution of singularities for 3-folds in positive characteristic. Amer. J. Math. **131** (2009), 59–127.

[dJ] A. J. de Jong, Smoothness, semi-stability and alterations, Inst. Hautes Études Sci. Publ. Math. **83** (1996), 51–93.

[EH] S. Encinas and H. Hauser, Strong resolution of singularities in characteristic zero, Comment. Math. Helv. **77** (2002), 821–845.

[Gir] J. Giraud, Sur la théorie du contact maximal, Math. Z. **137** (1974), 285–310.

[Hir] H. Hironaka, Resolution of singularities of an algebraic variety over a field of characteristic zero, I, II, Ann. of Math. (2) **79** (1964), 109–203; ibid. (2) **79** (1964), 205–326.

[Kol] J. Kollár, *Lectures on resolution of singularities*, Annals of Mathematics Studies **166**, Princeton University Press, Princeton, NJ, 2007.

[Vil1] O. Villamayor, Constructiveness of Hironaka's resolution, Ann. Sci. École Norm. Sup. (4) **22** (1989), 1–32.

[Vil2] O. Villamayor, Patching local uniformizations, Ann. Sci. École Norm. Sup. (4) **25** (1992), 629–677.

[Wło] J. Włodarczyk, Simple Hironaka resolution in characteristic zero, J. Amer. Math. Soc. **18** (2005), 779–822.

A Short Course on Multiplier Ideals

Robert Lazarsfeld

IAS/Park City Mathematics Series
Volume 17, 2008

A Short Course on Multiplier Ideals

Robert Lazarsfeld

Introduction

These notes are the write-up of my PCMI lectures on multiplier ideals. They aim to give an introduction to the algebro-geometric side of the theory, with an emphasis on its global aspects. Besides serving as warm-up for the lectures of Hacon, my hope was to convey to the audience a feeling for the sorts of problems for which multiplier ideals have proved useful. Thus I have focused on concrete examples and applications at the expense of general theory. While referring to [**21**] and other sources for some technical points, I have tried to include sufficient detail here so that the conscientious reader can arrive at a reasonable grasp of the machinery by working through these lectures.

The revolutionary work of Hacon–McKernan, Takayama and Birkar–Cascini–Hacon–McKernan ([**14**], [**15**], [**28**], [**3**]) appeared shortly after the publication of [**21**], and these papers have led to some changes of perspectives on multiplier ideals. In particular, the first three made clear the importance of adjoint ideals as a tool in proving extension theorems; these were not so clearly in focus at the time [**21**] was written. I have taken this new viewpoint into account in discussing the restriction theorem in Lecture 3. Adjoint ideals also open the door to an extremely transparent presentation of Siu's theorem on deformation-invariance of plurigenera of varieties of general type, which appears in Lecture 5.

Besides Part III of [**21**], I have co-authored an overview of multiplier ideals once before, in [**2**]. Those notes focused more on the local and algebraic aspects of the story. The analytic theory is surveyed in [**26**], as well as in other lecture series in this volume.

I wish to thank Eugene Eisenstein, Christopher Hacon, János Kollár and Mircea Mustaţă for valuable suggestions. I am particularly grateful to Sam Grushevsky, who read through in its entirety a draft of these lectures, and made copious suggestions.

Department of Mathematics, University of Michigan, Ann Arbor, MI 48109
E-mail address: rlaz@umich.edu
Research partially supported by NSF grant DMS 0652845

LECTURE 1

Construction and examples of multiplier ideals

This preliminary lecture is devoted to the construction and first properties of multiplier ideals. We start by discussing the algebraic and analytic incarnations of these ideals. After giving the example of monomial ideals, we survey briefly some of the invariants of singularities that can be defined via multiplier ideals.

Definition of multiplier ideals

In this section, we will give the definition of multiplier ideals.

We work throughout with a smooth algebraic variety X of dimension d defined over $\mathbf{C}$. For the moment, we will deal with two sorts of geometric objects on X: an ideal sheaf $\mathfrak{a} \subseteq \mathcal{O}_X$ together with a weighting coefficient $c > 0$, and an effective $\mathbf{Q}$-divisor D on X. Recall that the latter consists of a formal linear combination

$$D = \sum a_i D_i,$$

where the D_i are distinct prime divisors and each $a_i \in \mathbf{Q}$ is a non-negative rational number. We will attach to these data multiplier ideal sheaves

$$\mathcal{J}(\mathfrak{a}^c) \ , \ \mathcal{J}(D) \subseteq \mathcal{O}_X.$$

The intuition is that these ideals will measure the singularities of D or of functions $f \in \mathfrak{a}$, with "nastier" singularities being reflected in "deeper" multiplier ideals.

Although we will mainly focus on algebraic constructions, it is perhaps most intuitive to start with the analytic avatars of multiplier ideals.

Definition 1.1 (Analytic multiplier ideals). Given $D = \sum a_i D_i$ as above, choose local equations $f_i \in \mathcal{O}_X$ for each D_i. Then the (analytic) multiplier ideal of D is given locally by

$$\mathcal{J}_{\text{an}}(X, D) =_{\text{locally}} \left\{ h \in \mathcal{O}_X \ \Big| \ \frac{|h|^2}{\prod |f_i|^{2a_i}} \text{ is locally integrable} \right\}.$$

Similarly, if $f_1, \ldots, f_r \in \mathfrak{a}$ are local generators, then

$$\mathcal{J}_{\text{an}}(X, \mathfrak{a}^c) =_{\text{locally}} \left\{ h \in \mathcal{O}_X \ \Big| \ \frac{|h|^2}{\left(\sum |f_i|^2\right)^c} \text{ is locally integrable} \right\}.$$

(One checks that these do not depend on the choice of the f_i.) □

Equivalently, $\mathcal{J}_{\text{an}}(D)$ and $\mathcal{J}_{\text{an}}(\mathfrak{a}^c)$ arise as the multiplier ideal $\mathcal{J}(\phi)$, where ϕ is the appropriate one of the two plurisubharmonic functions

$$\phi = \sum \log |f_i|^{2a_i} \quad \text{or} \quad \phi = c \cdot \log \left(\sum |f_i|^2\right).$$

Note that this construction exhibits quite clearly the yoga that "more" singularities give rise to "deeper" multiplier ideals: the singularities of $f \in \mathfrak{a}$ or of D are reflected in the rate at which the real-vallued functions

$$\frac{1}{\prod |f_i|^{2a_i}} \quad \text{or} \quad \frac{1}{\left(\sum |f_i|^2\right)^c}$$

blow-up along the support of D or the zeroes of $\mathfrak{a}$, and this in turn is measured by the vanishing of the multipliers h required to ensure integrability.

Exercise 1.2. Suppose that $D = \sum a_i D_i$ has simple normal crossing support. Then

$$\mathcal{J}_{\text{an}}(X, D) \;=\; \mathcal{O}_X(-[D]),$$

where $[D] = \sum [a_i] D_i$ is the round-down (or integer part) of D. (HINT: This boils down to the assertion that if $z_1, \ldots, z_d$ are the standard complex coordinates in $\mathbf{C}^d$, and if $h \in \mathbf{C}\{z_1, \ldots, z_d\}$ is a convergent power series, then

$$\frac{|h|^2}{|z_1|^{2a_1} \cdot \ldots \cdot |z_d|^{2a_d}}$$

is locally integrable near the origin if and only if

$$z_1^{[a_1]} \cdot \ldots \cdot z_d^{[a_d]} \mid h$$

in $\mathbf{C}\{z_1, \ldots, z_d\}$. By separating variables, this in turn follows from the elementary computation that the function $1/|z|^{2c}$ of one variable is locally integrable if and only if $c < 1$.) □

Multiplier ideals can also be constructed algebro-geometrically. Let

$$\mu : X' \longrightarrow X$$

be a log resolution of D or of $\mathfrak{a}$. Recall that this means to begin with that μ is a proper morphism, with X' smooth. In the first instance we require that $\mu^* D + \text{Exc}(\mu)$ have simple normal crossing (SNC) support, while in the second one asks that

$$\mathfrak{a} \cdot \mathcal{O}_{X'} \;=\; \mathcal{O}_{X'}(-F)$$

where F is an effective divisor and $F + \text{Exc}(\mu)$ has SNC support. We consider also the relative canonical bundle

$$K_{X'/X} \;=\; \det(d\mu),$$

so that $K_{X'/X} \equiv_{\text{lin}} K_{X'} - \mu^* K_X$. Note that this is well-defined as an actual divisor supported on the exceptional locus of μ (and not merely as a linear equivalence class).

Definition 1.3 (Algebraic multiplier ideal)**.** The multiplier ideals associated to D and to $\mathfrak{a}$ are defined to be:

$$\begin{aligned} \mathcal{J}(D) \;&=\; \mu_* \mathcal{O}_{X'}(K_{X'/X} - [\mu^* D]) \\ \mathcal{J}(\mathfrak{a}^c) \;&=\; \mu_* \mathcal{O}_{X'}(K_{X'/X} - [cF]). \end{aligned}$$

(As in the previous Exercise, the integer part of a $\mathbf{Q}$-divisor is defined by taking the integer part of each of its corefficients.) □

Observe that these are subsheaves of

$$\mu_*\mathcal{O}_{X'}(K_{X'/X}) \;=\; \mathcal{O}_X,$$

i.e. they are indeed ideal sheaves.

One can rephrase the definition more concretely in terms of discrepancies. Write

$$\mu^*D \;=\; \sum r_i E_i \;\;,\;\; K_{X'/X} \;=\; \sum b_i E_i, \tag{1.1}$$

where the E_i are distinct prime divisors on X': thus the r_i are non-negative rational numbers and the b_i are non-negative integers. We view each of the E_i as defining a valuation ord_{E_i} on rational or regular functions on X. Then it follows from Definition 1.3 that

$$\mathcal{J}(D) \;=\; \big\{ f \in \mathbf{C}(X) \,\big|\, \mathrm{ord}_{E_i}(f) \geq [r_i] - b_i \,,\text{ with } f \text{ otherwise regular} \big\},$$

with a similar interpretation of $\mathcal{J}(\mathfrak{a}^c)$. (Note that we are abusing notation a bit here: $\mathcal{J}(D)$ is actually the sheaf determined in the evident manner by the recipe on the right.) Observe that $b_i > 0$ only when E_i is μ-exceptional, so the condition

$$\mathrm{ord}_{E_i}(f) \geq [r_i] - b_i$$

does not allow f to have any poles on X. Thus we see again that $\mathcal{J}(D)$ is a sheaf of ideals.

Remark 1.4. The definitions of $\mathcal{J}_{\mathrm{an}}(D)$ and $\mathcal{J}(D)$ may seem somewhat arbitrary or unmotivated, but they are actually dictated by the vanishing theorems that multiplier ideals satisfy. In the algebraic case, this will become clear for example in the proof of Theorem 2.4. □

Example 1.5. We work out explicitly one (artificially) simple example. Let $X = \mathbf{C}^2$, let $A_1, A_2, A_3 \subseteq X$ be three distinct lines through the origin, and set

$$D \;=\; \frac{2}{3}(A_1 + A_2 + A_3).$$

Then D is resolved by simply blowing up the origin:

$$\mu : X' \;=\; \mathrm{Bl}_0(X) \longrightarrow X.$$

Writing E for the exceptional divisor of μ, and A_i' for the proper transform of A_i, one has

$$\begin{aligned} \mu^*(A_1 + A_2 + A_3) \;&=\; (A_1' + A_2' + A_3') \;+\; 3E \\ \mu^*D \;&=\; \frac{2}{3}(A_1' + A_2' + A_3') + 2E \\ [\mu^*D] \;&=\; 2E. \end{aligned}$$

Moreover $K_{X'/X} = E$, and hence

$$\mathcal{J}(D) \;=\; \mu_*\mathcal{O}_{X'}(-E)$$

is the maximal ideal of functions vanishing at the origin. Observe that this computation also shows that rounding does not in general commute with pull-back of $\mathbf{Q}$-divisors. □

The algebraic construction of multiplier ideals started by choosing a resolution of singularities. Therefore it is important to establish:

Proposition 1.6. *The multiplier ideals $\mathcal{J}(D)$ and $\mathcal{J}(\mathfrak{a}^c)$ do not depend on the resolution used to construct them.*

In brief, using the fact that any two resolutions can be dominated by a third, one reduces to checking that if X is already a log resolution of the data at hand, then nothing is changed by passing to a further blow-up:

Lemma 1.7. *Assume that D has SNC support, and let $\mu : X' \longrightarrow X$ be a further log resolution of (X, D). Then*

$$\mu_* \mathcal{O}_{X'}\big(K_{X'/X} - [\mu^* D]\big) \;=\; \mathcal{O}_X\big(-[D]\big).$$

This in turn can be checked by an elementary direct calculation. We refer to [**21**, 9.2.19] for details.

The next point is to reconcile the analytic and algebraic constructions of multiplier ideals.

Proposition 1.8. *Let D be an effective $\mathbf{Q}$-divisor on X. Then*

$$\mathcal{J}_{\text{an}}(X, D) \;=\; \mathcal{J}(X, D),$$

and similarly $\mathcal{J}_{\text{an}}(X, \mathfrak{a}^c) = \mathcal{J}(X, \mathfrak{a}^c)$ for any ideal sheaf $\mathfrak{a}$.

(Strictly speaking, the analytic multiplier ideals are the analytic sheaves associated to their algebraic counterparts, but we do not dwell on this distinction.)

For the Proposition, the key point is that both species of multiplier ideals transform the same way under birational morphisms:

Lemma 1.9. *Let $\mu : X' \longrightarrow X$ be a proper birational map, and let D be an effective $\mathbf{Q}$-divisor on X. Then:*

$$\begin{aligned} \mathcal{J}_{\text{an}}(X, D) &= \mu_*\big(\mathcal{O}_{X'}(K_{X'/X}) \otimes \mathcal{J}_{\text{an}}(X', \mu^* D)\big) \\ \mathcal{J}(X, D) &= \mu_*\big(\mathcal{O}_{X'}(K_{X'/X}) \otimes \mathcal{J}(X', \mu^* D)\big). \end{aligned}$$

In the analytic setting this is a consequence of the change of variables formula for integrals, while the algebraic statement is established with a little computation via the projection formula. The Proposition follows at once from the Lemma. In fact, one is reduced to proving Proposition 1.8 when D or $\mathfrak{a}$ are already in normal crossing form, and this case is handled by Exercise 1.2.

Remark 1.10 (Multiplier ideals on singular varieties)**.** Under favorable circumstances, Definition 1.3 makes sense even when X is singular. The main point at which non-singularity is used in the discussion above is to be able to define the relative canonical bundle $K_{X'/X} = K_{X'} - \mu^* K_X$ of a log resolution

$$\mu : X' \longrightarrow X$$

of (X, D). For this it is enough that X is normal and that K_X is Cartier or even $\mathbf{Q}$-Cartier, so that $\mu^* K_X$ is defined. Thus Definition 1.3 goes through without change provided that X is Gorenstein or $\mathbf{Q}$-Gorenstein. For an arbitrary normal variety X, one can introduce a "boundary" $\mathbf{Q}$-divisor Δ such that $K_X + \Delta$ is $\mathbf{Q}$-Cartier, and define multiplier ideals

$$\mathcal{J}((X, \Delta); D) \;\subseteq\; \mathcal{O}_X.$$

These generalizations are discussed briefly in [**21**, §9.3.G]. DeFernex and Hacon explore in [**4**] the possibility of defining multiplier ideals (without boundaries) on an arbitrary normal variety. However in the sequel we will work almost exclusively with smooth ambient varieties X. □

We conclude this section with two further exercises for the reader.

Exercise 1.11. Assume that X is affine, and let $\mathfrak{a} \subseteq \mathbf{C}[X]$ be an ideal. Given $c > 0$, choose $k > c$ general elements

$$f_1, \ldots, f_k \ \in \ \mathfrak{a},$$

let $A_i = \mathrm{div}(f_i)$, and put $D = \frac{c}{k}(A_1 + \ldots + A_k)$. Then

$$\mathcal{J}(D) \ = \ \mathcal{J}(\mathfrak{a}^c).$$

(By a "general element" of an ideal, one means a general $\mathbf{C}$-linear combination of a collection of generators of the ideal.) □

Exercise 1.12. Let $D = \sum a_i D_i$ be an effective $\mathbf{Q}$-divisor on X. Assume that

$$\mathrm{mult}_x(D) \ =_{\mathrm{def}} \ \sum a_i \cdot \mathrm{mult}_x(D_i) \ \geq \ \dim X$$

for some point $x \in X$. Then $\mathcal{J}(X, D)$ is non-trivial at x, i.e.

$$\mathcal{J}(D)_x \ \subseteq \ \mathfrak{m}_x \ \subseteq \mathcal{O}_x X,$$

where $\mathfrak{m}_x \subseteq \mathcal{O}_X$ is the maximal ideal of x. (Compute the multiplier ideal in question using a resolution $\mu : X' \longrightarrow X$ that dominates the blow-up of X at x, and observe that

$$\mathrm{ord}_E\big(K_{X'/X} - [\mu^* D]\big) \ \leq \ -1,$$

where E is the proper transform of the exceptional divisor over x.) □

Monomial ideals

It is typically very hard to compute the multiplier ideal of an explicitly given divisor or ideal. One important class of examples that has been worked out is that of monomial ideals on affine space. These are handled by a theorem of Howald [**16**].

Let $X = \mathbf{C}^d$, and let

$$\mathfrak{a} \ \subseteq \ \mathbf{C}[x_1, \ldots, x_d]$$

be an ideal generated by monomials in the x_i. Observe that such a monomial is specified by an exponent vector $w = (w_1, \ldots, w_d) \in \mathbf{N}^d$: we write

$$x^w \ = \ x_1^{w_1} \cdot \ldots \cdot x_d^{w_d}.$$

The *Newton polyhedron*

$$P(\mathfrak{a}) \ \subseteq \ \mathbf{R}^d$$

of $\mathfrak{a}$ is the closed convex set spanned by the exponent vectors of all monomials in $\mathfrak{a}$. This is illustrated in Figure 1, which shows the Newton polyhedron for the monomial ideal

$$\mathfrak{a} \ = \ \langle x^4, x^2y, xy^2, y^5 \rangle. \qquad (1.2)$$

Finally, put $\mathbf{1} = (1, \ldots, 1) \in \mathbf{N}^d$.

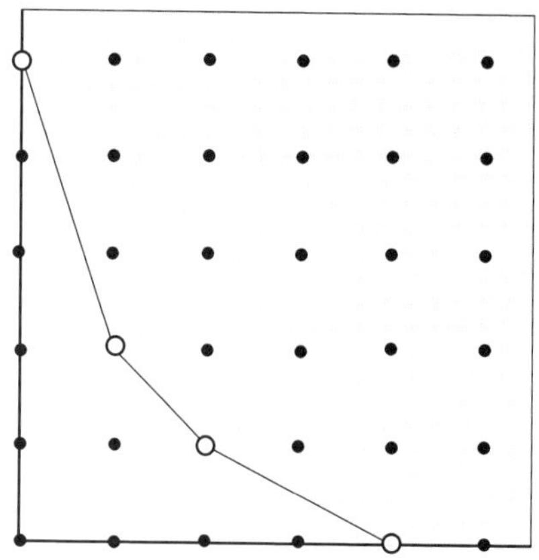

FIGURE 1. Newton Polyhedron of $\langle x^4, x^2y, xy^2, y^5\rangle$

Howald's statement is the following:

Theorem 1.13. *For any* $c > 0$,

$$\mathcal{J}(\mathfrak{a}^c) \subseteq \mathbf{C}[x_1, \ldots, x_d]$$

is the monomial ideal spanned by all monomials x^w *where*

$$w + \mathbf{1} \in \operatorname{int}\big(c \cdot P(\mathfrak{a})\big).$$

Once one knows the statement for which one is aiming, the proof is relatively straight-forward: see [**16**] or [**21**, §9.3.C].

Example 1.14. For $\mathfrak{a} = \langle x^4, x^2y, xy^2, y^5\rangle$, one has $\mathcal{J}(\mathfrak{a}) = (x^2, xy, y^2)$, while if $0 < c < 1$ then $(x, y) \subseteq \mathcal{J}(\mathfrak{a}^c)$. □

Example 1.15. Let

$$\mathfrak{a} = \big(x_1^{e_1}, \ldots, x_d^{e_d}\big) \subseteq \mathbf{C}[x_1, \ldots, x_d].$$

Writing $\xi_1, \ldots, \xi_d$ for the natural coordinates on $\mathbf{R}^d$ adapted to $\mathbf{N}^d \subseteq \mathbf{R}^d$, the Newton polyhedron $P(\mathfrak{a}) \subseteq \mathbf{R}^d$ of $\mathfrak{a}$ is the region in the first orthant given by the equation

$$\frac{\xi_1}{e_1} + \ldots + \frac{\xi_d}{e_d} \geq 1.$$

Hence $\mathcal{J}(\mathfrak{a}^c)$ is the monomial ideal spanned by all monomials x^w whose exponent vectors satisfy the equation

$$\frac{w_1 + 1}{e_1} + \ldots + \frac{w_d + 1}{e_d} > c. \quad \square$$

Invariants defined by multiplier ideals

Multiplier ideals lead to invariants of the singularities of a divisor or the functions in an ideal. The most important and well-known is the following:

Definition 1.16 (Log-canonical threshold). Let D be an effective $\mathbf{Q}$-divisor on a smooth variety X, and let $x \in X$ be a fixed point. The *log-canonical threshold* of D at x is

$$\mathrm{lct}_x(D) \ =_{\mathrm{def}}\ \min\big\{\, c > 0 \,\big|\, \mathcal{J}(cD) \ \text{ is non-trivial at } x \,\big\}.$$

The log-canonical threshold of an ideal sheaf $\mathfrak{a} \subseteq \mathcal{O}_X$ is defined analogously. □

Thus small values of $\mathrm{lct}_x(D)$ reflect more dramatic singularities. Definition 1.25 below explains the etymology of the term.

Exercise 1.17. Let $\mu : X' \longrightarrow X$ be a log resolution of D, and as in equation (1.1) write

$$\mu^* D \ = \ \sum r_i E_i \ \ , \ \ K_{X'/X} \ = \ \sum b_i E_i.$$

Then

$$\mathrm{lct}_x(D) \ = \ \min_{\mu(E_i) \ni x} \Big\{ \frac{b_i + 1}{r_i} \Big\},$$

the minimum being taken over all i such that x lies in the image of the corresponding exceptional divisor. In particular, $\mathrm{lct}_x(D)$ is rational.

Example 1.18 (Complex singularity exponent). One of the early appearances of the log-canonical threshold was in the work **[29]** of Varchenko, who studied the *complex singularity exponent* $c_0(f)$ of a polynomial or holomorphic function f in a neighborhood of $0 \in \mathbf{C}^d$. Specifically, set

$$(*) \qquad c_0(f) \ = \ \sup\Big\{ c > 0 \,\Big|\, \frac{1}{|f|^{2c}} \ \text{ is locally integrable near } 0 \,\Big\}.$$

Writing $\mathrm{lct}_0(f)$ for the log-canonical threshold of the divisor determined by f (or equivalently of the principal ideal generated by f), it follows from Proposition 1.8 that

$$c_0(f) = \mathrm{lct}_0(f).$$

In view of the previous exercise, this establishes the fact – which is certainly not obvious from (*) – that $c_0(f)$ is rational. (The rationality of $c_0(f)$ was proven in this manner by Varchenko, although his work pre-dates the language of multiplier ideals.) □

Exercise 1.19. If $D = \{x^3 - y^2 = 0\} \subseteq \mathbf{C}^2$, then $\mathrm{lct}_0(D) = \frac{5}{6}$. □

Exercise 1.20. Consider as in Example 1.15 the monomial ideal

$$\mathfrak{a} \ = \ (x_1^{e_1}, \ldots, x_d^{e_d}) \ \subseteq \ \mathbf{C}[x_1, \ldots, x_d].$$

Then $\mathrm{lct}_0(\mathfrak{a}) = \sum \frac{1}{e_i}$. □

The log-canonical threshold is the first of a sequence of invariants defined by the "jumping" of multiplier ideals. Specifically, observe that the ideals $\mathcal{J}(cD)$ become deeper as the coefficient c grows. So one is led to:

Proposition/Definition 1.21. *In the situation of Definition 1.16, there exists a discrete sequence of rational numbers $\xi_i = \xi_i(D; x)$ with*

$$0 = \xi_0 \ < \ \xi_1 \ < \ \xi_2 \ < \ \ldots$$

characterized by the property that (the stalks at x of) the multiplier ideals $\mathcal{J}(cD)_x$ are constant exactly for

$$c \in [\xi_i, \xi_{i+1}).$$

The ξ_i are called the jumping numbers of D at x.

Jumping numbers of an ideal sheaf $\mathfrak{a} \subseteq \mathcal{O}_X$ are defined similarly.

It follows from the definition that $\mathrm{lct}_x(D) = \xi_1(D;x)$. In the notation of (1.1), the ξ_i occur among the rational numbers $(b_i + m)/r_i$ for various $m \in \mathbf{N}$. First appearing implicitly in work of Libgober and Loeser-Vaquié, these quantities were studied systematically in [**9**]. In particular, this last paper establishes some connections between jumping coefficients and other invariants.

Exercise 1.22. Compute the jumping numbers of the ideal $\mathfrak{a} = (x_1^{e_1}, \ldots, x_d^{e_d})$. □

Exercise 1.23. Let $\xi_i < \xi_{i+1}$ be consecutive jumping coefficients of an ideal $\mathfrak{a} \subseteq \mathcal{O}_X$ at a point $x \in X$. Then

$$\sqrt{\mathfrak{a}} \cdot \mathcal{J}(\mathfrak{a}^{\xi_i})_x \subseteq \mathcal{J}(\mathfrak{a}^{\xi_{i+1}})_x$$

in $\mathcal{O}_x X$. In particular,

$$(\sqrt{\mathfrak{a}})^m \subseteq \mathcal{J}(\mathfrak{a}^{\xi_m})_x$$

for every $m > 0$. (This was pointed out to us by M. Mustaţă. See Example 3.19 for an application.) □

Remark 1.24. In an analogous fashion, one can define the log-canonical threshold $\mathrm{lct}(D)$ and $\mathrm{lct}(\mathfrak{a})$, as well as jumping numbers $\xi_i(D)$ and $\xi_i(\mathfrak{a})$, globally on X, without localizing at a particular point. We leave the relevant definitions – as well as the natural extension of the previous Exercise – to the reader. □

Finally, we note that multiplier ideals lead to some natural classes of singularities for a pair (X, D) consisting of a smooth variety X and an effective $\mathbf{Q}$-divisor D on X.

Definition 1.25. One says that (X, D) is *Kawamata log-terminal* (KLT) if

$$\mathcal{J}(X, D) = \mathcal{O}_X.$$

The pair (X, D) is *log-canonical* if

$$\mathcal{J}(X, (1-\varepsilon)D) = \mathcal{O}_X$$

for $0 < \varepsilon \ll 1$. □

These concepts (and variants thereof) play an important role in the minimal model program, although in that setting one does not want to limit oneself to smooth ambient varieties.

Remark 1.26 (Characteristic p analogues). Work of Smith, Hara, Yoshida, Watanabe, Takagi, Mustaţă and others has led to the development of theory in characteristic $p > 0$ that closely parallels the theory of multiplier ideals, and reduces to it for ideals lifted from characteristic 0. We refer to the last section of [**10**] for a quick overview and further references. □

LECTURE 2

Vanishing theorems for multiplier ideals

In this Lecture we discuss the basic vanishing theorems for multiplier ideals, and give some first applications.

The Kawamata–Viehweg–Nadel vanishing theorem

We start by recalling some definitions surrounding positivity for divisors. Let X be an irreducible projective variety of dimension d, and let B be a (Cartier) divisor on X. One says that B is *nef* (or *numerically effective*) if

$$(B \cdot C) \ \geq\ 0 \quad \text{for every irreducible curve } C \subseteq X.$$

Nefness means in effect that B is a limit of ample divisors: see [**21**, Chapter 1.4] for a precise account. A divisor B is *big* if the spaces of sections of mB grow maximally with m, i.e. if

$$h^0\big(X, \mathcal{O}_X(mB)\big) \ \sim\ m^d \quad \text{for } m \gg 0.$$

These definitions extend in the evident manner to $\mathbf{Q}$-divisors. We will first deal with divisors that are both nef and big: a typical example arises by pulling back an ample divisor under a birational morphism.

A basic fact is that for a nef divisor, bigness is tested numerically:

Lemma 2.1. *Assume that B is nef. Then B is big if and only if its top self-intersection number is strictly positive: $(B^d) > 0$.*

See [**21**, §2.2] for a proof. This Lemma also remains valid for $\mathbf{Q}$-divisors.

The fundamental result for our purposes was proved independently by Kawamata and Viehweg in the early 1980's.

Theorem 2.2 (Kawamata–Viehweg Vanishing Theorem)**.** *Let X be a smooth projective variety. Consider an integral divisor L and an effective $\mathbf{Q}$-divisor D on X. Assume that*

(i). *$L - D$ is nef and big; and*
(ii). *D has simple normal crossing support.*

Then

$$H^i\big(X, \mathcal{O}_X(K_X + L - [D])\big) \ = \ 0 \quad \textit{for } \ i > 0.$$

As in the previous Lecture, the integer part (or round-down) $[D]$ of a $\mathbf{Q}$-divisor D is obtained by taking the integer part of each of its coefficients.

When $D = 0$ and L is ample, this is the classical Kodaira vanishing theorem. Still taking $D = 0$, the Theorem asserts in general that the statement of Kodaira vanishing remains true for divisors that are merely big and nef: this very useful fact – also due to Kawamata and Viehweg – completes some earlier results of Ramanujam, Mumford, and Grauert–Riemenschneider. However the real power (and subtlety) of Theorem 2.2 lies in the fact that while the positivity hypothesis is tested for a **Q**-divisor, the actual vanishing holds for a round of this divisor. As we shall see, this apparently technical improvement vastly increases the power of the result: taking integer parts can significantly change the shape of a divisor, so in favorable circumstances one gets a vanishing for divisors that are far from positive.

The original proofs of the theorem proceeded by using covering constructions to reduce to the case of integral divisors. An account of this approach, taking into account simplifications introduced by Kollár and Mori in [**20**], appears in [**21**, §4.3.A, §9.1.C]. An alternative approach was developed by Esnault and Viehweg [**12**], and the L^2 $\overline{\partial}$-machinery gives yet another proof. In any event, it is nowadays not substantially harder to establish Theorem 2.2 than to prove the classical Kodaira vanishing.

The main difficulty in applying Theorem 2.2 is that in practice the normal crossing hypothesis is rarely satisfied directly. Given an arbitrary effective **Q**-divisor D on a variety X, a natural idea is to apply vanishing on a resolution of singularities and then "push down" to get a statement on X. Multiplier ideals appear inevitably in so doing, and this leads to the basic vanishing theorems for these ideals.

There are two essential results.

Theorem 2.3 (Local vanishing theorem). *Let X be a smooth variety, D an effective **Q**-divisor on X, and*

$$\mu : X' \longrightarrow X$$

a log resolution of D. Then

$$R^j \mu_* \mathcal{O}_{X'}\big(K_{X'/X} - [\mu^* D]\big) \;=\; 0$$

for $j > 0$.

The analogous statement holds for higher direct images of the sheaves computing the multiplier ideals $\mathcal{J}(\mathfrak{a}^c)$.

Theorem 2.4 (Nadel Vanishing Theorem). *Let X be a smooth projective variety, and let L and D be respectively an integer divisor and an effective **Q**-divisor on X. Assume that $L - D$ is nef and big. Then*

$$H^i\big(X, \mathcal{O}_X(K_X + L) \otimes \mathcal{J}(D)\big) \;=\; 0$$

for $i > 0$.

PROOF OF THEOREM 2.4 (GRANTING THEOREMS 2.2 AND 2.3). Let $\mu : X' \longrightarrow X$ be a log resolution of D, and set

$$L' \;=\; \mu^* L \;\;,\;\; D' \;=\; \mu^* D.$$

Thus $L' - D'$ is a nef and big **Q**-divisor on X', and by construction D' has SNC support. Therefore Kawamata–Viehweg applies on X' to give

$$H^i\big(X', \mathcal{O}_{X'}(K_{X'} + L' - [D'])\big) \;=\; 0 \tag{2.1}$$

for $i > 0$. Now note that

$$K_{X'} + L' - [D'] \;=\; K_{X'/X} - [\mu^* D] + \mu^*(K_X + L).$$

On the other hand, one finds using the projection formula and the definition of $\mathcal{J}(D)$:

$$\begin{aligned}\mu_* \mathcal{O}_{X'}\big(K_{X'/X} - [\mu^* D] + \mu^*(K_X + L)\big) &= \mu_* \mathcal{O}_{X'}\big(K_{X'/X} - [\mu^* D]\big) \otimes \mathcal{O}_X(K_X + L)\\ &= \mathcal{O}_X(K_X + L) \otimes \mathcal{J}(D).\end{aligned}$$

But thanks to Theorem 2.3, the vanishing (2.1) is equivalent to the vanishing

$$H^i\big(X, \mathcal{O}_X(K_X + L) \otimes \mathcal{J}(D)\big) \;=\; 0$$

of the direct image of the sheaf in question, as required. □

The proof of Theorem 2.3 is similar: one reduces to the case when X is projective, and applies the result of Kawamata and Viehweg in the global setting. See [**21**, §9.4.A] for an account.

Singularities of plane curves and projective hypersurfaces

As a first illustration, we will apply these theorems to prove some classical results about singularities of plane curves and their extensions to hypersurfaces of higher dimension. Starting with a singular hypersurface, the strategy is to build a **Q**-divisor having a non-trivial multiplier ideal. Then the vanishing theorems give information about the postulation of the singularities of the original hypersurface.

Consider to begin with a (reduced) plane curve $C \subseteq \mathbf{P}^2$ of degree m, and let

$$\Sigma \;=\; \operatorname{Sing}(C),$$

considered as a reduced finite subset of the plane. Our starting point is the classical

Proposition 2.5. *The set Σ imposes independent conditions on curves of degree $k \geq m-2$, i.e.*

$$H^1\big(\mathbf{P}^2, \mathcal{I}_\Sigma(k)\big) \;=\; 0 \quad \textit{for} \;\; k \geq m-2. \tag{2.2}$$

Here $\mathcal{I}_\Sigma$ denotes the ideal sheaf of Σ. We give a proof using Nadel vanishing momentarily, but first we discuss a less familiar extension due to Zariski.

Specifically, suppose that $C \subseteq \mathbf{P}^2$ has a certain number of cusps, defined in local analytic coordinates by an equation $x^3 - y^2 = 0$. (C may have other singularities as well.) Let

$$\Xi \;=\; \operatorname{Cusps}(C),$$

again regarded as a reduced finite subset of $\mathbf{P}^2$. Zariski [**30**] proved that one gets a stronger result for the postulation of Ξ. In fact:

Proposition 2.6. *One has*

$$H^1\big(\mathbf{P}^2, \mathcal{I}_\Xi(k)\big) \;=\; 0 \quad \textit{for} \;\; k > \tfrac{5}{6}m - 3.$$

Interestingly enough, Zariksi proved this by considering the irregularity of the cyclic cover of $\mathbf{P}^2$ branched along C. One can see the Kawamata–Viehweg–Nadel theorem as a vast generalization of this approach.

Proof of Proposition 2.6. This is a direct consequence of Nadel vanishing. In fact, consider the $\mathbf{Q}$-divisor $D = \frac{5}{6}C$. Since the log-canonical threshold of a cusp is $= \frac{5}{6}$, one has $\mathcal{J}(D) \subseteq \mathcal{I}_\Xi$. But as C is reduced the multiplier ideal $\mathcal{J}(D)$ is non-trivial only at finitely many points. Thus $\mathcal{I}_\Xi/\mathcal{J}(D)$ is supported on a finite set, and therefore the map

$$H^1\big(\mathbf{P}^2, \mathcal{O}_{\mathbf{P}^2}(k) \otimes \mathcal{J}(D)\big) \longrightarrow H^1\big(\mathbf{P}^2, \mathcal{O}_{\mathbf{P}^2}(k) \otimes \mathcal{I}_\Xi\big)$$

is surjective for all k. So it suffices to prove that the group on the left vanishes for $k > \frac{5}{6}m - 3$. But this follows immediately from Nadel vanishing upon recalling that $\mathcal{O}_{\mathbf{P}^2}(K_{\mathbf{P}^2}) = \mathcal{O}_{\mathbf{P}^2}(-3)$. □

Proof of Proposition 2.5. Here an additional trick is required in order to produce a $\mathbf{Q}$-divisor whose multiplier ideal vanishes on finite set including $\Sigma = \mathrm{Sing}(C)$. Specifically, fix $0 < \varepsilon \ll 1$, and let Γ be a reduced curve of degree ℓ, not containing any components of C, passing through Σ. Consider the $\mathbf{Q}$-divisor

$$D \ = \ (1-\varepsilon)C \ + \ 2\varepsilon\Gamma.$$

This has multiplicity ≥ 2 at each singular point of C, and hence $\mathcal{J}(D)$ vanishes on Σ thanks to Exercise 1.12. Moreover $\mathcal{J}(D)$ is again cosupported on a finite set since no component of D has coefficient ≥ 1. Therefore, as in the previous proof, it suffices to show that

$$(*) \qquad H^1\big(\mathbf{P}^2, \mathcal{O}_{\mathbf{P}^2}(k) \otimes \mathcal{J}(D)\big) \ = \ 0$$

for $k \geq m - 2$. But

$$\deg D \ = \ (1-\varepsilon)m + 2\varepsilon\ell \ < \ m+1$$

for $\varepsilon \ll 1$, and so (*) again follows from Nadel vanishing. □

Example 2.7. When C is the union of m general lines, the bound $k \geq m-2$ in (2.2) is the best possible. However we will see in Exercise 3.8 that one can take $k \geq m - 3$ when C is irreducible. □

Finally, we present a generalization of Proposition 2.5 to higher dimensional hypersurfaces.

Proposition 2.8. *Let $S \subseteq \mathbf{P}^r$ be a (reduced) hypersurface of degree $m \geq 3$ having only isolated singularities, and set*

$$\Sigma \ = \ \mathrm{Sing}(S).$$

Then Σ imposes independent conditions on hypersurfaces of degree $\geq m(r-1) - (2r-1)$, i.e.

$$H^1\big(\mathbf{P}^r, \mathcal{I}_\Sigma(k)\big) \ = \ 0 \ \text{ for } k \geq m(r-1) - (2r-1).$$

When $r = 3$ the statement was given by Severi. The general case, as well as the proof that follows, is due to Park and Woo [**24**].

PROOF OF PROPOSITION 2.8. We may suppose that $r \geq 3$, in which case the hypotheses imply that S is irreducible. Write

$$\Sigma = \{P_1, \ldots, P_t\},$$

and denote by $\Lambda \subseteq |\mathcal{O}_{\mathbf{P}^r}(m-1)|$ the linear series spanned by the partial derivatives of a defining equation of S; observe that every divisor in Λ passes through the points of Σ. For each $P_i \in \Sigma$, there exists a divisor $\Gamma_i \in \Lambda$ with $\mathrm{mult}_{P_i}(\Gamma_i) \geq 2$.[1] Then for $0 < \varepsilon \ll 1$ and $\ell \gg 0$, set

$$D = (1-\varepsilon)S + \varepsilon \cdot \sum_{i=1}^{t} \Gamma_i + \frac{\big(r-2-\varepsilon(t-1)\big)}{\ell} \cdot \sum_{j=1}^{\ell} A_j,$$

where $A_1, \ldots, A_\ell \in \Lambda$ are general divisors. As S is irreducible, none of the Γ_i or A_j occur as components of S, and therefore $\mathcal{J}(D)$ is cosupported on a finite set provided that $\varepsilon \ll 1$ and $t \gg 0$. One has

$$\mathrm{mult}_{P_i}(D) \geq (2-2\varepsilon) + \varepsilon(t+1) + \big((r-2) - \varepsilon(t-1)\big) \geq r,$$

which guarantees that $\mathcal{J}(D) \subseteq \mathcal{I}_\Sigma$. Moreover:

$$\begin{aligned} \deg(D) &= m(1-\varepsilon) + \varepsilon t(m-1) + \big((r-2) - \varepsilon(t-1)\big)(m-1) \\ &< m(r-1) - (r-2), \end{aligned}$$

and so the required vanishing follows from Theorem 2.4. □

Singularities of theta divisors

We next discuss a theorem of Kollár concerning the singularities of theta divisors.

Let (A, Θ) be a principally polarized abelian variety (PPAV) of dimension g. Recall that by definition this means that $A = \mathbf{C}^g/\Lambda$ is a g-dimensional complex torus, and $\Theta \subseteq A$ is an ample divisor with the property that

$$h^0\big(A, \mathcal{O}_A(\Theta)\big) = 1.$$

The motivating example historically is the polarized Jacobian (JC, Θ_C) of a smooth projective curve of genus g.

In their classical work [**1**], Andreotti and Meyer showed that Jacobians are generically characterized among all PPAV's by the condition that $\dim \mathrm{Sing}(\Theta) \geq g-4$.[2] In view of this, it is interesting to ask what singularities can occur on theta divisors. Kollár used vanishing for **Q**-divisors to prove a very clean statement along these lines.

Kollár's result is the following:

Theorem 2.9. *The pair (A, Θ) is log-canonical. In particular,*

$$\mathrm{mult}_x(\Theta) \leq g$$

for every $x \in A$.

---

[1]This uses that $m \geq 3$: see [**24**, Lemma 3.2].

[2]The precise statement is that the Jacobians form an irreducible component of the locus of all (A, Θ) defined by the stated condition.

PROOF. Suppose to the contrary that $\mathcal{J}((1-\varepsilon)\Theta) \neq \mathcal{O}_A$ for some $\varepsilon > 0$. We will derive a contradiction from Nadel vanishing. To this end, let $Z \subseteq A$ denote the subscheme defined by $\mathcal{J}((1-\varepsilon)\Theta)$. Then $Z \subseteq \Theta$: this is clear set-theoretically, but in fact it holds on the level of schemes thanks to the inclusion

$$\mathcal{O}_A(-\Theta) \;=\; \mathcal{J}(\Theta) \;\subseteq\; \mathcal{J}((1-\varepsilon)\Theta).$$

Now consider the short exact sequence

$$0 \longrightarrow \mathcal{O}_A(\Theta) \otimes \mathcal{J}((1-\varepsilon)\Theta) \longrightarrow \mathcal{O}_A(\Theta) \longrightarrow \mathcal{O}_Z(\Theta) \longrightarrow 0$$

The H^1 of the term on the left vanishes thanks to Theorem 2.4 and the fact that $K_A = 0$. Therefore the map

$$H^0\big(A, \mathcal{O}_A(\Theta)\big) \longrightarrow H^0\big(Z, \mathcal{O}_Z(\Theta)\big)$$

is surjective. On the other hand, the unique section of $\mathcal{O}_A(\Theta)$ vanishes on Z, and so we conclude that

(*) $$H^0\big(Z, \mathcal{O}_Z(\Theta)\big) \;=\; 0.$$

To complete the proof, it remains only to show that (*) cannot hold. To this end, let $a \in A$ be a general point. Then $\Theta + a$ meets Z properly, and hence $H^0\big(Z, \mathcal{O}_Z(\Theta + a)\big) \neq 0$. Letting $a \to 0$, if follows by semicontinuity that

$$H^0\big(Z, \mathcal{O}_Z(\Theta)\big) \;\neq\; 0,$$

as required. □

Remark 2.10. In the situation of the theorem, the fact that (A, Θ) is log-canonical implies more generally that the locus

$$\Sigma_k(\Theta) \;=_{\text{def}}\; \big\{x \in A \,\big|\, \text{mult}_x\Theta \geq k\big\}$$

of k-fold points of Θ has codimension $\geq k$ in A. Equality is achieved when

$$(A, \Theta) \;=\; (A_1, \Theta_1) \times \ldots \times (A_k, \Theta_k)$$

is the product of k smaller PPAV's. It was established in [**6**] that this is the only situation in which $\text{codim}_A \Sigma_k(\Theta) = k$. It was also shown in that paper that if Θ is irreducible, then Θ is normal with rational singularities. □

Uniform global generation

As a final application, we prove a useful result to the effect that sheaves of the form $\mathcal{O}(L) \otimes \mathcal{J}(D)$, where D is a $\mathbf{Q}$-divisor numerically equivalent to L, become globally generated after twisting by a fixed divisor. This was first observed by Esnault and Viewheg, and later rediscovered independently by Siu. The statement plays an important role in the extension theorems of Siu discussed in Lecture 5.

The theorem for which we are aiming is the following:

Theorem 2.11. *Let X be a smooth projective variety of dimension d. There exists a divisor B on X with the following property:*

- *For any divisor L on X; and*
- *For any effective $\mathbf{Q}$-divisor $D \equiv_{num} L$,*

the sheaf $\mathcal{O}_X(L + B) \otimes \mathcal{J}(D)$is globally generated.

Note that the hypothesis implies that L is $\mathbf{Q}$-effective, i.e. that $H^0(X, \mathcal{O}_X(mL)) \neq 0$ for some $m \gg 0$. The crucial point is that B is independent of the choice of L and D.

Corollary 2.12. *There is a fixed divisor B on any smooth variety X with the property that*

$$H^0(X, \mathcal{O}_X(L+B)) \neq 0$$

for any big (or even $\mathbf{Q}$-effective) divisor L on X. □

The Theorem is actually an immediate consequence of Nadel vanishing and the elementary lemma of Castelnuovo–Mumford:

Lemma 2.13 (Castelnouvo–Mumford). *Let $\mathcal{F}$ be a coherent sheaf on a projective variety X, and let H be a basepoint-free ample divisor on X. Assume that*

$$H^i(X, \mathcal{F} \otimes \mathcal{O}_X(-iH)) = 0 \quad \text{for } i > 0.$$

Then $\mathcal{F}$ is globally generated.

We refer to [**21**, §1.8] for the proof.

Proof of Theorem 2.11. As above, let $d = \dim X$. It sufficies to take $B = K_X + (d+1)H$ for a very ample divisor H, in which case Theorem 2.4 gives the vanishings required for the Castelnuovo–Mumford lemma. □

LECTURE 3

Local properties of multiplier ideals

In this Lecture we will discuss some local properties of multiplier ideals. First we take up the restriction theorem: here we emphasize the use of adjoint ideals, whose importance has lately come into focus. The remaining sections deal with the subadditivity and Skoda theorems. As an application of the latter, we give a down-to-earth discussion of the recent results of [**22**] concerning syzygetic properties of multiplier ideals.

Adjoint ideals and the restriction theorem

Let X be a smooth complex variety, let D be an effective $\mathbf{Q}$-divisor on X, and let $S \subseteq X$ be a smooth irreducible divisor, not contained in any component of D. Thus the restriction D_S of D to S is a well-defined $\mathbf{Q}$-divisor on S.

There are now two multiplier-type ideals one can form on S. First, one can take the multiplier ideal $\mathcal{J}(X, D)$ of D on X, and then restrict this ideal to S. On the other hand, one can form the multiplier ideal $\mathcal{J}(S, D_S)$ on S of the restricted divisor D_S. In general, these two ideals are different:

Example 3.1. Let $X = \mathbf{C}^2$, let S be the x-axis, and let $A = \{y - x^2 = 0\}$ be a parabola tangent to S. If $D = \frac{1}{2}A$, then

$$\mathcal{J}(X, D) \ = \ \mathcal{O}_X \ \ , \ \ \mathcal{J}(S, D_S) \ = \ \mathcal{O}_S(-P),$$

where $P \in S$ denotes the origin. □

However a very basic fact is that there is a containment between these two ideal sheaves on S.

Theorem 3.2 (Restriction Theorem). *One has an inclusion*

$$\mathcal{J}(S, D_S) \ \subseteq \ \mathcal{J}(X, D) \cdot \mathcal{O}_S.$$

This result is perhaps the most important local property of multiplier ideals. In the analytic perspective, it comes from the Osahawa–Takegoshi extension theorem: an element in the ideal on the left is a function on S satisfying an integrability condition, and Osahawa–Takegoshi guarantees that it is the restriction of a function satisfying the analogous integrability condition on X.

We will prove Theorem 3.2 by constructing and studying the *adjoint ideal* $\mathrm{Adj}_S(X, D)$ of D along S. This is an ideal sheaf on X that governs the multiplier ideal $\mathcal{J}(S, D_S)$ of the restriction D_S of D to S.

Theorem 3.3. *With hypotheses as above, there exists an ideal sheaf*

$$\mathrm{Adj}_S(X, D) \subseteq \mathcal{O}_X$$

sitting in an exact sequence:

$$0 \longrightarrow \mathcal{J}(X, D) \otimes \mathcal{O}_X(-S) \xrightarrow{\cdot S} \mathrm{Adj}_S(X, D) \longrightarrow \mathcal{J}(S, D_S) \longrightarrow 0. \tag{3.1}$$

Moreover, for any $0 < \varepsilon \leq 1$:

$$\mathrm{Adj}_S(X, D) \subseteq \mathcal{J}(X, D + (1-\varepsilon)S). \tag{3.2}$$

The sequence (3.1) shows that

$$\mathrm{Adj}_S(X, D) \cdot \mathcal{O}_S = \mathcal{J}(S, D_S).$$

Therefore (3.2) not only yields the Restriction Theorem, it implies that in fact

$$\mathcal{J}(S, D_S) \subseteq \mathcal{J}(X, D + (1-\varepsilon)S) \cdot \mathcal{O}_S$$

for any $0 < \varepsilon \leq 1$.

Before proving Theorems 3.2 and 3.3, we record some consequences.

Corollary 3.4. *Let $Y \subseteq X$ be a smooth subvariety not contained in the support of D, so that the restriction D_Y of D to Y is defined. Then*

$$\mathcal{J}(Y, D_Y) \subseteq \mathcal{J}(X, D) \cdot \mathcal{O}_Y.$$

(This follows inductively from the restriction theorem since Y is locally a complete intersection in X.) □

Corollary 3.5. *In the situation of the Theorem, assume that $\mathcal{J}(S, D_S)$ is trivial at a point $x \in S$. Then $\mathcal{J}(X, D + (1-\varepsilon)S)$ (and hence also $\mathcal{J}(X, D)$) are trivial at x.* □

Exercise 3.6. If D is an effective $\mathbf{Q}$-divisor on X such that $\mathrm{mult}_x(D) < 1$, then $\mathcal{J}(X, D)$ is trivial at x. (Using the previous corollary, take hyperplane sections to reduce to the case $\dim X = 1$, where the result is clear.) □

Proof of Theorem 3.3. Let $\mu : X' \longrightarrow X$ be a log resolution of $(X, D+S)$, and denote by $S' \subseteq X'$ the proper transform of S, so that in particular $\mu_S : S' \longrightarrow S$ is a log resolution of (S, D_S). Write

$$\mu^* S = S' + R,$$

and put $B = K_{X'/X} - [\mu^* D] - R$. We define:

$$\mathrm{Adj}_S(X, D) = \mu_* \mathcal{O}_{X'}(B).$$

To establish the exact sequence (3.1), note first that

$$K_S \equiv_{\mathrm{lin}} \left(K_X + S\right)_{|S} \quad , \quad K_{S'} \equiv_{\mathrm{lin}} \left(K_{X'} + S'\right)_{|S'},$$

and hence

$$K_{S'/S} = \left(K_{X'/X} - R\right)_{|S'}.$$

(One can check that this holds on the level of divisors, and not only for linear equivalence classes.) On X', where the relevant divisors have SNC support, rounding commutes with restriction. Therefore

$$\begin{aligned} \mathcal{J}(S, D_S) &= \mu_{S,*}\mathcal{O}_{S'}(K_{S'/S} - [\mu_S^* D_S]) \\ &= \mu_* \mathcal{O}_{S'}(B_{S'}). \end{aligned}$$

Observing that

$$B - S' = K_{X'/X} - [\mu^* D] - \mu^* S,$$

the adjoint exact sequence (3.1) follows by pushing forward

$$0 \longrightarrow \mathcal{O}_{X'}(B - S') \longrightarrow \mathcal{O}_{X'}(B) \longrightarrow \mathcal{O}_{S'}(B_{S'}) \longrightarrow 0$$

since the higher direct images of the term on the left vanish thanks to local vanishing. Finally, note that

$$\begin{aligned} B = K_{X'/X} - [\mu^* D] - R &\preccurlyeq K_{X'/X} - \big[\mu^* D + (1-\varepsilon) R\big] \\ &= K_{X'/X} - \big[\mu^* D + (1-\varepsilon)R + (1-\varepsilon)S'\big], \end{aligned}$$

which yields (3.2). □

Remark 3.7. We leave it to the reader to show that $\mathrm{Adj}_S(X, D)$ is independent of the choice of log resolution. □

If $\mathfrak{a} \subseteq \mathcal{O}_X$ is an ideal that does not vanish identically on S, then for $c > 0$ one can define in the analogous manner an adjoint ideal

$$\mathrm{Adj}_S(X, \mathfrak{a}^c) \subseteq \mathcal{O}_X$$

sitting in exact sequence

$$0 \longrightarrow \mathcal{J}(X, \mathfrak{a}^c) \otimes \mathcal{O}_X(-S) \longrightarrow \mathrm{Adj}_S(X, \mathfrak{a}^c) \longrightarrow \mathcal{J}(S, (\mathfrak{a}_S)^c) \longrightarrow 0,$$

where $\mathfrak{a}_S =_{\mathrm{def}} \mathfrak{a} \cdot \mathcal{O}_S$ is the restriction of $\mathfrak{a}$ to S. In particular, the analogue of the restriction theorem holds for the multiplier ideals associated to $\mathfrak{a}$.

Finally, Theorem 3.3 works perfectly well if S is allowed to be singular, as in Remark 1.10. Since in any event S is Gorenstein, when in addition it is normal there is no question about the meaning of the multiplier ideals on S appearing in the Theorem. In this case the statement and proof of 3.3 remain valid without change. For arbitrary S one can twist by $\mathcal{O}_X(K_X + S)$ and rewrite (3.1) as

$$\begin{aligned} (3.3)\quad 0 \longrightarrow \mathcal{O}_X(K_X) \otimes \mathcal{J}(X, D) &\longrightarrow \mathcal{O}_X(K_X + S) \otimes \mathrm{Adj}_S(X, D) \\ &\longrightarrow \mu_*\big(\mathcal{O}_{S'}(K_{S'}) \otimes \mathcal{J}(S', D'_S)\big) \longrightarrow 0, \end{aligned}$$

where $D_{S'} = \mu^* D_S$ denotes the pullback of D_S to S'. When $D = 0$ this is the adjoint exact sequence appearing in [**6**] and [**21**, Section 9.3.E].

We conclude with some exercises for the reader.

Exercise 3.8 (Irreducible plane curves). Let $C \subseteq \mathbf{P}^2$ be an irreducible (reduced) plane curve of degree m, and as in Proposition 2.5 put $\Sigma = \mathrm{Sing}(C)$. Then the points of Σ impose independent conditions on curves of degree $\geq m - 3$. (Let $f : C' \longrightarrow C$ be the desingularization of C, and use the adjoint sequence

$$0 \longrightarrow \mathcal{O}_{\mathbf{P}^2}(-3) \longrightarrow \mathcal{O}_{\mathbf{P}^2}(m-3) \otimes \mathrm{Adj}_C \longrightarrow f_* \mathcal{O}_{C'}(K_{C'}) \longrightarrow 0$$

coming from (3.3) with $D = 0$ to show that

$$H^1\big(\mathbf{P}^2, \mathcal{O}_{\mathbf{P}^2}(k) \otimes \mathrm{Adj}_C\big) \;=\; 0$$

when $k \geq m - 3$.) □

Example 3.9 (Condition for an embedded point)**.** It is sometimes interesting to know that a multiplier ideal has an embedded prime ideal. As usual, let X be a smooth variety of dimension d, and fix a point $x \in X$ with maximal ideal $\mathfrak{m}$. Consider an effective $\mathbf{Q}$-divisor D with integral multiplicity $s = \mathrm{mult}_x(D) \geq d$, and denote by $\overline{D}$ the effective $\mathbf{Q}$-divisor of degree s on $\mathbf{P}(T_xX) = \mathbf{P}^{d-1}$ arising in the natural way as the "projectivized tangent cone" of D at x. The following proposition asserts that if there is a hypersurface of degree $s - d$ on $\mathbf{P}^{d-1}$ vanishing along the multiplier ideal $\mathcal{J}(\mathbf{P}^{d-1}, \overline{D})$, then $\mathcal{J}(X, D)$ has an embedded point at x:

Proposition. *If*

$$H^0\Big(\mathbf{P}^{d-1}, \mathcal{J}(\mathbf{P}^{d-1}, \overline{D}) \otimes \mathcal{O}_{\mathbf{P}^{d-1}}(s-d)\Big) \;\neq\; 0,$$

then $\mathfrak{m}$ is an associated prime of $\mathcal{J}(X, D)$.

Observe that the statement is interesting only when $\mathcal{J}(X, D)$ is not itself cosupported at x.

For the proof, let $\pi : X' \longrightarrow X$ be the blowing up of x, with exceptional divisor $E = \mathbf{P}^{d-1}$, so that $K_{X'/X} = (d-1)E$. Write D' for the proper transform of D, so that $D' \equiv_{\mathrm{num}} \pi^*D - sE$, and note that $D_E = \overline{D}$. Now consider the adjoint sequence for D':

$$0 \longrightarrow \mathcal{J}(X', D') \otimes \mathcal{O}_{X'}(-E) \longrightarrow \mathrm{Adj}_E(X', D') \longrightarrow \mathcal{J}(\mathbf{P}^{d-1}, \overline{D}) \longrightarrow 0,$$

and twist through by $\mathcal{O}_{X'}\big((d-s)E\big)$. The resulting term on the left pushes down with vanishing higher direct images to $\mathcal{J}(X, D)$ thanks to Lemma 1.9. One finds an exact sequence having the shape

$$0 \longrightarrow \mathcal{J}(X, D) \longrightarrow \mathcal{A} \longrightarrow H^0\Big(E, \mathcal{J}(E, \overline{D}) \otimes \mathcal{O}_E((d-s)E)\Big) \longrightarrow 0,$$

where $\mathcal{A}$ is an ideal on X, and the vector space on the right is viewed as a skyscraper sheaf supported at x. But the hypothesis of the Proposition is exactly that this vector space is non-zero, and the assertion follows. □

The subadditivity theorem

The Restriction Theorem was applied in [**5**] to prove a result asserting that the multiplier ideal of a product of two ideals must be at least as deep as the product of the corresponding multiplier ideals. This will be useful at a couple of points in the sequel.

We start by defining "mixed" multiplier ideals:

Definition 3.10. Let $\mathfrak{a}, \mathfrak{b} \subset \mathcal{O}_X$ be two ideal sheaves. Given $c, e \geq 0$, the multiplier ideal

$$\mathcal{J}(\mathfrak{a}^c \cdot \mathfrak{b}^e) \;\subseteq\; \mathcal{O}_X$$

is defined by taking a common log resolution $\mu : X' \longrightarrow X$ of $\mathfrak{a}$ and $\mathfrak{b}$, with

$$\mathfrak{a} \cdot \mathcal{O}_{X'} = \mathcal{O}_{X'}(-A) \ , \ \mathfrak{b} \cdot \mathcal{O}_{X'} = \mathcal{O}_{X'}(-B)$$

for divisors A, B with SNC support, and setting

$$\mathcal{J}(\mathfrak{a}^c \cdot \mathfrak{b}^e) = \mu_* \mathcal{O}_{X'}\big(K_{X'/X} - [cA + eB]\big). \quad \square$$

The subadditivity theorem compares these mixed ideals to the multiplier ideals of the two factors.

Theorem 3.11 (Subadditivity Theorem). *One has an inclusion*

$$\mathcal{J}(\mathfrak{a}^c \cdot \mathfrak{b}^e) \subseteq \mathcal{J}(\mathfrak{a}^c) \cdot \mathcal{J}(\mathfrak{b}^e).$$

Similarly, $\mathcal{J}(X, D_1 + D_2) \subseteq \mathcal{J}(X, D_1) \cdot \mathcal{J}(X, D_2)$ for any two effective $\mathbf{Q}$-divisors D_1, D_2 on X

Sketch of Proof of Theorem 3.11. Consider the product $X \times X$ with projections

$$p_1 \, , \, p_2 \, : \, X \times X \longrightarrow X.$$

The first step is to show via the Künneth formula that

$$(*) \qquad \mathcal{J}(X \times X, (p_1^*\mathfrak{a})^c \cdot (p_2^*\mathfrak{b})^e) = p_1^*\mathcal{J}(X, \mathfrak{a}^c) \ \cdot \ p_2^*\mathcal{J}(X, \mathfrak{b}^e).$$

The one simply restricts to the diagonal $\Delta = X \subseteq X \times X$ using Corollary 3.4. Specifically,

$$\begin{aligned} \mathcal{J}(X, \mathfrak{a}^c \cdot \mathfrak{b}^e) &= \mathcal{J}(\Delta, ((p_1^*\mathfrak{a})^c \cdot (p_2^*\mathfrak{b})^e)_{|\Delta}) \\ &\subseteq \mathcal{J}(X \times X, (p_1^*\mathfrak{a})^c \cdot (p_2^*\mathfrak{b})^e)_{|\Delta} \\ &= \mathcal{J}(X, \mathfrak{a}^c) \cdot \mathcal{J}(X, \mathfrak{b}^e), \end{aligned}$$

the last equality coming from (*). $\square$

Exercise 3.12. Let $f : Y \longrightarrow X$ be a morphism, and let D be a $\mathbf{Q}$-divisor on X whose support does not contain the image of Y. Then

$$\mathcal{J}(Y, f^*D) \subseteq f^{-1}\mathcal{J}(X, D). \quad \square$$

Skoda's theorem

We now discuss Skoda's theorem, which computes the multiplier ideals associated to powers of an ideal.

Let X be a smooth variety of dimension d, and let $\mathfrak{a}, \mathfrak{b} \subseteq \mathcal{O}_X$ be ideal sheaves. In its simplest form, Skoda's theorem is the following:

Theorem 3.13 (Skoda's Theorem). *Assume that $m \geq d$. Then*

$$\mathcal{J}(\mathfrak{a}^m \cdot \mathfrak{b}^c) = \mathfrak{a} \cdot \mathcal{J}(\mathfrak{a}^{m-1} \cdot \mathfrak{b}^c)$$

for any $c \geq 0$.

Note that it follows that $\mathcal{J}(\mathfrak{a}^m\mathfrak{b}^c) = \mathfrak{a}^{m+1-d}\mathcal{J}(\mathfrak{a}^{d-1}\mathfrak{b}^c)$.

The algebraic proof of the theorem – as in [**7**] or [**21**, §10.6] – actually yields a more general statement that in turn has some interesting consequences. Before stating this, we record an elementary observation, whose proof we leave to the reader:

Lemma 3.14. *For any $\ell, k \geq 0$ there is an inclusion*

$$\mathfrak{a}^k \cdot \mathcal{J}(\mathfrak{a}^\ell \cdot \mathfrak{b}^c) \ \subseteq \ \mathcal{J}(\mathfrak{a}^{k+\ell} \cdot \mathfrak{b}^c). \quad \square$$

Theorem 3.13 will follow from the exactness of certain "Skoda complexes." Specifically, assume that X is affine, fix any point $x \in X$, and choose d general elements

$$f_1, \ldots, f_d \ \in \ \mathfrak{a}.$$

Theorem 3.15. *Still supposing that $m \geq d = \dim X$, the f_i determine a Koszul-type complex*

$$0 \longrightarrow \Lambda^d \mathcal{O}^d \otimes \mathcal{J}(\mathfrak{a}^{m-d}\mathfrak{b}^c) \longrightarrow \Lambda^{d-1}\mathcal{O}^d \otimes \mathcal{J}(\mathfrak{a}^{m+1-d}\mathfrak{b}^c) \longrightarrow \ldots$$
$$\ldots \longrightarrow \Lambda^2\mathcal{O}^d \otimes \mathcal{J}(\mathfrak{a}^{m-2}\mathfrak{b}^c) \longrightarrow \mathcal{O}^d \otimes \mathcal{J}(\mathfrak{a}^{m-1}\mathfrak{b}^c) \longrightarrow \mathcal{J}(\mathfrak{a}^m\mathfrak{b}^c) \longrightarrow 0$$

that is exact in a neighborhood of x.

The homomorphism $\mathcal{O}^d \otimes \mathcal{J}(\mathfrak{a}^{d-1}\mathfrak{b}^c) \longrightarrow \mathcal{J}(\mathfrak{a}^d\mathfrak{b}^c)$ on the right is given by multiplication by the vector $(f_1, \ldots, f_d)$. The surjectivity of this map implies that

$$\mathcal{J}(\mathfrak{a}^m\mathfrak{b}^c) \ = \ (f_1, \ldots, f_d) \cdot \mathcal{J}(\mathfrak{a}^{m-1}\mathfrak{b}^c).$$

But thanks to the Lemma one has

$$(f_1, \ldots, f_d) \cdot \mathcal{J}(\mathfrak{a}^{m-1}\mathfrak{b}^c) \ \subseteq \ \mathfrak{a} \cdot \mathcal{J}(\mathfrak{a}^{m-1}\mathfrak{b}^c) \ \subseteq \ \mathcal{J}(\mathfrak{a}^m\mathfrak{b}^c),$$

so Skoda's theorem follows.

Proof of Theorem 3.15. Let $\mu : X' \longrightarrow X$ be a log resolution of $\mathfrak{a}$ and $\mathfrak{b}$ as in Definition 3.10. We keep the notation of that definition, so that in particular $\mathfrak{a} \cdot \mathcal{O}_{X'} = \mathcal{O}_{X'}(-A)$. The d general elements $f_1, \ldots .f_d \in \mathfrak{a}$ determine sections

$$f_i' \ \in \ H^0\big(X', \mathcal{O}_{X'}(-A)\big),$$

and an elementary dimension count shows that after possibly shrinking X one can suppose that these sections actually generate $\mathcal{O}_{X'}(-A)$ (cf. [**21**, 9.6.19]). The f_i' then determine an exact Koszul complex

(3.4)

$$0 \longrightarrow \Lambda^d\mathcal{O}^d \otimes \mathcal{O}_{X'}(-(m-d)A) \longrightarrow \Lambda^{d-1}\mathcal{O}^d \otimes \mathcal{O}_{X'}(-(m-d-1)A) \longrightarrow \ldots$$
$$\ldots \longrightarrow \Lambda^2\mathcal{O}^d \otimes \mathcal{O}_{X'}(-(m-2)A) \longrightarrow \mathcal{O}^d \otimes \mathcal{O}_{X'}(-(m-1)A) \longrightarrow \mathcal{O}_{X'}(-mA) \longrightarrow 0$$

of vector bundles on X'. Now twist (3.4) through by $\mathcal{O}_{X'}\big(K_{X'/X} - [cB]\big)$. The higher direct images of all the terms of the resulting complex vanish thanks to Theorem 2.3. This implies the exactness of the complex on X obtained by taking direct images of the indicated twist of (3.4), which is exactly the assertion of the Theorem. $\square$

Remark 3.16. The functions $f_1, \ldots, f_r \in \mathfrak{a}$ occuring in Theorem 3.15 do not necessarily generate $\mathfrak{a}$. Rather (after shrinking X) they generate an ideal $\mathfrak{r} \subseteq \mathfrak{a}$ with the property that

$$\mathfrak{r} \cdot \mathcal{O}_{X'} = \mathfrak{a} \cdot \mathcal{O}_{X'},$$

which is equivalent to saying the $\mathfrak{r}$ and $\mathfrak{a}$ have the same integral closure. Such an ideal $\mathfrak{r}$ is called a "reduction" of $\mathfrak{a}$. See [**21**, §10.6.A] for more details. □

Following [**22**], we use Theorem 3.15 to show that multiplier ideals satisfy some unexpected syzygetic conditions. By way of background, it is natural to ask which ideals $\mathfrak{d} \subseteq \mathcal{O}_X$ can be realized as a multiplier ideal $\mathfrak{d} = \mathcal{J}(\mathfrak{b}^c)$ for some $\mathfrak{b}$ and $c \geq 0$. It follows from the definition that multiplier ideals are integrally closed, meaning that membership in a multiplier ideal is tested by order of vanishing along some divisors over X. However until recently, multiplier ideals were not known to satisfy any other local properties. In fact, Favre–Jonsson [**13**] and Lipman–Watanabe [**23**] showed that in dimension $d = 2$, any integrally closed ideal is locally a multiplier ideal.

The next theorem implies that the corresponding statement is far from true in dimensions $d \geq 3$. We work in the local ring $(\mathcal{O}, \mathfrak{m})$ of X at a point $x \in X$.

Theorem 3.17. *Let*

$$\mathfrak{j} = \mathcal{J}(\mathfrak{b}^c)_x \subseteq \mathcal{O}$$

be the germ at x of some multiplier ideal, and choose minimal generators

$$h_1, \ldots, h_r \in \mathfrak{j}$$

of $\mathfrak{j}$. Let $b_1, \ldots, b_r \in \mathfrak{m}$ be functions giving a minimal syzygy

$$\sum b_i h_i = 0$$

among the h_i. Then there is at least one index i such that

$$\operatorname{ord}_x(b_i) \leq d - 1.$$

To say that the h_i are minimal generators means by definition that they determine a basis of the $\mathcal{O}/\mathfrak{m} = \mathbf{C}$-vector space $\mathfrak{j}/\mathfrak{m} \cdot \mathfrak{j}$, the hypothesis on the b_i being similar. Note that there are no restrictions on the order of vanishing of *generators* of a multiplier ideal, since for instance $\mathfrak{m}^\ell = \mathcal{J}(\mathfrak{m}^{\ell+d-1})$ for any $\ell \geq 1$. On the other hand, the Theorem extends to statements for the higher syzygies of $\mathfrak{j}$, for which we refer to [**22**].

Example 3.18. Assume that $d = \dim X \geq 3$, choose two general functions $h_1, h_2 \in \mathfrak{m}^p$ vanishing to order $p \geq d$ at x, and consider the complete intersection ideal

$$\mathfrak{d} = (h_1, h_2) \subseteq \mathcal{O}.$$

Since $d \geq 3$, the zeroes of $\mathfrak{d}$ will be a reduced algebraic set of codimension 2, and therefore $\mathfrak{d}$ is integrally closed. On the other hand, the only minimal syzygy among the h_i is the Koszul relation

$$(h_2) \cdot h_1 + (-h_1) \cdot h_2 = 0.$$

In particular, the conclusion of the Theorem does not hold, and so $\mathfrak{d}$ is not a multiplier ideal. (When $d = 2$, the ideal $\mathfrak{d}$ is not integrally closed.) □

Idea of Proof of Theorem 3.17. The plan is to apply Theorem 3.15 with $\mathfrak{a} = \mathfrak{m}$. Specifically, assume for a contradiction that each of the b_i vanishes to order $\geq d$, and choose local coordinates $z_1, \ldots, z_d$ at x, so that $\mathfrak{m} = (z_1, \ldots, z_s)$. We can write

$$b_1 \;=\; z_1 b_{11} + \ldots + z_d b_{1d}\;, \quad \cdots \quad, \; b_r \;=\; z_1 b_{r1} + \ldots + z_d b_{rd},$$

for some functions b_{ij} vanishing to order $\geq d-1$ at x. Now put

$$G_1 \;=\; b_{11}h_1 + \ldots + b_{r1}h_r\;, \quad \cdots \quad, \; G_d = b_{1d}h_1 + \ldots + b_{rd}h_d.$$

Then $G_j \in \mathcal{J}(\mathfrak{m}^{d-1}\mathfrak{b}^c)$ thanks to Lemma 3.14, and

$$(*) \qquad z_1G_1 + \ldots + z_dG_d \;=\; 0$$

by construction. The relation (*) means that $(G_1, \ldots, G_d)$ is a cycle for the Skoda complex

$$(**) \qquad \mathcal{O}^{\binom{n}{2}} \otimes \mathcal{J}(\mathfrak{m}^{d-2}\mathfrak{b}^c) \longrightarrow \mathcal{O}^d \otimes \mathcal{J}(\mathfrak{m}^{d-1}\mathfrak{b}^c) \longrightarrow \mathcal{J}(\mathfrak{m}^d\mathfrak{b}^c)$$

and using the fact that the b_i are minimal one can show via some Koszul cohomology arguments that it gives rise to a non-trivial cohomology class in (**). But this contradicts the exactness of (**). □

Example 3.19 (Skoda's theorem and the effective Nullstellensatz)**.** One can combine Skoda's theorem with jumping numbers to give statements in the direction of the effective Nullstellensatz. Specifically, let $\mathfrak{a} \subseteq \mathcal{O}_X$ be an ideal. Then there is an integer $s > 0$ such that

$$(\sqrt{\mathfrak{a}})^s \;\subseteq\; \mathfrak{a},$$

and it is interesting to ask for effective bounds for s: see for instance [**19**] and [**7**], or [**21**, §10.5] for a survey and references. Now fix a point point $x \in X$, and consider the jumping numbers $\xi_i = \xi_i(\mathfrak{a}; x)$ of $\mathfrak{a}$ at x (Proposition/Definition 1.21). Let $\sigma = \sigma(\mathfrak{a}; x)$ be the least index such that $\xi_\sigma \geq d$. Then

$$\mathcal{J}(\mathfrak{a}^{\xi_\sigma})_x \;\subseteq\; \mathcal{J}(\mathfrak{a}^d)_x \;\subseteq\; (\mathfrak{a})_x,$$

thanks to Skoda's theorem, so it follows from Execise 1.23 that

$$(\sqrt{\mathfrak{a}})^{\sigma(\mathfrak{a};x)} \;\subseteq\; \mathfrak{a}$$

in a neighborhood of x. An analogous global statement holds using non-localized jumping numbers. It would be interesting to know whether one can recover or improve the results of [**19**] or [**7**] by using global arguments to bound σ. (The arguments in [**7**] also revolve around Skoda's theorem, but from a somewhat different perspective.) □

LECTURE 4

Asymptotic constructions

In this lecture we will study asymptotic constructions that can be made with multiplier ideals. It is important in many geometric problems to be able to analyze for example the linear systems $|mL|$ associated to arbitrarily large multiples of a given divisor. Unfortunately one cannot in general find one birational model of X on which these are all well-behaved. By contrast, it turns out that there is some finiteness built into multiplier ideals, and the constructions discussed here are designed to exploit this.

Asymptotic multiplier ideals

In this section we construct the asymptotic multiplier ideals associated to a big divisor. For the purposes of motivation, we start by defining their non-asymptotic parents, which we have not needed up to now.

Let X be a smooth projective variety, and L a divisor on X such that the complete linear series $|L|$ is non-trivial. Given $c > 0$ we construct a multiplier ideal

$$\mathcal{J}(c \cdot |L|) \subseteq \mathcal{O}_X$$

as follows. Take $\mu : X' \longrightarrow X$ to be a log resolution of $|L|$: this means that μ is a projective birational morphism, with

$$\mu^*|L| = |M| + F,$$

where $|M|$ is a basepoint-free linear series, and $F + \mathrm{Exc}(\mu)$ has SNC support. (This is the same thing as a log-resolution of the base-ideal $\mathfrak{b}(|L|) \subseteq \mathcal{O}_X$ of $|L|$.) One then defines

$$\mathcal{J}(c \cdot |L|) = \mu_* \mathcal{O}_{X'}\big(K_{X'/X} - [cF]\big).$$

One can think of these ideals as measuring the singularities of the general divisor $A \in |L|$.

The multiplier ideals attached to a linear series enjoy a Nadel-type vanishing:

Theorem 4.1. *Assume that B is a nef and big divisor on X. Then*

$$H^i\big(X, \mathcal{O}_X(K_X + L + B) \otimes \mathcal{J}(|L|)\big) = 0$$

for $i > 0$.

A proof is sketched in the following Exercise.

Exercise 4.2. Choose $k > c$ general divisors $A_1, \ldots, A_k \in |L|$, and let

$$D = \frac{1}{k} \cdot (A_1 + \ldots + A_k).$$

Then

$$\mathcal{J}(c \cdot |L|) = \mathcal{J}(cD).$$

In particular, Theorem 4.1 is a consequence of Theorem 2.4. □

Exercise 4.3. Let $\mathfrak{b} = \mathfrak{b}(|L|) \subseteq \mathcal{O}_X$ be the base-ideal of $|L|$. Then

$$\mathcal{J}(c \cdot |L|) = \mathcal{J}(\mathfrak{b}^c). \quad \square$$

Remark 4.4 (Incomplete linear series)**.** Starting with a non-trivial linear series $|V| \subseteq |L|$, one constructs in the similar manner a multiplier ideal $\mathcal{J}(c \cdot |V|)$. The analogue of Theorem 4.1 holds for these, as do the natural extensions of Exercises 4.2 and 4.3. The reader may consult [**21**, §9.2] for details. □

Exercise 4.5. (Adjoint ideals for linear series). With X as above, suppose that $S \subseteq X$ is a smooth irreducible divisor not contained in the base locus of $|L|$. Then, as in the previous Lecture, one constructs for $c > 0$ an adjoint ideal

$$\mathrm{Adj}_S(X, c \cdot |L|) \subseteq \mathcal{O}_X.$$

This sits in an exact sequence

$$0 \longrightarrow \mathcal{J}(X, c \cdot |L|) \otimes \mathcal{O}_X(-S) \longrightarrow \mathrm{Adj}_S(X, c \cdot |L|) \longrightarrow \mathcal{J}(S, c \cdot |L|_S) \longrightarrow 0,$$

where the term $|L|_S$ on the right involves the (possibly incomplete) linear series on S obtained as the restriction of $|L|$ from X to S. (One can use Exercise 4.2 to reduce to the case of divisors.) □

The ideals $\mathcal{J}(c \cdot |L|)$ reflect the geometry of the linear series $|L|$. But a variant of this construction gives rise to ideals that involve the asymptotic geometry of $|mL|$ for all $m \gg 0$.

Proposition/Definition 4.6. *Assume that L is big, and fix $c > 0$. Then for $p \gg 0$ the multiplier ideals*

$$\mathcal{J}(X, \tfrac{c}{p} \cdot |pL|)$$

all coincide. The resulting ideal, written $\mathcal{J}(X, c \cdot \| L \|)$, is the asymptotic multiplier ideal *of $|L|$ with coefficient c.*

Idea of Proof. It follows from the Noetherian property that as p varies over all positive integers, the family of ideals $\mathcal{J}(\frac{c}{p} \cdot |pL|)$ has a maximal element. But one checks that

$$\mathcal{J}(\tfrac{c}{p} \cdot |pL|) \subseteq \mathcal{J}(\tfrac{c}{pq} \cdot |pqL|)$$

for all $q \geq 1$, and therefore the family in question has a *unique* maximal element $\mathcal{J}(c \cdot \| L \|)$, which coincides with $\mathcal{J}(\frac{c}{p} \cdot |pL|)$ for all sufficiently large and divisible p. For the fact that these agree with $\mathcal{J}(\frac{c}{p} \cdot |pL|)$ for any sufficiently large p (depending on c), see [**21**, Proposition 11.1.18]. □

The asymptotic multiplier ideals associated to a big divisor L are the algebro-geometric analogues of the multiplier ideals associated to metrics of minimal singularities in the analytic theory. It is conjectured – but not known – that the two sorts of multiplier ideals actually coincide provided that L is big. (Ssee [**5**]).

The following theorem summarizes the most important properties of asymptotic multiplier ideals.

Theorem 4.7 (Properties of asymptotic multiplier ideals). *Assume that L is a big divisor on the smooth projective variety X.*

(i). *Every section of $\mathcal{O}_X(L)$ vanishes along $\mathcal{J}(\| L \|)$, i.e. the natural inclusion*

$$H^0\big(X, \mathcal{O}_X(L) \otimes \mathcal{J}(\| L \|)\big) \longrightarrow H^0\big(X, \mathcal{O}_X(L)\big)$$

is an isomorphism. Equivalently, $\mathfrak{b}\big(|L|\big) \subseteq \mathcal{J}(X, \| L \|)$.

(ii). *If B is nef, then*

$$H^i\big(X, \mathcal{O}_X(K_X + L + B) \otimes \mathcal{J}(\| L \|)\big) \;=\; 0$$

for $i > 0$.

(iii). *$\mathcal{J}(\| mL \|) = \mathcal{J}(m \cdot \| L \|)$ for every positive integer m.*

(iv). *(Subadditivity.) $\mathcal{J}(\| mL \|) \subseteq \mathcal{J}(\| L \|)^m$ for every $m \geq 0$.*

(v). *If $f : Y \longrightarrow X$ is an étale covering, then*

$$\mathcal{J}(Y, \| f^*L \|) \;=\; f^*\mathcal{J}(X, \| L \|).$$

Concerning the vanishing in (ii), note that it is not required that B be big. In particular, one can take $B = 0$.

Exercise 4.8. Give counterexamples to the analogues of properties (ii) – (v) if one replaces the asymptotic ideals by the multiplier ideals $\mathcal{J}(|L|)$ attached to a single linear series. (For example, suppose that L is ample, but that $|L|$ is not free. Then $\mathcal{J}(m \cdot |L|) \neq \mathcal{O}_X$ for $m \gg 0$, whereas $\mathcal{J}(|mL|) = \mathcal{O}_X$ for large m.) □

Exercise 4.9. In the situation of the Theorem, one has

$$\mathcal{J}(c \cdot \| L \|) \;\subseteq\; \mathcal{J}(d \cdot \| L \|)$$

whenever $c \geq d$. □

Indications of Proof of Theorem 4.7. We prove (ii) and (iii) in order to give the flavor. For (ii), choose $p \gg 0$ such that $\mathcal{J}(\| L \|) \;=\; \mathcal{J}(\frac{1}{p}|L|)$, and let $\mu : X' \longrightarrow X$ be a log-resolution of $|pL|$, with

$$\mu^*|pL| \;=\; |M_p| + F_p,$$

where $|M_p|$ is free. We can suppose that M_p is big (since L is). Arguing as in the proof of Theorem 2.4, one has

$$H^i\big(X, \mathcal{O}_X(K_X + L + B) \otimes \mathcal{J}(\| L \|)\big) \;=\; H^i\big(X', \mathcal{O}_{X'}(K_{X'} + \mu^*(L + B) - [\tfrac{1}{p}F_p])\big).$$

But

$$\mu^*(L + B) - \tfrac{1}{p}F_p \;\equiv_{\text{num}}\; \mu^*(B) + \tfrac{1}{p}M_p$$

is nef and big, so the required vanishing follows from the theorem of Kawamata and Viehweg. For (iii), the argument is purely formal. Specifically, for $p \gg 0$ one has thanks to Proposition/Definition 4.6

$$\begin{aligned} \mathcal{J}(\| mL \|) &= \mathcal{J}(\tfrac{1}{p} \cdot |pmL|) \\ &= \mathcal{J}(\tfrac{m}{mp} \cdot |pmL|) \\ &= \mathcal{J}(m \cdot \| L \|), \end{aligned}$$

where in the last equality we are using mp in place of p for the large index computing the asymptotic multiplier ideal in question. For the remaining statements we refer to [**21**, 11.1, 11.2]. □

Exercise 4.10 (Uniform global generation). Theorem 2.11 extends to the asymptotic setting. Specifically, there exists a divisor B on X with the property that $\mathcal{O}_X(L+B) \otimes \mathcal{J}(\| L \|)$ is globally generated for any big divisor L on X. □

Exercise 4.11 (Characterization of nef and big divisors). Let L be a big divisor on X. Then L is nef if and only if

$$(*) \qquad \mathcal{J}(\| mL \|) = \mathcal{O}_X \quad \text{for all } m > 0.$$

(Suppose (*) holds. Then it follows from the previous exercise that $\mathcal{O}_X(mL+B)$ is globally generated for all $m > 0$ and some fixed B. This implies that $\big((mL+B) \cdot C\big) \geq 0$ for every effective curve C and every $m > 0$, and hence that $(L \cdot C) \geq 0$.) □

Variants

We next discuss some variants of the construction studied in the previous section. To begin with, one can deal with possibly incomplete linear series. For this, recall that a *graded linear series* $W_\bullet = \{W_m\}$ associated to a big divisor L consists of subspaces

$$W_m \subseteq H^0\big(X, \mathcal{O}_X(mL)\big),$$

with $W_m \neq 0$ for $m \gg 0$, satisfying the condition that

$$R(W_\bullet) =_{\text{def}} \oplus W_m$$

be a graded subalgebra of the section ring $R(L) = \oplus H^0\big(X, \mathcal{O}_X(mL)\big)$ of L. This last requirement is equivalent to asking that

$$W_\ell \cdot W_m \subseteq W_{\ell+m},$$

where the left hand side denotes the image of $W_\ell \otimes W_m$ under the natural map $H^0\big(X, \mathcal{O}_X(\ell L)\big) \otimes H^0\big(X, \mathcal{O}_X(mL)\big) \longrightarrow H^0\big(X, \mathcal{O}_X((\ell+m)L)\big)$. Then just as above one gets an asymptotic multiplier ideal by taking

$$\mathcal{J}(c \cdot \| W_\bullet \|) = \mathcal{J}(\tfrac{c}{p} \cdot |W_p|)$$

for $p \gg 0$. Analogues of Properties (i) – (iv) from Theorem 4.7 remain valid in this setting; we leave precise statements and proofs to the reader.

Example 4.12 (Restricted linear series). An important example of this construction occurs when $Y \subseteq X$ is a smooth subvariety not contained in the stable base locus $\mathbf{B}(L)$ of the big line bundle L on X. In this case, one gets a graded linear series on Y by taking

$$W_m \;=\; \operatorname{Im}\Big(H^0\big(X, \mathcal{O}_X(mL)\big) \longrightarrow H^0\big(Y, \mathcal{O}_Y(mL_Y)\big)\Big).$$

We denote the corresponding multiplier ideals by $\mathcal{J}(Y, c\cdot \| L \|_Y)$. Note that there is an inclusion

$$\mathcal{J}(Y, c\cdot \| L\|_Y) \;\subseteq\; \mathcal{J}(Y, c\cdot \| L_Y\|),$$

but these can be quite different. For example if the restriction L_Y is ample on Y, then the ideal on the right is trivial. On the other hand, if for all $m \gg 0$ the linear series $|mL|$ on X have base-loci that meet Y, then the ideals on the left could be quite deep. □

One can also extend the construction of adjoint ideals to the asymptotic setting. Assume that $S \subseteq X$ is a smooth irreducible divisor which is not contained in the stable base locus $\mathbf{B}(L)$ of a big divisor L. Then one defines an asymptotic adjoint ideal $\operatorname{Adj}_S(X, \| L\|)$ that fits into an exact sequence:

$$0 \longrightarrow \mathcal{J}(X, \| L\|) \otimes \mathcal{O}_X(-S) \longrightarrow \operatorname{Adj}_S(X, \| L\|) \longrightarrow \mathcal{J}(S, \| L\|_S) \longrightarrow 0. \tag{4.1}$$

The ideal on the right is the asymptotic multiplier ideal associated to the restricted linear series of L from X to S, as in the previous example. This sequence will play a central role in our discussion of extension theorems in the next Lecture.

Finally, we discuss the asymptotic analogues of the ideals $\mathcal{J}(\mathfrak{a}^c)$. A *graded family of ideals* $\mathfrak{a}_\bullet = \{\mathfrak{a}_m\}$ on a variety X consists of ideal sheaves $\mathfrak{a}_m \subseteq \mathcal{O}_X$ satisfying the property that

$$\mathfrak{a}_\ell \cdot \mathfrak{a}_m \;\subseteq\; \mathfrak{a}_{\ell+m}.$$

We will also suppose that $\mathfrak{a}_0 = \mathcal{O}_X$ and that $\mathfrak{a}_m \neq (0)$ for $m \gg 0$. Then one defines:

$$\mathcal{J}(\mathfrak{a}_\bullet^c) \;=\; \mathcal{J}(c\cdot \mathfrak{a}_\bullet) \;=_{\text{def}}\; \mathcal{J}\big((\mathfrak{a}_p)^{c/p}\big)$$

for $p \gg 0$.

Example 4.13. The prototypical example of a graded system of ideals is the family of base-ideals

$$\mathfrak{b}_m \;=\; \mathfrak{b}\big(|mL|\big)$$

associated to multiples of a big divisor L when X is projective. In this case

$$\mathcal{J}(\mathfrak{b}_\bullet^c) \;=\; \mathcal{J}(c\cdot \| L\|). \quad \square$$

Example 4.14 (Symbolic powers). A second important example involves the symbolic powers of a radical ideal $\mathfrak{q} = \mathcal{I}_Z$ defining a (reduced) subscheme $Z \subseteq X$. Assuming as usual that X is smooth, define

$$\mathfrak{q}^{(m)} \;=\; \big\{ f \in \mathcal{O}_X \;\big|\; \operatorname{ord}_x(f) \geq m \;\text{ for general } x \in Z \big\}.$$

(If X is reducible, we ask that the condition hold at a general point of each irreducible component of X.) Observe that by construction, membership in $\mathfrak{q}^{(m)}$ is tested at a general point of Z, i.e. this is a primary ideal. The inclusion

$$\mathfrak{q}^{(\ell)} \cdot \mathfrak{q}^{(m)} \;\subseteq\; \mathfrak{q}^{(\ell+m)}$$

being evident, these form a graded family denoted $\mathfrak{q}_{(\bullet)}$. □

Exercise 4.15. Let $\mathfrak{a}_\bullet$ be a graded family of ideals. Then for every $m \geq 1$:

(i). $\mathfrak{a}_m \subseteq \mathcal{J}(\mathfrak{a}_\bullet^m)$;

(ii). $\mathcal{J}(\mathfrak{a}_\bullet^m) \subseteq \mathcal{J}(\mathfrak{a}_\bullet)^m$. □

Étale multiplicativity of plurigenera

As a first illustration of this machinery, we prove a theorem of Kollár concerning the behavior of plurigenera under étale covers.

Given a smooth projective variety X, recall that the mth plurigenus $P_m(X)$ of X is the dimension of the space of m-canonical forms on X:

$$P_m(X) =_{\text{def}} h^0\big(X, \mathcal{O}_X(mK_X)\big).$$

These plurigeneral are perhaps the most basic birational invariants of a variety.

Kollár's theorem is that these behave well under étale coverings:

Theorem 4.16. *Assume that X is of general type, and let $f : Y \longrightarrow X$ be an unramified covering of degree d. Then for $m \geq 2$,*

$$P_m(Y) \;=\; d \cdot P_m(X).$$

Kollár was led to this statement by the observation that it would (and now does) follow from the minimal model program.

Exercise 4.17. Show that the analogous statement can fail when $m = 1$. (Consider the case $\dim X = 1$.) □

Proof of Theorem 4.16. We use the various properties of asymptotic multiplier ideals given in Theorem 4.7. To begin with, 4.7 (i) yields:

$$\begin{aligned} H^0\big(X, \mathcal{O}_X(mK_X)\big) &= H^0\big(X, \mathcal{O}_X(mK_X) \otimes \mathcal{J}(\| mK_X \|)\big) \\ &= H^0\big(X, \mathcal{O}_X(mK_X) \otimes \mathcal{J}(\| (m-1)K_X \|)\big), \end{aligned}$$

the second equality coming from the includion $\mathcal{J}(\| (m-1)K_X \|) \subseteq \mathcal{J}(\| mK_X \|)$. But since X is of general type, when $m \geq 2$ the vanishing statement (ii) implies that

$$H^i\big(X, \mathcal{O}_X(mK_X) \otimes \mathcal{J}(\| (m-1)K_X \|)\big) \;=\; 0$$

when $i > 0$. Therefore

$$P_m(X) \;=\; \chi\Big(X, \mathcal{O}_X(mK_X) \otimes \mathcal{J}(\| (m-1)K_X \|)\Big),$$

and similarly

$$P_m(Y) \;=\; \chi\Big(Y, \mathcal{O}_Y(mK_Y) \otimes \mathcal{J}(\| (m-1)K_Y \|)\Big).$$

On the other hand,

$$\begin{aligned} f^*K_X &\equiv_{\text{lin}} K_Y \\ f^*\mathcal{J}(X, \| (m-1)K_X \|) &= \mathcal{J}(Y, \| (m-1)K_Y \|) \end{aligned}$$

thanks to 4.7 (v). The Theorem then follows from the fact that Euler characteristics are multiplicative under étale covers. □

Remark 4.18. It was suggested in [**21**, Example 11.2.26] that a similar argument handles adjoint bundles of the type $\mathcal{O}_X(K_X + mL)$. However this – as well as the reference given there – is erroneous. □

Exercise 4.19. Assume that X is of general type. Then for $m \geq 1$ and $i > 0$ the maps

$$H^0\big(X, \mathcal{O}_X(mK_X)\big) \otimes H^i\big(X, \mathcal{O}_X(K_X)\big) \longrightarrow H^i\big(X, \mathcal{O}_X((m+1)K_X)\big)$$

defined by cup product are zero. (Argue as in the proof of Kollár's theorem that the map factors through $H^i\big(X, \mathcal{O}_X((m+1)K_X \otimes \mathcal{J}(\|mK_X\|)\big)$.) □

A comparison theorem for symbolic powers

We start with a statement that follows formally from the subadditivity theorem in the form of Exercise 4.15. It shows that if a graded system of ideals has any non-trivial multiplier ideal, then that system must grow like the power of an ideal.

Proposition 4.20. *Let $\mathfrak{a}_\bullet$ be a graded family of ideals, and fix an index ℓ. Then for any m:*

$$\mathfrak{a}_\ell^m \ \subseteq \ \mathfrak{a}_{\ell m} \ \subseteq \ \mathcal{J}(\mathfrak{a}_\bullet^{\ell m}) \ \subseteq \ \mathcal{J}(\mathfrak{a}_\bullet^{\ell})^m.$$

In particular, if $\mathcal{J}(\mathfrak{a}_\bullet^\ell) \subseteq \mathfrak{b}$ for some ideal $\mathfrak{b}$, then

$$\mathfrak{a}_{\ell m} \ \subseteq \ \mathfrak{b}^m$$

for every $m \geq 0$. □

In spite of the rather formal nature of this result, it has a surprising application to symbolic powers. Specifically, recall from Example 4.14 that if $Z \subseteq X$ is a reduced subscheme with ideal $\mathfrak{q} \subseteq \mathcal{O}_X$, then the symbolic powers of $\mathfrak{q}$ are defined to be:

$$\mathfrak{q}^{(m)} \ = \ \big\{ f \in \mathcal{O}_X \ \big| \ \mathrm{ord}_x(f) \geq m \ \text{ for general } x \in Z \big\}.$$

Clearly $\mathfrak{q}^m \subseteq \mathfrak{q}^{(m)}$, and if Z is non-singular then equality holds. However in general the inclusion is strict:

Example 4.21. Let $Z \subseteq \mathbf{C}^3 = X$ be the union of the three coordinate axes, so that

$$\mathfrak{q} \ = \ (xy\,,\, yz\,,\, xz) \ \subseteq \ \mathbf{C}[x,y,z].$$

Then $xyz \in \mathfrak{q}^{(2)}$, but $xyz \notin \mathfrak{q}^2$.

A result of Swanson [**27**] (holding in much greater algebraic generality) states that there is an integer $k = k(Z)$ with the property that

$$\mathfrak{q}^{(km)} \ \subseteq \ \mathfrak{q}^m$$

for every m. It is natural to suppose that $k(Z)$ measures in some way the singularities of Z, but in fact it was established in [**8**] that there is a uniform result.

Theorem 4.22. *Assume that Z has pure codimension e in X. Then*

$$\mathfrak{q}^{(em)} \ \subseteq \ \mathfrak{q}^m$$

for every m. In particular, for any reduced Z one has

$$\mathfrak{q}^{(dm)} \ \subseteq \ \mathfrak{q}^m$$

for every m, *where as usual* $d = \dim X$.

PROOF. Consider the graded system $\mathfrak{q}_{(\bullet)}$ of symbolic powers. Thanks to the previous Proposition, it suffices to show that

$$\mathcal{J}(\mathfrak{q}^{e}_{(\bullet)}) \subseteq \mathfrak{q}.$$

As $\mathfrak{q}$ is radical, membership in $\mathfrak{q}$ is tested at a general point of (each component of) Z, so we can assume that Z is non-singular. But in this case

$$\mathcal{J}(\mathfrak{q}^{e}_{(\bullet)}) = \mathcal{J}(\mathfrak{q}^{e}) = \mathfrak{q},$$

as required. □

Example 4.23. Let $T \subseteq \mathbf{P}^2$ be a finite set, viewed as a reduced scheme, and denote by $I \subseteq \mathbf{C}[X, Y, Z]$ the homogeneous ideal of T. Let F be a form with the property that

$$\operatorname{mult}_x(F) \geq 2m \quad \text{for all } x \in T.$$

Then $F \in I^m$. (Apply the Theorem to the affine cone over T.) □

LECTURE 5

Extension theorems and deformation invariance of plurigenera

The most important recent applications of multiplier ideals have been to prove extension theorems. Originating in Siu's proof of the deformation invariance of plurigenera, extension theorems are what opened the door to the spectacular progress in the minimal model program. Here we will content ourselves with a very simple result of this type, essentially the one appearing in Siu's original article [**25**] (see also [**17**], [**18**]). As we will see, the use of adjoint ideals renders the proof very transparent.[1]

The statement for which we aim is the following.

Theorem 5.1. *Let X be a smooth projective variety, and $S \subseteq X$ a smooth irreducible divisor. Set*

$$L \;=\; K_X \,+\, S \,+\, B,$$

where B is any nef divisor, and assume that L is big and that $S \not\subseteq \mathbf{B}_+(L)$. Then for every $m \geq 2$ the restriction map

$$H^0\big(X, \mathcal{O}_X(mL)\big) \longrightarrow H^0\big(S, \mathcal{O}_S(mL_S)\big)$$

is surjective.

Recall that by definition $\mathbf{B}_+(L)$ denotes the stable base-locus of the divisor $kL - A$ for A ample and $k \gg 0$, this being independent of A provided that k is sufficiently large. The hypothesis that $S \not\subseteq \mathbf{B}_+(L)$ guarantees in particular that L_S is a big divisor on S.

Before turning to the proof of Theorem 5.1, let us see how it implies Siu's theorem on plurigenera in the general type case:

Corollary 5.2 (Siu's Theorem on Plurigenera). *Let*

$$\pi : Y \longrightarrow T$$

be a smooth projective family of varieties of general type. Then for each $m \geq 0$, the plurigenera

$$P_m(Y_t) \;=_{\text{def}} h^0\big(Y_t, \mathcal{O}_{Y_t}(mK_{Y_t})\big)$$

are independent of t.

[1]The utility of adjoint ideals in connection with extension theorems became clear in the work [**14**], [**28**] of Hacon-McKernan and Takayama on boundedness of pluricanonical mappings. The theory is further developed in [**15**] and [**11**].

PROOF. We may assume without loss of generality that $m \geq 2$ and that T is a smooth affine curve, and we write $K_t = K_{Y_t}$. Fixing $0 \in T$, one has $P_m(Y_0) \geq P_m(Y_t)$ for generic t by semicontinuity, so the issue is to prove the reverse inequality

$$(*) \qquad h^0\big(Y_0, \mathcal{O}_{Y_0}(mK_0)\big) \ \leq \ h^0\big(Y_t, \mathcal{O}_{Y_t}(mK_t)\big)$$

for $t \in T$ in a neighborhood of 0. For this, consider the sheaf $\pi_*\mathcal{O}_Y(mK_{Y/T})$ on T. This is a torsion-free (and hence locally free) sheaf, whose rank computes the generic value of the mth-plurigenus $P_m(Y_t)$. Moreover, the fibre of $\pi_*\mathcal{O}_Y(mK_{Y/T})$ at 0 consists of those pluri-canonical forms on Y_0 that extend (over a neighborhood of 0) to forms on Y itself. So to prove (*), it suffices to show that any $\eta \in H^0\big(Y_0, \mathcal{O}_{Y_0}(mK_0)\big)$ extends (after possibly shrinking T) to some

$$\tilde{\eta} \ \in \ H^0\big(Y, \mathcal{O}_Y(mK_{Y/T})\big).$$

We will deduce this from Theorem 5.1.

Specifically, we start by completing π to a morphism

$$\overline{\pi} : \overline{Y} \longrightarrow \overline{T}$$

of smooth projective varieties, where $T \subseteq \overline{T}$ and $Y = \overline{\pi}^{-1}(T)$. We view $Y_0 \subseteq \overline{Y}$ as a smooth divisor on $\overline{Y}$. Let A be a very ample divisor on $\overline{T}$, let $B = \pi^*(A - K_{\overline{T}})$, and set

$$L \ = \ K_{\overline{Y}} + Y_0 + B \ \equiv_{\text{lin}} \ K_{\overline{Y}/\overline{T}} + Y_0 + \pi^* A.$$

We can assume by taking A sufficiently positive that B is nef, and we assert that we can arrange in addition that L is big and that $Y_0 \not\subseteq \mathbf{B}_+(L)$. For the first point, it is enough to show that if D is an ample divisor on $\overline{Y}$, and if A is sufficiently positive, then $kL - D$ is effective for some $k \gg 0$. To this end, since Y_t is of general type for general t, we can choose k sufficiently large so that $kK_{\overline{Y}/\overline{T}} - D$ is effective on a very general fibre of $\overline{\pi}$. By taking A very positive, we can then guarantee that

$$\overline{\pi}_*\mathcal{O}_{\overline{Y}}(kL - D) \ = \ \overline{\pi}_*\mathcal{O}_{\overline{Y}}(kK_{\overline{Y}/\overline{T}} - D) \otimes \mathcal{O}_{\overline{T}}(kA + k0)$$

is non-zero and globally generated, which implies that $h^0\big(\overline{Y}, \mathcal{O}_{\overline{Y}}(kL - D)\big) \neq 0$. For the assertion concerning $\mathbf{B}_+$, observe first that since $|rY_0| = \overline{\pi}^*|r \cdot 0|$ is free for $r \gg 0$, the divisor Y_0 cannot not lie in the base locus of $|kL - D + qY_0|$ for arbitrarily large q. On the other hand, by making A more positive, we are free to replace L by $L + pY_0$, i.e. we may suppose that $Y_0 \not\subseteq \mathbf{B}_+(L)$, as claimed.

We may therefore apply Theorem 5.1 with $X = \overline{Y}$ and $S = Y_0$ to conclude that the restriction map

$$(*) \qquad H^0\big(\overline{Y}, \mathcal{O}_{\overline{Y}}(mL)\big) \longrightarrow H^0\big(Y_0, \mathcal{O}_{Y_0}(mL)\big)$$

is surjective for every $m \geq 2$. So provided that we take T sufficiently small so that $\mathcal{O}_{\overline{T}}(A + 0)|T$ is trivial, then

$$\mathcal{O}_Y(mL) \ \cong \ \mathcal{O}_Y(mK_{Y/T}),$$

and surjectivity of (*) implies the surjectivity of

$$H^0\big(Y, \mathcal{O}_Y(mK_{Y/T})\big) \longrightarrow H^0\big(Y_0, \mathcal{O}_{Y_0}(mK_0)\big),$$

as required. □

The proof of Theorem 5.1 basically follows [**25**] (and [**21**]), except that as we have mentioned the use of adjoint ideals substantially clarifies the presentation. Two pieces of vocabulary will be useful. First, if A is a divisor on a projective variety V, we denote by $\mathfrak{b}\big(|A|\big) \subseteq \mathcal{O}_V$ the base ideal of the complete linear series determined by A. Secondly, given an ideal $\mathfrak{a} \subseteq \mathcal{O}_V$, we say that a section $s \in H^0\big(V, \mathcal{O}_V(A)\big)$ *vanishes along* $\mathfrak{a}$ if it lies in the image of the natural map $H^0\big(V, \mathcal{O}_V(A) \otimes \mathfrak{a}\big) \longrightarrow H^0\big(V, \mathcal{O}_V(A)\big)$.

We also recall from the previous lecture a couple of facts about the asymptotic multiplier ideals associated to L and its restriction to S:

Remark 5.3. Assume that L is big. Then:

(i). For every $m \geq 0$,

$$\mathcal{J}(X, \|(m+1)L\|) \subseteq \mathcal{J}(X, \|mL\|) \quad , \quad \mathcal{J}(S, \|(m+1)L\|_S) \subseteq \mathcal{J}(S, \|mL\|_S)$$

(ii). (Nadel vanishing). If P is nef, then

$$H^i\Big(X, \mathcal{O}_X(K_X + mL + P) \otimes \mathcal{J}(X, \|mL\|)\Big) = 0 \ \text{ for } i > 0 \text{ and } m \geq 1.$$

(iii). (Subadditivity). For any $m, q > 0$,

$$\mathcal{J}(X, \|mqL\|) \subseteq \mathcal{J}(X, \|mL\|)^q \quad , \quad \mathcal{J}(S, \|mqL\|_S) \subseteq \mathcal{J}(S, \|mL\|_S)^q. \quad \square$$

Returning to the situation and notation of the Theorem, fix $m \geq 2$. Our analysis of extension questions revolves around the adjoint exact sequence:

(5.1)

$$0 \to \mathcal{J}(X, \|(m-1)L\|) \otimes \mathcal{O}_X(-S) \to \operatorname{Adj}_S(X, \|(m-1)L\|) \to \mathcal{J}(S, \|(m-1)L\|_S) \to 0.$$

We summarize what we will use in

Lemma 5.4. (i). *In order to prove the Theorem, it suffices to establish the inclusion*

$$\mathfrak{b}\big(|mL_S|\big) \subseteq \mathcal{J}(S, \|(m-1)L\|_S).$$

(ii). *If* $\mathfrak{a} \subseteq \mathcal{J}(S, \|(m-1)L\|_S)$ *is an ideal such that* $\mathcal{O}_S(mL_S) \otimes \mathfrak{a}$ *is globally generated, then*

$$\mathfrak{a} \subseteq \mathcal{J}(S, \|mL\|_S).$$

Proof. For both statements, we twist through in (5.1) by $\mathcal{O}_X(mL)$. Noting that

$$mL - S \equiv_{\text{lin}} (m-1)L + K_X + B,$$

it follows from Nadel vanishing and the hypotheses that

$$H^1\Big(X, \mathcal{O}_X(mL - S) \otimes \mathcal{J}(X, \|(m-1)L\|)\Big) = 0$$

provided that $m \geq 2$. (Note that this is where it is important that we use asyptotic multiplier ideals: we do not assume that B is more than nef, and so there is no "excess positivity" in the required vanishing.) Therefore the exact sequence (5.1) gives the surjectivity of the map

$$H^0\Big(X, \mathcal{O}_X(mL) \otimes \operatorname{Adj}_S(\|(m-1)L\|\Big) \longrightarrow H^0\Big(S, \mathcal{O}_S(mL_S) \otimes \mathcal{J}(\|(m-1)L\|_S)\Big).$$

The group on the left being a subspace of $H^0\big(X, \mathcal{O}_X(mL)\big)$, this means that any section of $\mathcal{O}_S(mL_S)$ vanishing along $\mathcal{J}(S, \|(m-1)L\|_S)$ lifts to a section of $\mathcal{O}_X(mL)$.

So if we know the inclusion in (i), this implies that all sections of $\mathcal{O}_S(mL_S)$ lift to X.

For (ii), remark that if M is a (Cartier) divisor on a projective variety V, and if $\mathfrak{a}, \mathfrak{b} \subseteq \mathcal{O}_V$ are ideals such that $\mathcal{O}_V(M) \otimes \mathfrak{a}$ is globally generated, then $\mathfrak{a} \subseteq \mathfrak{b}$ if and only if

$$H^0\big(V, \mathcal{O}_V(M) \otimes \mathfrak{a}\big) \ \subseteq\ H^0\big(V, \mathcal{O}_V(M) \otimes \mathfrak{b}\big)\big),$$

both sides being viewed as subspaces of $H^0\big(V, \mathcal{O}_V(M)\big)$. So in our situation it is enough to show that

$$H^0\Big(S, \mathcal{O}_S(mL_S) \otimes \mathfrak{a}\Big) \ \subseteq\ H^0\Big(S, \mathcal{O}_S(mL_S) \otimes \mathcal{J}(S, \| mL \|_S)\Big).$$

Suppose then that

$$\bar{t} \ \in\ H^0\big(S, \mathcal{O}_S(mL_S) \otimes \mathfrak{a}\big)$$

is a section of $\mathcal{O}_S(mL_S)$ vanishing along $\mathfrak{a}$. Since $\mathfrak{a} \subseteq \mathcal{J}(S, \| (m-1)L \|_S)$, it follows as above from (5.1) that $\bar{t}$ lifts to a section $t \in H^0\big(X, \mathcal{O}_X(mL)\big)$. By definition t vanishes along $\mathfrak{b}\big(|mL|\big) \subset \mathcal{O}_X$, and hence its restriction $\bar{t}$ vanishes along

$$\mathfrak{b}\big(|mL|\big) \cdot \mathcal{O}_S \ \subseteq\ \mathcal{J}(S, \| mL \|_S),$$

as required. □

In order to apply the statement (i) of the Lemma, the essential point is a comparison between the multiplier ideals of the restricted divisor pL_S and those of the restricted linear series of pL from X to S:

Proposition 5.5. *There exist a very ample divisor A on X, a positive integer $k_0 > 0$, and a divisor $D \in |k_0 L - A|$ meeting S properly, such that*

$$\mathcal{J}(S, \| pL_S \|) \otimes \mathcal{O}_S(-D_S) \ \subseteq\ \mathcal{J}(S, \| (p + k_0 - 1)L \|_S) \tag{5.2}$$

for every $p \geq 0$.

Granting this for the moment, we complete the proof of the Theorem. In fact, fix m, and apply equation (5.2) with $p = qm$ for $q \gg 0$. One finds:

$$\begin{aligned} \mathfrak{b}\big(|mL_S|\big)^q \otimes \mathcal{O}_S(-D_S) \ &\subseteq\ \mathfrak{b}\big(|mqL_S|\big) \otimes \mathcal{O}_S(-D_S) \\ &\subseteq\ \mathcal{J}(S, \| mqL_S \|) \otimes \mathcal{O}_S(-D_S) \\ &\subseteq\ \mathcal{J}(S, \| (mq + k_0 - 1)L \|_S) \\ &\subseteq\ \mathcal{J}(S, \| mqL \|_S) \\ &\subseteq\ \mathcal{J}(S, \| mL \|_S)^q, \end{aligned}$$

the last inclusion coming from the subadditivity theorem. But we assert that having the inclusion

$$\mathfrak{b}\big(|mL_S|\big)^q \otimes \mathcal{O}_S(-D_S) \ \subseteq\ \mathcal{J}(S, \| mL \|_S)^q \tag{+}$$

for all $q \gg 0$ forces $\mathfrak{b}\big(|mL_S|\big) \subseteq \mathcal{J}(S, \| mL \|_S)$ and hence also the inclusion in Lemma 5.4 (i). In fact, it follows from the construction of multiplier ideals that there are finitely many divisors E_α lying over S, together with integers $r_\alpha > 0$, such that a germ $f \in \mathcal{O}_S$ lies in $\mathcal{J}(S, \| mL \|_S)$ if and only if $\mathrm{ord}_{E_\alpha}(f) \geq r_\alpha$ for every α. But (+) implies that if $f \in \mathfrak{b}\big(|mL_S|\big)$, then

$$q \cdot \mathrm{ord}_{E_\alpha}(f) + \mathrm{ord}_{E_\alpha}(D_S) \ \geq\ q \cdot r_\alpha$$

for each α, and letting $q \to \infty$ one finds that $\mathrm{ord}_{E_\alpha}(f) \geq r_\alpha$, as required. Thus Lemma 5.4 applies, and this completes the proof of the Theorem granting Proposition 5.5.

It remains to prove the Proposition. Here the essential point is statement (ii) of the Lemma.

PROOF OF PROPOSITION 5.5. By Nadel vanishing and Castelnuovo-Mumford regularity, we can find a very ample divisor A so that for every $q \geq 0$ the sheaf

$$\mathcal{O}_S(qL_S + A_S) \otimes \mathcal{J}(S, \| qL \|_S)$$

is globally generated (cf Exercise 4.10). Next, since $S \not\subseteq \mathbf{B}_+(L)$, for $k_0 \gg 0$ we can take a divisor $D \in |k_0L - A|$ that does not contain S. We will show by induction on p that (5.2) holds with these choices of the data.

For $p = 0$ the required inclusion holds by virtue of the fact that $D + A \equiv_{\mathrm{lin}} k_0L$, which yields

$$\mathcal{O}_S(-D_S) \subseteq \mathcal{J}(S, \| k_0L \|_S) \subseteq \mathcal{J}(S, \| (k_0 - 1)L \|_S).$$

Assuming that (5.2) is satisfied for a given value of p, we will show that it holds also for $p + 1$. To this end, observe first that

$$\mathcal{O}_S\big((p + k_0)L_S - D_S\big) \otimes \mathcal{J}(S, \| pL_S \|)$$

is globally generated thanks to our choice of A. Applying Lemma 5.4 (ii) with $m = p + k_0$ and $\mathfrak{a} = \mathcal{J}(S, \| pL_S \|) \otimes \mathcal{O}_S(-D_S)$, it follows using the inductive hypothesis

$$\mathcal{J}(S, \| pL_S \|) \otimes \mathcal{O}_S(-D_S) \subseteq \mathcal{J}(S, \| (p + k_0 - 1)L \|_S)$$

that in fact

$$\mathcal{J}(S, \| pL_S \|) \otimes \mathcal{O}_S(-D_S) \subseteq \mathcal{J}(S, \| (p + k_0)L \|_S).$$

Therefore also

$$\mathcal{J}(S, \| (p + 1)L_S \|) \otimes \mathcal{O}_S(-D_S) \subseteq \mathcal{J}(S, \| (p + k_0)L \|_S),$$

which completes the induction. □

Bibliography

[1] Aldo Andreotti and Alan Mayer, On period relations for abelian integrals on algebraic curves, Ann. Scuola Norm. Sup. Pisa **21** (1967), pp. 189 – 238.

[2] Manuel Blickle and Robert Lazarsfeld, An informal introduction to multiplier ideals, in *Trends in Commutative Algebra*, MSRI Publ. **51**, Cambridge Univ. Press, 2004, pp. 87 – 114.

[3] Caucher Birkar, Paolo Cascini, Christopher Hacon and James McKernan, Existence of minimal models for varieties of log general type, to appear.

[4] Tommaso de Fernex and Christopher Hacon, Singularities on normal varieties, to appear.

[5] Jean-Pierre Demailly, Lawrence Ein and Robert Lazarsfeld, A subadditivity property of mulitplier ideals, Michigan Math. J. **48** (2000), pp. 137 – 156.

[6] Lawrence Ein and Robert Lazarsfeld, Singularities of theta divisors and the birational geometry of irregular varieties, J. Amer. Math. Soc. **10** (1997), pp. 243 – 258.

[7] Lawrence Ein and Robert Lazarsfeld, A geometric effective Nullstellensatz, Invent. Math. **137** (1999), pp. 427–488.

[8] Lawrence Ein, Robert Lazarsfeld and Karen Smith, Uniform bounds and symbolic powers on smooth varieties, Invent. Math. **144** (2001), pp. 241 – 252.

[9] Lawrence Ein, Robert Lazarsfeld, Karen Smith and Dror Varolin, Jumping coefficients of multiplier ideals, Duke Math. J. **123** (2004), pp. 469 – 506.

[10] Lawrence Ein and Mircea Mustaţă, Invariants of singularities of pairs, Int. Cong. Math., Zurich 2006, Vol. II, pp. 583–602.

[11] Lawrence Ein and Mihnea Popa, Extension of sections via adjoint ideals, to appear.

[12] Hélène Esnault and Eckart Viehweg, Logarithmic DeRham complexes and vanishing theorems, Invent. Math. **86** (1986), pp. 161 – 194.

[13] Charles Favre and Mattias Jonsson, Valuations and multiplier ideals, J. Amer. Math. Soc. **18** (2005), pp. 655–684.

[14] Christopher Hacon and James McKernan, Boundedness of pluricanonical maps of varieties of general type, Invent. Math. **166** (2006), pp. 1 – 25.

[15] Christopher Hacon and James McKernan, Existence of minimal models for varieties of log general type, II, preprint.

[16] Jason Howald, Multiplier ideals of monomial ideals, Trans. Amer. Math. Soc. **353** (2001), pp. 2665–2671.

[17] Yujiro Kawamata, Deformations of canonical singularities, J. Amer. Math. Soc. **12** (1999), pp. 85 – 92.

[18] Yujiro Kawamata, On the extension problem of pluricanonical forms, in *Algebraic Geometry: Hirzebruch 70*, Amer. Math. Soc., Providence, 1999, pp. 193 – 207.

[19] János Kollár, Sharp effective Nullstellensatz, J.Amer. Math. Soc. **1** (1988), pp. 963 – 975.

[20] János Kollár and Shigefumi Mori, *Birational Geometry of Algebraic Varieties*, Cambridge Tracts in Mathematics, vol. 134, Cambridge University Press, Cambridge, 1988.

[21] Robert Lazarsfeld, *Positivity in Algebraic Geometry, I & II*, Ergebnisse der Mathematik und ihrer Grenzgebiete, Vols. 48 & 49, Springer Verlag, Berlin, 2004.

[22] Robert Lazarsfeld and Kyungyong Lee, Local syzygies of multiplier ideals, Invent. Math. **167** (2007), pp. 409 – 418.

[23] Joseph Lipman and Kei-ichi Watanabe, Integrally closed ideals in two dimensional regular local rings are multiplier ideals, Math. Res. Lett. **10** (2003), pp. 423 – 434.

[24] Jihun Park and Youngho Woo, A remark on hypersurfaces with isolated singularities, Manuscripta Math. **121** (2006), pp. 451 – 456.

[25] Yum-Tong Siu, Invariance of plurigenera, Invent. Math. **134** (1998), pp. 661 – 673.

[26] Yum-Tong Siu, Multiplier ideal sheaves in complex and algebraic geometry, Sci. China Ser. A **48** (2005), suppl., pp. 1 – 31.

[27] Irena Swanson, Linear equivalence of ideal topologies, Math. Z. **234** (2000), pp. 755 – 775.

[28] Shigeharu Takayama, Pluricanonical systems on algebraic varieties of general type, Invent. Math. **165** (2006), pp. 551 – 587.

[29] A. N. Varchenko, Semicontinuity of the complex singularity exponent, Izv. Akad. Nauk SSSR Ser. Mat. **45** (1981), pp. 540 – 591.

[30] Oscar Zariski, On the irregularity of cyclic multiple planes, Ann. of Math. **32** (1931), pp. 485 – 511.

Exercises in the Birational Geometry of Algebraic Varieties

János Kollár

IAS/Park City Mathematics Series
Volume 17, 2008

Exercises in the Birational Geometry of Algebraic Varieties

János Kollár

The book [**KM98**] gave an introduction to the birational geometry of algebraic varieties, as the subject stood in 1998. The developments of the last decade made the more advanced parts of Chapters 6 and 7 less important and the detailed treatment of surface singularities in Chapter 4 less necessary. However, the main parts, Chapters 1–3 and 5, still form the foundations of the subject.

These notes provide additional exercises to [**KM98**]. The main definitions and theorems are recalled but not proved here. The emphasis is on the many examples that illustrate the methods, their shortcomings and some applications.

1. Birational classification of algebraic surfaces

For more detail, see [**BPVdV84**].

The theory of algebraic surfaces rests on the following three theorems.

Theorem 1. *Any birational morphism between smooth projective surfaces is a composite of blow-downs to points. Any birational map between smooth projective surfaces is a composite of blow-ups and blow-downs.*

Theorem 2. *There are 3 species of "pure-bred" surfaces:*

- (Rational): *For these surfaces the internal birational geometry is very complicated, but, up to birational equivalence, we have only $\mathbb{P}^2$. These frequently appear in the classical literature and in "true" applications.*
- (Calabi-Yau): *These are completely classified (Abelian, K3, Enriques, hyperelliptic) and their geometry is rich. They are of great interest to other mathematicians.*
- (General type): *They have a canonical model with Du Val singularities and ample canonical class. Although singular, this is the "best" model to work with. There are lots of these but they appear less frequently outside algebraic geometry.*

There are also two types of "mongrels":

- (Ruled): *Birational to $\mathbb{P}^1 \times$ (curve of genus ≥ 1).*
- (Elliptic): *These fiber over a curve with general fiber an elliptic curve.*

Department of Mathematics, Princeton University, Fine Hall, Washington Road Princeton NJ 08544-1000
E-mail address: kollar@math.princeton.edu

I thank Ch. Hacon, S. Mori and Ch. Xu and the participants of the OSU workshop organized by H. Clemens for many useful comments and corrections. Partial financial support was provided by the NSF under grant number DMS-0500198.

The "mongrels" are usually studied as an afterthought, with suitable modifications of the existing methods. In a general survey, it is best to ignore them.

Theorem 3. *Assume that S is neither rational nor ruled. Then there is a unique smooth projective surface S^{min} birational to S such that every birational map $S' \dashrightarrow S^{min}$ from a smooth projective surface S' is automatically a morphism.*

Some of these theorems are relatively easy, and some condense a long and hard story into a short statement.

> *The first aim of higher dimensional algebraic geometry is to generalize these theorems to dimensions three and up.*

In these notes we focus only on certain aspects of this project. Let us start with mentioning the parts that we will not cover.

The correct higher dimensional analogs of rational surfaces are *rationally connected varieties* and the ruled surfaces are replaced by *rationally connected fibrations.* We do not deal with them here. See [**Kol01**] for an introduction and [**Kol96**] for a detailed treatment.

The study of higher dimensional Calabi-Yau varieties is very active, with most of the effort going into understanding mirror symmetry rather than developing a general classification scheme.

It is known that any birational *map* between smooth projective varieties is a composite of blow-ups and blow-downs of smooth subvarieties [**Wło03, AKMW02**]. While it is very useful to stay with these easy-to-understand elementary steps, in practice it is very hard to keep track of geometric properties during blow-ups. It is much more useful to factor every birational *morphism* between smooth projective varieties as a composite of elementary steps. It turns out that smooth blow-ups do not work (22). From our current point of view, the natural question is to work with varieties with terminal singularities and consider the factorization of birational morphisms as a special case of the MMP. However, the following intriguing problem is still open.

Question 4. Let $f : X \to Y$ be a birational morphism between smooth projective 3-folds. Is f a composite of smooth blow-downs and flops?

2. Naive minimal models

This is a more technical version of my notes [**Kol07**].

Much of the power of affine algebraic geometry rests on the basic correspondence

$$\left\{ \begin{array}{c} \text{affine} \\ \text{schemes} \end{array} \right\} \quad \underset{\text{spectrum}}{\overset{\text{ring of regular functions}}{\rightleftarrows}} \quad \left\{ \begin{array}{c} \text{commutative} \\ \text{rings} \end{array} \right\}$$

Thus every affine variety is the natural existence domain for the ring of all regular functions on it.

Exercise 5. Let X be a $\mathbb{C}$-variety of finite type. Prove that X is affine iff the following 2 conditions are satisfied:

(1) (Point separation) For any two points $p \neq q \in X$ there is a regular function f on X such that $f(p) \neq f(q)$.

(2) (Maximality of domain) For any sequence of points $p_i \in X$ that does not converge to a limit in X, there is a regular function f on X such that $\lim f(p_i)$ does not exist.

Exercise 6. Reformulate and prove Exercise 5 for varieties over arbitrary fields.

7. As we move to more general varieties, this nice correspondence breaks down in two distinct ways.

Quasi affine varieties. Let $X := \mathbb{A}^n \setminus (\text{point})$ for some $n \geq 2$. Check that every regular function on X extends to a regular function on $\mathbb{A}^n$. Thus the function theory of X is rich but the natural existence domain for the ring of all regular functions on X is the larger space $\mathbb{A}^n$. Similarly, if

$$X = (\text{irreducible affine variety}) \setminus (\text{codimension} \geq 2 \text{ subvariety}),$$

then every regular function on X extends to a regular function on the irreducible affine variety.

Projective varieties. On a projective variety every regular (or holomorphic) function is constant, hence the regular (or holomorphic) function theory of a projective variety is not interesting.

On the other hand, a projective variety has many interesting *rational functions.* That is, functions that can locally be written as the quotient of two regular functions. At a point the value of a rational function f can be finite, infinite or undefined. The set of points where f is undefined has codimension ≥ 2. This makes it hard to control what happens in codimensions ≥ 2.

Rational functions on a k-variety X form a field $k(X)$, called the *function field* of X.

Exercise 8. Let $X = (xy - uv = 0) \subset \mathbb{A}^4$ and $f = x/u$. Show that X is normal and f is undefined only at the origin $(0, 0, 0, 0)$.

Exercise 9. Let X be a normal, proper variety over an algebraically closed field k. Prove that X is projective iff for any two points $p \neq q \in X$ and finite subset $R \subset X$, there is a rational function f on X such that $f(p) \neq f(q)$ and f is defined at all points of R.

Following the example of affine varieties we ask:

Question 10. *How tight is the connection between X and $k(X)$?*

Assume that we have $X_1 \subset \mathbb{P}^r$ with coordinates $(x_0 : \cdots : x_r)$, $X_2 \subset \mathbb{P}^s$ with coordinates $(y_0 : \cdots : y_s)$ and an isomorphism $\Psi : k(X_1) \cong k(X_2)$. Then $\phi_i := \Psi(x_i/x_0)$ are rational functions on X_2 and $\phi_j^{(-1)} := \Psi^{-1}(y_j/y_0)$ are rational functions on X_1. (Note that $\phi_j^{(-1)}$ is not the inverse of ϕ_j.) Moreover,

$$\Phi : (y_0 : \cdots : y_s) \mapsto \bigl(1 : \phi_1(y_0 : \cdots : y_s) : \cdots : \phi_r(y_0 : \cdots : y_s)\bigr)$$

defines a rational map $\Phi : X_2 \dashrightarrow X_1$ and

$$\Phi^{-1} : (x_0 : \cdots : x_r) \mapsto \bigl(1 : \phi_1^{(-1)}(x_0 : \cdots : x_r) : \cdots : \phi_s^{(-1)}(x_0 : \cdots : x_r)\bigr)$$

defines a rational map $\Phi^{-1} : X_2 \dashrightarrow X_1$ such that Ψ is induced by pulling back functions by Φ and Ψ^{-1} is induced by pulling back functions by Φ^{-1}. That is, X_1 and X_2 are *birational* to each other.

Exercise 11. Let C_1, C_2 be 1-dimensional, irreducible, projective with all local rings regular. Prove that every birational map $C_1 \dashrightarrow C_2$ is an isomorphism.

The situation is more complicated in higher dimensions. A map with an inverse is usually an isomorphism, but this fails in the birational case since Φ and Φ^{-1} are not everywhere defined. The simplest examples are blow-ups and blow-downs.

12 (Blow-ups)**.** Let X be a smooth projective variety and $Z \subset X$ a smooth subvariety. Let $B_Z X$ denote the *blow-up* of X along Z and $E_Z \subset B_Z X$ the exceptional divisor. We refer to $\pi : B_Z X \to X$ as a blow-up if we imagine that $B_Z X$ is created from X, and a *blow-down* if we start with $B_Z X$ and construct X later. Note that E_Z has codimension 1 and Z has codimension ≥ 2. Thus a blow-down decreases the Picard number by 1.

By blowing up repeatedly, starting with any X we can create more and more complicated varieties with the same function field. Thus, for a given function field $K = k(X)$, there is no "maximal domain" where all elements of K are rational functions. (The inverse limit of all varieties birational to X appears in the literature occasionally as such a "maximal domain," but so far with limited success.) On the other hand, one can look for a "minimal domain" or "minimal model."

As a first approximation, a variety X is a minimal model if the underlying space X is the "best match" to the rational function theory of X.

Example 13. Let S be a smooth projective surface which is neither rational nor ruled. Explain why it makes sense to say that S^{min} (as in (3)) is a "minimal domain" for the field $k(S)$.

Exercise 14. Let X be a projective variety that admits a finite morphism to an Abelian variety. Prove that every rational map $f : Y \dashrightarrow X$ from a smooth projective variety Y to X is a morphism.

Thus, if X is smooth, it makes sense to say that X is a "minimal domain" of its function field $k(X)$.

Not all varieties have a "minimal domain" with the above strong properties.

Example 15. Let $\mathbf{Q}^3 \subset \mathbb{CP}^4$ be the quadric hypersurface given by the equation $x^2 + y^2 + z^2 + t^2 = u^2$. Let

$$\pi : (x : y : z : t : u) \dashrightarrow (x : y : z : u - t)$$

be the projection from the north pole $(0 : 0 : 0 : 1 : 1)$ to the equatorial plane $(t = 0)$. Its inverse π^{-1} is given by

$$(x : y : z : u) \dashrightarrow (2xu : 2yu : 2zu : x^2 + y^2 + z^2 - u^2 : x^2 + y^2 + z^2 + u^2).$$

These maps show that the meromorphic function theory of $\mathbf{Q}^3$ is the same as that of $\mathbb{CP}^3$.

Show that π contracts the lines $(a\lambda : b\lambda : c\lambda : 1 : 1)$ to the points $(a : b : c : 0)$ whenever $a^2 + b^2 + c^2 = 0$, and π^{-1} contracts the plane at infinity $(u = 0)$ to the point $(0 : 0 : 0 : 1 : 1)$. Write π as a composite of blow ups and blow downs with smooth centers.

On the other hand, $\mathbf{Q}^3$ and $\mathbb{CP}^3$ are quite different as manifolds. Show that they have the same Betti numbers but they are not homeomorphic. Prove that $\mathbf{Q}^3$ and $\mathbb{CP}^3$ both have Picard number 1.

A more subtle example is the following.

Exercise 16. Let Y be a smooth projective variety of dimension 3 and f, g, h general sections of a very ample line bundle L on Y. Consider the hypersurface

$$X := (s^2 f + 2stg + t^2 h = 0) \subset Y \times \mathbb{P}^1_{s:t}.$$

Show that X is smooth and compute its canonical class.

Show that the projection $\pi : X \to Y$ has degree 2; let $\tau : X \dashrightarrow X$ be the corresponding Galois involution. Write it down explicitly in coordinates and decide where τ is regular.

Show that X contains (L^3) curves of the form (point) $\times \mathbb{P}^1$ and they are numerically equivalent to each other. (This may need the Lefschetz theorem on the Picard groups of hyperplane sections.)

Assume that Y admits a finite morphism to an Abelian variety. Prove that the following hold:

(1) Any smooth projective variety X' that is birational to X has Picard number at least $\rho(X)$.
(2) If X' has Picard number $\rho(X)$ then it is isomorphic to X.
(3) If $(L^3) > 1$ then there are nonprojective compact complex manifolds Z that are bimeromorphic to X, have Picard number $\rho(X)$, but are not isomorphic to X.

Exercise 17. Let X be a smooth projective variety such that K_X is nef. Let $f : X \dashrightarrow X'$ be a birational map to a smooth projective variety. Prove that the exceptional set $\operatorname{Ex}(f)$ has codimension ≥ 2 in X. Generalize to the case when X is canonical and X' is terminal (60).

Hint: You should find (107) helpful.

Definition 18. We say that a birational map $f : X_1 \dashrightarrow X_2$ *contracts* a divisor $D \subset X_1$ if f is defined at the generic point of D and $f(D) \subset X_2$ has codimension ≥ 2. The map f is called a *birational contraction* if f^{-1} does not contract any divisor.

A birational map $f : X_1 \dashrightarrow X_2$ is called *small* if neither f nor f^{-1} contracts any divisor.

The simplest examples of birational contractions are composites of blow-downs, but there are many, more complicated, examples.

Exercise 19. Let $f : S_1 \dashrightarrow S_2$ be a birational contraction between smooth projective surfaces. Show that f is a morphism.

Exercise 20. Let $L, M \subset \mathbb{P}^3$ two lines intersecting at a point. The identity on $\mathbb{P}^3$ induces a rational map $g : B_L B_M \mathbb{P}^3 \dashrightarrow B_L \mathbb{P}^3$. (With a slight abuse of notation, we also denote by L the birational transform of L on $B_M \mathbb{P}^3$, etc.) Show that g is a contraction but it is not a morphism. Describe how to factor g into a composite of smooth blow ups and blow downs.

There is essentially only one way to write a birational morphism between smooth surfaces as a composite of point blow ups. The next exercise shows that this no longer holds for 3-folds.

Exercise 21. Let $p \in L \subset \mathbb{P}^3$ be a point on a line. Let $C \subset B_L \mathbb{P}^3$ be the preimage of p. Show that the identity on $\mathbb{P}^3$ induces an isomorphism $B_C B_L \mathbb{P}^3 \cong B_L B_p \mathbb{P}^3$.

The next exercise shows that not every birational morphism between smooth 3-folds is a composite of smooth blow-ups.

Exercise 22. Let $C \subset \mathbb{P}^3$ be an irreducible curve with a unique singular point which is either a node or a cusp. Show that $B_C\mathbb{P}^3$ has a unique singular point; call it p. Check that $B_pB_C\mathbb{P}^3$ is smooth. Prove that $\pi : B_pB_C\mathbb{P}^3 \to \mathbb{P}^3$ can not be written as a composite of smooth blow-ups.

Write π as a composite of two smooth blow-ups and a flop (74).

Exercise 23. Let $f : X \dashrightarrow Y$ be a birational map between smooth, proper varieties. Show that

$$\rho(X) - \rho(Y) = \#\{\text{divisors contracted by } f\} - \#\{\text{divisors contracted by } f^{-1}\}$$

We are not yet ready to define minimal models. As a first approximation, let us focus on the codimension 1 part.

Temporary Definition 24. Let X be a smooth projective variety. We say that X is *minimal in codimension 1* if every birational map $f : Y \dashrightarrow X$ from a smooth variety Y is a birational contraction.

In particular, this implies that X has the smallest Picard number in its birational equivalence class.

Exercise 25. 1. Let X be a smooth projective variety such that K_X is nef. Prove that X is minimal in codimension 1.

2. $\mathbb{P}^3$ has the smallest Picard number in its birational equivalence class but it is not minimal in codimension 1.

3. Let $X \subset \mathbb{P}^4$ be a smooth degree 4 hypersurface. Then K_X is not nef but, as proved by Iskovskikh-Manin, X is minimal in codimension 1. (See [**KSC04**, Chap. 5] for a proof and an introduction to these techniques.)

Exercise 26. Set $X_0 := (x_1x_2 + x_3x_4 + x_5x_6 = 0) \subset \mathbb{A}^6$. Let $L \subset X_0$ be any 3-plane through the origin. Prove that, after a suitable coordinate change, L can be given as $(x_1 = x_3 = x_5 = 0)$. Prove that B_LX_0 is smooth.

Let Y be a smooth projective variety of dimension 3 and f_i, g_i are general sections of a very ample line bundle L on Y. Set

$$X' := \left(\sum_{i=1}^3 f_i(\mathbf{x})g_i(\mathbf{y}) = 0\right) \subset Y_{\mathbf{x}} \times Y_{\mathbf{y}},$$

where $\mathbf{x}$ (resp. $\mathbf{y}$) are the coordinates on the first (resp. second) factor.

Assume that Y admits a finite morphism to an Abelian variety. Show that X' is not birational to any smooth proper variety X that is minimal in codimension 1.

Exercise 27 (Contractions of products). [**KL07**] Let X, U, V be normal projective varieties and $\phi : U \times V \dashrightarrow X$ a birational contraction. Assume that X is smooth (or at least has rational singularities). Prove that there are normal projective varieties U' birational to U and V' birational to V such that $X \cong U' \times V'$.

In particular, $U \times V$ is minimal in codimension 1 iff U and V are both minimal in codimension 1

Hints to the proof. First reduce to the case when U, V are smooth.

Let $|H|$ be a complete, very ample linear system on X and $\phi^*|H|$ its pull back to $U \times V$. Using that ϕ is a contraction, prove that $\phi^*|H|$ is also a complete linear system.

If $H^1(U, \mathcal{O}_U) = 0$, then $\operatorname{Pic}(U \times V) = \pi_U^* \operatorname{Pic}(U) + \pi_V^* \operatorname{Pic}(V)$, thus there are divisors H_U on U and H_V on V such that $\phi_*|H| \sim \pi_U^* H_U + \pi_V^* H_V$. Therefore

$$H^0(U \times V, \mathcal{O}_{U\times V}(\phi^*|H|)) = H^0(U, \mathcal{O}_U(H_U)) \otimes H^0(V, \mathcal{O}_V(H_V)).$$

Let now U' be the image of U under the complete linear system $|H_U|$ and V' the image of V under the complete linear system $|H_V|$.

The $H^1(U, \mathcal{O}_U) \neq 0$ case is a bit harder. Replace H by a divisor $H^* := H + B$ where B is a pull back of a divisor from the product of the Albanese varieties of U and V. Show that for suitable B, there are divisors H_U on U and H_V on V such that $\phi_*|H^*| \sim \pi_U^* H_U + \pi_V^* H_V$. The rest of the argument now works as before.

3. The cone of curves

For details, see [**KM98**, Chap.3].

Definition 28. Let X be a projective variety over $\mathbb{C}$. Any irreducible curve $C \subset X$ has a homology class $[C] \in H_2(X, \mathbb{R})$. These classes generate a cone $NE(X) \subset H_2(X, \mathbb{R})$, called the *cone of curves* of X. Its closure is denoted by $\overline{\mathrm{NE}}(X) \subset H_2(X, \mathbb{R})$.

If X is over some other field, we can use the vector space $N_1(X)$ of curves modulo numerical equivalence instead of $H_2(X, \mathbb{R})$ to define the cone of curves $NE(X) \subset N_1(X)$.

Exercise 29. Show that every effective curve in $\mathbb{P}^{a_1} \times \cdots \times \mathbb{P}^{a_n}$ is rationally equivalent to a nonnegative linear combination of lines in the factors. Thus

$$\overline{\mathrm{NE}}\big(\mathbb{P}^{a_1} \times \cdots \times \mathbb{P}^{a_n}\big) \subset \mathbb{R}^n$$

is the polyhedral cone spanned by the basis elements corresponding to the lines.

Exercise 30. Assume that a connected, solvable group acts on X with finitely many orbits. Show that $\overline{\mathrm{NE}}(X)$ is the polyhedral cone spanned by the homology classes of the 1-dimensional orbits. (The same holds even for rational equivalence instead of homological equivalence.)

Hint. Use the Borel fixed point theorem: A connected, solvable group acting on a proper variety has a fixed point. Apply this to the Chow variety or the Hilbert scheme parametrizing curves in X.

Exercise 31. Let $S \subset \mathbb{P}^3$ be a smooth cubic surface. Show that every effective curve is linearly equivalent to a linear combination of lines. Thus $\overline{\mathrm{NE}}(X) \subset \mathbb{R}^7$ is a polyhedral cone spanned by the classes of the 27 lines. (Note that the Cone theorem implies this only with rational coefficients, not with integral ones. The proof is easiest using the basic theory of linear systems.)

Exercise 32. Let A be an Abelian surface. If Z is an ample $\mathbb{R}$-divisor, then $(Z \cdot Z) > 0$. Prove that, conversely, the condition $(Z \cdot Z) > 0$ defines a subset of $N_1(A)$ with 2 connected components, one of which consists of ample $\mathbb{R}$-divisors. Show that its closure is $\overline{\mathrm{NE}}(A)$.

Check that if $A = E \times E$ and E does not have complex multiplication then every curve is algebraically equivalent to a linear combination $aE_1 + bE_2 + cD$ where E_i are the two factors and D the diagonal. Thus

$$\overline{\mathrm{NE}}(E \times E) = \{aE_1 + bE_2 + cD : ab + bc + ca \geq 0 \text{ and } a + b + c \geq 0\} \subset \mathbb{R}^3$$

is a "round" cone.

Despite what these examples suggest, the cone of curves is usually extremely difficult to determine. For instance, we still don't know the cone of curves for the following examples.

(1) $C \times C$ for a general curve C. (See [**Laz04**, Sec.1.5] for the known results and references.)
(2) The blow up of $\mathbb{P}^n$ at more than a few points, cf. [**CT06**].

A basic discovery of [**Mor82**] is that the part of the cone of curves which has negative intersection with the canonical class is quite well behaved. Subsequently it was generalized to certain perturbations of the canonical class. The precise definitions will be given in Section 4. For now you can imagine that X is smooth and $\Delta = \sum a_i D_i$ is a $\mathbb{Q}$-divisor where $\sum D_i$ is a simple normal crossing divisor and $0 < a_i < 1$ for every i.

Theorem 33 (Cone theorem). *(cf.* [**KM98**, Thm.3.7.1–2]*) Let (X, Δ) be a projective klt pair with Δ effective. Then:*

(1) *There are (at most countably many) rational curves $C_j \subset X$ such that $0 < -(K_X + \Delta) \cdot C_j \le 2 \dim X$ and*

$$\overline{\mathrm{NE}}(X) = \overline{\mathrm{NE}}(X)_{(K_X+\Delta)\ge 0} + \sum \mathbb{R}_{\ge 0}[C_j],$$

where $\overline{\mathrm{NE}}(X)_{(K_X+\Delta)\ge 0}$ denotes the set of those elements of $\overline{\mathrm{NE}}(X)$ that have nonnegative intersection number with $K_X + \Delta$.

(2) *For any $\epsilon > 0$ and ample $\mathbb{Q}$-divisor H,*

$$\overline{\mathrm{NE}}(X) = \overline{\mathrm{NE}}(X)_{(K_X+\Delta+\epsilon H)\ge 0} + \sum_{\text{finite}} \mathbb{R}_{\ge 0}[C_j].$$

If $-(K_X + \Delta)$ is ample then taking $H = -(K_X + \Delta)$ and $\epsilon < 1$ in (33.2), the first summand on the right is trivial. Hence we obtain:

Corollary 34. *Let (X, Δ) be a projective klt pair with Δ effective and $-(K_X + \Delta)$ ample. There are finitely many rational curves $C_j \subset X$ such that*

$$\overline{\mathrm{NE}}(X) = \sum \mathbb{R}_{\ge 0}[C_j].$$

In particular, $\overline{\mathrm{NE}}(X)$ is a polyhedral cone. □

Warning 35. If the cone is 3-dimensional, the cone theorem implies that the $(K_X + \Delta)$-negative part of $\overline{\mathrm{NE}}(X)$ is locally polyhedral. This, however, fails for 4-dimensional cones.

Use (32) to show that such an example is given by $\overline{\mathrm{NE}}(E \times E \times \mathbb{P}^1)$ where E is an elliptic curve which does not have complex multiplication.

Definition 36. In convex geometry, a closed subcone $F \subset \overline{\mathrm{NE}}(X)$ is called an *extremal face* if $u, v \in \overline{\mathrm{NE}}(X)$ and $u+v \in F$ implies that $u, v \in F$. A 1-dimensional extremal face is called an *extremal ray.*

In algebraic geometry, one frequently assumes in addition that intersection product with K_X (or $K_X + \Delta$) gives a strictly negative linear function on $F \setminus \{0\}$.

Thus, extremal rays of $\overline{\mathrm{NE}}(X)$ are precisely those summands $\mathbb{R}_{\ge 0}[C_j]$ in (33.1) that are actually needed.

The next result shows that there are contraction morphisms associated to any extremal face.

Theorem 37 (Contraction theorem). *(cf.* [**KM98**, Thm.3.7.2–4]*) Let* (X, Δ) *be a projective klt pair with* Δ *effective. Let* $F \subset \overline{\mathrm{NE}}(X)$ *be a* ($(K_X + \Delta)$*-negative) extremal face. Then there is a unique morphism* $\mathrm{cont}_F : X \to Z$*, called the contraction of* F*, such that* $(\mathrm{cont}_F)_* \mathcal{O}_X = \mathcal{O}_Z$ *and an irreducible curve* $C \subset X$ *is mapped to a point by* cont_F *iff* $[C] \in F$*. Moreover,*

(1) $R^i(\mathrm{cont}_F)_* \mathcal{O}_X = 0$ *for* $i > 0$*, and*
(2) *if* L *is a line bundle on* X *such that* $(L \cdot C) = 0$ *whenever* $[C] \in F$ *then there is a line bundle* L_Z *on* Z *such that* $L \cong \mathrm{cont}_F^* L_Z$.

Exercise 38. Let Z be a smooth, projective variety and $W \subset X$ a smooth, irreducible subvariety of codimension ≥ 2. Show that $\pi : B_W Z \to Z$ is the contraction of an extremal ray on $B_W Z$.

Exercise 39. Let Z be an n-dimensional projective variety with a unique singular point p of the form

$$x_1^m + \cdots + x_{n+1}^m + (\text{higher terms}) = 0.$$

Show that $B_p Z$ is smooth and $\pi : B_p Z \to Z$ is the contraction of an extremal face on $B_p Z$ iff $m < n$. The exceptional divisor is the smooth hypersurface $(x_1^m + \cdots + x_{n+1}^m = 0) \subset \mathbb{P}^n$.

If $n \geq 4$, then by the Lefschetz theorem, π is the contraction of an extremal ray. Find examples with $n = 3$ where we do contract a face.

Exercise 40. Let $f_i(x_1, \ldots, x_4)$ for $i = m, m+1$ be homogeneous of degree i. Assume that

$$X := \bigl(x_0 f_m(x_1, \ldots, x_4) + f_{m+1}(x_1, \ldots, x_4) = 0\bigr) \subset \mathbb{P}^4$$

is smooth away from the origin. Prove that every Weil divisor on X is obtained by intersecting X with another hypersurface.

Exercise 41. Let Z be an n-dimensional projective variety with a unique singular point p of the form

$$x_1^m + \cdots + x_n^m + x_{n+1}^{m+1} + (\text{higher terms}) = 0.$$

Show that $B_p Z$ is smooth and $\pi : B_p Z \to Z$ is the contraction of an extremal ray on $B_p Z$ iff $m < n$ and $n \geq 3$. The exceptional divisor is the singular hypersurface $(x_1^m + \cdots + x_n^m = 0) \subset \mathbb{P}^n$.

Exercise 42. Let $f_m(x_1, \ldots, x_{n+1})$ be an irreducible, homogeneous degree m polynomial and $g_{m+1}(x_1, \ldots, x_{n+1})$ a general, homogeneous degree $m+1$ polynomial. Let Z be an n-dimensional projective variety with a unique singular point p of the form

$$f_m(x_1, \ldots, x_{n+1}) + g_{m+1}(x_1, \ldots, x_{n+1}) + (\text{higher terms}) = 0.$$

Use (51) and (67) to prove that $B_p Z$ has only canonical singularities (60).

Show that $\pi : B_p Z \to Z$ is the contraction of an extremal face on $B_p Z$ iff $m < n$.

Note that the exceptional divisor is the hypersurface $(f_m(x_1, \ldots, x_{n+1}) = 0) \subset \mathbb{P}^n$, which can be quite singular.

Exercise 43. Let $Z \subset \mathbb{P}^n$ be defined by $x_0 = f(x_1, \ldots, x_n) = 0$ where f is irreducible. Show that $B_Z \mathbb{P}^n \to \mathbb{P}^n$ is the contraction of an extremal ray on $B_Z \mathbb{P}^n$.

Show that Z has only cA-type singularities (67). When is Z canonical or terminal (60)?

Note that the exceptional divisor is a $\mathbb{P}^1$-bundle over Z, which can be quite singular.

Exercise 44. Let X be a smooth, projective variety, $D \subset X$ a smooth hypersurface and $C \subset D$ any curve. Assume that the Picard number of D is 1 and the conormal bundle $N^*_{D|X}$ is ample.

Prove that $[C]$ is an extremal ray of $\overline{\mathrm{NE}}(X)$ in the convex geometry sense (36). When is it a K_X-negative extremal ray?

Assume in addition that $-K_D$ is ample. Generalize the proof of Castelnuovo's theorem (for instance, as in [**Har77**, V.5.7]) to prove (37) in this case. (That is, there is a contraction $\pi : X \to X'$ that maps D to a point and is an isomorphism on $X \setminus D$.)

Exercise 45. With notation as in (44), assume that $D \cong \mathbb{P}^{n-1}$ and $N_{D|X} \cong \mathcal{O}_{\mathbb{P}^{n-1}}(-m)$. Set $x' := \pi(D)$. Prove that the completion of X' (at x') is isomorphic to the completion (at the origin) of the quotient of $\mathbb{A}^n$ by the $\mathbb{Z}/m$-action $(x_1, \dots, x_n) \mapsto (\epsilon x_1, \dots, \epsilon x_n)$ where ϵ is a primitive mth root of 1. (Hint: Use the methods of [**Har77**, Exrc.II.8.6–7].)

Exercise 46. Let Z be a smooth, projective variety and $X \subset Z \times \mathbb{P}^m$ a smooth hypersurface such that $X \cap (\{z\} \times \mathbb{P}^m)$ is a hypersurface of degree d for general $z \in Z$.

Show that the projection $\pi : X \to Z$ is the contraction of an extremal face on X iff $d < m + 1$ and $m \geq 2$.

If $m = 2$ and $\dim Z = 2$ then show that every fiber of $\pi : X \to Z$ is either a line (if $d = 1$) or a (possibly singular) conic (if $d = 2$). (This can fail if X has an ordinary double point.)

If $m = 2$ and $\dim Z = 3$ then find smooth examples where the general fiber of $\pi : X \to Z$ is a line or a conic but special fibers are $\mathbb{P}^2$.

Exercise 47. If you know some about the deformation theory and the Hilbert scheme of curves on smooth varieties, prove the following. (You will find (37.1) very helpful.)

Let $\pi : X \to Z$ be an extremal contraction with X smooth where every fiber has dimension ≤ 1. Then Z is smooth and we have one of the following cases:

(1) $X = B_W Z$ for some smooth $W \subset Z$ of codimension 2.
(2) X is a $\mathbb{P}^1$-bundle over Z.
(3) X is a hypersurface in a $\mathbb{P}^2$-bundle over Z and every fiber of $\pi : X \to Z$ is a (possibly singular) conic.

Exercise 48. Let $X \subset \mathbb{P}^4$ be a degree 3 hypersurface with a unique singular point that is an ordinary node. (That is, analytically isomorphic to $(xy - zt = 0)$.)

Let $\pi : Y \to X$ denote the blow up of the node. Prove that its exceptional divisor E is isomorphic to $\mathbb{P}^1 \times \mathbb{P}^1$ and its normal bundle is $\mathcal{O}_{\mathbb{P}^1 \times \mathbb{P}^1}(-1, -1)$.

Thus E looks like it could have been obtained by blowing up a curve $C \cong \mathbb{P}^1$ with normal bundle $\mathcal{O}_{\mathbb{P}^1}(-1) + \mathcal{O}_{\mathbb{P}^1}(-1)$ in a smooth 3-fold. Nonetheless, use (40) to show that there is no such projective 3-fold.

Example 49. Let X be the cE_7-type singularity $(x^2 + y^3 + yg_3(z,t) + h_5(z,t) = 0) \subset \mathbb{A}^4$, where g_3 and h_5 do not have a common factor. Show that X has an isolated singular point at the origin and its (3, 2, 1, 1)-blow up $Y \to X$ has only terminal singularities. (See [**KM98**, 4.56] or [**KSC04**, 6.38] for weighted blow-ups.) Conclude from this that X itself has a terminal singularity.

One of the standard charts on the blow up is given by the substitutions $x = x_1y_1^3, y = y_1^2, z = z_1y_1, t = t_1y_1$ and the exceptional divisor has equation

$$E = (g_3(z_1, t_1) + h_5(z_1, t_1) = 0)/\tfrac{1}{2}(1, 1, 1) \subset \mathbb{A}^3/\tfrac{1}{2}(1, 1, 1).$$

This gives examples of extremal contractions whose exceptional divisor E has quite complicated singularities.

(1) $x^2 + y^3 + yz^3 + t^5$. E is singular along $(z_1 = t_1 = 0)$, with a transversal singularity type $z^3 + t^5$, that is E_8.
(2) $x^2 + y^3 + y(z - at)(z - bt)(z - ct) + t^5$. E has triple self-intersection along $z_1 = t_1 = 0$.

Exercise 50. Let X be a smooth Fano variety, $\dim X \geq 4$. Let $Y \subset X$ be a smooth divisor in $|-K_X|$ (thus $K_Y = 0$). Show that the natural map $i_* : \overline{\mathrm{NE}}(Y) \to \overline{\mathrm{NE}}(X)$ is an isomorphism. Thus $\overline{\mathrm{NE}}(Y)$ is a polyhedral cone. (See [**Bor90, Bor91**] for many such interesting examples.)

Steps of the proof.

1. By a theorem of Lefschetz, i_* is an injection. Thus we need to show that for every extremal ray R of $\overline{\mathrm{NE}}(X)$ there is a curve $C_R \subset Y$ such that C_R generates R in $\overline{\mathrm{NE}}(X)$.

2. Let $f : X \to Z$ be the contraction morphism of R. If there is a fiber $F \subset X$ of f whose dimension is at least two then $Y \cap F$ contains a curve C_R which works.

3. If every fiber of f has dimension one then we use (47). We need to show that in these cases Y contains a fiber of f.

4. Prove the following lemma. Let $g : U \to V$ be a $\mathbb{P}^1$-bundle over a normal projective variety. Let $V' \subset U$ be an irreducible divisor such that $g : V' \to V$ is finite of degree one (thus an isomorphism). If V' is ample then $\dim V \leq 1$.

5. In the divisorial contraction case apply this lemma to $U :=$ the exceptional divisor of f.

6. In the $\mathbb{P}^1$-bundle case apply this lemma to $U :=$ normalization of the branch divisor of $Y \to Z$. (If there is no branch divisor, then to $X \times_Z Y \to Y$.)

7. In the conic bundle case there are two possibilities. If every fiber is smooth, this is like the $\mathbb{P}^1$-bundle case. Otherwise apply the lemma to $U :=$ normalization of the divisor of singular fibers of $Y \to Z$.

Exercise 51. Prove the following result of [**Kol97**, 4.4].

Theorem. Let X be a smooth variety over a field of characteristic zero and $|B|$ a linear system of Cartier divisors. Assume that for every $p \in X$ there is a $B(p) \in |B|$ such that $B(p)$ is smooth at p (or $p \notin B(p)$).

Then a general member $B^g \in |B|$ has only cA-type singularities (67).

Hint. By Noetherian induction it is sufficient to prove that for every irreducible subvariety $Z \subset X$ there is an open subset $Z^0 \subset Z$ such that a general member $B^g \in |B|$ has only cA-type singularities at points of Z^0.

If $Z \not\subset \mathrm{Bs}\,|B|$ then use the usual Bertini theorem.

If $Z \subset \mathrm{Bs}\,|B|$ and $\mathrm{codim}(Z, X) = 1$, then use the usual Bertini theorem for $|B| - Z$.

If $Z \subset \operatorname{Bs}|B|$ and $\operatorname{codim}(Z, X) > 1$ then restrict to a suitable hypersurface $Z \subset Y \subset X$ and use induction.

Exercise 52. Use the following examples to show that the conclusion of (51) is almost optimal:

Let $X = \mathbb{C}^n$ and $f \in \mathbb{C}[x_3, \dots, x_n]$ such that $(f = 0)$ has an isolated singularity at the origin. Consider the linear system $|B| = (\lambda x_1 + \mu x_1 x_2 + \nu f = 0)$. Show that at each point there is a smooth member and the general member is singular at $(0, -\lambda/\mu, 0, \dots, 0)$ with local equation $(x_1 x_2 + f = 0)$.

Consider the linear system $\lambda(x^2 + zy^2) + \mu y^2$. At any point $x \in \mathbb{C}^3$ its general member has a cA-type singularity, but the general member has a moving pinch point.

4. Singularities

For details, see [**KM98**, Chaps.4–5].

We already saw in several examples that even if we start with a smooth variety, the contraction of an extremal ray can lead to a singular variety. It took about 10 years to understand the correct classes of singularities that one needs to consider. Instead of going through this historical process, let us jump into the final definitions.

Remark 53. In the early days of the MMP, a lot of effort was devoted to classifying the occurring singularities in dimensions 2 and 3. While it is comforting to have some concrete examples and lists at hand, the recent advances use very little of these explicit descriptions. In most applications, we fall back to the definitions via log resolutions. The key seems to be an ability to work with log resolutions.

Definition 54. Let X be a normal scheme and Δ a $\mathbb{Q}$-divisor on X such that $K_X + \Delta$ is $\mathbb{Q}$-Cartier. Let $f : Y \to X$ be a birational morphism, Y normal. Let $E_i \subset \operatorname{Ex}(f)$ be the exceptional divisors. If $m(K_X + \Delta)$ is Cartier, then $f^*\mathcal{O}_X\big(m(K_X + \Delta)\big)$ is defined and there is a natural isomorphism

$$f^*\mathcal{O}_X\big(m(K_X + \Delta)\big)|_{Y \setminus \operatorname{Ex}(f)} \cong \mathcal{O}_Y\big(m(K_Y + f_*^{-1}\Delta)\big)|_{Y \setminus \operatorname{Ex}(f)}, \tag{54.1}$$

where $f_*^{-1}\Delta$ denotes the birational transform of Δ. Hence there are integers b_i such that

$$\mathcal{O}_Y\big(m(K_Y + f_*^{-1}\Delta)\big) \cong f^*\mathcal{O}_X\big(m(K_X + \Delta)\big)\big(\textstyle\sum b_i E_i\big). \tag{54.2}$$

Formally divide by m and write this as

$$K_Y + \Delta_Y \sim_{\mathbb{Q}} f^*(K_X + \Delta) \quad \text{where} \quad \Delta_Y := f_*^{-1}\Delta - \textstyle\sum (b_i/m) E_i.$$

The rational number $a(E_i, X, \Delta) := b_i/m$ is called the *discrepancy* of E_i with respect to (X, Δ).

The closure of $f(E_i) \subset X$ is called the *center* of E_i on X. It is denoted by $\operatorname{center}_X E_i$.

If $f' : Y' \to X$ is another birational morphism and $E_i' := \big((f')^{-1} \circ f\big)(E_i) \subset Y'$ is a divisor then $a(E_i', X, \Delta) = a(E_i, X, \Delta)$ and $\operatorname{center}_X E_i = \operatorname{center}_X E_i'$. Thus the discrepancy and the center depend only on the divisor up to birational equivalence, but not on the particular variety where the divisor appears.

Definition 55. Let X be a normal variety. An $\mathbb{R}$-*divisor* on X is a formal $\mathbb{R}$-linear combination $\sum r_i D_i$ of Weil divisors. We say that two $\mathbb{R}$-divisors A_1, A_2 are $\mathbb{R}$-*linearly equivalent*, denoted $A_1 \sim_{\mathbb{R}} A_2$, if there are rational functions f_i and real numbers r_i such that $A_1 - A_2 = \sum r_i (f_i)$.

One can pretty much work with $\mathbb{R}$-divisors as with $\mathbb{Q}$-divisors, but some basic properties need to be thought through.

Exercise 56. Prove the following about $\mathbb{R}$-divisors and $\mathbb{R}$-linear equivalence.

(1) Let A_1, A_2 be two $\mathbb{Q}$-divisors. Show that $A_1 \sim_{\mathbb{R}} A_2$ iff $A_1 \sim_{\mathbb{Q}} A_2$.

(2) Define the pull back of $\mathbb{R}$-divisors and show that it is well defined.

(3) Let A be an $\mathbb{R}$-divisor such that $A \sim_{\mathbb{R}} 0$. Prove that one can write $A = \sum r_i(f_i)$ such that $\operatorname{Supp}\bigl((f_i)\bigr) \subset \operatorname{Supp} A$ for every i.

Exercise 57. Let X be a normal scheme and Δ an $\mathbb{R}$-divisor on X such that $K_X + \Delta$ is $\mathbb{R}$-Cartier. Let $f : Y \to X$ be a proper birational morphism, Y normal. Show that there is a unique $\mathbb{R}$-divisor Δ_Y such that

(1) $f_*(\Delta_Y) = \Delta$, and
(2) $K_Y + \Delta_Y \equiv_f f^*(K_X + \Delta)$, where $\equiv_f$ denotes relative numerical equivalence, that is, $(K_Y + \Delta_Y \cdot C) = (f^*(K_X + \Delta) \cdot C)$ for every curve $C \subset Y$ such that $\dim f(C) = 0$. (Note that the latter is just 0.)

Use this to define discrepancies for $\mathbb{R}$-divisors.

Exercise 58. Formulate (54) in case $f : Y \dashrightarrow X$ is a birational map which is defined outside a codimension 2 set. (This holds, for instance if X is proper over the base scheme S.)

Exercise 59 (Divisors and rational maps). Let $f : X \dashrightarrow Y$ be a generically finite rational map between proper, normal schemes. Define the push forward $f_* : \operatorname{Div}(X) \to \operatorname{Div}(Y)$ of Weil divisors. Show that if f, g are morphisms then $(f \circ g)_* = f_* \circ g_*$ but this fails even for birational maps.

Let $f : X \dashrightarrow Y$ be a dominant rational map between normal schemes, Y proper. Define the pull back $f^* : \operatorname{CDiv}(Y) \to \operatorname{Div}(X)$ from Cartier divisors to Weil divisors. Show that if f is a morphism then we get $f^* : \operatorname{CDiv}(Y) \to \operatorname{CDiv}(X)$ but not in general. Find examples of birational maps between smooth projective varieties such that $(f \circ g)^* \neq f^* \circ g^*$.

Definition 60. Let (X, Δ) be a pair where X is a normal variety and $\Delta = \sum a_i D_i$ is a sum of distinct prime divisors. (We allow the a_i to be arbitrary real numbers.) Assume that $K_X + \Delta$ is $\mathbb{R}$-Cartier. We say that (X, Δ) is

$$\left.\begin{array}{c} \textit{terminal} \\ \textit{canonical} \\ \textit{klt} \\ \textit{plt} \\ \textit{dlt} \\ \\ \textit{lc} \end{array}\right\} \quad \text{if } a(E, X, \Delta) \text{ is} \quad \left\{\begin{array}{ll} > 0 & \forall\, E \text{ exceptional}, \\ \geq 0 & \forall\, E \text{ exceptional}, \\ > -1 & \forall\, E, \\ > -1 & \forall\, E \text{ exceptional}, \\ > -1 & \forall\, E \text{ such that } (X,\Delta) \text{ is not snc at} \\ & \quad \text{the generic point of } \operatorname{center}_X(E), \\ \geq -1 & \forall\, E. \end{array}\right.$$

Here klt is short for *Kawamata log terminal*, plt for *purely log terminal*, dlt for *divisorial log terminal*, lc for *log canonical* and snc for *simple normal crossing*. (The phrase "(X, Δ) has terminal etc. singularities" may be confusing since it could refer to the singularities of $(X, 0)$ instead.)

Each of these 5 notions has an important place in the theory of minimal models:

(1) *Terminal.* Assuming $\Delta = 0$, this is the smallest class that is necessary to run the minimal model program for smooth varieties. If $(X, 0)$ is terminal then $\operatorname{Sing} X$ has codimension ≥ 3. All 3-dimensional terminal singularities

are classified, see (71) for some examples. It is generally believed that already in dimension 4 a complete classification would be impossibly long. The $\Delta \neq 0$ case appears only infrequently.

(2) *Canonical.* Assuming $\Delta = 0$, these are precisely the singularities that appear on the canonical models of varieties of general type. Two dimensional canonical singularities are classified, see (66). There is some structural information in dimension 3 [**KM98**, 5.3]. This class is especially important for moduli problems.

(3) *Kawamata log terminal.* This is the smallest class that is necessary to run the minimal model program for pairs (X, Δ) where X is smooth and Δ a simple normal crossing divisor with coefficients < 1.

The vanishing theorems (cf. [**KM98**, 2.4–5]) seem to hold naturally in this class. In general, proofs that work with canonical singularities frequently work with klt. Most unfortunately, this class is not large enough for inductive proofs.

(4) *Purely log terminal.* This is useful mostly for inductive purposes. (X, Δ) is plt iff (X, Δ) is dlt and the irreducible components of $\lfloor \Delta \rfloor$ are disjoint.

(5) *Divisorial log terminal.* This is the smallest class that is necessary to run the minimal model program for pairs (X, Δ) where X is smooth and Δ a simple normal crossing divisor with coefficients ≤ 1.

By [**Sza94**], there is a log resolution $f : (X', \Delta') \to (X, \Delta)$ such that every f-exceptional divisor has discrepancy > -1 and f is an isomorphism over the snc locus of (X, Δ).

While the definition of this class is somewhat artificial looking, it has good cohomological properties and is much better behaved than general log canonical pairs.

If $\Delta = 0$ then the notions klt and dlt coincide and in this case we say that X has *log terminal* singularities (abbreviated as *lt*).

(6) *Log canonical.* This is the largest class where discrepancy still makes sense and inductive arguments naturally run in this class. There are three major complications though:
   (a) The refined vanishing theorems frequently fail.
   (b) The singularities are not rational and not even Cohen-Macaulay, hence rather complicated from the cohomological point of view; see, for example, (71).
   (c) Several tricks of perturbing coefficients can not be done since a perturbation would go above 1; see, for example, (95).

Exercise 61. Let $f : X \to Y$ be a birational morphism, Δ_X, Δ_Y $\mathbb{R}$-divisors such that $f_*\Delta_X = \Delta_Y$ and D an effective $\mathbb{R}$-divisor. Assume that $K_Y + \Delta_Y$ and D are $\mathbb{R}$-Cartier and

$$K_X + \Delta_X \sim_{\mathbb{R}} f^*(K_Y + \Delta_Y) + D.$$

Prove that for any E, $a(E, X, \Delta_X) \leq a(E, Y, \Delta_Y)$ and the inequality is strict iff $\operatorname{center}_X E \subset \operatorname{Supp} D$.

Exercise 62. Show that the assumptions of (61) are fulfilled (for suitable Δ_Y and D) if X is $\mathbb{Q}$-factorial, f is the birational contraction of a $(K_X + \Delta_X)$-negative extremal ray and $\operatorname{Ex}(f)$ has codimension 1.

The following exercise shows why log canonical is the largest class defined.

Exercise 63. Given (X, Δ) assume that there is a divisor E_0 such that $a(E_0, X, \Delta) < -1$. Prove that $\inf_E\{a(E, X, \Delta)\} = -\infty$.

Exercise 64. Show that if $(X, \sum a_i D_i)$ is lc (and the D_i are distinct) then $a_i \leq 1$ for every i.

Exercise 65. Assume that X is smooth and Δ is effective. Show that if $\operatorname{mult}_x \Delta < 1$ (resp. ≤ 1) for every $x \in X$ then (X, Δ) is terminal (resp. canonical).

Prove that the converse holds for surfaces but not in higher dimensions.

Exercise 66 (Du Val singularities). In each of the following cases, construct the minimal resolution and verify that its dual graph is the graph given. Check that these singularities are canonical. (One can see that these are all the 2-dimensional canonical singularities.) See [**KM98**, Sec.4.2] or [**Dur79**] for more information. (The equations below are correct in characteristic zero. The dual graphs are correct in every characteristic.)

A_n: $x^2 + y^2 + z^{n+1} = 0$, with $n \geq 1$ curves in the dual graph:

$$2 - 2 - \cdots - 2 - 2$$

D_n: $x^2 + y^2 z + z^{n-1} = 0$, with $n \geq 4$ curves in the dual graph:

$$\begin{array}{ccccccccc} & & 2 & & & & & & \\ & & | & & & & & & \\ 2 & - & 2 & - & \cdots & - & 2 & - & 2 \end{array}$$

E_6: $x^2 + y^3 + z^4 = 0$, with dual graph:

$$\begin{array}{ccccccccc} & & & & 2 & & & & \\ & & & & | & & & & \\ 2 & - & 2 & - & 2 & - & 2 & - & 2 \end{array}$$

E_7: $x^2 + y^3 + yz^3 = 0$, with dual graph:

$$\begin{array}{ccccccccccc} & & & & 2 & & & & & & \\ & & & & | & & & & & & \\ 2 & - & 2 & - & 2 & - & 2 & - & 2 & - & 2 \end{array}$$

E_8: $x^2 + y^3 + z^5 = 0$, with dual graph:

$$\begin{array}{ccccccccccccc} & & & & 2 & & & & & & & & \\ & & & & | & & & & & & & & \\ 2 & - & 2 & - & 2 & - & 2 & - & 2 & - & 2 & - & 2 \end{array}$$

Exercise 67 (cA-type singularities). Let $0 \in X$ a normal cA-type singularity. That is, either X is smooth at 0, or, in suitable local coordinates $x_1, \ldots, x_n$, the equation of X is $x_1 x_2 +$ (other terms) $= 0$.

Show that X is

(1) canonical near 0 iff $\dim \operatorname{Sing} X \leq \dim X - 2$, and
(2) terminal near 0 iff $\dim \operatorname{Sing} X \leq \dim X - 3$.

Hint. First show that being cA-type is an open condition. Then use a lemma of Zariski and Abhyankar (cf. [**KM98**, 2.45]) to reduce everything to the statements:

(3) The exceptional divisor(s) of $B_0 X \to X$ have discrepancy $\dim X - 2$, save when X is smooth.

(4) B_0X has only cA-type singularities.

Exercise 68 (Some simple elliptic singularities). In each of the following cases, construct the minimal resolution. Verify that the exceptional set is a single elliptic curve with self intersection $-k$.

$(k=3)$ $(x^3+y^3+z^3=0)$. (This is very easy)
$(k=2)$ $(x^2+y^4+z^4=0)$.
$(k=1)$ $(x^2+y^3+z^6=0)$. (This is a bit tricky.)

In general, prove that for any elliptic curve E and any $k\geq 1$ there is a normal singularity whose minimal resolution contains E as the single exceptional curve with self intersection $-k$.

Check that all of these are log canonical.

Use the methods of [**Har77**, Exrc.II.8.6–7] to prove that the completion of the singularity is uniquely determined by E.

Exercise 69. Construct the minimal resolutions of the following quotients of the singularities in (68). (See (72) for the notation.)

$(x^3+y^3+z^3=0)$: $\frac{1}{3}(1,0,0)$, $\frac{1}{3}(1,1,1)$.
$(x^2+y^4+z^4=0)$: $\frac{1}{2}(1,0,0)$, $\frac{1}{4}(0,0,1)$.
$(x^2+y^3+z^6=0)$: $\frac{1}{6}(0,0,1)$.

Exercise 70. Let $X\subset\mathbb{P}^n$ be a smooth variety and $C(X)\subset\mathbb{A}^{n+1}$ the cone over X. Show that $C(X)$ is normal iff $H^0(\mathbb{P}^n,\mathcal{O}_{\mathbb{P}^n}(m))\to H^0(X,\mathcal{O}_X(m))$ is onto for every $m\geq 0$.

Assume next that $C(X)$ is normal. Let Δ be an effective $\mathbb{Q}$-divisor on X. Prove that

(1) $K_{C(X)}+C(\Delta)$ is $\mathbb{Q}$-Cartier iff $K_X+\Delta\sim_{\mathbb{Q}} r\cdot H$ for some $r\in\mathbb{Q}$ where $H\subset X$ is the hyperplane class.
(2) If $K_X+\Delta\sim_{\mathbb{Q}} r\cdot H$ then $\big(C(X),C(\Delta)\big)$ is
   (a) terminal iff $r<-1$ and (X,Δ) is terminal,
   (b) canonical iff $r\leq -1$ and (X,Δ) is canonical,
   (c) klt iff $r<0$ and (X,Δ) is klt, and
   (d) lc iff $r\leq 0$ and (X,Δ) is lc.

Exercise 71. Notation as in (70). Prove that $C(X)$ has a rational singularity iff $H^i(X,\mathcal{O}_X(m))=0$ for every $i>0, m\geq 0$ and a Cohen-Macaulay singularity iff $H^i(X,\mathcal{O}_X(m))=0$ for every $\dim X>i>0, m\geq 0$. In particular:

(1) If X is an Abelian variety and $\dim X\geq 2$ then $C(X)$ is log canonical but not Cohen-Macaulay.
(2) If X is a K3 surface then $C(X)$ is log canonical, Cohen-Macaulay but not rational.
(3) If X is an Enriques surface then $C(X)$ is log canonical and rational.

72 (Quotient singularities). Let G be any finite group. A homomorphism $G\to GL_n$ is equivalent to a linear G-action on $\mathbb{A}^n$. The resulting quotient singularities $\mathbb{A}^n/G$ are rather special but they provide a very good test class for many questions involving log-terminal singularities.

One can always reduce to the case when the G-action on $\mathbb{A}^n$ is effective and fixed point free outside a codimension 2 set. (Unless you are into stacks.) Thus assume this in the sequel.

Show that any such $\mathbb{A}^n/G$ is log terminal.

Show that if $G \subset SL_n$ then the canonical class of $\mathbb{A}^n/G$ is Cartier. In particular, $\mathbb{A}^n/G$ is canonical.

Assume that $G = \langle g \rangle$ is a cyclic group. Any cyclic action on $\mathbb{A}^n$ can be diagonalized and written as

$$g : (x_1, \dots, x_n) \mapsto (\epsilon^{a_1} x_1, \dots, \epsilon^{a_n} x_n),$$

where $\epsilon = e^{2\pi i/m}$, $m = |G|$ and $0 \leq a_j < m$. Define the *age* of g as $\operatorname{age}(g) := \frac{1}{m}(a_1 + \cdots + a_n)$. As a common shorthand notation, we denote the quotient by this action by

$$\mathbb{A}^n / \tfrac{1}{m}(a_1, \dots, a_n).$$

The following very useful criterion tells us when $\mathbb{A}^n/G$ is terminal or canonical.

Reid-Tai criterion. $\mathbb{A}^n/G$ is canonical (resp. terminal) iff the age of every non-identity element $g \in G$ is ≥ 1 (resp. > 1).

(This is not hard to prove if you know some basic toric techniques. Otherwise, look up [**Rei87**].)

As a consequence, prove that the 3-fold quotients $\mathbb{A}^3/\frac{1}{m}(1, -1, a)$ are terminal if $(a, n) = 1$. (It is a quite tricky combinatorial argument to show that these are all the 3-dimensional terminal quotients, cf. [**Rei87**].)

By contrast, every "complicated" higher dimensional quotient singularity is terminal. By the results of [**KL07, GT08**], if the G-action on $\mathbb{A}^n$ is irreducible and primitive, then $\mathbb{A}^n/G$ is terminal whenever $n \geq 5$.

5. Flips

For more on flips, see [**KM98**, Chap.6], [**Cor07**] or [**HM05**].

The following is the most general definition of flips.

Definition 73. Let $f^- : X^- \to Y$ be a proper birational morphism between pure dimensional S_2 schemes such that the exceptional set $\operatorname{Ex}(f^-)$ has codimension at least two in X^-. Let H^- be an $\mathbb{R}$-Cartier divisor on X^- such that $-H^-$ is f^--ample. A pure dimensional S_2 scheme X^+ together with a proper birational morphism $f^+ : X^+ \to Y$ is called an H^-*-flip* of f^- if

(1) the exceptional set $\operatorname{Ex}(f^+)$ has codimension at least two in X^+.
(2) the birational transform H^+ of H^- on X^+ is $\mathbb{R}$-Cartier and f^+-ample.

By a slight abuse of terminology, the rational map $\phi := (f^+)^{-1} \circ f^- : X^- \dashrightarrow X^+$ is also called an H^--flip. We will see in (75) or (90) that a flip is unique and the main question is its existence. A flip gives the following diagram:

$$\begin{array}{ccccc} & X^- & \overset{\phi}{\dashrightarrow} & X^+ & \\ (-H^- \text{ is } f^-\text{-ample}) & f^- & \searrow \quad \swarrow & f^+ & (H^+ \text{ is } f^+\text{-ample}). \\ & & Y & & \end{array}$$

Warning 74. In the literature the notion of flip is frequently used in more restrictive ways. Here are the most commonly used variants that appear, sometimes without explicit mention.

(1) In older papers, flip refers to the case when X^- is terminal and $H = K_{X^-}$. These are the ones needed when we start the MMP with a smooth variety.

(2) In the MMP for pairs (X, Δ) we are interested in flips when (X^-, Δ^-) is a klt (or dlt or lc) pair and $H = K_{X^-} + \Delta^-$. In older papers this is called a log-flip, but more recently it is called simply a flip.
(3) Given (X^-, Δ^-), a $\big(K_{X^-} + \Delta^-\big)$-flip is frequently called a Δ^--flip.
(4) The statement "n-dimensional terminal (or canonical, klt, ...) flips exist" means that the H^--flip of $f^- : X^- \to Y$ exists whenever $\dim X^- = n$, $H^- = K_{X^-} + \Delta^-$ and (X^-, Δ^-) is terminal (or canonical, klt, ...).
(5) In many cases the relative Picard number of X^-/Y is 1. Thus, up to $\mathbb{R}$-linear equivalence, there is a unique f^--negative divisor and the choice of H^- is irrelevant; hence omitted. This variant is frequently used for nonprojective schemes or complex analytic spaces, when a relatively ample divisor may not exist.
(6) A flip is called a *flop* if K_{X^-} is numerically f^--trivial, or, if one has in mind a fixed (X^-, Δ^-), if $K_{X^-} + \Delta^-$ is numerically f^--trivial.
(7) Let X be a scheme and H an $\mathbb{R}$-divisor on X. Especially when studying sequences of flips, an H-flip could refer to any H^-*-flip* of $f^- : X^- \to Y$ if there is a birational contraction $g : X \dashrightarrow X^-$ and H^- is the birational transform of H.

Exercise 75. Prove the following result of Matsusaka and Mumford [**MM64**].

Let X_i be pure dimensional S_2-schemes and $X_i \to S$ projective morphisms with relatively ample divisors H_i. Let $U_i \subset X_i$ be open subsets such that $X_i \setminus U_i$ has codimension ≥ 2 in X_i. Let $\phi_U : U_1 \to U_2$ be an isomorphism such that $\phi_U(H_1|_{U_1}) = H_2|_{U_2}$.

Then ϕ_U extends to an isomorphism $\phi_X : X_1 \to X_2$.

Exercise 76. Notation as in (73). Prove that $f^-_*(H^-)$ is not $\mathbb{R}$-Cartier on Y.

We see in (96) that not all flips exist. Currently, the strongest existence theorem is the following.

Theorem 77. [**HM05, BCHM06**] *Dlt flips exist.*

Exercise 78. Let $\phi : X^- \dashrightarrow X^+$ be a $(K_{X^-} + \Delta^-)$-flip. Prove that for any E, $a(E, X^-, \Delta_{X^-}) \leq a(E, X^+, \Delta_{X^+})$ and the inequality is strict iff the center of E on X^- is contained in $\mathrm{Ex}(\phi)$.

Definition 79. Let (X, Δ) be an lc pair and $f : X \to S$ a proper morphism. A *sequence of flips over* S starting with (X, Δ) is a sequence of birational maps ϕ_i and morphisms f_i

$$\begin{array}{ccc} X_i & \overset{\phi_i}{\dashrightarrow} & X_{i+1} \\ f_i \searrow & & \swarrow f_{i+1} \\ & S & \end{array}$$

(starting with $X_0 = X$) such that for every $i \geq 0$, ϕ_i is a $\big(K_{X_i} + \Delta_i\big)$-flip where Δ_i is the birational transform of Δ on X_i.

The basic open question in the field is the following

Conjecture 80. *Starting with an lc pair* (X_0, Δ_0), *there is a no infinite sequence of flips* $\phi_i : (X_i, \Delta_i) \dashrightarrow (X_{i+1}, \Delta_{i+1})$.

This is known in dimension 3, almost known in dimension 4 and known in certain important cases in general; see [**BCHM06**] or (99) for more precise statements.

Exercise 81. Let $\phi_i : (X_i, \Delta_i) \dashrightarrow (X_{i+1}, \Delta_{i+1})$ be a sequence of flips. Prove that the composite $\phi_n \circ \cdots \circ \phi_0 : X_0 \dashrightarrow X_{n+1}$ can not be an isomorphism.

Problem 82. Let $\phi_i : (X_i, \Delta_i) \dashrightarrow (X_{i+1}, \Delta_{i+1})$ be a sequence of flips. Prove that (X_n, Δ_n) can not be isomorphic to (X_0, Δ_0) for $n > 0$. (I do not know how to do this, but it may not be hard.)

By contrast, show that the involution τ in (16) is a flop and even a flip for some $H = K_X + \Delta$ where (X, Δ) is klt. (Thus X_n could be isomorphic to X_0, but the isomorphism should not preserve Δ.)

Exercise 83 (Simplest flop). Let $L_1, L_2 \subset \mathbb{P}^3$ be two lines intersecting at a point p. Let $X_1 := B_{L_1} B_{L_2} \mathbb{P}^3$ and $X_2 := B_{L_2} B_{L_1} \mathbb{P}^3$. Set $Y := B_{L_1+L_2} \mathbb{P}^3$.

Show that the identity on $\mathbb{P}^3$ induces morphisms $f_i : X_i \to Y$ and a rational map $\phi : X_1 \dashrightarrow X_2$. We get a flop diagram

$$\begin{array}{ccc} X_1 & \overset{\phi}{\dashrightarrow} & X_2 \\ f_1 & \searrow \quad \swarrow & f_2 \\ & Y. & \end{array}$$

Show that neither ϕ nor ϕ^{-1} contracts divisors but neither is a morphism. Describe how to factor ϕ into a composite of smooth blow ups and blow downs.

Exercise 84 (Non-algebraic flops). Let $X \subset \mathbb{P}^4$ be a general smooth quintic hypersurface. It is know that for every $d \geq 1$, X contains a smooth rational curve $\mathbb{P}^1 \cong C_d \subset X$ with normal bundle $\mathcal{O}_{\mathbb{P}^1}(-1) + \mathcal{O}_{\mathbb{P}^1}(-1)$ [**Cle83**].

Prove that the flop of C_d exists if we work with compact complex manifolds. Denote the flop by $\phi_d : X \dashrightarrow X_d$ and let $H_d \in H^2(X_d, \mathbb{Z})$ be the image of the hyperplane class. Compute the self-intersection (H_d^3). Conclude that the X_d are not homeomorphic to each other and not projective.

Exercise 85 (Harder flops). Let $C_1, C_2 \subset \mathbb{P}^3$ be two smooth curves intersecting at a single point p where they are tangent to order m. Let $X_1 := B_{C_1} B_{C_2} \mathbb{P}^3$ and $X_2 := B_{C_2} B_{C_1} \mathbb{P}^3$. Set $Y := B_{C_1+C_2} \mathbb{P}^3$.

Show that the identity on $\mathbb{P}^3$ induces morphisms $f_i : X_i \to Y$, a rational map $\phi : X_1 \dashrightarrow X_2$ and we get a flop diagram as before. Describe how to factor f into a composite of smooth blow ups and blow downs.

Exercise 86 (Even harder flops). Consider the variety

$$X := (sx + ty + uz = sz^2 + tx^2 + uy^2 = 0) \subset \mathbb{P}^2_{xyz} \times \mathbb{A}^3_{stu}.$$

Show that X is smooth, the projection $\pi : X \to \mathbb{A}^3$ has degree 2 and $C := \operatorname{red} \pi^{-1}(0,0,0)$ is a smooth rational curve. Compute $(C \cdot K_X)$ and the normal bundle of C.

Let $Y \to \mathbb{A}^3$ be the normalization of $\mathbb{A}^3$ in $k(X)$. Determine the singularity of Y sitting over the origin.

As before, the Galois involution of $Y \to \mathbb{A}^3$ provides the flop of $X \to Y$.

It is quite tricky to factor f into a composite of smooth blow ups and blow downs.

Exercise 87 (Simplest flips). Fix $n \geq 3$ and consider the affine hypersurface

$$Z := (u^n - u^{n-1}y + x^{n-1}z = 0) \subset \mathbb{A}^4,$$

which we view as a degree n covering of the (x, y, z)-space.

Show that Z is not normal and its normalization has a unique singular point which lies above $(0,0,0)$.

Show that

$$X^+ := (s^n x - s^{n-1}ty + t^n x^{n-1}z = 0) \subset \mathbb{A}^3_{xyz} \times \mathbb{P}^1_{st}$$

is a small resolution of Z. Write down the morphism $X^+ \to Z$. It has a unique 1-dimensional fiber $C^+ \subset X^+$. Determine the normal bundle of C^+ in X^+ and the intersection number of C^+ with the canonical class.

Construct another small modification $X^- \to Z$ as follows. First blow up the ideal (z, u^{n-1}). We get the variety X_1 defined by equations

$$\big(s(y-u) - tx^{n-1} = sz - tu^{n-1} = u^n - u^{n-1}y + x^{n-1}z = 0\big) \subset \mathbb{A}^4_{xyzu} \times \mathbb{P}^1_{st}.$$

Show that the $s \neq 0$ chart is smooth and on the $t \neq 0$ chart we have a complete intersection

$$\big(w(y-u) - x^{n-1} = wz - u^{n-1} = 0\big) \subset \mathbb{A}^4_{xyzuw} \quad \text{with } w = s/t.$$

Setting $y' := y - u$ we have the local equations for X_1

$$wy' - x^{n-1} = wz - u^{n-1} = 0.$$

Write down a $\mathbb{Z}/(n-1)$-invariant finite morphism to the above local chart on X_1 from $\mathbb{A}^3_{pqr}$ with the $\mathbb{Z}/(n-1)$-action $(p,q,r) \mapsto (\epsilon p, \epsilon q, \epsilon^{-1} r)$, where ϵ is a primitive $(n-1)$-st root of unity. Let X^- be the normalization of X_1. Show that X^- has a single quotient singularity of the above form.

Write down the morphism $X^- \to Z$. It has a unique 1-dimensional fiber $C^- \subset X^-$. Determine the intersection number of C^- with the canonical class.

Exercise 88. Let now Y be any smooth 3-fold and L a very ample line bundle on Y with 3 general sections f, g, h. Fix $n \geq 3$ and consider the hypersurface

$$Z := (u^n - u^{n-1}g + f^{n-1}h = 0) \subset L^{-1}.$$

One small resolution is given by

$$X^+ := (s^n f - s^{n-1}tg + t^n h = 0) \subset Y \times \mathbb{P}^1_{st}.$$

Compute its canonical class in terms of K_Y and L.

Exercise 89 (Log terminal flips)**.** Work out the analog of (87) when we start with

$$X^+ := (s^n x - s^{n-i}t^i y + t^n z = 0) \subset \mathbb{A}^3_{xyz} \times \mathbb{P}^1_{st}.$$

Exercise 90. Let X be a Noetherian, reduced, pure dimensional, S_2-scheme and D a Weil divisor on X which is Cartier in codimension 1. Prove that the following are equivalent.

(1) $\sum_{m\geq 0} \mathcal{O}_X(mD)$ is a finitely generated sheaf of $\mathcal{O}_X$-algebras.

(2) There is a proper, birational morphism $\pi : X^+ \to X$ such that the exceptional set $\mathrm{Ex}(\pi)$ has codimension ≥ 2 and the birational transform $D^+ := \pi_*^{-1}(D)$ is $\mathbb{Q}$-Cartier and π-ample.

Hint of proof. (2) $\Rightarrow$ (1) is easy.

To see the converse, set $X^+ := \mathrm{Proj}_X \sum_{m\geq 0} \mathcal{O}_X(mD)$. We need to show that $X^+ \to X$ is small. Assume that $E \subset \mathrm{Ex}(\pi)$ is an exceptional divisor. Study the sequence

$$0 \to \mathcal{O}_{X^+}(mD^+) \to \mathcal{O}_{X^+}(mD^+ + E) \to \mathcal{O}_E\big((mD^+ + E)|_E\big) \to 0$$

to get, for some $m > 0$, a section of $\mathcal{O}_{X^+}(mD^+ + E)$ which is not a section of $\mathcal{O}_{X^+}(mD^+)$. By pushing forward to X, we would get extra sections of $\mathcal{O}_X(mD)$.

Exercise 91. Let (X, Δ) be klt. Let $f : X \to Y$ be a small $(K_X + \Delta)$-negative contraction. Show that there is a $\mathbb{Q}$-divisor D on X such that $(X, \Delta + D)$ is klt and $(K_X + \Delta + D) \sim_{\mathbb{Q},f} 0$.

Conclude from this that $\big(Y, f_*(\Delta + D)\big)$ is klt.

A consequence of the relative MMP is the following finite generation result, which we prove in (109). By (91), it formally implies the existence of dlt flips.

Theorem 92. *Let (X, Δ) be klt and D a $\mathbb{Q}$-divisor on X. Then $\sum_{m \geq 0} \mathcal{O}_X(\lfloor mD \rfloor)$ is a finitely generated sheaf of $\mathcal{O}_X$-algebras.*

Exercise 93. Show that $\lfloor A + B \rfloor \geq \lfloor A \rfloor + \lfloor B \rfloor$ for any divisors A, B, thus, for any divisor D, $R(X, D) := \sum_{m \geq 0} H^0(X, \mathcal{O}_X(\lfloor mD \rfloor))$ is a ring.

Give examples where $R^u(X, D) := \sum_{m \geq 0} H^0(X, \mathcal{O}_X(\lceil mD \rceil))$ is not a ring. Note, however, that $\lceil A + B \rceil \geq \lfloor A \rfloor + \lceil B \rceil$, thus $R^u(X, D)$ is an $R(X, D)$-module.

Exercise 94. Let X be normal and D an $\mathbb{R}$-divisor. Show that if $\sum_{m \geq 0} \mathcal{O}_X(\lfloor mD \rfloor)$ is a finitely generated sheaf of $\mathcal{O}_X$-algebras then D is a $\mathbb{Q}$-divisor.

The following example shows that (92) fails for lc pairs.

Exercise 95. Let $E \subset \mathbb{P}^2$ be a smooth cubic. Let S be obtained by blowing up 9 general points on E and let $E_S \subset S$ be the birational transform of E. Let H be a sufficiently ample divisor on S giving a projectively normal embedding $S \subset \mathbb{P}^n$. Let $X \subset \mathbb{A}^{n+1}$ be the cone over S and $D \subset X$ the cone over E_S.

Prove that (X, D) is lc yet $\sum_{m \geq 0} \mathcal{O}_X(mD)$ is not a finitely generated sheaf of $\mathcal{O}_X$-algebras.

Hints. First show that $H^0(X, \mathcal{O}_X(mD)) = \sum_{r \geq 0} H^0\big(S, \mathcal{O}_S(mE_S + rH)\big)$. Check that $\mathcal{O}_S(mE_S + rH)$ is very ample if $r > 0$ but $\mathcal{O}_S(mE_S)$ has only the obvious section which vanishes along mE_S. Thus the multiplication maps

$$\sum_{a=0}^{m-1} H^0\big(S, \mathcal{O}_S(aE_S + H)\big) \otimes H^0\big(S, \mathcal{O}_S((m-a)E_S)\big) \to H^0\big(S, \mathcal{O}_S(mE_S + H)\big)$$

are never surjective.

The next exercise shows that log canonical flops sometimes do not exist.

Exercise 96. Let E be an elliptic curve, L a degree 0 non-torsion line bundle and $S = \mathbb{P}_E(\mathcal{O}_E + L)$. Let $C_1, C_2 \subset S$ be the corresponding sections of $S \to E$. Note that $K_S + C_1 + C_2 \sim 0$. Let $0 \in X$ be a cone over S and $D_i \subset X$ the cones over C_i. Show that $(X, D_1 + D_2)$ is lc.

Following the method of (95) show that $\sum_{m \geq 0} \mathcal{O}_X(mD_i)$ is not a finitely generated sheaf of $\mathcal{O}_X$-algebras for $i = 1, 2$.

Let $F \subset S$ be a fiber of $S \to E$ and $B \subset X$ the cone over F. Show that $\sum_{m \geq 0} \mathcal{O}_X(mB)$ is a finitely generated sheaf of $\mathcal{O}_X$-algebras and describe the corresponding small contraction $\pi : Z \to X$.

Prove that the flip of $\pi : Z \to X$ does not exist (no matter what H we choose).

What happens if L is a torsion element in $\mathrm{Pic}(E)$?

Exercise 97. Let S be a Noetherian, reduced, 2-dimensional, S_2-scheme and D a Weil divisor on S. Prove that $\sum_{m\geq 0} \mathcal{O}_S(mD)$ is a finitely generated sheaf of $\mathcal{O}_S$-algebras iff $\mathcal{O}_S(mD)$ is locally free for some $m > 0$.

Use this to show that the following algebras are not finitely generated.

(1) S is a cone over an elliptic curve and $D \subset S$ a general line. State the precise generality condition.
(2) Let $C \subset \mathbb{P}^n$ be a projectively normal curve of genus ≥ 2 and $S \subset \mathbb{A}^{n+1}$ the cone over C. Assume that $\mathcal{O}_C(1)$ is a general line bundle and let $D = K_S$. Again, state the precise generality condition.
(3) Let S be the quadric cone $(xy - z^2 = 0) \subset \mathbb{A}^3$ and the (u, v)-plane glued together along the lines $(x = z = 0)$ and $(v = 0)$. (Show that this surface does not embed in $\mathbb{A}^3$ but realize it in $\mathbb{A}^4$ by explicit equations.) Set $D = K_S$.

The following conjecture is known if $x \in H$ is a quotient singularity [**KSB88**] or when $x \in H$ is a quadruple point [**Ste91**]. It is quite remarkable that, aside from the case when $x \in H$ is a quotient singularity, the conjecture seems unrelated to the minimal model program.

Conjecture 98. [**Kol91**, 6.2.1] *Let $x \in X$ be a 3-dimensional normal singularity and $x \in H \subset X$ a Cartier divisor. Assume that $x \in H$ is a (normal) rational surface singularity. Then $\sum_{m\geq 0} \mathcal{O}_X(mK_X)$ is a finitely generated sheaf of $\mathcal{O}_X$-algebras.*

6. Minimal models

For more details, see [**KM98**, 3.7–8] or [**BCHM06**].

Definition 99 (Running the MMP). Let (X, Δ) be a pair such that $K_X + \Delta$ is $\mathbb{Q}$-Cartier and $f : X \to S$ a proper morphism. Assume for simplicity that X is $\mathbb{Q}$-factorial. A *running of the* $(K_X + \Delta)$*-MMP* over S yields a sequence

$$(X, \Delta) =: (X_0, \Delta_0) \overset{\phi_0}{\dashrightarrow} (X_1, \Delta_1) \overset{\phi_1}{\dashrightarrow} \quad \cdots \quad \overset{\phi_{n-1}}{\dashrightarrow} (X_r, \Delta_r),$$

where each ϕ_i is either the divisorial contraction of a $(K_{X_i} + \Delta_i)$-negative extremal ray or the flip of a small contraction of a $(K_{X_i} + \Delta_i)$-negative extremal ray, $\Delta_{i+1} := (\phi_i)_* \Delta_i$ and all the X_i are S-schemes $f_i : X_i \to S$ such that $f_i = f_{i+1} \circ \phi_i$. We say the the $(K_X + \Delta)$-MMP *stops* or *terminates* with (X_r, Δ_r) if

(1) either $K_{X_r} + \Delta_r$ is f_r-nef (and there are no more extremal rays),
(2) or there is a Fano contraction $X_r \to Z_r$.

Sometimes we impose a stronger restriction:

(2') every extremal contraction of (X_r, Δ_r) is Fano.

Conjecturally, every running of the $(K_X + \Delta)$-MMP stops. This is known if $\dim X \leq 3$ [**Kaw92**], in many cases in dimension 4 [**AHK07**] or when the generic fiber of f is of general type [**BCHM06**] and at each step the extremal rays are chosen "suitably." Note that the latter includes the case when f is birational (or generically finite), since a point is a 0-dimensional variety of general type.

(Everything works the same if X is not $\mathbb{Q}$-factorial, except in that case it does not make sense to distinguish divisorial contractions and flips.)

Definition 100. Let (X, Δ) be a pair and $f : X \to S$ a proper morphism. We say that (X, Δ) is an

$$\left.\begin{array}{l} f\text{-weak canonical} \\ f\text{-canonical} \\ f\text{-minimal} \end{array}\right\} \text{model if } (X,\Delta) \text{ is} \left\{\begin{array}{c} \text{lc} \\ \text{lc} \\ \text{dlt} \end{array}\right\} \text{and } K_X + \Delta \text{ is} \left\{\begin{array}{c} f\text{-nef} \\ f\text{-ample} \\ f\text{-nef} \end{array}\right\}.$$

Warning 101. Note that a canonical model (X, Δ) has *log* canonical singularities, not necessarily canonical singularities. This, by now entrenched, unfortunate terminology is a result of an incomplete shift. Originally everything was defined only for $\Delta = 0$. When Δ was introduced, its presence was indicated by putting "log" in front of adjectives. Later, when the use of Δ became pervasive, people started dropping the prefix "log". This is usually not a problem. For instance, the canonical ring $R(X, K_X)$ is just the $\Delta = 0$ special case of the log canonical ring $R(X, K_X + \Delta)$.

However, canonical singularities are not the $\Delta = 0$ special cases of log canonical singularities.

Definition 102. Let (X, Δ) be a pair such that $K_X + \Delta$ is $\mathbb{Q}$-Cartier and $f : X \to S$ a proper morphism. A pair (X^w, Δ^w) sitting in a diagram

$$\begin{array}{ccccc} X & & \overset{\phi}{\dashrightarrow} & & X^w \\ f & \searrow & & \swarrow & f^w \\ & & S & & \end{array}$$

is called a *weak canonical model of* (X, Δ) over S if

(1) f^w is proper,
(2) ϕ is a contraction, that is, ϕ^{-1} has no exceptional divisors,
(3) $\Delta^w = \phi_* \Delta$,
(4) $K_{X^w} + \Delta^w$ is $\mathbb{Q}$-Cartier and f^w-nef, and
(5) $a(E, X, \Delta) \le a(E, X^w, \Delta^w)$ for every ϕ-exceptional divisor $E \subset X$. Equivalently, $(K_X + \Delta) - \phi^*(K_{X^w} + \Delta^w)$ is effective and ϕ-exceptional.

A weak canonical model $(X^m, \Delta^m) = (X^w, \Delta^w)$ is called a *minimal model of* (X, Δ) over S if, in addition to (1–4), we have

(5^m) $a(E, X, \Delta) < a(E, X^m, \Delta^m)$ for every ϕ-exceptional divisor $E \subset X$.

A weak canonical model $(X^c, \Delta^c) = (X^w, \Delta^w)$ is called a *canonical model of* (X, Δ) over S if, in addition to (1–3) and (5) we have

(4^c) $K_{X^c} + \Delta^c$ is $\mathbb{Q}$-Cartier and f^c-ample.

Exercise 103. Let (X, Δ) be a pair such that $K_X + \Delta$ is $\mathbb{Q}$-Cartier and $f : X \to S$ a proper morphism. Run the MMP:

$$(X, \Delta) =: (X_0, \Delta_0) \overset{\phi_0}{\dashrightarrow} (X_1, \Delta_1) \overset{\phi_1}{\dashrightarrow} \quad \cdots \quad \overset{\phi_{n-1}}{\dashrightarrow} (X_r, \Delta_r),$$

and assume that $K_{X_r} + \Delta_r$ is f-nef. Show that (X_r, Δ_r) is a minimal model of (X, Δ) over S.

Exercise 104. Let $f : (X, \Delta) \to S$ be a canonical model. Let $g : X' \to X$ be a proper birational morphism with exceptional divisors E_i. When is $f : (X, \Delta) \to S$ a canonical model of $(X', g_*^{-1}\Delta + \sum e_i E_i)$?

Exercise 105. Let $\phi : (X, \Delta) \dashrightarrow (X^w, \Delta^w)$ be a weak canonical model. Prove that

$$a(E, X^w, \Delta^w) \geq a(E, X, \Delta) \quad \text{for every divisor } E.$$

Hint. Fix E and consider any diagram

$$\begin{array}{ccccc} & & Y & & \\ g & \swarrow & & \searrow & h \\ X & & \stackrel{\phi}{\dashrightarrow} & & X^w \\ f & \searrow & & \swarrow & f^w \\ & & S & & \end{array}$$

where $\operatorname{center}_Y E$ is a divisor. Write K_Y in two different ways and apply (107).

Exercise 106. Let $\phi : (X, \Delta) \dashrightarrow (X^w, \Delta^w)$ be a weak canonical model. Prove that if a curve $C \subset X$ is not contained in $\operatorname{Ex}(\phi)$ then

$$C \cdot (K_X + \Delta) \geq \phi_*(C) \cdot (K_{X^w} + \Delta^w).$$

Exercise 107. Let $h : Z \to Y$ be a proper birational morphism between normal varieties. Let $-B$ be an h-nef $\mathbb{Q}$-Cartier $\mathbb{Q}$-divisor on Z. Then

(1) B is effective iff h_*B is.
(2) Assume that B is effective. Then for every $y \in Y$, either $h^{-1}(y) \subset \operatorname{Supp} B$ or $h^{-1}(y) \cap \operatorname{Supp} B = \emptyset$.

Hint. Use induction on $\dim Z$ by passing to a hyperplane section $H \subset Z$. Be careful: $h_*(B \cap H)$ need not be contained in h_*B.

Exercise 108 ($\mathbb{Q}$-factorialization)**.** Let (X, Δ) be klt. Let $f : Y \to X$ be a log resolution with exceptional divisor E. For $0 < \epsilon \ll 1$ run the $\bigl(Y, f_*^{-1}\Delta + (1-\epsilon)E\bigr)$-MMP over X and assume that it stops. (This is not a restriction by (99).)

Prove that the MMP stops at a small contraction $f_r : Y_r \to X$ such that Y_r is $\mathbb{Q}$-factorial.

It is called a *$\mathbb{Q}$-factorialization* of X.

More generally, prove that $\mathbb{Q}$-factorializations exist if (X, Δ) is dlt. Find lc examples without any $\mathbb{Q}$-factorialization.

Exercise 109. Notation as in (108). Let D be any Weil divisor on X. Prove that there is a $\mathbb{Q}$-factorialization $f_D : Y_D \to X$ such that the birational transform of D on Y_D is f_D-nef.

Use this to prove that $\mathbb{Q}$-factorializations are never unique, save when X itself is $\mathbb{Q}$-factorial.

Use this and the contraction theorem to prove (92).

Warning 110. You may have noticed already that we have not defined when a pair (X', Δ') is birational to another pair (X, Δ). The problem is: what should the coefficient of a divisor $D \subset X'$ be in Δ' when the center of D on X is not a divisor.

One approach is to insist that birational pairs have the same canonical rings. Then the next exercise suggests a definition.

It is, however, best to keep in mind that birational equivalence of pairs is a problematic concept.

Exercise 111. Let $f_1 : X_1 \to S$ and $f_2 : X_2 \to S$ be proper morphisms of normal schemes and $\phi : X_1 \dashrightarrow X_2$ a birational map such that $f_1 = f_2 \circ \phi$. Let Δ_1 and Δ_2

be $\mathbb{Q}$-divisors such that $K_{X_1}+\Delta_1$ and $K_{X_2}+\Delta_2$ are $\mathbb{Q}$-Cartier. Prove that

$$f_{1*}\mathcal{O}_{X_1}(mK_{X_1}+\lfloor m\Delta_1\rfloor)=f_{2*}\mathcal{O}_{X_2}(mK_{X_2}+\lfloor m\Delta_2\rfloor)\quad \text{for } m\geq 0$$

if the following conditions hold:

(1) $a(E,X_1,\Delta_1)=a(E,X_2,\Delta_2)$ if ϕ is a local isomorphism at the generic point of E,
(2) $a(E,X_1,\Delta_1)\leq a(E,X_2,\Delta_2)$ if $E\subset X_1$ is ϕ-exceptional, and
(3) $a(E,X_1,\Delta_1)\geq a(E,X_2,\Delta_2)$ if $E\subset X_2$ is ϕ^{-1}-exceptional.

Hints: Let Y be the normalization of the closed graph of ϕ in $X_1\times_S X_2$ and $g_i:Y\to X_i$ the projections. We can write

$$\begin{aligned} K_Y &\sim_{\mathbb{Q}} g_1^*(K_{X_1}+\Delta_1)+\textstyle\sum_E a(E,X_1,\Delta_1)E, \quad \text{and}\\ K_Y &\sim_{\mathbb{Q}} g_1^*(K_{X_2}+\Delta_2)+\textstyle\sum_E a(E,X_2,\Delta_2)E. \end{aligned}$$

Set $b(E):=\max\{-a(E,X_1,\Delta_1),-a(E,X_2,\Delta_2)\}$. Prove that

$$\sum_E \bigl(b(E)+a(E,X_i,\Delta_i)\bigr)E$$

is effective and g_i-exceptional for $i=1,2$. Conclude that

$$\begin{aligned} &(f_i\circ g_i)_*\mathcal{O}_Y(mK_Y+\textstyle\sum_E mb(E)E)\\ &\quad=f_{i*}g_{i*}\mathcal{O}_Y\Bigl(g_1^*(mK_{X_1}+m\Delta_1)+\textstyle\sum_E\bigl(mb(E)+ma(E,X_1,\Delta_1)\bigr)E\Bigr)\\ &\quad=f_{i*}\mathcal{O}_{X_i}(mK_{X_i}+m\Delta_i). \end{aligned}$$

Exercise 112. Let (X,Δ) be a lc pair with $\Delta\geq 0$, $f:X\to S$ a proper morphism and $f^w:(X^w,\Delta^w)\to S$ a weak minimal model. Prove the following:

(1) $f_*\mathcal{O}_X(mK_X+\lfloor m\Delta\rfloor)=f^w_*\mathcal{O}_{X^w}(mK_{X^w}+\lfloor m\Delta^w\rfloor)$ for every $m\geq 0$.
(2) If a canonical model (X^c,Δ^c) exists then

$$X^c=\operatorname{Proj}_S\textstyle\sum_{m\geq 0}f_*\mathcal{O}_X(mK_X+\lfloor m\Delta\rfloor),$$

and the right hand side is a sheaf of finitely generated algebras. In particular, a canonical model is unique.
(3) Any two minimal models of (X,Δ) are isomorphic in codimension one. (Hint: Prove this first when $\Delta=0$ and $(X,0)$ is terminal. The general case is more subtle.)

Exercise 113. Assume that X is irreducible,

$$R(X,K_X+\Delta):=\textstyle\sum_{m\geq 0}f_*\mathcal{O}_X(mK_X+\lfloor m\Delta\rfloor)$$

is a sheaf of finitely generated algebras and

$$\dim X=\dim\operatorname{Proj}_S R(X,K_X+\Delta).$$

Prove that the natural map $\phi:X\dashrightarrow\operatorname{Proj}_S R(X,K_X+\Delta)$ is birational and

$$(X^c,\Delta^c):=\bigl(\operatorname{Proj}_S R(X,K_X+\Delta),\phi_*\Delta\bigr)$$

is the canonical model of (X,Δ).

Hint: You should find (114) useful.

Exercise 114. Let X be an irreducible and normal scheme, L a Weil divisor on X and $f:X\to S$ a proper morphism, S affine. Write $|L|=|M|+F$ where $|M|$ is the moving part and F the fixed part. Assume that $R(X,L):=\sum_{m\geq 0}f_*\mathcal{O}_X(mL)$ is generated by $f_*\mathcal{O}_X(L)$. Set $Z:=\operatorname{Proj}_S R(X,L)$ with projection $p:Z\to S$ and let $\phi:X\dashrightarrow Z$ be the natural morphism. Prove that

(1) $Z \setminus \phi(X)$ has codimension ≥ 2 in Z.

(2) If ϕ is generically finite then it is birational and F is ϕ-exceptional.

(Hint: This is similar to (90).)

Exercise 115 (Chambers in the cone of big divisors)**.** Let X be a normal variety and D_i big $\mathbb{Q}$-divisors. Assume that the rings

$$R(D_i) := \sum_{m\geq 0} H^0\bigl(X, \mathcal{O}_X(\lfloor mD_i \rfloor)\bigr)$$

are finitely generated and the maps $X \dashrightarrow \operatorname{Proj} R(D_i)$ are birational and independent of i. Let $D = \sum a_i D_i$ be a nonnegative $\mathbb{Q}$-linear combination.

Prove that $R(D)$ is finitely generated and $X \dashrightarrow \operatorname{Proj} R(D)$ is the same map as before.

Conclude that the set of all big $\mathbb{Q}$-divisors with the same $X \dashrightarrow \operatorname{Proj} R(D)$ forms a convex subcone, called a *chamber* in the cone of big divisors.

Exercise 116. Develop a relative version of the notion of chambers of divisors for maps. (Note that for birational maps, every divisor is relatively big.)

Let $Y \to X$ be a $\mathbb{Q}$-factorialization of a klt pair (X, Δ) (108). Prove that there is a one-to-one correspondence between open chambers of $N^1(Y/X)$ and $\mathbb{Q}$-factorializations of X.

What kind of maps correspond to the other chambers?

Exercise 117. Let a_i be different complex numbers. Consider the singularity

$$X = X(a_1, \dots, a_n) := \bigl(xy - \prod_i (u - a_i v) = 0\bigr) \subset \mathbb{A}^4.$$

Find a small resolution of X by repeatedly blowing up planes of the form $(x = u - a_i v = 0)$.

Prove that the class group $\operatorname{Cl}(X)$ of X is generated by the planes $(x = u - a_i v = 0)$, with a single relation $\sum_i [x = u - a_i v = 0] = 0$.

Describe all small resolutions of X and the corresponding chamber structure on $\operatorname{Cl}(X)$.

(The same method can be used to describe the class group and the chamber structure for any cA-type terminal 3-fold singularity, see [**Kol91**, 2.2.7]. A similarly explicit description is not known for the cD and cE-type cases.)

Exercise 118. Let $S := (xy - z^3 = 0) \subset \mathbb{A}^3$ and $f : X \to S$ its minimal resolution with exceptional curves D_1, D_2. Let D_3, D_4 be the birational transforms of the lines $(x = z = 0)$ and $(y = z = 0)$. For $0 \leq a_i \leq 1$ describe minimal and canonical models of $(X, \sum a_i D_i)$ over S. Describe the chamber decomposition of $[0,1]^4$.

Exercise 119. Let S be one of the singularities in (69) and $f : X \to S$ its minimal resolution with exceptional curves D_i. For $0 \leq a_i \leq 1$ describe minimal and canonical models of $(X, \sum a_i D_i)$ over S and the corresponding chamber decomposition.

(This is pretty easy for the $\mathbb{Z}/2$-quotient. Some of the others have many curves to check.)

For the theory behind the next exercises, see [**KL07**].

Exercise 120. Let E be the projective elliptic curve with affine equation $(y^2 = x^3 - 1)$ and set $\tau : (x, y) \mapsto (x, -y)$. Check that

(1) $E/\tau \cong \mathbb{P}^1$.

(2) $(E \times E)/(\tau \times \tau)$ has Kodaira dimension 0. It is an example of a Kummer surface. If $u = y_1 y_2$ then it has affine equation

$$u^2 = (x_1^3 - 1)(x_2^3 - 1).$$

Find the singularities using this equation.

(3) For $n \geq 3$, $\big(E^n\big)/(\tau, \dots, \tau)$ has Kodaira dimension 0.

Exercise 121. Let E be the projective elliptic curve with affine equation $(y^3 = x^3 - 1)$ and set $\sigma : (x, y) \mapsto (x, \epsilon y)$ where $\epsilon = \sqrt[3]{1}$. Check that

(1) $E/\sigma \cong \mathbb{P}^1$.

(2) $(E \times E)/(\sigma, \sigma^2)$ has Kodaira dimension 0. It is an example of a K3 surface. If $u = y_1 y_2$ then it has affine equation

$$u^3 = (x_1^3 - 1)(x_2^3 - 1).$$

Find the singularities using this equation.

(3) $(E \times E)/(\sigma, \sigma)$. If $v = y_1 y_2^2$ then it has affine equation

$$v^3 = (x_1^3 - 1)(x_2^3 - 1)^2.$$

Find the singularities using this equation.

Prove that this surface is rational in two ways:

(a) Find many rational curves on it as preimages of rational curves of bi-degree $(2, 2)$ on $\mathbb{P}^1 \times \mathbb{P}^1$.

(b) Show that it is birational (even over $\mathbb{Z}$) to the cubic surface $y_1^3 - y_2^3 = x_1^3 - 1$.

(4) For $n \geq 3$, $\big(E^n\big)/(\sigma, \dots, \sigma)$ has Kodaira dimension 0.

Exercise 122. Let E be the projective elliptic curve with affine equation $\big(y^6 = x(x-1)^2(x+1)^3\big)$ and set $\rho : (x, y) \mapsto (x, \epsilon y)$ where $\epsilon = \sqrt[6]{1}$. Check that

(1) $E/\rho \cong \mathbb{P}^1$.

(2) For $2 \leq n \leq 5$, $\big(E^n\big)/(\rho, \dots, \rho)$ is uniruled, that is, it has a covering family of rational curves. Try to find explicitly such a family. (Such a family exists by [**KL07**], but I do not know how to construct one.) I don't know if these examples are rational or unirational.

(3) For $6 \leq n$, $\big(E^n\big)/(\rho, \dots, \rho)$ has Kodaira dimension 0.

Bibliography

[AHK07] Valery Alexeev, Christopher Hacon, and Yujiro Kawamata, *Termination of (many) 4-dimensional log flips*, Invent. Math. **168** (2007), no. 2, 433–448. MR MR2289869 (2008f:14028)

[AKMW02] Dan Abramovich, Kalle Karu, Kenji Matsuki, and Jarosław Włodarczyk, *Torification and factorization of birational maps*, J. Amer. Math. Soc. **15** (2002), no. 3, 531–572 (electronic). MR MR1896232 (2003c:14016)

[BCHM06] Caucher Birkar, Paolo Cascini, Christopher D. Hacon, and James McKernan, *Existence of minimal models for varieties of log general type*, http://www.citebase.org/abstract?id=oai:arXiv.org:math/0610203, 2006.

[Bor90] Ciprian Borcea, *Homogeneous vector bundles and families of Calabi-Yau threefolds*, Duke Math. J. **61** (1990), no. 2, 395–415. MR MR1074303 (91k:14029)

[Bor91] ———, *Homogeneous vector bundles and families of Calabi-Yau threefolds. II*, Several complex variables and complex geometry, Part 2 (Santa Cruz, CA, 1989), Proc. Sympos. Pure Math., vol. 52, Amer. Math. Soc., Providence, RI, 1991, pp. 83–91. MR MR1128537 (92k:14035)

[BPVdV84] W. Barth, C. Peters, and A. Van de Ven, *Compact complex surfaces*, Ergebnisse der Mathematik und ihrer Grenzgebiete (3) [Results in Mathematics and Related Areas (3)], vol. 4, Springer-Verlag, Berlin, 1984. MR MR749574 (86c:32026)

[Cle83] Herbert Clemens, *Homological equivalence, modulo algebraic equivalence, is not finitely generated*, Inst. Hautes Études Sci. Publ. Math. (1983), no. 58, 19–38 (1984). MR MR720930 (86d:14043)

[Cor07] Alessio Corti, *Flips for 3-folds and 4-folds*, Oxford Lecture Ser. Math. Appl., vol. 35, Oxford Univ. Press, Oxford, 2007. MR MR2359340

[CT06] Ana-Maria Castravet and Jenia Tevelev, *Hilbert's 14th problem and Cox rings*, Compos. Math. **142** (2006), no. 6, 1479–1498. MR MR2278756 (2007i:14044)

[Dur79] A. Durfee, *Fifteen characterizations of rational double points and simple critical points*, L'Ens. Math. **25** (1979), 131–163.

[GT08] Robert M. Guralnick and Pham Huu Tiep, *(in preparation)*, 2008.

[Har77] R. Hartshorne, *Algebraic geometry*, Springer, 1977.

[HM05] Christopher Hacon and James McKernan, *On the existence of flips*, http://www.citebase.org/abstract?id=oai:arXiv.org:math/0507597, 2005.

[Kaw92] Yujiro Kawamata, *Termination of log flips for algebraic 3-folds*, Internat. J. Math. **3** (1992), no. 5, 653–659. MR MR1189678 (93j:14019)

[KL07] János Kollár and Michael Larsen, *Quotients of Calabi-Yau varieties*, http://www.citebase.org/abstract?id=oai:arXiv.org:math/0701466, 2007.

[KM98] János Kollár and Shigefumi Mori, *Birational geometry of algebraic varieties*, Cambridge Tracts in Mathematics, vol. 134, Cambridge University Press, Cambridge, 1998, With the collaboration of C. H. Clemens and A. Corti, Translated from the 1998 Japanese original. MR MR1658959 (2000b:14018)

[Kol91] János Kollár, *Flips, flops, minimal models, etc.*, Surv. in Diff. Geom. **1** (1991), 113–199.

[Kol96] ———, *Rational curves on algebraic varieties*, Ergebnisse der Math., vol. 32, Springer Verlag, 1996.

[Kol97] ———, *Singularities of pairs*, Algebraic geometry—Santa Cruz 1995, Proc. Sympos. Pure Math., vol. 62, Amer. Math. Soc., Providence, RI, 1997, pp. 221–287. MR MR1492525 (99m:14033)

[Kol01] ———, *Which are the simplest algebraic varieties?*, Bull. Amer. Math. Soc. (N.S.) **38** (2001), no. 4, 409–433 (electronic). MR MR1848255 (2002f:14001)

[Kol07] ———, *What is ... a minimal model?*, Notices Amer. Math. Soc. **54** (2007), no. 3, 370–371. MR MR2292143 (2007k:14020)

[KSB88] János Kollár and N. Shepherd-Barron, *Threefolds and deformations of surface singularities*, Inv. Math. **91** (1988), 299–338.

[KSC04] János Kollár, Karen E. Smith, and Alessio Corti, *Rational and nearly rational varieties*, Cambridge Studies in Advanced Mathematics, vol. 92, Cambridge University Press, Cambridge, 2004. MR MR2062787 (2005i:14063)

[Laz04] Robert Lazarsfeld, *Positivity in algebraic geometry. I–II*, Ergebnisse der Mathematik und ihrer Grenzgebiete. 3. Folge., vol. 48, Springer-Verlag, Berlin, 2004. MR MR2095471 (2005k:14001a)

[MM64] T. Matsusaka and D. Mumford, *Two fundamental theorems on deformations of polarized varieties*, Amer. J. Math. **86** (1964), 668–684. MR 30 #2005

[Mor82] S. Mori, *Threefolds whose canonical bundles are not numerically effective*, Annals of Math. **116** (1982), 133–176.

[Rei87] M. Reid, *Young person's guide to canonical singularities*, Algebraic Geometry Bowdoin 1985, Proc. Symp. Pure Math., vol. 46, Amer. Math. Soc., 1987.

[Ste91] Jan Stevens, *Partial resolutions of rational quadruple points*, Internat. J. Math. **2** (1991), no. 2, 205–221. MR MR1094706 (91m:14006)

[Sza94] Endre Szabó, *Divisorial log terminal singularities*, J. Math. Sci. Univ. Tokyo **1** (1994), no. 3, 631–639. MR MR1322695 (96f:14019)

[Wło03] Jarosław Włodarczyk, *Toroidal varieties and the weak factorization theorem*, Invent. Math. **154** (2003), no. 2, 223–331. MR MR2013783 (2004m:14113)

Higher Dimensional Minimal Model Program for Varieties of Log General Type

Christopher D. Hacon

IAS/Park City Mathematics Series
Volume 17, 2008

Higher Dimensional Minimal Model Program for Varieties of Log General Type

Christopher D. Hacon

Introduction

Let X be a smooth complex projective variety. The minimal model program aims to show that if K_X is pseudo-effective (respectively if K_X is not pseudo-effective), then there exists a finite sequence

$$X = X_1 \dashrightarrow X_2 \dashrightarrow \ldots \dashrightarrow X_n = \bar{X}$$

of well understood birational maps known as flips and divisorial contractions such that $\bar{X}$ is a minimal model i.e. $K_{\bar{X}}$ is nef (respectively $\bar{X}$ has the structure of a Mori fiber space i.e. there is a morphism $f : \bar{X} \to Z$ such that $-K_{\bar{X}}$ is ample over Z).

The main features of the minimal model program had been understood in the 1980's by work of S. Mori, Y. Kawamata, J. Kollár, M. Reid, V. Shokurov and others. In order to produce the above sequence of birational maps $X_i \dashrightarrow X_{i+1}$, one proceeds as follows:

(1) If K_{X_i} is nef, then X_i is a minimal model and there is nothing to do.
(2) If K_{X_i} is not nef, then by the Cone Theorem, there exists a negative extremal ray and a corresponding morphism $f_i : X_i \to Z_i$ (of normal varieties, surjective with connected fibers and $\rho(X_i/Z_i) = 1$) such that $-K_{X_i}$ is ample over Z_i.
(3) If $\dim Z_i < \dim X_i$, then $X_i \to Z_i$ is a Mori fiber space and we are done.
(4) If $\dim Z_i = \dim X_i$ and $\dim \mathrm{Ex}(f_i) = \dim X_i - 1$, then this is a divisorial contraction and we let $X_{i+1} = Z_i$.
(5) If $\dim Z_i = \dim X_i$ and $\dim \mathrm{Ex}(f_i) < \dim X_i - 1$, then this is a flipping contraction. In this case, Z_i has bad singularities and we may not let $X_{i+1} = Z_i$. Instead we let X_{i+1} be the flip of f_i cf. 1.1 (assuming that this exists).

In order to complete the minimal model program, it is therefore necessary to prove that flips exist and to show that there is no infinite sequence of flips. In dimension 3, the existence of flips was proved by S. Mori [**Mori88**], and the termination of flips is straightforward. In higher dimensions, the existence and termination of flips have proven to be difficult problems. Recently, V. Shokurov proved the existence of

Department of Mathematics University of Utah 155 South 1400 East JWB 233 Salt Lake City, UT 84112, USA
E-mail address: hacon@math.utah.edu
The author was partially supported by NSF research grant no: 0456363 and an AMS Centennial fellowship.

flips in dimension 4 (cf. [**Shokurov03**]) and then C. Hacon and J. McKernan (cf. [**HM07**]) showed that assuming the minimal model program in dimension $n-1$ (for varieties of log-general type) then pl-flips exist in dimension n (cf. 1.3). Finally C. Birkar, P. Cascini, C. Hacon and J. McKernan (cf. [**BCHM06**]) showed that assuming that pl-flips exist in dimension n, then the minimal model program for varieties of log-general type holds in dimension n. In particular it follows that flips exist in all dimensions and that if X is a smooth complex projective variety, then its canonical ring $R(K_X) = \bigoplus_{m\geq 0} H^0(\mathcal{O}_X(mK_X))$ is finitely generated (this was also established for varieties of general type, using analytic methods, by Y.-T. Siu [**Siu06**]). We remark that in [**BCHM06**] it is shown that sequences of flips for the minimal model program with scaling terminate. This is sufficient to show that minimal models exist, however the problem of termination of an arbitrary sequence of flips remains open.

The main purpose of these lectures is to give a proof of the fact that assuming the minimal model program in dimension $n-1$ (for varieties of log-general type) then flips exist in dimension n cf. (1.10). This result was first proved in [**HM07**] using ideas of V. Shokurov and Y.-T. Siu. In these lectures we will follow the approach of [**HM08**]. In the last lecture, we will briefly review some of the results of [**BCHM06**]. In particular we will discuss the minimal model program with scaling.

Notation

We work over the field of complex numbers $\mathbb{C}$.

If X is a normal variety, then $\mathrm{WDiv}(X)$ will denote the group of Weil divisors on X. $\mathrm{WDiv}_{\mathbb{R}}(X) = \mathrm{WDiv}(X)\otimes_{\mathbb{Z}}\mathbb{R}$ is the group of $\mathbb{R}$ **divisors**. $D \in \mathrm{WDiv}_{\mathbb{R}}(X)$ is **$\mathbb{R}$-Cartier** if $D = \sum r_i D_i$ where $r_i \in \mathbb{R}$ and D_i are Cartier divisors. $D \in \mathrm{WDiv}_{\mathbb{R}}(X)$ is **effective** if $D = \sum p_i P_i$ where P_i are prime divisors and $p_i \geq 0$. If $D = \sum p_i P_i$ and $D' = \sum p_i' P_i$, then we let $\llcorner D \lrcorner = \sum \llcorner p_i \lrcorner P_i$ where $\llcorner p_i \lrcorner$ is the biggest integer $\leq p_i$, we let $\{D\} = D - \llcorner D \lrcorner$ and we let $D\wedge D' = \sum \min\{p_i, p_i'\} P_i$. X is **$\mathbb{Q}$-factorial** if for any prime divisor $P \subset X$, there is a positive integer $m > 0$ such that mP is Cartier.

If $f : X \to U$ is a morphism of normal varieties and $D_i \in \mathrm{WDiv}_{\mathbb{R}}(X)$, then $D_1 \sim_{\mathbb{R},U} D_2$ (i.e. D_1 and D_2 are **$\mathbb{R}$-linearly equivalent over** U) if $D_1 - D_2 = \sum r_i(q_i) + f^*C$ where $r_i \in \mathbb{R}$, q_i are rational functions on X and C is an $\mathbb{R}$-Cartier divisor on U. An $\mathbb{R}$-Cartier divisor D is **nef over** U if $D\cdot C \geq 0$ for any curve $C \subset X$ contracted by f.

An $\mathbb{R}$-Cartier divisor D is **ample over** U if $D \sim_{\mathbb{R},U} \sum r_i D_i$ where $r_i \in \mathbb{R}_{\geq 0}$ and D_i are ample Cartier divisors over U. An $\mathbb{R}$-Cartier divisor D is **semiample over** U if there is a morphism $g\colon X \to Y$ over U such that $D \sim_{\mathbb{R},U} g^*D'$ where D' is ample over U. An $\mathbb{R}$-Cartier divisor D is **big over** U if

$$\limsup \frac{\mathrm{rk} f_*\mathcal{O}_X(mD)}{m^{\dim f}} > 0$$

where $\dim f$ denotes the dimension of a general fiber of f. Note that D is big over U if and only if $D \sim_{\mathbb{R},U} A + B$ where A is ample over U and $B \geq 0$. An $\mathbb{R}$-Cartier divisor D is **pseudo-effective over** U if its numerical class belongs to the closure of the cone of big divisors over U.

The **real linear system** associated to an $\mathbb{R}$-divisor D on X is

$$|D/U|_{\mathbb{R}} = \{C \geq 0 | C \sim_{\mathbb{R},U} D\}.$$

The **stable base locus** of D over U is the Zariski closed subset

$$\mathbf{B}(D/U) := \bigcap_{C \in |D/U|_{\mathbb{R}}} \mathrm{Supp}(C).$$

When D is $\mathbb{Q}$-Cartier then $\mathbf{B}(D/U)$ is the usual stable base locus. The **stable fixed divisor** of D over U is the divisorial component of $\mathbf{B}(D/U)$. The **augmented stable base locus** of D over U is given by $\mathbf{B}(D - \epsilon A/U)$ for any divisor A ample over U and any $0 < \epsilon \ll 1$. This set is denoted by $\mathbf{B}_+(D/U)$.

If D is a Weil divisor and $V \subset |D|$ is a linear series, then the **base locus of** V is

$$\mathrm{Bs}(V) := \bigcap_{0 \leq C \in V} \mathrm{Supp}(C)$$

and the **fixed divisor** $\mathrm{Fix}(V)$ is the divisorial component of $\mathrm{Bs}(V)$. We will abuse notation and we will denote $\mathrm{Fix}(|D|)$ simply by $\mathrm{Fix}(D)$. The divisor $\mathrm{Mob}(D) = D - \mathrm{Fix}(D)$ is the **mobile part of** D.

A **log pair** (X, Δ) is a normal quasi-projective variety X and an $\mathbb{R}$-divisor $\Delta \geq 0$ such that $K_X + \Delta$ is $\mathbb{R}$-Cartier.

A **log resolution** of a log pair (X, Δ) is a projective birational morphism $f : Y \to X$ from a smooth quasi-projective variety Y such that $f^{-1}(\Delta) \cup \mathrm{Ex}(f)$ has simple normal crossings support. We will often write

$$(*) \qquad K_Y + \Gamma = f^*(K_X + \Delta) + E$$

where K_Y is a canonical divisor such that $f_* K_Y = K_X$, Γ and E are effective divisors with no common components such that $f_*\Gamma = \Delta$ and $f_*E = 0$. The divisors Γ and E are uniquely determined.

We also write $E - \Gamma = \sum a_F(X, \Delta) F$ where F runs over all prime divisors in Y. For all but finitely many divisors $F \subset Y$, we have $a_F(X, \Delta) = 0$. The numbers $a_F(X, \Delta) = 0$ are the **discrepancies** of (X, Δ) along F. They are independent of the choice of f.

If $a_F(X, \Delta) \geq -1$ (respectively $a_F(X, \Delta) > -1$) for all divisors F over X, then (X, Δ) is **log canonical** (respectively **kawamata log terminal**).

If $a_F(X, \Delta) \geq 0$ (respectively $a_F(X, \Delta) > 0$) for all exceptional divisors F over X, then (X, Δ) is **canonical** (respectively **terminal**).

If the coefficients of Δ are ≤ 1 and (X, Δ) admits a log resolution $f : Y \to X$ such that $a_F(X, \Delta) > -1$ for all exceptional divisors $F \subset Y$ over X, then (X, Δ) is **divisorially log terminal**. This is equivalent to requiring that there is a closed subset $Z \subset X$ such that $X - Z$ is smooth, $\Delta|_{X-Z}$ has simple normal crossings support and if F is an irreducible divisor with center contained in Z, then $a_F(X, \Delta) > -1$.

If $a_F(X, \Delta) > -1$ for all exceptional divisors F over X, then (X, Δ) is **purely log terminal**. If (X, Δ) is purely log terminal, then $\llcorner \Delta \lrcorner = S$ is a disjoint union of normal prime divisors. If (X, Δ) is purely log terminal and $\llcorner \Delta \lrcorner = S$ is irreducible, then the pair (S, Δ_S) defined by $K_S + \Delta_S = (K_X + \Delta)|_S$ is kawamata log terminal.

We will say that a $\mathbb{Q}$-divisor A is a **general ample $\mathbb{Q}$-divisor** if there is an integer $k > 0$ such that kA is a very ample Cartier divisor and $kA \in |kA|$ is general.

Cone and Base Point Free Theorems

For the convenience of the reader, we recall the Cone and Base Point Free Theorems. We refer the reader to [**KM98**] for a particularly clear exposition.

Theorem 0.1 (Base Point Free Theorem). *Let $f\colon X \to Z$ be a proper morphism and (X, Δ) be a kawamata log terminal pair. If D is a f-nef Cartier divisor such that $aD - (K_X + \Delta)$ is f-nef and f-big for some $a > 0$, then $|bD|$ is f-free for all $b \gg 0$ (i.e. $f^*(f_*\mathcal{O}_X(D)) \to \mathcal{O}_X(D)$ is surjective).*

Theorem 0.2 (Cone Theorem). *Let $f\colon X \to Z$ be a projective morphism and (X, Δ) be a kawamata log terminal pair. Then*

(1) *There exist countably many rational curves $C_j \subset X$ contracted by f such that $0 < -(K_X + \Delta) \cdot C_j \leq 2 \dim X$, and*
$$\overline{NE}(X/Z) = \overline{NE}(X/Z)_{(K_X+\Delta)\geq 0} + \sum \mathbb{R}_{\geq 0}[C_j].$$
(2) *For any $\epsilon > 0$ and any f-ample divisor H,*
$$\overline{NE}(X/Z) = \overline{NE}(X/Z)_{(K_X+\Delta+\epsilon H)\geq 0} + \sum_{\text{finite}} \mathbb{R}_{\geq 0}[C_j].$$
(3) *If $F \subset \overline{NE}(X/Z)_{(K_X+\Delta)<0}$ is a $(K_X + \Delta)$ negative extremal face, then there is a unique morphism over Z, $\mathrm{cont}_F\colon X \to Z$ such that $(\mathrm{cont}_F)_*\mathcal{O}_X = \mathcal{O}_Z$ and such that an irreducible curve $C \subset X$ is contracted by f if and only if $[C] \in F$.*
(4) *If L is a line bundle on X such that $L \cdot C = 0$ for any curve C with $[C] \in F$, then $L = \mathrm{cont}_F^* L_Z$ for some line bundle L_Z on Z.*

Recall that $N_1(X/Z) = \{\sum a_i C_i : a_i \in \mathbb{R}\}/\equiv_Z$ denotes the $\mathbb{R}$ vector space of 1-cycles modulo numerical equivalence over Z, that $NE(X/Z) = \{\sum a_i[C_i] : a_i \geq 0\} \subset N_1(X/Z)$, $\overline{NE}(X/Z)$ is the closure of $NE(X/Z)$ and that if D is an $\mathbb{R}$-Cartier divisor, then $\overline{NE}(X/Z)_{D\geq 0} = \{x \in \overline{NE}(X/Z) : D \cdot x \geq 0\}$.

Often one says that $\overline{NE}(X/Z)$ is locally polyhedral even though this is not actually the case.

Exercise 0.1. Show that if $D \geq 0$ is a Cartier divisor, then there is an integer $m > 0$ such that $\mathbf{B}(D) = \mathrm{Bs}(mD)$.

Exercise 0.2. Show that if (X, Δ) is a log pair and $a_F(X, \Delta) < -1$ for some divisor F over X, then $\min\{a_F(X, \Delta)\} = -\infty$.

Exercise 0.3. Show that if (X, Δ) is a purely log terminal pair and $S = \llcorner D \lrcorner$ is a prime divisor, then (S, Δ_S) is kawamata log terminal. Using the Connectedness Theorem, show that if (S, Δ_S) is kawamata log terminal, then (X, Δ) is a purely log terminal pair on a neighborhood of S.

Exercise 0.4. Show that if (X, Δ) is a divisorially log terminal pair, then there exists an $\mathbb{R}$-divisor Δ_0 such that (X, Δ_0) is a kawamata log terminal pair.

LECTURE 1

Pl-flips

We begin by recalling the definition of a flip and of a pl-flip.

Definition 1.1. Let (X, Δ) be a log canonical pair. A **flipping contraction** is a small projective morphism $f : X \to Z$ of normal quasi projective varieties such that $f_*\mathcal{O}_X = \mathcal{O}_Z$, X is $\mathbb{Q}$-factorial, $\rho(X/Z) = 1$ and $-(K_X + \Delta)$ is f-ample.

A small projective morphism $f^+ : X^+ \to Z$ of normal quasi-projective varieties is the **flip** of f if $f^+_*\mathcal{O}_{X^+} = \mathcal{O}_Z$, X^+ is $\mathbb{Q}$-factorial, $\rho(X^+/Z) = 1$ and $K_{X^+} + \Delta^+$ is f-ample where Δ^+ denotes the strict transform of Δ.

Recall that a **small** morphism is a birational morphism such that $\dim \operatorname{Ex}(f) < \dim X - 1$.

Remark 1.2. It is known that if the flip f^+ exists, then it is unique and given by

$$X^+ = \operatorname{Proj}_Z \bigoplus_{m \geq 0} f_*\mathcal{O}_X(m(K_X + \Delta)).$$

To prove the existence of the flip f^+, one may work locally over Z. If we assume that $Z = \operatorname{Spec}(A)$ is affine and $\Delta \in \operatorname{WDiv}_{\mathbb{Q}}(X)$, then it suffices to show that

$$R(K_X + \Delta) = \bigoplus_{m \geq 0} H^0(\mathcal{O}_X(m(K_X + \Delta)))$$

is a finitely generated A-algebra.

Definition 1.3. Let (X, Δ) be a purely log terminal pair with $S = \llcorner \Delta \lrcorner$. A **pl-flipping contraction** is a flipping contraction $f : X \to Y$ such that $\Delta \in \operatorname{WDiv}_{\mathbb{Q}}(X)$, S is irreducible and $-S$ is f-ample.

The flip of f is sometimes also called the **pl-flip** of f.

Remark 1.4. It is known that the existence of kawamata log terminal flips follows from the existence of pl-flips and special termination cf. [**Fujino07**]. Pl stands for pre-limiting.

Definition 1.5. Let (X, Δ) be a purely log terminal pair where $\llcorner \Delta \lrcorner = S$ is irreducible. The **restricted algebra** $R_S(K_X + \Delta)$ is given by

$$\bigoplus_{m \geq 0} \operatorname{Im}\left(H^0(\mathcal{O}_X(m(K_X + \Delta))) \to H^0(\mathcal{O}_S(m(K_X + \Delta)))\right).$$

We have the following result due to V. Shokurov.

Theorem 1.6. *Let $f : X \to Z$ be a pl-flipping contraction for a purely log terminal pair (X, Δ) where $\llcorner \Delta \lrcorner = S$. If $Z = \operatorname{Spec}(A)$ is affine, then the flip of f exists if and only if the restricted algebra $R_S(K_X + \Delta)$ is finitely generated.*

Theorem (1.6) is useful because it allows us to proceed by induction on the dimension of X. Before proving (1.6), we will need some results about finite generation.

Lemma 1.7. *Let $R = \bigoplus_{m\geq 0} R_m$ be a graded ring and $d > 0$ be an integer. If R is an integral domain, then R is finitely generated if and only if so is $R^{(d)} = \bigoplus_{m\geq 0} R_{md}$.*

PROOF. Suppose that R is finitely generated. Since $R^{(d)}$ is the ring of invariants of R by a $\mathbb{Z}/d\mathbb{Z}$ action, then $R^{(d)}$ is finitely generated by a Theorem of E. Noether cf. [**Eisenbud95**, 13.17].

Suppose that $R^{(d)}$ is finitely generated. Note that if $f \in R_m$, then f is integral over $R^{(d)}$ as it is a zero of the monic polynomial $x^d - f^d \in R^{(d)}[x]$. By E. Noether's Theorem on the finiteness of integral closures (cf. [**Eisenbud95**, 13.13]), it follows that R is finitely generated. □

PROOF OF (1.6). If $R(K_X+\Delta) = \bigoplus_{m\geq 0} H^0(\mathcal{O}_X(m(K_X+\Delta)))$ is finitely generated, then clearly $R_S(K_X+\Delta)$ is finitely generated.

Assume now that $R_S(K_X+\Delta)$ is finitely generated. Let $S' \sim S$ be an effective divisor not containing S (cf. (1.3)) and let g be a rational function such that $S - S' = (g)$. There are positive integers a and b such that $a(K_X+\Delta) - bS' \equiv_Z 0$. By the Base Point Free Theorem cf. (0.1), $a(K_X+\Delta) - bS'$ is semiample over Z and so we may assume that $a(K_X+\Delta) \sim_Z bS'$. After replacing Z by an open subset, we may assume that $a(K_X+\Delta) \sim bS'$. By (1.7), $R(K_X+\Delta)$ is finitely generated if and only if so is $R(S')$. By (1.7), $R_S(S')$ is finitely generated and so it suffices to show that the kernel of $\phi : R(S') \to R_S(S')$ is finitely generated. Let g_m be a rational function corresponding to a non-zero element of $R(S')_m$ so that $(g_m) + mS' \geq 0$. If $\phi(g_m) = 0$, then $(g_m) + mS' = S + B$ for some $B \geq 0$. But then $(g_m/g) + (m-1)S' = B$ so that g_m/g is a rational function corresponding to a non-zero element of $R(S')_{m-1}$. In particular the kernel of ϕ is the principal ideal generated by g and the lemma follows. □

In these lectures we will prove that a consequence of the existence of of log terminal models for $(n-1)$-dimensional log pairs of general type (cf. Theorem 1.9 below) implies the existence of pl-flips for log pairs of dimension n. Recall the following.

Definition 1.8. Let (X,Δ) be a kawamata log terminal pair and $f\colon X \to Z$ be a morphism to a normal variety. A **log terminal model** of (X,Δ) over Z is a birational morphism $\phi\colon X \dashrightarrow Y$ over Z such that ϕ^{-1} contracts no divisors, Y is normal and $\mathbb{Q}$-factorial, $-(K_Y+\phi_*\Delta)$ is nef over Z and $a_E(X,\Delta) \leq a_E(Y,\phi_*\Delta)$ for all divisors E over X where the strict inequality holds if $E \subset X$ is a ϕ-exceptional divisor.

Theorem 1.9. *Let $f\colon X \to Z$ be a projective morphism to a normal affine variety Z. Let (X,Δ) be a kawamata log terminal pair where $K_X+\Delta$ is pseudo-effective, $\Delta = A + B$ and $A \geq 0$ is an ample $\mathbb{Q}$-divisor and $B \geq 0$. Then*

(1) *(X,Δ) has a log terminal model $\mu\colon X \dashrightarrow Y$ over Z. In particular if $K_X+\Delta$ is $\mathbb{Q}$-Cartier, then $R(K_X+\Delta)$ is finitely generated.*
(2) *Let $V \subset \mathrm{WDiv}_{\mathbb{R}}(X)$ be a finite dimensional affine subspace of $\mathrm{WDiv}_{\mathbb{R}}(X)$ containing Δ which is defined over the rationals. There exists a constant*

$\delta > 0$ such that if P is a prime divisor contained in $\mathbf{B}(K_X + \Delta)$, then P is contained in $\mathbf{B}(K_X + \Delta')$, for any $\mathbb{R}$-divisor $\Delta' \in V$ with $\|\Delta - \Delta'\| \leq \delta$.

(3) *Let $W \subset \mathrm{WDiv}_{\mathbb{R}}(X)$ be the smallest affine subspace containing Δ which is defined over the rationals. Then there is a real number $\eta > 0$ and an integer $r > 0$ such that if $\Delta' \in W$, $\|\Delta - \Delta'\| \leq \eta$ and $k > 0$ is an integer such that $k(K_X + \Delta')/r$ is Cartier, then every component of $\mathrm{Fix}(k(K_X + \Delta'))$ is a component of $\mathbf{B}(K_X + \Delta)$.*

PROOF. See [**BCHM06**, §2] or (4.6). □

Theorem 1.10. *Assume Theorem 1.9 in dimension $n - 1$. Let $f : X \to Z$ be a pl-flipping contraction, then the flip $f^+ : X^+ \to Z$ exists.*

PROOF. See Lecture 3. □

Exercise 1.1. Show that if $f\colon X \to Z$ is a flipping contraction for a pair (X, Δ), then $K_Z + f_*\Delta$ is not $\mathbb{Q}$-Cartier.

Exercise 1.2. Show that if $f\colon X \to Z$ is a flipping contraction for a divisorially log terminal pair (X, Δ) such that there is a component S of $\llcorner \Delta \lrcorner$ where $-S$ is f-ample, then there is a pair (X, Δ') such that f is a pl-flipping contraction for (X, Δ').

Exercise 1.3. Let $f\colon X \to Z$ be a flipping contraction where Z is affine. Show that for any irreducible divisor $S \subset X$ we have $S \sim S'$ where S' is an effective divisor whose support does not contain S.

Exercise 1.4. Let $f\colon X \to Z$ be a flipping contraction and $f^+\colon X^+ \to Z$ be its flip. Show that X^+ is $\mathbb{Q}$-factorial and $a_E(X, \Delta) \leq a_E(X^+, \Delta^+)$ for any divisor E over X and that strict inequality holds if the center of E is contained in the flipping or flipped locus. In particular, if (X, Δ) is log canonical, kawamata log terminal or divisorially log terminal, then so is (X^+, Δ^+).

Exercise 1.5. Let $f\colon X \to Z$ be a divisorial contraction (i.e. f is a birational projective morphism of normal quasi-projective varieties such that X is $\mathbb{Q}$-factorial, $\dim \mathrm{Ex}(f) = \dim X - 1$, $\rho(X/Z) = 1$ and $-(K_X + \Delta)$ is f-ample). Show that Z is $\mathbb{Q}$-factorial and $a_E(X, \Delta) \leq a_E(Z, f_*\Delta)$ for any divisor E over X and that strict inequality holds if the center of E is contained in $\mathrm{Ex}(f)$. In particular, if (X, Δ) is log canonical, kawamata log terminal or divisorially log terminal, then so is $(Z, f_*\Delta)$.

Exercise 1.6. A $K_X + \Delta$ **MMP** is a sequence of flips and divisorial contractions $\phi_i : X_i \dashrightarrow X_{i+1}$ for (X_i, Δ_i) where $\Delta_i = (\phi_{i-1})_*\Delta_{i-1}$ and $(X_1, \Delta_1) = (X, \Delta)$. Use (1.4) and (1.5) to show that if Δ is a $\mathbb{Q}$-divisor, then $R(K_X + \Delta) \cong R(K_{X_i} + \Delta_i)$.

Exercise 1.7. Show that if $\pi : X \to U$ is a projective morphism of normal varieties and (X, Δ) is a kawamata log terminal pair such that Δ is big over U and $K_X + \Delta$ is nef over U, then $K_X + \Delta$ is semiample over U.

Exercise 1.8. Let (X, Δ) be a kawamata log terminal pair and $\pi : X \to Z$ be a projective morphism such that Δ is big over Z. Show that the divisors contracted by a log terminal model of (X, Δ) over Z are the divisors contained in $\mathbf{B}(K_X + \Delta/Z)$.

Exercise 1.9. Show that (1) of (1.9) implies (2) of (1.9).

LECTURE 2

Multiplier ideal sheaves

In this section we will recall the definition and the main properties of multiplier ideal sheaves. The standard reference for multiplier ideal sheaves is [**Lazarsfeld04**]. In what follows we will focus on a generalization of this notion known as adjoint ideals.

Definition 2.1. Let X be a smooth quasi-projective variety, Δ be a reduced divisor with simple normal crossings support and V a linear series such that $\mathrm{Bs}(V)$ contains no log canonical center of (X, Δ) (i.e. $\mathrm{Bs}(V)$ contains no strata of Δ). Let $f\colon Y \to X$ be a log resolution of V and of (X, Δ). We let

$$K_Y + \Gamma = f^*(K_X + \Delta) + E$$

as in $(*)$ of Notation.

For any real number $c > 0$, we define the **multiplier ideal sheaf**

$$\mathcal{J}_{\Delta, c\cdot V} := f_*\mathcal{O}_Y(E - \llcorner cF \lrcorner)$$

where $F = \mathrm{Fix}(f^*V)$.

If $D \geq 0$ is a $\mathbb{Q}$-divisor and $m > 0$ is an integer such that mD is Cartier, then we let $\mathcal{J}_{\Delta, c\cdot D} = \mathcal{J}_{\Delta, \frac{c}{m}\cdot V}$ where $V = \{mD\}$ is the linear series consisting of the unique divisor mD.

Remark 2.2. If $\Delta = 0$, then $\mathcal{J}_{\Delta, c\cdot V} = \mathcal{J}_{c\cdot V}$ is the usual multiplier ideal sheaf.

Lemma 2.3. *The definition given in* (2.1) *is independent of the log resolution.*

PROOF. Let $f\colon Y \to X$ and $f'\colon Y' \to X'$ be two log resolution. Since any two log resolutions can be dominated by a third log resolution, we may assume that there is a morphism $\mu\colon Y' \to Y$ such that $f' = f \circ \mu$. We let

$$K_Y + \Gamma = f^*(K_X + \Delta) + E \qquad \text{and} \qquad K_{Y'} + \Gamma' = (f')^*(K_X + \Delta) + E'$$

be as in $(*)$ of Notation. We have

$$K_{Y'} + \Gamma' = \mu^*(K_Y + \Gamma - E) + E'.$$

Let $F = \mathrm{Fix}(f^*V)$ and $F' = \mathrm{Fix}((f')^*V)$. Since $f^*V - F$ is base point free, it follows that $F' = \mu^*F$. We claim that

$$\mu_*\mathcal{O}_{Y'}(E' - \mu^*E - \llcorner \mu^*\{cF\} \lrcorner) = \mathcal{O}_Y.$$

Grant this for the time being, then

$$(f')_*\mathcal{O}_{Y'}(E' - \llcorner cF' \lrcorner) = (f')_*\mathcal{O}_{Y'}(E' - \mu^*E - \llcorner \mu^*\{cF\} \lrcorner + \mu^*(E - \llcorner cF \lrcorner)) =$$

$$f_*\left(\mu_*\mathcal{O}_{Y'}(E' - \mu^*E - \llcorner \mu^*\{cF\} \lrcorner) \otimes \mathcal{O}_Y(E - \llcorner cF \lrcorner)\right) = f_*\mathcal{O}_Y(E - \llcorner cF \lrcorner).$$

The Lemma now follows immediately. To see the claim, notice that $K_Y+\Gamma+\{cF\}$ is divisorially log terminal and its log canonical places coincide with those of $K_Y+\Gamma-E$. From the equation

$$K_{Y'}+\Gamma'+\mu^*(E+\{cF\})-E'=\mu^*(K_Y+\Gamma+\{cF\})$$

it follows that $\llcorner\mu^*(E+\{cF\})-E'\lrcorner\leq 0$ so that $E'-\mu^*E-\llcorner\mu^*\{cF\}\lrcorner$ is effective and μ-exceptional. The claim is now clear. □

Lemma 2.4. *Let X be a smooth quasi-projective variety, Δ be a reduced divisor with simple normal crossings support and $V\subset V'$ be linear series such that* $\mathrm{Bs}(V)$ *contains no log canonical center of (X,Δ). Let $c\geq c'$ be positive real numbers, $\Delta'\leq\Delta$ be reduced divisors and D and D' be effective $\mathbb{Q}$-Cartier divisors whose supports contain no log canonical centers of (X,Δ).*

Then

(1) $\mathcal{J}_{\Delta,c\cdot V}\subset\mathcal{J}_{\Delta',c'\cdot V'}$, *and in particular* $\mathcal{J}_{\Delta,c\cdot V}\subset\mathcal{J}_{c\cdot V}\subset\mathcal{O}_X$.

(2) *If Σ is an effective Cartier divisor such that $D\leq\Sigma+D'$ and $\mathcal{J}_{\Delta,D'}=\mathcal{O}_X$, then $\mathcal{I}_\Sigma\subset\mathcal{J}_{\Delta,D}$.*

Proof. Let $f\colon Y\to X$ be a common log resolution of V, V', (X,Δ) and (X',Δ'). We let

$$K_Y+\Gamma=f^*(K_X+\Delta)+E \qquad\text{and}\qquad K_Y+\Gamma'=f^*(K_X+\Delta')+E'$$

be as in $(*)$ of Notation. We have that $\Gamma'\leq\Gamma$ and $E'\geq E$. If we let $F=\mathrm{Fix}(f^*V)$ and $F'=\mathrm{Fix}(f^*V')$, then we also have $F\geq F'$. It follows that

$$\mathcal{J}_{\Delta,c\cdot V}=f_*\mathcal{O}_Y(E-\llcorner cF\lrcorner)\subset f_*\mathcal{O}_Y(E'-\llcorner c'F'\lrcorner)=\mathcal{J}_{\Delta',c'\cdot V'}.$$

Let $f\colon Y\to X$ be a log resolution of $(X,\Delta+\Sigma+D')$. We let

$$K_Y+\Gamma=f^*(K_X+\Delta)+E$$

be as in $(*)$ of Notation. Since $\mathcal{J}_{\Delta,D'}=\mathcal{O}_X$, we have $E-\llcorner f^*D'\lrcorner\geq 0$. Since Σ is Cartier, we have $\llcorner f^*D\lrcorner\leq f^*\Sigma+\llcorner f^*D'\lrcorner$. Therefore

$$E-\llcorner f^*D\lrcorner\geq E-\llcorner f^*D'\lrcorner-f^*\Sigma\geq -f^*\Sigma$$

and hence

$$\mathcal{I}_\Sigma=f_*\mathcal{O}_Y(-f^*\Sigma)\subset f_*\mathcal{O}_Y(E-\llcorner f^*D\lrcorner)=\mathcal{J}_{\Delta,D}.$$

□

The next lemma shows that multiplier ideals of the form $\mathcal{J}_{\Delta,D}$ may be viewed as being obtained by successive extensions of the usual multiplier ideal sheaves $\mathcal{J}_D$.

Lemma 2.5. *Let X be a smooth quasi-projective variety, Δ be a reduced divisor with simple normal crossings support and $D\geq 0$ be a $\mathbb{Q}$-Cartier divisor whose supports contains no log canonical center of (X,Δ). Let S be a component of Δ, then there is a short exact sequence*

$$0\to\mathcal{J}_{\Delta-S,D+S}\to\mathcal{J}_{\Delta,D}\to\mathcal{J}_{(\Delta-S)|_S,D|_S}\to 0.$$

Proof. Let $f: Y\to X$ be a log resolution of $(X,\Delta+D)$ which is an isomorphism at the generic point of each log canonical center of (X,Δ) (cf. [**Szabó94**]). We write

$$K_Y+\Gamma=f^*(K_X+\Delta)+E$$

as in $(*)$ of Notation. Let $T = (f^{-1})_* S$ and consider the short exact sequence

$$0 \to \mathcal{O}_Y(E - \llcorner f^*D \lrcorner - T) \to \mathcal{O}_Y(E - \llcorner f^*D \lrcorner) \to \mathcal{O}_T(E - \llcorner f^*D \lrcorner) \to 0.$$

By definition, we have $f_*\mathcal{O}_Y(E - \llcorner f^*D \lrcorner) = \mathcal{J}_{\Delta,D}$. Since

$$K_Y + \Gamma - T = f^*(K_X + \Delta - S) + E + f^*S - T$$

where $\Gamma - T$ and $E + f^*S - T$ are effective with no common components, it follows that

$$\mathcal{J}_{\Delta-S,D+S} = f_*\mathcal{O}_Y(E + f^*S - T - \llcorner f^*(D+S) \lrcorner) = f_*\mathcal{O}_Y(E - \llcorner f^*D \lrcorner - T).$$

Since

$$K_T + (\Gamma - T)|_T = (f|_T)^*(K_S + (\Delta - S)|_S) + E|_T,$$

it follows that

$$\mathcal{J}_{(\Delta-S)|_S, D|_S} = (f|_T)_*\mathcal{O}_T(E - \llcorner f^*D \lrcorner).$$

Finally, pick an effective and f-exceptional $\mathbb{Q}$-divisor F such that $-F$ is f-ample and $(Y, \Gamma - T + \{f^*D\} + F)$ is divisorially log terminal cf. (2.8). Since

$$E - \llcorner f^*D \lrcorner - T = K_Y + \Gamma - T + \{f^*D\} + F - F - f^*(K_X + \Delta + D),$$

by Kawamata-Viehweg vanishing (cf. (2.9)), we have that $R^1 f_*\mathcal{O}_Y(E - \llcorner f^*D \lrcorner - T) = 0$. □

Lemma 2.6. *Let X be a smooth quasi-projective variety, Δ be a reduced divisor with simple normal crossings support and $D \geq 0$ be a $\mathbb{Q}$-Cartier divisor whose support contains no log canonical center of (X, Δ). Let $\pi \colon X \to Z$ be a projective morphism to a normal affine variety Z and N be a Cartier divisor on X such that $N - D$ is ample. Then*

$$H^i(X, \mathcal{J}_{\Delta,D}(K_X + \Delta + N)) = 0 \qquad \text{for } i > 0,$$

and if S is a component of Δ, then the homomorphism

$$H^0(X, \mathcal{J}_{\Delta,D}(K_X + \Delta + N)) \to H^0(S, \mathcal{J}_{(\Delta-S)|_S, D|_S}(K_X + \Delta + N))$$

is surjective.

PROOF. Let $f : Y \to X$ be a log resolution of $(X, \Delta + D)$ which is an isomorphism at the generic point of each log canonical center of (X, Δ) (cf. [**Szabó94**]). We write

$$K_Y + \Gamma = f^*(K_X + \Delta) + E$$

as in $(*)$ of Notation. Pick an effective and f-exceptional $\mathbb{Q}$-divisor F such that $f^*(N - D) - F$ is ample and $(Y, \Gamma + \{f^*D\} + F)$ is divisorially log terminal. Since

$$E - \llcorner f^*D \lrcorner + f^*(K_X + \Delta + N) = K_Y + \Gamma + \{f^*D\} + F + f^*(N - D) - F,$$

by Kawamata-Viehweg vanishing (cf. (2.9)), we have that $R^j f_*\mathcal{O}_Y(E - \llcorner f^*D \lrcorner) = 0$ for all $j > 0$. Since $f_*\mathcal{O}_Y(E - \llcorner f^*D \lrcorner) = \mathcal{J}_{\Delta,D}$, by Kawamata-Viehweg vanishing it follows then that

$$H^i(X, \mathcal{J}_{\Delta,D}(K_X + \Delta + N)) = H^i(Y, \mathcal{O}_Y(E - \llcorner f^*D \lrcorner + f^*(K_X + \Delta + N))) = 0$$

for all $i > 0$ cf. (2.10).

The final claim follows immediately from (2.5). □

Asymptotic multiplier ideal sheaves

Definition 2.7. Let D be a divisor on a normal variety X and $V_i \subset |iD|$ be linear series, then $V_\bullet$ is an **additive sequence of linear series** if

$$V_i + V_j \subset V_{i+j}$$

for all $i, j \geq 0$.

Lemma 2.8. *Let X be a smooth quasi-projective variety and Δ be a reduced divisor with simple normal crossings support. If $V_\bullet$ is an additive sequence of linear series and c is a positive real number, k a positive integer such that $\mathrm{Bs}(V_k)$ contains no log canonical centers of (X, Δ), then*

$$\mathcal{J}_{\Delta, \frac{c}{p}\cdot V_p} \subset \mathcal{J}_{\Delta, \frac{c}{q}\cdot V_q}$$

for any integers p and q such that k divides p and p divides q.

In particular, since X is Noetherian, the set $\{\mathcal{J}_{\Delta, \frac{c}{p}\cdot V_p}\}$ has a unique maximal element $\mathcal{J}_{\Delta, \|c\cdot V\|}$. Note that $\mathcal{J}_{\Delta, \|c\cdot V\|} = \mathcal{J}_{\Delta, \frac{c}{p}\cdot V_p}$ for any $p > 0$ sufficiently divisible.

PROOF. This follows from (1) of (2.4), since $(q/p)V_p \subset V_q$. □

Definition 2.9. Let X be a smooth quasi-projective variety, Δ be a reduced divisor with simple normal crossings support and S an irreducible component of Δ. If c is a positive real number, D is a $\mathbb{Q}$-divisor on X such that $\mathbf{B}(D)$ contains no log canonical centers of (X, Δ) then we define the **asymptotic multiplier ideal sheaves**

$$\mathcal{J}_{\Delta, c\cdot\|D\|} = \mathcal{J}_{\Delta, c\cdot V_\bullet} \qquad \text{and} \qquad \mathcal{J}_{(\Delta - S)|_S, c\cdot\|D\|_S} = \mathcal{J}_{(\Delta-S)|_S, c\cdot W_\bullet}$$

where $V_m = |mD|$ and $W_m = |mD|_S$ is the restriction (or trace) of the linear series $|mD|$ to S.

Lemma 2.10. *Let X be a smooth quasi-projective variety, Δ be a reduced divisor with simple normal crossings support, S an irreducible component of Δ and D a $\mathbb{Q}$-divisor on X such that $\mathbf{B}(D)$ contains no log canonical centers of (X, Δ). If $\pi\colon X \to Z$ is a projective morphism to a normal affine variety. Then*

(1) *For any real numbers $0 < c' \leq c$ and any reduced divisor $0 \leq \Delta' \leq \Delta$, we have*

$$\mathcal{J}_{\Delta, \|c\cdot D\|} \subset \mathcal{J}_{\Delta', \|c'\cdot D\|}.$$

(2) *If D is Cartier, then*

$$\mathrm{Im}\,(\pi_*\mathcal{O}_X(D) \to \pi_*\mathcal{O}_S(D)) \subset \pi_*\mathcal{J}_{(\Delta-S)|_S, \|D\|_S}(D).$$

(3) *If D is Cartier, A is an ample Cartier divisor and H is a very ample Cartier divisor, then*

$$\mathcal{J}_{\|D\|}(K_X + D + A + nH)$$

is globally generated where $n = \dim X$.

(4) *If D is Cartier and $\mathbf{B}_+(D)$ contains no log canonical centers of (X, Δ), then the image of the homomorphism*

$$\pi_*\mathcal{O}_X(K_X + \Delta + D) \to \pi_*\mathcal{O}_S(K_X + \Delta + D)$$

contains

$$\pi_*\mathcal{J}_{(\Delta-S)|_S, \|D\|_S}(K_X + \Delta + D).$$

PROOF. (1) is immediate from (1) of (2.4).

To see (2), let $p > 0$ be an integer such that $\mathcal{J}_{(\Delta-S)|_S,\|D\|_S} = \mathcal{J}_{(\Delta-S)|_S,\frac{1}{p}\cdot|pD|_S}$ and consider a log resolution $f\colon Y \to X$ of (X,Δ) and $|pD|$. We let

$$K_Y + \Gamma = f^*(K_X+\Delta)+E$$

be as in $(*)$ of Notation. Let $F_p = \mathrm{Fix}(f^*|pD|)$, then $pF_1 \geq F_p$ so that

$$\pi_*\mathcal{O}_X(D) = (\pi\circ f)_*\mathcal{O}_Y(f^*D - F_1) \subset (\pi\circ f)_*\mathcal{O}_Y(f^*D + E - \llcorner\frac{1}{p}F_p\lrcorner)$$
$$\subset (\pi\circ f)_*\mathcal{O}_Y(f^*D+E) = \pi_*\mathcal{O}_X(D).$$

The assertion now follows as

$$\pi_*\mathcal{J}_{(\Delta-S)|_S,\|D\|_S}(D) = (\pi\circ f)_*\mathcal{O}_T(f^*D + E - \llcorner\frac{1}{p}F_p\lrcorner)$$

where $T = (f^{-1})_*S$.

(3) is immediate from (2.6) and (2.11).

To see (4), let $p > 0$ be an integer such that $pD \sim A + B$ where A is a general very ample divisor and B is an effective divisor containing no log canonical centers of (X,Δ). Replacing p by a multiple, we may assume that $\mathcal{J}_{\Delta,\|D\|_S} = \mathcal{J}_{\Delta,\frac{1}{p}\cdot|pD|_S}$. Let $f\colon Y\to X$ be a log resolution of (X,Δ) and $|pD|$ which is an isomorphism at the generic point of each log canonical center of (X,Δ). We let

$$K_Y + \Gamma = f^*(K_X+\Delta)+E$$

be as in $(*)$ of Notation. Let $F_p = \mathrm{Fix}(f^*|pD|)$, then $M_p := pf^*D - F_p$ is base point free and $f^*B \geq F_p$. Let $T = (f^{-1})_*S$ and F be an effective and exceptional divisor such that $f^*A - F$ is ample and $\delta > 0$ be a rational number such that $K_Y + \Gamma - T + \{\frac{1}{p}F_p\} + \delta(f^*B - F_p + F)$ is divisorially log terminal. We have

$$E - \llcorner\frac{1}{p}F_p\lrcorner - T + f^*(K_X+\Delta+D) = K_Y + \Gamma - T + f^*D - \llcorner\frac{1}{p}F_p\lrcorner \sim_{\mathbb{Q}}$$
$$K_Y + \Gamma - T + \{\frac{1}{p}F_p\} + \delta(f^*B - F_p + F) + (\frac{1}{p}-\delta)M_p + \delta(f^*A - F).$$

By Kawamata-Viehweg vanishing,

$$R^1(\pi\circ f)_*\mathcal{O}_Y(E - \llcorner\frac{1}{p}F_p\lrcorner - T + f^*(K_X+\Delta+D)) = 0.$$

Therefore, pushing forward the short exact sequence

$$0 \to \mathcal{O}_Y(E - \llcorner\frac{1}{p}F_p\lrcorner - T + f^*(K_X+\Delta+D)) \to \mathcal{O}_Y(E - \llcorner\frac{1}{p}F_p\lrcorner + f^*(K_X+\Delta+D))$$
$$\to \mathcal{O}_T(E - \llcorner\frac{1}{p}F_p\lrcorner + f^*(K_X+\Delta+D)) \to 0,$$

one sees that the homomorphism

$$(\pi\circ f)_*\mathcal{O}_Y(E - \llcorner\frac{1}{p}F_p\lrcorner + f^*(K_X+\Delta+D)) \to (\pi\circ f)_*\mathcal{O}_T(E - \llcorner\frac{1}{p}F_p\lrcorner + f^*(K_X+\Delta+D))$$

is surjective. (4) now follows as $(\pi\circ f)_*\mathcal{O}_Y(E - \llcorner\frac{1}{p}F_p\lrcorner + f^*(K_X+\Delta+D)) \subset \pi_*\mathcal{O}_X(K_X+\Delta+D)$ and

$$(\pi\circ f)_*\mathcal{O}_T(E - \llcorner\frac{1}{p}F_p\lrcorner + f^*(K_X+\Delta+D)) = \pi_*\mathcal{J}_{(\Delta-S)|_S,\|D\|_S}(K_X+\Delta+D).$$

□

Extending pluricanonical forms

Theorem 2.11. *Let* $\pi\colon X \to Y$ *be a projective morphism from a smooth quasi-projective variety to an affine variety,* $\Delta = \sum \delta_i \Delta_i$ *be a divisor with simple normal crossings support and rational coefficients* $0 \leq \delta_i \leq 1$ *and* S *an irreducible component of* $\llcorner \Delta \lrcorner$. *Let* $k > 0$ *be an integer such that* $D = k(K_X + \Delta)$ *is Cartier. If* $\mathbf{B}(D)$ *contains no log canonical centers of* $(X, \ulcorner \Delta \urcorner)$ *and if* A *is a sufficiently ample Cartier divisor, then*

$$(\star_m) \qquad \mathcal{J}_{\|mD|_S\|} \subset \mathcal{J}_{(\ulcorner \Delta \urcorner - S)|_S, \|mD+A\|_S} \qquad \text{for all } m > 0, \qquad \text{and}$$

$$\pi_* \mathcal{J}_{\|mD|_S\|}(mD + A) \subset \operatorname{Im}\left(\pi_* \mathcal{O}_X(mD + A) \to \pi_* \mathcal{O}_S(mD + A)\right)$$

for all $m > 0$.

PROOF. We proceed by induction on m. Since $\mathcal{J}_{(\ulcorner \Delta \urcorner - S)|_S, \|A\|_S} = \mathcal{O}_X$, $(\star_0)$ is clear. Therefore we must show that $(\star_{m+1})$ holds assuming that $(\star_m)$ holds. There is a unique choice of integral divisors

$$S \leq \Delta^1 \leq \Delta^2 \leq \ldots \leq \Delta^k = \Delta^{k+1} = \ulcorner \Delta \urcorner$$

such that $k\Delta = \sum_{i=1}^k \Delta^i$ (cf. (2.12)). We let

$$D_{\leq 0} = 0 \qquad \text{and} \qquad D_{\leq s} = K_X + \Delta^s + D_{\leq s-1} \quad \text{for } 0 < s \leq k+1.$$

In particular $D = D_{\leq k}$. Since $\mathcal{J}_{\|(m+1)D|_S\|} \subset \mathcal{J}_{\|mD|_S\|}$ (cf. (1) of (2.10)) and since $\mathcal{J}_{(\Delta^{k+1} - S)|_S, \|mD + D_{\leq k} + A\|_S} = \mathcal{J}_{(\ulcorner \Delta \urcorner - S)|_S, \|(m+1)D + A\|_S}$, to prove $(\star_{m+1})$, it suffices to show that

$$(\sharp_i) \qquad \mathcal{J}_{\|mD|_S\|} \subset \mathcal{J}_{(\Delta^{i+1} - S)|_S, \|mD + D_{\leq i} + A\|_S} \qquad \text{for } 0 \leq i \leq k.$$

We proceed by induction on $0 \leq i \leq k$. $(\sharp_0)$ holds since by $(\star_m)$ and (1) of (2.10), we have

$$\mathcal{J}_{\|mD|_S\|} \subset \mathcal{J}_{(\ulcorner \Delta \urcorner - S)|_S, \|mD+A\|_S} \subset \mathcal{J}_{(\Delta^1 - S)|_S, \|mD+A\|_S}.$$

Assume that $(\sharp_{i-1})$ holds for some $0 < i \leq k$. We have that

$$\begin{aligned} \pi_* \mathcal{J}_{\|mD|_S\|}(mD + D_{\leq i} + A) &\subset \pi_* \mathcal{J}_{(\Delta^i - S)|_S, \|mD + D_{\leq i-1} + A\|_S}(mD + D_{\leq i} + A) \\ &\subset \operatorname{Im}\left(\pi_* \mathcal{O}_X(mD + D_{\leq i} + A) \to \pi_* \mathcal{O}_S(mD + D_{\leq i} + A)\right) \\ &\subset \pi_* \mathcal{J}_{(\Delta^{i+1} - S)|_S, \|mD + D_{\leq i} + A\|_S}(mD + D_{\leq i} + A). \end{aligned}$$

The first inclusion follows as we are assuming $(\sharp_{i-1})$. The second inclusion follows by (4) of (2.10), since we may assume that $D_{\leq i-1} + A$ is ample and hence that $\mathbf{B}_+(mD + D_{\leq i-1} + A)$ contains no log canonical centers of (X, Δ^i). The third inclusion follows from (2) of (2.10) since we may assume that $D_{\leq i} + A$ is ample and hence that $\mathbf{B}(mD + D_{\leq i} + A)$ contains no log canonical centers of (X, Δ^{i+1}).

By (3) of (2.10)

$$\mathcal{J}_{\|mD|_S\|}(mD + D_{\leq i} + A) = \mathcal{J}_{\|mD|_S\|}(K_S + mD|_S + (\Delta^i - S + D_{\leq i-1} + A)|_S)$$

is generated by global sections since $(\Delta^i - S + D_{\leq i-1} + A)|_S$ is sufficiently ample. Therefore, $(\sharp_i)$ holds. This concludes the proof of $(\star_m)$ for all $m \geq 0$. The remaining claim is clear from what we have shown above. □

Theorem 2.12. *Let $\pi\colon X \to Z$ be projective morphism from a smooth quasi-projective variety to an affine variety. Let $\Delta = S + A + B$ be a $\mathbb{Q}$-divisor such that $S = \llcorner \Delta \lrcorner$ is irreducible and smooth, $A \geq 0$ is a general ample $\mathbb{Q}$-divisor, $B \geq 0$, (X, Δ) is purely log terminal, $(S, \Omega + A|_S)$ is canonical where $\Omega = (A + B)|_S$, and $\mathbf{B}(K_X + \Delta)$ does not contain S. For any sufficiently divisible integer $m > 0$ let*

$$F_m = \operatorname{Fix}(|m(K_X + \Delta)|_S)/m \qquad \text{and} \qquad F = \lim F_{m!}.$$

Let $\epsilon > 0$ be a rational number such that $\epsilon(K_X + \Delta) + A$ is ample. If Φ is a $\mathbb{Q}$-divisor on S and $k > 0$ is an integer such that $k\Delta$ and $k\Phi$ are Cartier and

$$\Omega \wedge (1 - \frac{\epsilon}{k})F \leq \Phi \leq \Omega$$

then

$$|k(K_S + \Omega - \Phi)| + k\Phi \subset |k(K_X + \Delta)|_S.$$

Proof. Since A is a general ample $\mathbb{Q}$-divisor, $\frac{k-1}{k}A$ is a multiple of a general very ample divisor and so $(X, \Delta + \frac{k-1}{k}A)$ is purely log terminal. By assumption $(S, \Omega + \frac{1}{k}A|_S)$ is canonical. Let $l > 0$ be a sufficiently big and divisible integer.

We will show (see (‡) below) that there exists an effective divisor H on X not containing S such that for all integers $m > 0$ divisible by l, we have

$$(\dagger) \qquad |m(K_S + \Omega - \Phi)| + m\Phi + (\frac{m}{k}A + H)|_S \subset |m(K_X + \Delta) + \frac{m}{k}A + H|_S.$$

Grant this for the time being. Then, for any $\Sigma \in |k(K_S + \Omega - \Phi)|$, we may choose a divisor $G \in |m(K_X + \Delta) + \frac{m}{k}A + H|$ such that $G|_S = \frac{m}{k}\Sigma + m\Phi + (\frac{m}{k}A + H)|_S$. If we define $\Lambda = \frac{k-1}{m}G + B$, then

$$k(K_X + \Delta) \sim_{\mathbb{Q}} K_X + S + \Lambda + (\frac{1}{k}A - \frac{k-1}{m}H)$$

where $\frac{1}{k}A - \frac{k-1}{m}H$ is ample as m is sufficiently big. By (2.6), we have a surjective homomorphism

$$H^0(X, \mathcal{J}_{S,\Lambda}(k(K_X + \Delta))) \to H^0(S, \mathcal{J}_{\Lambda|_S}(k(K_X + \Delta))).$$

Since (S, Ω) is canonical, $(S, \Omega + \frac{k-1}{m}H|_S)$ is kawamata log terminal as m is sufficiently big. Therefore $\mathcal{J}_{\Omega + \frac{k-1}{m}H|_S} = \mathcal{O}_S$. Since

$$\Lambda|_S - (\Sigma + k\Phi) = (\frac{k-1}{m}G + B)|_S - (\Sigma + k\Phi) \leq \Omega + \frac{k-1}{m}H|_S,$$

then by (2) of (2.4), we have $\mathcal{I}_{\Sigma + k\Phi} \subset \mathcal{J}_{\Lambda|_S}$ and so

$$\Sigma + k\Phi \in |k(K_X + \Delta)|_S$$

and the Theorem follows.

Let $f\colon Y \to X$ be a log resolution of $(X, \Delta + \frac{1}{k}A)$ and of $|l(K_X + \Delta + \frac{1}{k}A)|$. We write

$$K_Y + \Gamma = f^*(K_X + \Delta + \frac{1}{k}A) + E$$

as in $(*)$ of Notation. Define

$$\Xi = \Gamma - \Gamma \wedge \frac{\operatorname{Fix}(l(K_Y + \Gamma))}{l}.$$

We have that $l(K_Y + \Xi)$ is Cartier, $\operatorname{Fix}(l(K_Y + \Xi)) \wedge \Xi = 0$ and $\operatorname{Mob}(l(K_Y + \Xi))$ is free. Since $\operatorname{Fix}(l(K_Y + \Xi)) + \Xi$ has simple normal crossings support, it follows that $\mathbf{B}(K_Y+\Xi)$ contains no log canonical center of $(Y, \ulcorner\Xi\urcorner)$. Let $T=(f^{-1})_*S$. Let $\Gamma_T=(\Gamma-T)|_T$ and $\Xi_T=(\Xi-T)|_T$. Let $m>0$ be divisible by l. By (2.4)

$$\sigma \in H^0(T, \mathcal{O}_T(m(K_T+\Xi_T))) = H^0(T, \mathcal{J}_{\|m(K_T+\Xi_T)\|}(m(K_T+\Xi_T))).$$

By (2.11), there is an ample divisor H on Y such that if $\tau \in H^0(T, \mathcal{O}_T(H))$, then $\sigma\cdot\tau$ is in the image of the homomorphism

$$H^0(Y, \mathcal{O}_Y(m(K_Y+\Xi)+H)) \to H^0(T, \mathcal{O}_T(m(K_Y+\Xi)+H)).$$

Therefore

$$(\ddagger) \qquad |m(K_T+\Xi_T)| + m(\Gamma_T-\Xi_T) + |H|_T| \subset |m(K_Y+\Gamma)+H|_T.$$

We claim that

$$(\flat) \qquad \Omega + \frac{1}{k}A|_S \geq (f|_T)_*\Xi_T \geq \Omega - \Phi + \frac{1}{k}A|_S$$

and so, as $(S, \Omega+\frac{1}{k}A|_S)$ is canonical, we have

$$|m(K_S+\Omega-\Phi)| + m((f|_T)_*\Xi_T-\Omega+\Phi) \subset |m(K_S+(f|_T)_*\Xi_T)| = (f|_T)_*|m(K_T+\Xi_T)|$$

cf. (2.13). Pushing forward the inclusion ($\ddagger$), one sees that

$$|m(K_S+\Omega-\Phi)| + m\Phi + (\frac{m}{k}A+f_*H)|_S \subset |m(K_X+\Delta)+\frac{m}{k}A+f_*H|_S.$$

Equation ($\dagger$) now follows.

We will now prove the inequality ($\flat$) claimed above. We have $\Xi_T \leq \Gamma_T$ and $(f|_T)_*\Gamma_T = \Omega+\frac{1}{k}A|_S$ and so the first inequality follows.

In order to prove the second inequality, let P be any prime divisor on S and let P' be its strict transform on T. Assume that P is contained in the support of Ω, or equivalently, that P' is contained in the support of $(\Gamma-T)|_T$. Then there is a component G of the support of Γ such that

$$\mathrm{mult}_{P'}(\mathrm{Fix}|l(K_Y+\Gamma)|_T) = \mathrm{mult}_G(\mathrm{Fix}|l(K_Y+\Gamma)|)$$

and $\mathrm{mult}_{P'}(\Gamma_T)=\mathrm{mult}_G(\Gamma)$. It follows that

$$\mathrm{mult}_{P'}(\Xi_T) = \mathrm{mult}_G(\Xi) = \mathrm{mult}_G(\Gamma) - \min\{\mathrm{mult}_G(\Gamma), \ \frac{1}{l}\mathrm{mult}_G(\mathrm{Fix}|l(K_Y+\Gamma)|)\}.$$

Since $\mathrm{mult}_P(\Omega+\frac{1}{k}A|_S) = \mathrm{mult}_{P'}(\Gamma_T)$ and $\Omega\wedge(1-\frac{\epsilon}{k})F \leq \Phi$, it suffices to show that

$$\mathrm{mult}_{P'}(\mathrm{Fix}\,|l(K_Y+\Gamma)|_T) \leq l(1-\frac{\epsilon}{k})\mathrm{mult}_P(F).$$

Since $E|_T$ is exceptional, we have that

$$\mathrm{mult}_{P'}(\mathrm{Fix}\,|l(K_Y+\Gamma)|_T) = \mathrm{mult}_P(\mathrm{Fix}\,|l(K_X+\Delta+\frac{1}{k}A)|_S)$$

cf. (2.14). Let $\eta > \epsilon/k$ be a rational number such that $\eta(K_X+\Delta)+\frac{1}{k}A$ is ample. Since l is sufficiently big and divisible and

$$l(K_X+\Delta+\frac{1}{k}A) \cong l(1-\eta)(K_X+\Delta) + l(\eta(K_X+\Delta)+\frac{1}{k}A),$$

we have that

$$\mathrm{mult}_P(\mathrm{Fix}\,|l(K_X+\Delta+\frac{1}{k}A)|_S) \leq l(1-\frac{\epsilon}{k})\mathrm{mult}_P(F)$$

where we have used the fact that if $\mathrm{mult}_P(F)=0$, then $\mathrm{mult}_P(\mathrm{Fix}\,|l(K_X+\Delta+\frac{1}{k}A)|_S)=0$ for all $l \gg 0$ cf. (2.7). $\square$

Exercise 2.1. Notation as in (2.4). Show that if $c < 1$ and $D \in V$ is general, then $\mathcal{J}_{\Delta, c\cdot V} = \mathcal{J}_{\Delta, c\cdot D}$.

Exercise 2.2. Notation as in (2.4). Show that $\mathcal{J}_{\Delta, D} = \mathcal{O}_X$ if and only if $(X, D+\Delta)$ is divisorially log terminal and $\llcorner D \lrcorner = 0$.

Exercise 2.3. Notation as in (2.4). Show that $\mathcal{J}_{\Delta, c\cdot D} = \mathcal{O}_X$ if $c \ll 1$ (in fact $\mathcal{J}_{c\cdot D} = \mathcal{O}_X$ if $\operatorname{mult}_x(c \cdot D) < 1$ for all $x \in X$ cf. [**Lazarsfeld04**, 9.5.13]) and that $\mathcal{J}_{\Delta, D} \subset m_x$ if $\operatorname{mult}_x(D) \geq \dim X$.

Exercise 2.4. Let $\pi\colon X \to Z$ be a projective morphism from a smooth quasi-projective variety to an affine variety. Let $D \geq 0$ be a divisor on X. Show that $H^0(\mathcal{O}_X(D)) = H^0(\mathcal{J}_{||D||}(D))$.

Exercise 2.5. Let $D \geq 0$ be a Cartier divisor on a smooth variety X and $Z \subset X$ a closed subvariety such that $\lim \operatorname{mult}_Z(|m!D|)/m! = 0$. Show that $Z \not\subset \mathbf{B}_-(D)$.

Exercise 2.6. Use the arguments of [**Lazarsfeld04**, 9.2.19] to give an alternative proof of the claim in (2.3).

Exercise 2.7. Show that $\mathcal{J}_{\Delta, D} = \mathcal{J}_{\Delta, \{D\}} \otimes \mathcal{O}_X(-\llcorner D \lrcorner)$.

Exercise 2.8. Let $f\colon X \to Y$ be a birational morphism of normal varieties. If Y is $\mathbb{Q}$-factorial, show that there exists an effective f-exceptional divisor F such that $-F$ is ample over Y.

Exercise 2.9. The Kawamata-Viehweg Vanishing Theorem states the following: *If $f\colon Y \to X$ is a proper morphism of quasi-projective varieties with Y smooth, L is a line bundle on Y which is numerically equivalent to $M + \Delta$ where M is an f-nef and f-big $\mathbb{R}$-Cartier divisor and (X, Δ) is kawamata log terminal. Then*

$$R^j f_* \mathcal{O}_Y(K_Y + L) = 0 \qquad \text{for } j > 0.$$

Show that if $f\colon Y \to X$ is a projective morphism of quasi-projective varieties with Y smooth, (Y, Γ) is a divisorially log terminal pair and L a Cartier divisor such that $L - (K_Y + \Gamma)$ is f-ample, then

$$R^j f_* \mathcal{O}_Y(L) = 0 \qquad \text{for } j > 0.$$

Exercise 2.10. Let $f\colon Y \to X$ and $\pi\colon X \to Z$ be projective morphisms of normal quasi-projective varieties. Show that if F is a coherent sheaf on Y such that $R^j f_* F = 0$ for all $j > 0$, then $R^i(\pi \circ f)_* F \cong R^i \pi_*(f_* F)$ for all $i > 0$.

Exercise 2.11. Let $\pi\colon X \to Z$ be a projective morphism from a smooth variety to an affine variety. Let F be a coherent sheaf on X and H be a very ample divisor on X such that $H^i(X, F(mH)) = 0$ for all $i > 0$ and all $m \geq -\dim X$. Show that F is generated by global sections.

Exercise 2.12. Show that if $\Delta = \sum \delta_j \Delta_j$, then the divisors Δ^i in (2.11) are given by $\Delta^i = \sum_{j|\delta_j > (k-s)/k} \Delta_j$.

Exercise 2.13. Let $f\colon Y \to X$ be a proper birational morphism of normal varieties. Show that if $\Xi \geq 0$ is a $\mathbb{Q}$-divisor on Y such that $(X, f_*\Xi)$ is canonical and $m > 0$ is an integer such that $m(K_X + f_*\Xi)$ and $m(K_Y + \Xi)$ are Cartier, then $|m(K_X + f_*\Xi)| = f_*|m(K_Y + \Xi)|$.

Exercise 2.14. Let $f\colon Y \to X$ be a proper birational morphism of normal varieties, $S \subset X$ a divisor and $T = (f^{-1})_*S$. Let D be a Cartier divisor on X. Show that $|f^*D|_T = (f|_T)^*|D|_S$. Assume that T and S are normal and that P is a prime divisor on S and P' is its strict transform on T. Show that $\mathrm{mult}_{P'}(\mathrm{Fix}\,|f^*D|_T) = \mathrm{mult}_P(\mathrm{Fix}\,|D|_S)$.

LECTURE 3

Finite generation of the restricted algebra

Rationality of the restricted algebra

Theorem 3.1. *Assume Theorem 1.9_{n-1}.*

Let $\pi\colon X \to Z$ be projective morphism from a smooth quasi-projective variety of dimension n to a normal affine variety. Let $\Delta = S + A + B$ be a $\mathbb{Q}$-divisor such that $S = \llcorner \Delta \lrcorner$ is irreducible and smooth, $A \geq 0$ is a general ample $\mathbb{Q}$-divisor, $B \geq 0$, (X, Δ) is purely log terminal, $(S, \Omega + A|_S)$ is canonical where $\Omega = (A + B)|_S$, and $\mathbf{B}(K_X + \Delta)$ does not contain S. For any sufficiently divisible integer $m > 0$ let

$$F_m = \mathrm{Fix}(|m(K_X + \Delta)|_S)/m \qquad \text{and} \qquad F = \lim F_{m!}.$$

Then $\Theta = \Omega - \Omega \wedge F$ is rational and if $k\Delta$ and $k\Theta$ are Cartier, then

$$R_S(k(K_X + \Delta)) \cong R(k(K_S + \Theta)).$$

PROOF. Let $V \subset \mathrm{WDiv}_{\mathbb{R}}(S)$ be the sub-vector space generated by the components of Ω and $W \subset V$ be the smallest rational affine subspace containing Θ. By (1.9), there exists a constant $\delta > 0$ such that if $\Theta' \in V$ with $\|\Theta' - \Theta\| \leq \delta$, then any prime divisor contained in $\mathbf{B}(K_S + \Theta)$ is also contained in $\mathbf{B}(K_S + \Theta')$ and there exist a constant $\eta > 0$ and an integer $r > 0$ such that if $\Theta' \in W$ with $\|\Theta' - \Theta\| \leq \eta$ and $k > 0$ is an integer such that $k(K_S + \Theta')/r$ is Cartier, then every component of $\mathrm{Fix}(k(K_S + \Theta'))$ is contained in $\mathbf{B}(K_S + \Theta)$.

Note that if $l(K_X + \Delta)$ is Cartier and $\Theta_l = \Omega - \Omega \wedge F_l$, then $l(K_S + \Theta_l)$ is Cartier and

$$|l(K_X + \Delta)|_S \subset |l(K_S + \Theta_l)| + l(\Omega \wedge F_l).$$

It follows that no component of $\mathrm{Supp}(\Theta_l)$ is contained in $\mathrm{Fix}(l(K_S+\Theta_l))$ and hence in $\mathbf{B}(K_S+\Theta_l)$. For $l \gg 0$, we have that $\Theta_l \in V$, $\|\Theta_l - \Theta\| \leq \delta$ so that no component of $\mathrm{Supp}(\Theta_l)$ is contained in $\mathbf{B}(K_S + \Theta)$. Since $\mathrm{Supp}(\Theta) \subset \mathrm{Supp}(\Theta_l)$, no component of Θ is contained in $\mathbf{B}(K_S + \Theta)$.

Let $0 < \epsilon \ll 1$ be a rational number such that $\epsilon(K_X + \Delta) + A$ is ample. If Θ is not rational, then there exist a component P of Θ with $\mathrm{mult}_P(\Theta) \notin \mathbb{Q}$. Recall that $\Theta = \Omega - \Omega \wedge F$. Notice that if Q is a prime divisor on S such that $\mathrm{mult}_Q(\Omega \wedge F) \notin \mathbb{Q}$, then

$$\mathrm{mult}_Q(\Omega \wedge F) = \mathrm{mult}_Q(F)$$

and if $\mathrm{mult}_Q(\Omega \wedge F) \in \mathbb{Q}$, then $\mathrm{mult}_Q(\Phi) = \mathrm{mult}_Q(\Theta)$ for any $\mathbb{R}$-divisor $\Phi \in W$.

By Diophantine approximation, there exist an integer $k > 0$, an effective $\mathbb{Q}$-divisor $\Phi \in W$ such that $k\Phi/r$ and $k\Delta/r$ are Cartier, $\mathrm{mult}_P(\Phi) < \mathrm{mult}_P(\Omega \wedge F)$ and $\|\Phi - (\Omega \wedge F)\| \leq \min\{\eta, \gamma, \epsilon f/k\}$ where f is the smallest non-zero coefficient of F and γ is the smallest non-zero coefficient of $\Omega - \Omega \wedge F$. Then

$$\Omega \wedge (1 - \frac{\epsilon}{k})F \leq \Phi \leq \Omega.$$

By (2.12), we have that

$$(\flat) \qquad |k(K_S+\Omega-\Phi)|+k\Phi \subset |k(K_X+\Delta)|_S.$$

Since

$$\mathrm{mult}_P(\mathrm{Fix}(|k(K_X+\Delta)|_S)) \geq \mathrm{mult}_P(kF) = \mathrm{mult}_P(k(\Omega\wedge F)) > \mathrm{mult}_P(k\Phi),$$

it follows from ($\flat$) that P is contained in the support of $\mathrm{Fix}(k(K_S+\Omega-\Phi))$. Since $||(\Omega-\Phi)-\Theta|| \leq \eta$ and $k(K_S+\Omega-\Phi)/r$ is Cartier, P is contained in $\mathbf{B}(K_S+\Theta)$. This is a contradiction and therefore Θ is rational. The isomorphism $R_S(k(K_X+\Delta)) \cong R(k(K_S+\Theta))$ now follows from (2.12). □

Proof of (1.10)

PROOF OF (1.10). We may assume that Z is affine cf. (1.2). By (1.6), it suffices to show that the restricted algebra $R_S(K_X+\Delta)$ is finitely generated. By (1.7), this is equivalent to showing that the restricted algebra $R_S(k(K_X+\Delta))$ is finitely generated for some integer $k>0$. It follows that we may replace Δ by a $\mathbb{Q}$-linearly equivalent divisor Δ'. By (1.3), S is mobile. Since f is birational and Z is affine, $\Delta-S$ is big so that $\Delta-S\sim_{\mathbb{Q}} A'+B'$ where A' is ample, $B'\geq 0$ and $\mathrm{Supp}(B')$ does not contain S. Let $\epsilon>0$ be a sufficiently small rational number. We may replace Δ by

$$\Delta' = S+A+B \sim_{\mathbb{Q}} \Delta$$

where $A\sim_{\mathbb{Q}} \epsilon A'$ is a general ample $\mathbb{Q}$-divisor and $B=(1-\epsilon)(\Delta-S)+\epsilon B'$. Let $f\colon Y\to X$ be a log resolution of (X,Δ') and write

$$K_Y+\Gamma' = f^*(K_X+\Delta')+E'$$

as in $(*)$ of Notation. Let $T=(f^{-1})_*S$. We may assume that (Y,Γ') is purely log terminal and that $(T,(\Gamma'-T)|_T)$ is terminal. Let $F\geq 0$ be an effective exceptional $\mathbb{Q}$-divisor such that f^*A-F is ample, $(Y,\Gamma'+F)$ is purely log terminal and $(T,(\Gamma'-T+F)|_T)$ is terminal. We set $\Gamma=\Gamma'-f^*A+F+H$ where $H\sim_{\mathbb{Q}} f^*A-F$ is a general ample $\mathbb{Q}$-divisor. Then (Y,Γ) is purely log terminal and we may assume that $(T,(\Gamma-T+H)|_T)$ is terminal. Since $\mathbf{B}(K_X+\Delta)$ does not contain S, one sees that $\mathbf{B}(K_Y+\Gamma)$ does not contain T. By (3.1), it then follows that there is a $\mathbb{Q}$-divisor Θ on T such that $R_T(k(K_Y+\Gamma))\cong R(k(K_T+\Theta))$ for some $k>0$ sufficiently divisible. The Theorem now follows since for $k>0$ sufficiently divisible,

$$R_T(k(K_Y+\Gamma)) \cong R_S(k(K_X+\Delta)).$$

□

LECTURE 4

The minimal model program with scaling

In this section we will briefly explain the results of [**BCHM06**] which in particular show that the existence of pl-flips in dimension n implies Theorem 1.9 in dimension n. Therefore, by (1.10), pl-flips exist in all dimensions. We remark that:

(1) Theorem 1.9 is a consequence of the existence of log terminal models (for kawamata log terminal pairs (X, Δ) with Δ big) cf. (4.6).
(2) By a technique of V. Shokurov known as special termination, it is known that the existence of pl-flips in dimension n and the minimal model program in dimension $n-1$ implies the existence of flips in dimension n.

In what follows, we will therefore focus on explaining the main ideas behind showing that the existence of flips in dimension n (and the minimal model program in dimension $n-1$) implies the existence of log terminal models in dimension n.

The main idea is to use the **minimal model program with scaling**. We start with a $\mathbb{Q}$-factorial kawamata log terminal pair $(X, \Delta + H)$ and a projective morphism $\pi : X \to U$ such that $K_X + \Delta + H$ is nef over U and Δ is big over U. We let

$$\lambda = \inf\{t \geq 0 | K_X + \Delta + tH \text{ is nef over } U\}.$$

If $\lambda = 0$, then $K_X + \Delta$ is nef over U and we are done. If $\lambda > 0$, then there exists a $K_X + \Delta$ negative extremal ray R over U such that $(K_X + \Delta + \lambda H) \cdot R = 0$. We consider the corresponding contraction (over U) $\text{cont}_R : X \to Z$. If $\dim Z < \dim X$, then we have a $K_X + \Delta$ Mori fiber space and we are done. Otherwise (assuming the existence of the corresponding flips), we replace (X, Δ) by the corresponding flip or divisorial contraction $\phi : X \dashrightarrow X'$. Let $H' = \phi_* H$ and $\Delta' = \phi_* \Delta$. Since $K_X + \Delta + \lambda H$ is nef over U and $(K_X + \Delta + \lambda H) \cdot R = 0$, it follows that $K_{X'} + \Delta' + \lambda H'$ is nef over U and so we may repeat the process.

As in the usual minimal model program, this process terminates (yielding a log terminal model or a Mori fiber space) unless we get a sequence of flips $X_i \dashrightarrow X_{i+1}$. Let Δ_i and H_i denote the strict transforms of Δ and H on X_i. Then there is a sequence of real numbers $\lambda = \lambda_1 \geq \lambda_2 \geq \lambda_3 \geq \cdots > 0$ such that $K_{X_n} + \Delta_n + \lambda_n H_n$ is nef over U. In particular, $X \dashrightarrow X_n$ is a log terminal model for $(X, \Delta + \lambda_n H)$ over U. Notice that we have:

Lemma 4.1. *$X_n \dashrightarrow X_m$ is not an isomorphism for any $m > n$.*

PROOF. Let E be a divisor over X whose center is contained in $\text{Ex}(\text{cont}_{R_n})$, where $\text{cont}_{R_n} : X_n \to Z_n$ is the corresponding extremal contraction over U. Then $a_E(X_n, \Delta_n) < a_E(X_{n+1}, \Delta_{n+1})$ and $a_E(X_{n+i}, \Delta_{n+i}) \leq a_E(X_{n+i+1}, \Delta_{n+i+1})$ for all $i > 0$ cf. (1.4) and (1.5). It follows that $a_E(X_n, \Delta_n) < a_E(X_m, \Delta_m)$ an so $X_n \dashrightarrow X_m$ is not an isomorphism. □

Therefore, to show that the above sequence of flips $X_i \dashrightarrow X_{i+1}$ terminates, it suffices to show that: *For* $0 \leq \lambda \leq 1$, *the set of log terminal models for* $(X, \Delta + \lambda H)$ *over* U *is finite.* We will prove this fact (in dimension n) **assuming** the existence of log terminal models (in dimension n). I.e. assuming the following:

Theorem 4.2. *Let* (X, Δ) *be a* $\mathbb{Q}$*-factorial kawamata log terminal pair of dimension* n *and* $\pi : X \to U$ *a projective morphism such that* $K_X + \Delta$ *is pseudo-effective over* U *and* Δ *is big over* U. *Then there exists a log terminal model* $\phi : X \dashrightarrow Y$ *for* (X, Δ) *over* U.

we will show that

Theorem 4.3. *Let* X *be a* $\mathbb{Q}$*-factorial normal quasi-projective variety of dimension* n *and* $\pi : X \to U$ *a projective morphism. Let* A *be a general ample* $\mathbb{Q}$*-divisor over* U *and* $\mathcal{C} \subset \mathrm{WDiv}_{\mathbb{R}}(X)$ *be a rational polytope such that for any* $K_X + \Delta \in \mathcal{C}$, $\Delta = A + B$ *where* $B \geq 0$ *and* (X, Δ) *is kawamata log terminal.*

Then there exist finitely many $\phi_i : X \dashrightarrow Y_i$, $1 \leq i \leq k$, *birational maps over* U *such that if* $K_X + \Delta \in \mathcal{C}$ *and* $K_X + \Delta$ *is pseudo-effective over* U *then*

(1) *There exists an index* $1 \leq j \leq k$ *such that* ϕ_j *is a log terminal model of* (X, Δ) *over* U.
(2) *If* $\phi : X \dashrightarrow Y$ *is a log terminal model of* (X, Δ) *over* U, *then there exists an index* $1 \leq j \leq k$ *such that the rational map* $\phi_j \circ \phi^{-1} : Y \dashrightarrow Y_j$ *is an isomorphism.*

We will now prove the following:

Theorem 4.4. *Theorem 4.2 in dimension* n *implies Theorem 4.3 (1) in dimension* n.

PROOF. We proceed by induction on the dimension of $\mathcal{C}$. The case when $\dim \mathcal{C} = 0$ follows immediately from $(4.2)_n$.

Notice that $\mathrm{PSEF}(X/U)$ (the cone of pseudo-effective divisors over U) is closed. By compactness of $\mathcal{C} \cap \mathrm{PSEF}(X/U)$, it suffices to work locally around an $\mathbb{R}$-divisor $K_X + \Delta_0 \in \mathcal{C} \cap \mathrm{PSEF}(X/U)$. We are therefore free to replace $\mathcal{C}$ by an appropriate neighborhood of $K_X + \Delta_0$ inside $\mathcal{C} \cap \mathrm{PSEF}(X/U)$.

Step 1. *We may assume that* $K_X + \Delta_0$ *is nef over* U.

By $(4.2)_n$, there exists a log terminal model $\phi : X \dashrightarrow Y$ for $K_X + \Delta_0$ over U. In particular $K_Y + \phi_* \Delta$ is pseudo-effective over U. After shrinking $\mathcal{C}$, we have for all $K_X + \Delta \in \mathcal{C}$ that:

(1) $K_Y + \phi_* \Delta$ is kawamata log terminal cf. (4.1), and
(2) $a_F(X, \Delta) < a_F(Y, \phi_* \Delta)$ for any ϕ-exceptional divisor $F \subset X$.

It follows that if $\psi : Y \to Z$ is a log terminal model for $(Y, \phi_* \Delta)$ over U, then $\psi \circ \phi : X \to Z$ is a log terminal model for (X, Δ) over U cf. (4.2). We may therefore assume that $K_X + \Delta_0$ is nef over U.

Step 2. *We may assume that* $K_X + \Delta_0 \sim_{\mathbb{R}, U} 0$.

Since $K_X + \Delta_0$ is nef over U and Δ_0 is big over U, by the Base Point Free Theorem (cf. (1.7)), $K_X + \Delta_0$ is semiample over U and so there is a morphism $\psi : X \to Z$ over U and an $\mathbb{R}$-divisor H on Z ample over U such that $K_X + \Delta_0 \sim_{\mathbb{R}, U} \psi^* H$.

After further shrinking $\mathcal{C}$, we may assume that any log terminal model for (X, Δ) over Z is a log terminal model for (X, Δ) over U cf. (4.3). We may hence replace U by Z so that we may assume that $K_X + \Delta_0 \sim_{\mathbb{R}, U} 0$.

Step 3. Pick any $K_X+\Theta \in \mathcal{C}$ and let $K_X+\Delta$ be the point on the boundary of $\mathcal{C}$ such that

$$\Theta - \Delta_0 = \lambda(\Delta - \Delta_0) \qquad 0<\lambda<1.$$

Then we have that

$$K_X+\Theta = \lambda(K_X+\Delta)+(1-\lambda)(K_X+\Delta_0) \sim_{\mathbb{R},U} \lambda(K_X+\Delta).$$

It follows that $K_X+\Delta$ is pseudo-effective over U if and only if $K_X+\Theta$ is pseudo-effective over U and $\phi : X \dashrightarrow Y$ is a log terminal model for (X,Δ) over U if and only if it is a log terminal model for (X,Θ) over U. Since the boundary of $\mathcal{C}$ is a rational polytope of strictly smaller dimension, the induction is now complete. □

Remark 4.5. To gain some intuition for the fact that Theorem 4.4 (2) follows from Theorem 4.4 (1), one first notices that if $\phi : X \dashrightarrow Y$ is any log terminal model of (X,Δ) over U and $H \in \mathrm{WDiv}_{\mathbb{R}}(Y)$ is ample over U, then ϕ is a log terminal model over U for any kawamata log terminal pair (X,Δ') such that $\Delta' \equiv_U \Delta + \epsilon\phi_*^{-1}H$ and $0<\epsilon \ll 1$. Since $K_Y+\phi_*\Delta'$ is ample over U, ϕ is in fact the unique log terminal model over U for (X,Δ).

It suffices therefore to show that there exists a rational polytope $\mathcal{C} \subset \mathcal{C}'$ contained in a finite dimensional subspace of $\mathrm{WDiv}_{\mathbb{R}}(X)$ such that if $K_X+\Delta \in \mathcal{C}$ and H' is an $\mathbb{R}$-Cartier divisor on X, then there exists a kawamata log terminal pair $K_X+\Delta' \in \mathcal{C}'$ and a constant $\epsilon>0$ such that $\Delta' \equiv_U \Delta+\epsilon H'$. This can be achieved as follows: First, we translate $\mathcal{C}$ by $A'+B'-A$ where A' is a general ample $\mathbb{Q}$-divisor over U, $B' \geq 0$ is a $\mathbb{Q}$-divisor such that $A \sim_{\mathbb{R},U} A'+B'$, the components of $\mathrm{Supp}(B')$ generate $\mathrm{WDiv}_{\mathbb{R}}(X)/\equiv_U$, and for any $K_X+\Delta \in \mathcal{C}$, $\Delta - A + A' + B'$ is effective and $(X, \Delta - A + A' + B')$ is kawamata log terminal. We then let

$$\mathcal{C}' = \{K_X+\Delta' = K_X+\Delta+\epsilon C | K_X+\Delta \in \mathcal{C},\ \mathrm{Supp}(C) \subset \mathrm{Supp}(B') \text{ and } ||C|| < \epsilon\}$$

where $0<\epsilon \ll 1$.

Corollary 4.6. *Theorems 4.2 and 4.3 imply Theorem 1.9.*

PROOF. (1) is immediate from (4.2) and (1.7). We now prove (2). Notice that if $K_X+\Delta'$ is not pseudo-effective, then $\mathbf{B}(K_X+\Delta') = X$. Therefore, we may assume that $K_X+\Delta'$ is pseudo-effective. Let $\phi : X \dashrightarrow Y$ be a log terminal model for $K_X+\Delta$ over Z. We may pick $0<\delta$ such that if $||\Delta-\Delta'|| \leq \delta$, then $K_Y+\phi_*\Delta'$ is kawamata log terminal cf. (4.1) and $a_F(X,\Delta') < a_F(Y,\phi_*\Delta')$ for any ϕ-exceptional divisor $F \subset X$. By (4.2), if $\psi : Y \to Z$ is a log terminal model for $K_Y+\phi_*\Delta'$ over Z, then $\psi \circ \phi$ is a log terminal model for $K_X+\Delta'$ over Z. By (1.8), all ϕ-exceptional divisors are contained in $\mathbf{B}(K_X+\Delta')$.

We now prove (3). Let $\phi : X \dashrightarrow Y$ be a log terminal model for $K_X+\Delta$ over Z and let $\mathcal{C}$ be the set of all $\Delta' \in W$ such that $||\Delta-\Delta'|| \leq \eta$ and ϕ is a log terminal model for $K_X+\Delta'$ over Z. Since Δ is big, we may assume that there is $\mathbb{Q}$-divisor A on Y ample over Z such that if $||\Delta-\Delta'|| \leq \eta$, then $\phi_*\Delta' \geq A$. If η is sufficiently small, then $\phi_*(\Delta'-\Delta)+A$ is ample over Z for any $\Delta' \in W$ such that $||\Delta-\Delta'|| \leq \eta$ and moreover

$$\mathcal{C} = \{\Delta' \in W | K_Y+\phi_*\Delta' \text{ is nef over } Z, \text{ and } ||\Delta-\Delta'|| \leq \eta\}.$$

By the Cone Theorem cf. (0.2), there are finitely many negative extremal rays $R_1,\dots,R_k$ for $K_Y+\phi_*\Delta - A$. Since

$$K_Y+\phi_*\Delta' = (K_Y+\phi_*\Delta - A)+\phi_*(\Delta'-\Delta)+A,$$

if $K_Y + \phi_*\Delta'$ is not nef over Z, then there is an integer $1 \leq i \leq k$ such that R_i is a negative extremal ray for $K_Y + \phi_*\Delta'$ over Z. Therefore $\mathcal{C}$ is cut out by a subset of the finite collection of extremal rays $R_i = \mathbb{R}^+[C_i]$ and hence $\mathcal{C}$ is a rational polytope. Therefore the affine subspace of $\mathrm{WDiv}_{\mathbb{R}}(X)$ generated by $\mathcal{C}$ is W. We may then assume that $\mathcal{C} = \{\Delta' \in W : ||\Delta - \Delta'|| \leq \eta\}$.

Since Y is $\mathbb{Q}$-factorial, there is an integer $l > 0$ such that if $G \in \mathrm{WDiv}(Y)$, then lG is Cartier. By [**Kollár93**], there exists an integer $r > 0$ such that if $k(K_Y + \phi_*\Delta')/r$ is integral, then $k(K_Y + \phi_*\Delta')$ is generated by global sections. Let $p : W \to X$ and $q : W \to Y$ be a common resolution, then by definition of log terminal model, $p^*(K_X + \Delta') = q^*(K_Y + \phi_*\Delta') + E$ where $E \geq 0$ is q-exceptional and the support of $p_*(E)$ is the union of all ϕ-exceptional divisors on X. It follows that the support of $\mathrm{Fix}(k(K_X + \Delta'))$ is given by the divisors contained in $\mathbf{B}(K_X + \Delta')$ i.e. by the divisors contained the support of $p_*(E)$. Since ϕ is a log terminal model for $K_X + \Delta$ over Z, these divisors are also contained in $\mathbf{B}(K_X + \Delta)$.

□

Exercise 4.1. Let (X, Δ_0) be a $\mathbb{Q}$-factorial kawamata log terminal pair and $\Delta_0 \in V \subset \mathrm{WDiv}_{\mathbb{R}}(X)$ a finite dimensional vector space. Show that there exists an $\epsilon > 0$ such that $0 \leq \Delta \in V$ and $||\Delta|| \leq \epsilon$, then (X, Δ) is kawamata log terminal.

Exercise 4.2. Let (X, Δ) be a $\mathbb{Q}$-factorial kawamata log terminal pair $\pi : X \to U$ a projective morphism, $\phi : X \dashrightarrow Y$ be a rational map over U that extracts no divisors such that Y is normal, $\mathbb{Q}$-factorial and projective over U. If $a_F(X, \Delta) < a_F(Y, \phi_*\Delta)$ for any ϕ-exceptional divisor $F \subset X$ and if $\psi : Y \dashrightarrow Z$ is a log terminal model of $(Y, \phi_*\Delta)$ over U, then $\psi \circ \phi$ is a log terminal model of (X, Δ) over U.

Exercise 4.3. Let $\psi : X \to Z$ and $\eta : Z \to U$ be projective morphisms of normal varieties. If H is an ample $\mathbb{R}$-divisor on Z and (X, Δ) is a $\mathbb{Q}$-factorial kawamata log terminal pair such that Δ is big over U and $K_X + \Delta \sim_{\mathbb{R},U} \psi^* H$, show that there exists $0 < \epsilon \ll 1$ such that if $||\Delta - \Delta_0|| \leq \epsilon$, then any log terminal model for (X, Δ) over Z is a log terminal model for (X, Δ) over U.

Solutions to the exercises

Hint to (0.1). The inclusion $\mathbf{B}(D) \subset \mathrm{Bs}(mD)$ is clear. If $x \notin \mathbf{B}(D)$, then there is an $\mathbb{R}$-divisor $D' \geq 0$ such that $D \sim_{\mathbb{R}} D'$ and $x \notin \mathrm{Supp}(D')$. We have $D' - D = \sum r_i(f_i)$ where $r_i \in \mathbb{R}$ and f_i are rational functions on X. We may assume that D and D' have no common components and that $\mathrm{Supp}(f_i) \subset \mathrm{Supp}(D + D')$ (eg. by assuming that r_i are linearly independent over $\mathbb{Q}$). Then if q_i are rational numbers with $|q_i - r_i| \ll 1$, we have that $D'' = D + \sum q_i(f_i)$ is a $\mathbb{Q}$-divisor $\mathbb{Q}$-linearly equivalent to D with the same support as D'. □

Hint to (0.2). Blow up smooth divisors contained in F and its strict transforms. □

Solution to (0.3). [**KM98**, 5.48, 5.50]. □

Solution to (0.4). [**KM98**, 2.43]. □

Solution to (1.1). If $K_Z + f_*\Delta$ is $\mathbb{Q}$-Cartier, then as f is small, $K_X + \Delta = f^*(K_Z + f_*\Delta)$. This contradicts the fact that $-(K_X + \Delta)$ is ample over Z. □

Solution to (1.2). Let $\Delta = S + \sum \delta_i \Delta_i$ and set $\Delta' = S + \sum \delta_i' \Delta_i$ where $0 < \delta_i' < \delta_i$ are rational numbers such that $|\delta_i' - \delta_i| \ll 1$. It is easy to check that f is a pl-flipping contraction for (X, Δ'). □

Solution to (1.3). Since Z is affine, the sheaf $f_*\mathcal{O}_X(S)$ is generated by global sections. Let $S' \sim S$ be a divisor corresponding to a general section of $\Gamma(Z, f_*\mathcal{O}_X(S)) \cong \Gamma(X, \mathcal{O}_X(S))$. Since f is an isomorphism in codimension 1 and $\mathcal{O}_X(S)$ is invertible in codimension 1, S and S' have no common components. □

Solution to (1.4). See [**KM98**, 3.37, 3.38, 3.42]. □

Solution to (1.5). See [**KM98**, 3.36, 3.43]. □

Solution to (1.6). By (1.4) and (1.5), the X_i are $\mathbb{Q}$-factorial. It suffices to prove that $R(K_{X_i} + \Delta_i) \cong R(K_{X_{i+1}} + \Delta_{i+1})$. Let $p : W \to X_i$ and $q : W \to X_{i+1}$ be a common resolution that resolves the indeterminacies of ϕ_i. By (1.4) and (1.5), we have that

$$p^*(K_{X_i} + \Delta_i) = q^*(K_{X_{i+1}} + \Delta_{i+1}) + F$$

where $F \geq 0$ is a q-exceptional $\mathbb{Q}$-divisor. □

Solution to (1.8). Let $\phi \colon X \dashrightarrow Y$ be a log terminal model for (X, Δ) over Z. Then $(Y, \phi_*\Delta)$ is kawamata log terminal and Δ is big. By the Base Point Free Theorem, it follows that $K_Y + \phi_*\Delta$ is semiample over Z. Let $p : W \to X$ and $q : W \to Y$ be a common resolution that resolves the indeterminacies of ϕ. Then

$$p^*(K_X + \Delta) = q^*(K_Y + \phi_*\Delta) + F$$

where $F \geq 0$ is a q-exceptional $\mathbb{Q}$-divisor whose support contains the divisors contracted by ϕ. The assertion now follows easily. □

SOLUTION TO (1.9). Let $\phi\colon X \dashrightarrow Y$ be a log terminal model for $K_X + \Delta$ over Z, then $(Y, \phi_*\Delta)$ is kawamata log terminal and $\phi_*\Delta$ is big over Z therefore the same is true for $(Y, \phi_*\Delta')$. Therefore, there is a log terminal model for $K_Y + \phi_*\Delta'$ over Z, say $\psi\colon Y \dashrightarrow W$. It is easy to see that $\psi \circ \phi$ is a log terminal model for $K_X + \Delta'$ over Z. The claim now follows from (1.8). □

SOLUTION TO (2.1). Let $f\colon Y \to X$ be a log resolution of (X, Δ) and V, then $M = f^*D - \mathrm{Fix}(f^*V)$ is a reduced irreducible smooth divisor intersecting $F = \mathrm{Fix}(f^*V)$ transversely. Since $c < 1$, we have

$$\mathcal{J}_{\Delta, c\cdot V} = f_*\mathcal{O}_Y(E - \llcorner c \cdot F \lrcorner) = f_*\mathcal{O}_Y(E - \llcorner c \cdot (M + F)\lrcorner) = \mathcal{J}_{\Delta, c\cdot D}.$$

□

SOLUTION TO (2.2). Let $f\colon Y \to X$ be a log resolution of $(X, \Delta + D)$ that is an isomorphism at the general point of each log canonical center of (X, Δ). If $\mathcal{J}_{\Delta, D} = \mathcal{O}_X$, then $E - \llcorner f^*D \lrcorner \geq 0$ and in particular $\llcorner D \lrcorner = 0$. For any exceptional prime divisor F in Y, we have $a_F(X, \Delta + D) = \mathrm{mult}_F(E - f^*D) > -1$ so that $(X, \Delta + D)$ is divisorially log terminal. The reverse implication is similar. □

SOLUTION TO (2.4). Let $p > 0$ be an integer such that $\mathcal{J}_{||D||} = \mathcal{J}_{\frac{1}{p}\cdot|pD|}$. Let $f : Y \to X$ be a log resolution of $|pD|$. Then $\mathcal{J}_{||D||} = f_*\mathcal{O}_Y(K_{Y/X} - \llcorner \frac{1}{p} F_p \lrcorner)$ where $F_p = \mathrm{Fix}(|pf^*D|)$. If $F_1 = \mathrm{Fix}(|f^*D|)$, then $pF_1 \geq F_p$ and so

$$H^0(\mathcal{O}_X(D)) = H^0(\mathcal{O}_Y(f^*D - F_1)) \subset H^0(\mathcal{O}_Y(f^*D + K_{Y/X} - \llcorner \frac{1}{p} F_p \lrcorner))$$
$$= H^0(\mathcal{J}_{||D||}(D)) \subset H^0(\mathcal{O}_X(D)).$$

□

SOLUTION TO (2.3). Let $f\colon Y \to X$ be a log resolution of $(X, \Delta + D)$. Since $E \geq 0$, we have that $E - \llcorner c \cdot f^*D \lrcorner$ is effective and f-exceptional for $0 < c \ll 1$. Therefore $\mathcal{J}_{\Delta, c\cdot D} = \mathcal{O}_X$.

If $\mathrm{mult}_x(D) > 1$, let $\mu\colon X' \to X$ be the blow up of $x \in X$ with exceptional divisor F. Then $\mathrm{mult}_F(K_{X'} + (\mu^{-1})_*D - \mu^*(K_X + D)) \geq 1$. It follows that for any log resolution $f\colon Y \to X$ dominating μ, we have that $E - \llcorner f^*D \lrcorner$ is not effective and in fact contained in $\mathcal{O}_Y(-F')$ where F' is the strict transform of F. Therefore $\mathcal{J}_{\Delta, D} \subset f_*\mathcal{O}_Y(-F') = m_x$. □

SOLUTION TO (2.5). For any $m > 0$, there exists a number $k > 0$ such that $\mathrm{mult}_Z(|mkD|)/k < 1$. By [**Lazarsfeld04**, 9.5.13], we have that $\mathcal{J}_{\frac{1}{k}\cdot|mkD|} = \mathcal{O}_X$ on a neighborhood of a general point of Z. Let H be an ample $\mathbb{Q}$-divisor on X. For any $m > 0$ sufficiently big and divisible, $\mathcal{O}_X(m(D + H)) \otimes \mathcal{J}_{\frac{1}{k}\cdot|mkD|}$ is globally generated cf. (3) of (2.10). □

SOLUTION TO (2.7). Let $f\colon Y \to X$ be a log resolution of $(X, \Delta + D)$. By the projection formula, we have

$$\mathcal{J}_{\Delta, D} = f_*\mathcal{O}_Y(E - \llcorner f^*(\llcorner D \lrcorner + \{D\})\lrcorner) =$$
$$f_*\mathcal{O}_Y(E - \llcorner f^*\{D\}\lrcorner) \otimes \mathcal{O}_X(-\llcorner D \lrcorner) = \mathcal{J}_{\Delta, \{D\}} \otimes \mathcal{O}_X(-\llcorner D \lrcorner).$$

□

SOLUTION TO (2.8). Let H be an ample divisor on X and $H' = f_*H$. Then $F = f^*H' - H$ is effective and $-F$ is ample over Y. □

SOLUTION TO (2.9). Let $A = L - (K_Y + \Gamma)$. There exists an effective divisor $\Gamma' \sim_{\mathbb{R},X} \Gamma + A/2$ such that (Y, Γ') is kawamata log terminal and $L - (K_Y + \Gamma') \sim_{\mathbb{R},X} A/2$ is f-ample. □

SOLUTION TO (2.10). Since the spectral sequence $E^1_{p,q} = R^p\pi_* R^q f_* F$ degenerates at the E^1-term since $E^1_{p,q} = 0$ for $q > 0$. □

SOLUTION TO (2.11). See [**Lazarsfeld04**, 1.8.5]. □

SOLUTION TO (2.13). Since $(X, f_*\Xi)$ is canonical, $K_Y + \Xi = f^*(K_X + f_*\Xi) + E$ where E is effective and exceptional. □

SOLUTION TO (4.1). Let G be a divisor whose support contains the support of any effective divisor in V and let $f : Y \to X$ be a log resolution of (X, G). It follows that f is a log resolution of (X, Δ) for any $0 \leq \Delta \in V$. We let $K_Y + \Gamma = f^*(K_X + \Delta)$. (X, Δ) is kawamata log terminal if $\operatorname{mult}_F(\Gamma) < 1$ where F is any divisor in the support of G or any f-exceptional divisor on Y. Since (X, Δ_0) is kawamata log terminal and $\operatorname{mult}_F(\Gamma)$ are continuous functions of Δ, the assertion is clear. □

SOLUTION TO (4.2). By the Negativity Lemma, it suffices to check that $a_F(X, \Delta) < a_F(X, \psi_*\phi_*\Delta)$ for all $(\psi \circ \phi)$-exceptional divisors $F \subset X$. Let $p : W \to X$, $q : W \to Y$ and $r : W \to Z$ be a common resolution, then we have that $q^*(K_Y + \phi_*\Delta) = r^*(K_Z + \psi_*\phi_*\Delta) + E$ where $E \geq 0$ is r-exceptional and that $K_X + \Delta = p_*q^*(K_Y + \phi_*\Delta) + G$ where $G \geq 0$ is ϕ-exceptional. Therefore $K_X + \Delta = p_*r^*(K_Z + \psi_*\phi_*\Delta) + p_*E + G$ where $p_*E + G \geq 0$. Since $\operatorname{Supp}(q_*E)$ contains all ψ-exceptional divisors and $\operatorname{Supp}(G)$ contains all ϕ-exceptional divisors, $\operatorname{Supp}(G + p_*E)$ contains all $(\phi \circ \psi)$-exceptional divisors. □

SOLUTION TO (1.7). Since Δ is big over U, we may write $\Delta \sim_{\mathbb{R},U} A + B$ where A is ample over U and $B \geq 0$. Let $\Delta' = (1 - \epsilon)\Delta + \epsilon B'$ where $0 < \epsilon \ll 1$. Then (X, Δ') is kawamata log terminal and $K_X + \Delta - (K_X + \Delta')$ is ample over U. The assertion now follows from (0.1). □

SOLUTION TO (4.3). Let $\phi : X \dashrightarrow Y$ be a log terminal model for (X, Δ) over Z and $\zeta : Y \to Z$ be the corresponding morphism. We must show that if $||\Delta - \Delta_0|| \ll 1$, then ϕ is a log terminal model for (X, Δ) over U. Replacing Δ by an appropriate ($\mathbb{R}$-linearly equivalent over U) divisor, we may assume that $\Delta = A + B$ where A is ample. By the Cone Theorem (0.2), there are finitely many extremal rays R_i over U corresponding to curves $C_i \subset Y$ such that if $K_Y + \phi_*\Delta$ is not nef over U, then $(K_Y + \phi_*\Delta) \cdot C_i < 0$. Since $(K_Y + \phi_*\Delta) \sim_{\mathbb{R},U} \phi_*(\Delta - \Delta_0) + \zeta^*H$ and $||\Delta - \Delta_0|| \ll 1$, we may assume that $\zeta^*H \cdot C_i = 0$ and hence that the R_i are extremal rays over Z. By assumption $K_Y + \phi_*\Delta$ is nef over Z and hence it is nef over U. □

Bibliography

[BCHM06] C. Birkar, P. Cascini, C. D. Hacon, J. McKernan, *Existence of minimal models for varieties of log general type.* ArXiv:math.AG/0610203.

[Eisenbud95] D. Eisenbud, *Commutative algebra. With a view toward algebraic geometry.* Graduate Texts in Mathematics, 150. Springer-Verlag, New York, 1995. xvi+785 pp.

[Fujino07] O. Fujino, *Special termination and reduction to pl flips*, in: Flips for 3-folds and 4-folds, A. Corti editor. Oxford lecture series in mathematics and its applications, **35**.

[FM00] O. Fujino, S. Mori, *A canonical bundle formula.* J. Differential Geom. **56** (2000), no. 1, 167–188.

[HM07] C. D. Hacon, J. McKernan, *Extension Theorems and the existence of Flips*, in: Flips for 3-folds and 4-folds, A. Corti editor. Oxford lecture series in mathematics and its applications, **35**.

[HM08] C. D. Hacon, J. McKernan, *Existence of minimal models for varieties of log general typeII: pl-flips.* arXiv:0808.1929

[Kollár93] J. Kollár, *Effective base point freeness.* Math. Ann. **296** (1993), no. 4, 595–605.

[KM98] J. Kollár, S. Mori, *Birational geometry of algebraic varieties.* With the collaboration of C. H. Clemens and A. Corti. Translated from the 1998 Japanese original. Cambridge Tracts in Mathematics, 134. Cambridge University Press, Cambridge, 1998.

[Lazarsfeld04] R. Lazarsfeld, *Positivity in algebraic geometry.* Ergebnisse der Mathematik und ihrer Grenzgebiete. 3. Folge. A Series of Modern Surveys in Mathematics [Results in Mathematics and Related Areas. 3rd Series. A Series of Modern Surveys in Mathematics], 48-49. Springer-Verlag, Berlin, 2004.

[Mori88] S. Mori, *Flip theorem and the existence of minimal models for 3-folds.* J. Amer. Math. Soc. 1 (1988), no. 1, 117–253.

[Shokurov03] V. V. Shokurov, *Prelimiting flips.* Tr. Mat. Inst. Steklova **240** (2003), Biratsion. Geom. Linein. Sist. Konechno Porozhdennye Algebry, 82–219; translation in Proc. Steklov Inst. Math. 2003, no. 1 (240), 75–213

[Siu06] Y.-T. Siu, *A General Non-Vanishing Theorem and an Analytic Proof of the Finite Generation of the Canonical Ring.* math.AG/0610740

[Szabó94] E. Szabó, *Divisorial log terminal singularities.* J. Math. Sci. Univ. Tokyo 1 (1994), no. 3, 631–639.

Lectures on Flips and Minimal Models

Alessio Corti, Paul Hacking, János Kollár,
Robert Lazarsfeld, and Mircea Mustaţă

IAS/Park City Mathematics Series
Volume 17, 2008

Lectures on Flips and Minimal Models

Alessio Corti, Paul Hacking, János Kollár, Robert Lazarsfeld, and Mircea Mustaţă

These are notes from the lectures of Corti, Kollár, Lazarsfeld, and Mustaţă at the workshop "Minimal and canonical models in algebraic geometry" at the Mathematical Sciences Research Institute, Berkeley, CA, April 16-20, 2007. The lectures give an overview of the recent advances on canonical and minimal models of algebraic varieties obtained in the papers [**HM1**], [**BCHM**].

We have tried to preserve the informal character of the lectures, and as a consequence we have kept the subsequent changes to a minimum. Lecture 1, by Lazarsfeld, gives an introduction to the extension theorems used in the proof. For a detailed introduction to multiplier ideals, extension theorems, and applications, see Lazarsfeld's PCMI lectures, in this volume. Lectures 2 and 3 describe the proof from [**HM1**] of the existence of flips in dimension n assuming the MMP in dimension $n-1$. Lecture 2, by Mustaţă, gives a geometric description of the restriction of the log canonical algebra to a boundary divisor (it is a so called adjoint algebra). Lecture 3, by Corti, proves that this algebra is finitely generated, following ideas of Shokurov. For a slightly different approach to the proof of existence of flips, and for more details, see Hacon's PCMI lectures in this volume. Lecture 4, by Kollár, gives an overview of the paper [**BCHM**] on the existence of minimal models for varieties of log general type.

Videos of the lectures are available at

www.msri.org/calendar/workshops/WorkshopInfo/418/show_workshop

Acknowledgements. We thank MSRI for hosting the conference and financial support. Hacking, Kollár, Lazarsfeld, and Mustaţă were partially supported by

---

AC: Department of Mathematics, Imperial College London, Huxley Building, 180 Queen's Gate, London SW7 2AZ, United Kingdom
E-mail address: A.Corti@imperial.ac.uk
PH: Department of Mathematics, University of Washington, Box 354350, Seattle, WA 98195-4350, USA
E-mail address: hacking@math.washington.edu
JK: Department of Mathematics, Princeton University, Fine Hall, Washington Road, Princeton, NJ 08544-1000, USA
E-mail address: kollar@math.princeton.edu
RL: Department of Mathematics, University of Michigan, East Hall, 530 Church Street, Ann Arbor, MI 48109-1043, USA
E-mail address: rlaz@umich.edu
MM: Department of Mathematics, University of Michigan, East Hall, 530 Church Street, Ann Arbor, MI 48109-1043, USA
E-mail address: mmustata@umich.edu

NSF grants DMS-0650052, DMS-0500198, DMS-0139713, and DMS-0500127, respectively.

LECTURE 1

Extension theorems

These are very condensed notes of Lazarsfeld's lecture on the Hacon-McKernan extension theorem. No attempt has been made to flesh them out. As mentioned above, some of the essential ideas are discussed in detail in Lazarsfeld's PCMI lectures elsewhere in this volume.

The theorem in question is this:

Theorem 1.1 (Hacon–McKernan). *Let X be a smooth projective variety over $\mathbb{C}$. Let $T+\Delta$ be an effective $\mathbb{Q}$-divisor having textnormalSNC support, with T irreducible and $\lfloor \Delta \rfloor = 0$. Assume that*

$$\Delta \sim_{\mathbb{Q}} A + B,$$

where A is ample, B is effective, and $T \not\subseteq \operatorname{Supp} B$. Suppose furthermore that the stable base locus of $K_X + T + \Delta$ does not contain any intersection of irreducible components of $T+\Delta$. Choose k such that $k\Delta$ is integral and set $L = k(K_X+T+\Delta)$. Then the restriction map

$$H^0(X, mL) \to H^0(T, mL_T)$$

is surjective for all $m \geq 1$.

Here and in what follows we write D_T for the restriction of a divisor D to a subvariety T.

Remark 1.2. Hacon and McKernan prove an analogous result for a projective morphism $X \to Z$ with Z affine.

The proof draws on an idea of Siu which first appeared in his work on deformation invariance of plurigenera. There are also related works by Kawamata, Takayama, and others. Ein and Popa have given a generalization of the theorem.

1.1. Multiplier and adjoint ideals

For an effective $\mathbb{Q}$-divisor D on X, we have the multiplier ideal $\mathcal{J}(X, D) \subseteq \mathcal{O}_X$. Roughly speaking it measures the singularities of the pair (X, D) — worse singularities correspond to deeper ideals.

Let L be a big line bundle on X. Set

$$\mathcal{J}(X, \|L\|) = \mathcal{J}(X, \frac{1}{p} D_p)$$

where $D_p \in |pL|$ is general and $p \gg 0$.

Proposition 1.3. Multiplier ideals satisfy the following properties.

(1) Every section of L vanishes on $\mathcal{J}(\|L\|)$, i.e. the map
$$H^0(L \otimes \mathcal{J}(X, \|L\|)) \xrightarrow{\sim} H^0(L)$$
is an isomorphism.

(2) If $M = K_X + L + P$ where P is nef then
$$H^i(X, M \otimes \mathcal{J}(\|L\|)) = 0$$
for all $i > 0$.

(3) If $M = K_X + L + (\dim X + 1)H$ where H is very ample then $M \otimes \mathcal{J}(\|L\|)$ is globally generated.

Here, given an ideal $\mathfrak{a} \subseteq \mathcal{O}_X$, line bundle M, and section $s \in \Gamma(X, M)$, we say s *vanishes on* $\mathfrak{a}$ if $s \in \operatorname{Im}(\Gamma(M \otimes \mathfrak{a}) \to \Gamma(M))$. Statement (2) is a formulation of the Kawamata-Viehweg-Nadel vanishing theorem. Assertion (3) follows from (2) plus Castelnuovo–Mumford regularity: if a sheaf $\mathcal{F}$ on projective space satisfies $H^i(\mathcal{F}(-i)) = 0$ for all $i > 0$ then $\mathcal{F}$ is globally generated.

Consider a smooth irreducible divisor $T \subset X$, $T \not\subset \operatorname{Supp} D$. We can define the *adjoint ideal* $\operatorname{Adj}_T(X, D) \subseteq \mathcal{O}_X$, which sits in an exact sequence.
$$0 \to \mathcal{J}(X, D) \otimes \mathcal{O}_X(-T) \to \operatorname{Adj} \to \mathcal{J}(T, D_T) \to 0.$$
Similarly, suppose that T is not contained in the stable base locus of L. We have
$$\mathcal{J}(T, \|L\|_T) := \mathcal{J}(T, \frac{1}{p} D_p|_T) \subseteq \mathcal{J}(T, \|L_T\|)$$
where $D_p \in |pL|$ is general and $p \gg 0$. We get $\operatorname{Adj}_T(X, \|L\|) \subseteq \mathcal{O}_X$, with
$$0 \to \mathcal{J}(X, \|L\|) \otimes \mathcal{O}_X(-T) \to \operatorname{Adj} \to \mathcal{J}(T, \|L\|_T) \to 0.$$
Now apply $(\cdot) \otimes M$. The essential idea is this: If $s \in \Gamma(T, M_T)$ vanishes on $\mathcal{J}(T, \|L\|_T)$ and if $M - (K_X + L + T)$ is nef, then s extends to a section of $\mathcal{O}_X(M)$ (we get vanishing of H^1 by Kawamata–Viehweg vanishing).

Consider as above $L = k(K_X + T + \Delta)$, $\Delta = A + B$, etc.

Lemma 1.4. (Main Lemma) There exists a very ample divisor H (independent of p) such that for every $p \geq 0$, every section
$$\sigma \in \Gamma(T, \mathcal{O}(pL_T + H_T) \otimes \mathcal{J}(\|pL_T\|))$$
extends to $\hat{\sigma} \in \Gamma(X, \mathcal{O}_X(pL + H))$.

(0) We may assume (T, B_T) is klt. Take $h \in \Gamma(X, H)$ general.

(1) Let $s \in \Gamma(T, mL_T)$. Consider $\sigma = s^l \cdot h \in \Gamma(T, lmL_T + H_T)$. The section σ vanishes on $\mathcal{J}(T, \|lmL_T\|)$. So there exists $\hat{\sigma} \in \Gamma(X, lmL + H)$ such that $\hat{\sigma}|_T = \sigma$ (by the Main Lemma).

(2) Let $F = \frac{mk-1}{mlk} \operatorname{div}(\hat{\sigma}) + B$. We find (using $\Delta = A + B$) that $mL - F - T = K_X + (\text{ample})$ if $l \gg 0$.

(3) Using (T, B_T) klt, we check that $\mathcal{O}_T(-\operatorname{div}(s)) \subseteq \mathcal{J}(T, F_T)$. So s vanishes on $\mathcal{J}(T, F_T)$.

(4) Finally, consider the sequence
$$0 \to \mathcal{J}(X, F) \otimes \mathcal{O}_X(-T) \to \operatorname{Adj} \to \mathcal{J}(T, F_T) \to 0$$
tensored by mL. We have $s \in H^0(\mathcal{O}_T(mL_T) \otimes \mathcal{J}(T, F_T))$ and $H^1(\mathcal{O}_X(mL - T) \otimes \mathcal{J}(X, F)) = 0$ by vanishing. So s extends.

1.2. Proof of the Main Lemma

We will only prove the special case $k = 1$, $L = K_X + T$ (so $\Delta = 0$ — we don't need $\Delta = A + B$ here), T not contained in the stable base locus of L. (Note: $L_T = K_T$). We prove

$(*)_p$: If $H = (\dim X + 1)(\text{very ample})$, then

$$H^0(T, \mathcal{O}(pL_T + H_T) \otimes \mathcal{J}(\|(p-1)L_T\|))$$

lifts to $H^0(X, pL + H)$.

The proof is by induction on p. The case $p = 1$ is OK by vanishing. Assume $(*)_p$ holds.

Claim 1. $\mathcal{J}(T, \|(p-1)L_T\|) \subseteq \mathcal{J}(T, \|pL + H\|_T)$.

PROOF. By Castelnuovo–Mumford regularity and vanishing

$$\mathcal{O}_T(pL_T + H_T) \otimes \mathcal{J}(\|(p-1)L_T\|) \qquad (**)$$

is globally generated.

$(*)_p$ implies sections of $(**)$ lift to X. So

$$\mathcal{J}(T, \|(p-1)L_T\|) \subseteq \mathfrak{b}(X, |pL + H|) \cdot \mathcal{O}_T \subseteq \mathcal{J}(T, \|pL + H\|_T).$$

(here $\mathfrak{b}(X, |D|)$ denotes the ideal defining the base locus of $|D|$). □

Consider the adjoint sequence

$$0 \to \mathcal{J}(\|pL + H\|)(-T) \to \text{Adj} \to \mathcal{J}(T, \|pL + H\|_T) \to 0.$$

Apply $(\cdot) \otimes \mathcal{O}_X((p+1)L + H)$. By the Claim

$$H^0(\mathcal{O}_T((p+1)L_T + H_T) \otimes \mathcal{J}(\|pL_T\|)) \subseteq H^0(\mathcal{O}_T((p+1)L_T + H_T) \otimes \mathcal{J}(T, \|pL + H\|_T))$$

and

$$H^1(\mathcal{O}_X((p+1)L + H - T) \otimes \mathcal{J}(\|pL + H\|)) = 0.$$

This gives the desired lifting.

LECTURE 2

Existence of flips I

This chapter is an exposition of work of Hacon and McKernan from [**HM1**], that build on the extension results from [**HM2**], and on ideas and results of Shokurov from [**Sho**].

2.1. The setup

Let $f\colon (X,D) \to Z$ be a birational projective morphism, where X is $\mathbb{Q}$-factorial, D is an effective $\mathbb{Q}$-divisor, Z is normal, $\rho(X/Z) = 1$, and $-(K_X + D)$ is f-ample. Assume also that f is small, i.e., that the exceptional locus has codimension ≥ 2. We consider two cases:

(1) klt flip: (X, D) klt.
(2) pl flip: (X, D) plt, $D = S + \Delta$ with $\lfloor D \rfloor = S$ irreducible, and $-S$ is f-ample.

It is well-known that the flip of f exists iff the $\mathcal{O}_Z$-algebra

$$\oplus_{m\geq 0} f_*\mathcal{O}_X\left(\lfloor m(K_X + D)\rfloor\right)$$

is finitely generated. This is a local question on Z, hence we may and will assume that Z is affine.

It is a result of Shokurov that MMP in dimension $(n-1)$, plus existence of pl flips in dimension n implies existence of klt flips in dimension n. Therefore we need only consider the case of pl flips.

Remark 2.1. Let $R = \oplus_{i\geq 0} R_i$ be a graded domain such that R_0 is a finitely generated $\mathbb{C}$-algebra. Then the algebra R is finitely generated iff the *truncation*

$$R_{(k)} := \oplus_{i\geq 0} R_{ki}$$

is finitely generated. Indeed, we have an obvious action of $\mathbb{Z}/k\mathbb{Z}$ on R such that the ring of invariants is $R_{(k)}$, hence R finitely generated implies $R_{(k)}$ finitely generated. To see the converse, it is enough to note that for every $0 < j < k$, if $s \in \oplus_{i\geq 0} R_{ki+j}$ is a nonzero homogeneous element, then multiplication by s^{k-1} embeds $\oplus_{i\geq 0} R_{ki+j}$ as an ideal of $R_{(k)}$.

¿From now on, we assume that we are in the pl flip setting.

Remark 2.2. Since $\rho(X/Z) = 1$, and since we work locally over Z, it follows from our assumptions that we may assume that there are positive integers p and q such that $p(K_X + D)$ and qS are linearly equivalent Cartier divisors.

Remark 2.3. Since Z is affine and f is small, it follows that S is linearly equivalent to an effective divisor not containing S in its support. In particular, it follows from the previous remark that there is a positive integer k and $G \in |k(K_X + D)|$ such that $S \not\subset \operatorname{Supp}(G)$.

A key remark due to Shokurov is that the algebra

$$\mathcal{R}\colon = \oplus_{m\geq 0} H^0(X, \mathcal{O}_X(\lfloor m(K_X + S + \Delta)\rfloor))$$

is finitely generated iff the *restricted algebra* $\mathcal{R}|_S$, given as

$$\oplus_{m\geq 0} \operatorname{Im}\left(H^0(X, \mathcal{O}_X(\lfloor m(K_X + S + \Delta)\rfloor)) \to H^0(S, \mathcal{O}_X(\lfloor m(K_X + S + \Delta)\rfloor)|_S)\right)$$

is finitely generated. *Sketch of proof*: replacing X by a suitable $U \subset X$ such that $\operatorname{codim}(X - U) \geq 2$, we may assume that S is Cartier. Since

$$p(K_X + S + \Delta) \sim qS$$

for some positive integers p and q, it follows from Remark 2.1 that it is enough to show that the algebra $\mathcal{R}' = \oplus_{m\geq 0} H^0(X, \mathcal{O}(mS))$ is finitely generated. Using the fact that $\mathcal{R}|_S$ is finitely generated and Remark 2.1, we deduce that the quotient $\mathcal{R}'/h\mathcal{R}'$ is finitely generated, where $h \in \mathcal{R}'_1$ is an equation for S. Therefore $\mathcal{R}'$ is finitely generated.

The above discussion shows that existence of pl flips in dimension n (and so by Shokurov's result, existence of klt flips in dimension n) follows from MMP in dimension $(n-1)$ and the following Main Theorem, due to Hacon and McKernan.

Theorem 2.4. *Let $f\colon (X, S+\Delta) \to Z$ be a projective birational morphism, where X is $\mathbb{Q}$-factorial, $S+\Delta$ is an effective $\mathbb{Q}$-divisor with $\lfloor S+\Delta\rfloor = S$ irreducible, and Z is a normal affine variety. Let k be a positive integer such that $k(K_X + S + \Delta)$ is Cartier. Suppose*

(1) *$(X, S+\Delta)$ is plt.*
(2) *S is not contained in the base locus of $|k(K_X + S + \Delta)|$.*
(3) *$\Delta \sim_{\mathbb{Q}} A + B$, where A is ample, B is effective, and $S \not\subset \operatorname{Supp} B$.*
(4) *$-(K_X + S + \Delta)$ is f-ample.*

If the MMP (with $\mathbb{R}$-coefficients) holds in dimension $\dim X - 1$, *then the restricted algebra*

$$\mathcal{R}|_S = \oplus_{m\geq 0} \operatorname{Im}(H^0(X, L^m) \to H^0(S, L^m|_S))$$

is finitely generated, where $L = \mathcal{O}_X(k(K_X + S + \Delta))$.

2.2. Adjoint algebras

In the setting of Theorem 2.4, we have by adjunction $(K_X + S + \Delta)|_S = K_S + \Delta|_S$. Moreover, the pair $(S, \Delta|_S)$ is klt. The trouble comes from the fact that the maps

$$H^0(X, mk(K_X + S + \Delta)) \to H^0(S, mk(K_S + \Delta|_S)) \tag{2.1}$$

are not surjective in general.

The goal is find a model $T \to S$ such that (some truncation of) the restricted algebra $\mathcal{R}|_S$ can be written as

$$\oplus_{m\geq 0} H^0(T, B_m),$$

where $\{B_m\}_m$ is an additive sequence of Cartier divisors on T. Recall that an *additive sequence* is a sequence of divisors $\{B_m\}_m$ on a normal variety T such that $B_i + B_j \leq B_{i+j}$ for all i, j. Note that in this case $\oplus_{m\geq 0} H^0(T, \mathcal{O}(B_m))$ has a natural algebra structure. A typical example of additive sequence: start with a divisor D such that $|D|$ is nonempty, and let $B_m = \operatorname{Mob}(mD) := mD - \operatorname{Fix}|mD|$. More generally, if $\{D_m\}_m$ is an additive sequence such that $|D_m| \neq \emptyset$ for every m, and if we put $B_m = \operatorname{Mob}(D_m)$, then $\{B_m\}_m$ forms an additive sequence.

Given an additive sequence $\{B_m\}_m$, the *associated convex sequence* is given by $\{\frac{1}{m}B_m\}_m$. If each $\frac{1}{m}B_m$ is bounded above by a fixed divisor, set

$$B := \sup \frac{1}{m} B_m = \lim_{m\to\infty} \frac{1}{m} B_m.$$

Remark 2.5. If B is semiample and if there exists i such that $B = \frac{1}{i}B_i$ (hence, in particular, B is a $\mathbb{Q}$-divisor), then $\oplus_{m\geq 0} H^0(T, \mathcal{O}(B_m))$ is finitely generated. Indeed, in this case $B = \frac{1}{m}B_m$ whenever $i \mid m$, and it is enough to use the fact that if L is a globally generated line bundle, then $\oplus_{m\geq 0} H^0(X, L^m)$ is finitely generated.

Suppose that $f\colon T \to Z$ is a projective morphism, where T is smooth and Z is affine. An *adjoint algebra* on T is an algebra of the form

$$\oplus_{m\geq 0} H^0(T, \mathcal{O}(B_m)),$$

where $\{B_m\}$ is an additive sequence and $B_m = mk(K_T + \Delta_m)$ for some $k \geq 1$ and $\Delta_m \geq 0$ such that $\Delta := \lim_{m\to\infty} \Delta_m$ exists and (T, Δ) is klt.

Our goal in what follows is to show that under the hypothesis of Theorem 2.4 (without assuming $-(K_X + S + \Delta)$ ample or MMP in dimension $\dim(X) - 1$), the algebra $\mathcal{R}|_S$ can be written as an adjoint algebra. It is shown in Lecture 3 how one can subsequently use the fact that $-(K_X + S + \Delta)$ is ample to deduce that $\mathcal{R}|_S$ is "saturated", and then use MMP in dimension $\dim(X) - 1$ to reduce to the case when the limit of the above Δ_m is such that $K_T + (\text{this limit})$ is semiample. This is enough to give the finite generation of $\mathcal{R}|_S$ (Shokurov proved this using diophantine approximation).

2.3. The Hacon–McKernan extension theorem

As we have already mentioned, the difficulty comes from the non-surjectivity of the restriction maps (2.1). We want to replace X by higher models on which we get surjectivity of the corresponding maps as an application of the following Extension Theorem, also due to Hacon and McKernan.

Theorem 2.6. *Let $(Y, T + \Delta)$ be a pair with Y smooth and $T + \Delta$ an effective $\mathbb{Q}$-divisor with SNC support, with $\lfloor T + \Delta \rfloor = T$ irreducible. Let k be a positive integer such that $k\Delta$ is integral and set $L = k(K_Y + T + \Delta)$. Suppose*

(1) *$\Delta \sim_{\mathbb{Q}} A + B$, where A is ample, B is effective, and $T \not\subset \operatorname{Supp}(B)$.*
(2) *No intersection of components of $T + \Delta$ is contained in the base locus of L.*

Then the restriction map

$$H^0(Y, L) \to H^0(T, L|_T)$$

is surjective.

This theorem was discussed in Lecture 1.

2.4. The restricted algebra as an adjoint algebra

Let $f\colon (X, S + \Delta) \to Z$ be a projective morphism, where X is $\mathbb{Q}$-factorial, $S + \Delta$ is an effective $\mathbb{Q}$-divisor with $\lfloor S + \Delta \rfloor = S$ irreducible, $(X, S + \Delta)$ is plt, and Z is an affine normal variety. Let k be a positive integer such that $k(K_X + S + \Delta)$ is Cartier and S is not contained in the base locus of $|k(K_X + S + \Delta)|$. Assume

$$\Delta \sim_{\mathbb{Q}} A + B \qquad (*)$$

where A is ample, B is effective, and $S \not\subset \operatorname{Supp} B$.

We will replace $(X, S + \Delta)$ by a log resolution (in fact, a family of resolutions) on which we can apply the extension theorem and use this to exhibit the restricted algebra as an adjoint algebra. Consider a birational morphism $f \colon Y \to X$, and let T be the strict transform of S. Write

$$K_Y + T + \Delta_Y = f^*(K_X + S + \Delta) + E,$$

where Δ_Y and E are effective and have no common components, $f_*\Delta_Y = \Delta$, and E is exceptional.

Since E is effective and exceptional, we have $H^0(X, mk(K_X + S + \Delta)) \simeq H^0(Y, mk(K_Y + T + \Delta_Y))$ for every m. Therefore we may "replace" $K_X + S + \Delta$ by $K_Y + T + \Delta_Y$.

Step 1. After replacing Δ by the linearly equivalent divisor $A' + B'$, where $A' = \epsilon A$ and $B' = (1 - \epsilon)\Delta + \epsilon B$, with $0 < \epsilon \ll 1$, we may assume that we have equality of divisors $\Delta = A + B$ in $(*)$. We may also assume (after possibly replacing k by a multiple) that kA is very ample and that

$$A = \frac{1}{k} \cdot (\text{very general member of } |kA|).$$

Since A is general in the above sense, if $f \colon Y \to X$ is a projective birational morphism, we may assume $f^*A = \tilde{A}$ is the strict transform of A on Y. In this case, $(*)$ will also hold for $(Y, T + \Delta_Y)$: indeed, there exists an effective exceptional divisor E' such that $f^*A - E'$ is ample, hence

$$\Delta_Y = (f^*A - E') + E' + (\cdots),$$

where $(\cdots)$ is an effective divisor that does not involve T.

Caveat: we will construct various morphisms f as above starting from $(X, S + \Delta)$. We will then modify Δ to satisfy $f^*A = \tilde{A}$, and therefore we need to check how this affects the properties of f.

Step 2. Let $f \colon Y \to X$ be a log resolution of $(X, S + \Delta)$. After modifying Δ as explained in Step 1, we write $\Delta_Y = \tilde{A} + \sum a_i D_i$ (note that f remains a log resolution for the new Δ).

Claim 2. After blowing up intersections of the D_i (and of their strict transforms). we may assume $D_i \cap D_j = \emptyset$ for all $i \neq j$.

The proof of the claim is standard, by induction first on the number of intersecting components and then on the sum of the coefficients of intersecting components. Note also that as long as we blow up loci that have SNC with $\tilde{A}$, the condition $f^*(A) = \tilde{A}$ is preserved.

Step 3. We need to satisfy condition 2) in Theorem 2.6. The hypothesis implies that $T \not\subset \operatorname{Bs}|k(K_Y + T + \Delta_Y)|$. Since A is general, it follows that we only need to worry about the components D_i, and the intersections $D_i \cap T$ that are contained in $\operatorname{Bs}|k(K_Y + T + \Delta_Y)|$.

Canceling common components, we may replace Δ_Y by $0 \leq \Delta'_Y \leq \Delta_Y$ such that no component of Δ'_Y appears in the fixed part of $|k(K_Y + T + \Delta'_Y)|$. We put $\Delta'_Y = \tilde{A} + \sum_i a'_i D_i$.

Step 4. We now deal with intersections $T \cap D_i$ that are contained in the base locus of $|k(K_Y + T + \Delta'_Y)|$. Let $h \colon \overline{Y} \to Y$ be the blowup of $T \cap D_i$. Note that since T is smooth and $T \cap D_i \subset T$ is a divisor, the strict transform $\widetilde{T}$ of T maps isomorphically to T. Let $F \subset \overline{Y}$ be the exceptional divisor. We write

$$K_{\overline{Y}} + (h^*(T + \Delta'_Y) - F) = h^*(K_Y + T + \Delta'_Y).$$

Note that $h^*(T) = \widetilde{T} + F$, hence $h^*(T + \Delta'_Y) - F$ is effective (the coefficient of F being a'_i).

On $\overline{Y}$ the divisors $\widetilde{T}$ and $\widetilde{D_i}$ are disjoint. We need to blow up again along $F \cap \widetilde{D_i}$, but this gives an isomorphism around $\widetilde{T}$. We repeat this process; however, we can only continue finitely many times because

$$\mathrm{ord}_{T\cap F}|\mathrm{Mob}\,(k(K_{\overline{Y}} + h^*(T + \Delta'_Y) - F))\,|_T \leq \mathrm{ord}_{T\cap D_i}|\mathrm{Mob}(k(K_Y+T+\Delta'_Y))|_T - 1.$$

Step 5. We change notation to denote by $(Y, T + \Delta'_Y)$ the resulting pair (we emphasize that T hasn't changed starting with Step 3). We can now apply the extension theorem for $(Y, T + \Delta'_Y)$. Consider the commutative diagram

$$\begin{array}{ccc} H^0(X, k(K_X + S + \Delta)) & \to & H^0(S, k(K_X + S + \Delta)|_S) \\ \downarrow & & \downarrow \\ H^0(Y, k(K_Y + T + \Delta'_Y)) & \to & H^0(T, k(K_Y + T + \Delta'_Y)|_T) \end{array}$$

The left vertical arrow is an isomorphism and the right vertical arrow is injective by construction, while the bottom arrow is surjective by the extension theorem. Hence writing $\Theta_1 = \Delta'_Y|_T$, we see that the component of degree k in $\mathcal{R}|_S$ is isomorphic to $H^0(T, k(K_T + \Theta_1))$.

Step 6. In order to prove further properties of the restricted algebra, one needs to combine the previous construction with "taking log mobile parts". For every $m \geq 1$ write

$$mk(K_Y + T + \Delta_Y) = M_m + (\text{fixed part}).$$

Let $0 \leq \Delta_m \leq \Delta_Y$ be such that Δ_m has no common component with the above fixed part. After possibly replacing k by a multiple, $\{mk(K_Y + T + \Delta_m)\}$ is an additive sequence. We now apply Steps 3-5 for each $(Y, T + \Delta_m)$. Note that T remains unchanged. We get models $Y_m \to Y$ and divisors Θ_m on T such that the kth truncation of $\mathcal{R}|_S$ is isomorphic to

$$\oplus_{m\geq 0} H^0(T, km(K_T + \Theta_m)).$$

Moreover $\Theta_m \leq \Theta := \Delta_Y|_T$ implies $\Theta' := \lim_{m\to\infty} \Theta_m$ exists and (T, Θ') is klt. This proves that the restricted algebra is an adjoint algebra.

LECTURE 3

Existence of flips II

This lecture is an exposition of work of Shokurov.

Recall (from Lecture 2)

Definition 3.1. Let $Y \to Z$ be a projective morphism with Y smooth and Z affine. An *adjoint algebra* is an algebra of the form

$$\mathcal{R} = \oplus_{m \geq 0} H^0(Y, N_m)$$

where $N_m = mk(K_Y + \Delta_m)$, the limit $\Delta := \lim_{m\to\infty} \Delta_m$ exists, and (Y, Δ) is klt.

Remark 3.2. Note that Δ can be an $\mathbb{R}$-divisor.

In Lecture 2, it was shown that the "restricted algebra" is an adjoint algebra.

Definition 3.3. Given an adjoint algebra $\mathcal{R}(Y, N_\bullet)$, set $M_i = \operatorname{Mob} N_i$, the mobile part of N_i, and $D_i = \frac{1}{i}M_i$. We say $\mathcal{R}$ is *a-saturated* if there exists a $\mathbb{Q}$-divisor F on Y, $\lceil F \rceil \geq 0$, such that $\operatorname{Mob}\lceil jD_i + F\rceil \leq jD_j$ for all $i \geq j > 0$.

Remark 3.4. In applications, F is always the discrepancy of some klt pair (X, Δ), $f\colon Y \to X$. That is

$$K_Y = f^*(K_X + \Delta) + F.$$

We sometimes write $F = \mathbb{A}(X, \Delta)_Y$.

Example 3.5. An a-saturated adjoint algebra on an affine curve is finitely generated.

Let $Y = \mathbb{C}$ and $P = 0 \in \mathbb{C}$. Let $N_i = m_i \cdot 0$, where $m_i + m_j \leq m_{i+j}$, and $D_i = \frac{1}{i}m_i \cdot 0 = d_i \cdot 0$. By assumption $d = \lim_{i\to\infty} d_i \in \mathbb{R}$ exists. In this context, a-saturation means there exists $b < 1$, $F = -b \cdot 0$, such that

$$\lceil jd_i - b\rceil \leq jd_j$$

for all $i \geq j > 0$.

We want to show $d \in \mathbb{Q}$, and $d = d_j$ for j divisible. Passing to the limit as $i \to \infty$ we get

$$\lceil jd - b\rceil \leq jd_j.$$

Assume $d \notin \mathbb{Q}$. Then

$$\{\langle jd\rangle \mid j \in \mathbb{N}\} \subset [0,1]$$

is dense (here $\langle\cdot\rangle$ denotes the fractional part). So, there exists j such that $\langle jd\rangle > b$, and then

$$jd_j \leq jd < \lceil jd - b\rceil \leq jd_j,$$

a contradiction. So $d \in \mathbb{Q}$. The same argument shows that $d_j = d$ if $j \cdot d \in \mathbb{Z}$.

Definition 3.6. An adjoint algebra $\mathcal{R}(Y, N_\bullet)$ is *semiample* if the limit $D = \lim_{i\to\infty} \frac{1}{i}M_i$ is semiample, where $M_i := \operatorname{Mob}(N_i)$.

Remark 3.7. We say that an $\mathbb{R}$-divisor D on Y is *semiample* if there exists a morphism $f\colon Y \to W$ with W quasiprojective such that D is the pullback of an ample $\mathbb{R}$-divisor on W. We say an $\mathbb{R}$-divisor is *ample* if it is positive on the Kleiman–Mori cone of curves.

Theorem 3.8. *If an adjoint algebra $\mathcal{R} = \mathcal{R}(Y, N_\bullet)$ is a-saturated and semiample then it is finitely generated.*

The proof is a modification of the one dimensional case, based on the following

Lemma 3.9. Let $Y \to Z$ be projective with Y smooth and Z affine and normal. Let D be a semiample $\mathbb{R}$-divisor on Y, and assume that D is not a $\mathbb{Q}$-divisor. Fix $\epsilon > 0$. There exists a $\mathbb{Z}$-divisor M and $j > 0$ such that

(1) M is free.
(2) $\|jD - M\|_{\sup} < \epsilon$
(3) $jD - M$ is not effective.

Theorem 3.10. *Assume the MMP in dimension n (precisely, MMP with scaling for klt pairs with $\mathbb{R}$-coefficients. For the definition of MMP with scaling see Lecture 4). Let $\mathcal{R} = \mathcal{R}(Y, N_\bullet)$ be an adjoint algebra. Let $N_i = ik(K_Y + \Delta_i)$, $\Delta = \lim_{i\to\infty} \Delta_i$, and assume $K_Y + \Delta$ big. Then there exists a modification $\pi\colon Y' \to Y$ and $N'_\bullet$ such that $\mathcal{R} = \mathcal{R}(Y', N'_\bullet)$ is semiample.*

Remark 3.11. In the context of flips, the condition $K_Y + \Delta$ big is not an issue.

The proof of the theorem is based on the following

Lemma 3.12. Let (X, Δ) be a klt pair, where X is $\mathbb{Q}$-factorial, Δ is an $\mathbb{R}$-divisor, and $K_X + \Delta$ is big. Assume the MMP in dimension n. Let $\Delta \in V \subset \operatorname{Div}_{\mathbb{R}} X$ be a finite dimensional vector space. There exist $\epsilon > 0$ and finitely many $g\colon X \dashrightarrow W_i$ (birational maps) such that if $D \in V$, $\|D - \Delta\| < \epsilon$, then for some i the pair $(W_i, g_{i*}D)$ is a log minimal model of (X, D).

For the proof of the lemma see Lecture 4.

That the restricted algebra is a-saturated can be proved by a straightforward application of Kawamata–Viehweg vanishing as follows.

Theorem 3.13. *If $-(K_X + S + \Delta)$ is big and nef, then the restricted algebra is a-saturated.*

PROOF. Let $(X, S + \Delta) \to Z$ be a pl-flipping contraction, and consider a birational morphism $f\colon Y \to X$. Let T denote the strict transform of S. Write

$$K_Y + T + \Delta_Y = f^*(K_X + S + \Delta) + E,$$

where Δ_Y and E are effective (with no common components) and E is exceptional. Assume $k(K_X + S + \Delta)$ is integral and Cartier. Let Δ_m be the largest divisor such that $0 \leq \Delta_m \leq \Delta$ and

$$M_m := \operatorname{Mob}(mk(K_Y + T + \Delta_m)) = \operatorname{Mob}(mk(K_Y + T + \Delta)).$$

We may assume M_m is free (for some f). Write $M_m^0 = M_m|_T$. For simplicity we only consider the case $i = j$, i.e., we prove

$$\operatorname{Mob}\lceil M_j^0 + F\rceil \leq M_j^0,$$

where $F = \mathbb{A}(S, \Delta|_S)_T$.

If $\lceil E \rceil$ is an f-exceptional divisor, then $\mathrm{Mob}\lceil M_j + E \rceil \leq M_j$. Choose $E = \mathbb{A}(X, S + \Delta)_Y + T$, so $E|_T = F$, and consider the short exact sequence

$$0 \to \mathcal{O}_Y(M_j + \mathbb{A}(X, S + \Delta)_Y) \to \mathcal{O}_Y(M_j + E) \to \mathcal{O}_T(M_j^0 + F) \to 0.$$

By the Kawamata-Viehweg vanishing theorem and our assumption, we know $H^1(\mathcal{O}_Y(M_j + \mathbb{A}(X, S+\Delta)_Y)) = 0$ (because $\mathbb{A}(X, S+\Delta)_Y = K_Y - f^*(K_X + S + \Delta)$). Now the a-saturated condition $\mathrm{Mob}\lceil M_j^0 + F \rceil \leq M_j^0$ is implied by $\mathrm{Mob}\lceil M_j + E \rceil \leq M_j$ and this extension result. □

LECTURE 4

Notes on Birkar-Cascini-Hacon-McKernan

By the previous 3 lectures, we can start with the:

Assumption: We proved the existence of flips in dimension n, using minimal models in dimension $n-1$.

The main result is the following:

Theorem 4.1 (Minimal models in dimension n). *Assume that Δ is a big $\mathbb{R}$-divisor, (X, Δ) is klt and $K + \Delta$ is pseudo-effective. Then (X, Δ) has a minimal model.*

Let us first see some corollaries.

Corollary 4.2. Δ is a big $\mathbb{Q}$-divisor, (X, Δ) is klt and $K + \Delta$ is pseudo-effective. Then the canonical ring

$$\sum_{m \geq 0} H^0(X, \mathcal{O}_X(mK_X + \lfloor m\Delta \rfloor)) \quad \text{is finitely generated.}$$

PROOF. Get minimal model for (X, Δ), then use base point freeness. □

Corollary 4.3. X smooth, projective and K_X is big. Then the canonical ring

$$\sum_{m \geq 0} H^0(X, \mathcal{O}_X(mK_X)) \quad \text{is finitely generated.}$$

PROOF. Pick some effective $D \sim mK_X$. Then $(X, \epsilon D)$ is klt (even terminal) for $0 < \epsilon \ll 1$ and ϵD is big. So $(X, \epsilon D)$ has a minimal model. It is automatically a minimal model for X. □

Corollary 4.4. X smooth, projective. Then the canonical ring

$$\sum_{m \geq 0} H^0(X, \mathcal{O}_X(mK_X)) \quad \text{is finitely generated.}$$

PROOF. As in Kodaira's canonical bundle formula for elliptic surfaces, Fujino-Mori reduces the ring to a general type situation in lower dimension. □

Corollary 4.5. Let X be a Fano variety. Then the Cox ring

$$\sum_{D \in \mathrm{Pic}(X)} H^0(X, \mathcal{O}_X(D)) \quad \text{is finitely generated.}$$

PROOF. See original. □

Corollary 4.6. If $K + \Delta$ is not pseudo-effective, then there exists a birational map $X \dashrightarrow X'$ and a Mori fiber space $X' \to Z'$.

PROOF. Fix H ample and the smallest $c > 0$ such that $K + \Delta + cH$ is pseudo-effective. (Note that a priori, c may not be rational.) After MMP, we get $X \dashrightarrow X'$ such that $K + \Delta + cH$ is nef on X'. It cannot be big since then $K + \Delta + (c - \epsilon)H$ would still be effective. So base point freeness gives $X' \to Z'$. □

4.1. Comparison of 3 MMP's

In a minimal model program (or MMP) we start with a pair (X, Δ) and the goal is to construct a minimal model $(X, \Delta)^{min}$. (Note that minimal models are usually not unique, so $(X, \Delta)^{min}$ is a not well defined notational convenience.)

There are 3 ways of doing it.

4.1.1. Mori-MMP (libertarian)

This is the by now classical approach. Pick *any* extremal ray, contract/flip as needed. The hope is that eventually we get $(X, \Delta)^{min}$. This is known only in dimension ≤ 3 and almost known in dimension 4.

Note that even if (X, Δ) is known to have a minimal model, it is not at all clear that every Mori-MMP starting with (X, Δ) has to end (and thus yield a minimal model).

4.1.2. MMP with scaling (dictatorial)

Fix H and $t_0 > 0$ such that $K + \Delta + t_0 H$ is nef.

Let $t \to 0$. For a while $K + \Delta + tH$ is nef, but then we reach a critical value $t_1 \leq t_0$. That is, there exists an extremal ray R_1 such that $R_1 \cdot \big(K + \Delta + (t_1 - \eta)H\big) < 0$ for $\eta > 0$. Contract/flip this R_1 and continue.

We will show that MMP with scaling works in any dimension, provided that Δ is big and $K + \Delta$ is pseudo-effective.

4.1.3. Roundabout MMP of BCHM

Instead of directly going for a minimal model, we start by steps that seem to make things more complicated. Then we aim for the minimal model on a carefully chosen path. There are 5 stages:

(1) Increase Δ in a mild manner to Δ^+.
(2) Increase Δ^+ wildly to $\Delta^+ + M$.
(3) Construct $(X, \Delta^+ + M)^{min}$.
(4) Remove the excess M to get $(X, \Delta^+)^{min}$.
(5) Prove that $(X, \Delta^+)^{min}$ is also $(X, \Delta)^{min}$.

You can imagine the process as in the picture:

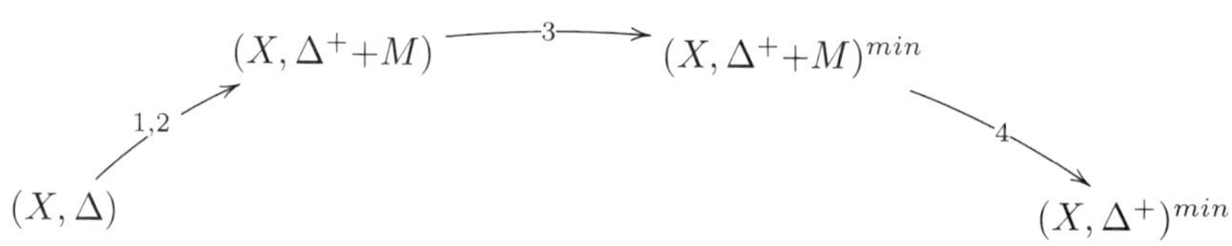

Bending it like BCHM

(Note that the "C" of Cascini is pronounced as a "K".) We will show that bending works in any dimension, provided that Δ is big and $K + \Delta$ is pseudo effective.

(Major side issues). My presentation below ignores 3 important points.

(1) Difference between klt/dlt/lc. The main results are for klt pairs but many intermediate steps are needed for dlt or lc. These are purely technical points but in the past proofs collapsed on such technicalities.
(2) Relative setting. The induction and applications sometimes need the relative case: dealing with morphisms $X \to S$ instead of projective varieties. The relevant technical issues are well understood.
(3) I do not explain how to prove the *non-vanishing theorem*: If $K + \Delta$ is pseudo-effective (that is, a limit of effective divisors) then it is in fact effective.

4.1.4. Spiraling induction

We prove 4 results together as follows:

MMP with scaling in dim. $n - 1$

$\Downarrow$ Section 4.3

Termination with scaling in dim. n near $\lfloor \Delta \rfloor$

$\Downarrow$ Section 4.4

Existence of $(X, \Delta)^{min}$ in dim. n

$\Downarrow$ Section 4.5

Finiteness of $(X, \Delta + \sum t_i D_i)^{min}$ in dim. n for $0 \leq t_i \leq 1$

$\Downarrow$ Section 4.2

MMP with scaling in dim. n

(Finiteness questions). While we expect all 3 versions of the MMP to work for any (X, Δ), the above finiteness is quite subtle. Even a smooth surface can contain infinitely many extremal rays, thus the very first step of the Mori-MMP sometimes offers an infinite number of choices.

By contrast, if Δ is big, then there are only finitely many possible models reached by Mori-MMP. This is, however, much stronger than what is needed for the proof.

It would be very useful to pin down what kind of finiteness to expect in general.

4.2. MMP with scaling

Begin with $K + \Delta + t_0 H$ nef, $t_0 > 0$.

(1) Set $t = t_0$ and decrease it.
(2) We hit a first *critical value* $t_1 \leq t_0$. Here $K + \Delta + t_1 H$ nef but $K + \Delta + (t_1 - \eta)H$ is not nef for $\eta > 0$.
(3) This means that there exists an extremal ray $R \subset \overline{NE}(X)$ such that
$$R \cdot (K + \Delta + t_1 H) = 0 \quad \text{and} \quad R \cdot H > 0.$$
Thus $R \cdot (K + \Delta) < 0$ and R is a "usual" extremal ray.
(4) Contract/flip R to get $X_0 \dashrightarrow X_1$ and continue.

The problem is that we could get an infinite sequence
$$X_1 \dashrightarrow X_2 \dashrightarrow X_3 \dashrightarrow \cdots .$$

*A*dvantage of scaling:

In the MMP with scaling, each X_i is a minimal model for some $K + \Delta + tH$. So, if we know finiteness of models as t varies then there is no infinite sequence. Thus we have proved the implication

$$\boxed{\text{Finiteness of } (X, \Delta + tH)^{min} \text{ in dim. } n \text{ for } 0 \leq t \leq t_0}$$
$$\Downarrow$$
$$\boxed{\text{MMP with scaling in dim. } n}$$

Remark 4.7. In the Mori-MMP, the X_i are *not* minimal models of anything predictable. This makes it quite hard to prove termination as above since we would need to control all possible models in advance.

4.3. MMP with scaling near $\lfloor \Delta \rfloor$

Start with $(X, S + \Delta)$, S integral. Here $(X, S + \Delta)$ is assumed dlt and in all interesting cases $S \neq 0$, so $(X, S + \Delta)$ is not klt.

We run MMP with scaling to get a series of contractions/flips
$$\cdots \dashrightarrow (X_i, S_i + \Delta_i) \overset{\phi_i}{\dashrightarrow} (X_{i+1}, S_{i+1} + \Delta_{i+1}) \dashrightarrow \cdots$$

Instead of termination, we claim only a much weaker result, traditionally known as *special termination.*

Proposition 4.8. There are only finitely many i such that $S_i \cap \mathrm{Ex}(\phi_i) \neq \emptyset$.

PROOF. This is all old knowledge, relying on 6 basic observations.

(1) We can concentrate on 1 component of S and by a perturbation trick we can assume that S is irreducible.
(2) The discrepancy $a(E, X_i, S_i + \Delta_i)$ weakly increases with i and strictly increases iff $\mathrm{center}_{X_i}(E) \subset \mathrm{Ex}(\phi_i)$.
(3) There are very few E with $a(E, X_i, S_i + \Delta_i) < 0$.
(4) If $S_i \dashrightarrow S_{i+1}$ creates a *new* divisor $F_{i+1} \subset S_{i+1}$, then there exists E_{i+1} with $a(E_{i+1}, X_i, S_i + \Delta_i) < a(E_{i+1}, X_{i+1}, S_{i+1} + \Delta_{i+1}) \leq 0$.
(5) Combining these we see that $S_i \dashrightarrow S_{i+1}$ creates no new divisors for $i \gg 1$.
(6) Only finitely many $S_i \dashrightarrow S_{i+1}$ contracts a divisor, since at such a step the Picard number of S_i drops.

Thus, for $i \gg 1$, each $S_i \dashrightarrow S_{i+1}$ is an isomorphism in codimension 1. (Maybe such a map could be called a *traverse*.)

Therefore, if $X_0 \dashrightarrow X_1 \cdots$ is an MMP with scaling, then

$$S_N \overset{\phi_N|_{S_N}}{\dashrightarrow} S_{N+1} \dashrightarrow S_{N+1} \dashrightarrow \cdots$$

is also an MMP with scaling for $N \gg 1$, except that $\phi_i|_{S_i}$ is an isomorphism when $S_i \cap \mathrm{Ex}(\phi_i) \neq \emptyset$. By induction, $\phi_i|_{S_i}$ is an isomorphism for $i \gg 1$, hence $S_i \cap \mathrm{Ex}(\phi_i) = \emptyset$. □

Thus we have shown that

$$\boxed{\text{MMP with scaling in dim. } n-1}$$
$$\Downarrow$$
$$\boxed{\text{Termination with scaling in dim. } n \text{ near } \lfloor \Delta \rfloor}$$

4.4. Bending it like BCHM

This is the hardest part. We assume termination with scaling near $\lfloor \Delta \rfloor$ (for many different X and Δ), and we prove that the roundabout MMP also works.

The proof uses 2 basic lemmas. The first one shows that under certain conditions, every flip we have to do involves $\lfloor \Delta \rfloor$. The second one shows how to increase $\lfloor \Delta \rfloor$ without changing $(X, \Delta)^{min}$.

Lemma 4.9 (Scaling to the boundary). Assume that

(1) $K + \Delta \sim cH + F$ for some $c \geq 0$ and $F \geq 0$.
(2) $K + \Delta + H$ is nef.
(3) $\mathrm{Supp}(F) \subset \lfloor \Delta \rfloor$.

Then $(X, \Delta + t \cdot H)^{min}$ exists for every $0 \leq t \leq 1$.

PROOF. (To accomodate $\mathbb{R}$-divisors, $\sim$ can stand for $\mathbb{R}$-linear equivalence.) Start scaling. At the critical value, get a ray R such that $R \cdot H > 0$ and $R \cdot (K + \Delta) < 0$. Thus $R \cdot (cH + F) < 0$ and $R \cdot F < 0$, so $\mathrm{locus}(R) \subset \mathrm{Supp}(F) \subset \lfloor \Delta \rfloor$.

Thus every flip encountered in the scaling MMP intersects the boundary. Thus we have termination. □

If we start with a klt pair (X, Δ), then $\lfloor \Delta \rfloor = 0$. From condition (3) thus $F = 0$ and hence $K + \Delta \sim cH$. Therefore $K + \Delta \sim \frac{c}{c+1}(K + \Delta + H)$ is nef and we have nothing to do. Conclusion: *We need a way to increase Δ without changing the minimal model!*

Lemma 4.10 (Useless divisor lemma). If $\Delta' \subset$ stable base locus of $(K + \Delta)$, then $(X, \Delta + \Delta')^{min} = (X, \Delta)^{min}$.

PROOF. Note that

$$\begin{array}{ccc} (\Delta')^{min} & \subset & \text{stable base locus of } (K+\Delta)^{min} \\ & & \| \\ & & \text{stable base locus of } (K+\Delta+\Delta')^{min} \end{array}$$

However, $(K + \Delta + \Delta')^{min}$ is base point free, thus $(\Delta')^{min} = 0$, and so $(X, \Delta + \Delta')^{min} = (X, \Delta)^{min}$. □

Corollary 4.11. We can always reduce to the case when the stable base locus of $K+\Delta^+$ is in $\lfloor\Delta^+\rfloor$.

PROOF. First we can take a log resolution to assume that (X,Δ) has simple normal crossings only. Write the stable base locus as $\sum_{i=1}^r D_i$ and $\Delta=\sum_{i=1}^r d_iD_i+$ (other divisors), where $d_i=0$ is allowed. Set $\Delta' := \sum_{i=1}^r (1-d_i)D_i$. Then $\Delta^+ := \Delta+\Delta' = \sum_{i=1}^r D_i +$ (other divisors) and $\lfloor\Delta+\Delta'\rfloor \supset \sum_{i=1}^r D_i$. □

4.12 (Bending I: $K+\Delta$ is a $\mathbb{Q}$-divisor).
We proceed in 6 steps:

(0) Write $K+\Delta\sim rM+F$ where M is mobile, irreducible and F is in the stable base locus.
(1) Take log resolution. With Δ' as above, set $\Delta^+ := \Delta+\Delta'$ and $F^+ := F+\Delta'$. Then $K+\Delta^+\sim rM+F^+$ and $\mathrm{Supp}(F^+)\subset\lfloor\Delta^+\rfloor$.
(2) Add M to Δ^+ and pick an ample H to get
  (a) $K+\Delta^+ + M\sim 0\cdot H+\big((r+1)M+F^+\big)$
  (b) $K+\Delta^+ + M+H$ is nef, and
  (c) $\mathrm{Supp}(M+F^+)\subset\lfloor\Delta^+ + M\rfloor$.
(3) Scale by H to get $(X,\Delta^+ + M)^{min}$. Now we have:
  (a) $K+\Delta^+\sim rM+F^+$.
  (b) $K+\Delta^+ + M$ is nef.
  (c) $\mathrm{Supp}(F^+)\subset\lfloor\Delta^+\rfloor$.
(4) Scale by M to get $(X,\Delta^+)^{min}$.
(5) By the useless divisor lemma, $(X,\Delta^+)^{min}=(X,\Delta)^{min}$.

4.13 (Bending II: $K+\Delta$ is an $\mathbb{R}$-divisor).
We follow the same 6 steps, but there are extra complications.

(0) Write $K+\Delta\sim r_iM_i+F$, where the M_i are mobile, irreducible and F is in the stable base locus. (This is actually not obvious.)
(1–3) goes as before and we get
  (a) $K+\Delta^+\sim\sum r_iM_i+F^+$
  (b) $K+\Delta^+ +\sum M_i$ is nef, and
  (c) $\mathrm{Supp}(F^+)\subset\lfloor\Delta^+\rfloor$.
(4) Let me first describe 2 attempts that do not work.

(First try): Scale $\sum M_i$. The problem is that $\sum r_iM_i\neq c\sum M_i$.

(Second try): Scale M_1. This works, but the support condition (c) fails at next step when we try to scale M_2.

(Third try): Do not take *all* of M_1 away at the first step.

Reorder the indices so that $r_1\le r_2\le\cdots\le r_k$. We then construct inductively minimal models for

$$\big(X,\Delta^+ + \tfrac{1}{r_j}(r_1M_1+\cdots+r_{j-1}M_{j-1})+M_j+\cdots+M_k\big)$$

We already have the $j=1$ case. Let us see how to do $j\to j+1$.

By assumption $K+\Delta^+ + \frac{1}{r_j}(r_1M_1+\cdots+r_{j-1}M_{j-1})+(M_j+\cdots+M_k)$ is nef. Move M_j from the right sum to the left. Thus now we have:

(a) $K+\Delta^+\sim(r_1M_1+\cdots+r_jM_j)+(r_{j+1}M_{j+1}+\cdots+r_kM_k+F^+)$
(b) $K+\Delta^+ + \frac{1}{r_j}(r_1M_1+\cdots+r_jM_j)+(M_{j+1}+\cdots+M_k)$ is nef, and
(c) $\mathrm{Supp}(r_{j+1}M_{j+1}+\cdots+r_kM_k+F^+)\subset\lfloor M_{j+1}+\cdots+M_k+\Delta^+\rfloor$.

Scale $r_1M_1+\cdots+r_jM_j$ with scale factor r_j/r_{j+1}. This gives the $j+1$ case.

At the end (and after moving M_k into the sum) we have:
(a) $K + \Delta^+ \sim \sum r_i M_i + F^+$
(b) $K + \Delta^+ + \frac{1}{r_k}(r_1 M_1 + \cdots + r_k M_k)$ is nef.
(c) $\mathrm{Supp}(F^+) \subset \lfloor \Delta^+ \rfloor$.
Now we can scale by $r_1 M_1 + \cdots + r_k M_k$ to get $(X, \Delta^+)^{min}$.
(5) By the useless divisor lemma, $(X, \Delta^+)^{min} = (X, \Delta)^{min}$.

Thus we have proved that

Termination with scaling in dim. n near $\lfloor \Delta \rfloor$

$\Downarrow$

Existence of $(X, \Delta)^{min}$ in dim. n

4.5. Finiteness of models

More generally, we claim that the set of models $(X, \Delta_w)^{min}$ is finite as Δ_w moves in a compact set of $\mathbb{R}$-divisors satisfying 3 conditions:

(1) Every Δ_w is big. Note that being big is not a closed condition and it would be very good to remove this restriction. Without it one could get minimal models for non-general type (X, Δ) by getting $(X, \Delta + \epsilon(\text{ample}))^{min}$ and then letting $\epsilon \to 0$.
(2) Every $K + \Delta_w$ is effective. By definition, being pseudo-effective is a closed condition. It is here that non-vanishing comes in: it ensures that being effective is also a closed condition.
(3) $K + \Delta_w$ is klt. This is preserved by making Δ_w smaller, which is what we care about.

By compactness, it is enough to prove finiteness locally, that is, finiteness of the models $(X, \Delta' + \sum t_i D_i)^{min}$ for $|t_i| \le \epsilon$ (depending on Δ'). (Important point: Even if we only care about $\mathbb{Q}$-divisors, we need this for $\mathbb{R}$-divisors Δ'!)

PROOF. Induction on r for $D_1, \ldots, D_r$. Let $(X^m, \Delta^m) := (X, \Delta')^{min}$. By the base point free theorem, we have $g : X^m \to X^c$ such that $K_{X^m} + \Delta^m \sim g^*(\text{ample})$.

So, for $|t_i| \ll 1$, $g^*(\text{ample}) \gg \sum t_i D_i$, *except* on the fibers of g. (This is more delicate than it sounds, but not hard to prove for extremal contractions.)

Thus the MMP to get $(X^m, \Delta^m + \sum t_i D_i)^{min}$ is relative to X^c. We can switch to working locally over X^c, we thus assume that $K_{X^m} + \Delta^m \sim 0$. So

$$K_{X^m} + \Delta^m + c \sum t_i D_i \sim c\big(K_{X^m} + \Delta^m + \sum t_i D_i\big)$$

Therefore $(X^m, \Delta^m + \sum t_i D_i)^{min} = (X^m, \Delta^m + c\sum t_i D_i)^{min}$. For $(t_1, \ldots, t_r)$ choose c such that $\max_i |ct_i| = \epsilon$, that is, $(ct_1, \ldots, ct_r)$ is on a face of $[-\epsilon, \epsilon]^r$. Ths shows that we get all possible $(X^m, \Delta^m + \sum t_i D_i)^{min}$ by computing $(X^m, \Delta^m + \sum t_i D_i)^{min}$ only for those $(t_1, \ldots, t_r)$ which are on a face of the r-cube $[-\epsilon, \epsilon]^r$. The faces are $2r$ copies of the $(r-1)$-cube. So we are done by induction on r.

The notation $(X, \Delta)^{min}$ helped us skirt the issue whether there may be infinitely many minimal models for a given (X, Δ). (This happens in the non-general type cases, even for smooth minimal models.) This is, however, no problem here. We found one $\mathbb{Q}$-factorial model $g : X^m \to X^c$ such that K_{X^m} is numerically g-trivial.

Let $D_1, \ldots, D_r$ be a basis of the Néron-Severi group. Every other model where $K + \Delta$ is nef lives over X^c and some $\sum t_i H_i$ is g-ample on it. We can thus find every such model as a canonical model for $(X, \Delta + \epsilon \sum t_i H_i)$ for $0 < \epsilon \ll 1$. □

We have now proved that

$$\boxed{\text{Existence of } (X, \Delta)^{min} \text{ in dim. } n}$$
$$\Downarrow$$
$$\boxed{\text{Finiteness of } (X, \Delta + \textstyle\sum t_i D_i)^{min} \text{ in dim. } n \text{ for } 0 \le t_i \le 1}$$

and the spiraling induction is complete.

Bibliography

[BCHM] C. Birkar, P. Cascini, C. Hacon, J. McKernan, Existence of minimal models for varieties of log general type, preprint arXiv:math/0610203v1, J.Amer. Math. Soc., to appear.

[HM1] C. Hacon, J. McKernan, Extension theorems and the existence of flips, in *Flips for 3-folds and 4-folds*, 76–110, Oxford Lecture Ser. Math. Appl., 35, Oxford Univ. Press, Oxford, 2007.

[HM2] C. Hacon, J. McKernan, Boundedness of pluricanonical maps of varieties of general type, Invent. Math. **166** (2006), 1–25.

[Sho] V. V. Shokurov, Prelimiting flips, Proc. Steklov Inst. Math. **240** (2003), 75–213.

Titles in This Series

17 **Jeffery McNeal and Mircea Mustaţă, Editors,** Analytic and Algebraic Geometry, 2010

16 **Scott Sheffield and Thomas Spencer, Editors,** Statistical Mechanics, 2009

15 **Tomasz S. Mrowka and Peter S. Ozsváth, Editors,** Low Dimensional Topology, 2009

14 **Mark A. Lewis, Mark A. J. Chaplain, James P. Keener, and Philip K. Maini, Editors,** Mathematical Biology, 2009

13 **Ezra Miller, Victor Reiner, and Bernd Sturmfels, Editors,** Geometric Combinatorics, 2007

12 **Peter Sarnak and Freydoon Shahidi, Editors,** Automorphic Forms and Applications, 2007

11 **Daniel S. Freed, David R. Morrison, and Isadore Singer, Editors,** Quantum Field Theory, Supersymmetry, and Enumerative Geometry, 2006

10 **Steven Rudich and Avi Wigderson, Editors,** Computation Complexity Theory, 2004

9 **Brian Conrad and Karl Rubin, Editors,** Arithmetic Algebraic Geometry, 2001

8 **Jeffrey Adams and David Vogan, Editors,** Representation Theory of Lie Groups, 2000

7 **Yakov Eliashberg and Lisa Traynor, Editors,** Symplectic Geometry and Topology, 1999

6 **Elton P. Hsu and S. R. S. Varadhan, Editors,** Probability Theory and Applications, 1999

5 **Luis Caffarelli and Weinan E, Editors,** Hyperbolic Equations and Frequency Interactions, 1999

4 **Robert Friedman and John W. Morgan, Editors,** Gauge Theory and the Topology of Four-Manifolds, 1998

3 **János Kollár, Editor,** Complex Algebraic Geometry, 1997

2 **Robert Hardt and Michael Wolf, Editors,** Nonlinear Partial Differential Equations in Differential Geometry, 1996

1 **Daniel S. Freed and Karen K. Uhlenbeck, Editors,** Geometry and Quantum Field Theory, 1995